A
DOCUMENTARY
HISTORY OF
BIOCHEMISTRY
1770–1940

A
DOCUMENTARY
HISTORY OF
BIOCHEMISTRY
1770–1940

Mikuláš Teich
with Dorothy M. Needham

LEICESTER UNIVERSITY PRESS
Leicester and London

© Mikuláš Teich 1992

First published in Great Britain in 1992 by Leicester University Press
(a division of Pinter Publishers Ltd)

Editorial offices
Fielding Johnson Building, University of Leicester,
University Road, Leicester, LE1 7RH

Trade and other enquiries
25 Floral Street, London, WC2E 9DS

British Library Cataloguing in Publication Data
A CIP cataloguing record for this book is available
from the British Library

ISBN 0 7185 1341 X

Typeset by Best-Set Typesetter Ltd, Hong Kong
Printed and bound in Great Britain by
Biddles Ltd, Guildford and King's Lynn

To the memory of two good people who gave me life and lost theirs
in the concentration camps of Oranienburg and Ravensbrück (1945)

and

of J.D. Bernal (1901–71) and J.B.S. Haldane (1892–1964) who
made me aware of the historical and social context of science

Contents

SECTION I: ENZYMES

SECTION II: PHOTOSYNTHESIS

SECTION III: RESPIRATION

SECTION IV: CARBOHYDRATES

SECTION V: PROTEINS

SECTION VII: CONCEPTUAL AND DISCIPLINARY ISSUES

List of illustrations

Foreword

This volume contains a selected collection of reprints, most of them extracted from papers and books about the chemistry of life. Well over half of them are translated into English for the first time. They have been brought together – in a few cases the whole paper is reprinted – in an attempt to make available source materials on the evolution of the study of the chemistry of life into modern biochemistry. An evolution, that is, in which the chemistry of life, initially seen as an integral part of a still unified science of the chemistry of the three kingdoms (mineral, animal and vegetable), was eventually recognized as an independent branch of the biological sciences.

The implication here is that there was a time when the chemistry of life was not a science. As to the meaning of the term 'science', the view adopted in this work has been that it relates to *the* human activity, historically evolved, of a systematic practical examination of nature along qualitative and quantitative lines, and the production of correlative theoretical knowledge about it. Intellectual curiosity and utilitarian necessity were equally involved and intertwined, and it is not always easy to disentangle the 'pure' and 'applied' impulses and motives which were pushing scientific advance forward.

The case for starting with the years 1770–1800

But, equally, it is not easy to determine precisely the point at which a non-scientific chemistry of life yielded to a scientific one. In this volume it is presumed that the change-over occurred during 1770–1800 and the following brief remarks may help to clarify the reasons for maintaining that position. Those were the years when the doctrine of phlogiston was replaced by the oxidation theory of combustion, leading up to Lavoisier's formulation of the principle of conservation of matter (1789): the principle, that is, governing the chemical combination and recombination of thirty-three elementary substances which then, according to Lavoisier, made up the ultimate non-decomposable constituents of matter. Effectively, it became the basis for chemical bookkeeping in terms of elements, not only with respect to inanimate but also animate nature. Growing, as it partly did, out of concerns with the phenomena of fermentation, respiration and of what became known as photosynthesis, the transition from the doctrine of phlogiston to the oxidation theory resulted in the inversion of the method of their study and interpretation. That has had a profound and lasting effect on the unification of the chemical approach to the worlds of non-life and life.

What had to be explained was why the same elements formed inorganic compounds under certain circumstances and organic compounds – that is, compounds produced by plants and animals – under others. The explanation was found in the existence of a 'vital force', peculiar to living things. It appears that originally the term was introduced and discussed in 1774 (C.F. Medicus). But it was not until its extensive treatment in 1795 in the first issue of the *Archiv für die Physiologie* (J.C. Reil), that it began to be widely invoked as *the* principle of life. Soon it came to be viewed as belonging to *animas, entelechies*, souls, *archaei*, which 'dance processionally through the history of European thinking because a *deus* always had to be found for a *machina*.'[a] It is for its religious as well as for its philosophically ambiguous connotation (idealist and/or materialist) that vital force remained pivotal to thought focused on the chemistry of life for another century.

This, then, is the reasoning behind the assumption that the last three decades of the eighteenth century form a watershed which serves as a dividing line between the non-scientific and scientific chemistry of life. Hence almost all the material heading the Sections in this volume is taken from works produced between 1770 and 1800.

The case for stopping at around 1940

In the light of the assembled material, it took more than a century and a half for the chemistry of life to attain the status of an independent branch of science. It was partly this consideration that prompted the decision to leave off with texts published around 1940. Concerned with the chemical nature of substances occurring in living things and the dynamic relations between them, the chemistry of life had become by then 'biochemistry' and 'biological chemistry' respectively (though the term 'physiological chemistry' continued to be used). Looked at in the context of scientific history, the making of the chemistry of life into a separate discipline was a long drawn out process; and it is the function of the seven thematic Sections of the volume to offer guidance in understanding it. Before listing them, it may be useful to look more closely at the reasons why the years around 1940 mark the tail-off for this selection of texts.

Since the mid-1930s biochemists had come to appreciate the part X-ray crystallography could play in the study of the structure of biologically interesting molecules. It was the lack of familiarity with this technique that induced them to link up with X-ray crystallographers interested in the structural organization of living matter. Thus X-ray crystallography was made use of in the historic assault on nucleoprotein from plants infected with tobacco mosaic virus (1936). It was the joint work of a plant pathologist (F.C. Bawden), a biochemist (N.W. Pirie) and two crystallographers (J.D. Bernal and I. Fankuchen).[b]

Indeed, there was a particular proposal – which proved to be abortive – to further this sort of interdisciplinary approach on an institutional basis by creating an Institute for Mathematico-Physico-Chemical Morphology at Cambridge. In 1935 it was put forward by a biochemist researching into embryonic development (J. Needham) on behalf of a group that included a crystallographer with biological concerns (J.D. Bernal), an embryologist with genetical concerns (C.H. Waddington), a biologist engrossed in philosophy (J.H. Woodger) and a mathematician probing the molecular structure of chromosomes (Dorothy M. Wrinch). The proposal and its failure to obtain the support of Cambridge University and the Rockefeller Foundation have become a subject of discussion although, clearly, this is not the place to join it.[c]

It is, however, appropriate to draw attention to Needham's bold vision of bridging the gap between biochemistry and morphology which was to have governed the programme of work at the institute. If this task was to be fulfilled, according to Needham, biochemists had to change their attitude to large molecules by coming to terms with them. Needham was led to this consideration by X-ray findings which pointed to molecular organization in plant and animal fibres based on the presence of carbohydrate and protein crystals forming themselves into polymeric chains. He thought that this approach could provide an avenue for the study of morphogenesis, for example during embryonic development.[d]

All this contrasted with the hitherto prevailing conception, originally pioneered by F.G. Hopkins (1913),[e] that the key to the chemistry of life was through small molecules undergoing comprehensible reactions. But was Hopkins's theory negated by that? Needham himself made it plain that the large molecules of the living world could not be but a part of metabolism and it would, therefore, be a task of biochemistry to trace their formation and breakdown.

Actually, the question of the underlying unity of the structure and dynamic sides of biochemistry, raised by Needham, was brought into sharper relief long after 1940: after, that is, the potential of the two methods that entered the biochemical scene in the mid-1930s realized itself more fully. One of those methods, the application of X-ray crystallography to the study of molecular organization of biological structures, has already been touched upon. The isotope technique was the other one. Its relevance to the field of intermediate metabolism was demonstrated through the use of deuterium.[f]

We have now reached the eve of the Second World War. Although a good deal has been written about its effects on scientific research and scientists, more will have to be done before we have a fuller picture. The problem is to differentiate between the continuous and discontinuous components of what is actually a single historical process. This has become conspicuous, in relation to the study of the chemistry of life, when scientists have tried to reach agreement on the demarcation between 'biochemistry' and 'molecular biology'.[g]

An historian cannot but conclude that the roots of the latter, however the term is defined, lie in the years around 1940 when this *Documentary History* is brought to an end.

Sections and Selections

At the heart of the chemistry of life lies the comprehension of the chemical changes maintaining life. The maintenance of life depends on the intake of food materials which with plants is, as a rule, an invisible process.

Historically, the division of foodstuffs into carbohydrates, proteins and fats goes back to 1827.[h] This, coupled with the recognition of their basic part in the composition of living organisms and in the chemical changes proceeding in them, has led to the arrangement of the contents of the book. They provide the principal themes of Sections I–VI: I Enzymes, II Photosynthesis, III Respiration, IV Carbohydrates, V Proteins, VI Lipids. Broadly the material in them focuses on observations, experiments and their theoretical interpretation. In contrast, in the final Section, VII, entitled 'Conceptual and disciplinary issues', the emphasis is on the growth of biochemistry's self-awareness, under the impact of questions and answers on how to approach and understand life at the chemical level.

The selected material embraces particular topics which are dealt with in an 'earlier' Subsection, (a) – up to c. 1880–1900 – and a 'later' Subsection, (b) – up to c. 1940 – into which each Section is subdivided. The heading of the Subsection indicates, in the main, the topic(s) covered. Thus in Ia the accent is on fermentation and digestion, and in Ib on the nature of enzymatic action and the chemical nature of enzymes.

As a rule, Selections in each Section are set out chronologically. They are numbered from 1 to 180 and, leaving out Selections 72, 88, 107, 128, 151 and 171, each relates to a single source. Apart from Selection 98, originally written in Russian, the texts are taken from English, German and French publications; an effort has been made to include contributions by scientists from different countries. The translation of scientific texts from the past is a venture fraught with peril. There is always the danger of producing an anachronistic version which distorts the historical picture. That the language of science has its own history is, as it were, attested in this volume by the inclusion of a few contemporary translations. In our own translations of material for the present work, our intention has been to remain as close as possible to the original, without necessarily following the latter's punctuation.

In a work such as this, there is an additional risk that scientific history will be distorted by the process of selection itself. Almost inescapably the materials give the impression of a continuous movement forward, scarcely, if at all, affected by zigzaggedness. That impression should be neutralized, at least to some extent, by the Introduction which precedes each Selection and places it broadly in its historical setting 'within' and 'without' the study of the chemistry of life. Here, inevitably, the treatment can be neither in depth nor uniform. Nevertheless, it is hoped that the Introductions with their conjoining Selections will help to make readers aware that the division into the 'internal' and 'external' history of biochemistry is useful, but cannot be regarded as absolute. A picture of the past mechanically put together on the basis of the two histories will inevitably be distorted.

'Internal' and 'external' history of biochemistry: the case of fermentation[i]

Here the example of fermentation, possibly the earliest biochemical process made use of by man in daily life, is apposite. Assumptions about fermentation and explanations of its nature, already important in the history of natural, technical and medical knowledge before chemistry became fully a science at the end of the eighteenth century, played an even more decisive role afterwards. Indeed, Lavoisier formulated and applied the principle of the conservation of matter (one of the corner-stones of modern chemistry) in a discussion of alcoholic fermentation, which he considered one of the most extraordinary operations in chemistry.[j]

Identifying chemical change with fermentation was common among sixteenth- and seventeenth-century writers trying to explain chemical reactions. As will be shown, a similar idea, but restricted to the chemistry of life, was developed by T. Schwann, who in 1835 published the results of his work on fermentation showing that yeast was growing during alcoholic fermentation and this was probably the cause of fermentation. Referring to this biological interpretation of alcoholic fermentation in his classical book on the cell, *Mikroskopische Untersuchungen über die Uebereinstimmung in der Struktur und dem Wachsthum der Thiere und Pflanzen* (1838–9), Schwann pointed out (in a footnote) that the metabolic changes in the yeast cell represented the best known and the simplest example of a process which repeated itself in every living cell.[k]

The further development of the study of the chemistry of life confirmed Schwann's important observations but the historical development of the understanding of the nature of chemical changes and the means by which they operate was not straightforward. Whilst the study of fermentation was gradually taken up in a systematic manner, the chemistry of the cell essentially was not until after Pflüger's substantiation of the idea of cell respiration (1872, 1875).[l] Chemists, like Berzelius and Liebig, rejected completely the idea that fermentation had anything to do with life. Liebig maintained that it was a purely chemical process, related to putrefaction and decay of those organic bodies which had already ceased to live.[m] Whether fermentation constituted a pure chemical reaction in which the yeast ferment decomposed sugar into alcohol or whether it had something to do with the life of yeasts played a part in the mechanist–vitalist controversy. Liebig and Pasteur were perhaps the most prominent participants in the later stages of the controversy, but there were others, and it became important not only because of the scientific and philosophical issues involved. It is possible, for example, to draw attention to Pasteur's findings that besides alcohol and carbon dioxide, alcoholic fermentation also yielded succinic acid, glycerol and other products.[n] We can also point to Emil Christian Hansen's work on pure yeasts in the Carlsberg Laboratory (1882).[o] There can be no doubt that this work of Pasteur and Hansen was linked, in both its origin and its far-reaching technical and economic consequences, to the fermentation industries.

The disclosure of cell-free fermentation by Eduard Buchner (1897) furnishes another example of the difficulty of separating the 'internal' and 'external' aspects of scientific development. The discovery found itself at the point where science, medicine, philosophy, technology and industry met. The demonstration of fermentation outside living cells, proving that both Liebig and Pasteur had been partly right and partly wrong, indicated that the mechanist and vitalist approaches were not as widely separated as it had seemed. Indeed, it could be argued on the basis of what they wrote that the 'mechanist' Liebig adopted at times a 'vitalist' position and the 'vitalist' Pasteur a 'mechanist' one. Following Buchner's discovery, the systematic study of ferments (enzymes)[p] and their role in cellular chemistry was taken up. It brought to fruition the ideas which had been in the mind of Schwann and of later investigators. Prominent among them was the wine-merchant cum scientist, Moritz Traube. 'In so far as the chemical processes in the living organism mostly depend on ferment activities,' Traube maintained, 'an understanding of the chemistry of life without a correct theory of fermentation is altogether impossible.' (1861)[q]

Some of the more important developments in enzyme chemistry at the turn of the century originated in work on the environment in which fermentational changes were taking place. Among these developments was the recognition of the phosphate buffer system by the French brewing chemists, A. Fernbach and L. Hubert (1900). Possibly their perception sprang from an understanding of the role of phosphates in agriculture, and was reinforced by the idea of the constancy of the internal environment which Claude Bernard had identified as a necessary condition for normal activities in the living organism.[r]

Much of research along these lines was carried out in the laboratories of the Carlsberg Foundation in Copenhagen, established by the brewer J.C. Jacobsen in 1876 for fundamental studies of chemical and physiological problems related to brewing. It was here that Hansen demonstrated the obnoxious effects of 'wild' yeast, which eventually led Jacobsen to introduce important changes in brewing practice because pure yeast cultures began to be used in his breweries. There is no doubt that in research conducted in the Chemical Department of the Carlsberg Laboratory, first headed by Johannes Kjeldahl, and after his death by S.P.L. Sørensen, relations between science and the brewing industry existed, though not necessarily on a linear basis.[s] Sørensen's name is indissolubly linked with his studies of the influence of the hydrogen ion concentration on enzyme reactions, which he had likened to temperature effects. In the course of this research he had developed the immensely useful concept of the hydrogen ion exponent, pH (1909). Although the work was not ostensibly undertaken in direct response to the needs of brewing, the available evidence suggests that the connections were not as remote as they would seem. The mediating link can be found in the studies of some other Carlsberg workers which were more specifically related to scientific problems of brewing and which were referred to by Sørensen in his highly seminal paper. Thus Fr. Weis was concerned with enzymes during the germination of barley. He published a paper (1903) in which he described, as Fernbach and Hubert had done, the phosphate buffer's marked effect on proteolytic enzymes from malt. At that time, according to Sørensen, two other Carlsberg workers, Petersen and Sollied, confirmed these findings in relation to proteolytic enzymes from yeast.[t] In the light of Sørensen's work on the hydrogen ion concentration the physico-chemical nature of the constancy maintained by buffer solutions found its explanation. Sørensen's work, geared as it had been to the needs of the fermentation industries, nevertheless acquired a much wider significance because it provided the key for the understanding and control of those chemical, biological and industrial processes which require a specific optimum range of acidity and alkalinity.

The previous paragraph illustrates some of the multiform reciprocal links which make the historical study of the phenomena of fermentation as much a part of the history of biochemistry (and other sciences) as a part of the technological and economic history of several industries. Another feature of the history of fermentation is that it was intimately connected with endeavours to answer the perennial question of the origin of living matter and the nature of the processes underlying it.[u]

This does not mean that the pursuit of 'internal' and 'external' aspects of the history of biochemistry respectively is meaningless. Both are justified provided we understand that the questions asked and the answers given are, by their very nature, partial and limited. The point at issue for an historian is how to amalgamate them because history is in reality an integral process, not divided into 'internal' and 'external' compartments, but governed by a multitude of circumstances, deriving from and belonging inseparably to spheres 'inside' and 'outside' science.

Notes

a. J. Needham, *Science and Civilisation in China* (Cambridge, 1956), II, p. 302.
b. F.C. Bawden, N.W. Pirie, J.D. Bernal and I. Fankuchen, 'Liquid crystalline substances from virus infected plants', *Nature, 138*, 1051, 1936.
c. Cf. Abir-Am, 'Discourse', Introduction to Selection 134, note b.
d. Selection 180, Section VII.
e. Selection 174, Section VII.
f. Selection 133, Section V; Selection 156, Section VI.
g. See J.N. Davidson, 'Chairman's introduction'; J.C. Kendrew, 'Some remarks on the history of molecular biology' in T.W. Goodwin (ed.), *British Biochemistry Past and Present* (London–New York, 1970), pp. 3–4, 5–10.
h. Selection 160, Section VII.
i. This draws on M. Teich, 'From "enchyme" to "cyto-skeleton": the development of ideas on the chemical organization of living matter' in M. Teich and R. Young (eds), *Changing Perspectives in the History of Science* (London, 1973), pp. 439–71.
j. Selection 1, Section I.
k. Selection 12, Section I; Selection 163, Section VII.
l. Selection 53, Section III.
m. Introduction to Selection 13, Section I, note b; Selection 14, Section I.
n. L. Pasteur, 'Mémoire sur la fermentation alcoolique', *Ann.Chim.* [3] *68*, 323–63, 354–426, 1860; cf. also Selection 16, Section I.
o. Selection 19, Section I, note d; also M. Teich, 'Fermentation theory and practice: the beginnings of pure yeast cultivation and English brewing, 1883–1913', *History of Technology, 18*, 119–33, 1983.
p. Cf. M. Teich, 'Ferment or enzyme: what's in a name', *Hist.Phil. Life.Sci.*, *3*, 193–215, 1981.
q. Selection 51, Section III.
r. Introduction to Selection 80, Section IV, note e; Selection 169, Section VII.
s. H. Holter and K. Max Møller (eds), *The Carlsberg Laboratory 1876/1976* (Copenhagen, 1976).
t. Selection 23, Section I.
u. A full historical treatment of the connection between intra-scientific and extra-scientific influences on fermentation theory and practice would require, for example, consideration

of such a motive as patriotism. This certainly prompted Pasteur's work on beer after France's defeat by Prussia in the Franco-Prussian war. Cf. L. Pasteur, 'L'idée de ces recherches m'a été inspirée par nos malheurs. Je les ai entreprises aussitôt après la guerre de 1870 et pursuivies sans relâche depuis cette époque, avec la résolution de les mener assez loin pour marquer d'un progrès durable une industrie dans laquelle l'Allemagne nous est supérieure' (*Études sur la bière*, Paris, 1876, p. vii).

Acknowledgements

Habent sua fata libelli! Looking back I am very aware of the length of time it has taken to put together the ingredients of this *Documentary History* since the search for material for it began in the autumn of 1972. What was hoped to be concluded in three years took four times as long and perhaps a word or two of explanation is in order here. The work on the *Documentary History* proved to be more complex and, therefore, more time-consuming than expected. Also, economic circumstances made it difficult to give uninterrupted attention to the work.

That the *Documentary History* appears in print is primarily due to the sustained encouragement and backing which I have received from my family. Above all I owe a measureless debt of gratitude to my wife, Professor Alice Teichova, who has been my closest adviser and helper, in matters great and small – material and spiritual. Nor can I refrain from acknowledging that I owe a special debt of gratitude to my brother, Dr Paul Campbell-Tiech, for extending a generous helping hand at a time when it was most needed.

Next, my warmest thoughts go to the memory of Dr Dorothy Moyle Needham, FRS, whose name appears on the title page; she sadly died on 22 December 1987 at the age of 91. Her knowledge of the inter-war developments of biochemistry. in which she participated with great distinction as an active researcher, has been invaluable in preparation of the volume in draft form. Moreover, she has made a vital contribution to this volume by translating the selected German and French texts.

Biochemists of recent years have been remarkable among scientists both in their concern for the history of their discipline and in the body of historical work they have produced. I have been fortunate and privileged to obtain helpful comments and suggestions from Professor J.T. Edsall on Section V, Professor J.S. Fruton on Section I and IV, Dr J.C. Gray on Section II, Dr R. Hill on Section II, Professor A. Kleinzeller on Section VI, the late Professor Sir Hans Krebs on Section III and from Mr N.W. Pirie on Section VII. In recording my gratitude to them, it goes without saying that they bear no responsibility for the choice of Selections or for the views expressed in Introductions and annotations.

My thanks are also due to Ms Alison Hennegan, MA, for reading the Introductions and annotations from the language point of view and to Mrs Valerie Striker for the technical preparation of the book. Nor can I omit to say that the suggestion to publish it came from Professor Philip Cottrell, a member of the Editorial Committee of Leicester University Press. For accepting it, as well as getting the manuscript to press, and helping with the proofs, I should like to enter my special debt to Alec McAulay of Leicester University Press and Pinter Publishers.

The work on the *Documentary History* began while I was a member of Gonville and Caius College (Cambridge) between 1972 and 1975. To its then Master, Dr Joseph Needham, FRS, FBA, and Fellows for making me Visiting Scholar, and to the Society for the Protection of Science and Learning for assisting me financially, I am deeply grateful. Since 1976 it has been my gratifying association with the newly founded Robinson College which has underpinned the resolve to persevere with the enterprise. The appointment as Senior Research Fellow in 1983 provided opportunities for continuing with the project. For this and for other forms of support shown to me throughout I should like to record my indebtedness to the Trustees of Robinson College, presided over by Professor Charles Brink, FBA, and to the Fellowship headed by Professor Lord Jack Lewis, FRS.

The completion of the book was aided by a Wellcome Research Fellowship; it allowed me to probe more deeply into material regarding biochemistry's search for identity which has been incorporated, in particular, into Section VII.

I wish to thank the following for permission to reproduce copyrighted material, from the sources indicated. Author(s), titles and dates of publications, including page references, are specified in Selections. Académie des Sciences for material from *Comptes Rendus de l'Académie des Sciences de Paris*; Akademie der Wissenschaften Göttingen for material from *Nachrichten von der Königlichen Gesellschaft der Wissenschaften zu Göttingen Mathematisch-Physikalische Klasse*; Akademische Verlagsgesellschaft Geest & Portig for material from *Zeitschrift für physikalische Chemie*; American Association for the Advancement of Science and J.S. Fruton for materials from *Science*; American Chemical Society for material from *The Journal of the American Chemical Society*; American Physiological Society and W.C. Rose for material from *Physiological Reviews*; American Society of Biological Chemists for material from *The Journal of Biological Chemistry*; Francis Aprahamian, Literary Executor (J.D. Bernal); The Biochemical Society and the late J.H. Quastel for material from *The Biochemical Journal*; Blackwell Scientific Publications Ltd. for material from *The Journal of the Society of Chemical Industry*; Gebrüder Bornträger for material from *Chemie der Zelle und Gewebe*; British Association for the Advancement of Science for material from its *Report* (1913); Cambridge University Press for material from J. Needham and D.E. Green (eds), *Perspectives in Biochemistry* (1937); and from J. Needham, *Biochemistry and Morphogenesis* (1942); Trustees of the Carlsberg Laboratory for material from *Comptes Rendus des Travaux du Laboratoire Carlsberg*; Cold Spring Harbor Laboratory for material from its *Symposia on Quantitative Biology*; R.P. Cook for material from C. Bernard, *Phenomena of Life Common to Animals and Vegetables* (1974); Elsevier Science Publishers BV for material from *Nobel Lectures Physiology or Medicine 1922–41*; Hoppe-Seyler's *Zeitschrift für Physiologische Chemie*; Dr W. Junk Publishers for material from *Enzymologia*; Karolinska Institutet for material from *Skandinavisches Archiv für Physiologie*; McGraw Hill (UK) Ltd. for material from J. Loeb, *Proteins and the Theory of Colloidal Behaviour* (1924); Macmillan Journals Ltd. for material from *Nature*; The Physiological Society for material from *The Journal of Physiology*; The Rockefeller University Press for material from *The Journal of General Physiology*; The Royal Society and the late A.C. Chibnall for material from its *Proceedings*; The Royal Society of Chemistry for material from *Transactions of the Faraday Society*; Société de Biologie et de ses Filiales for material from its *Comptes Rendus*; Society for Experimental Biology and Medicine for material from its *Proceedings*; Springer-Verlag for material from *Archiv für experimentelle Pathologie und Pharmakologie, Biochemische Zeitschrift, Deutsches Archiv für klinische Medizin, Klinische Wochenschrift, Pflügers Archiv*; Dr Dietrich Steinkopff Verlag for material from *Kolloid-Zeitschrift*; Verlag Chemie for material from *Berichte der Deutschen chemischen Gesellschaft*; John Wiley and Sons Ltd. for material from *Advances of Enzymology*.

M.T.

Note on the text

Annotations, cross-references, a select bibliography and biographical data are offered to enquiring readers in the hope that they will find them useful.

There are two kinds of annotations: numerical and alphabetical. The numerical notes, which may have been renumbered (starting with the numeral 1), refer to the original text while the styles of citation have remained unaltered. Tables and illustrations accompanying the original text have also been renumbered.

The alphabetical notes are editorial and here the citation of periodicals is uniform. An attempt has been made to produce a comprehensive list of serial publications met with in this volume (see pages xxxi–xxxvii). As to book references, as a rule in each Selection they are given in full where they occur first, followed by a shortened version in subsequent citations. Occasionally, the shortened title of a book or journal article is used when cross-referenced to the full citation in another Selection.

There is a seven-part bibliography, broadly supplementing the references in the notes to the Foreword, Selections in respective Sections, and the Afterword to about 1985.

On the whole, available dates of birth (and death) of individuals in each Section appear once when their name is mentioned first. Here J.S. Fruton's *A Bio-Bibliography for the History of the Biochemical Sciences since 1800* (1982) has been an invaluable source of information.

Square brackets enclose editorial insertions and interpolations. A number within square brackets preceding the volume number of a scientific periodical indicates that the latter was published in more than one series.

Abbreviations of serial publications

Adv.Carb.Chem.	Advances in Carbohydrate Chemistry
Adv.Enzym[ol].	Advances in Enzymology
Adv.Protein Chem.	Advances in Protein Chemistry
Agric.Hist.	Agricultural History
Am.Chem.J.	American Chemical Journal
Am[er].J.Path.	American Journal of Pathology
Am.J.Physiol.	American Journal of Physiology
Am.Natur.	American Naturalist
Ann.	Annalen der Pharmazie (1832–9); also as *Ann.Chem.*; *Ann.Chem.Pharm.*
Ann.Bot.	Annals of Botany
Ann.[d.]Chem[ie].	Justus Liebigs Annalen der Chemie (1873–)
Ann.Chem.Pharm.	Annalen der Chemie und Pharmacie (1840–73)
Ann.[de]Chim.[Phys.]	Annales de Chimie [et de Physique]
Ann.de l'Inst.Pasteur	Annales de l'Institut Pasteur
Ann.N.Y.Acad.Sci.	Annals of the New York Academy of Sciences
Ann.Phil.	Annals of Philosophy
Ann.Phys.	Poggendorffs Annalen der Physik und Chemie
Ann.Rep.Prog.Chem.	Annual Reports on the Progress of Chemistry
Ann.Rev.Biochem.	Annual Review of Biochemistry
Ann.Rev.Plant Physiol.	Annual Review of Plant Physiology
Ann.Sci.	Annals of Science
Ann.Sci.Nat.(Zool.)	Annales de Sciences Naturelles (Zoologie)
Arch.Anat.Physiol.	as *Müllers Arch.*
Arch.f.Anat.Phys.u.wiss.Medizin	Ibid.
Arch.Biochem.Biophys.	Archives of Biochemistry and Biophysics
Arch.[f.]exp.Path.[und]Pharm.	Archiv für experimentelle Pathologie und Pharmakologie
Arch.f.Heilk.	Archiv für Heilkunde
Arch.Int.Hist.Sci.	Archives Internationales d'Histoire des Sciences
Arch.klin.Med.	Deutsches Archiv für klinische Medizin

Arch.mikr.Anat.	Archiv für mikroskopische Anatomie
Arch.Mikrobiol.	Archiv für Mikrobiologie
Arch.Néerl.Sci.Ex.Nat.	Archives Néerlandaises des Sciences Exactes et Naturelles
Arch.Path.Anat.	as *Virchows Arch.*
Arch.Phys.Biol.	Archives de Physique Biologique
Arch.Physiol.	Archiv für die Physiologie; as *Müllers Archiv*
B.	Berichte der Deutschen chemischen Gesellschaft
B.H.	Botanische Hefte
Beitr.chem.Physiol.u.Path.	as *Hofmeisters Beitr.*
Ber.bot.Ges.	Berichte der Deutschen botanischen Gesellschaft
Ber.chem.Ges.	as *B.*
Berichte d.d.chem.Gesellsch.	Ibid.
Biochem.J[l].	Biochemical Journal
Biochem.Z[s].	Biochemische Zeitschrift
Bio.Z.	Ibid.
Biochem.Zeitschr.	Ibid.
Biog.Mem.F.R.S.	Biographical Memoirs of Fellows of the Royal Society
Biol.Revs.	Biological Reviews
Bot.Centrlb.	Botanisches Centralblatt
Bot.Ztg.	Botanische Zeitung
Brit.For.Med.Chir.Rev.	British and Foreign Medico-Chirurgical Review
Brit.J.Hist.Sci.	British Journal for the History of Science
Brit.J.Phil.Sci.	British Journal for the Philosophy of Science
Brit.Med.Bull.	British Medical Bulletin
Brit.Med.J.	British Medical Journal
Bull.Agr.Chem.Soc.Japan	Bulletin of the Agricultural Chemical Society of Japan
Bull.Biol.et Med.Exper.Moscou	Bulletin de biologie et de médicine expérimentale d l'U.R.S.S. Moscou
Bull.Chim.Biol.	Bulletin de la Société de Chimie Biologique
Bull.Hist.Med.	Bulletin of the History of Medicine
Bull.Johns Hopkins Hospital	Bulletin of Johns Hopkins Hospital
Bull.Pharm.	Bulletin de Pharmacie; as *J. Pharm.*
Bull.Soc.Chim.	Bulletin de la Société Chimique de France
Bull.Soc.Chim.Biol.	as *Bull.Chim.Biol.*
C.R. [de l'Acad. des Sciences]	Comptes Rendus Hebdomadaires

	de Séances de l'Académie des Sciences
Centralbl.f.Bakt. und Parasit.	Centralblatt für Bakteriologie, Parasitenkunde u. Infektions-Krankheiten
Ch.J.Physiol.	Chinese Journal of Physiology
Chem.News	Chemical News and Journal of Physical Science
Chem.Revs.	Chemical Reviews
Chem.Soc.Proc.	Proceedings of the Chemical Society
Chem.Zelle Gew.	Chemie der Zelle und Gewebe
Chem.Ztg.	Chemiker-Zeitung
Cold Spr.Harb.Symp.Quant.Biol.	Cold Spring Harbor Symposia on Quantitative Biology
Comp[t].Rend [de l'Acad.des Sciences]	as *C.R.*
Comp.Rend.Carlsberg	Comptes Rendus des Travaux du Laboratoire de Carlsberg
Comp.Rend.Soc.Biol.	Comptes Rendus des Séances et Mémoires de la Société de Biologie (et de ses filiales)
Deutsch.Arch.klin.Med.	as *Arch.klin.Med.*
Deutsch.Med.Wochenschr.	Deutsche medizinische Wochenschrift
Erg[eb].Enzymf[orsch].	Ergebnisse der Enzymforschung
Ergeb.Physiol.	Ergebnisse der Physiologie
Enz.	Enzymologia
Fed.Proc.	Federation [of American Societies for Experimental Biology] Proceedings
Geneesk.Tijdschr.Nederland Indië	Geneeskundig Tijdschrift voor Nederlandsch-Indië
Ges.Wiss.Göttingen, Math.physik. Klasse	as *Nachr.Ges.Wiss.(Göttingen)*
Gött.Nachr.	as *Nachr.Ges.Wiss.(Göttingen)*
Helv.Chim.Acta	Helvetica Chimica Acta
Hist.Acad.Sci.	Histoire de l'Académie des Sciences; as *Mem.Acad.Sci.*
Hist.Phil.Life Sci.	History and Philosophy of Life Sciences
Hist.Sci.	History of Science
Hofmeisters Beitr.	Beiträge zur chemischen Physiologie und Pathologie. Zeitschrift für die gesamte Biochemie
Ind.Eng.Chem.	Industrial and Engineering Chemistry
Interdiscipl.Sci.Revs.	Interdisciplinary Science Reviews

J[ourn].Am[er].Chem.Soc.	Journal of the American Chemical Society
J.Biochem., Japan	Journal of Biochemistry, Japan
J[l].Biol.Chem.	Journal of Biological Chemistry
J.Canad.Biochem.	Journal of Canadian Biochemistry
J.Chem.Ed.	Journal of Chemical Education
J.Chem.Soc.	Journal of the Chemical Society, London
J.Electrochem.	Journal of Electrochemistry
J.Exper.Med.	Journal of Experimental Medicine
J.Gen.Microbiol.	Journal of General Microbiology
J.Gen.Physiol.	Journal of General Physiology
J.Hist.Biol.	Journal of the History of Biology
J.Hist.Med.	Journal of the History of Medicine
J.Lab. and Clin.Med.	Journal of Laboratory and Clinical Medicine
J.Med.Chir.Pharm.	Journal de Médicine, Chirurgie, Pharmacie, etc.
J.Pharm.	Journal de Pharmacie et Chemie; also Journal de Pharmacie et des Sciences Accessoires
J.Phys.	Journal de Physique Théorique et Appliquée
J.Phys.Chem.	Journal of Physical (and Colloid) Chemistry
J.Physiol.	Journal of Physiology
J.de Physiol. et de Path.Gen.	Journal de Physiologie et de Pathologie Générale
J[ourn.f.]prakt.Chem.	Journal für praktische Chemie
J.Roy.Statist.Soc.	Journal of the Royal Statistical Society
J.Soc.Chem.Ind.	Journal of the Society of Chemical Industry
J.Tokyo Chem.Soc.	Journal of the Tokyo Chemical Society
Jahrb.wiss.Kritik	Jahrbücher für wissenschaftliche Kritik
Jahresb.	Jahres-Bericht über die Fortschritte der physischen Wissenschaften [über die Chemie und Mineralogie]
Jahres-Ber.	Ibid.
Johns Hopkins Hosp.Bull.	as *Bull.Johns Hopkins Hospital*
Jour.Pharm. and Exper.Therap.	Journal of Pharmacology and Experimental Therapeutics
Kazansk.Mediz.J.	Kazansky meditzinsky zhurnal
Klin.Wochensch.	Klinische Wochenschrift
Koll.Z[eitsch].	Kolloid-Zeitschrift

Landw.Versuch-St.	Die Landwirtschaftlichen Versuchs-Stationen
Malys Berichte	Jahres-Bericht über die Fortschritte der Thier-Chemie
Med.Chem.Unt.	Medicinisch-chemische Untersuchungen (Tübingen)
Med.Chir.Trans.	Medico-Chirurgical Transactions of the Royal Medical and Chirurgical Society of London
Meded.Dienst Volksgezdh.Nederl. -Indië	Mededeelingen van den Dienst der Volksgezondheid in Nederlandsch-Indië
Med.Hist.	Medical History
Mém.Acad.Sci.	Histoire de l'Académie Royale des Sciences avec les Mémoires de Mathematique et de Physique, also Mémoires de l'Académie des Sciences de Paris
Mem.Lit.Phil.Soc.(Manchester)	Memoires of the Literary and Philosophical Society of Manchester
Monatsh.[f.]Chem.	Monatshefte für Chemie
Müllers Arch.	Archiv für Anatomie, Physiologie und wissenschaftliche Medizin
Münch.Med.Woch.	Münchener medizinische Wochenschrift
Nachr.Ges.Wiss.(Göttingen)	Nachrichten von der Königlichen Gesellschaft der Wissenschaften zu Göttingen Mathematische-Physikalische Klasse
Naturwiss[ensch].	Naturwissenschaften
Nicholson's J.	Journal of Natural Philosophy, Chemistry and the Arts
Nord.Med.Ark.	Nordiskt Medicinskt Arkiv
Not.Rec.R.S.	Notes and Records of the Royal Society of London
Obit.Not.F.R.S.	Obituary Notices of Fellows of the Royal Society
Observ.Phys.	Observations sur la Physique, sur l'Histoire Naturelle et sur les Arts
Persp.Biol. and Med.	Perspectives in Biology and Medicine
Pflügers Arch.	Pflügers Archiv für die gesam[m]te Physiologie des Menschen und der T[h]iere
Pharm.Journal and Trans.	Pharmaceutical Journal and Transactions of the Pharmaceutical Society

Phil. Mag.	Philosophical Magazine and Journal of Science
Phil. Trans.	Philosophical Transactions of the Royal Society of London
Phys. Zeitsch.	Physikalische Zeitschrift
Physiol. Revs.	Physiological Reviews
Plant Physiol.	Plant Physiology
Proc. Am. Phil. Soc.	Proceedings of the American Philosophical Society
Proc. (Amsterdam)	Proceedings of the Section of Sciences, Koninklijke Akademie van Wetenschappen te Amsterdam
Proc. Cam. Phil. Soc. Biol. Sci.	Proceedings of the Cambridge Philosophical Society Biological Sciences; as *Biol. Revs.*
Proc. Natl. Acad. Sci.	Proceedings of the National Academy of Sciences (US)
Proc. Physiol. Soc.	Proceedings of the Physiological Society
Proc. Roy. Soc.	Proceedings of the Royal Society of London
Proc. Soc. Exp. Biol. Med.	Proceedings of the Society of Experimental Biology and Medicine
Quart. J. Med.	Quarterly Journal of Medicine
Quart. J. Microscop. Sci.	Quarterly Journal of Microscopical Science
Quart. Rev. Biol.	Quarterly Review of Biology
Rec. Trav. Bot. Néerl.	Recueil des Travaux Botaniques Néerlandais
Rec. Trav. Chim.	Recueil des Travaux Chimiques
Rep. Brit. Ass.	Report of the British Association for the Advancement of Science
Roy. Soc. Proc.	as *Proc. Roy. Soc.*
Schmiedebergs Arch.	as *Arch. [f.] exp. Path. [und] Pharm.*
Schweiggers J.	Neues Journal für Chemie und Physik
Sitzber. (Wien)	Sitzungsberichte der Kaiserlichen Akademie der Wissenschaften Mathematisch-Naturwissenschaftliche Klasse
Skand. Arch. Physiol.	Skandinavisches Archiv für Physiologie
Soc. de Biol.	as *Comp. Rend. Soc. Biol.*
Stud. Hist. Phil. Sci.	Studies in History and Philosophy of Science
Sudhoffs Arch.	Sudhoffs Archiv für Geschichte der Medizin und der Naturwissenschaften

Symp.Soc.Exp.Biol.	Symposia of the Society of Experimental Biology
TIBS	Trends in Biochemical Science
Trans.Am.Phil.Soc.N.S.	Transactions of the American Philosophical Society New Series
Trans.Bose Res.Inst.	Transactions of the Bose Research Institute
Trans.Farad.Soc.	Transactions of the Faraday Society
Verhandelingen (Amsterdam)	Verhandelingen der Koninklijke Akademie van Wetenschappen te Amsterdam (Tweede Sectie)
Vierteljs. (Zürich)	Vierteljahrschrift der Naturforschenden Gesellschaft in Zürich
Virchows Arch.	Archiv für pathologische Anatomie und Physiologie und für klinische Medizin
Yale J.Biol. and Med.	Yale Journal of Biology and Medicine
Z.anal.Chem.	Zeitschrift für analytische Chemie
Z.Biol.	Zeitschrift für Biologie
Z.[eitsch.f.]Elektrochem.	Zeitschrift für Elektrochemie
Z.f.Gärungsphysiologie	Zeitschrift für Gärungsphysiologie
Z.Morph.Anthr.	Zeitschrift für Morphologie und Anthropologie
Z.phys[ik].Chem.	Zeitschrift für physikalische Chemie
Z[s].physiol.Chem.	Zeitschrift für physiologische Chemie
Z.Unters.Nahr.Genuss.	Zeitschrift für Untersuchung der Nahrungs- und Genussmittel
Zeitschr.f.Phys.	Zeitschrift für Physik
Zeitschr.[f.]physiol.Chem.	as *Z[s].physiol.Chem.*
Zentralbl.Physiol.	Zentralblatt für Physiologie

Section I:
ENZYMES

a. From the 1780s to *c.* 1880: fermentation and digestion

Selection 1

When and how early man became aware of fermentation will probably never be known. What is certain is that the practice of fermentation offered constant opportunities to improve upon our understanding of the chemistry of the animate and inanimate. That is true even of an age before chemistry was transformed into a truly scientific discipline, but it can be particularly effectively shown by the study of Lavoisier's epoch-making contributions. That is why his results on alcohol fermentation and his explanation of the processes have been placed at the beginning of the material assembled in this volume.[a]

The passages on alcoholic fermentation have been extracted from *Elements of Chemistry*, the English version, by the Scottish surgeon Robert Kerr, of Lavoisier's *Traité élémentaire de Chimie*, published first in Edinburgh in October 1790, nineteen months after the appearance of the French original (1789). It is one of the great texts in the history of scientific thought and indeed it can be looked upon as the forerunner of the comprehensive textbooks of chemistry written by later authors. Chemists delving into the book today can soon find their way about in it, whereas the same cannot be said of many of the pre-Lavoisierian publications.

The attitude of Antoine-Laurent Lavoisier (1743–94) was to treat of living and non-living things from the unified standpoint of the central role assigned to oxygen. If combustion was the union of oxygen with combustible substances, respiration was slow combustion.[b] Oxygenation (oxidation) products of metallic and non-metallic bodies were marked off as oxides. Their non-oxygen component was known as 'base' or 'radical'. When it came to bodies of biological origin, they were regarded as oxides of compound bases. Thus plants provided substances thought of as vegetable oxides having mostly two bases in varying proportions: hydrogen and charcoal (carbon). On the other hand, animals produced materials, that is animal oxides, containing up to five bases: hydrogen, charcoal, azote (nitrogen), phosphorus and sulphur.

On analysis sugar was shown to be a vegetable oxide. Its conversion into alcohol and carbonic acid (carbon dioxide) during fermentation was conceived of as a chemical decomposition involving a mutual process. One of the products, carbonic acid, was thought to have reached a higher stage of combination with oxygen, whereas the other substance, alcohol, was seen to represent a compound at a lower stage of oxidation and therefore combustible. The reasoning in favour of this conclusion was linked to the principle of conservation of matter, here formulated by Lavoisier for the first time. Paradoxically, the inaccuracies of analytical results submitted by Lavoisier did not invalidate the principle because seemingly they cancelled themselves out.

Although Lavoisier did not claim to have originated the ideas of the principle, neither the form in which he enunciated it nor the fact that he came to it through the study of the process of fermentation was accidental. It reflected a major shift in his thinking on the nature of the elementary constitution of substances of biological origin. There existed a good deal of uncertainty whether water, carbonic acid and oil obtained from plant and animal material on dry distillation were present in that form before distillation. It was through the study of alcoholic fermentation in particular that Lavoisier concluded that carbon and hydrogen in the alcohol were not present as oil, and hydrogen and oxygen in sugar not as water. Referring to these findings Lavoisier stated:

It may be readily conceived, that it must have cost me a good deal to abandon my first notions, but by several years reflection, and after a great number of experiments and observations upon vegetable substances, I have [changed] my ideas . . .[c]

Here it is of more than passing interest to note that even before he published the *Traité élémentaire de Chimie*, Lavoisier had been greatly stimulated by contemporary work on the significance of sunlight in plant growth. He had investigated the mechanism of interaction between carbonic acid and water and wrote:

It is necessary to know that there is no vegetation without water and carbonic acid; these two substances mutually decompose each other during the process of vegetation . . . hydrogen leaves oxygen to unite with charcoal to produce oils, resins and to develop the plant; simultaneously oxygen of water and of the carbonic acid are abundantly given off, as Messrs. Priestley, Inguenhouz [sic] and Sennebier [sic][d] have observed, and combines with light to form oxygen gas.[e]

It is beyond the scope of these comments to deal in more detail with the 'fermentational' context of Lavoisier's pronouncement on the principle of conservation of matter. Suffice it to say that not until his detailed study of the relationship between the end products – the carbonic acid and alcohol, resulting from the fermentation of the starting material, sugar – did Lavoisier become really aware of the reciprocal relation as regards the quantity and quality of identical elements in chemical reactions to be established by analysis and synthesis.[f]

[A.-L.] Lavoisier, *Elements of Chemistry in a New Systematic Order, Containing All the Modern Discoveries* **(Edinburgh, 1790)**

pp. 129–32

Of the Decomposition of Vegetable Oxyds by the Vinous Fermentation.
The manner in which wine, cyder, mead, and all the liquors formed by the spiritous fermentation, are produced, is well known to every one . . .

When the fermentation is completed, the juice of grapes is changed from being sweet, and full of sugar, into a vinous liquor which no longer contains any sugar, and from which we procure, by distillation, an inflammable liquor, known in commerce under the name of Spirit of Wine. As this liquor is produced by the fermentation of any saccharine matter whatever diluted with water, it must have been contrary to the principles of our nomenclature to call it spirit of wine rather than spirit of cyder, or of fermented sugar; wherefore, we have adopted a more general term and the Arabic word *alkohol*[a] seems extremely proper for the purpose.

This operation is one of the most extraordinary in chemistry: We must examine whence proceed the disengaged carbonic acid and the inflammable liquor produced, and in what manner a sweet vegetable oxyd becomes thus converted into two such opposite substances, whereof one is combustible, and the other eminently the contrary. To solve these two questions, it is necessary to be previously acquainted with the analysis of the fermentable substance, and of the products of the fermentation. We may lay it down as an incontestible axiom, that, in all the operations of art and nature, nothing is created; an equal quantity of matter exists both before and after the experiment; the quality and quantity of the elements remain precisely the same; and nothing takes place beyond changes and modifications in the combination of these elements. Upon this principle the whole art of performing chemical experiments depends: We must always suppose an exact equality between the elements of the body examined and those of the products of its analysis . . .

From these considerations, it became necessary accurately to determine the constituent elements of the fermentable substances; and, for this purpose, I did not make use of the compound juices of fruits, the rigorous analysis of which is perhaps impossible, but made choice of sugar, which is easily analysed, and the nature of which I have already explained. This substance is a true vegetable oxyd with two bases, composed of hydrogen and charcoal brought to the state of an oxyd, by a certain proportion of oxygen; and these three elements are combined in such a way, that a very slight force is sufficient to destroy the equilibrium of their connection. By a long train of experiments, made in various ways, and often repeated, I ascertained that the proportion in which these ingredients exist in sugar, are nearly eight parts of hydrogen, 64 parts of oxygen, and 28 parts of charcoal, all by weight, forming 100 parts of sugar.

Sugar must be mixed with about four times its weight of water, to render it susceptible of fermentation; and even then the equilibrium of its elements would remain undisturbed, without the assistance of some substance, to give a commencement to the fermentation. This is accomplished by means of a little yeast from beer; and, when the fermentation is once excited, it continues of itself until completed. I shall, in another place, give an account of the effects of yeast, and other ferments, upon fermentable substances.

p. 139

The effects of the vinous fermentation upon sugar is thus reduced to the mere separation of its elements into two portions; one part is oxygenated at the expence of the other, so as to form carbonic acid, whilst the other part, being disoxyginated in favour of the former, is converted into the combustible substance alkohol; therefore, if it were possible to reunite alkohol and carbonic acid together, we ought to form sugar.

Selection 2

From the above account by Lavoisier it is evident that he did not ignore the role of yeast in fermentation, but we have no information about his views on the subject.

As an example of the attempt to elucidate the nature of the substance initiating fermentation we can consult Fabbroni's account.[a] This many-sided and capable investigator represented his native Tuscany at the international deliberations on the metric system, which began in Paris in September 1798 and continued well into 1799.[b] In view of his previous work[c] and certain new features of the way he explained the decomposition of the sugar during fermentation, Fabbroni's communication, read in his absence, received a good deal of attention as Fourcroy's remarks reveal.

Fabbroni's concern was with vinous fermentation which he chemically likened to the decomposition of carbonates by acids with evolution of carbonic acid. He thought it was induced by a particular kind of matter (called 'vegeto-animal substance') present in plant structure, which he described as 'utricles' (plant cells?). This term, vegeto-animal substance, was applied by H.M. Rouelle (1716–78) to gluten, the substance obtained by I.M. Beccari (1682–1766) by washing out the starch from wheat flour.[d] The younger Rouelle clearly expressed the notion that all plants contained not only a substance resembling the 'glutinous matter' obtained from wheat but also the 'caseous portion' of milk. This view belongs to the early history of albuminous or protein chemistry before the nineteenth-century development of exact procedures of ultimate (elementary) analysis of organic compounds.[e]

For a considerable time fermentation was associated with changes affecting substances of plant origin whereas putrefaction was assumed to be allied to the decomposition of animal substances. Gluten attracted considerable attention, as a plant substance containing nitrogen. It was also crucial to the much debated question of whether fermentation was or was not spontaneous. Antoine François Fourcroy (1759–1809), well known for his work on animal chemistry, emphasized that it was the identification of the fermenting agent with gluten that made the view of Fabbroni novel, whereas the finding that fermentation could take place in a vacuum (evidently earlier studied by Lavoisier) was not. Although he inclined to a chemical view of fermentation, Fourcroy was not satisfied that it was caused by a reaction analogous to the action of acids on carbonates.

Fabbroni's paper and Fourcroy's critical comments are valuable because they give a detailed picture of varied arguments for the chemical interpretation of the nature of fermentation advanced at the end of the eighteenth century. These selections not only illuminate an historical facet of a particular biochemical problem, they also give the reader a vivid sense of the 'scientific workshop' atmosphere of the time.

[A.F.] Fourcroy, *Ann.Chim.*, *31*, 299. 1799

'D'un mémoire du cit. Fabroni, sur les fermentations vineuse, putride, acéteuse, et sur l'éthérification; lu à la société philomatique le 3 fructidor an 7; et Réflexions sur la nature et les produits de ces phénomènes' (Concerning a memoir by *citoyen* Fabroni on vinous, putrid, acetic fermentations, and on etherification; read at the Philomatic Society on the 3rd of fructidor in the year 7;[a] and reflections on the nature and the products of these phenomena)

p. 299

The *citoyen* Fabroni, celebrated physicist of Florence, sent to Paris by his government to take part in the definitive work on weights and measures to which he effectively contributed . . . on departing left a memoir on fermentation and etherification, which has been read in the meeting of the Philomatic Society on the 3rd of fructidor in the year 7, and on which some observations have been made by members of the Society.

pp. 300–3

§. I *New propositions extracted from the memoir of citoyen Fabroni*
  1. It is on the occasion of a work with which he was charged on vinous fermentation that *citoyen* Fabroni has published in 1787, and after the experiments and much reading which he did on this subject that he thought of remarking that fermentation was not as well understood as the actual state of science would allow.
  2. The terms fermentation, and particularly spirituous or alcoholic, are according to him susceptible of error.
  3. It is not a spontaneous movement, since there is no spontaneity in dead matter.
  4. Sugary matter is the necessary element for this fermentation; it decomposes there; it ferments only with the help of another substance capable of reacting upon it and of disengaging an elastic fluid.[b]
  5. Fermentation is only a decomposition of one substance by another, like that of a carbonate by an acid, or of the sugar by nitric acid; it gives, like this last case especially, a slow effervescence. Fermentation is then an effervescence which one must name *vinous effervescence*.
  6. The material which decomposes sugar in vinous effervescence is the vegeto-animal substance; it is situated in particular utricles, in the grape as in the wheat. On crushing the grape one mixes this glutinous matter with the sugar, as if one poured an acid and a carbonate into a vessel; as soon as the two substances are in contact, the effervescence or the fermentation begins there, as this takes place in any other chemical operation.

7. The author considers the reciprocal action of these two bodies as that of an oxide with two bases, the sugar, and a kind of carbonate with two bases, the glutinous material, which because of the slight attraction of these for carbon has, he says, a very powerful attraction for oxygen. When they find themselves liquid the carbon of the latter is carried on to the oxygen of the sugar, is burnt and released in the form of gas; the sugar, partly deoxygenated, forms a new manner of combination with hydrogen and nitrogen, which gives it its piquant flavour and its intoxicating property; a combination analogous to that of opium. One finds no more of vegeto-animal matter nor of sugar in the bread nor in the wine after fermentation, if the respective proportion of these substances was that necessary for their total decomposition. Then the wine is perfect, and as much wine as there can be.

8. The contact with air is not necessary for the formation of wine; *citoyen* Fabroni has obtained it in a barometric vacuum; Collier,[c] the English chemist, similarly made beer without this contact.

9. The intoxicating property of the wine is not due to ready-formed alcohol, since opium and ammonia which also intoxicate do not contain alcohol. Chemists have deceived themselves in regarding this *product* of distillation as a simple *educt*, or in believing in the release just by heat of the alcohol all contained in the wine. This is a manifest error, according to *citoyen* Fabroni, since one cannot remake the wine by mixing the alcohol obtained with its residue ... It is then to the action of heat that the production of alcohol is due; it is a true product of the fire.

11. The author concludes from these experiments and these data that alcohol is not one of the principal constituents of wine, nor a product of fermentation; that one must not name this phenomenon spiritous or alcoholic fermentation, but *vinous effervescence*; and that it is due, without being a spontaneous movement, to the reciprocal action of two agents, of two reactants, put artificially in contact.

12. According to him the same applies to what is named putrid fermentation ...

13. The acetification is also due, in the opinion of the author, to the reaction of the fluid substances which undergo it, to the decomposition of a true mucilage. He proves it by the nature of the mucous wines tending to turn sour more quickly than the others, [and] by that of the wines which flow all the better for their containing more vegeto-animal matter. Nevertheless there is not here any evolution of gas; it is then no more an effervescence than a fermentation; the absorption of air is not necessary and acetification is not due to atmospheric oxygen; but indeed to the simple decomposition of the very much oxygenated mucous matter contained in the wine, since this liquid into which one has put mucilage becomes very good vinegar when one exposes it for a long time to a mild temperature in well-closed vessels. The scum, employed by the vinegar makers under the name of *mother of vinegar*, is only a mucilage which easily acetifies the wine into which it is plunged ...

p. 307

§.II *Reflections on the 14 preceding propositions*

pp. 309–15

II. Since the epoch of the new nomenclature, established at the end of the summer of 1787,[d] I have raised my voice against the expression *spiritous* fermentation, since the word *spirit* must henceforth be banished from science. I had proposed the names of vinous or alcoholic fermentation; and one will soon see that these words can still be adopted and do not express an error as *citoyen* Fabroni thinks. In fact, the word *fermentation* properly signifies only an internal or intestine movement, and one cannot deny that this happens in fact in the substances which ferment, by whatever cause this movement is produced. The word *vinous* gives unequivocally the nature of the product which follows the movement. As to the word *alcoholic*, I have proposed and employed it because (although I think like *citoyen* Fabroni on several points, as I shall have occasion to show in a moment) it is none the less true that alcohol is a necessary product of the decomposition of the wine, and consequently real, although remote from the vinous fermentation . . .

III. When *citoyen* Fabroni raises his voice against the definition of *spontaneous movement*, applied to fermentation, under the pretext that there is no spontaneity in dead matter, he manifestly credits the chemists, who gave or admitted this definition, with an idea which they have not had. Not one of them has meant that there was in the fermenting must an action comparable to what happens in living animals; they have not meant anything else by this expression of spontaneous movement than that the agitation, the reciprocal jostling of the molecules[e] which are brought into being in the fermentation liquor (without any cause strange to their real nature, without any addition to their composition) were alone necessary to cause their generation. Now it is a fact well acknowledged and recognized that spontaneous generation, that the conditions indispensable for this change depend on the nature of the vegetal liquids which experience it. It is a fact that the molecules of these liquids agitate themselves, jostle, move the one against the other; it is a fact that this molecular movement, which accompanies and precedes the change in these liquids, is brought about spontaneously, without the chemist needing to be involved, to add any reactant. Plant life has so disposed the nature and the composition of the vegetal substances that it has placed there the necessary production of this movement and of this change which is its result. It is quite evident that it is thus that chemists have understood the words *spontaneous movement*, applied to fermentation and that, in the sense stated, one cannot find in the employment of this expression, any error prejudicial to the clearness and precision of the language of science.

IV. The fourth proposition of *citoyen* Fabroni expresses a truth recognized for a long time by the chemists, that of the necessity of a sugary material for the production of the vinous fermentation, and above all of another material capable of acting on the first. This last has been named *ferment* by the chemists. Thus a solution of sugar, rather concentrated but not to the consistency of syrup, placed alone on a sand-bath, shows no movement nor other change but evaporation of part of the water; whilst as soon as one adds a small proportion of beer yeast and mixes well, one sees the whole mass become agitated, rise up, become covered with a thick froth, rapidly undergo in a word the movement of fermentation. It is thus that the juices, sugary as well as mucous, starchy, glutinous, of the ripe fruits, the must of grape, of plums, of pears, etc. are susceptible of passing alone and without addition into vinous fermentation. Up to this point *citoyen* Fabroni is in accord with all the chemists who have preceded him in the field: but this fourth proposition contains something more than has been said until now, a statement more precise than that which has up to the present made part of the chemical theory of vinous fermentation. In speaking of this matter different from sugar which is necessary for developing in it the fermentative movement, *citoyen* Fabroni represents it as capable of reacting on the sugary substance, and of *disengaging from it an elastic fluid*. It is this last statement which really begins to make known the particular ideas of the author; one sees there an action of the ferment on the sugary body made precise, whilst until now it had only been enunciated vaguely and in an indeterminate manner.

V.   The 5th proposition offers the development of the preceding; vinous fermentation is shown there as a simple decomposition of one material by another, compared to that of a carbonate on an acid, or to that of sugar by nitric acid, when it is slow or feeble or moderated; it is presented as a real and simple effervescence, and it is proposed to name it *vinous effervescence*; one cannot fail to recognize here a very ingenious idea, which can only be put forward by a man well-versed in profound consideration and in the precise comparison of the phenomena of nature, very clever in the art of bringing facts together and of finding there connections, relations, which escape ordinary minds. But the more an idea is new and ingenious, the more it offers to the imagination this display which pleases and is so fit to seduce, the more without doubt it merits being examined with care, weighed with prudence, studied deeply at leisure. Then one recognizes that the theory proposed by *citoyen* Fabroni has not all the correctness and all the solidity that one was so eager and even so disposed to find there at the first glance. In fact, can one really compare as analogous the effervescence produced by acid poured on carbonate, and the action excited by the ferment in the sugar solution; the gas which is given off in the first is an acid all formed in advance, which needs to take the fluid-elastic form only in proportion as the stronger acid added reacts with the base by which it is more attracted; it is the immediate effect of a chemical attraction between three substances, all settled, all complete from the beginning, and which retain afterwards the same nature which they had before. Now there are not the same circumstances nor similar agents in the case of the ferment and the fermenting [substance]; the first is not an acid, the second is not a carbonate; the carbonic acid set free is formed from all the constituents and did not exist before the fermentation.

p. 316

VI and VII. The 6th and 7th propositions contained in the memoir of the physicist from Florence are only the continuation of the 5th, and the precise statement both of the nature of the material added to the sugar as ferment, and of the decomposition which the two fermenting substances experience the one by the other.

pp. 318–20

It is more difficult still to conceive exactly what the author wished to say, when in order to explain the reciprocal action of the vegeto-animal matter and the sugary matter, he presents the former as a carbonate with two bases – at least as if there is no mistake in the manuscript which has been communicated to me. For the rest, by whatever manner he describes the glutinous body, the explanation of its decomposition as well as that of the sugar which accompanies it is not less clear and ingenious. One sees there the sugar abandoning its oxygen to the carbon of the glutinous body, taking from it its nitrogen, and becoming thus a body more hydrogenated and nitrogenous, whose piquant flavour and intoxicating property depend on this new combination; in adopting the greater part of this theory, in admitting the removal of oxygen from the sugar and its nitrogenation in the vinous liquor, although no experiment has yet proved the presence of the nitrogen, I think nevertheless that the carbon of the carbonic acid which is formed and given off belongs rather perhaps to the sugar than to the glutinous body. For the rest, it is easy to see that in this respect scientific knowledge is lacking for the exact proportions of sugary and vegeto-animal substances necessary for their reciprocal decomposition; and that it is in the appreciation of these proportions that the solution lies for the problem which is occupying me.

VIII. It is very well known and confirmed today that vinous fermentation can take place without contact with air, and the latter has no use in large-scale wine production except to serve as the recipient of the carbonic acid gas whose release accompanies this natural movement, and termination of this release announces that the wine is completely formed. Lavoisier had proved very exactly existence of vinous fermentation without contact with air when he thought of making a solution of sugar ferment with yeast in a well-closed apparatus; this was so arranged that nothing was lost of the fermenting material and all the carbonic acid gas formed was exactly recovered. Thus in this 8th proposition the skilful physicist from Florence is entirely in accord with the French chemists who had recognized, already a long time ago, the truth of this assertion.

IX and X. As for the opinion that alcohol is not ready formed in the wine, that it does not constitute one of the materials of the wine, that in obtaining it by the distillation it is a real *product* of the action of fire and not a simple educt: in spite of all the confidence which the vast and profound knowledge of *citoyen* Fabroni merits and has inspired me, I cannot adopt this opinion.

pp. 322–6

Citoyen Fabroni has not brought together enough proofs, particularly proofs strong and convincing enough, to establish unequivocally and without any room for doubt, his opinion on the non-existence of alcohol ready formed in the wine ...

XI. Thus the 11th proposition of the author in which, briefly summing up, he puts forward as the principal conclusion of his work, that since alcohol is not an elementary constituent of the wine but a product of its decomposition, the name of alcoholic fermentation which I gave to it ten years ago does not suit him; this proposition, if I am not deceived, is not free from all reproach and superior to all objection: and even if alcohol should be a product of wine decomposed by fire, the necessity of taking wine to procure this body and the ease with which one extracts the alcohol from it in the emission to which it is so readily disposed, would suffice to make legitimate this designation of alcoholic fermentation; for I believe that I have already made clear that the word *effervescence* cannot be substituted for that of fermentation with which it had been confused a century ago, to the point where chemists said that the alkalis entered into fermentation with the acids . . .

XII.   I have almost nothing to say on the new manner in which he explains the phenomenon of putrefaction, which he regards no longer as a spontaneous movement, like a fermentation, but as a true effervescence . . .

XIII. The author has nothing particular on acetification except the influence of mucilage, which he regards as a ferment in the wines which turn sour, and to which he attributes the formation of vinegar. The facts which he cites in this regard are both exact and well considered. He no longer sees this conversion of wine into vinegar as an effervescence; neither does he adopt acetic fermentation; he sees in this formation only the reaction of two liquids, and a change of form due to the effect of two reactants on one another. He would have been able to extend the history of acetification, for one knows today an abundance of circumstances where acetic acid is formed without a fermentation, and without necessity for pre-existence of wine − such as the action of fire, that of sulphuric acid, of nitric acid, of oxymuriatic acid.[f] Nevertheless it is quite evident that wine in turning sour, and the different vegetal materials which are susceptible without being vinous, experience a true fermentation since they show an internal and spontaneous movement; this, following the sense which I give to these words in the chemical language, modifies their nature and makes them become acetic acid.

p. 327

I have wished to show to the learned public, in delivering this discussion and in seeking to prove that the chemical expressions of vinous and alcoholic fermentation must continue to be employed in chemistry, however much value I set upon the work and opinions of *citoyen* Fabroni.

Selection 3

In order to see how the knowledge of fermentation developed we consider now the investigations of Louis Jacques Thenard (1777–1857). When he started his enquiry into the subject of alcoholic fermentation Thenard had seemingly not been acquainted with Fabbroni's work. When he did familiarize himself with it he did not think that fermentation could be associated with the action of gluten on sugar. He asserted that well-washed gluten did not promote the fermentation of sugar.

Indeed, Thenard turned his attention back to the action of yeast on sugar believing that its study could throw light on the nature of the fermenting agent. It is the effort to trace the origin of the true ferment and its relationship to and differentiation from the yeast that makes Thenard's account historically significant. He considered the real ferment to be a substance analogous or closely related to the beer yeast. The latter could be produced in the course of fermentation and then deposited. Thenard also speculated about the chemical nature of the ferment and the way it interacted with sugar and converted it into carbonic acid and alcohol. The ferment belonged to animal substances, its carbon became to a certain extent a component part of carbonic acid and its nitrogen was possibly absorbed by the alcohol.

[L.J.] Thenard, *Ann.Chim.*, *46*, 294. 1803

'Sur la fermentation vineuse' (On vinous fermentation)

pp. 296–7

All our knowledge of fermentation is confined effectively to knowing that sugary matter is changed into alcohol and carbonic acid by means of an intermediate body. But what is the nature of this body? How does it act on the sugar? These are two great questions which form the subject of this memoir and that has been often approached without reaching a resolution.

p. 301

I made my first observations on the juice of gooseberries.

pp. 302–9

I expressed into a piece of linen of close texture the juice of a kilogram of gooseberries; it was cloudy and held in suspension a slightly sticky matter which I separated by filtering, and that I washed with plenty of water. As nothing is to be neglected in the observational sciences, and as often the smallest fact leads to great results, I submitted this matter to the following examination. My first care was to put it with sugar and water, to know whether it would be capable of making it ferment; soon, in fact, I saw disengage themselves many bubbles of an elastic fluid[a] which I recognized as being carbonic acid. The effervescence lasted eight days, and at the end of this time the liquor, very agreeable to drink, was only slightly sugary and resembled (in a way to deceive one) a wine which is not yet completely made. One feels that I had to redouble my zeal and attention in examining a substance which offered me that which I was seeking; it was natural to see first whether as a whole it was fitted to decompose sugar. Scarcely a sixth of its weight being able to operate this decomposition, I concluded that it contained only in small quantity the fermenting principle, which I tried vainly by all sorts of means to separate and obtain apart. It remained to me after that to examine it in a comparative way before and after it had served for the fermentation. This appeared not to have changed it: this substance was always insipid, insoluble in water and in alcohol, without action on tincture of sunflower or on syrup of violets; but on distilling it, it no longer gave any trace of volatile alkali.[b] Confirmed by a second experiment this result, which did not astonish me, was nevertheless a beam of light which reassured me in the journey I was making. It made me see that the germ of fermentation was of animal nature in accord with the ideas which I had conceived and gave an appearance of reality to my suspicions.

I searched then with great care in the gooseberry juice for this animal matter which already I regarded as the true ferment . . . It was in a word a matter entirely analogous to the yeast of beer. I hastened to see if this phenomenon was general; it ought to be according to my manner of reasoning. Experience soon taught me in fact that it belonged to all fermenting juices; and the must of the grape, the juice of the cherry, that of the pear, the peach, the apple, the decoction of barley, wheat, gave yeast on fermenting. That of the grape gave more than the others but less than the gooseberry juice; also it ferments less promptly than the latter; the cherry and the peach deposit almost the same quantity of it. The pear and the apple give very little of it, and that is why their fermentation is so slow. I should have liked to have a greater number of fruits at my disposal, in order to vary my trials more; they sufficed however to prove that wherever alcohol is formed usually a deposit of yeast is formed . . . another body capable of exciting fermentation it has the greatest analogy with it; that it differs from it very little; that like it, it is composed of nitrogen, oxygen, carbon and hydrogen, and in fact has without doubt the same manner of acting on the sugar. I am going to set forth with the greatest care the properties of this material which I shall henceforward call ferment, and above all to consider its action on the sugary principle in order to be able to establish the theory of the fermentation. This theory, even supposing that the yeast is not formed but produces itself in the juices which ferment, will always be useful and will find a number of applications, as one will see below.

I shall not go back to the physical properties of the ferment. I have spoken of these several times in this memoir: I shall occupy myself only with its chemical properties which alone are essentially interesting. It has no taste; it does not redden sunflower tincture nor turn syrup of violets green. The putrid fermentation which it experiences with time is in everything similar to that of animal matter. On drying it loses three quarters of its weight; this loss is entirely due to the water which it contains and which volatilises. Thus dried, it is always ready to initiate fermentation; it is not decomposed; and in this state it can be conserved indefinitely without alteration. One could even profit by this property in sending it to places far from breweries, and with which communications are so difficult that fresh ferment would not arrive there, especially in summer, without becoming bad.

pp. 310–12

But of all the properties of the ferment none is so remarkable and at the same time so useful, and none in consequence merits so much study, as its action on sugar; this interests man in all orders of society, from the artisan to the philosopher, both by its products; and the latter furthermore because it can be a fruitful source of reflections and new truths . . .

In order to obtain the solution to this problem I put together different quantities of ferment and of sugar; I observed in all cases what became of the one and the other; and I confirmed by other observations what the first had suggested. Sixty grams of ferment, not dried, and three hundred grams of sugar promptly entered into fermentation; the temperature was 15°; in the space of 4 to 5 days all the sugary material disappeared; 51.5 litres of carbonic acid were liberated.

p. 313

... the quantity of matter arising from 300 grams of sugar and sixty of ferment amounts only to twelve grams; and neither of the two gives ammonia on distillation, while the ferment gives much of it. If these observations are well made, if I have well observed all the phenomena, if nothing has imposed itself on me, one is forced to conclude that nitrogen must exist in the alcohol; but I have sought its presence in this liquid in vain, and in ether and acetic acid too ... These results in any case enlighten us enough to see what happens in the act of fermentation: I do not share in this respect the opinion of Lavoisier. I do not believe with him that all the carbonic acid formed arises from the sugar. How would one conceive then the action of the ferment on it? I think that the first portions of acid are due to a combination of carbon of the ferment and oxygen of the sugar, and that it is in removing from the latter a portion of this principle[c] that the ferments starts the fermentation.

One can then, I repeat, announce as a demonstrated proposition that in all spirituous fermentation an animal matter is deposited resembling in every way that provided by the wort of beer; possessing absolutely the same properties above all that of decomposing sugar and converting it into carbonic acid and spirit of wine.[d] This proposition raises a new question which naturally presents itself and which must now occupy us. Does the yeast produce itself in the act of fermentation, or is it already formed. Does it serve as a ferment?

I must admit that there are still no experiments which prove directly that nature makes use only of this material in operating the transformation of sugar into alcohol and carbonic acid. For why should it deposit itself when the fermentation takes place? One could say, in truth, that the sugar holds it in solution, that it can dissolve more of it than it needs for its decomposition, and that then the excess is precipitated. But this theory is only feebly confirmed by experience. However, is this a strong enough reason to reject it totally? Has one not several examples of compounds which demand much time for forming themselves; and is it perhaps what happens in the juices of the fruits where the ferment and the sugar are a long time in contact. What is certain, or at least appears probable is that if the yeast is a product of the fermentation, as all the liquors which ferment deposit it, it owes its origin without doubt to a similar material from which it probably differs little, and which by its reaction on the sugar produces it.

Many more researches [will have to show] which of these two should be given preference. As I do not doubt that the yeast is not a proximate principle of plants[e] and plays in consequence not a great role in the phenomena of nature and art.

pp. 317–18

It appears then, according to this, that the ferment removes oxygen from the sugar, not only by means of a part of its carbon but also by means of a part of its hydrogen. For the quantity of carbon given up by the ferment is too little to be the only germ of fermentation, nitrogen disappears and enters perhaps into the composition of the alcohol; the other principles of the ferment form acetic acid and a particular white material which precipitates.

pp. 319–20

I do not know if I shall succeed in seeing what becomes of the nitrogen of the ferment. But I shall determine without difficulty whether the residual matter that one obtains, and which I regard as particular, is a product of the fermentation, which I believe; whether sugar contributes to its formation, which is possible; or whether it is already formed and only deposits itself, which is against all probability.

Selection 4

Another significant advance in the study of fermentation arose from public concern in France with the art of conserving food. In 1810 a confectioner and distiller, Nicolas Appert (1750–1841), published in Paris a book known under the shorter title *L'art de conserver, pendant plusieurs années, tous les substances animales et végétales.*[a] In it he showed that food could be preserved provided it was bottled in such a way that air had no contact with it.[b] The bottles had to be heated in boiling water. After confirming Appert's findings, the eminent French scientist Joseph Louis Gay-Lussac (1778–1850) tried to explain them by assuming that the process of fermentation was conditioned by the presence of air (free oxygen).[c]

[J.L.] Gay-Lussac., *Ann.Chim.*, **76**, 245. 1810

'Memoire sur la fermentation' (Memoir on fermentation)

pp. 245–8

It is well shown, according to the experiments of Lavoisier and those of MM. Fabroni and Thenard, that in order to bring about alcoholic fermentation it is necessary to have a sugary material combining with a particular ferment of animal nature. Circumstances favourable for fermentation have been observed for a long time; and there appears to be agreement today that it can begin and continue without the help of any extraneous body, particularly of oxygen gas. One is assured, in fact, that when beer yeast is introduced with sugar and water into a vessel which is filled entirely, fermentation develops in the same manner as in free air; and from this it is concluded that the fermentation of must of grapes, of sugared fruits and of cereals must take place, like that of sugar and beer yeast, without contact of oxygen gas. But in order that this conclusion may be legitimate it is necessary to suppose that the ferment, contained in the fermentable substances, is of the same nature as that of the beer yeast. M. Thenard, to whom we owe a very fine memoir on fermentation,[1] has also reached the opinion that that ferment was always one identical substance. Some experiments that I have made have led me to a different opinion, and the principal object of this memoir will be to prove that fermentation of the must of grapes cannot begin without the help of oxygen gas. The result of this proposition will be that the ferment of grapes is not of the same nature as the beer yeast, or rather that they are not, the one and the other, in the same state.

I have been led to these researches on examining the methods of M. Appert for the conservation of vegetable and animal substances.[2] I had noticed with surprise that must of grapes which had been kept without alteration for an entire year, entered into fermentation some days after having been decanted. It is just as M. Appert prepared sparkling wines at all seasons of the year. This fact made me suspect that the air had a certain influence on the fermentation and led me to the following experiments.

I took a bottle of must of grapes conserved for a year and perfectly limpid; I decanted the must into another bottle that I corked exactly, and I exposed it to a temperature of 15 to 30°. Eight days after the must has lost its transparency; fermentation is established, and soon it [the must] is found to be changed into a vinous drink, sparkling like the best Champagne. A second bottle of must conserved for a year, like the preceding one, but which had not had contact with air, showed no sign of fermentation, although placed in the most favourable circumstances for developing it.

I then took this bottle of must of grapes and, after having made a deep mark on the neck with a file, I inverted it on a bath of mercury and easily detached the neck without the must having contact with air. I made a portion of it pass on to mercury in a bell-jar containing a small quantity of oxygen gas, and a second portion into another perfectly free from air. The first fermented in a few days; but the second showed no sign of fermentation after forty days. On absorption, with potash, of the carbonic acid gas which was set free during the fermentation of the first portion, only an extremely small residue remained: consequently the oxygen gas which I had added had been absorbed in very great part.

These results obviously prove that the must conserved for a long time cannot ferment without contact with oxygen gas.

Selection 5

The preceding selections have shown something of the work at the close of the eighteenth and the beginning of the nineteenth centuries which conceived of fermentation as a chemical process. During the course of explaining how to evaluate the 'goodness' and the 'strength' of beer the idea was put forward, possibly for the first time, that fermentation was a chemical operation associated with plant life.

It was suggested in 1818 by Christian Polykarp Friedrich Erxleben (1765–1831), a pharmacist and industrial entrepreneur of German origin, settled in Bohemia. Among other interests he occupied himself with brewing, and in particular with the application of hydrometry (saccharometry) to it. There was at this time considerable uncertainty, among both brewers and consumers, about what constituted the strength of beer and how to gauge it. Erxleben believed that the strength of beer was determined by the amount of alcohol contained in it and he strove hard to demonstrate that the latter could be estimated quantitatively only after separating it from beer by distillation, and not by the hydrometer. This instrument, he thought, was to be employed in detecting adulteration of beer to which water was added. Clearly, Erxleben was aware of 'attenuation', that is, the fall in the specific gravity of the fermenting liquid due to formation of alcohol during fermentation. His efforts to come to terms with this phenomenon belong to the history of making use of the scientific method in brewing practice with which he was thoroughly familiar. It would seem that it was his detailed knowledge of brewing which enabled Erxleben to break new ground in his study of fermentation. Explicit in his approach was the assumption that fermentation has to be seen as much in chemical as in biological terms.

How did Erxleben arrive at this novel notion of the nature of fermentation? He started from the supposition that the preparation of malt was only partly under the control of man, with the result that the brewer had to make use of diverse malts. Erxleben traced the cause of this variability to plant growth: the germination of the barley used resulted in non-uniform brewers' malts. Accordingly, Erxleben maintained, worts brewed from them would necessarily be of changeable composition so the specific gravity of the beer could not be expected to be uniform, even if the fermenting procedures used were always the same.[a]

[Ch.P.F.] Erxleben, *Uiber Guete und Staerke des Biers, und die Mittel, diese Eigenschaften richtig zu wuerdigen* (Prague, 1818)

(*On Goodness and Strength of Beer, and the Means of Correctly Appreciating These Properties*)

p. 69

As a rule one can indeed state in advance with certainty the *result* of any once known chemical *operation*, but here an exception occurs. Because the fermentation, although until now always considered as such, appears in no way a mere chemical *operation* but much rather in part a *process* by which *plants grow*, and must be considered as the link in the great chain in nature which brings about union of the activities which we call chemical processes with those of *plantlike growth*.

Selection 6

Since the early days of the nineteenth century the conversion of starch into sugar received increasing attention. It was felt that the scientific understanding of the change would benefit brewing and distilling and also the sugar and starch industries. It was this concern with the application of scientific findings to industrial practice that in 1833 led the French chemists Anselme Payen (1795–1871) and Jean François Persoz (1805–68) to their discovery and isolation of a thermolabile substance obtained from germinating barley and capable of transforming starch into sugar: they named it 'diastase'.

It was thought that the starchy material in plants, known as 'fecula'[a] was composed of globular structures covered by a sheath. Payen and Persoz advanced the view that diastase was a constant plant component involved in the saccharification of starch and also in the piercing of the covering of the starch particles.[b] They likened the process in the living plants to the changes produced on heating starch to a moderate temperature or subjecting it to the action of dilute sulphuric acid.

The authors suggested that purified diastase obtained by repeatedly dissolving the material in water and precipitating it with alcohol was devoid of nitrogen.

[A.] Payen and [J.F.] Persoz, *Ann.Chim.*, *53*, 73. 1833

'Mémoire sur la diastase, les principaux produits de ses reactions, et leurs applications aux arts industriels' (Memoir on diastase, the principal products of its reactions, and their applications to the industrial arts)

pp. 74–5

It seems to us, in fact, that one does not yet possess any economic means of extracting from fecula the interior substance which is characterised by a new optical phenomenon recently observed by M. Biot.[a] That after several years of research, far from knowing the active principle developed on germination, one had attributed its reactions first to hordein, then to a sort of *soluble gluten*, which we have separately found to be inert.

That one admitted the transformation of fecula into sugar under this influence without having perceived the *dextrin* liberated, which becomes today the source of numerous applications.

That in consequence the circumstances, the phenomena of the saccharification of fecula in presence of germinated barley were not made precise; that the volumes written on this subject by the English brewers and distillers, and by our authors, leave an abundance of practical anomalies unforeseen and inexplicable.

That in short, none of the consequences of the discovery of *diastase* for organic chemistry, physiology and the industrial arts had been able to be foreseen.

pp. 77–8

... one extracts it from germinating barley by the following procedures...

After having macerated in cold water for some instants the mixture of water and germinated barley, one throws it into a filter, or better one submits it to a strong pressure and filters the solution; the clear liquid is heated in a water bath to 70 degrees. This temperature coagulates the greatest part of the nitrogenous matter, which one must separate then by a new filtration. The filtered liquid contains the active principle together with a little nitrogenous matter, some colouring matter and a quantity of sugar depending on the progress of the germination. In order to separate this last one pours alcohol into the liquid up to cessation of precipitation, and the diastase being insoluble disposes itself in the form of flakes which one can recover and dry at a low temperature. Finally, in order not to alter it, it is necessary above all to avoid heating it damp to 90 or 100°. To obtain it still more pure, one must dissolve it in water and precipitate it anew with alcohol, and even repeat these solutions and precipitations twice. One recovers the diastase free from nitrogenous matter without coagulating it by rise of temperature, but only by several precipitations with alcohol. After each precipitation less of this substance dissolves and the diastase becomes more and more white and pure.

Selection 7

The desire to understand what happens to food in the alimentary canal and the belief in the fermentational nature of digestion had played a major role in the development of the chemistry of life. A novel approach useful in the interpretation of digestion in the stomach and fermentation resulted from a discovery published in 1834 by a practising physician, Johann Nepomuk Eberle (1798–1834). As shown below, by treating gastric mucosa with dilute hydrochloric acid he obtained a fluid showing digestive properties outside the stomach *in vitro* ('artificial digestion').

J.N. Eberle, *Physiologie der Verdauung nach Versuchen auf natürlichem und künstlichem Wege* (Würzburg, 1834)

(*Physiology of the Digestion According to Experiments in the Natural and Artificial Manner*)

pp. 78–9

But if one takes only the mucous membrane for such experiments, then one obtains the whole acid mucus of the stomach of the living animals, which chymifies as well in the artificial way as does the stomach of an animal. Mucus from other organs, from the nose, air tubes or also from the stomach of fasting animals, does not chymify, but in the warm causes rapid putrefaction in the foodstuff. If one combines it however with hydrochloric acid or acetic acid the chymification is successful.

It follows from this that the mucus plays a chief role in the chymifaction and the associated changes in the food material, and that without it neither the acids nor the rest of the substances contained in the stomach fluid are capable of chymifying the foodstuffs. It is now clear why we meet in no system of the animal body such an amount of mucus as just in the system of the digestive organs. All simple animal substances may indeed participate to a great degree in the formation and breakdown of organic substances in the different life processes, but they are certainly not of such great importance to the common metabolism [*gegenseitigen Stoffwechsel*] as is the mucus.

One can make the acid mucus by means of hydrochloric acid, acetic acid or a mixture of these two acids. Probably other acids too, e.g. butyric acid etc. are suitable for mucilage formation; so far I have made no experiments concerning this.

For the mucus which I had artificially prepared I used the mucus membrane of the calf stomach. It was detached from the rest of the stomach membrane and then washed with cold water until it no longer had an acid reaction. Then it was dried in the open air. Now whenever I needed mucus I took a piece of this, cut it into several small pieces and placed it in a medicine bottle. Water was poured on and the bottle was closed with a cork. As I did not always have the opportunity to maintain the bottle and its contents at a measured and uniform temperature, I held it in my hand which when closed always developed a heat of 28° R. In this way the little pieces of mucous membrane soften before long, swell up and suck in all the water. They now show, except for the reddish colour which they have lost, just the appearance of a well-organized [organisirten] mucous membrane, and it has also already formed more or less mucus, which however is not acid. If one omits to add acid the mixture very quickly becomes putrid.

Selection 8

Kirchhof's decomposition of starch with sulphuric acid[a] furnished one of the early examples of a chemical operation which apparently contradicted the known laws of chemical combination. The significant feature was that in order for the change to succeed the acid had to be present even though it was not used up in the reaction. The suggestion that sulphuric acid was in effect serving as a 'contact substance', without actually participating in the chemical change, came from Eilhard Mitscherlich (1794–1863). As shown in the excerpt below Mitscherlich supposed that this type of 'contact reaction' might not only be involved in the preparation of ether from alcohol and sulphuric acid, but might also provide a clue to the understanding of fermentative chemical change in general.

E. Mitscherlich,[1] *Ann.Phys.*, [2] *1*, 279. **1834**

'Ueber die Aetherbildung' (On the formation of ether)

p. 279

The decomposition of alcohol into ether and water is not only important because ether is formed thereby, but chiefly because it provides an example of a specific chemical breakdown which one can follow so well with no other substance, and which takes place in the formation of some very important substances, for example of alcohol itself.

It has been tried to explain the ether formation through the formation of sulphovinic acid[a] which arises through the action of sulphuric acid on alcohol. It was believed at that time that this acid consists of sulphuric acid together with hydrocarbon in which are contained one part of carbon with two parts of hydrogen. Through exact researches however it has been shown that it consists of sulphuric acid and alcohol,[2] and that if one distils it without setting free sulphuric acid, alcohol is formed. For example, if one mixes potassium sulphovinate with lime, there are found at a temperature going finally to over 200°, sulphuric acid and alcohol, which contains only some oil of wine.[b]

Thus from the facts quoted it follows that alcohol in contact with sulphuric acid at a temperature of about 140° breaks down into ether and water. Decomposition and combinations which are brought about in this way appear very frequently.

We wish to term them decomposition and combination by contact. Oxidized water provides the most beautiful example; the smallest trace of manganese superoxide, of gold, of silver and other substances brings about a breakdown of the compound into water and oxygen gas which evolves, without these bodies suffering the slightest alteration. The breakdown of the sugars into alcohol and carbonic acid, the oxidation of alcohol when it is converted into acetic acid, the breakdown of urea and water into carbonic acid and ammonia belong here. By themselves these substances suffer no alteration, but through the addition of a very small amount of ferment which thereby is the contact substance, and at a definite temperature, this immediately takes place. The transformation of starch into starch sugar,[c] when one boils starch with water and sulphuric acid, is quite similar to ether formation, only that in reverse, with this sugar formation water is split, and the constituents of the same unite themselves with those of the starch in a new compound. When compounds of ether with acids, for example acetic ether, are treated with potassium solution then potassium acetate and alcohol are formed; so that here the opposite takes place, as with sulphuric acid.

Selection 9

To account for chemical reactions started by substances which do not themselves undergo any change Jöns Jacob Berzelius (1779–1848) in 1835 propounded the term 'catalysis', which eventually came into use. He did this in the course of writing his widely read and respected annual report on scientific advances in chemistry and mineralogy, presented to the Swedish Academy of Sciences: in it he also mentioned Mitscherlich's work. It says much for Berzelius' imagination that, while reflecting on the catalytic participation of diastase in the conversion of starch in living plants, he visualized catalytic reactions playing an essential part in the chemistry of living systems. Being a firm believer in the electrical interpretation of chemical combination Berzelius assumed that the hidden nature of catalysis would be one day explained on an electrochemical basis.

J. Berzelius, *Jahres-Ber.*, *15*, 237. 1836

'Einige Ideen über eine bei der Bildung organischer Verbindungen in der lebenden Natur wirksame, aber bisher nicht bemerkte Kraft' (A few ideas about a force active in the formation of organic compounds in living nature but hitherto not observed)

pp. 243–5

This is a new force for developing chemical activity: belonging both to inorganic and to organic nature, which certainly must be more widely distributed than has been thought till now and whose nature is still hidden from us. If I call it a new force it is in no way my view to explain it as a property independent of the electrochemical relationships of matter; on the contrary I can only guess that it may be a specific variety of the expression of these. Meanwhile so long as their reciprocal relationship remains hidden from us, it lightens our researches to consider it as a force in itself; similary our discussions are made easier if we have a specific name for it. Availing myself of a well-known etymology in chemistry, I will therefore name it the *catalytic force* of bodies and breakdown caused by it *catalysis*, in the same way as we understand with the word analysis the separation of the constituents of bodies by virtue of the ordinary chemical affinity. The catalytic force appears to consist intrinsically in this: that bodies through their mere presence, and not through their affinity, may awaken affinities slumbering at this temperature. So that as a result of this the elements in a complex body arrange themselves in altered relations through which a greater electrochemical neutralization is called forth. In this way they work on the whole in the same manner as does heat; and here the question can arise whether an unequal degree of catalytic force can excite in dissimilar bodies the same inequality in catalytic products as often results from heat or unequal temperatures; and so whether dissimilar catalyzing bodies can produce different catalytic products from a certain compound body? It is not yet possible to decide whether this question should be answered by yes or no. Another question is whether bodies possessing catalytic force exert this on a larger number of compound bodies or whether, as still seems at present, they catalyze certain ones without acting on others? The answer to this and other questions must be left for future investigation. Here it suffices already to have demonstrated the presence of catalytic force through a sufficient number of examples. If we turn now with this idea to the chemical processes of living nature, a quite new light dawns on us.

If nature, for example, has laid down the diastase in the eyes of the potato (Jahresb. 1835, p. 283), and this moreover is not contained in the root nodules and in the germs sprouting from them, we are led to the manner in which the insoluble starch changes through catalytic force into gum[a] and sugar; and the neighbourhood of the eyes becomes a secretory organ for soluble bodies from which the sap in the growing germs will be formed. It does not indeed follow from this that this catalytic process must be the only one in plant life; we have well-grounded reason to conjecture the opposite – that in the living plants and animals thousands of catalytic processes go on between the tissues and the fluids, and produce the amount of dissimilar chemical syntheses for whose formation from the common raw material, the plant sap or the blood, we could never see acceptable cause, which perhaps in the future we shall discover in the catalytic force of the organic tissue of which the organs of the living body consist.

Selection 10

Eberle provided a more convenient method of studying the chemical changes incident to digestion than the one employed by William Beaumont (1785–1853).[a] This American army-surgeon made use of the gastric fistula which a trapper, Alexis St Martin, acquired after being accidentally wounded by a gunshot. In view of previous findings by René Antoine Ferchault Réaumur[b] (1683–1757), Lazzaro Spallanzani[c] (1729–99), William Prout[d] (1785–1850), Friedrich Tiedemann (1781–1861) and Leopold Gmelin[e] (1788–1853), to mention a few who made important studies which advanced the understanding of the process of digestion, it is somewhat surprising that in the early 1830s the existence of gastric juice seems still to have been doubted. Beaumont's collection of gastric juice from a living man and its utilization in experimental studies of digestion removed, as it were, the last vestiges of doubt about the existence of gastric juice but it did not clear up what it was that made the stomach digest. Summarizing his results Beaumont made the suggestion, among others, that gastric juice contained in addition to free hydrochloric acid some other active chemical principles.

By 1834 the idea that gastric juice might be exclusively related to hydrochloric acid (or a combination of acids) could hardly be upheld. The presence of another constituent responsible for digestion was suspected but its nature remained undetermined. A year earlier Payen and Persoz reported the discovery of diastase.[f] It occurred to the great German physiologist Johannes Müller (1801–58), while preparing his *Handbuch der Physiologie des Menschen*, that the decomposition of starch by diastase represented a prototype of the way the unknown organic principle of the gastric juice could effect the breakdown of meat or egg white.

Müller in collaboration with Theodor Schwann (1810–82), who worked as his assistant at that time, largely confirmed the results of Eberle.[g] As shown below, continuing his researches on 'artificial digestion' Schwann established that, in addition to hydrochloric acid, gastric fluid contained another consitituent with a digestive function which he named 'pepsin' (Greek *pepsis* = digestion).

Originally Schwann's paper was published in *Poggendorffs Annalen der Physik und Chemie* [*38*, 358–64. 1838]. It was reprinted in *Annalen der Pharmacie*, with a note by Liebig of considerable historical interest. In it he emphasized the need to reconcile the concept of catalytic (contact) action with organic analysis and the laws of chemical combination.

T. Schwann, *Ann., 20*, 28. 1836

'Ueber das Wesen des Verdauungsprocesses' (On the nature of the digestion process)

pp. 29–32

Through J. Müller's and other experiments it was shown that the solution of egg white is not brought about simply by dilute acids, thus either that the acid is not active at all or that besides it something else is active. I now first observed that digestive fluid lost its activity on neutralization, that thus the acid really plays an important role in the digestion of egg white; however besides it still another substance is necessary. The experiments on the manner of working of the acid give the following facts: 1. Nearly complete neutralization of the digestive fluid, during which nothing yet is precipitated from it, puts an end to its digesting power. 2. The digestive fluid, diluted to a very high degree with acid water digests very well, but not if it is diluted merely with water. The necessary amount of acid is thus not determined by the amount of digestive principle, but by the amount of water in which it must amount to approximately 2¾ per cent (commercial hydrochloric acid). 3. The amount of free acid remains unaltered during the digestion. From these facts it can be concluded that the free acid serves not merely for the formation of the other digestive principle, nor for mere solution or for a chemical combination, but that just as with the change of starch meal into sugar it is active through contact.

The properties of the other active digestive principle besides the acid were now to be studied. It first follows from this that also the filtered, quite clear digestive fluid dissolves egg white, the latter being then soluble in dilute hydrochloric and acetic acid. As the digestive fluid, when it is neutralized and then filtered and the appropriate quantity of acid is added again, retains its digestive power, the digestive principle must have remained dissolved in the neutral fluid. If one evaporates down the neutralized digestive fluid at a low temperature (whereby the digestive power is not lost) in order to investigate its solubility in spirit of wine, and treats the residue with spirit of wine, the digestive power is completely done away with. The digestive principle is thus destroyed by spirit of wine.[a] If the digestive fluid is heated to boiling point the digestion activity is likewise lost. In order to test the behaviour of the same towards customary reagents I added these to the acid or neutralized digestive fluid, separated the precipitate by filtration from the unprecipitated constituents of the fluid, washed the former thoroughly, mixed it then again with water which contained the requisite amount (2¾ per cent) of hydrochloric acid, and added to this a reagent whose action might completely or partly abolish the action of the first reagent, e.g. hydrogen sulphide and the like. According as how the digestive power manifested itself either in the latter, the precipitate-containing fluid, or in the filtrate of the digestive fluid which contained the unprecipitated constituents, the digestive principle had to be precipitated or not by the reagent. In this way it was ascertained that lead acetate precipitates the same from the acid and still more completely from the neutral digestive fluid; that it also is precipitated by sublimate from the neutral solution, but not by potassium ferrocyanide from the acid digestive fluid. However the most characteristic reaction is the precipitation of casein or the curdling of milk.[b] The following facts show that this is caused by the digestive principle: 1. The digestive fluid causes curdling of the milk in the warm even if its quantity amounts only to 0.42 per cent, whilst of a fluid containing only hydrochloric acid at the same degree of dilution more than 3.3 per cent is necessary. 2. Also the neutralized digestive fluid causes curdling of milk. 3. At boiling heat this capacity of the neutralized digestive fluid is abolished, which agrees with the fact that such treatment, as described above, destroys the digestive principle. (The last two facts [2 and 3] taken together show that the digestive fluid and the dissolved casein can be used reciprocally as reagents on one another. A fluid which contained only 0.0625 per cent casein was still precipitated by the neutralized digestive fluid.) By all these reactions the digestive principle is characterized as an individual substance, to which I have given the name *pepsin*. Its behaviour with casein alone suffices to show its difference from other substances, especially from mucus. Mucus appears however to be the substance from which pepsin is formed by a specific transformation on treatment with dilute hydrochloric acid. At least, pure mucus prepared from saliva showed in one experiment after treatment with dilute hydrochloric acid digestive activity on protein, although very slight.

Concerning now the nature of the action of the digestive principle on egg white, it seems that this must be counted as catalytic or contact action. At least the extremely small quantity of pepsin which suffices to dissolve a large quantity of egg white speaks for this.

p. 33

The solution of coagulated egg white and fibrin by the action of pepsin together with acids is no simple solution, but at the same time a breakdown and indeed from the egg white there appears: a. a substance closely related to the coagulated egg white, which is merely dissolved in the acid and precipitates from it on neutralization of the same, b. osmazôme,[c] c. ptyalin.[d] The digestion of the fibrin yields the same products; but besides the fluid in which fibrin has been digested contain uncoagulated albuminous body [*Eiweisstoff*], which can be precipitated from it by boiling heat. Muscle flesh, as well as raw and boiled and roasted is dissolved in the same way as the pure fibrin, only with rather more difficulty.[1]

Selection 11

From the end of the eighteenth century the efforts to find a solution to the fermentation problem (also backed up by prize money) bear witness to the attempt to approach and explain it purely chemically. However from the mid-1830s evidence began to accumulate which pointed to biological aspects of fermentation. With no apparent knowledge of Erxleben's earlier reference to the part played by plant life in fermentation,[a] three independently written articles appeared during 1837 and 1838; all of them confirmed Erxleben's suggestion although they approached the subject in different ways.

Let us first consider the work of Charles Cagniard-Latour[b] (1777–1859), presented to the French Academy of Sciences on 12 June 1837 and published in 1838. More than a quarter of a century earlier Cagniard-Latour, occupied with researches on how best to prepare alcohol, began to examine fresh yeast microscopically. Due to the imperfect optical design of microscopes before the 1830s, yeast appeared to Cagniard-Latour as fine sand composed of crystalline particles. Later, with improved instruments which magnified the image 300 to 400 times and more, he observed at regular intervals the changes in samples of yeast collected from fermenting vessels. He concluded that the yeast ferment was composed of globules displaying reproduction, a characteristic feature of living matter. He described it as budding, but he seemed to differentiate this way of multiplication from an elongation of the body of the organism and its division into two, three or more globules. He also measured the size of globules by a micrometer incorporated into the microscope. From the absence of locomotion he deduced that the yeast organism was a living plant rather than an animal capable of breaking up sugar into alcohol and carbonic acid. He also showed that neither drying nor the absence of carbonic acid nor cooling (to −60°C with the aid of 'carbonic acid snow') affected the fermenting activity of yeast.

[C.] Cagniard-Latour, *Ann.Chim.*, *68*, 206. 1838

'Mémoire sur la fermentation vineuse' (Memoir on vinous fermentation)

pp. 206–7

In the year VIII[a] the class of physical and mathematical sciences of the *Institut*[b] had proposed for the subject of a prize the following question: What are the characters in vegetable and animal matter which distinguish those which are used as ferment from those which are made to submit to fermentation. The prize was a medal worth a kilogram of gold, that is to say a little more than three thousand francs. This prize was proposed anew in the year X;[c] but it was withdrawn in the year XII,[d] like all those of other classes, as the result of an event which deprived the *Institut* of the funds from which these prizes must be paid.

The question concerning fermentation, having remained without solution, can be considered as interesting now as in the time when it was the object of competition; with this in mind and believing that the competition had principally in view the most important fermentation (that of which the effect is the conversion of sugary matter to alcohol and carbonic acid, in a word vinous fermentation) I have undertaken a series of researches but proceeding otherwise than had been done. That is by studying the phenomena of this activity by the aid of a microscope.

pp. 220–2

I have examined the principal works which treat of vinous fermentation; in none of them have I seen that anyone proposes to try the use of the microscope to study the phenomena on which it depends.[1]

This attempt, as one can judge by the researches just described, was useful since it has supplied several new observations with the following principal results: 1. That the yeast of beer (this ferment of which one makes so much use and which for this reason was suitable for examination in a particular manner) is a mass of little globular bodies able to reproduce themselves, consequently organized, and not a substance simply organic or chemical, as one supposed. 2. That these bodies appear to belong to the vegetable kingdom and to regenerate themselves into two different ways. 3. That they seem to act on a solution of sugar only as long as they are living. From which one can conclude that it is very probably by some effect of their vegetable nature that they disengage carbonic acid from this solution and convert it into a spirituous liquor.

I will remark besides that the yeast, considered as organized material, perhaps merits the attention of physiologists in this sense: 1. That it can begin to grow and develop in certain circumstances with great promptitude, even in carbonic acid as in the vats of the brewers. 2. That its mode of regeneration presents pecularities which had not been observed with other microscopic productions composed of isolated globules. 3. That it is not destroyed by very considerable cold nor by deprivation of water.

In ending I will add that the question formerly proposed by the *Institut* appears now to be solved according to the results which I have just reported and several others that I have communicated to the Philomatic Society[2] during the years 1835 and 1836, for they lead to this conclusion that generally the ferments, at least those that produce vinous fermentation in the manner of the yeast, are composed of very simple organized microscopic bodies, and that the materials which they submit to this fermentation are purely chemical substances since they are, as one knows, sugar and compounds related to it.

Selection 12

The direct effect of improved microscopy after 1830 was the growing interest in 'infusoria'. These were living creatures observed by the microscope in preparations obtained by infusions of organic material, for instance hay. They were thought to belong to the animal kingdom.[a] This work revived experimental explorations undertaken in the hope of finding whether the microscopic animate world could be the product of 'equivocal' or 'spontaneous' generation. It was from the study of spontaneous generation that Theodor Schwann obtained a clue to the understanding of fermentation.

Schwann reported first on his experiments concerning spontaneous generation to the Annual Assembly of the Society of German Naturalists and Physicians, held in Jena in September 1836. No infusoria appeared in a vessel containing infusion of organic material if it was placed in boiling water.[b]

Following this, Schwann concentrated on the role of air entering into vessels containing organic infusion. He demonstrated that provided the air was heated neither mould nor infusoria appeared in the infusion of meat and, indeed, the organic material did not decompose and become putrid. Schwann perceived that these experiments did not support those who accepted the view of spontaneous generation. They could be explained on the basis that air normally contained the germs (*Keime*) of moulds and infusoria which on heating had been destroyed. Further, he comprehended that what was called putrefaction was brought about by these germs feeding on the organic substance and in the process decomposing it.

It was generally held that the phenomena of putrefaction and fermentation were closely related. Therefore Schwann decided to extend his experimental approach, which led him to connect organic decay, due to the activity of tiny organisms, to the study of respiration and fermentation. Thus he showed that a frog could make use of previously heated air.

The extract below indicates how Schwann concluded that alcoholic fermentation was promoted by proliferating yeast organisms, classified by him as sugar fungi.[c] Their feeding on sugar (and a nitrogenous body) induced the fermentation process and the liberation of alcohol and carbonic acid.

Th. Schwann, *Ann.Phys.*, [2] *11*, 184. 1837

'Vorläufige Mittheilung, betreffend Versuche über die Weingärung und Fäulniss' (Preliminary communication concerning experiments on vinous fermentation and putrefaction)

pp. 187–90

With fermentation I made the experiment in the following manner. A solution of cane sugar was mixed with beer yeast, and four small flasks were completely filled with this and corked. The flasks were then placed for the same length of time (about 10 minutes) in boiling water, so that all the fluid in them reached boiling heat. Then they were taken out, inverted under mercury, and after cooling atmospheric air was led into all four flasks amounting to about ⅓ to ¼ of the volume of the whole liquid. With two this happened through a thin glass tube which at one place was heated to redness, with the two others through a similar glass tube, but not heated. An analysis, by the aid of a platinum pellet, showed that atmospheric air, which had been led through a glowing glass tube, still contains about 19.4 per cent of oxygen. The objection that could be brought on account of the small oxygen diminution was prevented in this way: that into one of the tubes that contained heated air rather more of this was led than into the others. The flasks were then corked and were placed inverted at a temperature of 10 to 14°R. After 4 to 6 weeks fermentation appeared in both flasks which contained air not heated, and showed itself in this way that the flasks, because they were inverted, were flung away [*weggeschleudert*]. The two other flasks, even after double the time, remain quite undisturbed.[1]

It is thus with vinous fermentation, as with putrefaction, not oxygen, at least not oxygen alone of the atmospheric air, which brings these about, but a substance in the atmospheric air that can be destroyed by heat.

The thought immediately presents itself that perhaps also the fermentation may be a breakdown of sugar which is caused by the development of infusoria or some kind of plant. Since *Extr. Nucis vom. spir.*[a] is a poison for infursoria but not for mould, whilst arsenic kills not only infusoria but most kinds of mould, these substances were first used in order to ascertain in a preliminary way whether I should direct my attention more to infusoria or to plants. It happened that not *Extr. Nucis vom. spir.* but indeed some drops of potassium arsenite stopped vinous fermentation. Thus a plant was probably to be expected.

With microscopic examination of beer yeast there appear the known granules [*Körnchen*] which form the ferment; but I saw at the same time most of them hanging together in rows. They are partly round, for the most part however oval granules of a yellowish-white colour, which are present partly singly, for the most part however in rows of two to eight or still more hanging together. On such a row there stand ordinarily one or several other rows obliquely. Frequently also one sees, between two particles of a row, a small granule seated sideways as the foundation of a new row, and there is usually to be found on the last granule of a row a small corpuscle sometimes drawn out in length. In short the whole has great similarity to many organized moulds, and is without doubt a plant.

Hr. Prof. Meyen,[b] who had the kindness to examine this substance at my request, was quite of the same opinion. He expressed himself in this way, that one would only be doubtful whether it should be considered more as an alga or as a hyphomyces, which latter seemed to him more correct on account of the lack of green pigment.

Beer yeast is made up almost entirely of these fungi. In freshly pressed-out grape juice nothing of the kind is present. But if one places the latter at a temperature of about 20°R one finds after 36 hours some such plants therein, which however consist of only a few such granules. These grow visibly under the microscope, so that already after ½ to 1 hour one can observe the increase in volume of a very small granule which sits on a larger one. Only some hours later than one observes the first of these plants does gas development appear since [*weil*] the first carbonic acid remains dissolved in the water. The formation of such plants increases now very much in the course of fermentation and after its end they settle in great quantity as a yellowish-white powder on the bottom. They usually show some small differences from the fungi in beer yeast. Only some conform completely to the latter; with most others the granules approximate more to the round form and do not lie so regularly in straight lines. Finally, the number of the single granules and of those where out of a single granule only a second small granule grows, is far greater than is the case with beer yeast. Observation of its growth leaves however no doubt of its plant nature.[2]

p. 192

The connection between vinous fermentation and the development of the sugar fungus cannot be misconstrued, and it is highly probable that the latter by its development causes the phenomena of fermentation. As however for the fermentation besides sugar a nitrogen-containing body is necessary, it appears that the latter also is a condition for the life of that plant as it is already probable that the fungus contains nitrogen. One must therefore picture the vinous fermentation as the decomposition which is so brought about that the sugar fungus draws from the sugar and a nitrogen-containing body the substances necessary for its own nutrition and growth. Whereby the elements of these bodies not entering the fungus (probably amongst several other substances) combine preferentially to alcohol. Most of the observations made on vinous fermentation follow very naturally from this explanation.

Selection 13

The third contribution which directed attention in the 1830s to the biological side of fermentation came from the algologist Friedrich Traugott Kützing (1807–93).[a] In principle supporting Cagniard-Latour and Schwann, Kützing did not doubt that alcoholic and acetic fermentation were chemical activities of low plant life, associated with the formation of yeasts and an acetic fungus known as the 'mother of vinegar'. He was not prepared to say whether yeasts should be classified as algae or fungi. Kützing's description of the microscopical features of yeasts differed, in some respects, from that of the other two workers. He recognized nucleation in yeasts and suggested that precisely because of the solid nature of the 'inner nucleus' the reported outflowing from the parent yeast globules (budding) could not take place.

Thus besides linking fermentation to the life of yeasts, Kützing's article is historically significant for indicating the way their microscopic image was interpreted. But beyond that what attracts attention is the author's determination to distinguish between the inanimate and the animate. Convinced that yeasts and the 'mother of vinegar' were organic products of spontaneous generation, Kützing contended that the expression 'organic' should not be applied to the so-called organic acids, alkaloids, oils, resins. They were no more 'organic' than the inorganic materials, containing calcium compounds or silica, detected in plants.

Without denying the existence of a chemical force (affinity) in living bodies, he considered that its role was necessarily subsidiary to the dominating vital force. It was the latter which was responsible for organizing the non-living chemical components into life. Working from this assumption, Kützing discerningly recognized in fermentation a sustained process which was both biological and chemical.

The suggestion resulting from the investigations of Cagniard-Latour, Schwann and Kützing that yeast was an organism accountable for alcoholic fermentation was severely criticized by Berzelius. He admitted that Kützing's microsopical work could have contributed to the understanding of low plants but completely rejected the latter's view about which compounds had to be regarded as organic. He saw in it the detrimental expression of philosophical influence on science, thought to have been overcome a long time ago (*Naturphilosophie*).[b]

F. Kützing, *J.prakt.Chem.*, **2**, 385. 1837

'Microscopische Untersuchungen über die Hefe und Essigmutter, nebst mehreren andern dazu gehörigen vegetabilischen Gebilden' (Microscopic researches on yeast and mother of vinegar, besides several other relevant plant formations)

pp. 385–6

When I occupied myself three years ago with the microscopic investigation of the lower plant organisms forming in different kinds of fluid, I examined also yeast and mother of vinegar, products which as is well known form during vinous and sour [acetic] fermentation. At that time I was already convinced that yeast is a purely plant formation, and has not the same meaning as several other so-called chemical-organic compounds as hitherto supposed in the chemistry of the yeast.

The researches on this subject, together with some other physiological researches, I communicated already in the autumn of 1834 to Professors Horkel[a] and Ehrenberg[b] in Berlin, and mentioned them also incidentally by word of mouth to *Herr* Alexander von Humboldt.[c]

A journey to Dalmatia, Italy and Switzerland which I undertook shortly after that time was the reason that my researches were not made known. For I had it in view to incorporate these discoveries in the physiological part of an algological work which I have undertaken to prepare.

Since my return from the journey my work on this subject increased; and I had just begun to collect together the results when I was apprized by my friend *Herr* Hofrat Wallroth of the same discovery by *Herr* Cagniard-Latour in connection with yeast. Shortly afterwards (in September of this year) I received the *Annalen der Physik und Chemie* and saw the paper of *Herr* Schwann in Berlin, who had got similar results in connection with experiments on yeast.

Since now by three of us one and the same observation was made on the truly organic[1] nature of the yeast without any one having knowledge of the investigations of the others, this is to me more gratifying as I see my observations confirmed by other research workers. I willingly renounce my priority right in a discovery as it is all the same for science who first made it; and I am convinced that the communication of my observations should still not be superfluous as it still contains some facts which are not contained in the communications of *Herr* Cagniard-Latour and *Herr* Schwann.

p. 392

It is obvious that now chemistry must erase yeast from the number of the chemical compounds, because it is not a chemical compound but an organic body, an organism. Unfortunately, too many truly organized formations are being listed among the chemical compounds, where they do not belong at all.

pp. 408–9

If now we put together all the phenomena just mentioned *which in the whole of organic life repeat and run parallel,*[2] the understanding of fermentation must appear to us quite different from our understanding until now.

Fermentation is a battle between organic and inorganic life (chemism), which continues till both have reached equilibrium. Then after the organic and inorganic products formed in it have separated in pure state and set apart from one another, rest begins. In so far now as fermentation is equivalent to a reciprocal effect of generating organic and inorganic formations on the constituents of a given fluid, which can be considered as food material in relation to the organic product, so it is also necessarily synonymous with any organic life process. Therefore organic life = fermentation.

Those processes, on the other hand, which lead to the formation of acetic acid by means of platinum black or in other similar ways, cannot be compared with fermentation; they are *pure chemical processes* whilst fermentation is an *organo-chemical* process like the life process of any organic body.

Although in recent times, more than earlier, it happened that in chemical researches of organic products their organization has been considered, yet these researches give anew the proof that so-called organic chemistry cannot so much adhere to this point of view. At the same time one will always increasingly feel the need to distinguish the terms *organic* and *inorganic*, as I have just mentioned, always more exactly from one another; and indeed to limit the concept of *organic* in chemistry as much as it is absolutely necessary, in order to prevent conceptual confusions.

If I should have provided here the first inducement, I am recompensed adequately by the conviction of having been useful to science; and I gladly renounce priority rights in a discovery which in the end had to be made by the daily advance of science.

Selection 14

Berzelius was not alone in rejecting the biological interpretation of alcoholic fermentation. Thus the leading German chemist, Justus Liebig (1803–73), published in 1839 an article seeking to demonstrate from first principles that fermentation could be satisfactorily explained as a chemical transformation.

First, he pointed out that formation and decomposition of organic compounds outside the living body (and therefore not under the influence of the still accepted vital force) were governed by the unequal affinities of carbon, hydrogen, oxygen and nitrogen for each other. Then he stressed the widely recognized notion that fermentation was related to decay and putrefaction. They were regarded as spontaneous chemical changes, akin to slow combustion, affecting nitrogenous organic substances separated from the living organisms.

Decay appeared to thrive in the presence of water and air. According to Liebig what constituted decay in an organic substance was the regrouping of its elements due to their unequal attraction for the atmospheric oxygen. A rearrangement of carbon, hydrogen and nitrogen composing the organic substance could occur also, in the absence of the atmospheric oxygen, as a result of their interaction with the oxygen contained in the substance itself. It was termed putrefaction if it was accompanied by fetid odour and fermentation in the case of vegetable substances, yielding no smell, or at any rate not one with a disagreeable effect.

Pursuing this line of approach Liebig developed a special view of the nature of the ferment, the substance responsible for fermentation. Its characteristic feature was that in the process of formation it concentrated all nitrogen from the nitrogenous (albuminous) plant substances. This made the ferment particularly prone to putrefaction and therefore internally unstable. Liebig believed that the inner mechanism of fermentation was connected with the transfer of the unstable condition of the ferment to the substance to be fermented.

It should be added that neither Friedrich Wöhler (1800–82), Liebig's closest scientific companion, nor Berzelius received favourably what had come from Giessen on the subject of fermentation. In view of the fact that Liebig rejected Berzelius's concept of catalysis, Liebig was criticized for his failure to explain how the substance undergoing putrefaction actually acts as a ferment.[a]

As will be seen below, Liebig contrived to deal with the suggestion that alcoholic fermentation resulted from the biological activity of yeasts. Combatting this idea he maintained that globules of yeast, for instance, appearing during the fermentation of brewer's wort were precipitated from it. They were not living entities, but non-crystalline globular solids.

J. Liebig, *Ann.*, *30*, 250. 1839

'Ueber die Erscheinungen der Gährung, Fäulniss und Verwesung und ihre Ursachen' (On the phenomena of fermentation, putrefaction and decay and their causes).

pp. 272–3

Many nitrogen-containing materials which constitute ingredients of animals and plants suffer a progressive alteration from the moment when they cease to belong to the living organism, if they are brought into contact with air and water; they go spontaneously into decay, into putrefaction. Blood and plant juices cannot be put into contact with air without alteration of their state; oxygen is absorbed, decay begins and following this putrefaction.

Putrefaction of this material falls into several periods; the compounds which are formed in the beginning vanish towards the end of the metamorphosis; carbonic acid, ammonia, water and a body similar to humus are the final products.

Nitrogen-free organic compounds in a state of purity do not, with few exceptions, spontaneously undergo putrefaction. This metamorphosis only sets in if they are exposed to substances engaged in putrefaction, thus as a rule to nitrogen-containing substances. Rotting meat, wine, isinglass, osmazome, egg white, cheese, gliadin, gluten, legumin, blood added to water containing sugar bring about the putrefaction of sugar (fermentation).

A body that possesses this property in a remarkable degree, just on this account, has been given the name *ferment*. The ferment is considered as a chemical compound which through its contact with water containing sugar brings about a breakdown.

The manner of action was compared with that of silver and platinum on hydrogen peroxide, neither of which suffers the slightest alteration. But this opinion, on more exact consideration, only reflects the phenomenon, it gives no elucidation of the cause by means of which it is brought about.

The so-called *ferment* evolves following a metamorphosis which begins in sugar-containing plant juices on admission of air, and on shutting off the latter continues without interruption up to a certain point. In the ferment is found all the nitrogen of the nitrogen-containing constituents of the plant juice which one knows as *vegetable albumen, gluten, vegetable gluten*; it arises therefore as the result of a change which takes places in these materials; it possesses in most cases the same nature.

pp. 278–9

From the foregoing it appears that on the fermentation of the pure sugar by the ferment both together suffer a breakdown as a sequel to which they vanish completely. Their elements are arranged in new compounds. Of the sugar one knows with positive certainty that through this change carbonic acid and alcohol are formed, compounds which, judging by their elements, were contained in it except for 1 atom of water[a] which is found again in the alcohol. What the products are which are formed on the transformation of the ferment has not been investigated, only with regard to its nitrogen content one knows that it is found again as ammonia in the fermented fluid.

Thus the ferment is a decomposing, putrefying body, its ability [to act] results from contact with oxygen, through decay . . .

A certain amount of ferment is necessary in order to bring a portion of sugar into fermentation. However its action is no mass action but its influence is limited solely by its presence up to the point in time when the last atom of sugar has been broken down. The ferment as stimulant of fermentation therefore does not exist; the insoluble part does not possess this property; the soluble part which arises on its decomposition lacks it likewise. However, both materials excite fermentation from the moment when through the influence of air and water they suffer a change, the last result of which is their own destruction. There is therefore no specific body, no substance or matter which brings about breakdown, *but these are only carriers of an activity which extends beyond the sphere of the decomposing body.*

pp. 285–6

The fermentation of sugar in contact with ferment is essentially different from the fermentation of a plant juice or of beer wort. In the first the ferment vanishes with the sugar, in the other it is *formed* as an accessory to or in the metamorphosis which the sugar undergoes.

The form and nature of this insoluble precipitate has misled many physiologists to a very peculiar view about the fermentation.

Beer and wine yeast dispersed in water, examined under a good magnifying glass, display flattened transparent globules which sometimes hanging together in rows assume plant form; in the eyes of others they resemble some infusoria.

It would be certainly a highly remarkable phenomenon if vegetable gluten and albumen which separate themselves in altered state on fermentation of beer and plant juices, on this separation assumed a geometric form since these bodies never have been observed in a crystalline state. This is now not the case, they separate out like all substances which possess no crystalline nature, in the form of globules which either swim around free or hang together with one another.

These investigators of nature were misled through this form to declare the ferment as living organic beings, as plants and animals which, in order to develop, acquire the components of sugar and give [these] off in the form of carbonic acid and alcohol as waste matter. In this way they explain the breakdown of the sugar and the increase in the mass of the added ferment during beer fermentation.

Selection 15

The two interpretations of fermentation, as propounded by Liebig and Schwann, remained in dispute until the end of the nineteenth century and played a major part in the conflict between mechanists and vitalists. Under its impetus much theoretical and experimental work was undertaken by supporters of both views without, apparently, contributing to the solution of the problem of the chemical and biological nature of fermentation. In fact, as a result of focusing attention of the contradictory aspects of fermentation, real progress was made, even though the problem proved to be much more complex than the adherents of both approaches had imagined.

Among those who attacked the problem of fermentation and recognized the crucial importance of ferments in the chemistry of life was Moritz Traube (1826–94). Not the least remarkable thing about this German wine merchant was that his scientific pursuits were a side line to his business activities. The extract below is taken from Traube's earliest paper on the topic, in which he really summarized the work which he dealt with at greater length in a separate study published in the same year (1858).[a]

Traube accepted the notion that putrefaction, decay and fermentation are chemical processes related to oxidation. Compared with Liebig his work reflected clearer insight into the problem of fermentation and the action of ferments. They were not mere carriers of motion, interacting with and rearranging the fermentable substances. They were defined compounds of proteinic origin with the property of transmitting oxygen and participating in a sequence of oxidation and reduction. The designation of three groups of ferments appears to be an early attempt at their classification.

Historically important is Traube's endeavour to come to terms with Schwann's findings. While accepting fermentation as underlying (most) biochemical activities (*vital-chemische Prozesse*), he turned Schwann's approach upside down in the sense that he claimed ferments were chemical and not living entities. Moreover, thinking that they were proteins he, effectively, connected their ability to bring about fermentation with their chemical structure.

M. Traube, *Ann.Phys.*, [4] *103*, 331. 1858

'Zur Theorie der Gährungs- und Verwesungserscheinungen, wie der Fermentwirkungen überhaupt' (On the theory of fermentation and decay phenomena, also of ferment activity in general)

pp. 332–3

The true cause of the phenomena of fermentation lies in the principles developed in the following, principles partly found on direct experiment partly irrefutably concluded from already known facts:

1. The putrefaction and decay ferments are *definite chemical compounds* arising from the reaction of the protein substances with water (perhaps with the co-operation of oxygen), arising thus from a chemical process, which we are accustomed to designate shortly *putrefaction*.
   Little as with the mutability of the ferments a purification is possible, all the facts however show that they in their composition can diverge only little from the protein substances, from the reaction of which with water they originated.
2. The ferments present in the organism likewise have arisen *highly probably* from the reaction of the protein substances with water (perhaps with the co-operation with oxygen). Only, because formed under special conditions provided in the living organism, they have also other properties than those of the putrefaction ferments formed outside the organism. The Schwann hypothesis, which considers putrefaction and decay as conditioned by lower organisms, by vital processes, must be *reversed*. That is to say the power depending on the atomic composition [*atomistische Zusammensetzung*] of the protein substances to decompose water and to form ferments is also in the organisms the cause of most fermentation processes, of most vital-chemical processes altogether.
3. Amongst the ferments formed inside and outside the organism there are:
   a. Such which are capable to take up merely free oxygen with ease and to hold it only loosely bound. (*Decay ferments*);
   b. Such which also take up already bound oxygen, that is to say are

capable easily to remove oxygen from other bodies. The process of oxygen removal is in most cases the following: the ferment attracts the oxygen of the water to itself, whilst the passive substance, for example indigo or indigo-sulphuric acid, takes up the hydrogen. The water is thus through the action of two mutually supporting affinities broken down to oxygen and hydrogen respectively. (*Reduction ferments*);

c. Such ferments which even without participation of a second affinity for the hydrogen are in the state directly to split the water, whereby the hydrogen is freely evolved. These ferments, or this ferment develops in the advanced stage of putrefaction of gluten and casein. (We call it the . *highest putrefaction ferment*).

4. All these ferments have the power of transferring to other bodies the oxygen taken up in one or the other manner, that is to say to become again reduced by them and to be put into the state to take up new quantities of oxygen, again to transfer it and so on. In this way all ferments may transfer free or bound oxygen to other substances in almost endless amounts, that is to *bring about fermentation and decay.*

Selection 16

From the latter part of the 1850s a strong counter current in favour of the biological explanation of fermentation made itself felt. It arose from the work of Louis Pasteur (1822–95), a remarkable fabric woven from intellectual, experimental and social threads.[a]

Regarding alcoholic fermentation, Pasteur furnished renewed experimental evidence that it was an adjunct of yeast (cellular) life. But this is only one side of the picture. The other side that led Pasteur to support the biological view of fermentation was his discovery that racemic compounds could be resolved biologically. He drew an analogy between the action of yeast on sugar and the action of *Penicillium glaucum* on the optically inactive form of ammonium tartrate. From this followed his suggestion that ferments resembled lower plants and the specificity of the chemistry of life had to be considered as related to the asymmetric character of substances naturally occurring in the living world and derived from it.[b]

We have seen that from the end of the eighteenth century the study of fermentation had been marked by attempts to regard it in terms of oxygen participation. But not until Pasteur considered the nature of the relationship between fermentation by yeast and oxidation was the subject put on a firmer basis. As indicated below, Pasteur argued that under aerobic conditions yeast behaved more like a plant and under anaerobic conditions it acted more like a ferment. Further, he was able to show that, in fact, a quantitative interdependent relationship existed between the two processes: the amount of sugar decomposed by a unit quantity of yeast depended on whether fermentation took place in the absence of air oxygen.[c]

L. Pasteur, *Comp.Rend.*, *52*, 1260. 1861

'Experiences et vues nouvelles sur la nature des fermentations' (Experiments and new views on the nature of fermentations)

pp. 1262–4

It results from this that beer yeast has two ways of living, essentially distinct. Free oxygen gas can be totally absent, or it can be present in any volume whatever. In the second case it is used by the plant, the life of which is singularly activated. The little plant thus lives then in the manner of the lower plants; and as I have previously recognized with regard to the assimilation of carbon, of phosphates and of nitrogen, beer yeast does not offer essential differences from the moulds [*Mucédinées*]. It is well established that the yeast, placed in circumstances where it respires free oxygen gas, has a mode of life comparable in every respect with that of plants and lower animalcules. Now experience proves that the analogy goes much further, and that it extends to the disposition to ferment [*caractère ferment*]. In fact, if one determines the fermenting power of the yeast when it is assimilating free oxygen gas, one finds that this fermenting power of the yeast has almost completely disappeared.

I have little doubt that I could succeed in suppressing it completely; but what is certain is that I have already made it nearly twenty times less than it is in ordinary conditions, that is to say that for a development of yeast equal to 1 part, only 6 to 8 parts of sugar are transformed. Let us notice besides that the beer yeast that was just developing in contact with air absorbing oxygen gas, and which, under this influence and by this mode of special life, loses its disposition to ferment, has not however changed its nature. On the contrary: for if one transfers it to sugared water, protected from air, it immediately brings about there the most energetic fermentation. I have never known alcoholic yeast more active, no doubt because all the globules are budded and turgid. It is impossible to see a yeast more homogenous and more remarkable in shapes, and in health, if I may express myself thus.

To sum up, the little cellular plant, commonly called beer yeast, can develop without free oxygen gas and it is a ferment; a dual [*double*] property which separates it from all the lower beings. Or else it can develop while assimilating free oxygen gas and with such activity that one can say it is its normal life, and it loses its ferment character; a dual property which brings it near to all the low beings. But let us not forget to notice that if the yeast loses its disposition to ferment while it is multiplying under the influence of the oxygen of the air, it nevertheless puts itself back into the very same state for acting as a ferment, if one suppresses the free oxygen gas.

These are the facts in all their simplicity. Now what is their immediate consequence? Is it necessary to admit that the yeast so avid of oxygen that [the former] removes it from atmospheric air with great activity, has no more need of it and does without it when denied this gas in the free state, while provided with it in profusion in the combined form in the fermentable matter? There is all the mystery of fermentation. For if one replies to the question that I have just posed by saying: since beer yeast assimilates oxygen gas with energy when it is free, that proves that it has need of it for living, and consequently it must take it from the fermentable matter if one refuses it this gas in the free state; immediately the plant appears to us as an agent of decomposition of sugar. When with each respiration movement of its cells, there will be molecules of sugar the equilibrium of which will be destroyed by the removal of a part of their oxygen. A phenomenon of decomposition will follow, and from this the ferment character, which on the contrary will be absent when the plant assimilates free oxygen gas.

To sum up, besides all the beings known until the present day, and which, without exception (at least so one believes) can respire and nourish themselves only in assimilating free oxygen gas, there would be a class of beings whose respiration would be active enough for them to be able to live outside the influence of air in taking up oxygen from certain compounds, whence there would result for the latter a slow and progressive decomposition. This second class of organized beings would be made up of the ferments, at all points similar to the beings of the first class, living like them assimilating in their way carbon, nitrogen and phosphates, and like them having need of oxygen. But differing from them in that they would be able, in absence of free oxygen gas, to respire with the oxygen gas removed from little stable compounds.

Selection 17

By the eighth decade of the nineteenth century, studies of alcoholic fermentation and digestion had established the existence of two types of ferment. They became known as unformed (unorganized) and formed (organized) ferments. The terms reflected the prevailing view that the two groups differed fundamentally in the conditions under which they could promote or participate in chemical changes performed in living systems. Pepsin was typical of the unformed group of soluble ferments. Derived from living cells, it was looked upon as a definite chemical compound, possibly akin to a catalyst, capable of achieving chemical changes associated with gastric digestion outside the living body. Yeast embodied the class of insoluble organized ferments. It was envisaged that their fermenting activity coincided with the life of the organism itself and could not be separated from it.

In connection with the study of pancreatic digestion, as well as giving the name 'trypsin' to the already known proteolytic ferment present in the pancreatic juice, the German physiologist, Willy Kühne (1837–1900), referred in 1876 to the pepsin type of ferments as 'enzymes' (see below). Two years later, in view of some opposition[a] to his giving unorganized ferments a new name, he explained his reasons for doing so. The particular term was chosen to indicate that fermentation was due to a constituent to be found in yeast (en zyme, in Greek). The existence implied that fermentative activities known to occur in multicellular organisms (from which pepsin, trypsin and other unorganized ferments could be obtained) did not essentially differ from those of unicellular organisms.[b] Thus Schwann starting from the unicellular organisms and Kühne from the multicellular organisms arrived at a similar conclusion regarding the nature of chemical processes in the cell. They associated them with the kind of phenomena encountered in fermentations.[c]

W. Kühne, *Verhandlungen des naturhistorisch-medicinischen Vereins,* ***1*, 190. 1877**

'Ueber das Verhalten verschiedener organisirter und sog. ungeformter Fermente' (On the behaviour of different organized and so-called unformed ferments)

p. 190

Herr W. Kühne reports on the different organized and so-called unformed ferments. In order to avoid misunderstandings and burdensome transcribing the lecturer proposes to designate as *enzymes* the unformed or not organized ferments, whose activity can occur without presence of the organisms and outside the latter. More exactly investigated was especially the protein-digesting [*Eiweiss verdauende*] enzyme of the pancreas, for which, as it at the same time causes splitting of the albumin bodies [*Albuminkörper*], the name *trypsin* was chosen. The trypsin first prepared by the lecturer, and indeed free from the protein substances [*Eiweissstoffen*] which itself digests and breaks down, digests only in alkaline, neutral or very weakly acid solution.

b. From *c.* 1880 to *c.* 1940: nature of action and chemical nature

Selection 18

To students of the subject in the 1880s the nature of enzyme action appeared to be a puzzle. Though products and agents of life, enzymes were capable of acting outside living systems. Was this capability attributable to a certain amount of 'vital force' imparted by the living organs or cells to enzymes, or was it comparable with and not unlike a chemical reaction in the non-living world? In order to help to elucidate this knotty point the Irish-born, brewing-oriented Cornelius O'Sullivan (1842–1907) undertook together with Frederick William Tompson (1859–1930) to examine in detail the action of invertase[a] on cane sugar. Allowing the mutarotation to come to completion, they investigated the rate at which the inversion takes place and in what manner the various factors (such as amounts of invertase and cane sugar, temperature, alkalinity and acidity) influence it. The long paper which took more than three years to elaborate[b] also includes much information on the preparation, purification and properties of invertase.

In their work the authors utilized the earlier findings of A. Vernon Harcourt (1834–1919) who studied the quantitative aspects of the reaction between hydrogen peroxide and hydrogen iodide. They found that the rate of inversion of cane sugar by invertase gave the same kind of curve as that of Harcourt's for the chemical reaction between the organic compounds.[c] In this way they indicated that chemical changes in living and non-living matter were governed by the same laws, thus discounting the idea that a 'vital force' came into play in enzymic reactions. They also raised the suggestion at this early stage that a complex was formed between enzyme and the substrate cane sugar, the invert sugar remaining in combination with the enzyme. This was to account for the protective action of the substrate against a temperature rise of 25° above that inactivating the enzyme without substrate. A part of their summary is given below.

C. O'Sullivan and F.W. Tompson, *J.Chem.Soc.*, **57, 834. 1890**

'Invertase: a contribution to the history of an enzyme or unorganized ferment'

pp. 926–8

1. The rate of inversion of cane-sugar by means of invertase may always be expressed by a definite time-curve; this curve is practically that given by Harcourt as being the one expressing a chemical change 'of which no

condition varies, excepting the diminution of the changing substance.'
Whatever the conditions may be under which inversion is taking place, as
long as these conditions remain unchanged, this curve is adhered to.
There are, however, some slight, but apparently constant, deviations from
the theoretical curve.

2. When the degree of acidity is that most favourable for the action of
   invertase, the rapidity of the action is in proportion to the amount of
   invertase present.
3. The most favourable concentration of the sugar solution at a temperature
   of 54° is about 20 per cent. Below that, there is a rapid decline in the speed
   of inversion. Greater concentrations are only slightly less favourable until
   about 40 grams per 100 c.c. is reached. In saturated solutions inversion
   only proceeds with extreme slowness.
4. The speed of inversion increases rapidly with the temperature until 55–60°
   is reached. At 65°, the invertase is slowly destroyed, and at 73°, it is
   immediately destroyed. At the lower temperatures; the speed of the action
   increases with rise of temperature in accordance with Harcourt's law, the
   rate being about doubled for 10° rise, but above 30° the increase is not
   nearly as rapid. Elevated temperatures have no permanent effect on the
   activity of invertase, so long as they are not sufficiently high to destroy it.
5. The caustic alkalis, even in very small proportions, are instantly and
   irretrievably destructive of invertase.
6. Minute quantities of sulphuric acid are exceedingly favourable to the
   action, but a slight increase of acidity beyond the most favourable point is
   very detrimental. The most favourable amount of acid increases to some
   extent with the proportion of invertase, and decreases with rise of
   temperature, but we have not been able to discover on what it depends.
   We find that in studying the action of invertase, it is of the utmost
   importance that the most favourable amount of acid should be employed,
   otherwise correct results cannot be obtained. At a temperature of 60°, the
   action is almost stopped, unless exactly the right amount of acid is used,
   whilst if this factor is properly adjusted, inversion proceeds at (probably)
   the maximum speed.
7. The influence of alcohol varies in direct proportion with the amount
   present. 5 per cent of alcohol decreases the speed of the action by about
   one half.
8. The dextrose formed by the action of invertase is initially in the birotary
   state, and, therefore, the optical activity of a solution undergoing inversion
   is no guide to the amount of inversion that has taken place.
9. If a caustic alkali be added to a solution undergoing inversion, and the
   optical activity be allowed sufficient time to become constant, it is a true
   indicator of the amount of inversion that had taken place at the moment of
   adding the alkali.
10. A sample of invertase which had induced inversion of 100,000 times its
    own weight of cane-sugar was still active; and we have shown that
    invertase itself is not injured or destroyed by its action on cane-sugar.
    Under these circumstances there is evidently no limit to the amount of
    sugar which can be hydrolysed by a given amount of invertase.
11. The inversion of cane-sugar by means of invertase is a simple chemical
    change, differing in no important way from those which inorganic
    substances undergo.
12. The products of inversion have no influence on the rate of the action.

13. A solution of invertase will withstand a temperature 25° higher in the presence of cane-sugar than in its absence. From this fact we are of opinion that when invertase hydrolyses cane-sugar, combination takes place between the two substances, and the invertase remains in combination with the invert-sugar. The combination breaks up in the presence of molecules of cane-sugar.

Selection 19

Within ten years of his first publication on sugars (1884)[a], during which time he greatly illuminated the configurative relationship between them, Emil Fischer (1852–1919) turned his attention to the question of their fermentability.[b] Apart from the base phenylhydrazine for the isolation and identification of sugars, Fischer's most powerful tool in the elucidation of their configuration had been the concept of the asymmetric carbon atom. Originating in Pasteur's observations on the connection between crystallographic properties and optical activity, it was propounded in 1874 independently by Joseph Achille Le Bel (1847–1930) and Jacobus Henricus van't Hoff (1852–1911).[c]

Familiar with the work of Pasteur and others, who demonstrated that microorganisms were capable of acting specifically by consuming one or other optical antipode, Fischer undertook to investigate whether yeast also possessed specific fermentative activities. In this he was aided by the pure yeast cultivation method introduced by Emil Christian Hansen (1842–1909) in 1883.[d] Working in the Physiological Laboratory of the Carlsberg Laboratory, it was this Danish botanist who supplied Fischer and his co-worker, H. Thierfelder (1858–1930), with eight different pure yeasts. Having altogether twelve species of yeasts at their disposal, Fischer and Thierfelder obtained in 1894 an indication of a relation between the stereochemical configuration of the fermentative agent (formed and made use of by the yeast cell) and the fermentable sugar.

Following this up, Fischer examined the mode of action of enzymes, that is unorganized ferments, on glucosides.[e] For this purpose he employed a freshly prepared yeast extract, which he called invertin,[f] and the commercially available emulsin. He found that the invertin decomposed the α-methylglucoside and emulsin the β-methylglucoside only. As shown below, this led Fischer to assume that the specificity of enzyme action was determined stereochemically. In order to underline its restrictive character he employed an analogy that became renowned – that the interaction between enzyme and substrate depended on whether they correlated in the lock-and-key manner.

E. Fischer, *Ber.chem.Ges.*, 27, 2985. 1894

'Einfluss der Configuration auf die Wirkung der Enzyme' (Influence of the configuration on the activity of the enzyme)

pp. 2985–6

The differing behaviour of the stereoisomeric hexoses towards yeast has led Thierfelder and me to the hypothesis that the active chemical agents of the yeast cell can only attack those sugars with which they posses a related configuration.[1]

This stereochemical understanding of the fermentation process had to gain in probability if it were to be possible to establish similar differences also with the ferments separable from the organism, the so-called enzymes.

I have now succeeded in this in an unambiguous manner first for two glucoside-splitting enzymes, invertin and emulsin. The means for this were provided by the artificial glucosides, which according to the procedure found by me can be prepared from the different sugars and alcohols in great number.[2] For comparison however also several natural products of the aromatic series and some polysaccharides, which I considered as *glucosides of the sugars themselves* were drawn into the sphere of the research. The result of the same can be summarized in the proposition that the action of both the enzymes is in a striking way dependent on the configuration of the glucoside molecule.

pp. 2992–3

The observations suffice to prove in principle that the enzymes are as particular concerning the objects of their attack as are the yeast and other microorganisms. The analogy of the two phenomena appears in this point so complete that one must assume for it the same cause, and with this I turn back to the previously mentioned hypothesis of Thierfelder and myself. Invertin and emulsin have, as known, some similarities with the protein substances and no doubt possess like them an asymmetrically built molecule. Their restricted action on the glucosides could thus be explained also by the assumption that only with similar geometrical structure that approach of the molecules, which is necessary for the chemical process, can take place. Making use of an image, I will say that enzyme and glucoside must fit one another like lock and key in order to be able to exert on one another a chemical action. This idea has in any case acquired probability and value for stereochemical research after the phenomenon itself is transferred out of the biological into the purely chemical realm. If forms a widening of the theory of asymmetry without however being a direct consequence of the same; for the conviction that the geometrical structure of the molecule even with mirror image formations, exerts such a great influence on the game [*Spiel*] of the chemical affinities could, according to my opinion, only be gained by means of new factual observations. Experience hitherto, that salts formed from two asymmetric components can be distinguished by their solubility and melting point, certainly did not suffice for it. That one will soon find the fact first established only for the complicated enzymes also with simpler asymmetric agents, I doubt as little as the usefulness of the enzymes for the detection of the configuration of asymmetric substances.

The experience that the activity of the enzyme is limited to so high a degree by molecular geometry could also be of some use in physiological research. But more important for the latter seems to me the proof that the difference, earlier frequently accepted, between the chemical activity of the living cell and the activity of chemical agents in regard to molecular asymmetry, in fact does not exist. By this means is the analogy of 'living and non-living ferments' so frequently expressed by Berzelius, Liebig and others, in one not unimportant point again established.

Selection 20

The substitution of 'enzyme' for 'unorganized ferment' might be seen as a simple change in name. Actually it has to be seen against the background of a continuing uncertainty as to how to reconcile the chemical and biological sides of fermentative processes.

In the past, efforts to decide between the chemical and biological conception of alcoholic fermentation, by making use of crushed yeast or yeast dehydrated at high temperatures, were inconclusive.[a] A critical contribution to overcoming the contrariety grew out of an interplay of the experimental and theoretical interests of two brothers, Eduard Buchner (1860–1917) in fermentation, and Hans Buchner (1850–1902) in bacteriology and immunochemistry.

Eduard Buchner's interest in fermentation can be traced to his first-hand experience of it when he worked in a preserve and canning factory before becoming a full-time undergraduate at Munich University in 1884. There he studied chemistry under Liebig's successor, Adolf (von) Baeyer (1835–1917), and botany under Carl Wilhelm (von) Nägeli (1817–91),[b] both of whom have a place in the history of fermentational aspects of biochemistry during the period from 1870 to 1900.

Baeyer, who was awarded the Nobel Prize for his work on indigo and triphenylmethane dyes in 1905, published in 1870 an influential paper explaining fermentation as a chemical action based on the alternative process of hydration and dehydration, whereby the sugar molecule is eventually decomposed into alcohol and carbonic acid.[c]

Nägeli's treatise on fermentation, appearing in 1879, also exercised a great deal of influence on those working in the field. Although accepting that chemical (molecular) action occurs in fermentaion, Nägeli stressed that it could not be separated from the plasma, i.e. from the substance of the living cell.[d] Thus the Nägeli view represents an effort to overcome the contradiction between the chemical (Liebig) and biological (Schwann–Pasteur) approaches to fermentation.

Among problems of great theoretical and practical interest at the time was the controversy arising from Pasteur's observation on the nature of the relationship between fermentaion by yeast and oxidation. What was disputed was Pasteur's claim that the amount of sugar decomposed by a unit quantity of yeast depended on whether fermentation took place in the presence or absence of air oxygen. Under aerobic conditions yeast behaved more like a plant; under anaerobic conditions yeast acted more like a ferment.[e] It was clearly under the impetus of Nägeli (who opposed Pasteur) that Buchner undertook to examine this question by using bacteria instead of yeast. The results of the investigation, published in 1885, however, did not dislodge Pasteur's position. They showed that in the presence of free oxygen bacterial propagation was favoured, and fermentative activity was reduced (calculated per cell).[f]

In 1891 Eduard Buchner qualified as *Privatdozent*, entitling him to give courses of lectures of his own choosing at the Institute of Chemistry, headed by Baeyer. He lectured there on the problem of fermentation, which also became the subject of his research. Maintaining his interest in the still unsolved nature of the phenomenon of fermentation, Baeyer obtained a research grant for Buchner to study it, and also provided him with special laboratory facilities for this purpose.

Appreciating that the problem was how to get at the cellular contents of yeast without materially affecting their chemical state, Eduard Buchner employed sand-grinding in order to disrupt the yeast cells in 1893. Then the experiments were discontinued, partly because the method was assessed not to be new and promising – hence the financial support ceased – and partly because Eduard left to take up a post at the University of Kiel. They were resumed three years later at the Munich Institute of Hygiene, directed since 1894 by his brother Hans.

At the suggestion of Martin Hahn who held there the position of *Universitätsassistent*, yeast juice was eventually obtained by a combined method of grinding and pressing yeast cells by the employment of a hydraulic press. In order to protect the yeast juice, sugar was added as a means of preserving it.[g] The ensuing phenomenon of fermentation was recognized as such by Eduard, who meanwhile had become *ausserordentlicher Professor* of analytical and pharmaceutical chemistry at Tübingen University and, in fact, made it a subject of his *Antritts-Rede* (Inaugural Lecture) on 4 February 1897.[h]

This was almost certainly the first public report on the observation that alcoholic fermentation need not depend on living yeast cells but could be promoted by press-juice obtained from them. It looked as if this was due to a doubtless proteinic substance contained in the yeast juice possessing the property of an enzyme, which Eduard named 'zymase'.[i]

Without going into detail, Eduard Buchner covered essentially the same ground in the Inaugural Lecture as in the renowned 'Preliminary communication' that he sent off from Tübingen earlier on 9 January 1897, reaching the *Berichte der Deutschen chemischen Gesellschaft* two days later, from which the Selection is taken.

These two texts constitute authentic historical records with respect to Eduard Buchner's view of the whole problem of cell-free fermentation at the time of its discovery, and they ought to receive appropriate attention in writing on the background and early phases of the subject. From Eduard Buchner's presentation of the problem of cell-free fermentation in 1897 in its historical, conceptual and contemporary scientific context, it emerges unambiguously that he then regarded his achievement as falling within the ambit of the fermentation theme.

Kohler's suggestion that the issue of the nature of fermentation was quiescent in the 1880s and 1890s is not accurate. On the theoretical side it largely focused on the idea held by Nägeli and others that fermentation was a chemical process residing in the plasma, i.e. living matter of the yeast cell.

This in turn gave rise to experiments with the view of extracting from fresh or dried yeast, substances capable of ferment (enzyme) action. As brought out by Eduard Buchner, the work of Emil Fischer fell into this category. In 1895, collaborating with a well-known brewing microbiologist, Paul Lindner (1861–1945), Fischer published a paper describing how, by grinding with glass powder, they obtained from yeast an inverting agent ('invertin'), believed to be not a stable water-soluble enzyme but to be a part of living protoplasm.[j] It is significant that at the time of composing his Inaugural Lecture and his 'Preliminary communication', Buchner assumed that zymase was a true protein and more closely related[k] to the living protoplasm of the yeast cell than invertin.

E. Buchner, *Ber.chem.Ges.*, *30*, 117. 1897

'Alkoholische   Gährung   ohne   Hefezellen.   Vorläufige   Mittheilung'[a] (Alcoholic fermentation without yeast cells. Preliminary communication)

pp. 117–21

A separation of the fermentative action of living yeast cells has not so far been successful; in the following a procedure is described which solves this problem.

1,000 grams of brewers' yeast,[1] purified for the preparation of pressed yeast but still not treated with potato starch, is carefully mixed and then rubbed with an equal weight of quartz sand[2] and 250 grams of kieselguhr, until the mass becomes moist and plastic. The paste is then treated with 100 grams of water and, wrapped in a press cloth, is placed gradually under a pressure of 4 to 500 atmospheres: 300 cubic centimeters of press juice result. The residual mass is then again rubbed, sieved and treated with 100 grams of water. Treated once more in the hydraulic press with the same pressure, it gives another 150 cubic centimeters of press juice. Thus from 1 kilo of yeast 500 cubic centimeters of press juice are gained containing about 300 cubic centimeters of substances which are cell contents. To remove traces of cloudiness, the press juice is finally shaken with 4 grams of kieselguhr and filtered through a paper filter with repeated refiltrations for the first portions.

The press juice thus obtained is a clear, slightly opalescent, yellow liquid with a pleasant yeastlike odour. The specific gravity was once found to be 1.0416 (17°C). When this is boiled, a strong separation of coagulum occurs, so that the liquid almost completely solidifies: the formation of insoluble flakes begins at about 35–40 degrees. Even before this bubbles of gas, demonstrably carbonic acid, are observed rising and consequently saturating the liquid with this gas.[3] The press juice contains over 10 per cent dry substance. In an earlier, less well-prepared press juice there were 6.7 per cent dry substance, 1.15 per cent ash, and according to nitrogen content, 3.7 per cent protein substances.

The most interesting property of the press juice lies in this, that it can promote fermentation of carbohydrates. By mixing it with an equal volume of a concentrated cane sugar solution, there occurs after one quarter to one hour a regular evolution of carbonic acid which lasts for days. Glucose, fructose and maltose behave in the same way, but no appearance of fermentation occurs in mixtures of the press juice with saturated lactose as well as mannite solutions, just as these bodies are also not fermented by living brewers' yeast cells. Mixtures of press juice and sugar solution which have been fermenting for several days, when set in the icebox, usually grow turbid without the appearance of microscopic organisms, but at 700 times magnification they show a rather considerable number of protein curds, whose separation probably depends upon the acids resulting from the fermentation. Saturation of the mixture of press juice and saccharose solution with chloroform does not hinder the fermentation but leads to an early slight precipitation of proteins. Filtration of the press juice through a sterilized Berkefeldt kieselguhr filter, which safely holds back all yeast cells, has just as slight a destructive effect on the fermenting power; the mixture of the entirely clear filtrate with sterilized cane sugar solution even at the temperature of the icebox undergoes fermentation, albeit somewhat delayed, after about a day. If a parchment-paper bag is filled with press juice and hung in a 37 per cent cane sugar solution the surface of the bag after some hours is covered with numerous minute gas bubbles; naturally lively gas evolution was also observed inside the bag, due to diffusion of the sugar solution inward. Further experiments must determine whether the bearer of the fermenting action can actually diffuse through the parchment paper, as it seems. The fermenting power of the press juice is gradually lost with time; press juice kept five days in ice water in a half-filled flask showed itself as inactive towards saccharose. However, it is noteworthy that when treated with cane sugar solution, this fermentatively active press juice retains its fermenting power at least two weeks in the icebox. To begin with, a favourable action must be assumed from this for the action of the carbonic acid formed in the reaction in holding off the oxygen of the air; but the easily assimilable sugar could also contribute to preservation of the agent.

Too few experiments have yet been made to enable a conclusion to be drawn as to the *nature of the active substance* in the press juice. When the press juice is warmed to 40–50 degrees, carbonic acid evolution first occurs, then general separation of coagulated protein. After an hour this was filtered, with numerous refilterings. The clear filtrate still had weak fermenting power towards cane sugar in one experiment, but in a second, not any more; the active substance therefore appears either to lose its action at this strikingly low temperature or to coagulate and precipitate. Further, 20 cubic centimeters of the press juice were put into three times as great volume of absolute alcohol and the precipitate sucked off and dried over sulphuric acid in a vacuum; 2 grams of dry substance resulted, and upon digestion of this with 10 cubic centimeters of water, only the smallest part again dissolved. The filtrate from this had no fermenting action on cane sugar. These experiments must be repeated; especially also, the isolation of the active substance by ammonium sulphate will be attempted.

Up to now, the following conclusions have to be drawn for the *theory of fermentation*. First, it is proved that to bring about the fermentation process, such a complicated apparatus as represented by the yeast cell is not required. It is considered that the bearer of the fermenting action of the press juice is more truly a dissolved substance, doubtless a protein; this will be designated as *zymase*.

The view that a specially formed protein descended from the yeast cells causes fermentation was expressed as long ago as 1858 by M. Traube as the *enzyme* or *ferment* theory,[b] and it was later especially defended by F. Hoppe-Seyler.[c] The separation of such an enzyme from yeast cells, however, had not so far been successful.

It still remains questionable also whether the zymase can be numbered among the enzymes already longer known. As C. v. Nägeli[4] has already stressed, there are important differences between fermenting action and the action of ordinary enzymes. The latter is solely hydrolytic and can be imitated by the simplest chemical means. Even though A. von Baeyer[5] has recently increased our understanding of the chemical processes in alcoholic fermentation when he reduced them to relatively simple principles, the decomposition of sugar to alcohol and carbonic acid still belongs to the more complicated reactions; the loosening of carbon bonds in this manner has not been accomplished by any other method. There is also an important difference in the heat of the reaction.[6]

Invertin can be extracted with water from yeast cells killed by dry heat (heated one hour at 150 degrees), and by precipitation with alcohol it can be isolated as easily water-soluble powder. The substance active in fermentation cannot be obtained in a similar way. It may altogether no longer be present in yeast cells heated so high; alcoholic precipitation changes it to a water-insoluble modification, provided an experiment such as the above permits of a conclusion. One can hardly go wrong in assuming, therefore, that zymase belongs to the true proteins and stands closer to the living protoplasm of the yeast cell than does invertin.

The French bacteriologist, Miguel,[d] has expressed similar views as to urease, the enzyme secreted [ausgeschieden] by bacteria of so-called urea fermentation; he denotes it straightforwardly as protoplasm which lacking the protection of the cell wall acts outside, and which differs from that of the cell contents only in this.[7] Also the experiences of E. Fischer and P. Lindner[8] regarding the action of yeast-fungus [Hefepilz] Monilia candida on cane sugar belong here. This yeast-fungus ferments saccharose; but neither Ch. E. Hansen nor the authors mentioned were successful in extracting from fresh or dried yeast with water an enzyme like invertin which, from what has already been stated, could perform the splitting into glucose and fructose. The experiment went entirely otherwise when Fischer and Lindner used fresh Monilia yeast in which, by careful grinding with glass powder, first a part of the cells was opened. The inverting action was now unmistakable. The inverting agent seems here indeed not to be a stable water-soluble enzyme but to be a part of the living protoplasm.

The fermentation of sugars by zymase can occur within the yeast cells;[9] but more probably the yeast cells secrete this protein into the sugar solution, where it causes fermentation.[10] The process in alcoholic fermentation is, then, perhaps only to be considered a physiological action in so far as it is the living yeast cells which secrete the enzyme. Nägeli[11] and Löw have shown that from yeast cells in an originally weakly alkaline nutrient solution (from K_3PO_4), which later becomes neutral at 30 degrees, already after 15 hours a considerable amount of protein coagulable by heat has diffused [herausdiosmiren] out. Actually, it appears, as the above experiment shows, that zymase can pass through parchment paper.

Selection 21

Among the remarkable features of enzyme chemistry during the first two decades of the century was the development of enzyme kinetics. As mentioned earlier, O'Sullivan and Tompson demonstrated the value of the quantitative approach. Not only did it emerge from this research that the inversion of cane sugar by invertase proceeded in agreement with the Mass Law but evidence was also provided for the combination of the enzyme with the substrate and also with the products of the hydrolysis.[a]

Further light was thrown on these questions by Edward Frankland Armstrong (1878–1945). He suggested that the enzyme and the sugar, linked together in the manner of the 'lock-and-key' relationship envisaged by E. Fischer, formed an intermediate active system.[b]

As shown in the extract below, he published fresh evidence for this view in 1904. He examined the rate of hydrolytic change by certain sugar-splitting enzymes after additions of hexoses. He found that in those cases where the hexoses were decomposition products of hydrolysis they inhibited the progress of the enzyme reaction. Although the possibility of reversibility of enzyme action was perceived, the subject was in its infancy at that time.[c]

E.F. Armstrong, *Proc.Roy.Soc.*, *73*, 516. 1904

'Studies on enzyme action. III. The influence of the products of change on the rate of change conditioned by sucroclastic enzymes'

p. 516

The experiments to be described have been made with the object of ascertaining by direct observation whether and to what extent the action of a given enzyme is affected by one or more of the products formed under its influence. They establish very clearly the existence of a close relationship between the configuration of the hexose and the enzyme in those cases in which a retarding influence is apparent: it is difficult to explain such a result except on the assumption that the enzyme and hexose combine together in some intimate manner.

p. 520–1

Correlation of Differential Action with Configuration of Hydrolyte. Combining my results with EmilFischers[a] earlier observations the following table is arrived at.

The compounds entered in the second column of the table are those which, according to Emil Fischer[a], are alone hydrolysed by the particular enzymes indicated in the first column. My own observations on the specific retarding influence of the various hexoses are entered in the remaining columns. It is clear that the only hexoses which retard hydrolysis by any given enzyme are those derived from the hexosides[1] which undergo hydrolysis under the influence of that enzyme.

A more absolute proof of the close correlation in configuration between enzyme and hydrolyte cannot well be imagined: it is difficult to interpret such behaviour in any other way than as evidence that the enzyme combines with the hexose in some special, peculiarly intimate manner and is thereby withdrawn from the sphere of action. The retardation cannot well be due to reversion, as in the case of milk sugar the retardation is effected chiefly by glucose when emulsin is the active agent but by galactose alone when lactase is used to effect hydrolysis.

Enzyme	Corresponding hydrolyte	Effect of hexose on rate of change		
		Glucose	Galactose	Fructose
Lactase	β-Galactosides (i.e., milk sugar, β-alkyl galactosides)	No influence	Retards	No influence
Emulsin	β-Glucosides (i.e., most natural glucosides, β-alkyl glucosides) β-Galactosides (as above)	Retards considerably	Retards slightly	No influence
Maltase	α-Glucosides (i.e., maltose, α-alkyl glucosides) α-Galactosides (i.e., α-alkyl galactosides)	Retards considerably	Retards slightly	No influence
Invertase	Fructosides[2] (i.e., cane-sugar, raffinose, gentianose, manneotetrose)	No influence	—	Retards

Selection 22

That in addition to enzyme and substrate the presence of another substance, the so-called 'coferment', could be required for the enzymic reaction to occur had been known since at least 1897.[a] A more general interest in the subject grew out of the research of Arthur Harden (1865–1940), in collaboration with William John Young (1878–1942), at the Lister Institute during the first decade of this century. It originated in attempts by the head of the Institute, Allan Macfadyen (1860–1907), to produce anti-ferments by injecting animals with yeast juice. More particularly, it could be traced back to Harden's studies of the effect of adding serum of the treated animals to the mixture of yeast juice and sugar. It was then that Harden observed that about 60 to 80 per cent more sugar was fermented in the absence of the serum.[b]

Their first extensive paper in 1906[c] comprised experimental details regarding the effect of boiled and filtered yeast juice on the fermentation of glucose, including investigations with fractions obtained by dialysis of yeast juice. Important features of their experimental technique were the use of a porcelain filter saturated with gelatine[d] and the volumetric measurement of the amount of carbon dioxide evolved.

Harden and Young showed that yeast juice could be separated by dialysis into a residue and a dialysate, both by themselves inactive but capable of fermenting the glucose when brought together. They established that the dialysate contained a thermostable substance promoting fermentation. In attempting to isolate this component of boiled juice they found that an increase in fermentation was linked to the presence of soluble phosphates.

Also if these were added to yeast juice which was fermenting glucose, a rapid evolution of carbon dioxide followed.

It was in the second paper (in 1906) that Harden and Young employed the term 'coferment' for the heat-stable dialysable constituent of the boiled yeast juice without which alcoholic fermentation could not take place. They observed that the coferment disappeared from yeast juice during fermentation or when the latter underwent autolysis. It was also in this publication that they tackled the relationship of the coferment and phosphates in alcoholic fermentation. As shown in the excerpt below Harden and Young demonstrated that the addition of inorganic phosphates to the inactive residue did not restore fermentation. In other words these phosphates were not to be considered in the same light as the chemically as yet undetermined coferment of alcoholic fermentation appearing to be an organic phosphate.

A. Harden and W.J. Young, *Proc.Roy.Soc.*, (B), *78*, 369. 1906

'The alcoholic ferment of yeast-juice. Part II. The coferment of yeast-juice'

pp. 373–5

In view of the fact that soluble phosphates, as described in the previous communication,[a] exert a remarkable effect on the fermentation of glucose by yeast-juice, experiments were made to ascertain whether the addition of a soluble phosphate to a solution of the inactive residue in glucose is sufficient to set up fermentation. All the attempts hitherto made to effect this have yielded entirely negative results, although both the kind of phosphate and the amount added have been varied.

Dipotassium hydrogen phosphate, a mixture of this with the dihydrogen phosphate, disodium hydrogen phosphate, diammonium hydrogen phosphate, microsmic salt, and a mixed phosphate of potassium and magnesium obtained by boiling a solution of potassium dihydrogen phosphate with magnesium carbonate, were employed, all of which are capable of producing the characteristic effect of phosphates on the fermentation of glucose by yeast-juice.

Although these substances did not set up fermentation when added to an inactive filtered residue and glucose, they did not affect the potential activity of the residue. This is evident from the fact that the subsequent addition of boiled juice produced an immediate fermentation.

In every case the solution of the phosphate was saturated with carbon dioxide at 26° and added to the solution of the inactive residue in glucose solution also saturated with carbon dioxide at 26°, and in no case was any evolution of gas observed after the cessation of the slight disturbance which inevitably occurs when the solutions are mixed. The phosphate solutions were all of $3/10$ molar strength, with the exception of the mixed phosphate of magnesium and potassium, 5 c.c. of which yielded 0.1594 gramme $Mg_2P_2O_7$.

In those cases in which the residue employed was slightly active, incubation was continued until all fermentation had ceased before the phosphate solution was added, and the amount of carbon dioxide evolved in this preliminary

	Amount of residue	Preliminary fermentation	Phosphate added	Carbon dioxide produced	Subsequent evolution of carbon dioxide after the addition of boiled juice
1	25 c.c.	41.2	2.5 c.c. KMg phosphate	0	18.4
2	25 "	47.7	5 " "	0	—
3	25 "	0.6	3 " KHPO₄	0	268.8
4	1.6 gr.	21.5	3 " KMg phosphate	0	159.2
5	1.2 "	10.7	5 " "	0	—
6	0.5 "	0	2 " mixture of K_2HPO_4 and KH_2PO_4	0	3 c.c. per hour
7	0.5 "	0	2 " $(NH_4)_2HPO_4$	0	4.2 " "
8	0.5 "	0	2 " $(NH_4)NaHPO_4$	0	3.8 " "
9	0.5 "	0	2 " Na_2HPO_4	0	3.6 " "

period is given in the table. The amount of glucose, both in the solution of the residue and in that of the phosphate, was throughout 10 grammes to 100. The foregoing conclusion is confirmed by the observation, made in the experiments on the disappearance of the coferment from yeast-juice, that boiled autolysed juice does not set up fermentation in a mixture of the inactive residue with glucose, although it itself contains a large amount of phosphate precipitable by magnesia mixture. The following numbers were obtained by the analysis of three specimens of boiled autolysed juice employed in those experiments, two of which were quite inactive, whilst the other only produced a fermentation of 2.6 c.c.

No. of experiment	Volume of juice	Carbon dioxide evolved	Phosphate present in grammes of $Mg_2P_2O_7$
	c.c.	c.c	
1. b	20	0	0.3100
2. b	16	2.6	0.3011
3. c	15	0	0.1893

These experiments throw no light on the actual chemical nature of the coferment, but show that most probably it does not consist of a phosphate precipitable by yeast-juice. They also indicate that substances, which, like phosphates, increase the total fermentation produced by yeast-juice, are not necessarily capable of setting up fermentation when added to a mixture of inactive residue and glucose.

Selection 23

Following the developments in electrochemistry at the turn of the century, the significance of the acid-base equilibrium began to be more fully appreciated and received attention.[a] The use of the theory of ionization made it possible to link acidity (alkalinity) of aqueous solutions to the concentration of hydrogen ions.

Considerations such as these, taken in connection with the development of the concept of buffers,[b] led to a profounder study of the influence of acidity on enzyme action. Of all the work done on this subject during the first decade of the century none became more influential than the paper of Søren Peter Lauritz Sørensen (1868–1939), published in 1909. Carrying out a critical examination of methods of determination of hydrogen ion concentration, Sørensen pointed out suggestively that the part played by the latter in enzyme reactions was comparable to that of temperature (see below). His designation of the hydrogen ion concentrations in terms of their negative logarithms to the base 10 (pH) made this idea tangible. From it the study of the conditions under which the watery medium of living systems operates received a decisive impulse.

S.P.L. Sörensen, *Biochem.Z.*, *21*, 130. 1909

'Enzymstudien. II. Mitteilung. Über die Messung und die Bedeutung der Wasserstoffionenkonzentration bei enzymatischen Prozessen' (Enzyme studies. II. Communication. On the measurement and the significance of the hydrogen ion concentration in enzymatic processes)

pp. 133–5

If one indicates the concentration of hydrogen ions, of hydroxyl ions and of water by $C_{H^.} \times C_{OH'}$ and C_{H_2O} respectively, then as is known, in virtue of the law of mass action, the following equation holds:

$$\frac{C_{H^.} \times C_{OH'}}{C_{H2O}} = \text{constant.}$$

As C_{H_2O} is to be regarded as constant also for solutions diluted only to some degree, also the product

$$C_{H^.} \times C_{OH'} = \text{constant.}$$

This product, which is usually, also in this paper, called *the dissociation constant of water* is set at 18° as equal to 0.64×10^{-14}; in a series of measurements which were carried out here in the laboratory . . . we have found the mean value 0.72×10^{-14}, or otherwise written $10^{-14.14}$. The value makes it possible, as may easily be seen, to calculate the hydrogen ion concentration of an aqueous solution if the concentration of the hydroxyl ions is known and *vice versa*. Because naturally the dissociation constant of water is subject to errors, and because further the concentration of hydrogen ions can usually be estimated more exactly and easily than that of the hydroxyl ions: it is rational, as H. Friedenthal[1] has proposed, as far as possible always to estimate and calculate the hydrogen ion concentration of a solution also then when the solution has an alkaline reaction. This procedure is therefore also used in the following, where for example a solution of which the normality referred to hydrogen ions is found equal to 0.01, or is 0.01 n, with neglect of the normality notation is shortly labelled 10^{-2}. In the same way the hydrogen ion concentration of a solution (which is 0.01 n referred to hydroxyl ions) is given by $10^{-12.12}$, since $10^{-12.14} \times 10^{-2} = 10^{-14.14}$. Completely pure water and really neutral solutions with the use of this manner of expression, have a hydrogen ion concentration of $10^{-7.07}$, since $10^{-7.07} \times 10^{-7.07} = 10^{-14.14}$.

The value of the hydrogen ion concentration is accordingly expressed by the normality factor of the hydrogen ions of the solution concerned, and this factor is written in *the form of a negative power of 10* . . . I will here only add *that I use the name 'hydrogen ion exponent' and the symbol pH for the numerical value of the exponent of this power.* In the three examples quoted above pH is respectively 2, 12.14 and 7.07 . . .

The hydrogen ion concentration plays in enzymatic processes a role quite similar to that of temperature. By the *temperature curve of an enzyme* one usually understands the curve which has as ordinate the substrate amounts split under given conditions in the time interval whilst the experimental temperatures serve as abscissae. Such a temperature curve thus gives information on the rate at which the substrate is split at different temperatures but with otherwise similar conditions; and shows further, as is well known, that there is for each single enzyme a definite temperature, the optimal temperature, at which the splitting reaches its greatest rate. In the neighbourhood of the optimal temperature – in the optimal zone – the splitting goes with about the same high rate as at the optimal temperature itself; with a temperature above or below the optimal temperature, however, the rate usually falls on both sides especially with rising temperature, so that the substrate is practically not split if the temperature is sufficiently low or high. That the rate of splitting decreases with falling temperature has certainly no foundation in the destruction of the enzyme; on the other hand one can scarcely doubt that the great fall in the temperature curve at temperatures above that of the optimal zone is due to destruction of the enzyme happening under these conditions. The temperature curve is thus to be regarded as the difference between two curves: the true temperature curve which not only below but probably also above the optimal zone will run uniformly rising; and on the other hand the destruction curve of the enzyme which at low temperatures below the optimal has smaller ordinates;[2] whilst with higher temperatures it rises greatly, soon to take a course almost parallel with the ordinate axis.[3]

It is easy to see that a curve which has common abscissae with these two curves – the 'true temperature curve' and the destruction curve of the enzyme – whilst its ordinates are equal to the difference of the ordinates of these curves will run quite like the usual temperature curve of an enzyme. Quite similar relations hold also for the *hydrogen ion concentration curve* of an enzyme; by this is meant the curve which one obtains if one uses as ordinates the substrate amounts split in the time interval under the given experimental conditions while the hydrogen ion component of the experimental fluid functions as abscissae. Such a hydrogen ion concentration curve which makes clear the splitting rate at different hydrogen ion concentrations, if all else remains unchanged . . . has quite like the temperature curve its optimal point, its optimal zone and its falling off both with increasing and decreasing hydrogen ion concentration. Whether the analogy between the influence of temperature and that of hydrogen ion concentration goes still further: so that for example the descent of the one curve branch depends on a slower action of the enzyme at the corresponding ion concentration, whilst the fall of the other branch has its cause in an increasing destruction of the enzyme at the corresponding ion concentration; or whether, perhaps, with both branches of the curve cause of sinking lies in the latter circumstance. This has not yet been experimentally investigated by us, but investigations in this connection are being undertaken in the laboratory.

By this comparison between the influence of temperature and that of the hydrogen ion concentration – a comparison which on consideration of the hydrogen ion concentration curves obtrudes itself – I have wished to emphasize as strongly as possible that knowledge of the hydrogen ion concentration of the fluid in which an enzymatic process takes place is of as decisive significance as knowledge of any other factor, for example the temperature, which influences the rate of the process.

Selection 24

Sørensen's concept of the influence of pH was used effectively in the development of a rate equation for enzyme reactions. This is recognizable when the work of Victor Henri (1872–1940), published in 1903[a] is compared with that of Leonor Michaelis (1875–1949) and Maud Leonora Menten (1879–1960), which appeared in 1913.[b]

In developing mathematical expressions describing the course of enzyme changes Henri treated enzymes as catalysts. He utilized existing suggestions about the union of the enzyme with its substrate and the dependence of enzyme catalyzed processes upon the concentration of the enzyme and its substrate.

Henri's work influenced the investigations of the kinetics of invertase action undertaken by Michaelis and Menten. By taking the hydrogen ion concentration factor and the glucose mutarotation effect, which Henri disregarded, into consideration, they developed a mathematical theory of enzyme processes by considering the action of invertase on cane sugar.

They assumed that free invertase and sucrose are in equilibrium with a labile sucrose-invertase compound. By using a rather unwieldy graphical method they established a value of 0.0167 or $\frac{1}{60}$ for the dissociation constant of the sucrose-invertase compound. Further, they held that the invertase also formed compounds with the products of inversion and gave the values of $0.088 = \frac{1}{11}$ and $0.058 = \frac{1}{17}$ for the dissociation constants of the glucose-invertase and fructose-invertase compounds respectively.

Picturing the breakdown of the sucrose-invertase compound as a monomolecular reaction, they took the rate of breakdown of the sucrose at any time to be simply proportional to the concentration of the sucrose-invertase compound.

These, then, were the assumptions which led, as shown below, to the derivation of a differential equation for the fermentative splitting of cane sugar, the integral of which was tested with observed values of the initial amount of cane sugar, and the amount that had disappeared.

L. Michaelis and M.L. Menten, *Biochem. Z.*, *49*. 333, 1913

'Die Kinetik der Invertinwirkung' (The kinetics of invertase activity)

pp. 364–8

3. *The equation of the reaction of fermentative splitting of cane sugar*
... we are now in a better position to solve the old problem of the reaction
equation of invertase in a rational manner and without help of more than one
arbitrary constant. Until now V. Henri has come nearest of all authors to this
solution, and we can consider our derivation as a further modification, on the
grounds of newly acquired knowledge, of the Henri derivation.
    The fundamental assumption of this deduction is this, that the rate of
breakdown at any moment is proportional to the concentration of the sucrose-
invertase compound; and that the concentration of this compound at any
moment is determined by the concentration of the ferment, of the sucrose, and
also of the split products capable of binding of the ferment. However whilst
Henri brought into consideration an 'affinity constant of the split products', we
operated particularly with the dissociation constant of the sucrose-ferment
combination, $k = \frac{1}{60}$, with that of the fructose-ferment combination, $k_1 = \frac{1}{17}$,
and with that of the glucose-ferment combination, $k_2 = \frac{1}{11}$.
    We apply besides the following symbols:

Φ = the total ferment concentration
φ = the concentration of the sucrose-ferment combination
ψ_1 = the concentration of the fructose-ferment combination
ψ_2 = the concentration the glucose-ferment combination

S = the concentration of sucrose ⎫ that is to say, the concentration of the
F = the concentration of fructose ⎬ sugar in question in the *the free state*
G = the concentration of glucose ⎭ which however is practically equal to
  that of the total sugar in question.

Since upon splitting just as much fructose as glucose is formed, G is always
equal to F.
    Now in any moment according to the law of mass action

$$S(\Phi - \varphi - \Psi_1 - \Psi_2) = k\varphi \tag{1}$$

$$F(\Phi - \varphi - \Psi_2 - \Psi_2) = k_1\psi_1 \tag{2}$$

$$G(\Phi - \varphi - \Psi_1 - \Psi_2) = k_2\psi_2 \tag{3}$$

From (1) it follows

$$\varphi = \frac{S(\Phi - \Psi_1 - \Psi_2)}{S + k} \tag{4}$$

We can eliminate φ_1 and φ_2, if we first find by division of (2) and (3):

$$\psi_2 = \frac{k_1}{k_2} \cdot \psi_1,$$

and further by division of (1) and (3)

$$\psi_1 = \frac{k}{k_1} \cdot \varphi \cdot \frac{F}{S},$$

so that

$$\psi_1 + \psi_2 = k \cdot \varphi \cdot \frac{F}{S}\left(\frac{1}{k_1} + \frac{1}{k_2}\right).$$

Let us refer next to the abbreviation

$$\frac{1}{k_1} + \frac{1}{k_2} = q,$$

so that

$$\psi_1 + \psi_2 = k \cdot q \cdot \varphi \frac{F}{S}.$$

This gives, substituted in (4) and solved for φ,

$$\varphi = \Phi \cdot \frac{S}{S + k(1 + qF)} \tag{4}$$

Now we can arrive at the differential equation. If a is the initial amount of sucrose, t the time, x the amount of fructose *or* glucose present at time t, that is $a - x$ the amount of sucrose still present at time t, then the breakdown rate at time t is defined by

$$v_t = \frac{dx}{dt}.$$

According to assumption this is proportional to φ, so that our differential equation with use of equation (4) runs:

$$\frac{dx}{dt} = C \cdot \frac{a - x}{a + k - x(1 - kq)} \tag{5}$$

where C represents the single arbitrary constant; and indeed this is proportional to the amount of ferment.

The general integral of the equation is given without mathematical difficulty:

$$C \cdot t = (1 - kq)x - k(1 + aq)\ln(a - x) + \text{constant}.$$

For the elimination of the integration constants we give below the corresponding equation for the initial state of the process, for which $x = 0$ and $t = 0$:

$$0 = k(1 + aq)\ln a + \text{constant},$$

and find finally by subtraction of the two last equations the definite integral:

$$C \cdot t = k(1 + aq) \cdot \ln \frac{a}{a - x} + (1 - kq)x \tag{6}$$

or with introduction of the value for q:

$$\frac{k}{t}\left(\frac{1}{a} + \frac{1}{k_1} + \frac{1}{k_2}\right) \cdot a \cdot \ln \frac{a}{a - x} + \frac{k}{t}\left(\frac{1}{k} - \frac{1}{k_1} - \frac{1}{k_2}\right)x = C.$$

Now one can as well include k in the constant on the right side and obtain:

$$\frac{1}{t}\left(\frac{1}{a} + \frac{1}{k_1} + \frac{1}{k_2}\right) \cdot a \cdot \ln \frac{a}{a - x} + \frac{1}{t}\left(\frac{1}{k} - \frac{1}{k_1} - \frac{1}{k_2}\right)x = \text{constant} \tag{7}$$

It is characteristic for this function as for that of Henri, that it gives by superposition a linear and a logarithmic function of the type

$$m \cdot \ln \frac{a}{a - x} + n \cdot x = t \cdot \text{constant} \tag{8}$$

where m and n have the meaning apparent from the previous equation; they are factors whose magnitude is determined by the particular dissociation constants and the initial amount of sugar.

If we use the values of k, k_1 and k_2 obtained by us at the experimental temperature of 25°, we have

$$\frac{1}{t}(1 + 28a) \cdot 2{,}303 \log^{10} \frac{a}{a - x} + \frac{1}{t} \cdot 32 \cdot x = \text{constant} \tag{9}$$

Instead of $\log \dfrac{a}{a - x}$ we write more simply for the calculation $-\log\left(1 - \dfrac{x}{a}\right)$.

This constant must be proportional to the ferment concentration. That it is so appears from all earlier investigations, and especially it was shown by L. Michaelis and H. Davidson[a] . . . that an equation of the form

$$\text{amount of ferment} \times \text{time} = f(a, x) \tag{10}$$

strictly proved correct. The function left undetermined on the right-hand side of this equation is given a definite form by means of our equation (8), otherwise nothing is altered and one sees forthwith by comparison of (8) and (10) that the constant of equation (8) must be proportional to the ferment concentration.

Thus it is superfluous to test the correctness of equation (9) with changing amounts of ferment; but it must be tested once more whether this constant remains the same with the same ferment amount but changing sugar amounts; and above all whether during a single experiment it is independent of the time. We use now . . . the values of x, which hitherto we described in arbitrary polarimetric units of measure, and have to convert them into moles. We make use in this of the assumption that the theoretical final rotation of a sucrose solution, which at first rotates m°, amounts to $-0.313\,m°$.[b]

Time (t)	$\frac{x}{a}$	K	Mean
	I. Sucrose 0,333 n.		
7	0,0164	0,0496	
14	0,0316	0,0479	
26	0,0528	0,0432	
49	0,0923	0,0412	
75	0,1404	0,0408	
117	0,2137	0,0407	
1052	0,9834	[0,0498]	0,0439
	II. Sucrose 0,1667 n.		
8	0,0350	0,0444	
16	0,0636	0,0446	
28	0,1080	0,0437	
52	0,1980	0,0444	
82	0,3000	0,0445	
103	0,3780	0,0454	0,0445
	III. Sucrose 0,0833 n.		
49,5	0,352	0,0482	
90,0	0,575	0,0447	
125,0	0,690	0,0460	
151,0	0,766	0,0456	
208,0	0,900	0,0486	0,0465

Time (t)	$\frac{x}{a}$	K	Mean
	IV. Sucrose 0,0416 n.		
10,25	0,1147	0,0406	
30,75	0,3722	0,0489	
61,75	0,615	0,0467	
90,75	0,747	0,0438	
112,70	0,850	0,0465	
132,70	0,925	0,0443	
154,70	0,940	0,0405	
1497,00	0,972	[0,0514]	0,0445
	V. Sucrose 0,0208 n.		
17	0,331	0,0510	
27	0,452	0,0464	
38	0,611	0,0500	
62	0,736	0,0419	
95	0,860	[0,0388]	
1372	0,990	[0,058]	0,0474

Average mean value 0,0454.

The constant is in all experiments so consistent and, apart from slight fluctuations, shows no 'slant' either with time or with sugar concentration, that we can consider it as satisfactorily constant.

Selection 25

The progress in enzyme kinetics between 1900 and 1920 was not paralleled by quantitative work in other branches of enzyme chemistry. This was, among other things, due to difficulties (the First World War, conceptual uncertainties) in obtaining pure preparations, with the consequence that no secure experimental basis existed for the elucidation of the chemical nature of enzymes. At that time most investigators inclined to think of them as proteins, especially as their colloidal nature was not in doubt. Among those who came to oppose the belief in the protein nature of enzymes was the great German scientist Richard Willstätter (1872–1942), known already for his investigations on chlorophyll.[a]

Among the characteristic properties attributed to colloids was the display of adsorption phenomena. Willstätter recognized in adsorption techniques a viable method of purifying enzymes and developed its employment during the 1920s. In his investigations of enzymes Willstätter was guided by the assumption that an enzyme was composed of a colloidal carrier (of perhaps slightly varying composition) and a specific active group. The isolation of enzymes from non-enzymatic substances depended on their relatedness in their colloidal and chemical properties. Bodies from which the enzymes were first formed or products resulting from their breakdown were thought to resemble enzymes most closely in their properties and therefore to be difficult to separate from them.

By paying attention to conditions of dilution and acidity, to suitable adsorbents and also to fractional adsorption Willstätter and his associates obtained preparations of higher purity and activity, compared with previous work. For instance, on a dry weight basis, they were able to increase the degree of purification of horse-radish peroxidase about 12,000 times and that of yeast invertase about 4,000 times.[b] The purity of enzymic preparations was judged by negative tests, i.e. the vanishing reactions identifying proteins, carbohydrates and other substances assumed to accompany enzymes as admixtures. Preparations of this degree of purity seemed to support the view promoted by Willstätter that enzymes were, in fact, chemical individuals of a particular as yet unknown kind.

The extract below is from a review of the subject given by Willstätter at the commemoration of the 100th anniversary of the Annual Assembly of the Society of German Naturalists and Physicians held at Leipzig in 1922. It presents Willstätter's characteristic point of view on adsorption and enzymes to which he remained faithful until the end of his life.[c]

R. Willstätter, *Ber. chem. Ges.*, 55, 3601. 1922

'Über Isolierung von Enzymen' (On isolation of enzymes)

pp. 3610–11

Significance of the adsorption methods for isolation

For their isolation from the mixtures with large amounts of foreign substances, namely of proteins, carbohydrates and salts, the enzymes offer, so far as one knows, no points of attack by chemical means. Disregarding the reaction capability of their specifically active groups they are chemically indifferent substances. It does indeed frequently happen that they are separated in precipitation reactions, for example invertin by lead acetate or uranyl acetate, but the enzymes are in general only adsorbed by precipitates.[1] As the precipitate formation is to be ascribed not to the enzyme itself but to the accompanying substances, it thus depends too much on the changing, more accidental nature of the enzyme containing solution. Further, making the isolation difficult is the solubility behaviour of the enzymes unfavourable for the separation of foreign bodies, since the former are insoluble in many organic solvents, particularly in those not miscible with water.

There is therefore only one *single general method for the isolation of the enzymes, the use of the adsorption processes depending on small amounts of affinity residues*. It is capable of adjustment and development and manifold, as is the nature of the adsorption phenomena themselves. In the first more exactly investigated applications of this method some essentially different enzymes have been brought to a high degree of purity, e.g. the cane sugar-splitting (*yeast-saccharase*), a fat-splitting (*pancreas-lipase*), a starch-splitting (*pancreas-amylase*), an oxidation enzyme (*plant peroxidase*). In this way three tasks have become detectable the discharge of which will be decisive for our knowledge of the enzymes.

One task consists in this, to shape the adsorptions more selectively. The adsorbent generally takes up an enzyme together with a considerable amount of foreign bodies, which are partly simply mixed with it, for the other part tightly bound with it. The smaller the necessary amount of adsorbent, the higher the degree of purity will be in the adsorbate.

The second task, *separation of enzymes*, is set by e.g. the investigation of the *pancreas*. The gland produces chiefly a fat-, starch- and protein-splitting enzyme. In order to learn to recognize the behaviour of one of them it does not suffice to describe the mixture out of the pancreas by means of only *one* activity and according to this to name it. More correctly one will consider as a preliminary condition for the investigation of one of the pancreatic enzymes, the separation of the accompanying enzymes (according to their action quite different) and develop the adsorption methods for this.

The increase of the degree of purity of the enzyme is in the beginning easily followed with the disappearance of the sensitive reactions for proteins, purines, carbohydrates and the like. Then appears the third task which is still almost totally unsolved and which until now only in the example of peroxidase was taken up with some success. This is the *separation of the enzymatically inactive portion* which does not betray itself through chemical reactions, colour or precipitation phenomena. It is still not certainly known how far the adsorption means of our methods may select between the active enzymes on the one hand, their pre-existing stages and the products obtained on their decomposition on the other hand. Probably the most obstinate companions are the transformation products, in the colloid properties most closely related to the enzyme, distinguishing themselves from it by the lack of the active specific group.

Selection 26

Despite improvements in the purification of enzymes brought about by adsorption techniques and also by attempts to apply other methods (precipitation, dialysis, cataphoresis), progress was slow. Among the few early successful examples, as shown below, belongs the purification of xanthine oxidase of milk by Malcolm Dixon (1898–1985) in collaboration with Keizo Kodama in 1926. The work is noteworthy for the way the precipitation and adsorption methods were combined. In order to separate the principal protein in milk from the xanthine oxidase it was coagulated by rennin. Adsorption was employed in further efforts to concentrate the enzyme.

M. Dixon and K. Kodama, *Biochem. J.*, *20*, 1104. 1926

'On the further purification of xanthine oxidase'

p. 1104

Dixon and Thurlow[a] obtained an active and stable oxidase preparation by precipitating the enzyme, together with the caseinogen and fat, by half-saturation of the milk with ammonium sulphate. The fat was removed from the dried precipitate by thorough extraction with ether. The resulting preparation is soluble, contains the whole amount of the enzyme, and is free from other milk enzymes such as peroxidase and catalase.

pp. 1105–7

The problem of further purification resolves itself into that of eliminating the protein without loss of enzyme. Up to the present attempts to purify Dixon and

Thurlow's preparation further have not been successful. Precipitation and adsorption methods proved unsatisfactory, cataphoresis experiments at various p_H's showed that the oxidase always migrated with the caseinogen, and it cannot be dialysed away from the protein. We found, however, that, by making use of the fact discovered by Dixon and Thurlow that when the milk proteins are coagulated by the action of rennin a considerable part of the oxidase remains in the whey, it was possible to get rid of the greater part of the protein at the outset, and so obtain very much more active preparations. The procedure we adopted is as follows.

Rennin method of preparation

The fresh milk, after its activity had been tested, was warmed to 35° and mixed with a small amount of rennin preparation ("Birks' Junket Powder"). The mixture was allowed to stand for 30 minutes, by which time it had set to a firm clot. This clot was then broken up by stirring, and the insoluble portion or curd was centrifuged off. The whey thus obtained was mixed with an equal volume of saturated ammonium sulphate solution and allowed to stand. A small amount of precipitate formed, and quickly rose to the surface of the solution, where it formed a compact layer. The liquid was then siphoned off, and the precipitate transferred to a filter-paper allowed to drain, and squeezed as dry as possible between filter-papers. It was then thoroughly extracted with ether to remove traces of fat. It was again squeezed between filter-papers, and finally dried in a vacuum desiccator. We shall refer to the resulting preparation as "the whey preparation."

It was found that when the milk clots the whole of the oxidase passes into the whey, and the curd, after washing with water, is quite inactive. Thus by this method the caseinogen and fat are got rid of without loss of enzyme.

The whole of the enzyme is precipitated by the half-saturation with ammonium sulphate.

The process of preparation should be carried out as quickly as possible, as the enzyme is much more easily destroyed by oxidation than in the case of the caseinogen preparation. In particular, the drying must be carried out *in vacuo*. Allowing the preparation to dry in the air is sometimes sufficient to destroy the activity almost entirely, and it always considerably reduces it. The enzyme is perfectly stable in this form once the preparation is completely dry, and it can be kept in the vacuum desiccator for long periods without loss of activity.

The whey preparation is very readily soluble in water, giving a practically clear solution. The yield is about 10 g. from 1.5 l. of milk, and the activity was found by means of the methylene blue technique to be from 500 to 700 times that of the original milk...

Methods for further purification

(a) *Charcoal*. It was stated by Dixon and Thurlow that charcoal does not adsorb the oxidase from a solution of the caseinogen preparation. This observation was confirmed with the whey preparation. It was found, however, that in this case the charcoal did remove a considerable proportion of the protein...

The charcoal method, therefore, enables us to concentrate the whey preparation a further five times. The filtrate after four charcoal adsorptions is a clear, colourless solution of very high activity; and by half-saturation of this with ammonium sulphate, filtering off the precipitate, and drying *in vacuo*, we obtained the whole of the enzyme in the form of a preparation which was, as we expected, five times as active as the whey preparation, or 2500 times as active as milk.

(*b*) *Kaolin.* It was found that kaolin could also be used for the further purification of the whey preparation.

p. 1109

We may say, then, that the charcoal method gives a means of purifying the whey preparation without loss of enzyme, but can only be used for the removal of 80% of the protein. On the other hand, a single kaolin adsorption and extraction is sufficient to give this degree of purification, and a repeated application of this treatment would undoubtedly give preparations of a very much higher activity, but unfortunately the 50% loss of enzyme at each stage renders the method unsatisfactory.

The most active preparations we have obtained have been made by a combination of both methods.

Selection 27

Throughout the period from 1900 to around 1940 the advance of knowledge as regards the nature of enzymes was in many ways tentative. This was partly due to workers like Willstätter who, exerting a commanding influence in the field, did not appreciate sufficiently the significance of crystallization of proteins for the understanding of the physico-chemical properties of enzymes. At the same time, it is necessary to recognize the existing laboratory predicaments regarding the handling of protein preparations characterized as colloidal.

By 1900 a number of proteins (haemoglobins, globulins, albumins) had been obtained in crystalline state.[a] Among the workers – few in number – stimulated to apply methods employed in the crystallization of proteins to attack the chemistry of enzymes was the American James Batchellor Sumner (1887–1955). As shown below, in 1926 he announced that he had obtained a new crystalline protein 'whose solutions possess to an extraordinary degree the ability to decompose urea into ammonium carbonate'. The fact that it took nearly nine years to achieve this result is an indication of the complexities encountered by those early investigators engaged in purification of enzymes.

J.B. Sumner, *J. Biol. Chem.*, *69*, 435. 1926

'The isolation and crystallization of the enzyme urease. Preliminary paper'

pp. 437–9

I undertook the task of isolating urease in the fall of 1917 with the idea that it might be found to be a crystallizable globulin, in which case the proof of its isolation would be greatly simplified. Other reasons for choosing urease were that the quantitative estimation of urease is both rapid and accurate, that urease can be reasonably expected to be an individual enzyme, rather than a mixture of enzymes, and that the jack bean appears to contain a very large amount of urease, if it is permissible to draw a parallelism between the urease content of the jack bean and the amounts of other enzymes found in other plant and animal materials.

In previous work in collaboration with Graham and Noback[1,2] and in unpublished work of my own it has been found that urease is very completely precipitated, together with the jack bean globulins, by cooling its 35 per cent alcoholic solution to −5 to −10°C, provided the reaction is sufficiently acid. We have found that urease can be precipitated by neutral lead acetate and neutralized cadmium chloride and that most of the urease can be reextracted by decomposing the precipitate with potassium be oxalate; that urease can be precipitated by tannic acid without very much inactivation and that urease can be rendered insoluble, with loss of a part of its activity, by the action of dilute alcohol or very dilute acid.

Although the literature contains numerous references to coenzyme of urease, I believe that no specific coenzyme exists. My evidence rests upon the fact that the loss of activity that occurs when the octahedral crystals are separated from a jack bean extract is almost exactly equal to the activity obtained when these crystals are washed with dilute acetone and then dissolved in water. If anything could separate an enzyme from its coenzyme crystallization might be expected to do so. The proteins in impure urease solutions doubtless exert a protective action as buffers and both proteins and polysaccharides may exert protective colloidal action.

I present below a list of reasons why I believe the octahedral crystals to be identical with the enzyme urease.

1. The fact that the crystals can be seen by the microscope to be practically uncontaminated by any other material.
2. The great activity of solutions of the crystals.
3. The fact that solvents which do not dissolve the crystals extract little or no urease and that to obtain solutions of urease one must dissolve the crystals.
4. The fact that the other crystallizable jack bean globulins, concanavalin A and B, carry with them very little urease when they are formed from solutions that are comparatively rich in urease.
5. The unique crystalline habit of the octahedra and their ready denaturation by acid.
6. The fact that the crystals are purely protein in so far as can be determined by chemical tests, combined with evidence from previous work to the effect that urease behaves like a protein in its reactions towards heavy metals, alkaloid reagents, alcohol, and acids.
7. The fact that the crystals are nearly free from ash and the fact that we have previously prepared solutions of urease that contained neither iron, manganese, nor phosphorus.

The method which I have used to obtain the crystals is extremely simple. It consists in extracting finely powdered, fat-free jack bean meal with 31.6 per cent acetone and allowing the material to filter by gravity in an ice chest. After standing overnight the filtrate is centrifuged and the precipitate of crystalline urease is stirred with cold 31.6 per cent acetone and centrifuged again. The crystals can be now dissolved in distilled water and centrifuged free from insoluble and inactive matter that has passed through the filter during the filtration. Of the urease extracted from the meal as much as 47 per cent may be present in the crystals. If one uses coarsely ground jack bean meal that has not been freed from fat the crystals are still obtained, but in traces only. I have carried out the process described above about fifteen times since first discovering the crystals and have always had success.

Selection 28

During the first three decades of this century a good deal of experimental and theoretical work had been conducted on the assumption that an enzyme combines with its substrate to form a complex. However, the problem of how to picture the pattern of the union remained largely speculative as experimental evidence was not easily forthcoming.

In the second half of the 1920s a notable but hardly noted attempt was made to deal with this question.[a] It arose from investigations of the mechanism of enzyme action by bacteria pursued by Juda Hirsch Quastel (1899–1987) at Cambridge. In 1926 he published observations showing that in the presence of *B.coli*, 56 substances out of 103 examined either reduced methylene blue or oxidized the leuco form.[b] In order to explain these chemical reactions promoted by living material Quastel adopted a combined approach evolved from two contemporary developments in physical chemistry. They were the electronic theory of valence applied to oxidation (interpreted as dehydrogenation)[c] and the recognition of the role played by surfaces in catalytic phenomena.[d] Quastel envisaged that specific areas of the bacterial surface, due to their electronic structure, would act as polarizing fields. Under their influence the substrates (substrate molecules) could be activated to donate or accept hydrogen.

What emerged from this work was a conception of enzyme action structurally localized in 'active centres' electronic in character. The active sites were integral portions of the cell surface or of an intracellular surface. As shown in he following excerpts from a paper published in 1927 by Quastel in collaboration with Walter R. Wooldridge, the proposed mechanism seemed to suggest that it was unnecessary to postulate a large number of highly specific enzymes.

J.H. Quastel and W.R. Wooldridge, *Biochem. J.*, *21*, 1224. 1927

'Experiments on bacteria in relation to the mechanism of enzyme action'

pp. 1246–7

Formation of active centres at biological surfaces

There is, naturally, great vagueness at present as to the mechanism of formation of enzymes and their relationship to the remaining constituents of the cells, but it seems to be fairly widely held that the cell is able to elaborate at least two distinct classes of molecules, the very highly specialised molecules which exhibit enzymic behaviour, and the enzymically inert substances which together make up the protoplasmic and histological structures of the cell. This view, which calls for a sharp line of demarcation between the architectural units of the cell, proteins, nucleotides, etc., and the specific enzymes, which are not only being synthesised themselves in the cell but which are regulating the course of metabolism and the growth of the cell itself, seems to be greatly strengthened by the fact that a large number of enzymes can be secreted by the cell. This gives the impression that enzymes are simply products of the cell in much the same way as, let us say, adrenaline.

When it is considered (a) that the number of highly specialised enzyme molecules, which the cell is presumed to contain, must be very large indeed and that they must vary considerably in their type and constitution, (b) that the evidence, presented in this paper, concerning the dehydrogenations effected by bacteria is contrary to the supposition that activity is due to the presence of many highly specific enzyme molecules, (c) that it is extremely difficult to understand how a cell is able to cope with material to which hitherto it has not been accustomed, if its content of specific enzymes, though large, is yet limited, it will be granted that the view stated above stands in need at any rate of some emendation.

The hypothesis we put forward, that enzymic activity may be regarded as the property of the active centres of cellular and intracellular structures (and this includes the smaller structures capable of extraction from or secretion by the cell) leads to a considerable simplification of the above view. Precisely what enzymic behaviour a particular structure or colloidal aggregate in the cell may possess depends on the nature of the active centres which form a part of the structures or of the colloidal aggregates. Enzymes, on this view, are themselves part of and cannot be dissociated from the architectural units of the cell. This does not imply, of course, that only the histological structures are involved; the smaller colloidal aggregates are just as much part of the architecture of the organism. The conditions existing in the cell are such as to bring about just that arrangement or juxtaposition of molecules which makes for the formation of active centres on the normal material of the cell. Thus we may imagine that the protein, nucleotides, etc., are not only so arranged as to form the various substances of the cell but that the arrangement is such that active centres are formed on these particular substances. Enzymes, therefore, and cellular structures are inseparably connected.

p. 1249

The active centre we would imagine in the molecule –A –CO–S–B– to be
$$NH_2$$
made up at least of the groups – NH_2, –CO– and –S– and each of these groups will play its part in rendering a substrate accessible to the centre. The field due to these groupings may not be homogeneous and hence the orientation of a substrate at the centres, so that activation may occur, becomes a highly important consideration.

We have given this illustration of a possible formation of an active centre in order to make clear the difference between the centre and the usual conception of an enzyme. The actual composition of an active centre will in all probability be much more complex than in the illustration given. We may regard the entire aggregate as the enzyme, or the particular centre (at the link) as the enzyme. Each view is equally legitimate. But the residue of the aggregate, distinct from the region occupied by the centre, may be the seat of other active centres, so that the aggregate as a whole may have a much wider range of specificity than were the residue inert in this respect. Such an aggregate would be a relatively large colloidal particle and it would certainly be difficult to regard it as a *specific* enzyme. On the other hand, the residue may be of comparatively small dimensions and contain no other active centres. The specificity of action would be determined by the single centre and the range of specificity may be so small as to make the particle a highly specific enzyme.

Selection 29

Sumner's announcements concerning the protein nature of urease was largely not accepted at its face value and it remained an isolated case of its kind for four years. The problem was to find crystallization techniques without destroying the enzyme. Then in 1930 John Howard Northrop (1891–1987) of the Rockefeller Institute for Medical Research discovered how to crystallize pepsin from commercial material[a] and produced evidence that the enzyme could be protein.

In some general tests the pepsin crystals behaved like proteins. They also did not change their composition nor did they alter their optical and proteolytic activities on repeated recrystallization. Further, solubility measurements done on recrystallized material also seemed to indicate that it was pure protein or possibly a mixture of proteins.

The excerpt from Northrop's paper demonstrates that his approach was in accord with the view linking enzymes to proteins. At the same time it reveals the persisting uncertainties concerning the properties of enzymes. Should enzymes be characterized wholly as proteins or did they also possess non-protein portions and if so what relative part did these components play in enzyme action?

J.H. Northrop, *J.Gen.Physiol.*, *13*, 739. 1930

'Crystalline pepsin. I. Isolation and tests of purity'

pp. 763–4

The preceding experiments have shown that no evidence for the existence of a mixture of active and inactive material in the crystals could be obtained by recrystallization, solubility determinations in a series of solvents, inactivation by either heat or alkali, or by the rate of diffusion. It is reasonable to conclude therefore that the material is either a pure substance or a solid solution of two very closely related substances. If it is a solid solution of two or more substances it must be further assumed that these substances have about the same degree of solubility in the various solvents used, as well as the same diffusion coefficient and rate of inactivation or denaturization by heat. It must also be assumed that both substances are changed by alkali at the same rate and to the same extent. This could hardly be true with the possible exception of two closely related proteins. It is conceivable that two proteins might be indistinguishable by any of the tests applied in this work. But in this case it would follow that the enzyme itself was a protein and this, after all, is the main point. It does not necessarily follow even if the material represents the pure enzyme that it is the most active preparation that can be obtained nor that it is the only compound which has proteolytic activity. There is some evidence that the activity of the preparation may depend on its physical state as is known to be the case with the catalytic activity of colloidal metals. It is possible, on the other hand, that hemoglobin is the type structure for the enzymes and that they consist of an active group combined with a protein as suggested by Pekelharing. The active group may be too unstable to exist alone, but it is quite conceivable that a series of compounds may exist containing varying numbers of active groups combined with the protein, and that the activity of the compound would depend on the number of these active groups. This hypothetical complex would not differ much from that assumed by Willstätter[1] and his coworkers, except that it supposes a definite chemical compound with the protective group in place of an adsorption complex. It is of course possible that both types of complex may be formed under suitable conditions. The reactivation of enzymes as reported in the literature also suggests their protein nature since the conditions for this reactivation are similar to those found by Anson and Mirsky[2] to be suitable for the formation of native from denatured protein. The fact that the crystalline urease prepared by Sumner is also a protein and that the temperature coefficient for the rate of inactivation of enzymes in general is that characteristic for the denaturization of proteins, suggests that the protein fraction in the purification of enzymes be given special attention even though it may not be the most active fraction.

Selection 30

Since the time of Emil Fischer's enzyme investigations on the α- and β-glucosides in the 1890s the doctrine of specificity for substrate has been an integral part of the study of enzymes. It may, therefore, come as a surprise that in the case of the long established proteolytic enzyme pepsin[a] the knowledge of specificity was not put on a firmer foundation until the late 1930s.

The problem was how to picture the action of the enzymes hydrolyzing in the light of the peptide hypothesis proposed independently by Fischer and Hofmeister early in the century.[b] Following Fischer's own efforts,[c] synthetic advances have been made giving support for the peptide link in joining amino acids in proteins. Much important work was done by Max Bergmann (1886–1944), a pupil of Emil Fischer. By 1932, while still heading the Kaiser Wilhelm Institute for Leather Research in Dresden, Bergmann developed, in collaboration with Leonidas Zervas (1902–80), the so-called 'carbobenzoxy method' of peptide synthesis.[d]

By drawing upon this improved technique of synthetic organic chemistry M. Bergmann, J.S. Fruton and H. Pollok, all at the Rockefeller Institute for Medical Research, were able to supply in 1937 synthetic protein substrates of known composition for each of the three trypsins occurring in the pancreas and to study their specific activities. In 1938 Joseph Stewart Fruton (*b.* 1912) and Bergmann applied the same approach to demonstrate for the first time the specificity of pepsin action (see the report below).

J.S. Fruton and M. Bergmann, *Science*, *87*, 557. 1938

'The specificity of pepsin action'

p. 557

The various enzymes which attack genuine proteins and which therefore are designated proteinases exhibit striking differences in the specificity of their chemical action. The clearest demonstration of these differences in enzymatic specificity has been obtained by means of synthetic substrates. Such substrates have recently been described[1] for all the known types of proteinases, with the exception of pepsin.

In this communication we wish to report the finding of a synthetic substrate for swine pepsin. Carbobenzoxy-*l*-glutamyl-*l*-tyrosine is extensively hydrolyzed in the presence of pepsin with the formation of carbobenzoxy glutamic acid and tyrosine; under our conditions the hydrolysis attained 70 per cent. in 3 days. This enzymatic hydrolysis occurs at pH 4. At the generally accepted pH optimum of pepsin – pH 2 –, a hydrolysis of only 10 per cent. of the synthetic substrate was observed. Once recrystallized pepsin is more effective than a good commercial preparation.

The availability of synthetic substrates for pepsin makes possible a study of the specificities, homogeneity and kinetics of pepsin preparations from various animal species.[a]

Notes to Section I

Selection 1

Introduction:

a. Clearly, the impetus to learn about fermentation came with the evolution of bread-making and the preparation of alcoholic beverages. Derived from *fermentum* (leaven) the term 'fermentation' served over centuries as a prototype of chemical changes occurring in the mineral, vegetable and animal worlds. Following the work of Johannes Baptista van Helmont (1579–1644) and of other investigators like Johann Joachim Becher (1635–1682 [1685]) and Georg Ernst Stahl (1659–1734), all of whom played a major role in the development of chemical practice and thought before Lavoisier, the word become associated with the production of alcohol and vinegar (vinous and acetic fermentation) and putrefaction (putrid fermentation). This division into three kinds of fermentation was adopted by Lavoisier and later chemists, although it had been acknowledged that the phenomena could include other transformations such as the leavening of dough (panary fermentation). In the present subsection 'fermentation' will refer to alcoholic fermentation, unless otherwise stated.
b. Selection 46, Section III.
c. Cf. Lavoisier, *Elements of Chemistry* (Edinburgh, 1790), pp. 139–40.
d. See Selections 32, 33, Section II.
e. Lavoisier, 'Réflexions sur la décomposition de l'eau par les substances végétales et animales', *Mém.Acad.Sci.*, 590–605. 1786 (1788) (p. 605).

f.   This draws on M. Teich, 'Circulation, transformation and conservation of matter and the balancing of the biological world in the eighteenth century', *Ambix, 29,* 17–28. 1982. Also published in W. Bernardi and La Vergata (eds), *Lazzaro Spallanzani e la biologia del settecento* (Firenze, 1982), pp. 363–80. For a detailed treatment of Lavoisier's investigation of fermentation, see F.L. Holmes's major study *Lavoisier and the chemistry* of life *An exploration of scientific activity* (Madison, 1985).

Text:

a.   This term was employed, in the form of 'al-kohl', for the black eyebrow paint used by the Arabic peoples and consisting of very finely powdered antimony sulphide. The transfer of the name by Paracelsus (*c.* 1493–1541) to our liquid alcohol was in reference to its subtlety and volatility, which reminded him of the fineness of the impalpable powder. See J. Needham, *Science and Civilisation in China* (Cambridge, 1974), V, pt 2, pp. 267–8; and J.R. Partington, *A History of Chemistry* (London, 1961), II, p. 149.

Selection 2

Introduction:

a.   For the correct name, Giovanni Valentino Mattia Fabbroni, see Mario Gliozzi, *Dictionary of Scientific Biography* (New York, 1971), IV, p. 503. In the older encyclopaedias and biographical dictionaries the name is frequently spelled as Fabroni.
b.   For a discussion of the circumstances of this meeting see M. Crosland, 'The Congress on definite metric standards, 1798–1799: the first international scientific conference', *Isis, 60,* 226–31. 1969.
c.   The work in question, *Dell'arte di fare il vino*, was published in Florence in 1787. As W. Bulloch has already pointed out, *The History of Bacteriology* (London, 1960), pp. 42–3, the copy of this book in the British Museum bears the name of Adamo Fabbroni.
d.   Iacopo Bartolomeo Beccari should not be mistaken for Giambatista Beccaria (1716–87), mainly known for his contribution to knowledge of electrical phenomena. On Beccari, see E.F. Beach, 'Beccari of Bologna, the discoverer of vegetable protein', *J.Hist.Med., 16,* 354–73. 1961. For a dissenting view on Beccari's discovery of gluten, see R. Savelli, 'Iacopo Bartolomeo Beccari n'a pas decouvert le gluten', *Actes du VIIIe Congrès International d'Histoire des Sciences*, 588–9. 1956.
e.   See Introduction to Selection 101 and Introduction to Selection 103, Section V, note b.

Text:

a.   20 August 1789.
b.   Term applied to gas.
c.   It has not been possible to establish the identity of Collier. Perhaps Fourcroy refers to Joseph Collier who experimented on 'what effect an atmosphere of some factitious airs would have on fermentation'. Experimenting with 'hydrogen, oxygen, and a mixture of the two', he concluded that these experiments were not 'worth repeating as objects of profit; but they serve to confirm our opinion against the admission of air to fermenting liquids . . . we cannot suppose that the latter is ever freely admitted at all, by any of the common processes of fermentation'. See J. Collier, 'Experiments and observations on fermentation and the distillation of ardent spirits', *Mem.Lit.Phil.Soc. (Manchester), 5,* 243–74. 1798 (pp. 263–5).
d.   Reference to the publication by de Morveau, Lavoisier, Bertholet [*sic*] and de Fourcroy, *Méthode de nomenclature chimique* (Paris, 1787) in which the principles of the new (Lavoisierian) nomenclature were formulated. On Guyton de Morveau see G. Bouchard, *Guyton Morveau Chimiste et Conventionnel (1737–1816)* (Paris, 1938); W.A. Smeaton, 'L.B. Guyton de Morveau (1737–1816): a bibliographical study', *Ambix, 6,* 18–34. 1957 and 'Guyton de Morveau and the phlogiston theory' in *L'Aventure de la Science (Mélanges Alexandre Koyré)*, introduced by I.B. Cohen and R. Taton (Paris, 1964), pp. 522–40. On Berthollet, see M. Sadoun-Goupil, *Le Chimiste Claude-Louis Berthollet (1748–1822), Sa Vie, Son Oeuvre* (Paris, 1977). On Fourcroy, see W.A. Smeaton, *Fourcroy Chemist and Revolutionary 1755–1809* (Cambridge, 1962).

e. The term 'molecule', derived from the Latin word *moles* (mass), meant a minute particle of matter. Although Amadeo Avogadro (1776–1856) distinguished between an 'atom' and a 'molecule' (1811) it was not till 1858 that the confusion over what they represented was resolved. See also Introduction to Selection 108, Section V.

f. Chlorine.

Selection 3

Text:

a. Term applied to gas; cf. Selection 2, note b.
b. Ammonia.
c. Element.
d. Alcohol.
e. From the end of the eighteenth century the study of the chemistry of plants grew in conjunction with the so-called 'proximate analysis'. It involved the separation and identification of sugar, starch, gum, a few acids and certain other substances to be found frequently as constant constituents of plants. They were considered to be common to all plants and termed 'proximate principles'.

Selection 4

Introduction:

a. Actually, the book carries two titles. The longer title reads: *Le livre de tous les ménages ou l'art de conserver, pendant plusieurs années, tous les substances animales et végétales.*
b. That organic substances could be preserved provided they were not in contact with air (in a vacuum) was recorded in 1782, if not before. Cf. de Fourcroy, *Elements of Natural History and of Chemistry*, trans. W. Nicholson, 2nd edn, (London, 1788), IV, p. 137.
c. On Gay-Lussac see M. Crosland, *Gay-Lussac Scientist and Bourgeois* (Cambridge, 1978).

Text:

1. Ann. de chim., vol. XLVI, p. 294.
2. These procedures are extremely simple, consisting in putting the substances which one wishes to conserve into bottles which one corks very exactly; afterwards one exposes these bottles to the temperature of boiling water during a more or less considerable time. See the instruction which M. Appert has published on this subject.

Selection 5

Introduction:

a. It is not easy to determine the influence of Erxleben's work outside Bohemia. Here his writings were studied by the remarkable reformer of Bohemian brewing František Ondřej Poupě, known also as Franz Andreas Paupie (1753–1805), and by Carl Josef Nepomuk Balling (1805–68), who taught at the Prague Polytechnic. Balling substantially contributed to the understanding of attenuation. His *Attenuationslehre* enabled brewers and distillers to control the course of fermentation. The topic is dealt with by M. Teich, *Bier, Wissenschaft und Wirtschaft in Deutschland 1800–1914* (in preparation) and more briefly in 'The origins of carbohydrate chemistry in Bohemia (A study in the social relations of science)', *Acta historiae rerum naturalium necnon technicarum Special issue, 1,* 85–102. 1965 and 'Essay review' of G. Böhme et al., *Die gesellschaftliche Orientierung des wissenschaftlichen Fortschritts* (1978), *Ann.Sci., 37,* 101–8. 1980.

Selection 6

Introduction:

a. During the latter part of the eighteenth century and even early in the nineteenth century 'fecula' was identified with vegetable constituents bearing a resemblance to animal matter. Thus the great French chemist, Joseph Louis Proust (1754–1826), wrote 'fecula then has something in its nature analogous to wool, silk, etc.: it is gluten.' See Proust, 'An essay on the fecula of green plants', *Phil.Mag.*, *16*, 128. 1803. Later the term denoted starch, though apparently in France a distinction was made between cereal starch (*amidon*) and potato starch (*fécule*). Cf. *La grande encylopédie*, (Paris, n.d.), II, p. 748; ibid. (Paris, n.d.), XVII, p. 110. Cf. also Introduction to Selection 101, Section V, note d.

b.   *Diastasis* (Greek) means 'separation' but also 'breach in a barrier'.

Text:

a.   This presumably refers to Biot's discovery of cane sugar undergoing inversion. Cf. Biot, 'Sur un caractère optique à l'aide duquel on reconnaît immédiatement les sucs végétaux qui peuvent donner du sucre analogue au sucre de cannes, et ceux qui ne peuvent donner que du sucre semblable au sucre de raisin', *Ann.Chim.*, *52*, 58–72. 1833.

Selection 8

Introduction:

a.   Selection 74, Section IV.

Text:

1.   Taken from the second edition of the author's textbook published in the beginning of November last year. P.[d]
2.   Liebig and Wöhler in these Annalen, vol. XXII p. 186; Magnus, the same, vol. XXVII p. 367.
a.   Ethyl hydrogen sulphate.
b.   A name given to a substance obtained, under certain conditions from a mixture of alcohol and sulphuric acid.
c.   Glucose.
d.   P. = Johann Christian Poggendorff (1796–1877), the editor of *Annalen der Physik und Chemie*.

Selection 9

Text:

a.   Dextrin.

Selection 10

Introduction:

a.   W. Beaumont, *Experiments and Observations on the Gastric Juice, and the Physiology of Digestion* (Edinburgh, 1838). Reprinted from the Plattsburgh edition (1833), with notes by A. Combe.
b.   de Réamur, 'Sur la digestion des oiseaux', *Mém.Acad.Sci.*, 1752, 266–307, 461–95.
c.   L. Spallanzani, *Dissertations relative to the Natural History of Animals and Vegetables* (London, 1784), I.
d.   W. Prout, 'On the nature of the acid and saline matters usually existing in the stomachs of animals', *Phil. Trans.*, 45–9, 1824. See also Selection 160, Section VII.
e.   F. Tiedemann and L. Gmelin, *Die Verdauung nach Versuchen* (Heidelberg and Leipzig, 1826–7), I, II. See also Selection 159, Section VII.
f.   See Selection 6, this Section.
g.   J. Müller and Th. Schwann, 'Versuche über die künstliche Verdauung des geronnenen Eiweisses', *Müllers Arch.*, 1836, 69–89; also Th. Schwann, 'Ueber das Wesen des Verdauungsprocesses', ibid., 90–138.

Text:

1.   Schwann's observations must lead to noteworthy and interesting results, I will, however, not leave unmentioned that they can only be perceived from the right point of view when

the substances which the hydrochloric acid dissolves, are prepared and the alterations which through its action on egg white, etc. are produced, are followed up by means of elementary analysis and studied. The name *pepsin* is for the present only the representative of an idea, and before we determine to bring the name *catalysis* into these researches all means towards solving the riddle by way of analysis must be exhausted. By catalytic actions we explain something only by the name, and at this moment osmazôme and ptyalin are likewise only designations for a certain behaviour; these bodies then exist for us only in actuality if elementary analysis protects us from confusion, if it allows us again to recognize those bodies which we meet in research. The task is not light, it is however soluble, and he who has carried it through will be looked upon as the discoverer. J.L.

a.   Among the investigators who continued the work of Eberle, Müller and Schwann were the Czech physiologist Jan Evangelista Purkyně (Purkinje) (1787–1869) and his collaborator S. Pappenheim (1811–1882). They reported (1837) that pepsin was soluble in spirit of wine (alcohol). Schwann later examined the solubility of swine pepsin in alcohol and found it depended on the concentration of alcohol. See J.E. Purkyně, *Opera omnia* (Prague, 1837) II, pp. 89–90; *Opera omnia* (Prague, 1939), III, pp. 49–51; *Opera omnia* (Prague, 1973), XII, pp. 183–94; commentary by V. Kruta, ibid., pp. 194–9. See also Schwann's footnote on pp. 127–8 in his article 'Versuche, um auszumitteln, ob die Galle im Organismus eine für das Leben wesentliche Rolle spielt', *Müllers Arch.*, 1844, 127–59.

b.   The origins of the subsequent prolonged discussion on the milk clotting power of pepsin and its distinctness from rennin may be traced to this observation by Schwann. It is not without interest that Schwann envisaged modifications of pepsin due to the presence or absence of rennet activity. Calf pepsin clotted milk but swine pepsin did not. Cf. footnote on p. 128 of Schwann's article quoted above.

c.   The term *osmazôme* was introduced by Thenard to denote the constituent of the aqueous extract of meat soluble in alcohol, supposed to be responsible for its taste and smell. It would seem that Thenard was not sure whether the material was really a definable chemical substance. See L.J. Thenard, *Traité de chimie élémentaire, théorique et pratique* (Paris, 1815), III, pp. 447–9. Later the name was attributed to all alcohol-soluble extracts of nitrogenous materials derived from animals and plants. See J.J. Berzelius, *Lehrbuch der Chemie*, 4th edn, (Dresden and Leipzig, 1840), 9, pp. 569, 575–6, 584.

d.   'Ptyalin' was the name given by Berzelius to an 'animal matter' obtainable from saliva (*Speichelstoff*). See Berzelius, ibid., p. 218. To Schwann it represented one of the cleavage products yielded by the action of pepsin on albuminous bodies (proteins). It should be noted that the designation 'ptyalin' for the ferment (enzyme) in saliva converting starch into sugar did not come into usage until later.

Selection 11

Introduction:

a.   See Selection 5.
b.   Spelled variously de Latour, (de) La Tour.

Text:

1.   Leuwenhoek, in 1680, had already seen by the aid of the microscope that the yeast of beer was composed of globules, the origin of which he attributed to the meals used in the making of the wort but this observation, of which I have had knowledge only a year after the presentation of my Memoir to the Academy, did not bring its author to the most important point, which was to know that the globules are capable of budding [*germer*] and of growing [*végéter*] in the wort of beer during its fermentation. See the Memoir of M. Turpin, in the Comptes rendus de l'Academie des Sciences, 20 August 1838, p. 396.

2.   See the journal *l'Institut*, numbers 158, 159, 164, 165, 166, 167, 168, 185 and 199.

a.   1799–1800.
b.   *Institut de France* – the leading French institution of learning, since 1832 composed of five Academies.
c.   1801–2.
d.   1803–4.

Selection 12

Introduction:

a. The term *Infusionstierchen* (infusion animalcules) appears to have been introduced in 1763 by the German naturalist Martin Frobenius Ledermüller (1719–69). By the mid-1830s the name referred to one of four classes of *Protozoa*.
b. Cf. 'Dr Schwann theilte ferner Versuche über *generatio aequivoca* mit', *Isis*, (1837), 524.
c. The term *Saccharomyces* derives clearly from the German *Zuckerpilz* or sugar fungus.

Text:

1. Later repetitions of this experiment showed me that it did not always succeed so well, and sometimes fermentation appeared in none of the flasks (if that is one had boiled them too long), sometimes also in the flasks which contained heated air the liquid ferments. For all that, this is easily explicable by the way in which the experiments were set up, since from the surface of the mercury (although this had been directly before strongly heated) and while airing and replacing the stopper, some unboiled organic substance could easily enter. The method used with the putrefaction experiments was not usable here because for it long boiling is necessary. On this account I would not state the above result, had it not been highly probable from the analogy to putrefaction and mould formation, once the existence of a plant had shown itself. The matter meanwhile will be decided by another safer method.
2. If sugar solution is kept with muscle, urine or glue, then similar plants appear therein, however in reduced number, mostly smaller and to some extent deformed.
a. *Extractum Nucis vomicae spirituosum.* The seeds of the Strychnine tree (*Strychnos nux vomica* L.) – a source of strychnine.
b. Franz Julius Ferdinand Meyen (1804–40) – German botanist and plant physiologist, active as a microscopist.

Selection 13

Introduction:

a. Kützing's algological work is discussed but his contribution to fermentation studies is completely left out in the *Dictionary of Scientific Biography* (New York, 1973), VII, pp. 533–4.
b. J. Berzelius, 'Weingährung', *Jahres-Ber.*, *18*, 400–3. 1839.

Text:

1. I call it truly organic in order to differentiate it from the falsely so-called chemo-organic compounds, as for example sugar, alcohol, ether, etc.
2. I point out hereby that in any organic body where organic substance is formed which is assimilated by the organism, also *every time* one or more inorganic products are produced, which partly remain in the organism (oils, fats, resins), partly are given out (urine, milk, etc.).
a. Johann Horkel (1769–1846) taught medical subjects in the universities of Halle and Berlin.
b. Christian Gottfried Ehrenberg (1795–1876) – author of the influential book on microscopic organisms: *Die Infusionsthierchen als vollkommene Organismen* (Leipzig, 1838).
c. Alexander von Humboldt (1769–1859) – naturalist and explorer of great authority in the world of science of his time.

Selection 14

Introduction:

a. A.W. Hofmann (ed.), *Aus Justus Liebig and Friedrich Wöhlers Briefwechsel in den Jahren 1829–1873*, (Braunschweig, 1888), I. pp. 148–9.

Text:

a.   At that time the term 'atom' applied to 'elementary' as well as to 'compound' atoms
     (molecules).

Selection 15

Introduction:

a.   The rare *Theorie der Fermentwirkungen* (Berlin, 1858) is to be found in M. Traube's
     Gesammelte Abhandlungen (Berlin, 1899) containing practically all his scientific
     publications.

Selection 16

Introduction:

a.   The impetus to work on fermentation came to Pasteur from industry around Lille when in
     1856 he was approached to investigate difficulties befalling the local manufacturers of
     alcohol by the fermentation of beetroot. He entered the field knowing from his previous
     optical studies that crude amyl alcohol, a by-product of alcoholic fermentation, contained
     two isomeric forms, an optically active and an optically inactive amyl alcohol. His
     knowledge of crystallographic properties of matter lay at the root of his stress on the role
     of asymmetric forces in the animate world. See Pasteur, *Oeuvres* (ed. Pasteur Vallery-
     Radot), I (*Dissymétrie moleculaire*), II *Fermentations et générations dites spontanées*)
     (Paris, 1922).
b.   Pasteur, 'Mémoire sur la fermentation de l'acide tartarique', *Comp.Rend.*, *46*, 615–18.
     1858. Also *Oeuvres*, II, pp. 25–8. For a stimulating discussion of Pasteur's ideas on the
     connection between molecular assymetry and life, see N. Roll-Hansen, 'Louis Pasteur – a
     case against reductionist historiography', *Brit.J.Phil. Sci.*, *23*, 347–61. 1972.
c.   This finding had important repercussions on industries employing or producing yeast
     (brewing, distilling and the manufacture of compressed yeast). The phenomenon (termed
     later by Warburg the 'Pasteur reaction') became the subject of continuous study. See O.
     Warburg, *Über den Stoffwechsel der Tumoren* (Berlin, 1926), p. 241; also E. Racker,
     'History of the Pasteur effect and its pathology', *Proceedings of the Conference on the
     Historical Development of Bioenergetics* (sponsored by the American Academy of Arts
     and Sciences) (Boston, 1973), pp. 47–53; idem, 'From Pasteur to Mitchell: a hundred
     years of bioenergetics', *Fed.Proc.*, *39*, 210–15. 1980.

Selection 17

Introduction:

a.   Forcibly voiced by F. Hoppe-Seyler, 'Ueber Gährungsprozesse', *Z.physiol.Chem.*, *2*, 3–4.
     1878–9. See Selection 54, Section III.
b.   W. Kühne, 'Erfahrungen und Bemerkungen über Enzyme und Fermente',
     Untersuchungen aus dem physiologischen Institut Heidelberg, *1*, 291–324. 1878.
c.   Some time ago during a controversy regarding the replacement of the concept 'ferment' by
     the term 'enzyme' R. Kohler assumed, it seems to me rightly, that there was more to the
     change than meets the eye. However, his suggestion that compared with 'enzyme' the
     term 'ferment' was broader and less specific as regards the nature of its functions, relates
     only to one aspect of the problem. Namely, it can be also argued that 'ferment' was a
     narrow and specific term because it historically developed out of and was applied to the
     specific process of alcoholic fermentation or processes allied to it. Equally, it can be
     maintained that 'enzyme' is a broad and non-specific term in the sense that it concerns not
     only fermentation but all catalytic chemical changes taking place in biological systems.
     Gradually, it was realized that chemical changes in the cell, in fact, could not be identified
     with fermentation. I suggested that this was the reason, ultimately, why 'enzyme' was
     adopted and 'ferment' relinquished. Cf. M. Teich, 'Ferment or enzyme: what's in a

name?', *Hist.Phil.Life Sci.*, *3*, 193–215. 1981. Professor Fruton doubts the validity of this suggestion because it leaves out of consideration other factors such as national tradition. He points out that Carl Oppenheimer, the widely known writer on the subject, or the eminent Otto Warburg continued with the use of 'ferment' in the 1930s (private correspondence, 1982).

Selection 18

Introduction:

a.   A soluble ferment [*ferment glucosique*] first obtained 1860 from yeast by M. Berthelot (1827–1907). Cf. 'Sur la fermentation glucosique du sucre de canne', *Comp.Rend.*, *50*, 980–4. 1860.

b.   See H.D. O'Sullivan, *The Life and Work of C. O'Sullivan, F.R.S.* (Guernsey, n.d.), p. 75.

c.   See A. Vernon Harcourt, 'On the observation of the course of chemical change', *J.Chem.Soc.*, *20*, 460–92. 1867. Among others it seems that Harcourt was one of the first to point out 'that for every 10°C the rate of change is doubled'. Harcourt did not refer to the earlier work by Ludwig Wilhelmy (1812–64). By measuring the change in optical activity of a hydrolyzed solution of cane sugar, Wilhelmy obtained the first expression for the rate at which inversion takes place. It emerged that the amount of sugar undergoing change decreased proportionally with time. The proportionality of the rate constant was designated as 'transformation coefficient of the sugar in unit time' (*Umwandlungscoefficient des Zuckers für die Zeiteinheit*). Cf. L. Wilhelmy, 'Ueber das Gesetz, nach welchem die Einwirkung der Säuren auf den Rohrzucker stattfindet', *Ann.Phys.* [3] *21*, 413–28; 499–526. 1850. Coincident in time with Harcourt's efforts was the work of the Norwegians Cato Maximilian Guldberg (1836–1902) and Peter Waage (1833–1900) which led to the birth of the Law of Mass Action (1864, 1867). *The Law of Mass Action. A Centenary Volume* (Oslo, 1964), published by the Norwegian Academy of Sciences, contains a certain amount of historical information on the subject. It is examined in more detail by M. Christine King, 'Experiments with time: progress and problems in the development of chemical kinetics', *Ambix*, *28*, 70–82. 1981; 'Part 2', *Ambix*, *29*, 49–61. 1982.

Selection 19

Introduction:

a.   Selection 78, Section IV.

b.   Fischer's interest in fermentation can be traced to his student days in Strasbourg when he became acquainted with Pasteur's *Études sur la bière* (1876). Under the impact of this work he did inform himself thoroughly about yeasts, moulds and bacteria. As Fischer pointed out in his reminiscences, he was probably the first chemist in Germany to apply Pasteur's lessons industrially, in a Dortmund brewery with which his father was associated. Cf. E. Fischer, *Aus meinem Leben* (Berlin, 1922), pp. 18–20.

c.   J.A. le Bel, 'Sur les relations qui existent entre les formules atomiques des corps organiques et le pouvoir rotatoire de leur dissolutions', *Bull.Soc.Chim.*, *22*, 337–47. 1874; J.A. van't Hoff, 'Sur les formules de structure dans l'espace', *Arch.Néerl.Sci.Ex.Nat.*, *9*, 445–45. 1874.

d.   E.C. Hansen, *Untersuchungen aus der Praxis der Gärungsindustrie* (Munich and Leipzig, 1888, 1892), I, II. The English version, revised by the author, appeared as *Practical Studies in Fermentation Being a Contribution to the Life History of Micro-Organisms* (London–New York, 1896).

e.   See note 2, this Selection and Selection 79, Section IV.

f.   This yeast infusion was not identical with invertase. See Selection 18. Compare also J.S. Fruton, *Molecules and Life*, (New York, 1972), p. 81.

Text:

1.   These Berichte *27*, 2031.

2.   These Berichte *26*, 2400.

Selection 20

Introduction:

a. F.W. Lüdersdorff, 'Ueber die Natur der Hefe', *Ann.Phys.*, *67*, 408–16. 1846; Marie Mannassein, 'Beiträge zur Kenntnis der Hefe und zur Lehre von der alkoholischen Gärung' in J. Wiesner (ed.), *Mikroskopische Untersuchungen* (Stuttgart, 1872) pp. 116–28. Cf. M. Teich, 'A little known chapter in the history of enzymes', *Acta historiae rerum naturalium necnon technicarum Special issue*, *2*, 69–74. 1966.

b. *Nobel Lectures Chemistry 1901–1921* (Amsterdam–London–New York, 1966), p. 121.

c. A. Baeyer, 'Ueber die Wasserentziehung und ihre Bedeutung für das Pflanzenleben und die Gährung', *Ber.chem.Ges.*, *3*, 63–78. 1870. This paper contained the idea regarding the formation of carbohydrates on the basis of the polymerization of formaldehyde. It went back to some experiments by Alexander Mikhailovich Butlerov (1828–86) in 1861 and persisted long into the 1930s. See A. Butlerov, 'Bildung einer zuckerartigen Substanz durch Synthese', *Ann.Chem.*, *120*, 295–8. 1861; P. Karrer, *Organic Chemistry* (Amsterdam and New York, 1938), p. 147.

d. C. v. Nägeli, *Theorie der Gärung, Abhandlungen der Mathematisch-Physikalischen Classe der Königlichen Bayerischen Akademie der Wissenschaften*, *13*, 1878–80.

e. Selection 16.

f. E. Buchner, 'Ueber den Einfluss des Sauerstoffs auf Gährungen', *Z.physiol.Chem.*, *9*, 389–415. 1885.

g. E. Buchner, H. Buchner and M. Hahn, *Die Zymasegärung* (Munich and Berlin, 1903); E. Buchner, 'Cell-free fermentation', *Nobel Lectures*, pp. 103–20.

h. E. Buchner, *Fortschritte in der Chemie der Gährung. Antritts-Rede bei Übernahme der ausserordentlichen Professur für analytische und pharmaceutische Chemie an der Hochschule zu Tübingen* am 4. Februar 1897 (Tübingen, 1897). Hans Buchner gave his paper to the *Society for Morphology and Physiology in Munich* on 16 March 1897, see 'Die Bedeutung der activen löslichen Zellproducte für den Chemismus der Zelle' in *Sitzungsberichte der Gesellschaft für Morphologie und Physiologie in München*. Robert E. Kohler considers this to be 'The first announcement of the discovery of zymase . . .', see 'The reception of Eduard Buchner's discovery of cell-free fermentation', *J.Hist.Biol.*, *5*, 327–53. 1972 (p. 328). In view of Eduard Buchner's earlier Inaugural Lecture this is clearly incorrect. Consult also, by the same author, 'The background to Eduard Buchner's discovery of cell-free fermentation', *J.Hist.Biol.*, *4*, 35–61. 1971.

i. Analogously with diastase, the term was first used in 1864 by the French microbiologist, Antoine Béchamp (1816–1909), to denote a soluble sugar-inverting ferment produced by low plant organisms to be known as invertin and later invertase. Curiously, Béchamp did not regard cane sugar as a sugar. A. Béchamp, 'Sur de nouveaux ferments solubles', *Comp.Rend.*, *59*, 496–500. 1864. See also Introduction to Selection 18, note a.

j. E. Fischer and P. Lindner, 'Ueber die Enzyme eniger Hefen', *Ber.chem.Ges.*, *28*, 3034–9. 1895 (p. 3038).

k. This draws on M. Teich, '*A* History of Biochemistry', *Hist.Sci.*, *18*, 46–67. 1980.

Text:

1. This is freed from superficially adhering water to the extent that at a pressure of 25 atmospheres no more water is given off.

2. Glass powder is less suitable on account of its action as a weak alkali.

3. Plant physiologists may decide whether this carbonic acid perhaps originates in the oxidation processes connected with respiration.

4. Theorie der Gärung. Munchen 1879. p.15.

5. These Berichte *3*, 73.

6. The heat development appearing on alcoholic fermentation by yeast has recently been estimated again by A. Bouffard, Compt.rend., *121*, 357.

7. It must be noted however that the so-called urea fermentation, the decomposition of urea into ammonia and carbonic acid, is chemically very different from the proper fermentation processes and therefore many do not regard it at all as fermentation. It is a simple hydrolysis, obtained even with water at 120 degrees.

8. These Berichte *28*, 3037 [*sic*].

9. The diosmotic properties make this seem possible. Cf. v. Nägeli, l.c. p. 39.
10. This also probably clarifies the experiences of J. de Rey-Pailhade (Compt. rend., 118, 201), who prepared a weak alcoholic extract (22 per cent) from fresh brewers' yeast with the addition of some grape sugar. After being freed from micro-organisms by filtering through a sterile Arsonval candle, this sugar-containing extract spontaneously developed carbonic acid in the absence of oxygen.
11. loc.cit. p. 94. The experiments were repeated with the same result; only it was shown that they proceeded in the same way in lactose as in saccharose solutions. The diffusion processes are therefore not bound to the carrying on of the fermentative activity, as the named authors assumed.
a. The translation follows in the main that published by H.M. Leicester and H.S. Klickstein, *A Source Book in Chemistry 1400–1900* (New York–Toronto–London, 1952), pp. 506–15.
b. Traube did not employ the term 'enzyme' in 1858 as it was not introduced until 1876. See Selection 15.
c. Felix Hoppe-Seyler (1825–95), the influential German biochemist, deplored the introduction of the term 'enzyme'. He certainly supported the chemical view of fermentation. However he seemed to be less concerned with the chemical nature of ferments than with the chemical nature of processes accompanying putrefaction. See Selection 54, Section III.
d. The name of the French bacteriologist is misspelt and should read Miquel (Pierre, 1850–1922).

Selection 21

Introduction:

a. See Selection 18.
b. E.F. Armstrong, 'Studies on enzyme action. II. The rate of the change conditioned by sucroclastic enzymes and its bearing on the Law of Mass Action', *Proc.Roy.Soc.*, *73*, 500–16. 1904.
c. E.F. Armstrong expressed in 1905 the view that for the most part the enzyme involved in synthesis of certain bodies could not hydrolyze them. See his 'Studies on enzyme action. VII. The synthetic action of acids contrasted with that of enzymes. synthesis of maltose and isomaltose', *Proc.Roy.Soc.*, *(B)*, *76*, 592–9. 1905. Cf. also Selection 149, Section VI.

Text:

1. It is probable that these compounds are not derivatives of fructose of the -OR type corresponding to the simple glucosides; probably the linkage is of a peculiar character, two centres being concerned. In this connection, it may be mentioned that invertase has no action on methyl fructoside, a substance in every way analogous to methylglucoside. This point will be more fully dealt with in a later communication.
2. The term 'hexoside' is used as a general expression to include all compounds of a glucoside character derived from a hexose.
a. Cf. E. Fischer, 'Bedeutung der Stereochemie', *Z.physiol.Chem.*, *26*, 60–87. 1898–9.

Selection 22

Introduction:

a. The term 'coferment' (later replaced by the expression 'co-enzyme') was coined by G. Bertrand in 'Sur l'intervention du manganèse dans les oxydations provoquées par la laccase', *Comp.Rend.*, *124*, 1032–5. 1897.
b. A. Harden, 'Über alkoholische Gärung mit Hefepresssaft (Buchner's 'Zymase') bei Gegenwart von Blutserum', *Ber.chem.Ges.*, *36*, 715–16. 1903. See also E.F. Korman, 'The discovery of fructose-1, 6-diphosphate (Harden-Young ester) in the molecularization of fermentation and of bioenergetics', *Proceedings of the Conference of the Historical Development of Bioenergetics* (sponsored by the American Academy of Arts and Sciences)

(Boston, 1973), pp. 190–3; R. Kohler, 'The background to Arthur Harden's discovery of cozymase', *Bull.Hist.Med.*, *48*, 22–40. 1974.

c.   A. Harden and W.J. Young, 'The alcoholic ferment of yeast-juice', *Proc.Roy.Soc.*, *(B)*, 77, 405–20. 1906.
d.   Devised by Charles James Martin (1866–1955), who headed the Lister Institute, and communicated the paper to the Royal Society.

Text:

a.   A. Harden and W.J. Young, 'The alcoholic ferment of yeast-juice', *Proc.Roy.Soc.*, *(B)*, 77, 405–22. 1906.

Selection 23

Introduction:

a.   Apparently, among the first to perceive the marked effect of acidity and alkalinity on chemical reactions in physiological systems was Claude Bernard, in collaboration with C.L. Barreswil (1845). See Introduction to Selection 139, Section VI.
b.   Introduction to Selection 80, Section IV.
c.   Sørensen headed the Department of Chemistry of the Carlsberg Laboratory at Copenhagen. On the relations of his work to industrial practice see the obituary by K. Lindenstrøm-Lang, reprinted in H. Holter and K. Max Møller (eds), *The Carlsberg Laboratory 1876/1976* (Copenhagen, 1976), pp. 63–81.

Text:

1.   Zeitsch. f. Elektrochem. *10*, 114. 1904.
2.   Obviously such a comparison is only possible if the ordinates of the two curves are expressed in the same units of weight. The ordinates of the destruction curve of the enzyme must therefore give the difference between the amount of substrate split in the experiment and the amount which would have been split if the enzyme had lost none of its activity.
3.   See for example T. Madsen and L. Walbum, Recherches sur l'affaiblissement de la présure, Festschrift tillegnad Olof Hammarsten 1906, Abh. Nr 10; and L.W. Famulener und Thorwald Madsen, Die Abschwächung der Antigene durch Erwärmung. This Zeitsch. *11*, 186, 1908.

Selection 24

Introduction:

a.   V. Henri, *Lois générales de l'action des diastases*, (Paris, 1903).
b.   For a valuable review in this field, see H.L. Segal's 'The development of enzyme kinetics' in P.D. Boyer, H. Lardy, K. Myrbäck (eds), *The Enzymes*, 2nd edn (New York and London, 1959), I, pp. 1–48.

Text:

a.   A reference to L. Michaelis and H. Davidson, 'Die Wirkung der Wasserstoffionen auf den Invertin', *Biochem.Z.*, *35*, 398–400. 1911.
b.   The authors refer here to S.P.L. Sørensen's 'Enzymstudien. II. Mitteilung', *Biochem.Z.*, *21*, 262. 1909. See Selection 23.

Selection 25

Introduction:

a.   Selection 38, Section III.
b.   R. Willstätter and A. Pollinger, 'Über Peroxydase, [Dritte Abhandlung]', *Ann.*, *430*, 269–319. 1923.

c. See also his Faraday Lecture, 'Problems and methods in enzyme research', *J.Chem.Soc.*, 1359–81. 1927. Willstätter considered his investigations on enzymes as his chief contribution to science. Cf. R. Willstätter, *Aus meinem Leben* (Weinheim/Bergstr., 1949), pp. 360f.

Text:

1. Until now only the precipitation of the peroxide by tannin appears to be a reaction of the enzyme itself (compare R. Willstätter and A. Pollinger, III. Abh. über Peroxydase [*Ann.*, *430*, 269–319. 1923].

Selection 26

Text:

a. M. Dixon and S. Thurlow, 'Studies on xanthine oxidase. I. Preparation and properties of the active material', *Biochem.J.*, *18*, 971–5. 1924.

Selection 27

Introduction:

a. See Selection 112, Section V.

Text:

1. Sumner, J.B., Graham, V.A., and Noback, C.V., *Proc. Soc. Exp. Biol. and Med.*, 1924, xxi, 551.
2. Sumner, J.B. and Graham, V.A. *Proc. Soc. Exp. Biol. and Med.*, 1925, xxii, 204.

Selection 28

Introduction:

a. Cf. 'These ideas seemed to make very little impact at the time and I felt very discouraged.' See J.H. Quastel, 'Fifty years of biochemistry. A Personal account', *J.Canad.Biochem.*, *52*, 71–82, 433. 1974 (p. 74).
b. J.H. Quastel, 'Dehydrogenations produced by resting bacteria. IV. A theory of the mechanisms of oxidations and reductions *in vivo*', *Biochem.J.*, *20*, 166–94. 1926.
c. Selection 60, Section III.
d. See I. Langmuir, 'Chemical reactions on surfaces', *Trans.Farad.Soc.*, *17*, 607–20. 1921–2.

Selection 29

Introduction:

a. Northrop improved on the precipitation method by dialysis from acid solution, originally proposed in 1896 by the Dutch worker, Cornelis Adrianus Pekelharing (1848–1922). Cf. C.A. Pekelharing, 'Ueber eine neue Bereitungsweise des Pepsins', *Z.physiol.Chem.*, *22*, 233–44. 1896. Northrop observed 'that the precipitate which formed in the dialysing sac appeared in more or less granular form and filtered rather easily, as though on the verge of crystallization. This precipitate dissolved on warming the suspension and it was eventually found that it could be induced to crytallize by warming to 45°C, filtering and allowing filtrate to cool slowly.'

Text:

1. Willstätter, R., *Naturwissensch.*, 1927, *15*, 585.
2. Anson, M.L. and Mirsky, A.E., *J.Gen.Physiol.*, 1929, *13*, 121.

Selection 30

Introduction:

a. On the history of pepsin, see E. Hickel, 'Pepsin ein Veteran der Enzymchemie', *Naturwissenschaftliche Rundschau.*, *28*, 14–18, 1975.
b. See Selections 115 and 116, Section V.
c. See Introduction to Selection 124, Section V, note b.
d. Named after the reagent carbobenzoxy chloride (benzyloxy-carbonyl chloride) $C_6H_5CH_2COCl$, used in the synthesis of peptides in order to introduce the carbobenzoxy radical into the amino group of an amino acid. The free carboxyl group of the amino acid was chlorinated to give an acid chloride or alternatively heated to produce an azide. In either case these products could be made to react with another amino acid to form a peptide with the carbobenzoxy radical still attached. The latter was removed by reduction brought about by catalytic hydrogenation. Cf. M. Bergmann and L. Zervas, 'Über ein allgemeines Verfahren der Peptid-Synthese', *Ber.chem.Ges.*, 65, 1192–1201. 1932. For background and the development of the method, see J.S. Fruton, 'The carbobenzoxy method of peptide synthesis', *TIBS*, 7, 37–9. 1982.

Text:

1. M. Bergmann, J.S. Fruton and H. Pollok, Science, 85: 410, 1937.
a. Cf. also their later paper, J.S. Fruton and M. Bergmann, 'The specificity of pepsin', *J.Biol.Chem.*, *127*, 627–41. 1939. Here it was shown that all the peptides they hydrolyzed by swine pepsin contained the aromatic amino acid residues tyrosine or phenyl alanine in the bond. The sequence of these amino acids was also important; presence of free carboxyl groups in the neighbourhood of the bond was favourable to enzyme activity. Further details of the effects of substrate structure on the sensitivity of the enzyme towards it were described.

Section II: PHOTOSYNTHESIS

a. From the 1770s to *c.* 1880: discovery and groundwork

Selection 31

The origins of the knowledge of 'photosynthesis'[a] go back to the last quarter of the eighteenth century, when Joseph Priestley (1733–1804) embarked on the systematic investigation of 'airs', as the gases were then called. The word 'gas' was coined much earlier by Johannes Baptista van Helmont (1579–1644) whose experiments and ideas initiated the study of gases.[b] However, it took some time for the implications of his work to sink in, including the use of the term 'gas' which was re-introduced by Lavoisier. The delay was due, among other reasons, to the inadequacy of contemporary manipulative techniques, and also to the dominance of the four-element theory which, incidentally, the Flemish chemist had rejected. It was consonant with this view to think of the gaseous state of matter – largely independent of the nature of the gas – in terms of the behaviour of common air and to consider different gases as modified 'airs'.

Priestley received his first impetus to study gases from observations of the evolution of 'fixed air' (carbon dioxide) in a Leeds brewery which stood next to the place where he stayed (1767). By making full use of the pneumatic trough, Priestley developed the qualitative chemical study of gases.[c] In the course of this work he discovered a whole range of novel 'airs', among them in 1774 a remarkably respirable variety, which he called 'dephlogisticated air'. In view of the reigning doctrine of phlogiston it was an appropriate name for the gas which, following Lavoisier, became known as oxygen. The doctrine of phlogiston assumed that calcination (heating) of metals, burning of candles and also animal breathing were so-called phlogistic processes involving the release of an inflammable principle (phlogiston) from combustible bodies into air.[d]

Accordingly, the contamination of the air in a confined space through burning or respiration was ascribed to phlogistication and ways were sought of improving or dephlogisticating it. It was the popular topic of 'goodness' of air which attracted Priestley's scientific attention,[e] especially Count Saluzzo's[f] attributing to cold the capacity for renewing the quality of air. While being unable to confirm this claim of the Italian investigator, Priestley discovered restorative powers in growing plants. Coming as it did before his preparation of 'dephlogisticated air' this find surprised him considerably, because according to the commonly-held opinion there should have been no difference in the way common air was required by plants and animals to sustain life.

J. Priestley, *Phil.Trans.*, *62*, 147. 1772

'Observations on different kinds of air'

pp. 166–7

One might have imagined that, since common air is necessary to vegetable, as well as to animal life, both plants and animals had affected it in the same manner, and I own I had that expectation, when I first put a sprig of mint into a glass-jar, standing inverted in a vessel of water; but when it had continued growing there for some months, I found that the air would neither extinguish a candle, nor was it at all inconvenient to a mouse, which I put into it.

Finding that candles burn very well in air in which plants had grown a long time, and having had some reason to think, that there was something attending vegetation, which restored air that had been injured by respiration, I thought it was possible that the same process might also restore the air that had been injured by the burning of candles.

Accordingly, on the 17th of August, 1771, I put a sprig of mint into a quantity of air, in which a wax candle had burned out, and found that, on the 27th of the same month, another candle burned perfectly well in it. This experiment I repeated, without the least variation in the event, not less than eight or ten times in the remainder of the summer. Several times I divided the quantity of air in which the candle had burned out, into two parts, and putting the plant into one of them, left the other in the same exposure, contained, also, in a glass vessel immersed in water, but without any plant; and never failed to find, that a candle would burn in the former, but not in the latter. I generally found that five or six days were sufficient to restore this air, when the plant was in its vigour; whereas I have kept this kind of air in glass vessels, immersed in water many months, without being able to perceive that the least alteration had been made in it. I have also tried a great variety of experiments upon it, as by condensing, rarefying, exposing to the light and heat, &c. and throwing into it the effluvia of many different substances, but without any effect.

Experiments made in the year 1772, abundantly confirmed my conclusion concerning the restoration of air, in which candles had burned out by plants growing in it. The first of these experiments was made in the month of May; and they were frequently repeated in that and the two following months, without a single failure.

For this purpose I used the flames of different substances, though I generally used wax or tallow candles. On the 24th of June the experiment succeeded perfectly well with air in which spirit of wine had burned out, and on the 27th of the same month it succeeded equally well with air in which brimstone matches had burned out, an effect of which I had despaired the preceding year.

This restoration of air I found depended upon the vegetating state of the plant; for though I kept a great number of the fresh leaves of mint in a small quantity of air in which candles had burned out, and changed them frequently, for a long space of time, I could perceive no melioration in the state of the air.

This remarkable effect does not depend upon any thing peculiar to mint, which was the plant that I always made use of till July 1772; for on the 16th of that month, I found a quantity of this kind of air to be perfectly restored by sprigs of balm, which had grown in it from the 7th of the same month.

Selection 32

Although Priestley's interest continued he did not throw substantially more light on the purification of contaminated air by plants. This was largely due to uncertainty regarding the biological nature of the microscopic algal layer called by him 'green matter', which formed on the walls of his jars and produced bubbles of dephlogisticated air. He inclined to the belief that the green substance was neither vegetable nor animal in nature, but 'a thing *sui generis*' constituted, probably, under the influence of light.[a]

It was the Dutch investigator, Jan Ingen-Housz[b] (1730–99), who when in London took the essential step further (1779). On the basis of extensive investigations he pointed out that the elaboration of dephlogisticated air took place only in the green parts of the plants during daytime under the influence of sun.[c] He also originated the widely influential idea that as plants were unable to improve the quality of air in the dark, they could, in a closed space or at night, even seriously impair it.

J. Ingen-Housz, *Experiments upon Vegetables, Discovering Their Great Power of Purifying the Common Air in the Sun-shine, and of Injuring it in the Shade and at Night* (London, 1779).

pp. xxxiii–xxxviii

I observed, that plants not only have a faculty to correct bad air in six or ten days, by growing in it, as the experiments of Dr Priestley indicate, but that they perform this important office in a compleat manner in a few hours; that this wonderful operation is by no means owing to the vegetation of the plant, but to the influence of the light of the sun upon the plant. I found that plants have, moreover, a most surprising faculty of elaborating the air which they contain, and undoubtedly absorb continually from the common atmosphere, into real and fine dephlogisticated air; that they pour down continually, if I may so express myself, a shower of this depurated air, which, diffusing itself through the common mass of the atmosphere, contributes to render it more fit for animal life; that this operation is far from being carried on constantly, but begins only after the sun has for some time made his appearance above the horizon, and has, by his influence, prepared the plants to begin anew their beneficial operation upon the air, and thus upon the animal creation, which was stopt during the darkness of the night; that this operation of the plants is more or less brisk in proportion to the clearness of the day, and the exposition of the plants more or less adapted to receive the direct influence of that great luminary; that plants shaded by high buildings, or growing under a dark shade of other plants, do not perform this office, but, on the contrary, throw out an air hurtful to animals and even contaminate the air which surrounds them; that this operation of plants diminishes towards the close of the day, and ceases entirely at sun-set, except in a few plants, which continue this duty somewhat longer than others; that this office is not performed by the whole plant, but only by the leaves and the green stalks that support them; that acrid, ill-scented and even the most poisonous plants perform this office in common with the mildest and the most salutary; that the most part of leaves pour out the greatest quantity of this dephlogisticated air from their under surface, principally those of lofty trees; that young leaves, not yet come to their full perfection, yield dephlogisticated air less in quantity and of an inferior quality, than what is produced by full-grown and old leaves; that some plants elaborate dephlogisticated air better than others; that some of the aquatic plants seem to excell in this operation; that all plants contaminate the surrounding air by night, and even in the day time in shaded places; that, however, some of those which are inferior to none in yielding beneficial air in sun-shine, surpass others in the power of infecting the circumambient air in the dark, even to such a degree that in a few hours they render a great body of good air so noxious, that an animal placed in it loses its life in a few seconds; that all flowers render the surrounding air highly noxious, equally by night and by day; that the roots removed from the ground do the same, some few, however, excepted; but that in general fruits have the same deleterious quality at all times, though principally in the dark, and many to such an astonishing degree, that even some of those fruits which are the most delicious, as, for instance, peaches, contaminate so much the common air as would endanger us to lose our lives, if we were shut up in a room in which a great deal of such fruits are stored up; that the sun by itself has no power to mend air without the concurrence of plants, but on the contrary is apt to contaminate it.

Selection 33

At this stage (the 1780s) the phlogiston cycle, envisaged already by Stahl, principally influenced notions about the mutual relationship of the non-living and living worlds. Underlying it was the conjecture that the phlogiston of the air was built into plants during their growth, then taken up via vegetable food by animals and passed by them back into the air through breathing. Unless we take into consideration this view about the circulation of phlogiston it is hardly possible to understand Jean Senebier's (1742–1809) early contribution to the study of the subject (1782) which focused attention on 'fixed air' (carbon dioxide), previously not taken into account by Priestley and Ingen-Housz. Senebier showed that 'fixed air' – which he believed dissolved in aerial water and resulted from the interaction of aerial phlogiston with 'dephlogisticated air' – was indispensable to the latter's production by plants.

J. Senebier, *Mémoires physico-chymiques* (Geneva, 1782), I

(*Physico-chemical Memoirs*)

pp. 373–6

It is evident, from my experiments, that the leaves of plants provide much air; that this air, mixing in the atmosphere with the phlogiston which is contained there, precipitates a rather great quantity of fixed air. And that it thus diminishes the quantity of phlogistic matters which are always mingled there, which would accumulate there without this precipitation, and which would finally make the ordinary air the cause of our death in removing from it the means it has of prolonging our life.

But if one gains by rendering the air respirable by depriving it of the phlogistic matters which give it deadly qualities, is one not in the position of losing this advantage, since the fixed air which is formed, by the combination of the pure air with the phlogistic matters is also dangerous? It is here that one will be able to judge if I have found the way of NATURE, in seeing the reciprocal relationship [*liaison*] which it appears to have in all its operations.[a]

This fixed air precipitated by the presence of the pure air provided by the plants, and which is formed by the union of this pure air with the phlogiston always contained in the common air. This fixed air is itself absorbed by the plants which draw it off from the atmosphere with the humidity it contains, in which it is dissolved and for which the leaves are very avid, as M. Bonnet[b] has so well demonstrated, and as my experiments prove equally by another means.

I do not actually know, whether one cannot affirm that the plants do not receive other air than that which penetrates with the water they suck in. And I believe so much the more that they have an indispensable need of this fixed air to yield their air, as I have proved, when I demonstrated that waters distilled and boiled are only suitable for rendering air to leaves of plants which one has exposed to the sun on account of the fixed air which they had dissolved. It would result then from this, that the plants only grow well in phlogisticated air because it furnishes the leaves with an abundant quantity of fixed air by its combination with the pure air which they allow to escape; and that this air is improved only because in the precipitation there is always a part of phlogiston which enters into the leaf with the fixed air, and which settles there, or which is lost in the water with the fixed air absorbed with it. These two causes produce the great diminution of air where plants are growing, as one sees in experiments of this kind made in close vessels: it results from this that plants which grow in the atmosphere do not draw directly into themselves the phlogiston which is contained there, but that they secure it only by means of the fixed air which transmits it to them while decomposing in their vessels by the act of the plant growth.

Selection 34

The fundamental transformation of the theoretical basis of chemistry due to the replacement of the doctrine of phlogiston by the theory of oxidation failed to affect Priestley, but was registered in the later writings of Ingen-Housz and Senebier.[a] By the turn of the century the new chemistry had definitely been extended to plant studies, as shown by Nicolas Théodore de Saussure (1767–1845). Among other things, he attempted to ascertain the quantitative side of the decomposition of carbonic acid by the green parts of seven periwinkle plants placed in an artificial atmosphere over mercury. In effect, his analytical results provided the first figures for the relation between the volume of carbon dioxide absorbed and that of oxygen given out during photosynthesis.[b] De Saussure also traced the origin of carbon in plants to the carbon dioxide of the air by especially determining the carbon content of mint plants grown in distilled water.

Th. de Saussure, *Recherches chimiques sur la végétation* (Paris, 1804) (*Chemical Researches on Vegetation*)

pp. 42–4

It results from the eudiometric observations described above, that the mixture of common air and acid gas contained before the experiment:
4199 cub. centim. or (211.92 cub. inches) of nitrogen gas.
1116 . . . . (56.33 . . . .) of oxygen gas.
 431 . . . . (21.75 . . . .) of carbonic acid gas.
5746 . . . . (290   . . . .)

The same air contained after the experiment:

4338 cub. centim. or (218.95 cub. inches) of nitrogen gas.
1408 . . . . (7.05 . . . .) of oxygen gas.
   0 . . . . (0   . . . .) of carbonic acid gas.
5746 . . . .

The periwinkle plants have then elaborated or made to disappear 431 cubic centimetres (21¾ cubic inches) of carbonic acid gas, if they had eliminated all the oxygen gas from it they would have produced a volume equal to that of the acid gas which has disappeared; but they have released only 292 cubic centimetres (14¾ cubic inches) of oxygen gas; they have then assimilated 139 cubic centimetres (7 cubic inches) of oxygen gas in the decomposition of the acid gas and they have produced 139 cubic centimetres (7 cubic inches) of nitrogen gas.

One comparative experiment proved to me that the seven periwinkle plants which I had used weighed dry, before the decomposition of the acid gas, 2707 grams (51 grains) and that they yielded on carbonization by fire in a closed vessel, 528 milligrams (9.95 grains) of carbon. The plants which had decomposed the acid gas have been dried and carbonized by the same procedure, and they have yielded 649 milligrams (12.23 grains) of carbon. The decomposition of the acid gas has thus caused one to obtain 120 milligrams or 2.28 grains of carbon.

I have in the same way carbonized periwinkle plants which had grown in atmospheric air deprived of acid gas and I found that the proportion of their carbon had rather diminished than augmented during their stay under the vessel.

pp. 51–2

The hundred parts of mint, after two and a half months growth in free air, have weighed green 216 parts; but up to the present this increase in weight teaches us nothing, since perhaps it is due to addition of water which always increases in plants when they are transplanted into a more humid place than that where they were growing before. They are reduced, by dessication at the temperature of the atmosphere, to a weight equivalent to 62 parts. The plant had then augmented, by the cooperation of air and water, their dry vegetable material by 21.71 parts. These 62 parts have given, on carbonization, 15.78 parts of carbon, or 4.82 more than they would have given before having grown in the distilled water. When I have made the same plants grow in similar circumstances in a feebly illuminated place, I have found that they had lost a small quantity of their carbon.

Selection 35

During the succeeding years the scientific interest shifted to the green substance of the leaves. The French chemists, Pierre Joseph Pelletier (1788–1842) and Joseph Bienaimé Caventou (1795–1877), sought to determine its composition and found in 1817 that it contained a large amount of hydrogen.[a] They recognized that its usual classification among 'resins' had been unsatisfactory and suggested that it should be considered as a 'proximate principle', that is as a chemically distinct and constant constituent of plants. The provisional name they gave to it – chlorophyll (leaf-green) – had been retained, but the chemistry of the green colouring matter of plants proved to be difficult to elucidate.

As a result of technical developments in microscopy the other major step was the discovery of the granular structure of chlorophyll and also the association of starch with chlorophyll by Hugo von Mohl (1805–72) in 1837.[b] His observation that starch grains were frequently embedded in the chlorophyll granules was subsequently (1858) confirmed by Carl Wilhelm Nägeli (1817–91), who stated that the absence of starch from chlorophyll structures was exceptional.[c] As to the relation of chlorophyll to starch, opinions were divided, but some suspected, e.g. Jöns Jacob Berzelius (1779–1848), that a chemical process, perhaps of a catalytic nature, was involved.[d]

It was to the great merit of the German plant physiologist, Julius Sachs (1832–87), that he took into account the implications of the earlier investigations of the gaseous exchange and the role of sunlight for the understanding of the connection between chlorophyll and the production of starch in the life of plants. In 1862 he provided experimental evidence that only light-growing plants assemble starch in the chlorophyll grains of the leaves. He also suggested that starch, as a reserve substance of plants, was the organic end product of a possible chain of chemical transformations, involving initially the inorganic compounds of carbon dioxide and water.

J. Sachs, *Bot.Ztg.*, *44*, 365. 1862.

'Ueber den Einfluss des Lichtes auf die Bildung des Amylum in den Chlorphyllkörnern' (On the influence of light on the formation of starch in the chlorophyll bodies)

p. 368

For the purpose before us the most important property of the etiolated chlorophyll bodies is this, that they, so far as my researches go, *never contain starch granules*. If one allows green leaves of the plants just enumerated[a] (with the exception of *Allium*) lying in strong alcohol to stand in the sun, until the green pigment is completely extracted and destroyed, if one then allows fine sections of the same to lie in caustic alkali for some days or warms them in it for some time, then washes with water and neutralizes with acetic acid, one then obtains on addition of dilute iodine solution an extremely clear starch reaction in the cells containing chlorophyll. In place of the chlorophyll bodies one now sees starch grains, each of which usually consists of a number of individual granules; and if the potash has not worked too strongly one recognizes the form of the chlorophyll body itself, in which lies the starch, the violet-blue colour of which stands out very clearly.... Whilst one in this way can most clearly demonstrate the smallest starch granules contained in the chlorophyll bodies, this on the other hand *never* succeeds in the etiolated chlorophyll bodies of the leaves cultivated in the dark in showing even the slightest trace of starch; and I have no hesitation, supported by the exactitude of the method, in expressing the conviction *that the yellow chlorophyll bodies formed in the dark contain no starch*. I lay such emphasis on this result because on it especially rest the results mentioned below and because the fact given here is the basis for further research . . .

A greater number of seeds was placed in different flower pots in earth and these were put in a roomy dark cupboard. The germinating plants remained here until the total reserve substances of the cotyledons and endosperm were consumed and until they, as a result, ceased to form new leaves in the dark. These etiolated and ready germinated plants were now used in 3–4 groups for research and further experiment. One lot was left in darkness where they still remained unaltered for some time to go then to pot; another lot in this state was placed in very strong alcohol for further examination; a third lot was on the other hand put against or in front of a sunny window in order to turn green there. Of these last some were examined after a few days when they had just become green, others only after a longer time when they had begun to develop further. Also these plants were placed in alcohol and bleached in the light. Of all parts, leaves, roots, stalks, buds of the etiolated, the green and the already further grown plants thin longitudinal and cross sections were examined, partly on the fresh, partly on the alcohol treated specimens; the presence of starch was particularly investigated and the effort was made to obtain a clear picture of its distribution.

pp. 370–1

Through experiments of a similar kind it can now be shown with the same certainty *that for the production of starch in the chlorophyll bodies light of higher intensity is necessary*, that for instance light in the interior of an ordinary living-room is not sufficient, while on the other hand this light intensity suffices completely for the turning green of chlorophyll. By diminished light one is therefore in a position to retain chlorophyll bodies long in a green state without formation of starch in their interior; and the end result consists here also in that the plants, thus turned green, do not grow further than if they stood in the dark, obviously just for the reason that in the chlorophyll no starch is formed . . .

The phenomena just mentioned now raise the question: in what way the first starch granules originate in the chlorophyll of the leaves under the influence of light? One can here make two valid hypotheses: one can assume on the one hand that the starch stored in the chlorophyll simply arises through *conversion* of an organic substance already present in the plant; but as with the continued influence of the light the starch formation continually increases one would have to assume a continued new formation of this substance through the metamorphosis of which the starch in the chlorophyll bodies must originate, and the origin of this substance would presuppose an assimilation process which is only stimulated by light; and indeed this process necessarily would have to take place in the chlorophyll-containing cells themselves ... Or one could answer the above question by assuming that the green chlorophyll bodies under the influence of light were in the state to form starch by a specific activity from *inorganic* substances (carbonic acid, water in the presence of mineral salts coming from the ground). I do *not* wish with these words to say that the starch in the chlorophyll so arises that from carbonic acid and water by elimination of oxygen immediately ready starch is formed. Rather the possibility remains open that here, inside the chlorophyll bodies themselves a longer series of chemical reactions sets in. With this assumption the single characteristic circumstance must be emphasized that here the process begins with *inorganic* substances and ends with production of starch, so that one can consider the starch produced here to be the first [state of synthesis], formed from inorganic substances.

Selection 36

The fact that it is the chlorophyllous parts of the plant cell which are concerned with the production of oxygen when under light was imaginatively demonstrated by the German investigator Theodor Wilhelm Engelmann (1843–1909). Under the microscope bacteria were observed to move towards the location where oxygen was evolved when plant-like microscopic organisms like *Euglena* were illuminated. Of considerable interest are Engelmann's grounds for regarding bacteria as animal-like living things.

T.W. Engelmann, *Bot.Ztg.*, *39*, 441. 1881.

'Neue Methode zur Untersuchung der Sauerstoffausscheidung pflanzlicher under tierischer Organismen' (A new method for the investigation of oxygen liberation by plant and animal organisms)

pp. 443–4

If one adds to a drop (enclosed between two pieces of glass) in which all bacteria at first swarming have now come to rest, a drop of defibrinated blood (oxygen rich through shaking with air) so that it flows from the edge of the cover glass, then movement starts up soon again at the boundary of the two liquids. This does not happen (or at most quite sporadically and for a very short time) if instead of arterial blood one uses blood through which immediately before has been led a strong stream of carbon monoxide.

Now if one takes a drop richly supplied with bacteria capable of movement and adds some green cells, for example *Euglena*, pieces of filamentous algae or some brown diatoms (e.g. *Navicula*) – protected with the cover-slip – and places one or more of these cells in the illuminated field of view of the microscope at 200–300 times magnification, one sees how in a short time lively swarming bacteria pile up around these cells. The former remain in the most lively movement when in all other parts of the drop there is already complete standstill.

If one now suddenly darkens the field of vision so far that the swarming bacteria still remain clearly visible (often much less darkening suffices), the latter soon suspend their movement and either remain still in their place or scatter gradually by molecular movement into the surrounding liquid.

Now if one allows complete light to fall again, the hither and thither movements round the chlorophyll-containing cells begin again immediately, the swarming bacteria gradually piling up again.

These experiments could be repeated many times over a short period on the identical object always with the same result.

The nearest explanation of the phenomena described, and as more exact testing teaches, the only permissible one is this: the chlorophyll-containing cells give off oxygen in the light and it is this which allows the bacteria to move and assemble at the source of oxygen.[1] In the dark liberation of oxygen ceases and the oxygen lack resulting from the rapid oxygen consumption by the bacteria now brings the movements to an end.

b. *From c. 1900 to c. 1940: elaboration*

With the turn of the century the study of the pigments in the leaf and also of the photosynthetic activity entered upon a new stage. During the preceding half-century there was no lack of attempts to solve the chemical nature of chlorophyll, and many of them resulted in conflicting and unsatisfactory data, summarized, for instance, in the writings of one of the leaders in the field, Leon Marchlewski (1869–1946).[a] Nevertheless, the historical value of this work cannot be discarded. It not only pointed to a close chemical relationship between the green colouring matter of plants and the red colouring matter of blood but, as interpreted by another Polish scientist, Marceli Nencki (1847–1901), this chemical information could be of significance for the understanding of the evolutionary aspects of biological processes.[b]

Actually, the most fruitful impetus to identification of leaf pigments came from people who were applying physico-chemical methods to the needs of biological study. Thus the distinguished physicist, George Gabriel Stokes (1819–1903), announced in 1864 that the chlorophyll of land plants was a mixture of two green and two yellow components. But he did not indicate how he obtained this information except that he had been 'for a good while engaged at intervals with an optico-chemical examination of chlorophyll'.[c] Following this, Henry Clifton Sorby (1826–1908)[d] confirmed the non-homogeneous nature of the colouring matter of plants. He did this by combining the separation method based on the partition of substances between immiscible solvents with the use of the microspectroscope he devised. By means of carbon disulphide, alcohol and water he obtained evidence for three substances belonging to the chlorophyll group: blue chlorophyll, yellow chlorophyll and chlorofucine. He also identified at least six forms of the xanthophyll group of pigments.[e]

For a considerable time neither Stokes' suggestion or Sorby's confirmation regarding the compound nature of the green colouring matter of plants produced any effect on the study of these subjects. Among the few investigators who appreciated Sorby's paper and indeed called it 'exemplary' (*musterhaft*) was the Russian botanist Mikhail Semenovich Tsvet (1872–1919).[f] Thoroughly conversant with solvent partition, he supplemented it with the adsorption technique. By running a carbon disulphide extract of the pigments through a column of calcium carbonate he was able to separate them in coloured zones (chromatograms). In demonstrating the existence of two chlorophyll components, chlorophyllin α and chlorophyllin β, he corroborated Sorby's earlier findings.

M. Tswett, *Ber.bot.Ges.*, *24*, 385. 1906

'Adsorptionsanalyse und chromatographische Methode. Anwendung auf die Chemie des Chlorophylls' (Adsorption analysis and chromatographic method. Application to the chemistry of chlorophyll)

p. 385

If a mixed solution (e.g. a chlorophyll solution in CS_2) filters through a column of adsorbent, the pigments precipitate in the manner [similar] to adsorption, mutually repel each other however and arrange themselves according to the adsorption series in the direction of the stream. Substances which do not enter into any undissociable adsorption compounds with the adsorption medium used wander away more or less quickly through the column. Subsequent filtration of the pure solution medium will understandably make the separation of the substances more complete. It can however be supposed that two substances in a solvent might be adsorbed to the same degree. Relative differences in concentration of the two substances would however not allow the formation of a unitary mixed zone. Also the equal potency of two substances in *different* solvents can scarcely be imagined. In spite of all this, although the number of adsorption zones will correspond to the number of substances, it can happen that any one zone is not absolutely pure, as is to be concluded from what is said above. By extraction of the substance of one zone and renewed adsorption one will reach the desired degree of purity.

We see thus that the laws of mechanical affinity may by used for the most complete physical separation of the substances soluble in certain fluids.

p. 386

As means of adsorption any one of the powdered substances insoluble in the solute concerned can serve. As however very many substances do not remain without chemical effect on the adsorbed substances, the choice of the analyst will fall in general on such bodies as are chemically indifferent and at the same time can be brought into as fine a form as possible . . . Amongst the adsorption means I can provisionally recommend precipitated $CaCO_3$ which gives the most beautiful chromatograms.

pp. 388–90

The green pigment of the leaves, the chlorophyll, is known to be a mixture of pigments, the complexity of which was differently estimated by different investigators. Chromatographic analysis is called upon to settle finally this degree of complexity. Compared to the other methods, it behaves like spectral analysis of the colour of a substance in comparison with analysis by means of tinted glass specimens . . .

The chromatograms obtained from a CS_2 solution have the following form:

I.   (Top) Zone. Colourless . . .
II.   Zone, especially less sharply separated from the next. Yellow due to xanthophyll β^1 . . .
III. Zone. Dark olivegreen. Chlorophyllin β.

IV. Zone. Dark bluegreen. Due to chlorophyllin α (Sorby's blue chlorophyll).
V. Zone. Yellow (xanthophylls α' and α'').
VI. Zone. Colourless.
VII. Zone. Orangeyellow (xanthophyll α).

Selection 38

Although Tsvet's work on chlorophyll was well known to his contemporaries and, indeed, at times he defended it forcefully, few chemists were prepared to accept his opinion about the existence of more than one form of chlorophyll. What was lacking was an unambiguous analysis giving the composition and formula of the chlorophyll pigments.

The challenge was met by Richard Willstätter (1872–1942) in collaboration with a host of co-workers, including his foremost disciple Arthur Stoll (1887–1971), with whom he produced *Untersuchungen über Chlorophyll* (Berlin, 1913). This contains the account of an extensive series of skilful investigations begun in 1906[a] which substantially removed the uncertainties regarding chlorophyll chemistry. Willstätter reviewed his work in 1914 in a lecture to the *Deutsche chemische Gesellschaft* and the extract below is taken from its published version.

The Willstätter investigations of chlorophyll involved analysis of leaf-green from more than 200 plant species, the relating of its optimal extraction to the presence of a certain amount of water in solvents, the judicious use of liquid–liquid partition methods and a sensitive application of acid and alkaline degradation methods. It turned out that chlorophyll of different species did not vary and, indeed, was a mixture of two components named by Willstätter chlorophyll *a* and chlorophyll *b*. As to previous suggestions that iron, phosphorus or magnesium might be a part of chlorophyll, only magnesium's presence has been demonstrated, together with the discovery that the metal could be removed from the chlorophylls by weak acids. It has been further established that both compounds possessed the nature of diesters derived from the interaction of dibasic acids with methyl alcohol and phytol, a hitherto unknown unsaturated, aliphatic alcohol discovered in the Willstätter laboratory ($C_{20}H_{39}OH$).

Due to the success and extensive application of various forms of chromatography – often regarded as a low-grade chemical activity in the past – its history has attracted attention recently. Among the topics touched upon has also been the historical relationship of the work of Tsvet and Willstätter on chlorophyll.[b] This is not the place for a detailed historical examination of the subject except to the extent that it throws some light on the nomenclature of the two chlorophylls. Ostensibly, the well-known Russian plant physiologist, Kliment Arkadevich Timiryazev (1843–1920), already employed in 1871 the term 'chlorophyllin' for the designation of a chlorophyll derivative. This emerges from Willstätter's second paper, also published in 1906,[c] where he adopted that name for the magnesium products obtained by alkaline hydrolysis of chlorophyll. Two years later,[d] Willstätter returned to the question of the origins of his terminology and defended his position against Tsvet's objections. To all appearances this was the first time that Willstätter referred to Tsvet in his chlorophyll publications. This is interesting in view of the way he had taken great care to acknowledge the contributions of his predecessors and contemporaries in the field, including Polish and Russian investigators whose scientific output in this area was notable.

R. Willstätter, *Ber.chem.Ges.*, *47*, 2831. 1914

'Über Pflanzenfarbstoffe'. (On plant pigments)

pp. 2846−7

The isolation of chlorophyll, in which we succeeded in the year 1911,[1] was based on the colorimetric estimation of the degree of purity of the solutions and rested on systematic increase in purity through partition methods [*Entmischungsmethoden*]. The distribution between several solvents of the substances contained in the extracts is used here in a special way in order to separate the yellow and especially the colourless substances accompanying the chlorophyll. From extracts which, because of the large amount of colourless accompanying substances, contain only 8−16 per cent of chlorophyll, by the partition operations solutions result containing 70 per cent chlorophyll. Then finally an unexpected observation helps the solution of the problem. When the chlorophyll has reached a certain degree of purity, it is indeed still easily soluble in pure petroleum ether. If one removes the ethyl or methyl alcohol by washing, the chlorophyll separates, and can be purified by reprecipitation from ether by means of petroleum ether.

At first this procedure was troublesome and the yield small. In newer experiments of Willstätter and Stoll the method was improved through alterations in the methods of extraction[2] and partition.[3]

The material for our large-scale work is mostly dried and chopped leaves. It has appeared that a considerable water content of solvents makes substantially easier and hastens the extraction of the total leaf pigment. The pigments are found in the chloroplasts also after drying in the colloidal state and are difficultly soluble. By means of a solvent which dissolves salts out of the leaf substance they are flocculated and in this way made more easily soluble.

Besides the amount of accompanying substances going into solution is increased; not any more the solvent itself but its mixture with the accompanying substances is the real means of extraction for the green pigment of the leaf, and indeed so excellent that thereby the pigment is quickly and easily almost quantitatively extracted. It appears as if just the whole chloroplast substance was carried off by the solvent with suitable water content.

The best means of solution are 85−90 percent alcohol and 80−85 percent acetone; . . .

pp. 2848−9

With the partition processes, which serve for the isolation of the chlorophyll, we observe that the two components divide themselves unequally between methyl alcohol and petroleum ether. By systematic fractionation two white homogeneous components are finally obtained from the mixture.[4] The one component, chlorophyll a, is blue-green the second, chlorophyll b, yellow-green.[5]

Their constitution is very similar in spite of their optical difference. The distinction consists in a different stage of oxidation ... corresponding to the formulae:

Chlorophyll a
$C_{55}H_{72}O_5N_4Mg$ i.e. $[C_{32}H_{30}ON_4Mg]$ (CO_2CH_3) $(CO_2C_{20}H_{39})$
Chlorophyll b
$C_{55}H_{70}O_6N_4Mg$ i.e. $[C_{32}H_{28}O_2N_4Mg]$ (CO_2CH_3) $(CO_2C_{20}H_{39})$[a]

Selection 39

At about the same time the study of the nature of photosynthesis received fresh impetus from important experimental data and theoretical considerations, thanks to the Cambridge-based plant physiologist, Frederick Frost Blackman (1866–1947). It was his earlier work[a] finalizing knowledge about the passage of carbon dioxide through the stomata of leaves which led him to raise questions about factors influencing the rate of carbon assimilation or photosynthesis. As a result of experimental evidence, accumulated in his laboratory by Gabrielle Matthaei (b. 1904), on the influence of temperature on carbon assimilation, Blackman came to appreciate that in order to understand photosynthesis, the interaction of environmental factors like carbon dioxide, light intensity and temperature must be considered.

The carbon dioxide-temperature curves indicated that the rate of assimilation was merely in part a direct function of the temperature. At any given temperature, Matthaei concluded, the process ran at its greatest rate ('maximal assimilation for that temperature') only in the presence of adequate light and a sufficient amount of carbon dioxide.[b] The analysis of experimental data provided by Matthaei led Blackman to conceive the 'principle of limiting factors' outlined and illustrated in the following extract from his theoretical paper.

F.F. Blackman, *Ann.Bot.*, 74, 281. 1905

'Optima and limiting factors'

pp. 289–91

We start this section with the following axiom.
    When a process is conditioned as to its rapidity by a number of separate factors, the rate of the process is limited by the pace of the 'slowest' factor.
    I think one may fairly express surprise at the extent to which this principle has been overlooked by those who have proposed to work out the relation between a function and some *single* one of the various factors that control it.

This desirable end often cannot be really accomplished without taking deliberate thought to the other factors, lest surreptitiously one of *them*, and not the factor under investigation, becomes the real limiting factor to an increase of functional activity.

We will consider in some detail the application of this axiom to assimilation, and briefly its application to respiration and growth.

Carbon assimilation furnishes the most instructive case for the consideration of the inter-relation of conditioning factors, because these factors are largely external ones, whereas in growth they are internal and less under control.

Let us then consider first the case of assimilation. We can recognize five obvious controlling factors in the case of a given chloroplast engaged in photosynthesis.

(1) The amount of CO_2 available,
(2) the amount of H_2O available,
(3) the intensity of available radiant energy,
(4) the amount of chlorophyll present,
(5) the temperature in the chloroplast.

In theory any one of these five might be the limiting factor in the total effect, and it is comparatively easy to experiment with (1), (3), or (5) successively as limiting factors . . .

When the rate of a function exhibits, in experiment, a sudden transition from rapid increase to a stationary value, it becomes at once probable that a 'limiting factor' has come into play. The form of curve obtained is then like the curve ABC in Figure 1, where the limiting factor has soon come into play. If the factor in question only becomes 'limiting' when the function is near its high values, then the curve ABFG represents the result attained. If the factor only 'limits' when the function is close to its highest values we may get a curve recalling the conventional optimum curve with the top cut off . . .

Suppose a leaf in a glass chamber to have enough light falling upon it to give energy equal to decomposing 5 c.c. of carbon dioxide per hour. Then, as one gradually increases the carbon dioxide in the air current through the chamber from the amount (or pressure) that causes 1 c.c. to diffuse into the leaf through its stomata up to five times that pressure, so steadily the assimilation will increase from 1 c.c. to fivefold. After that, further increase of the carbon dioxide will produce no augmentation of the assimilation, but will

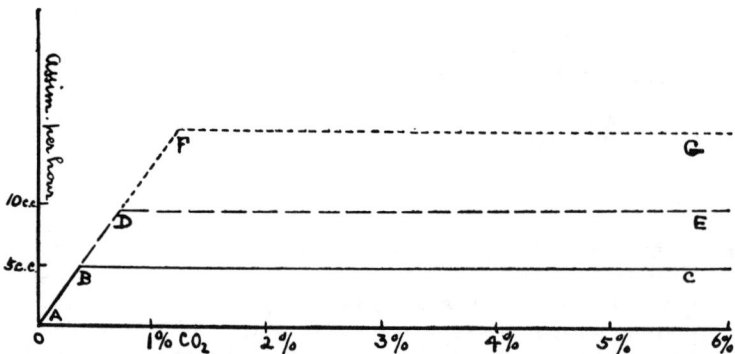

Figure 1 Diagram illustrating the 'principle of limiting factors'

give continually an effect of 5 c.c. of carbon dioxide assimilated – the light being now the limiting factor. The curve obtained will be of the form ABC. Ultimately, if the supply of carbon dioxide in the air current be increased up to 30, 50, or 70 per cent, the carbon dioxide will have a general depressing effect on the whole vitality, and before suspension of all function a diminution of assimilation undoubtedly occurs; this is, however, quite a separate process. Now, secondly, suppose the light falling on the leaf to be sufficient for the decomposition of 10 c.c. of carbon dioxide per hour, then twice the external pressure of carbon dioxide will be required to reach the limit and the angle of the curve, which will now be ABDE. With still stronger light we should get ABDFG. Those who would be prepared to admit that a curve like ABC shows an optimum, only with a very long drawn-out top, would have to further admit that for *each intensity of light falling on a leaf there is a different optimum amount of carbon dioxide*. This is not to be entertained.

The light-energy available fixes an upper limit to the carbon dioxide that can be decomposed, and when that amount is attained, which even for direct sunlight could be provided with a current of air containing less than 1 per cent. if the current were sufficiently fast, the limit of effect of carbon dioxide is reached: any more provided is wasted, and has no further effect till many times that concentration is reached and a general depressing effect comes in. Just as little can one speak of an optimum amount of carbon dioxide required to use up a fixed amount of radiant energy (i.e. a given intensity of light) as one can speak seriously of the 'optimum amount' of water required to fill a litre flask, while to attempt to speak of an optimal amount of carbon dioxide for assimilation in general is like speaking of 550 c.c. as the optimal amount of water *to fill flasks*, when the two flasks in question happen to be the one a litre flask and the other a 100 c.c. flask.

Selection 40

The concept of the 'limiting factor', developed as a by-product of studies on the effect of temperature on carbon assimilation, exerted seminal influence on subsequent work on the nature of the photosynthetic process. By pointing out the complex interplay of factors in photosynthesis it led to investigations of their actual relations. In the course of these researches something of the photosynthetic mechanism was revealed by showing that it entailed more than one step. The application of knowledge emanating from chemical kinetics and concerning the influence of temperature on the rate of chemical and photochemical change yielded this insight. It was recognized that the rate of chemical reaction doubled or trebled when temperature rose whereas the rate of photochemical reaction changed little. In other words, the temperature coefficients of chemical reaction velocity (Q_{10}) were found to have the value between 2 and 3 and those of photochemical reactions approached low values of about 1. As measurements in the laboratory yielded evidence that a rise in temperature would not always augment the rate of assimilation it became increasingly probable that photosynthesis involved a chemical step, probably conditioned by the presence of enzymes, and a photochemical step. Among those studying photosynthesis on these lines were Willstätter and Stoll. They assembled the fruit of their labours in the book *Untersuchungen über die Assimilation der Kohlensäure* (1918) consisting of seven studies from which the conclusion summarizing their findings is published below.

While Blackman's concern was the relations of external factors to assimilatory activity, the German-Swiss pair concentrated on internal factors which they related to chlorophyll and to an enzyme present in the protoplasm. That they regarded chlorophyll as the most conspicuous internal factor is not surprising. They associated the chemical stage of photosynthesis with the formation of a chlorophyll-carbon dioxide compound and the subsequent photochemical stage with its isomerization, resulting in the production of a compound belonging to the group of peroxides. They suggested an enzymic decomposition of the peroxide into oxygen, formaldehyde and re-constituted chlorophyll. Willstätter and Stoll appear to have accepted the then widely held view regarding the formation of carbohydrates on the basis of the polymerization of formaldehyde, despite the lack of any hard evidence that it was actually an intermediate assimilatory product from which the carbohydrates were formed in green plants.[a]

R. Willstätter and A. Stoll, *Untersuchungen über die Assimilation der Kohlensäure* **(Berlin, 1918)**

(*Investigations into the Assimilation of Carbonic Acid*)

pp. 433–7

Our work was devoted to the question by what chemical means the decomposition of carbonic acid happens through sunlight in the chlorophlast. It was investigated whether and in what way the chlorophyll reacts chemically in the assimilation process, whether a role of the carotenoids can be shown in the life processes of the plant, and in what manner constituents of the colourless stroma, which are to be determined more precisely, act together with the chlorophyll.

A function of the yellow pigments could be shown neither with assimilation nor in respiration. The chlorophyll, on the other hand, combines with [its] importance (which obviously is conditioned by its pigment nature) a more difficultly recognizable function which depends on its power of reacting chemically. The pigment is decomposed by carbonic acid with splitting off of magnesium; a dissociable carbonic acid compound is an intermediate of the reaction. The behaviour towards carbonic acid was tested with the pigment in the state most similar to its dispersion in the chloroplast, that is in its hydrosol.

On the observation that chlorophyll, indeed both components a and b, form dissociable addition products with carbonic acid a theory of assimilation is founded. The absorbed light performs its chemical work on the chlorophyll molecule itself (of which the carbonic acid becomes a constituent) through its addition to the magnesium complex, while through regrouping of the valencies the carbonic acid molecule isomerizes into a form appropriate to spontaneous breakdown (fourth study). On the addition of carbonic acid to the light absorbent the reaction is distinct from the effect of other sensitizing agents. This consideration must leave undecided whether the carbonic acid as such adds to the chlorophyll, of which it is quite capable, or whether a carbonic acid derivative is attached. Not the chlorophyll alone, but the unilluminated leaf, that is constituents of the leaf substance (which are not individually determined) combine with the carbonic acid to form loose, dissociable addition-products. It is probable that in this way the transfer of the carbonic acid from the air to the chloroplasts is mediated, the rate of the carbonic acid uptake is increased and the form of the carbonic acid is altered (third study).

This explanation of the action of the chlorophyll through addition and rearrangement of the carbonic acid has nothing in common with the conception that in the assimilation process the chlorophyll is destroyed and again built up. Such assumptions are contradicted by the demonstration (first study) that the chlorophyll in its amount and also in the relation of its components remains unaltered during assimilation, also with any arbitrarily increased and long lasting performance. The connection between assimilatory performance and the amount of chlorophyll could, as this remains constant, be followed under such conditions that the external factors – carbonic acid partial pressure, illumination and temperature – were without influence on the performance. The quotient of the assimilated carbonic acid and the chlorophyll amount, the photosynthetic number [*Assimilationszahl*] undergoes great fluctuations according to the concentration of chlorophyll in the leaves, further with growth and time of year. From the more exact investigation of the cases in which the photosynthetic number deviates furthest from the norm it was to be concluded (second study) that, apart from the pigment, a second internal factor of enzymic nature is determining for the assimilative process, and indeed probably an enzyme active in the breakdown of the intermediate product form from chlorophyll and carbonic acid. This result is in agreement with the observation that a very small oxygen content of the leaf is indispensable for the assimilation process. An agent acting together with chlorophyll in assimilation appears to react as a dissociating oxygen compound (sixth study).

The question concerning the reduction product which is condensed to carbohydrate is closely connected with consideration of the process in which oxygen is split off from the carbonic acid. Von Baeyer's explanation that formaldehyde is the intermediate stage in sugar formation is much disputed and, in a manner not permissible, it is often attempted to prove this still hypothetical assumption, for example by the demonstration of formaldehyde in the leaves.

Unequivocally, without hypothesis, it is proved that the carbonic acid is deoxygenated to the reduction stage of carbon itself or, which is quite identical, to the formaldehyde stage, when it is shown that in assimilation exactly and irreversibly the total oxygen is removed from carbonic acid. There has been much concern over the total gas exchange of the plant, but there are only isolated and incomplete estimations of the purely assimilatory gas exchange.

Our research (fifth study) deals with the assimilation gas exchange at highly increased assimilation performance. Thus the influence of respiration is excluded and a rigorous estimation of the photosynthetic [*assimilatorischen*] coefficient is made possible. At the same time this arrangement followed the aim of enforcing, with the increased performance under different conditions (it may be in the beginning or after longer time), deviations of the coefficient if this is at all possible. The result was: the coefficient amounts to 1 and is constant. An intermediate stage of reduction like oxalic acid, formic acid and the like therefore does not become free. If the reduction on the chlorophyll follows step by step no carbon compound is liberated before the complete removal of oxygen from the chlorophyll.

As it is the formaldehyde stage to which carbonic acid decomposition leads, it is an assumption of great probability that not only is the stage attained but that formaldehyde itself is formed. For it is the single carbon compound of this substitution grade with only one carbon atom in the molecule. All organic compounds of the same composition are derivatives of formaldehyde, indeed its further condensation products.

As one can demonstrate formaldehyde in the greatest dilution, many workers have already undertaken experiments to obtain its formation from carbonic acid outside the living cell by the action of chlorophyll. But the traces of aldehyde which were often observed in such experiments have originated through photooxidation and, indeed, generally from substances accompanying the chlorophyll. Now the possibility of working with the pure pigment and of adapting the experimental conditions to the relations in the chloroplasts better than happened earlier led us to do this: also in the experiments in light to test the carbonic acid decomposition or also to search for the formation of a peroxide compound (seventh study). All these experiments were undoubtedly unequivocal and completely negative. They are, on that account, not without value as in a field, which was yielding a sham harvest, they wipe the slate clean. A step forward will only be possible after realizing that the illumination of chlorophyll in an atmosphere of carbonic acid does not suffice, and that in this experimental arrangement still essential conditions are lacking for imitation of the assimilation process.

The investigation of the pigments in the green plants has advantage over that of the constituents of the colourless protoplasm though these are also essential for assimilation. Here it is up to chemical analysis to describe more fully the assimilation apparatus. In the leaf the chlorophyll, being in the state of pure hydrosol, is completely protected against photooxidation to which it is subjected. In the leaf the chlorophyll is protected from decomposition by carbonic acid observed in the pure colloid, without hindrance of its uptake. On the contrary in the leaf the carbonic acid is absorbed with far greater rapidity than with the action of even undiluted carbonic acid on the hydrosol.

Thus regarding the state of the chlorophyll in the chloroplast, regarding the form into which the carbonic acid changes and with regard to the enzymes active in the assimilation process new questions have emerged, because the work has furnished a deeper insight into the differences between the conditions of the assimilation experiment and the relations in the living cell.

Selection 41

The beginning of the third decade of the twentieth century saw the consolidation of the idea that photosynthesis was not a single process. One of Blackman's pupils, George Edward Briggs (1893–1985) adopted in 1920 the suggestive word 'dark' for the chemical stage of photosynthesis,[a] which has remained in use. Two years later Otto Warburg (1883–1970) described the slowly progressing dark reaction depending on high illumination intensity as the 'Blackman reaction' and he believed that under these conditions this reaction determined the rate of decomposition of carbon dioxide, that is the rate of photosynthesis.[b]

Among the work produced during the 1920s Warburg's novel methodological approach to the study of photosynthesis occupies a prominent place. By modifying the Haldane-Barcroft arrangement for the estimation of blood gases he developed manometry as a means for the study of gas exchange in photosynthesis (and other biochemical processes). The pressure of a gas at constant volume and constant temperature was measured. Changes in pressure, gauged by means of the Warburg apparatus (as the device became known), made it possible to measure rates of gas exchange and thus to obtain a more precise picture of biochemical reactions accompanied by the production or consumption of gas.[c]

His other innovatory device had been the replacement of leaves with the unicellular alga *Chlorella* as the experimental simple prototype of a green plant. Extending the work of Blackman and Willstätter, Warburg investigated in detail the relations of the photochemical and chemical stages in photosynthesis by also making use of intermittent illumination. Closely allied to these questions were his researches into the effects of substances like hydrocyanic acid and urethane on photosynthesis.[d]

By combining manometry with bolometry (measurement of small amounts of radiant heat) and by working with light of various wavelengths (436–660 μμ), together with E. Negelein (*b*.1897) he enquired into the efficiency of the photosynthetic process. From the relation between the energy used, (W), and the energy absorbed by the plant, (E), they derived the values for efficiency: $\phi = \dfrac{W}{E}$ and the limiting value ϕ_0, where E = O. They found about 60 per cent of the absorbed radiation was used. Stimulated by the developments in the field of quantum theory as applied to light, Warburg calculated how many quanta of light were required to decompose, that is to assimilate, one molecule of carbon dioxide. From the start the figures obtained – of about 4–5 quanta – were disputed.

O. Warburg and E. Negelein, *Z.Physik.Chem.*, *106*, 191. 1923

'Über den Einfluss der Wellenlänge auf den Energieumsatz bei der Kohlensäureassimilation' (On the influence of the wavelength on the energy exchange during carbonic acid assimilation)

pp. 205–206

The relation $\dfrac{W}{E}$ – the chemical effect brought about by one calorie of absorbed radiation – which we according to E. Warburg[1] label ϕ, depends with carbonic acid assimilation on the intensity of the radiation used. With increasing intensity ϕ becomes smaller and with decreasing intensity approaches a limiting value ϕ_0, which we designate as the 'yield' (*Ausbeute*) with carbonic acid assimilation.

In order to find ϕ_0, we have earlier measured two ϕ-values and then calculated ϕ_0 by extrapolation to zero intensity. As indeed the course of the ϕ-curve is unknown, we have given up the calculation of the yield by extrapolation. We now measure at intensities as low as possible. If ϕ does not change considerably with the intensity in the sphere in which we are measuring, then we consider the value measured at the lowest intensity as the yield of 63.5 per cent.

Thus we find the yield lower than perhaps it is, and lower than it was calculated in the previous work by extrapolation. As against the mean value of 70 per cent which was earlier obtained by *calculation*, there now stands – in red light – a *measured* value of 59 per cent and a measured maximum value of 63.5 per cent.

pp. 210–12

The result of our experiments is *that the yield with carbonic acid assimilation decreases with decreasing wavelength* – in the direction of red towards blue . . .

It is known that the quantum theory anticipates a decrease of the photochemical yield with decreasing wavelength (photochemical equivalent law of Einstein) and Emil Warburg has found this decrease in the yield demanded by the theory with photolysis of hydrobromic and hydriodic acids.[2]

We must now explain by the quantum theory the make of the yield with wavelength also in our case.

As one quantum in none of our spectral regions suffices for the breakdown of *one* carbonic acid molecule, the expression

$$n = \frac{Q}{h\nu}$$

must from the start be excluded, if we understand by n the number of decomposed carbonic acid molecules, by Q the absorbed irradiation energy. However

$$n = k\frac{Q}{h\nu} \tag{1}$$

would be possible (k being a proportionality factor), a statement which affirms: the number of carbonic acid molecules is *proportional* to the number of absorbed quanta and which includes the assumption that each absorbed quantum produces the chemical effect . . .

There remains still the calculation of the factor k in equation (1); k means the number of carbonic acid molecules which are decomposed by *one* quantum, or $\frac{1}{k}$ the number of quanta which are necessary for the breakdown of one carbonic acid molecule.

For the calculation of k we divide both sides of the equation by Avogadro's number N_o and obtain

$$\frac{n}{N_o} = k\frac{Q}{h\nu N_o}$$

from which

$$k = \phi_0 \cdot h\nu \cdot N_o, \; \phi_0 \text{ expressed in moles/calorie.}^a \tag{2}$$

If we put in equation (2) the *mean values* of ϕ_0 in moles/calorie we find

Wavelength $\mu\mu$	$\phi_0 \left[\dfrac{moles}{cal}\right]$	$N_0 h\nu$ [cal]	k	$\dfrac{1}{k}$
660	$5.25 \cdot 10{-}6$	43,000	0.226	4.4
578	$4.75 \cdot 10{-}6$	49,200	0.234	4.3
436	$3.01 \cdot 10{-}6$	65,100	0.196	5.1

If we put in equation (2) the *maximum values* of ϕ_0 in moles/calorie we find

Wavelength $\mu\mu$	$\phi_0 \left[\dfrac{moles}{cal}\right]$	$N_0 h\nu$ [cal]	k	$\dfrac{1}{k}$
660	$5.67 \cdot 10{-}5$	43,000	0.244	4.1
578	$5.42 \cdot 10{-}5$	49,200	0.267	3.8
436	$3.26 \cdot 10{-}5$	65,100	0.213	4.7

One sees from the last column of the table that in the red and yellow about four quanta, in the blue about five quanta are necessary for the decomposition of *one* carbonic acid molecule. Whether now the reduction of one carbonic acid molecule is not possible with less than four quanta, or whether with improvement of cultivation methods higher yields will appear, which means smaller values of $\dfrac{1}{k}$, is a question the answer to which we must leave to the future.

Selection 42

Following Warburg's studies with intermittent illumination, Robert Emerson (1903–59) and William Arnold (*b.* 1904) employed the same approach a decade later, in order to provide more information about the nature and relations of the photochemical and dark stages of photosynthesis. Although they assumed that the photochemical stage, involving a partial reduction of carbon dioxide, preceded the nonphotochemical phase during which the reduction was completed, they were prepared to visualize the process as a cycle. In that case the question of precedence or sequence of the light and dark reactions became immaterial, but not so the problem of their duration. The results of the American workers showed that whereas for the light reaction to take place a flash lasting about 10^{-5} seconds was necessary, the maximum yield of oxygen per flash required a dark interval of about 0.03 to 0.4 of a second depending on the temperature.

R. Emerson and W. Arnold, *J.gen.Physiol.*, **15, 391. 1932**

'A separation of the reactions in photosynthesis by means of intermittent light'

pp. 392–3

The experiments described in this paper indicate, we think, that the steps in photosynthesis which proceed in the dark involve what has hitherto been known as the Blackman reaction. Probably the reduction of carbon dioxide is not completed during the photochemical part of the process. A more correct way of representing the sequence of events in intermittent light would be as follows. Two steps are involved in the reduction of carbon dioxide: a reaction in which light is absorbed, followed by a reaction not requiring light – the so called Blackman reaction. If the light intensity is high the photochemical reaction is capable of proceeding at great speed, but in continuous light it can go no faster than the Blackman reaction. We suppose that the product formed in the photochemical reaction is converted to some other substance by the Blackman reaction, and at the same time the chlorophyll is set free to take part again in the photochemical reaction. If a green cell is illuminated, we think that the photochemical reaction proceeds rapidly until an equilibrium concentration of its product is formed. After this the photochemical reaction proceeds only as fast as the Blackman reaction removes the intermediate product. If the cell is now darkened, the photochemical reaction stops at once, but the Blackman reaction continues until its raw material, the product formed by the photochemical reaction, is exhausted. After this nothing further happens until the cell is again illuminated. Higher efficiency of the light would be obtained if each light flash lasted only long enough to build up the equilibrium concentration of the intermediate product, and each dark period were long enough to allow the Blackman reaction time to use up all the intermediate product present at the moment the light period ended. In Warburg's flicker experiments the light and dark periods were always of equal length. He found that the amount of work done by the light could be increased by shortening both the light and the dark periods. This indicates that his light periods were too long for maximum efficiency. In the latter part of each light period the photochemical reaction must have been brought down to near the speed of the Blackman reaction.

Using 133 light flashes per second, Warburg obtained an improvement of 100 per cent over the continuous light yield. We were able to improve the continuous light yield 300 per cent to 400 per cent by using only 50 flashes per second and making the light flashes much shorter than the dark periods. This opened the possibility of determining the length of the dark period necessary for the complete removal of the intermediate product formed in a light flash of given intensity and duration.

p. 417

The experiments described in this paper show that photosynthesis involves a light reaction not affected by temperature, and capable of proceeding at great speed, and a dark reaction dependent on temperature, which requires a relatively long time to run its course. The light reaction can take place in about a hundred-thousandth of a second. The dark reaction requires less than 0.04 second for completion at 25°C., and about 0.4 second at 1.1°C. The light reaction is dependent on carbon dioxide concentration and is inhibited by narcotics. The dark reaction is not noticeably inhibited by narcotics, is independent of carbon dioxide concentration, and is strongly inhibited by cyanide.

Selection 43

The fruitful idea that photosynthesis consisted of light and dark stages led to a significant accumulation of experimental and theoretical knowledge. Nevertheless, in the early 1930s, it could hardly be maintained that the chemical aspects of carbon dioxide assimilation were really understood. It was at this time that research on photosynthesis began to benefit from the combined impact of experimental work on photosynthesis of bacteria and its interpretation in terms of the removal and addition of hydrogen, derived from the new thinking on oxidation and reduction. The complexion of photosynthesis started to change, primarily due to Cornelis Bernardus van Niel (*b.* 1897), a product of the influential Dutch school of microbiology at Delft, which by subscribing to the idea that biochemically life constitutes a unity advanced the subject matter of comparative biochemistry.[a] As shown below, van Niel drew attention to the essential similarity of the way green plants and light dependent bacteria made use of the hydrogren transport mechanism for the reduction of carbon dioxide. At the same time he pointed out the difference in respect to the evolution of oxygen, which accompanied photosynthesis of green plants but not that of photosynthetic bacteria. What was even more astounding was that he attributed the origin of oxygen to water, the latter also acting as the hydrogen donor.

C.B. van Niel, *Cold Spr.Harb.Symp.*, *3*, 138. 1935.

'Photosynthesis of bacteria'

p. 139

The biological conversion of CO_2 into organic matter dependent upon illumination is photosynthesis. Inasmuch as this conversion is carried out by the green and purple sulphur bacteria we must, therefore, consider these organisms photosynthetic.

On the other hand, a study of the photosynthetic activity as displayed by the green plants has revealed that fundamentally this process can be expressed by the equation

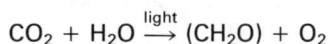

$$CO_2 + H_2O \xrightarrow{\text{light}} (CH_2O) + O_2$$

The absence of any evidence of O_2-production by the bacteria thus seems a considerable obstacle to the acceptance of such a process in these organisms. Yet, if one compares the metabolism of the green bacteria – and also the early stages of the metabolism of the Thiorhodaceae in H_2S containing media, during which there is a rapid disappearance of H_2S while the cells store up elementary sulphur – with photosynthesis of the green plants, the relationship at once becomes more clear. For the quantitative connection between the participants in the metabolism of the bacteria can be expressed by the equation:

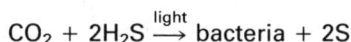

$$CO_2 + 2H_2S \xrightarrow{\text{light}} \text{bacteria} + 2S$$

This equation is a first approximation and implies a composition of the bacteria corresponding to that of a carbohydrate. But then one might also write this equation as follows:

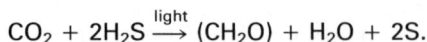

$$CO_2 + 2H_2S \xrightarrow{light} (CH_2O) + H_2O + 2S.$$

In doing so the above-mentioned relation can be paraphrased in the following terms: whereas in the green-plant-photosynthesis H_2O, interacting with CO_2, gives rise to O_2-production, the bacterial photosynthesis, starting with H_2S instead of H_2O, leads to S.

It is well known that in 1913 H. Wieland began to bring forth evidence in favor of the viewpoint that the mechanism of respiration processes can ultimately be reduced to a dehydrogenation of the substrate[1]. In 1925 this idea was extended by Kluyver and Donker to cover all metabolic processes[2] The most general form in which a metabolic process might then be expressed is

$$H_2A + B \rightarrow A + H_2B.$$

On the basis of this hypothesis the two types of photosynthetic reactions can then be regarded as special representatives of a generalized photosynthetic process:

$$CO_2 + 2H_2A \xrightarrow{light} CH_2O + H_2O + 2A$$

The experimental evidence so far presented in connection with the metabolism of the green and purple sulphur bacteria fits in with this formulation which implies that for different photosynthetically active organisms different hydrogen donors for the final reduction of CO_2 may be required.

This formulation also shows at least the possibility of the occurrence of photosynthetic processes in which the hydrogen donor H_2A is neither H_2O nor H_2S.

pp. 142–3

If one tries to understand the meaning of the generalized equation for photosynthesis it becomes clear that all those mechanisms proposed for the photosynthetic reaction which imply the formation of a carbonic acid-chlorophyll complex which is subsequently transformed into a formaldehyde peroxide are not quite in accordance with the formulation of photosynthesis as an oxidation-reduction process. Such schemes fail to give a satisfactory explanation for the photosynthetic process carried out by the green and purple bacteria (See also note 3).

From a unified point of view, as laid down in the generalized equation, green plant photosynthesis should be considered as a reduction of CO_2 with hydrogen obtained from H_2O, and the oxygen produced during illumination as dehydrogenated H_2O.

Selection 44

Van Niel's picture of photosynthesis as a process of oxidoreduction was confirmed by Robert Hill (*b.* 1899) in a series of striking investigations performed during the second half of the 1930s until 1940. Hill's approach was independent of the earlier work on van Niel as it derived from his prior interest in blood pigments. He had previously devised a spectroscopic method, based on the recognized affinity of myoglobin for oxygen, for the measurement of oxygen evolved by illuminated isolated chloroplasts. Later, partly in collaboration with R. Scarisbrick, he showed that the evolution of oxygen by preparations of isolated chloroplasts in the light took place in the presence of ferric oxalate and could be hindered by urethane. As will be seen, these observations were interpreted as a photosynthetic light reaction created *in vitro* in the absence of carbon dioxide. Under these circumstances it was concluded that the production of molecular oxygen went together with the reduction of ferric oxalate acting as a hydrogen acceptor. This type of 'chloroplast reaction' became known as the 'Hill reaction', a term apparently introduced by the American researchers C.S. French (*b.* 1907) and M.L. Anson (1901–68) in 1941.[a]

R. Hill and R. Scarisbrick, *Proc.Roy.Soc.(B).*, *129*, 238. 1940.

'The reduction of ferric oxalate by isolated chloroplasts'

pp. 238–9

The photosynthesis of green plants is characterized by the production of molecular oxygen when the living cells containing chlorophyll are illuminated. The oxygen, however, is only produced when carbon dioxide is taken up, and no other substance at present known will cause the photosynthetic reaction on the living cell. The chloroplasts, which are organized subcellular units containing chlorophyll, will themselves evolve traces of oxygen in light after removal from the cells. It was shown (Hill 1939)[a] that, in light, chloroplasts would reduce ferric oxalate to ferrous oxalate and that an equivalent amount of oxygen was set free. This reaction, here referred to as 'the ferric oxalate reaction', was capable of measurement and is probably the first indication of any measurable activity apart from the whole cell which has any bearing on photosynthesis. The chloroplasts, after removal from cells, have this property of evolving oxygen from certain hydrogen acceptors, but carbon dioxide will not act as a hydrogen acceptor. Moreover, the oxygen produced by isolated chloroplasts is maintained only at a very low pressure of 2–4 mm. Hg; a possible explanation of the low pressure of oxygen is the re-oxidation of some hydrogen acceptor. In spite of this, however, with the help of the respiratory pigment, haemoglobin, it is shown that the chloroplasts over this low range of oxygen pressure have a high activity when compared with other biological systems. The fact that the chloroplasts will produce molecular oxygen from substances other than carbon dioxide, gives a possibility of analysing the photosynthetic mechanism from a more chemical point of view. Hence it is of primary importance to establish, by varying external and internal factors, the behaviour of the chloroplast outside the living cell. The present paper is concerned with the effect of varying light intensity and the inhibiting effect of urethane on the ferric oxalate reaction.

p. 254

The ferric oxalate reaction with chloroplasts resembles photosynthesis in producing molecular oxygen but differs essentially in that carbon dioxide does not act as a hydrogen acceptor. The inhibition by urethane is closely similar for photosynthesis in chlorella cells and isolated chloroplasts of Stellaria; cyanide and hydroxylamine, while they inhibit photosynthesis, do not prevent the production of oxygen with ferric oxalate. If, then, the ferric oxalate reaction corresponds with the photochemical reaction these facts are in agreement with the conclusions of Warburg[b] from his work on Chlorella: urethane affected the photochemical reaction while cyanide affected the dark reaction in which carbon dioxide was involved. The new conclusion that can be drawn from the work on isolated chloroplasts is that oxygen itself is formed in a photochemical reaction during which there is no reaction involving carbon dioxide.

Selection 45

Although the work of van Niel and Hill undermined the long prevailing notion that carbon dioxide is the source of the photosynthetic oxygen, few of their contemporaries were conditioned to recognize water immediately as the alternative source. It should be added that, apart from van Niel's conjecture, this view was expressed before by Stoll who propounded in 1932 that the photolysis of water could provide the molecular oxygen and the hydrogen needed for the reduction of carbox dioxide.[a] Valid evidence supporting this suggestion had to wait until the introduction of the isotope technique in the analysis of photosynthesis, mainly under the influence of Samuel Ruben (1913–43).[b]

S. Ruben, M. Randall, M. Kamen and J.L. Hyde, *J.Am.Chem.Soc.*, *63*, 877. 1941

'Heavy oxygen (O^{18}) as a tracer in the study of photosynthesis'

pp. 877–8

It is generally agreed that the net reaction for green plant photosynthesis can be represented by the equation

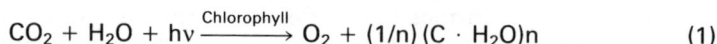

$$CO_2 + H_2O + hv \xrightarrow{\text{Chlorophyll}} O_2 + (1/n)(C \cdot H_2O)n \qquad (1)$$

and also that very little is known about the actual mechanism. It would be of considerable interest to know how and from what substance the oxygen is produced. Using O^{18} as a tracer we have found that the oxygen evolved in photosynthesis comes from water rather than from the carbon dioxide.

   The heavy oxygen water used in these experiments was prepared by fractional distillation[1] and was distilled from alkaline permanganate before use. The isotopic oxygen content was determined by the method of Cohn and Urey[2] using carbon dioxide and a mass spectrometer. Heavy oxygen carbonate was prepared by allowing a solution of potassium acid carbonate ($KHCO_3$) in heavy oxygen water to come to approximate isotopic equilibrium, adding a nearly equivalent quantity of potassium hydroxide and distilling off the water, finally drying in an oven at 120°. Isotopic analysis of this carbonate or of the carbonate in a solution, was performed by rendering the solution sufficiently alkaline to prevent exchange[3] and precipitating calcium carbonate. The calcium carbonate after filtering, washing and drying at 120°, was calcined at red heat in an evacuated platinum bulb connected to the gas handling system of the mass spectrometer, and the evolved carbon dioxide analyzed for heavy oxygen.

Young active Chlorella cells were suspended in heavy oxygen water (0.85% O^{18}) containing ordinary potassium bicarbonate and carbonate. Under these conditions the oxygen exchange between the water and bicarbonate ion is slow and readily measurable.[3] The isotopic ratio in the evolved oxygen was measured with a mass spectrometer. In other experiments the algae were allowed to carry on photosynthesis in ordinary water and heavy oxygen potassium bicarbonate and carbonate. The results of these experiments are summarized in Table 1.

It is apparent that the O^{18}/O^{16} ratio of the evolved oxygen is identical with that of the water. Since the oxygen in OH, COOH, O—O, C=O, etc., groups exchanges but very slowly[4] with water at room temperature and moderate pH, it seems reasonable to conclude that the oxygen originates solely from the water. While this conclusion makes it possible to reject many of the suggestions proposed in the past[5] it does not enable a choice to be made between the several more recent hypotheses. However it is of interest to note that van Niel[6] has specifically suggested that the oxygen may arise by a dehydrogenation of water.

Table 1 Isotopic ratio in oxygen evolved in photosynthesis by *Chlorella*[a]

Expt.	Substrate	Time between dissolving $KHCO_3$ + K_2CO_3 and start of O_2 collection, minutes	Time at end of O_2 collection, minutes	Percent. O^{18} in — H_2O	HCO_3^- + CO_2^-	O_2
1	0.09 M	0		0.85	0.20	..
	$KHCO_3$	45	110	0.85	0.41[b]	0.84
	+0.09 M	110	225	0.85	0.55[b]	0.85
	K_2CO_3	225	350	0.85	0.61	0.86
2	0.14 M	0		0.20		..
	$KHCO_3$	40	110	0.20	0.50	0.20
	+0.06 M	110	185	0.20	0.40	0.20
	K_2CO_3					
3	0.06 M	0	..	0.20	0.68	..
	$KHCO_3$	10	50	0.20		0.21
	+0.14 M	50	165	0.20	0.57	0.20
	K_2CO_3					

[a] The volume of evolved oxygen was large compared to the amount of atmospheric oxygen present at the beginning of the experiment. [b] These are calculated values.

Notes to Section II

Selection 31

Introduction:

a.   The term 'photosynthesis' which gradually replaced 'carbon assimilation' after the turn of the century, apparently, originated in American botanical circles in the 1890s. See Ch. A. Barnes, 'So-called assimilation', *Bot.Centrbl.*, 76, 257–9. 1898.

b.   [J.B.v.] Helmont, *Oriatrike or Physick refined* (London, 1662). At times not always comprehensible translation of Helmont's posthumous *Ortus medicinae* (Amsterdam, 1648).

c.   Cf. J. Parascandola amd A.J. Ihde, 'History of the pneumatic trough', *Isis*, 60, 351–61. 1969.

d.   Georg Ernst Stahl (1659–1734) deriving the term from *phlogistos* ('inflammable' in Greek) conceptualized it by the end of the second decade of the eighteenth century. The events in this and the following two selections have received attention from M. Teich, 'Circulation', Introduction to Selection 1, Section I, note f.

e.   The method of estimating the 'goodness' of air by means of vessels called 'eudiometers', elaborated by Priestley, became widely adopted. Equal volumes of common air and 'nitrous air' (nitric oxide) were mixed over water and the resulting diminution of volume (due to the absorption of nitric peroxide in water) was measured. Although at times he estimated the proportion of 'dephlogisticated air' in common air to be one-fifth, Priestley also gave other figures showing lesser and higher proportions. The diverse degrees of 'goodness' of air, determined by Priestley and others by means of eudiometry and indicating a varying composition of air, were already contested at the time and are of historic interest as an example of early gasometry.

f.   On Giuseppe Angelo, Count of Saluzzo (1734–1810) consult J.R. Partington, *A History of Chemistry* (London, 1962), III, pp. 127, 252.

Selection 32

Introduction:

a.   J. Priestley, *Experiments and Observations Relating to Various Branches of Natural Philosophy* (London, 1779), I, pp. 338f.; *Experiments and Observations Relating to Various Branches of Natural Philosophy* (London, 1781), II, pp. 16f.

b.   Known also as John Ingenhousz.

c.   Carl Wilhelm Scheele (1742–1786) observed that plants removed from a dark cellar produced a 'green resin' in sunlight (1777). He did not pursue this aspect of the subject further. Cf. C.W. Scheele, *Chemische Abhandlung von der Luft und dem Feuer. Nebst einem Vorbericht von Torbern Bergmann*, Facsimile of the original edition (Uppsala and Leipzig, 1777; Stockholm, 1970), p. 126.

Selection 33

Text:

a.   For Senebier this correlativity of natural processes stood out as a further proof of the Creator's teleological sway.

b.   Charles Bonnet (1720–93) is remembered for his influential experimental and methodological work in natural history and biology, which included problems related to photosynthesis. Cf. *Recherches sur l'usage des feuilles dans les plantes etc.* (Göttingen and Leiden, 1754).

Selection 34

Introduction:

a.   A more thorough analysis of the contribution of the Dutchman and the Swiss is called for. Accounts tend to emphasize the flair of Ingen-Housz and the verbosity of Senebier. So, for example, Senebier's pioneering investigations into the influence of light of different colours on the growing plants are hardly recognized.

b.   The historically surprising suggestion has been made by a leading scientist in the modern field of photosynthesis that de Saussure could have deduced from his experimental results that the source of oxygen was water. E.I. Rabinowitch, *Photosynthesis and Related Processes* (New York, 1945), I, p. 24.

Selection 35

Introduction:

a.   [P.J.] Pelletier and [J.B.] Caventou, 'Sur la matière verte des feuilles', *J.Pharm.*, *33*, 486–91. 1817.

b.   H. von Mohl, 'Untersuchungen uber die anatomischen Verhältnisse des Chlorophylls (Dissertation vom Jahre 1837)', reprinted in: *Vermischte Schriften botanischen Inhalts* (Tübingen, 1845), pp. 349–61.

c.   C. Nägeli, 'Die Stärkekörner' in C. Nägeli and C. Cramer (eds), *Pflanzenphysiologische Untersuchungen* (Zurich, 1858), fasc. 2, p. 383.

d.   J.J. Berzelius, *Lehrbuch der Chemie*, 5th edn (Dresden and Leipzig, 1847), IV, p. 118.

Text:

a.   *Zea Mais*, *Helianthus annus*, *Apium graveolans*, *Allium Cepa*, *Beta vulgaris*, *Phaseolus multiflorus* and *vulgaris*, *Cucurbita*.

Selection 36

Text:

1.   The last fact seems to me, like the flocking to *Euglena* and similar green organisms in the light to be explicable only on the assumption of a capability of sensation. Whoever studies exactly the movements of the bacteria, especially with stronger magnification, will not be able to dismiss from one's thought that they arouse so decidedly the appearance of voluntary intelligent movements like any such movements of microorganisms of decidedly animal type. This, taken together with the enormous oxygen requirement and the strong carbon dioxide emission, assigns the mobile bacteria with greater certainty to a place amongst the animal beings 'endowed with soul', rather than their morphological connections place them with plant organisms.

Selection 37

Introduction:

a.   L. Marchlewski, *Die Chemie des Chlorophylls* (Hamburg and Leipzig, 1895); *Die Chemie der Chlorophylle und ihre Beziehung zur Chemie des Blutfarbstoffes* (Braunschweig, 1909).

b.   M. Nencki, 'Uber die biologischen Beziehungen des Blatt- und des Blutfarbstoffes' (1896) in *Opera omnia* (Braunschweig, 1904), II, pp. 573–8. M.H. Bickel published a brief account of Nencki's work (including a biographical sketch), *Marceli Nencki (1847–1901)* (Bern, Stuttgart, Vienna, 1972) in *Berner Beiträge zur Geschichte der Medizin und Naturwissenschaften* (NF, 5).

c.   G.G. Stokes, 'On the supposed identity of biliverdin with chlorophyll with remarks on the constitution of chlorophyll', *Proc.Roy.Soc.*, *13*, 144–5. 1864.

d.   A brief readable account of the life and work of Sorby who pioneered studies of the microconstituents of steel and also of biological pigments by the aid of microspectroscopy, will be found in N. Higham's *A very scientific gentleman* (Oxford–New York, 1963), with a Foreword by C.S. Smith.

e.   H.C. Sorby, 'On comparative vegetable chromatology', *Proc.Roy.Soc.*, *21*, 442–83. 1872–3.

f.   Known also as Michael Tswett, see T. Robinson, 'Michael Tswett', *Chymia*, *6*, 146–61. 1960.

Text:

1.   My xanthophyll β (obviously identical with Sorby's yellow xanthophyll) . . .

Selection 38

Introduction:

a.   R. Willstätter, *Untersuchungen uber Chlorophyll*; R. Willstatter and W. Mieg, 'I. Über eine Methode der Trennung und Bestimmung von Chlorophyll-Derivaten', *Ann.*, *350*, 1–47. 1906.
b.   R.L.M. Synge, 'Tsvet, Willstätter, and the use of adsorption for purification of proteins', *Arch.Biochem.Biophys.*, *Suppl. 1*, 1–6. 1962.
c.   R. Willstätter, 'II. Zur Kenntniss der Zusammensetzung des Chlorophylls', *Ann.*, *350*, 48–82. 1906.
d.   R. Willstätter, 'Untersuchungen über Chlorophyll'; R. Willstätter and A. Pfannenstiel, 'V. Über Rhodophyllin', *Ann.*, *358*, 205–65. 1907.

Text:

1.   R. Willstätter and E. Hug, *Ann.*, *380*, 177 [1911].
2.   R. Willstätter and A. Stoll, *Untersuchungen über Chlorophyll* [Berlin, 1913], Chapter III.
3.   *Untersuchungen*, Chapter VI.
4.   R. Willstätter and M. Isler, *Ann.*, *390*, 269, 327 [1912]; R. Willstätter and A. Stoll, *Untersuchungen über Chlorophyll*, Chapter VI, 3 and 4.
5.   On the absorption spectra of the components and first derivatives of chlorophyll, see R. Willstätter, A. Stoll and M. Utzinger, *Ann.*, *385*, 156 [1911], as well as R. Willstätter and A. Stoll, *Untersuchungen über Chlorophyll*, Chapter XXV.
a.   The square brackets in the formulae are Willstätter's.

Selection 39

Introduction:

a.   F.F. Blackman, 'Experimental Researches on Vegetable Assimilation and Respiration, I. On a new method for investigating the carbonic acid exchanges of plants; II. On the paths of gaseous exchanges between aerial leaves and the atmosphere', *Phil.Trans. (B).*, *186*, 485–502; 503–62. 1895.
b.   G.L.C. Matthaei, 'Experimental researches on vegetable assimilation and respiration. III. On the effect of temperature on carbon-dioxide assimilation', *Phil.Trans. (B).*, *197*, 47–105. 1904.

Selection 40

Introduction:

a.   See Introduction to Selection 20, Section I, note c.

Selection 41

Introduction:

a.   G.E. Briggs, 'Experimental researches on vegetable assimilation and respiration, xiii. The development of photosynthetic activity during germination', *Proc.Roy.Soc. (B).*, *91*, 255–68, 1919–20.
b.   O. Warburg and E. Negelein, 'Über den Energieumsatz bei der Kohlensäureassimilation', *Z.physik.Chem.*, *102*, 248. 1922.
c.   See P. Oesper, 'The history of the Warburg apparatus. Some reminiscences on its use', *J.Chem.Ed.*, *41*, 294–6. 1964; A. Kleinzeller (ed.), *Manometrische Methoden und ihre Anwendung in Biologie und Biochemie* (Jena and Prague, 1965). Warburg's life and work is examined by H. Krebs. See Introduction to Selection 61, Section III, note c.

d. O. Warburg, 'Über die Geschwindigkeit der photochemischen Kohlensäurezersetzung in lebenden Zellen', *Biochem. Z.*, *100*, 230–70. 1919. The second part was published in *Biochem.Z.*, *103*, 188–217. 1920.

Text:

1. Quantentheoretische Grundlagen der Photochemie, *Zeitschr.f.Elektrochemie*, *26*, 54 (1920).
2. Ibid.
a. Warburg and Negelein did not elaborate on how the values of ϕ_0 in moles/cal were obtained. It would appear that they were arrived at on the basis that

$$\phi_0\left[\frac{moles}{cal}\right] = 4.48 \times 10^{-8} \times \phi_0\left[\frac{cmm}{cal}\right],$$

Where cmm denoted the volume of evolved oxygen in cubic millimetres measured manometrically (see p. 205). Thus the results of Warburg and Negelein for the mean values of ϕ_0 in moles/cal (5.25×10^{-6}, 4.75×10^{-6}, 3.01×10^{-6}) can be obtained by inserting into the above expression the experimentally based values of ϕ_0 in cubic millimetres/cal with light of different wavelengths, as given by the authors (p. 206):

Wavelength $\mu\mu$	$\phi_0\left[\dfrac{cmm}{cal}\right]$
660	117
578	106
436	67

Selection 43

Introduction:

a. See Selection 176, Section VII.

Text:

1. Wieland, H., *On the mechanism of oxidation*. Yale Univ. Press. 1932.
2. Kluyver, A.J. and Donker, H.J.L., *Chem. Zelle Gewebe*, *13*, 134, 1926; Kluyver, A.J., *Chemical Activities of Microorganisms*, London Univ. Press, 1931.
3. van Niel, C.B. and Muller, F.M., *Rec.Trav.Bot.Néerl.*, *28*, 245, 1931; Muller, F.M., *Arch.Mikrobiol.*, *4*, 131, 1933.

Selection 44

Introduction:

a. J. Myers, 'Conceptual developments in photosynthesis, 1924–1974', *Plant.Physiol.*, *54*, 420–6. 1974.

Text:

a. R. Hill, 'Oxygen produced by isolated chloroplasts', *Proc.Roy.Soc. (B).*, *127*, 192–210. 1939.
b. See Introduction to Selection 41, note d.

Selection 45

Introduction:

a. A. Stoll, 'Über den chemischen Verlauf der Photosynthese', *Naturwiss.*, *20*, 957–8. 1932.
b. According to Martin D. Kamen (*b*. 1913), who was closely associated with these

investigations: 'Ruben was responsible, almost single-handed, for the growth of interest in tracer methodology which occurred at Berkeley in the years 1937–1938.' See his article 'The early history of carbon–14' in A.J. Ihde and W.F. Kieffer (eds), *Selected Readings in the History of Chemistry* (Easton, 1965), p. 217. Kamen makes no reference to R. Schoenheimer's pioneering efforts to employ isotopic tracers in the biochemical field (cf. Selection 133, Section V and Selection 156, Section VI). It is of interest that in 1941 the Soviet researcher, A.P. Vinogradov (1896–1975), in collaboration with R.V. Teis, also showed by the use of isotopes that the oxygen originates in water. See C.B. van Niel, 'The present status of the comparative study of photosynthesis', *Ann.Rev.Plant Physiol.*, *13*, 2–3. 1962.

Text:

1.   Randall and Webb, *Ind.Eng.Chem.*, 31, 227 (1939).
2.   Cohn and Urey, THIS JOURNAL, *60*, 679 (1938).
3.   Mills and Urey, ibid., *62*, 1019 (1940).
4.   For a review of oxygen exchange reaction see Reitz, *Z. Elektrochem.*, *45*, 100 (1939).
5.   For an excellent review of this subject up to 1926 see H.A. Spoehr 'Photosynthesis', Chem. Cat. Co., New York, N.Y. 1926.
6.   Van Niel, *Cold Spring Harbor Symposia on Quant. Biol.*, *3*, 138 (1935).

Section III:
RESPIRATION

a. From the 1770s to *c*. 1880: the concept of slow combustion. Location

Selection 46

As conveyed in the two previous Sections, the replacement of the doctrine of phlogiston by the oxidation theory had a profound impact on the development of the understanding of fermentation and photosynthesis. Also closely connected with this, respiration became related to the chemical change sustaining calcination (heating) of metals and to combustion, eventually interpreted by Lavoisier as oxygenation (oxidation).

Despite a continuous interest in the subject, reflected in a considerable body of work, the historical picture is still incomplete. This is partly due to the fact that while there is historical knowledge of particular aspects of Lavoisier's work on respiration, combustion and the oxidation theory, exploration of the interrelation between them is still in its early stages.[a]

The object of these paragraphs is merely to emphasize the need for an approach, at once differentiated and integrated, to Lavoisier's work which progressed in stages. It now seems feasible that Lavoisier could have received his initial impetus from the prize-winning essay (1770) on fermentation in wine and the best way of obtaining alcohol, written by Abbé François Rozier (1734–94). The essay contained the suggestion that common air played a part in the souring of wine, a process that must have worried man ever since he became involved in its preparation. It was perhaps this idea that provided the first clue that led to Lavoisier's subsequent interest in the aeriform state, the physical and chemical properties of 'elastic fluids' or 'airs', including phenomena of heat and the composition of common air and water.[b] Through investigations of these problems Lavoisier was eventually brought to new conceptions of acidity, calcination and reduction of metals, combustion and respiration based on oxygen, all of them turning on the principle of conservation of matter.

Lavoisier's opinion on the chemistry of respiration was presented for the first time in a paper read to the Academy in May 1777 and published in 1780. At this stage he primarily raised the question of what happened to common air, at once the agent and subject matter of respiration, as he put it, during breathing. He also came up with two tentative answers.

According to one argument, a conversion occurred in the lungs, attended by the simple change of the essential respiratory constituent of common air (*partie ou portion éminemment respirable*) into a noxious elastic fluid (*partie ou résidue méphitique*). Lavoisier designated it as 'aeriform chalky acid' (*acide crayeux aëriforme*) thus replacing the currently accepted 'fixed air' because, as he pointed out in a footnote, he was keen to emphasize the provenance (from limestone) of this 'elastic fluid'.

According to the other reasoning, a chemical process took place in the blood of the lung, akin to the formation of the red-coloured metallic calces of mercury, lead and iron. It should be pointed out that this paper neither includes the analogy to combustion (slow or otherwise) nor contains a reference to caloric theory, as is frequently maintained even by knowledgable students of the subject.[c] By the same token, in order to avoid anachronism, it is necessary to bear in mind that there is no mention of 'oxygen' and 'carbonic acid'. They were products of later developments embodied in Lavoisier's *Traité élémentaire de Chimie* of 1789.[d]

The analogy between respiration and slow combustion which yields carbonic acid and water was recognized earlier (1780) by Lavoisier in the quantitative work on a guinea pig breathing in an ice-calorimeter. These experiments were conducted jointly with Pierre Simon Laplace (1748–1827).[e] But it was almost a decade later that this analogy including the generation of animal heat was explained, in the light of the transformed chemical thinking, by Lavoisier in collaboration with Armand Seguin (1767–1835).

The following excerpts from two papers give an indication of the growth of the understanding of the chemistry of respiration which resulted from Lavoisier's researches between 1777 and 1789.

[A.-L.] Lavoisier, *Mém.Acad.Sci.*, *185*. [1777] (1780)

'Expériences sur la respiration des animaux et sur les changements qui arrivent à l'air en passant par leur poumon' (Experiments on the respiration of animals and on the changes which occur in the air on passing through their lungs)

pp. 191–2

In fact, according to what one has just seen, one can conclude that for the purpose of respiration one of two things happens: either the amount of eminently respirable air contained in the atmospheric air is converted into aeriform chalky acid gas in passing through the lung; or an exchange takes place in this internal organ:[a] on the one hand the eminently respirable air is absorbed and, on the other, the lung receives in its place a portion of aeriform chalky acid gas almost equal in volume.

The first of these two opinions is supported by an experiment which I have already communicated to the Academy. I have shown, in a memoir read to the public sitting at Easter 1775,[b] that the eminently respirable air could be totally converted into aeriform chalky acid gas by addition of charcoal powder, and I shall prove in other memoirs that there are several other means of operating this same conversion. It is then possible that the respiration may have this same property, and that the eminently respirable air which has entered into the lung, may come out again as aeriform chalky acid; but on the other hand some strong analogies seem to militate in favour of the second opinion, and lead one to believe that a portion of eminently respirable air remains in the lung and combines there with the blood. One knows that it is a property of eminently respirable air to impart the colour red to the substances, and above all to the metallic substances with which it is combined: the mecury, the lead and the iron supply examples of this. These metals form, with eminently respirable air, calces of a beautiful red, the first, known under the name of *mercure precipité per se* or *red precipitate of mercury*,[c] the second, under the name of *minium*; finally the third, under the name of *colcothar*. The same effects, the same phenomena are found again, as one has just seen in the calcination of the metals and in the respiration of animals; all the circumstances are the same, up to the colour of the residues: will one not be able to deduce from this that the red colour of the blood is due to the combination of the eminently respirable air, or more exactly, as I shall show in a subsequent memoir, to the combination of the base of the eminently respirable air with an animal liquid, in the same manner as the red colour of the red precipitate of mercury and of the minium is due to the combination of this same air with a metallic substance?

[A.] Seguin and [A.-L.] Lavoisier, *Hist.Acad.Sci.*, *566*. [1789] (1793)

'Premier mémoire sur la respiration des animaux'. (First memoir on the respiration of animals)

pp. 570–1

Setting out from the knowledge acquired, and reducing it to simple ideas which everyone can easily grasp, we shall say first in general, that respiration is only a slow combustion of carbon and hydrogen, which is altogether similar to that which operates in a lamp or in a lighted candle, and that, from this point of view, animals which respire are real combustible bodies which burn and are consumed.

In respiration, as in combustion, it is the atmospheric air which supplies oxygen and caloric. But, as in respiration it is the substance itself of the animal, it is the blood which supplies the combustible material, if the animals do not habitually restore from their nourishment what they lose by respiration, the oil would soon be lacking in the lamp, and the animal would perish, as a lamp goes out when it lacks nourishment.

The proofs of this identity of effects between respiration and combustion are deduced immediately from experience. In fact, the air which has served for respiration no longer contains, on coming out from the lung, the same quantity of oxygen: it contains not only carbonic acid gas, but also very much more water than it contained before inspiration. Now, as the vital air[a] can be converted into carbonic acid only by an addition of carbon; as it can be converted into water only by addition of hydrogen; this double combination cannot operate unless the vital air loses a part of its specific caloric, it results that the effect of the respiration is to extract from the blood a portion of carbon and hydrogen, and to put in its place a portion of its specific caloric, which, in the circulation, is distributed with the blood in all parts of the animal economy, and keeps this temperature nearly constant, as one observes in all animals which respire.

Selection 47

Till the beginning of the nineteenth century it was widely understood that animal respiration (breathing) could be solely related to lungs. Unknown to others at that time, the first systematic steps in advancing knowledge on Lavoisierian lines about other systems of respiration had been taken by Lazzaro Spallanzani (1729–99). At the close of his life, this inventive and productive Italian investigator found, with the aid of the newly developed eudiometry,[a] that organs other than lungs could effect the taking up of oxygen and the giving off of carbonic acid. This conclusion followed from extensive observation and experimentation on cold-blooded animals (worms, insects, fishes, amphibians) and warm-blooded animals (birds, mammals).

Spallanzani's findings on animal respiration were brought to light by his Swiss scientific friend, Jean Senebier, who was known for his work on the relation of carbonic acid to oxygen formation by plants.[b] From Spallanzani's unedited manuscripts in Italian Senebier prepared three volumes in French which appeared between 1803 and 1807.[c] However, it would seem that Spallanzani's insight into aspects of animal respiration other than those connected with the work of the lungs had no immediate impact.[d]

All the extracts used here come from the portion of the first volume entitled 'Lettre de Spallanzani au citoyen Senebier, relative à la respiration'. They also contain Spallanzani's demonstration of carbonic acid production when the organs had no access to oxygen and of cutaneous respiration.

L. Spallanzani, *Mémoires sur la respiration* (Geneva, 1803)

(Memoirs on respiration)

pp. 62–3

I enclosed in a given measure of ordinary air different species of worm; it is with this class of animals that I have begun my researches. I learnt thus that those which had organs for respiration, like those which were deprived of them, absorbed all the oxygen from ordinary air, at least as much as the phosphorus of Kunckel[a] absorbs of it. I perceived that in these last animals the organ of the skin replaced the lungs; this innovation made me look for another. I wished to know if this organ ceased to absorb oxygen when the worms ceased to live, or if it then retains this property: to resolve this problem, when these animals were dead I confined them in closed vessels, placing them in the same circumstances where they were during their life. But the oxygen was just the same entirely absorbed.

pp. 66–8

To disembarrass myself of these apparent vexations, and to lighten this darkness,[b] I had recourse to an expedient which ought to be decisive: it was to place the dead animals in a medium completely deprived of oxygen gas. Because, either carbonic acid would not be engendered in this gas, which would have furnished me with an unanswerable proof that the production of this gas depended on atmospheric oxygen. Or what is the same thing, that it was the effect of the combination of this principle with the carbon exhaled by the animal. Or indeed I shall have had this carbonic acid gas almost as when the animals are confined in ordinary air, and then it was demonstrated that it did not depend on the oxygen of the air, and consequently that it is exhaled immediately from the body of these animals in an acid form, or in the state of carbonic acid, combined with caloric and become gaseous.

I confined then different species of worms freshly killed in pure nitrogen gas . . . but in these experiments carbonic acid gas was manifest. I confirmed this experiment by another, enclosing the animals in pure hydrogen gas, and more than once I had then a quantity of carbonic acid gas, produced in these noxious gases, greater than when these animals were confined in ordinary air. I was then forced to conclude that the carbonic acid gas produced in these two cases is not dependent on atmospheric oxygen, and consequently that the oxygen gas destroyed by the presence of these dead animals has its base[c] absorbed by these animals themselves.

p. 70

I extended these experiments to dead fishes from fresh water and from sea water, enclosed in ordinary air. Their size permits me to make these experiments on their separated internal parts, on the intestines, the stomach, the liver, the heart, the ovaries: but all these parts completely absorbed the oxygen of the air, like the insects and the worms.

p. 73

I have however been able to establish the unambiguous absorption of oxygen made by the cutaneous organ, without removing their lungs from the amphibians. I confined their bodies to vessels in such a manner than they were without communication with the exterior air, whilst they had their heads outside in the air, where they breathed without restraint. I thus informed myself clearly that the absorption made by these animals when dead is only a continuation of that they were making during their life.

Selection 48

As we have already seen, Lavoisier suggested that blood colouration could relate to the chemical process later interpreted as oxidation. Curiosity about the chemistry of colour phenomena in living beings was carried a stage further early in the nineteenth century. It was stimulated by findings of colour changes which guaiacum underwent in light of different colours and in nitric acid and oxymuriatic acid (chlorine).[a] Since the introduction by the Spaniards in the early sixteenth century guaiacum bark and resin (*lignum vitae* tree) had been widely used as drugs and had been subjected to adulteration. It was in the course of investigation of a suspected consignment of guaiacum resin that the noted French scientific pharmacist, Louis-Antoine Planche (1776–1840), discovered in 1810 in the blueing of guaiacum by fresh roots of horseradish a suitable method of distinguishing between the original and the adulterated substance.[b] On returning to the same topic a decade later Planche recognized that the production of the blue colour was due to a thermolabile constituent of plant roots and milk.

[L.-A.] Planche, *J.Pharm.*, *6*, 16. 1820

'Sur les substances qui développent le couleur bleue dans la résine de gaïac' (On the substances which develop blue colour in guaiacum resin)

p. 17

1st EXPERIMENT

If one plunges some slices of the fresh root of wild horseradish (*Cochlearia armoracia*, L.) into a glass containing alcoholic tincture of guaiacum resin one sees that this root becomes coloured blue here and there, and that the tincture participates in this colour, which afterwards vanishes rather promptly. The liquid becomes green again, but paler than before the experiment.

p. 20

4th EXPERIMENT

On the influence of light on the development of the blue colour

In order to know up to what point light can modify the colouration of guaiacum resin, I put a piece of *Pastinaca sativa* root in a brown earthenware saucer, covered it with a small bell jar (with a tube) wrapped in black paper; and I immediately poured through the tube several drops of tincture of guaiacum on to the root. After half an hour I took off the bell jar, and found the parsnip coloured a beautiful blue, as regards tint in every way similar to a piece of the same root which had been watered with tincture of guaiacum and exposed to diffuse light. One sees from this that the theory of the refrangibility of light rays being the cause of the colouration cannot apply.[a]

p. 22

9th EXPERIMENT

Action of milk on the tincture of guaiacum

The tincture of guaiacum and cold milk, agitated together in a flask offer a liquid of a colour celestial blue . . .

13th EXPERIMENT

Milk and tincture of guaiacum,[1] the one and the other in ebullition, then introduced into a really dry flask, heated during half an hour at a temperature of more than eighty degrees to expel the air, have given rise to no action.

p. 24

In considering the effect of heat on the bodies which contain the principle coloured by guaiacum resin I first thought that this principle was of volatile nature, and that it would be possible to drive it off by heating one of these bodies to different degrees up to that of boiling water, in an apparatus arranged to collect the liquid and gaseous products. The milk which I had chosen as the subject of experiment has given me no satisfactory result . . .

I have been obliged from this moment to renounce the idea that this principle was volatile. It seems then to me more probable that this kind of *cyanogen,*[b] whatever may be its nature, is absorbed by the bodies which, in the ordinary state, permit it to exercise its action as soon as they are exposed to a certain temperature, that it then obeys other laws and forms new combinations which specify its colouring properties.

Selection 49

Originating from the observations of the Italian anatomist, Luigi Galvani (1737–98), at the close of the eighteenth century, electrophysiology developed as a distinctive research field during the following century. It concentrated largely on the study of electrical activities arising from the stimulation of a muscle or nerve. It played a major part in the quest to understand the interconnection between functions of certain animal organs and systems and their cessation. In this way it was hoped to obtain a greater insight into the living and non-living state of the animal organism.

An important step in this direction had been taken by the many-sided naturalist, Alexander von Humboldt (1769–1859). By combining the electrophysiological method with the chemical approach (in the light of the new Lavoisierian system) Humboldt investigated in 1795 the 'vitality' of excised muscles from frogs, and hearts from frogs and other animals in different gases. He found that in oxygen the muscle retained its ability to twitch and the heart its power to beat for a longer time than in air, hydrogen, nitrogen or carbonic acid. Humboldt was not prepared to offer an explanation of these phenomena but, on observing a bright red colour of the muscle in the presence of oxygen, he suspected that it was due to blood undergoing oxidation.[a]

Although not ignored, these observations had, apparently, little immediate impact on work dealing with the problem of respiration. They were repeated and expanded by Georg Liebig (1827–1903), the son of the illustrious chemist, in the middle of the nineteenth century. Extending the electrophysiological experimentation, he compared the production of carbonic acid in muscles from frogs supplied with blood with its production in muscles where the blood was replaced by distilled water. As shown in the next excerpt, he suggested in 1850 that the muscle tissue constituted the site of respiration and not the capillaries, as it was currently held.[b]

G. Liebig, *Müllers Arch.*, 393. 1850

'Ueber die Respiration der Muskeln' (On the respiration of the muscles)

pp. 409–13

If we compare the results so far obtained, it is first established that a muscle in an atmosphere of oxygen or in air containing oxygen maintains its twitching capability longer or, what is the same thing, death rigor occurs later than in an atmosphere containing no oxygen, thus e.g. in pure nitrogen, hydrogen and carbonic acid.

Further it appears that an atmosphere of carbonic acid besides not maintaining the life of the muscle, indeed hastens its end by bringing about alteration in its structural elements [*Formtheile*] incompatible with the continuance of the properties of life, which does not happen with nitrogen.

Then from these experiments emerges that a muscle, during the prolonged period of time of its capability to twitch in an oxygen-containing atmosphere, simultaneously takes up oxygen and gives it off in the form of carbonic acid.

If we now name as respiration the taking up of oxygen by an animal organism whose survival is linked with this and with the simultaneous giving out of carbonic acid, then according to these experiments we must assume that muscle separated from the body still respires.

Already according to the absorption theory, the results contained here could be conjectured. [That is the theory] which considers the blood (in connection with respiration or the taking up of oxygen and giving out of carbonic acid in the lungs) merely as means of absorption for those gases, which transfers the oxygen towards the capillaries without chemically combining with it, and instead of this from here brings back carbonic acid. If the blood with respect to the exchange of both gases really performs only this role, and the greatest part of the inspired oxygen will not be used to change chemically the blood on its way to the capillaries and through this to make it only possible for the formation of carbonic acid in the capillaries to take place, as the chemical theories demand it, then all these experiments have to give the same results on removing all blood from the muscles used in the investigation.

In order to test this conclusion experiments were made with muscles of frogs, which, through injection of distilled water into the aortic bulb until only pure water flowed out of the opening in the heart, were freed from their blood and their vessels filled instead with water. In order to make this injection possible a firm ligature was made around the neck of the living frog, then the head half cut off and the spinal cord destroyed. Each time on injection, when a syringe-full of water went through the capillaries, the whole frog twitched . . .

The temperature was from 19°–24°. The behaviour here shows that the muscles emptied of blood lose in the same time in oxygen and air their power of twitching; . . . we find that also the muscles which contain no trace of blood live a much shorter time in an atmosphere in which there is no oxygen than in an oxygen-containing one, and that during the longer period of their twitching power they take up oxygen and again give out carbonic acid. From this it is clear that the blood in the process of respiration indeed serves only as means to accomplish the transport of the gases towards the capillaries and back, and that in the capillaries the *formation* of carbonic acid cannot go on, but only the *exchange* of that already formed, through the vessel walls, for the oxygen of the blood.

p. 414

According to this consideration there is nothing more likely than to assume that the formation of carbonic acid by means of a part of the inspired oxygen goes on also in the body not in the capillary vessels but outside them in the tissues of the muscles. Just as it happened with the surviving properties of life of the muscles outside the body.

Selection 50

The blue colouration obtained on bringing biological material (plant roots, milk) into contact with an alcoholic solution of guaiacum (as demonstrated by Planche) had become valuable in the development of work on biological oxidation during the mid-1850s. Thus in the hands of Christian Friedrich Schönbein (1799–1868) the use of the reaction led to historically important investigations of oxidation under physiological conditions and eventually to the fruitful suggestion that it occurred in a sequence. Schönbein envisaged that it was incorrect to ascribe to atmospheric oxygen the position of the oxidizing agent in living material. Instead he devised an oxidation system utilizing information on ozone, which he discovered in 1839.[a] According to this view certain substances in living material, on interaction with atmospheric oxygen (O), had the power to change it into the active ozonized form (O) and simultaneously to combine with it. In this union the ozonized oxygen was loosely held and when split off made available for oxidation purposes.

The point of view was developed in a letter written in 1856 by Schönbein to Michael Faraday (1797–1867), with whom he was on friendly terms, and from which the following excerpt is taken.

[C.F.] Schönbein, *Phil.Mag.*, *11*, 137. 1856

'On ozone and ozonic actions in mushrooms'

pp. 139–141

... what the botanists tell me to be called '*Boletus luridus*,' with some other sorts of mushroom, have the remarkable property of turning rapidly blue when their head and stem happen to be broken and exposed to the action of the atmospheric air. On one of my ramblings I found a specimen of the said *Boletus*, perceived the change of colour alluded to, and being struck with the curious phænomenon, took the bold resolution to ascertain, if possible, its proximate cause. I carried home the part, set to work, and found more than I looked for, which luckily enough happens now and then. Being, by the short space allotted even to the longest letter, prevented from entering into the details of the subject, I confine myself to stating the principal results obtained from my mushroom researches. *Boletus luridus* contains a colourless principle, easily soluble in alcohol; and in its relations to oxygen, bearing the closest resemblance to guaiacum, as appears from the fact, that all the oxidizing agents which have the power of bluing the alcoholic solution of guaiacum, also enjoy the property of colouring blue the alcoholic solution of our mushroom principle; and all the deoxidizing substances by which the blue solution of guaiacum is decolorized also discharge the colour of the blued solution of the *Boletus* matter. From this fact and others, I infer that this mushroom principle, like guaiacum, is capable of combining with Ȯ, and is not affected by O. Now the occurrence of a matter so closely related to guaiacum in a mushroom is a fact pretty enough of itself, but as to scientific importance far inferior to what I am going to tell you.

The fact that the resinous *Boletus* principle, after having been removed from the mushroom (by the means of alcohol), is not able to colour itself spontaneously in the atmospheric air, whilst it seems to have that power so long as it happens to be deposited in the parenchyma of the *Boletus*, led me to suspect that there exists in the *Boletus luridus*, besides the guaiacum-like substance, another matter, endowed with the property of exalting the chemical power of common oxygen, and causing that element in its Ȯ condition to associate itself to the resinous principle of the mushroom. The conjecture was correct; for I found that in the juice obtained by pressure from a number of mushrooms belonging to the genera *Boletus* and *Agaricus*, and notably from *Agaricus sanguineus* (upon which I principally worked), an organic matter is contained which enjoys the remarkable power of transforming O into Ȯ, and forming with the latter a compound from which Ȯ may easily be transferred to a number of oxidable matters, both of an inorganic and organic nature; and I must not omit to state, that the peculiar agaricus matter, after having been deprived of its Ȯ, may be charged with it again by passing through its solution a current of air. The easiest way of ascertaining the presence of Ȯ in the said agaricus juice, is to mix that liquid with an alcoholic solution of guaiacum, or the resinous matter of the *Boletus luridus*. If the juice happens to be deprived of Ȯ, the resiniferous solutions will not be coloured blue; but if it contains Ȯ, the solutions will assume a blue colour, just as if they were treated with peroxide of lead, permanganic acid, hyponitric acid, &c. From the facts stated, it appears that the organic matter in question is a true carrier of active oxygen, and therefore, when charged with it, an oxidizing agent. Indeed, that matter may in many respects be compared to NO^2, which, as is well known, enjoys to an extraordinary extent the power of instantaneously transforming O into Ȯ, and forming a compound ($NO^2 + 2$Ȯ) with that Ȯ, from which the latter may easily be transferred to a multitude of oxidable matters. Now in a physiological point of view, the existence of such an organic substance is certainly an important fact, and seems to confirm an old opinion of mine, according to which the oxidizing effects of the atmospheric oxygen (of itself inactive) produced upon organic bodies, such as blood, &c., are brought about by means of substances having the power both of exciting and carrying oxygen.

Before dropping this subject, I must not omit to mention a fact or two more. The peculiar matter contained in the juice of the *Agaricus sanguineus*, &c., and charged with Ȯ, gives up that oxygen to guaiacum, and the latter transfers it to the resinous matter of the *Boletus luridus*; thus the different organic matters capable of uniting with Ȯ as such, exhibit different affinities for that oxygen, a fact not without physiological importance. Another fact worthy of remark, is the facility with which the nature of our agaricus matter may be changed. On heating the aqueous solution which has the power of deeply bluing the guaiacum solution to the boiling-point, it not only loses the property, but also the capacity of again becoming an oxidizing agent, i.e. carrier of oxygen, however long it may be kept in contact with atmospheric air.

Selection 51

Schönbein's observations on the blueing of guaiacum by plant and animal material and also his supposition that respiration depended on oxygen transfer (albeit in the ozonized form) by intermediate carriers greatly influenced M. Traube (1826–94). He began to work out his ideas on the nature of ferments, and the fundamental role they played in the chemical changes in living organisms, in the late 1850s and early 1860s. As previously indicated, Traube considered ferments to be substances of definite composition derived from proteins. He treated them as oxygen carriers and he suggested that the oxidation and reduction processes in living bodies were fermentative in nature.[a]

As the following extracts show, Traube applied this conception to the elucidation of the chemical relationship between respiration and muscle work. Published in 1861, Traube's paper was notable in more than one way. He showed that Justus Liebig's still influential views on the nitrogenous (proteinic) source of muscular work were open to grave criticism.[b] Traube enunciated clearly that muscle work is a respiratory act, indeed he regarded the maintenance of muscle activity as the main function of respiration (the others being cell formation and heat production). Finally, he propounded a theory of muscle action which related both to his views on the oxygen-carrying function of ferments (in this particular case the ferment was thought to be contained in the muscle fibre) and to the role played by respiration in muscular work.

When it came to visualizing the mechanism of the chemical reaction in muscle respiration Traube resorted to processes drawn from knowledge of laboratory chemistry and industrial practice, among these, particularly, dyeing with indigo.[c] This process attracted early scientific attention because it did not support the original Lavoisierian claim of relating acidic properties exclusively to oxygen. Long before the acidic properties of phenols could be reasonably explained theoretically, the acidic nature of the colourless indigo (a phenolic derivative) had been accepted and the loss and recovery of the colour had been interpreted as reduction and oxidation respectively.[d]

M. Traube, *Virchows Arch.*, *21*, 386. 1861

'Ueber die Beziehung der Respiration zur Muskeltätigkeit und die Bedeutung der Respiration überhaupt' (On the relation of respiration to muscle activity and on the meaning of respiration in general)

p. 399

The free oxygen enters in the dissolved state through the capillary walls and *unites with the muscle fibre in a loose chemical combination*, which is capable of giving up this oxygen again to other substances dissolved in the muscle fluid and endowed with higher affinity for oxygen; and then it can take up new oxygen . . .

The muscle fibre according to this behaves on the one hand towards
reducing substances, and on the other hand towards oxygen in just the same
way as do indigo, indigosulphuric acid, nitric oxide, copper sulphate, acetic
ferment, and many other substances.

pp. 403–4

As one sees, the analysis given here of the chemical processes of muscle
respiration is essentially an application of the theory of slow combustion as I
have developed it in the place mentioned.[a] It arises from the assumption that
the muscle fibre, or rather the fibrous body contained in it is a *vital decay
ferment*,[1] which transfers the oxygen taken up out of the blood on to the
substances dissolved in the muscle fluid, itself suffering no damage in this
process.

Selection 52

When it transpired that muscle could contract in the absence of oxygen, (A.V.
Humboldt, G. Liebig), the question arose how carbonic acid was formed in
muscle respiration. Of the conjectural answers the so-called 'inogen'
conception of muscular contraction had the most lasting effect. It was
worked out by the German investigator, Ludimar Hermann (1838–1914),[a]
in the later 1860s and earlier 1870s and its influence extended into the first
decade of the twentieth century.[b]

Apart from drawing on his own results, Hermann took into account
contemporary experimental achievements and theoretical discussions.
William Kühne (1837–1900)[c] had discovered a muscle protein named myosin
(1859, 1864); Hermann had applied Traube's idea of the oxygen-carrying
ferment to myosin; and it was generally recognized that the criticisms and
results produced by Traube, Adolf Fick (1829–1901), Johannes Wislicenus
(1835–1902) and others tended to invalidate J. Liebig's supposition that
proteins were metabolized during muscular exertion.[d]

At the time when Hermann embarked on his studies of contraction under
vacuum conditions, the highest degree of evacuation was obtained by means
of the Geissler mercury air-pump. Hermann must have been one of the first
physiologists to make use of this technique to investigate the chemical
processes taking place in resting and active muscle. Among other discoveries, he found that the resting muscle exposed to vacuum and deprived
of oxygen produced a hardly noticeable amount of carbonic acid which
increased on contraction. Hermann concluded that in muscle respiration the
taking up of oxygen and the giving off of carbonic acid were not essentially
interdependent. He imagined that oxygen passed from the flowing blood to
the muscle-fibre and became bound to myosin. As for carbonic acid, he
suggested that its formation resulted not from an oxidative process but from
decomposition of a complex substance called 'inogen'. Myosin acted as an
organic catalyst in a cycle of chemical changes that involved the restitution
of the inogen but not the contraction of the muscle.

The following excerpts contain a summary of Hermann's theoretical view of the respiratory process in muscle as propounded in his *Grundriss der Physiologie des Menschen*, translated into a number of European languages. The extracts are taken from the version rendered into English by Arthur Gamgee (1841–1909), known for his work on haemoglobin, who based his translation on the fifth edition of the German text (1874).

L. Hermann, *Elements of Physiology* (London, 1875)

p. 253

The following may in all probability be regarded as the simplest theory of the chemical processes occurring during contraction and rigor: Muscle contains at any moment a store of a complicated nitrogenous substance dissolved in the contents of the muscle tubes and in the muscle-plasma, which may be described, for the sake of brevity, as the 'energy generating',[a] or 'inogene, substance.' This 'inogen substance' is capable of undergoing a decomposition in which energy is evolved, and the following products yielded, viz. carbonic acid, sarco-lactic acid, probably glycerin-phosphoric acid . . . and a gelatinous body, myosin, of an albuminous nature, which separates spontaneously, and afterwards contracts firmly, becoming probably concentrated. This decomposition occurs spontaneously, but slowly, while the muscle is at rest, the rapidity of its occurrence being determined by the height of the temperature. It takes place instantaneously at the temperature of heat-rigor. It is, moreover, at once accelerated by stimulation; and this acceleration is essentially what occurs during the active condition. When the substance is entirely used up, muscular activity is no longer possible.

The 'inogene substance' has not hitherto been isolated, as in every method of chemical investigation yet devised the characteristic decomposition occurs. The latter may, indeed, be prevented by subjecting the muscle suddenly to a strong heat (scalding) or to the action of mineral acids, but both these methods destroy the substance. As regards its composition, it would seem to resemble haemoglobin . . . as both yield an albuminous body on decomposition. On account of the analogy between the chemical processes of muscular activity and of rigor we must suppose an expenditure of glycogen . . . to occur during the former as during the latter.

p. 254

The whole of the products of the decomposition of 'inogene substance' do not leave the muscle; for, as the excretion of nitrogenous material is not increased by muscular exertion, it must be concluded that the myosin remains within the muscle. In consequence of this, and of the fact that it is not the prepared substance, but only the materials necessary to form it, which are conveyed to the muscle by the blood, it is most probable that the recovery of muscle after exhaustion, apart from the mere removal of the effete materials, consists in a synthesis of the 'inogene substance,' in which myosin plays a part, and for which the blood supplies oxygen and some non-nitrogenous organic body hitherto undiscovered (Hermann). Myosin, therefore, according to this theory, undergoes in muscle a complete cycle of chemical changes.

Selection 53

Respiratory physiologists conspicuously disregarded the cell during the three decades following the publication of Theodor Schwann's seminal piece on the cellular theory in 1838–9. Yet he very clearly stated in it:

Oxygen, or carbonic acid, in gaseous form, is essentially necessary to the metabolic phenomena of the cells. The oxygen disappears and carbonic acid is formed, or *vice versa*, carbonic acid disappears and oxygen is formed. The universality of respiration is based entirely upon this fundamental condition for the metabolic phenomena of the cells.[a]

This neglect in effect reflected and was a part of the continuing division of physiology into relatively separate components concerned with physiological functions, morphological structures and chemical processes underlying life.

It is therefore of some historical significance that Eduard Pflüger (1829–1910) of Bonn, who deplored and denounced this splitting,[b] was also the first physiologist to formulate explicitly the theory that the cells of tissues were the site of the respiration process. He elaborated this in two papers published in 1872 and 1875.

In the first paper he was mainly concerned with the taking up and giving off of oxygen by asphyxiated blood and with haemodynamics. He found that on saturating asphyxiated blood with oxygen it could still be pumped off from the blood under vacuum conditions. The argument was that the oxygen essentially did not enter into chemical combination in the blood with oxidizable substances that were presumably present. As to the circulation of blood, Pflüger was focusing critically on the investigations into respiration carried out by Carl Ludwig (1816–95) at Leipzig, where particular attention was paid to the analysis of blood gases, including the estimation of the velocity of blood flow[c] and the development of the perfusion technique.[d] To Ludwig and his collaborators, the experimental evidence indicated that the blood, rather than the tissues, was the site of respiration under control of the rate of blood flow. Denying the regulatory role of the velocity of blood, in opposition to the Leipzig school, Pflüger stressed the directive role of the cell demand for oxygen. He proclaimed that oxygen diffused from the blood – unfettered by the amount contained in it – to the cells in accordance with their requirements and that therefore the cell had to be the locus of the oxidation process (see first extract).

In his second more extended article, among others, Pflüger attempted to generalize by suggesting that cellular respiration was a universal property of the plant and animal world, including insects and embryos (see second extract).[e]

E. Pflüger, *Pflügers Arch.*, **6**, 43. 1872

'Ueber die Diffusion des Sauerstoffs, den Ort und die Gesetze der Oxydationsprocesse im thierischen Organismus' (On the diffusion of oxygen, the site and the laws of the oxidation processes in the animal organism)

p. 52

If, as the facts of lung diffusion clearly show, a so extraordinarily small motive force provides in so short a time by the way of diffusion a so great amount of oxygen through the wall of the capillaries, then it is clear that each *small* variation of the *extremely low* partial pressure of oxygen in the tissues exerts immediately a quite enormous influence on the rate of the oxygen stream. Only through the extreme *lowness* of the motive force, which suffices for the diffusion of the oxygen, the tissue regulates, I say, *the animal cell itself regulates so finely and easily the intensity of the oxygen stream.* For as soon as the tissue in unit of time, because of increased vital activity, uses more oxygen, that is it reduces the partial pressure of the oxygen in itself if only very little (perhaps not detectable by our method), the diffusion stream immediately increases vastly. For the small variation is a great part of a small whole. *Here lies, as I will once and for all declare, the intrinsic secret for the regulation of the oxygen amounts used by the whole organism, which only the cell itself determines, not the oxygen content of the blood, not the tension of the aortic system, not the rate of the blood stream, not the mode of the work of the heart, not the mode of the respiration.* All these factors are secondary and subordinate. They combine only in their action in the service of the *cells* which perform the *real animal work* and themselves again stand in a system of subordination to one another, so that a special class of small cells, I mean certain *nerve cells*, exert *supremacy* over the intensity of living processes of almost all cells and, that is to say, *according to the feeling of comfort* which the normal temperature relations of the blood produce.

E. Pflüger, *Pflügers Arch.*, *10*, **251. 1875**

'Beiträge zur Lehre von der Respiration. I. Ueber die physiologische Verbrennung in den lebendigen Organismen' (Contributions to the doctrine of respiration. I. On the physiological combustion in living organisms)

p. 270

The absolute necessity by living matter, respectively by the cell, to take up oxygen and to form carbonic acid is a fundamental property of all organic kingdoms. It belongs not only to animals but equally so, indispensably, to plants. No cell can grow without oxygen. All parts of plants, probably the green-coloured ones during the day as well as the roots, stems, flowers take up oxygen and exhale carbonic acid. Under conditions of daylight radiation the oxidation process in the green plants will no doubt be masked owing to simultaneous much more powerful reduction of carbonic acid. The plant continuously respires like the animal but takes in its organic food, carbonic acid, from time to time like the animal, by assimilating it by dint of solar radiation.

pp. 272–4

No group in the animal kingdom however gives to those who doubt the predominant significance of the cell for oxidation processes a more instructive example than the tracheates, and particularly the insects. The development of the circulation apparatus is here at a very low stage, for there still exists no capillary system and no vein but only a contractile heart, provided with entries, continues itself into an artery, so that the often colourless blood flows through the body cavity, only washes the organs and either does not penetrate into them, or by such scanty routes comes into connection with them, that one cannot remotely think of an intimate and lively connection between blood and tissues as with the vertebrates. With these animals endowed with intensive oxidation on this account, the air does not transfer itself to the blood but directly into the interior of the organ with the help of the air passages or tracheae, always branching more finely and coming closely into contact with the cell. Here one sees clearly how the blood, on account of its too slow movement through the body and its too little intimate contact with the interior of the organ, is avoided so that the air, that is the oxygen is led directly to the cell without intervention of the blood . . .

What comparative physiology makes known to us clearly, ascending from the simplest forms of the animal kingdom to the higher creatures, that teaches us accordingly of necessity also the study of foetal respiration . . . that with the first moment of the development of the embryo the oxygen absorption and formation of carbonic acid begin, thus at a time when *neither blood nor blood vessels* exist, when thus only cells can consume the oxygen and form carbonic acid.

Selection 54

We have already noted that, before Pflüger, physiologists and chemists largely ignored the cell theory in considering the chemistry of respiration. The identification of the morphological unity at the microscopic level was not paralleled by an acceptance of a chemical unity underlying events in the cells of plants and animals. The exphasis was rather on their chemical diversity in a tradition going back to earlier ideas such as those formulated by the German chemist Leopold Gmelin (1788–1853) in lectures in 1827 and 1828. According to him, plants had the capacity to synthesize substances by deoxidation (reduction) and animals possessed the ability to degrade them by oxidation.[a] This helps to explain why knowledge of respiration since Lavoisier, as outlined in the preceding pages, essentially grew out of the study of animal respiration. The latter also involved the issue of the activation of atmospheric (molecular) oxygen, its transfer to enable it to participate in the oxidation reactions in the animal organism (including the formation of carbonic acid) and the question of the site where it occurred.

Although this distinction between the plant and animal worlds was shown to be questionable by the 1850s, its acceptance lingered on during the second half of the century. By the 1870s, however, it was recognized that, in addition to the oxidation reactions, reduction processes also occurred in the animal body. Thus in 1876 Felix Hoppe-Seyler (1825–95) of Strasbourg (at that time in Germany), who did much to promote the view of biochemistry as a separate discipline at that time,[b] identified the formation of the bile pigment and urobilin, and the production of succinic acid from ingested asparagine as reductive action. He then proceeded to develop the tentative hypothesis that oxido-reduction reactions underlay animal and plant metabolism (excepting chlorophyll).[c]

What he did was to make use of the concept of 'nascent hydrogen', thought to be chemically more active than ordinary hydrogen. He suggested that hydrogen evolved during putrefaction (traditionally allied to fermentation) was 'nascent hydrogen'. As shown in the excerpt from the paper published in 1878, according to Hoppe-Seyler 'nascent hydrogen' participated in oxido-reduction in rendering molecular oxygen (O_2) active by splitting it into atomic oxygen (O) with formation of water.[d]

F. Hoppe-Seyler, *Z.physiol.Chem.*, **2**, 1. 1878–9

'Ueber Gährungsprozesse' (On fermentation processes)

pp. 25–6

It cannot appear unjustified to compare the phenomena called forth by palladium hydride with those of the putrefactive processes in that also here on the hydrogen, which demonstrably evolves from these compounds, with this transformation into other compounds will be conferred the same capability, which is so decisively recognizable in the palladium. The known facts stand in complete agreement with this. So long as the putrefying fluid contains absorbed oxygen, through the putrefaction is developed neither free hydrogen nor is a reduction of other substances than that of the free oxygen recognizable.

If one adds to a putrefying fluid some oxyhaemoglobin solution, methaemoglobin is soon formed. The same substance appears with the action on pure oxyhaemoglobin solution by a palladium plate loaded with hydrogen. The process here is as follows. The active hydrogen removes from the O_2 molecule, which is bound with the haemoglobin, one O atom to form water. The other O atom becomes active and passes immediately into firm combination with the haemochromogen-atom group of the haemoglobin and hereby methaemoglobin is formed.

b. From *c*. 1880 to *c*. 1940: intracellular respiration and oxidizing-reducing systems. Citric acid cycle

Selection 55

Another problem which affected the chemical enquiry into (animal) respiration was the colour of blood. Lavoisier opened the way in 1777 when he surmised that the underlying process in the formation of the red colour of the blood and of the red precipitate, red lead, and colcothar are the same.[a] The discovery by Antoine François Fourcroy (1759–1809) in 1790 that blood contained iron tended to substantiate such opinion.[b] There were, however, men like William Wells who in 1797 opposed the idea that the colouring matter of the blood came from iron: instead he associated it with 'the peculiar organization of the animal matter of one of its parts'.[c]

Though curiosity about the blood colouration and the significance of its changes persisted, a clearer knowledge of the chemical nature of the substance which gave blood its red colour did not begin until 1838. In this year Louis-René Lecanu (1800–71), who taught at the Paris school of pharmacy, published his thesis submitted to the medical faculty. In it chemical procedures for obtaining the red colouring matter from the blood globules were described. Moreover, it was confirmed that this substance ('globuline') termed 'haematosine', probably contained iron in the metallic state. It was also shown that it could be split into two parts – an albuminous substance and the colouring matter proper.[d]

Three years later the Swedish arbiter *in rebus chemicis* Jöns Jacob Berzelius (1779–1848) assigned to the albuminous portion the name 'globulin' (in order to differentiate it from other substances thought to be present in the blood globules and termed respectively albumin and fibrin) and insisted that 'haematosine' should be replaced by 'haematin'.[e] Thus the iron-containing blood pigment became known as 'haematoglobulin'.

A further change in the name occurred in 1864 when Hoppe-Seyler called it haemoglobulin or haemoglobin. Hoppe-Seyler's attention was drawn to the subject in 1857 when he was asked to investigate coal-gas poisoning in coal mines. Almost immediately after Gustav Robert Kirchoff (1824–87) and Robert Wilhelm Bunsen developed the spectroscope for analytical purposes, Hoppe-Seyler began to apply this method to his study of the blood-colouring matter in the early 1860s.[f]

Drawing on the evolutionary and comparative approach, studies of plant and animal pigments, and of their chemical relationship and distribution, were taken up in Britain around the same time.[g] Thus in the mid-1880s Charles Alexander MacMunn (1852–1911), a practising physician and gifted scientific investigator, was able to describe pigments in the tissues of vertebrate and invertebrate animals named 'histohaematins', and 'myohaematin' in the case of the muscle pigment. As shown below, these pigments changed their absorption hands on reduction and oxidation and MacMunn concluded that they were concerned with tissue respiration.[h]

[C.A.] MacMunn, *J.Physiol.* (Proc. Physiol. Soc.), 5, xxiv. 1884

'On myohaematin, an intrinsic muscle pigment of vertebrates and invertebrates, on histohaematin, and on the spectrum of the supra-renal bodies'.

pp. xxiv–xxvi

The absorption spectra to be described were detected by means of the microspectroscope, and most of them are only fully visible in it, as the dispersion of the chemical spectroscope is too great for the detection of some of the very feeble bands. A binocular microscope provided with a substage achromatic condenser to which is fitted two diaphragms was specially made for this kind of work. Its objectives are so adapted as to enable both fields to be fully illuminated when any power up to the $\frac{1}{8}$th is used. The left hand tube is used as a 'finder', and as a means of getting any required portion of the object into the centre of the field so that its spectrum may be obtained in the spectrum eyepiece of the right hand tube. In this way the various portions of a very small bit of tissue or organ may be readily differentiated from each other and their spectra observed. Moreover by the use of the iris diaphragm which is placed below the substage condenser the marginal part of the field can be readily cut off. Another piece of apparatus is indispensable namely the *compressorium*, as by its aid the section is squeezed out thin enough to allow the spectrum to be observed.

No reagent whatever is required for the detection of the spectra to be described so that the substances present cannot be altered in any way.

Myohaematin. Physiologists have accepted Kühne's statement[a] that muscle owes its colour to haemoglobin, but although the majority of voluntary muscles do owe their colour to it, it is accompanied by myohaematin in most cases, and sometimes entirely replaced by it, while in other cases it entirely replaces myohaematin. The *heart* muscle of every vertebrate animal which I have examined yields myohaematin which gives a very beautifully diffused spectrum totally distinct from any decomposition product of haemoglobin, e.g. methaemoglobin, acid or alkaline haematin, or haematoporphyrin. All one has to do in order to detect myohaematin is to cut off a bit of heart muscle, put it while fresh in the compressorium, press it down and observe the spectrum. *No reagent whatever is required.* The spectrum consists of three bands, two of which are very narrow and persist after the haemoglobin bands have gone when the tissue has been squeezed out to great thinness in the compressorium. The bands have been missed by other observers simply because when the oxyhaemoglobin bands are well marked they cover and are merged into the myohaematin bands. The first band of myohaematin occurs just before *D*, the next two (of great narrowness) are placed between *D* and *E*, and two other faint bands may be present near violet, of which the first covers *E* and *b* and the other is between *G* and *F*, close to latter line. Their wavelengths are 1st band λ 613–596.5, 2nd band λ 569–563, 3rd band λ 556–549 (heart of dog) and they have been measured in all cases with the same result. I find myohaematin in the heart muscles and some voluntary muscles of the following *Mammals*: man, dog, cat, rabbit, guinea-pig, hedgehog, sheep, cow, pig, rat and hare. In *Birds*: in pigeon, owl, duck, goose, turkey and fowl. In *Reptiles*: in green lizard, common ringed snake and fresh-water tortoise. In *Batrachians*: in toad, frog, salamander and tree frog. In *Fishes*: in herring, mackerel, tench, roach, eel, plaice, whiting and codfish.[1] But it is also found in Invertebrates, in which I first detected it. It is found in the muscle from thorax and in leg muscles of the following insect genera: Dytiscus, Hydrophilus, Lucanus, Cerambyx, Creophilus, Staphylinus, Geotrupes, Coccinella, Musca (3 species), Tipula, Gryllus, Blatta, Vespa, Apis, Bombus, Pieris, Enuomos, etc. It also occurs in the cephalo-thoracic muscles of spiders, in the heart of the crab, lobster and crayfish (and not in their voluntary muscles); in the heart and buccal muscles of Arion, Limax, Helix and other pulmonate mollusks, . . .

Two attempts have been made to isolate it. In the first it was got out of the muscle by digesting in pepsine solution and was slightly changed in the process; in the second it was got out of the frozen heart muscle of a rabbit by pressing out the plasma,[2] here it was mixed with traces of haemoglobin but could be differentiated from it: hence it probably occurs in muscle *plasma* like muscle-haemoglobin.

Histohaematin. This name has been given by me to a class of pigments or modifications of the same pigment, which are found widely distributed in the Animal Kingdom. Myohaematin belongs to them, as can easily be shown. They are found in Mollusks, Arthropods, Echinoderms, and, modified peculiarly, in Coelenterata. The bands are carefully measured and compared with spectra yielded by various organs and tissues of Vertebrates, and no difference is found between those of Vertebrates and Invertebrates. In order to see these spectra in the higher animals the blood vessels are washed out with salt solution thoroughly, and then the organs and tissues examined in the manner described. It is not possible to go into this subject in an abstract, as the facts are too numerous to be compressed into such a small space; it will suffice to say that the histohaematins are respiratory pigments as can be proved by oxidising and reducing them in the solid organs. Their bands occupy almost the same place as those of myohaematin, except that the second and third bands of the myohaematin spectrum appear compressed into one in some cases.

Myohaematin itself is also undoubtedly a respiratory substance.

C.A. MacMunn, *Phil.Trans.*, **177**, **267**. **1886**

'Researches on myohaematin and the histohaematins'

pp. 294–5

I need not repeat the results obtained by the action of oxidising and reducing agents on these pigments, it will suffice to say that they are capable of oxidation and reduction, and are hence respiratory. Just as in *Actiniæ* – as I have shown in a former paper[a] – actiniohæmatin is concerned in the respiration of the tissues, so these pigments are also concerned in it. They combine with the oxygen conveyed to them in the blood, and hold it for purposes of metabolism, parting with the carbon dioxide in exchange for the oxygen. This is the only conclusion which anyone who has gone over the same ground can come to. These observations appear to me to point to the fact that the formation of CO_2 and the absorption of oxygen takes place *in the tissues* themselves (Pflüger and Oertmann)[b] and not in the blood. Hence these observations are of value in helping to decide a difficult point.

Selection 56

Among the first to enquire into the location of the oxidation and reduction processes in the cell was the great German medical scientist, Paul Ehrlich (1854–1915). He did this in his publication *Das Sauerstoff-Bedürfniss des Organismus. Eine Farbenanalytische Studie* (1885). There were two major, and connected, impulses which impelled Ehrlich to investigate the problem. One came from the experimental and theoretical work of Pflüger. The other stemmed from Ehrlich's lasting interest in the theory and practice of dyeing and was aided by developments in the manufacture of synthetic dyes in Germany.

Adopting Pflüger's idea of the protoplasmic giant molecule,[a] Ehrlich put forward for consideration the suggestion that it consisted of two parts: a central chemical structure (nucleus) and side-chains. This arrangement recalled very strongly the configuration of aromatic hydrocarbons with side-chains.[b] Whereas the chemical nucleus was, according to Ehrlich, involved in activities specialized for particular cell (liver, kidney, etc.) functions, the side-chains were concerned with general cell activities. Ehrlich suggested that, by donating oxygen, some of the side-chains were responsible for the physiological process of combustion while others were consumed in the process.

Having thus localized cellular respiration at the chemical level and satisfied himself that, in effect, it can be related to the alternate changes in the acid – alkaline states of the protoplasm – Ehrlich proceeded to test the scheme by the use of alizarine blue and indophenol. The intensity of decolorization of these vat dyes when in contact with different kinds of tissue served as a measure of the reducing power of animal organs.[c] The reconversion of the leuco-compound to the original dye in the alkaline solution led Ehrlich to imagine that oxidative reactions (uptake of oxygen) in tissues were favoured by alkaline conditions.

The following extracts, submitted as *Habilitationsschrift*,[d] are from the historically important biochemical publication.

P. Ehrlich, *Das Sauerstoff-Bedürfniss des Organismus. Eine Farbenanalytische Studie* (Berlin, 1885)

(*The Requirement of the Organism for Oxygen. An Analytical Study with the Aid of Dyes*)

p. 19

From the considerations mentioned here it follows that dyes which should serve as a measure of the reducing power of animal organs must satisfy the following three conditions, which are dependent exclusively on their chemical nature. Firstly, they have to be capable of vat dyeing, secondly they have not to offer any maximal resistance to reduction, and thirdly they have to be introduced into the cell in granular form, that is to say they have to be insoluble in the blood. With these three demands, which are exclusively conditioned by the constitution, a fourth is associated in practice. Namely the possibility to provide the cells with the dye in the highest state of subdivision, I might say molecular division. This last factor limits to a high degree the number of utilizable materials, and I have not succeeded, for example, with indigo to satisfy the last requirement.

pp. 133–4

If we assume that the protoplasm, during its activity, is bathed now with alkaline, now with acid fluid, then we can assume without difficulty that we have to set the formation of oxygen in the alkaline phase, that of the consumption in the phase of acidification, or of diminished alkalinity respectively. We would therefore, quite naturally, without assuming any kind of life expression of an unknown nature, trace back the continual balance between two opposing functions, which precisely forms the basis of life, to the regularly and cyclically occurring alterations of alkalinity.

Taking into account that the products that arise through the combustion (especially the CO_2 hereby appearing) can at times call forth an acid reaction, which under the influence of the blood alkalinity is turned back again into the normal alkalinity, it will be easy to imagine a kind of self-regulation of the protoplasm.

Let us take the case that in a normal cell reacting alkaline a certain high number of unsaturated oxygen affinities is present, then the oxygen molecules pressing on first saturate the sites of the highest oxygen affinity, then, falling off, those of less. The lower the saturation extends so much the more loosely is the oxygen bound, so much the more easily will it be given up. A point will thus finally come where the oxygen is so loosely bound that it can, at a given moment under the prevailing conditions, oxidize any side-chains whatever [and] form carbonic acid. When this point is reached there begins the phase of acidification in which, as already mentioned, the real combustion takes place. Obviously this lasts until, with the always progressing reduction of the oxygen sites of the protoplasm, the sites are reached which bind the oxygen so firmly that they no longer give it up under the prevailing conditions to the combustible side-chains present. At this moment naturally the acid formation in the protoplasm ceases and under the influence of the alkaline blood the alkaline reaction begins to re-appear. With this the process of oxygen uptake begins anew.

In order to avoid misunderstanding, I should like immediately to emphasize that the size of the variation of the vital reaction must be different in the different organs, and that in some of them it will come perhaps not indeed to a direct acidification but only to a more or less significant decrease in alkalinity. I should like therefore to make precise my views regarding the physiological regulation in general that the higher degrees of alkalinity favour or condition oxygen formation, the lower the actual combustion process.[a]

Selection 57

As the nineteenth century was coming to the close, it was apparent that the nature of the oxidation process (slow combustion) in the living body was far from clear. What was particularly puzzling was the readiness of carbohydrates, proteins and fats, making up the foods, to undergo oxidation inside the animal body, in contrast to the resistance of these substances to oxidation by oxygen outside it. In order to account for this difference, several theories of the activation of the atmospheric oxygen under living conditions were advanced. Thus we saw that Schönbein connected the activation of oxygen with its passing into ozonized state and Hoppe-Seyler with the splitting of the oxygen molecule by means of 'nascent hydrogen'. This was to result in the production of one atom of oxygen in active state and the other uniting with hydrogen to form water.[a]

It was for this view that Hoppe-Seyler came under heavy attack from Traube. Referring to his own previous work, Traube reminded Hoppe-Seyler and others that the oxygen activation could not be separated from the mediating action of 'oxidation ferments'. As to the splitting of molecular oxygen by 'nascent oxygen', experimental evidence pointed in another direction, namely, to the formation of hydrogen peroxide.[b]

Another investigator who raised objection to Hoppe-Seyler's scheme was the Russian Aleksei Nikolaevich Bach (1857–1946).[c] Without mentioning Traube, he outlined in 1897 a peroxide theory of oxidation processes in the animal body occurring with the aid of ferments (enzymes). In fact, Bach suggested that the latter were made up of peroxides.[d]

A. Bach, *Comp. Rend.*, *124*, 951. 1897

'Du rôle des peroxydes dans les phénomènes d'oxydation lente' (On the role of the peroxides in the phenomena of slow oxidation)

p. 951

The energetic oxidations of which the animal organism is the seat necessarily imply the preliminary transformation of the passive oxygen of the blood into active oxygen. Amongst the different hypotheses stated on the mechanism of this transformation, the hypothesis of Hoppe-Seyler is that which has gained most credence. According to Hoppe-Seyler, the nascent hydrogen splits the molecule of passive oxygen, of which it fixes one of the atoms to form water, whilst the other atoms is set free and becomes capable of producing the most energetic oxidations. Analogous reactions happen in the organism where, under the action of certain ferments, there can be release of hydrogen or formation of substances easily oxidizable, which function like nascent hydrogen.

The study of the phenomena of slow oxidation has led me to the conclusion that the transformation of passive oxygen into active oxygen can be effected through the intermediation of the peroxides which arise in the oxidation of easily oxidizable substances. By peroxides I mean oxygenated substances functioning like oxygenated water and characterized by the presence of at least one group –O–O– of which the two free valencies are saturated by electropositive or electronegative, monovalent or bivalent radicals.

p. 953

Containing active oxygen, the peroxides formed in the oxidation of the easily oxidizable substances can provoke energetic oxidations. If one causes a current of air to pass into a solution of indigo to which essence of turpentine or benzaldehyde has been added, the indigo is rapidly oxidized to isatine. The nascent hydrogen set free by hydrogenated palladium produces the same effect. It is this last fact which Hoppe-Seyler considers as a decisive proof in support of his theory. In order to determine whether the oxidation of indigo is provoked by the atoms of free oxygen or by the peroxides resulting from the oxidation of nascent hydrogen, I have repeated the experiment of Hoppe-Seyler separating it into two phases: 1, oxidation of nascent hydrogen; 2, oxidation of the indigo by the products obtained, but *in the absence of nascent hydrogen*.

I have made pure air pass into a well-cooled test-tube containing 15 cc. of acidulated water and a sheet (12 × 4) of hydrogenated palladium. The air arrived through a tube drawn out to a capillary point. After one hour I withdrew the sheet of palladium, added to the liquid 1 cc. of a 1 per cent solution of indigo and noted the time elapsed until complete oxidation of the indigo. In a series of experiments, the time varied from thirty minutes to one hour twenty minutes. In order to obtain the same result with oxygenated water, it was necessary to use solutions containing 0.675 g. to 0.3275 g. per litre. However, on titrating by means of the permanganate the active oxygen in the product of oxidation of the hydrogen, I found only 0.012 g. to 0.018 g. per litre.

These experiments show that the theory of Hoppe-Seyler is void of foundation.

p. 954

All that has just been said on the mode of formation and on the oxidizing action of the peroxides can be applied equally to the processes of oxidation which go on in the animal organism. On being oxidized in presence of an excess of oxygen, the easily oxidizable substances, which arise in the blood, form peroxides which oxidize difficultly oxidizable substances in the same manner as the peroxide formed by the turpentine oxidizes indigo. The oxidizing ferments which exist in the blood are probably nothing else than these easily oxidizable substances and are eminently apt to form peroxides.

Selection 58

Notwithstanding vital staining, the twentieth century opened without any noticeable change in the method employed in the study of the chemical processes associated with respiration. It relied on the measurement of the two gases central to respiration: oxygen absorbed and carbon dioxide excreted by tissues. One new development was the tissue-mincing technique which made it possible to enlarge the diffusion area (an important factor in governing the exchange of respiratory gases). It was found reasonable to sacrifice the intactness of organs, obtained under perfusion, in favour of studying diffusion from and to the liquid saturated with air or oxygen, in which the minced tissue was suspended.

It was from such work on minced muscle, carried out by means of a microrespirometer and begun by the Swedish investigator, Torsten Thunberg (1873–1952), during the first decade of the new century,[a] that novel relationships began to emerge which threw light on the intermediate steps in the biochemical oxidation reactions.

Thunberg investigated the respiration of minced muscle under the influence of various substances. He tested a large number of organic acids and found an increase of oxygen consumption of the muscle on administering oxalic acid, malonic acid and especially succinic acid.[b] As shown below he then concluded that investigations of this kind would help to elucidate the nature of intermediate products of biological oxidation.

T. Thunberg, *Skand.Arch.Physiol.*, *24*, 23. 1911

'Studien über die Beeinflussung des Gasaustausches des überlebenden Froschmuskels durch verschiedene Stoffe. Vierte Mitteilung' (Studies of the gas exchange in the surviving frog muscle influenced by various substances. Fourth communication)

p. 60

The researches communicated here have given the result that no single one of the 38 organic acids investigated here exert the same action on the gas exchange which I have found earlier for the oxalic acid, malonic acid and succinic acid. The action of these acids is thus to be seen as specific for them. In the following I call this action the specific succinic acid-action because succinic acid shows it to the most marked degree. The question now arises how one should picture this action.

The possibility, which appears to me as lying nearest, proceeds from the thought that the normal oxidation processes run in such a manner that an intermediate process is interposed, whereby an unknown substance breaks down into succinic acid and carbonic acid, perhaps also into several other substances. If succinic acid is present in larger amount it acts inhibitively on this process but not hindering, at least not primarily, the other oxidation processes. The oxygen uptake, which is an independent partial process, is thus not influenced, only the carbonic acid formation. This conception, which one must apply also to the malonic acid and the oxalic acid if one wishes to be consistent, is bound up with several difficulties. It has however the advantage that it can be tested experimentally. It points also, if it is valid, to a method of work until now scarcely considered, as far as I know, for analyzing complicated chemical, especially biochemical processes, a method which one can name the method of partial resistances. It would consist in this, that one exposes the system, which shows the reaction under investigation, to the influence of a number of substances which can be considered as possible intermediate products. If one finds in this way that a certain substance exerts a specific slowing down action, one has found a possibility to follow up further.

The tendency of several organic acids, especially the di- and poly-carbonic acids, to split off CO_2, appears to be important for the theory of elementary respiration. It shows a possible, until now scarcely considered source of carbonic acid being formed under anaerobic conditions.

Selection 59

Concurrently with the work initiated by Thunberg, the Italian, Federico Battelli (1867–1941), and the Russian-born Lina Stern (1878–1968) began to develop a new hypothesis of respiration. They assumed that respiration in isolated living tissues consisted of two distinct processes which existed side by side: main respiration (*Hauptatmung*) and accessory respiration (*akzessorische Atmung*). They suggested that the two forms of respiration differed in both their function and nature and also in their relations to the structure of the cell.

The chief function of main respiration was the transformation of the chemical energy of the organic compounds, whereas accessory respiration was concerned with the oxidation of certain potentially noxious substances, such as uric acid by uricase. In contrast to main respiration, accessory respiration was considered to be fermentative (enzymatic) in nature and able to proceed in the absence of cells or cell fragments. It was claimed that after the death of the animal main respiration gradually decreased and finally stopped, but accessory respiration could continue, not being cell-bound.[a]

Under the impact of Thunberg's work it seemed to Battelli and Stern that the study of the oxidation of succinic acid could elucidate the obscure features of main respiration. Indeed, as shown below, they thought they had obtained evidence that the oxidation of succinic acid had much in common with the main respiratory process. They stated that most animal tissues were capable of oxidizing succinic acid into malic acid which, it was claimed, had been isolated as barium salt. Although it had been thought that conversion might be due to the presence of fermentative (enzymic) oxidizing agents, the experiments afforded no proof for such a view.

F. Battelli and L. Stern, *Biochem. Z.*, 30, 172. 1911

'Die Oxydation der Bernsteinsäure durch Tiergewebe' (The oxidation of succinic acid by animal tissues)

p. 172

We have most recently made the observation[1] that animal tissues possess the capability of oxidizing succinic acid with uptake of molecular oxygen to malic acid according to the formula:

$$C_4H_6O_4 + O = C_4H_6O_5$$

The study of the oxidation of succinic acid seems to us to be of quite special importance. We believe that a closer knowledge of this oxidation will put us in the position to win a glance into the dark and complicated processes of the main respiration of the tissues. Besides the oxidation of succinic acid offers a special interest, because it shows several characteristics which distinguish it from the oxidation of the other substances so far investigated.

p. 173

We have ... found that the succinic acid increases the oxygen uptake of the tissue very energetically whilst, according to the experimental conditions, the carbonic acid formation remains unaltered or is diminished. Malic, fumaric and citric acids can, under certain conditions to which we shall return in a later work,[2] cause an increase in the carbonic acid formation and the oxygen uptake, yet the latter is incomparably smaller than that caused by addition of succinic acid.

The increase in oxygen consumption caused by addition of succinic acid is not to be ascribed to activation of the intrinsic respiration of the muscle but, as we have already said, to oxidation of the succinic acid to malic acid.

pp. 191–2

Our knowledge concerning the mechanism of the main respiration is very limited. We have found that for the taking place of the main respiration the co-operation of two substances or substance groups is necessary, of which those of the one kind are soluble in water whilst those of the other kind remain adhering to the cell fragments. The first we have named *pnein* and the latter designated as *fundamental respiration process*. If for example we add water to a ground muscle and press the same out through a cloth, we obtain a residue (fundamental respiration process) and an extract (pnein). Separated, neither the residue nor the extract shows respiration activity but after bringing together the two the respiration again resumes. The fundamental respiration process plays the most important role in the mechanism of the main respiration, the pnein works only as activator or perhaps as direct fuel.

The oxidation of succinic acid now shows several characteristics closely related to the main respiration. We recall above all the extremely important fact that in the different tissues a very close parallelism is to be noted between the energy of the main respiration and the intensity of the succinic acid oxidation. Besides, the succinic acid oxidation like the main respiration requires the assistance of substances which are not soluble in water. Amongst the other characters common to both processes the following may be named. The main respiration and the oxidation of succinic acid are characteristic of all tissues so far investigated. The optimum temperature is for both 40°, and at 55° both processes sink down to a minimum. The treatment of the tissues with alcohol or acetone completely destroys the main respiration and at the same time also the oxidation of succinic acid.

Yet there are also clear differences between the process of the main respiration and of succinic acid oxidation. The most important among these are the following. The intensity of the main respiration gradually falls off after death and in some organs, such as the liver or the heart, this falling off is extremely rapid. Besides, the main respiration shows itself as very sensitive to all influences which encroach on the vitality of the cell, and it is for example impossible to wash the minced muscle 3 or 4 times without completely destroying the main respiration. The energy of the succinic acid oxidation on the other hand remains in all tissues unaltered for a longer time after death, and one can wash the muscle numberless times without markedly diminishing its capability to oxidize succinic acid. The main respiration is more energetic in slightly alkaline medium that in neutral medium; the succinic acid oxidation on the other hand carries on most energetically in neutral medium. Finally, the main respiration cannot take place without the cooperation of water-soluble substances (pnein), whilst for the oxidation of succinic acid the action only of substances insoluble in water suffices.

In consideration of the common characteristics as well as of the differences which can be noted between the main respiration and succinic acid oxidation the assumption can be made that the process operating the oxidation of succinic acid makes up a perhaps not inconsiderable part of the mechanism of the main respiration. In the main respiration still other processes naturally come into consideration which are distinguished essentially by a great lability. This hypothesis cannot claim precision, however, at the present time we do not possess the necessary data to be able to formulate our hypothesis more exactly.

It would naturally be of the greatest interest to know the nature of the process oxidizing succinic acid. One could assume that it is a question of a special oxidizing ferment, which possesses the characteristic of attaching itself very strongly to the water-insoluble substances of the tissue. However, no direct experimental evidence supports such an assumption and one can assume with just as much probability that substances are concerned which are quite different from the usual oxidizing ferments.

Selection 60

Early in the second decade of the twentieth century the study of biological oxidation took a new turn. It grew out of the work of the German investigator Heinrich Wieland (1877–1957), in connection with his interest in catalytic dehydrogenation and hydrogenation as developed by the Russian Vladimir Nikolaevich Ipatiev (1857–1952) and the French Paul Sabatier (1854–1941), working independently at about the same time. The use of these methods began to play an important role in the industrial and theoretical areas.

In his first paper (1912) Wieland pointed out that it was possible, for example, to assume a system in equilibrium consisting of quinone and hydroquinone. In the presence of palladium, quinone could be reduced to hydroquinone and conversely hydroquinone could be dehydrogenated to quinone. These considerations led him to demonstrate successfully *Dehydrierungen*, as he called the catalytic dehydrogenations of organic compounds at ordinary temperatures in the presence of palladium and other catalysts.[a]

On the basis of these results Wieland began to enquire whether the underlying mechanism of biological oxidation could be interpreted as catalytic dehydrogenation. It should be added that in considering this kind of reaction with respect to biochemical systems Wieland recognized that the oxidation process was accompanied by a reduction process involving enzymes termed 'dehydrases'. The first application of this approach is to be found in Wieland's second paper on this theme (1913) from which the following excerpts are taken.

H. Wieland, *Ber.chem.Ges.*, 46, 3327. 1913

'Über den Mechanismus der Oxydationsvorgänge' (On the mechanism of the oxidation processes)

pp. 3329–30

Of the oxidation processes, which go on in the interior of the cells, the slow combustion of grape sugar to carbon dioxide and water, the respiration process, is the most important. It has always rightly excited the wonder of chemists and physiologists that the not specially easily oxidizable molecule of glucose is burnt by the inert oxygen of the atmosphere in the tissues, already at low temperatures, smoothly to carbon dioxide and water. Already on grounds of definition there can be no doubt that catalysts of some kind must take part in the acceleration of this reaction which in itself runs with unending slowness; only on account of its sensitivity, probably it was as yet not possible to separate it from the life process. From its manner of working one has generally the conception that it consists in an activation of oxygen, which is taken up into a peroxide form of combination. I have now with oxygen-free palladium black already at blood temperature (about 40°) been able to bring about a rather rapidly running dehydrogenation of grape sugar during which (and this is the important thing) *already from the beginning an abundant formation of carbon dioxide appeared*. With the increasing saturation of the metal by hydrogen its activity weakens. One reaches however a further step in the reaction if one binds the hydrogen split off to quinone or methylene blue. *It is thus possible to burn the grape sugar extensively at low temperatures, with exclusion of oxygen, with the aid of palladium black alone or with it and quinoid compounds as hydrogen acceptors.*

pp. 3339–40

If one considers the oxidation processes as dehydrogenations, as according to results up to the present time has been shown exactly at least for some important cases, then they contain enclosed within themselves at the same time a *reduction process*, as indeed the hydrogen activated by the ferment must be taken up by some acceptor. Thus naturally the hydrogenation of the molecule of oxygen to water $O = O + 4H \rightarrow 2H_2O$ is just as much a reduction as the hydrogenation of methylene blue, or quinone, of plant dyes,[1] or of the nitrates[2] to nitrites and ammonia and so on. Also I think of the participation of hydrogen peroxide in oxidation reactions in such a way that, as an easily hydrogenizable substance, it has the function to take up hydrogen:

$$HO - OH \xrightarrow{2H} 2H_2O,$$

to dehydrogenize. Hydrogen peroxide is indeed also, as known, the first hydrogenation stage of the oxygen molecule.

In the light of the relation just set forth, the so-called *reduction ferments*, often treated in the literature, lose their special place if one can bring the proof that their obvious reduction activity, for example, decolourization of a dye by means of a substrate may also be used for the hydrogenation of the oxygen molecule. If one, in the sense of the view previously stated, can show that the 'reductase' can function also at the same time as 'oxidase'. The most thoroughly investigated reduction ferment, the enzyme discovered by Schardinger[a] in milk allows this proof. Its typical activity consists in the fact that in its presence methylene blue is rather rapidly decolourized by aldehydes, i.e. is reduced. The aldehydes go naturally in this way into the corresponding acids. If one replaces the dye by molecular *oxygen*, the ferment mediates in the same manner the dehydrogenation of the aldehyde; we have the action of an oxidase. The theory given here comprises, as need not be gone into more closely, reduction and oxidation as the two expressions of one process, the dehydrogenation.

Selection 61

Following the discovery of cell-free fermentation[a], it seemed natural to imagine that this approach was likely to prove significant for the comprehension of oxidation reactions in living tissues. Investigations undertaken on these lines, however, threw doubt on such assumptions. Thus Arthur Harden (1865–1940), in collaboration with Hugh MacLean (1879–1957), in 1911 found no definite evidence for respiration taking place in press juice obtained from mammalian muscles and livers.[b] Observations of this kind fostered notions that cell respiration was not accomplished by means analogous to the fermenting action of yeast juice on sugar.

Among investigators particularly impressed by the observations made by Harden and MacLean was Otto Warburg (1883–1970), who at that time began to make his first important contribution to science by researching into the chemistry and physical chemistry of cellular respiration.[c] Starting in 1908[d], Warburg wrote a series of papers reporting and discussing observations – in part conducted with Otto Meyerhof (1884–1951) – on respiratory activities of various materials (sea-urchin eggs, avian erythroctyes, liver tissue). The papers received attention, which led to Warburg writing a critical survey of the contemporary work done by himself and others on the subject. It was published in 1914.[e]

A major idea that suggested itself to Warburg was that cellular respiration was conspicuously connected with the structural elements of the cell. To support this view he could demonstrate the existence of special granules (*Körnchen*) in mammalian liver cells by centrifugation, and intimate that they took up oxygen and formed carbon dioxide. This made Warburg query the evidence for the existence of the two types of respiration (main and accessory) propounded, as we have seen, by Battelli and Stern. Warburg suggested – as did Harden and MacLean – that their experimental results were vitiated by faulty techniques (bacterial action) and, therefore, they had been deceived into believing that accessory respiration, in contrast to main respiration, did not require structurally localized water-insoluble substances.[f]

Now, Warburg accepted that cellular respiration involved catalysis. As to how it occurred, he pointed out that chemical catalysts present in the cell entered into contiguity with its structural components and thus accelerated the oxidation reaction.

Accordingly, the cell's oxidative faculty depended on its structure. When working, however, with disrupted unfertilized sea-urchin eggs, Warburg found that the granular suspensions obtained from them retained almost unchanged the capacity to take up oxygen and produce carbon dioxide. He then concluded:

The unfertilized sea-urchin egg is therefore – in contrast to many other cells – a machine, in which the rate of the chemical reaction that yields work is to a large extent not dependent on the structure of the machine.

Discussing the significance of this finding, Warburg thought that it opened up the possibility of investigating cellular respiration as an ordinary chemical reaction proceeding in glassware. On the basis of several thousand manometric[g] measurements Warburg was led to assume that the first step in the oxidative process, very probably, constituted a simple and known chemical reaction: catalysis brought about by iron.

Warburg regarded his own evidence for this supposition as tenable but not adequate. He argued that the conjecture that iron participated in the mechanism of cellular respiration was not new but lacked sufficient experimental support.[h] He pointed this out in a short historical note at the end of the paper from which the following excerpts are taken. Taken in conjunction with his experimental results, they contain the theoretical considerations that led him to interpret cell respiration in terms of a reaction catalyzed by iron.

O. Warburg, *Z.physiol.Chem.*, 92, 231. 1914

'Über die Rolle des Eisens in der Atmung des Seeigeleis nebst Bemerkungen über einige durch Eisen beschleunigten Oxydationen' (On the role of iron in the respiration of the sea-urchin egg, together with remarks on some oxidations accelerated by iron)

pp. 253–5

On the ground of the facts communicated I advance the theory *that the oxygen respiration in the egg is an iron catalysis; that in the respiration process the oxygen consumed is primarily taken up by the dissolved or adsorbed ferrous ion.*

A proof of the correctness of this conception is not provided by the factual material. Even if being pressed and assuming contingencies, the supposition can be defended that though the respiration is accelerated by iron addition, yet is itself no iron catalysis.

That the theory however gives a singularly probably and simple explanation of the different facts emanates perhaps best from a short summary of the experimental results.

1. *The respiring fluid from sea-urchin eggs contains 0.02–0.03 mg. of iron per 100 mg. of nitrogen.*
2. *The acetone precipitate of the fluid gives the iron [ferric] ion reaction with potassium thiocyanate and hydrochloric acid.*
3. *If one adds small amounts of iron salt to the freshly prepared fluid the oxidation rate increases, most probably also the carbon dioxide production; and indeed the increase in oxidation rate amounts to 70–100 per cent if one adds one hundredth of a milligram per 100 mg. nitrogen. Larger amounts of iron do not work much more strongly; considerably smaller amounts of iron do not act. The order of magnitude of the iron amounts active on addition and that of the iron amounts naturally present in the egg is thus the same.*
4. *If one adds the iron salt only after the respiration has become very weak, the increase in oxygen consumption is much smaller than if the iron is added at the time of unweakened respiration. The substance on to which the added iron is transferred is thus obviously used in the respiratory process.*
5. *The increase in oxygen use following on iron addition is inhibited by the narcotic ethylurethane to almost exactly the same degree as the respiration itself.*
6. *If one adds to the fluid substances the oxidation of which is accelerated by iron, one observes an increase in oxygen uptake; the fluid thus behaves as a catalyst like iron salts; or the naturally occurring iron in the fluid is in a state to accelerate oxidations.*

pp. 255–6

1. It speaks in favour of the theory that a fundamental fact of respiration chemistry is immediately explained by it: the inhibiting effect of small amounts of hydrocyanic acid. Hydrocyanic acid would form with Fe·· ion the complex and catalytically inactive ferrocyanide ion. It would act in minimal amounts because the reactant amounts of iron were minimal.
2. It often happens that oxidation in the cell shows great similarity to oxidation by H_2O_2 in presence of iron salts.[1] Now with the autoxidation there is probably formed first from ferrous oxide a superoxide and we would have then *in the cell superoxide[2] as the oxidizing agent in the presence of ferrous salt.*

The different peroxide and superoxide theories of respiration acquire from this standpoint a quite solid look.

Selection 62

Battelli's and Stern's conclusion about the conversion of succinic acid into malic acid in animal tissues was criticized by the German investigator, Hans Einbeck (*b.*1873), in 1913 and 1914. In the first of two papers on this theme Einbeck pointed out that Battelli and Stern omitted to analyze the barium salt of the oxidation product and only assumed that it was a malate, without considering that it could also be a fumarate.[a] In the subsequent paper Einbeck was able to show that succinic acid was oxidized to fumaric acid in fresh meat, just as in the extract of fresh meat.

Einbeck's findings led him to state his views as follows.

H. Einbeck, *Z.physiol.Chem.*, *90*, 301. 1914

'Über das Vorkommen der Fumarsäure im frischen Fleische' (On the occurrence of fumaric acid in fresh meat)

pp. 306–7

The result of the qualitative working up of the oxidation mixture (according to Batelli and Stern) which besides the originally added succinic acid, it is true, yielded not *malic acid* but, in fact, *fumaric acid*, is indeed just as surprising as the finding of fumaric acid in extracts of fresh beef. It agrees however throughout with the physicochemical observations of Batelli and Stern. For in order to convert succinic acid into fumaric acid one molecule of acid needs one atom of oxygen, exactly as much as Batelli and Stern have found in their oxidation experiments, and as likewise would be necessary to convert succinic acid into malic acid. On the course of the conversion of succinic into fumaric acid nothing could be found out for the present. Perhaps later experiments to be carried out *quantitatively* will yield closer elucidation of this, through which it must be established above all how the amounts of the acids recovered from the reaction mixture relate to the weight of the originally added succinic acid.

Selection 63

We have already noted Wieland's attempt to interpret biological oxidation as dehydrogenation taking place under the influence of enzymes called 'dehydrases'. Important evidence for this was obtained by Thunberg, who by 1916 had begun to investigate the oxidation of succinic acid by animal tissues with the aid of a simple technique.

Thunberg devized a glass test tube (holding about 10 cc) provided with a side-tube as well as a ground-in glass stopper with a hole. On turning the stopper the hole could communicate with the side-tube which allowed evacuation of the tube (hence 'Thunberg's vacuum tube').

He used these tubes to investigate the influence of a large number of organic substances (acids, alcohols, carbohydrates, amino acids, uric acid and others) on decoloration of methylene blue preparations of washed animal tissue (principally muscle) *in vacuo*. Thunberg found that in the presence of succinic acid the sample of methylene blue tissue was reduced. By taking into account his own earlier observations as well as the results obtained by Battelli and Stern, and Einbeck, and in the light of Wieland's concepts, Thunberg regarded the oxidation of succinic acid by tissues as a process of hydrogenation catalyzed by the enzyme 'succino-dehydrogenase'.

In the following will be found the summary of Thunberg's reasoning at the time, including the examination of the influence of cyanide on the reduction of the dye, and the discussion of the nomenclature of the succinic enzyme involved in dehydrogenation.

T. Thunberg, *Skand.Arch.Physiol.*, *35*, *163*. *1918*[a]

'Zur Kenntnis der Einwirkung tierischer Gewebe auf Methylenblau' (On the knowledge of the action of animal tissues on methylene blue)[1]

pp. 169–70

The experiments communicated above show that methylene blue, in presence of succinic acid and the muscle preparation described above, is decolorized. This suggests that the explanation of the process is in agreement with Wieland's conception of oxidation. According to this, the muscle ferment would first dehydrogenate the succinic acid with the formation of fumaric acid. If oxygen is present, the hydrogen then reacts with the oxygen. But the hydrogen can also react with other hydrogen-accepting substances. And in the absence of oxygen methylene blue can in this manner be dehydrogenated into its leuco compound ...

One could, however, regard the process also in another manner. That is to say, one could consider that the enzyme transfers oxygen to succinic acid and that, in absence of free oxygen, it provides oxygen for itself from water by splitting it, if a hydrogen acceptor is present. The enzyme would thus in presence of methylene blue split water, transfer the oxygen to succinic acid oxidizing this to fumaric acid, and at the same time change the methylene blue to the leuco compound by means of the hydrogen split off from the water. The behaviour of the reaction on addition of cyanide should permit a decision between these two possibilities, the Wieland dehydrogenation hypothesis and the 'hydroclastic' splitting of water hypothesis, if one compares the above behaviour with the action of potassium cyanide on the oxidation of succinic acid in presence of muscle ferment and oxygen.

pp. 171–3

Now it is interesting that the methylene blue decolorisation by muscle enzyme and succinic acid is not affected by potassium cyanide ... Thus we draw the conclusion that the inactivity of potassium cyanide towards methylene blue decolorisation and its activity towards the aerobic succinic acid oxidation rests in this: that the potassium cyanide, it is true, does not inhibit the dehydrogenation of succinic acid or its transport to methylene blue, but does inhibit the process thanks to which the oxygen oxidizes hydrogen.

pp. 185–6

As has emerged from the description above, we perceive the process in the system succinic acid-enzyme-oxygen or in the system succinic acid-enzyme-methylene blue in the way that the enzyme robs the succinic acid of two hydrogen atoms, which hydrogen atoms it then passes over to oxygen or methylene blue. The question now arises whether we insert an enzyme of this kind into the enzyme groups already named and can name it in agreement with its place in the system.

This however comes up against difficulties. It is clear at once that the enzyme is to be called 'succinase' if one follows the principle of denoting the enzyme with the suffix 'ase', according to the substance preferably attacked by it. But this name says nothing as to how the enzyme attacks the succinic acid. In order to characterize the mode of action more closely one can indeed name the enzyme 'succino-oxydase'. One can indeed regard the fumaric acid, as it is poorer in hydrogen than the succinic acid, as arising through oxidation of the latter. This name however arouses the impression that the enzyme exposes succinic acid to the action of oxygen, which indeed is not the case. In such circumstances the name 'succino-dehydrogenase' would be better. But with this last name one has not taken into consideration that the enzyme can transfer hydrogen to oxygen or to methylene blue or perhaps to other physiologically important substances. It would be best if one could find a name which expresses these properties of the enzyme.

The name 'hydrogen-transportase' is indeed not short and sounds at first perhaps rather unfamiliar. I believe however that it has the advantage of being easily understandable and giving no occasion for error. Besides, it is not my intention to put forward this name as the only designation to be used. Especially the term 'succino-dehydrogenase' is also a good designation. According as one wishes to emphasize the action on succinic acid or the possibility of transferring hydrogen, one can use one or the other term.

The designation 'transportase' which I have used here in a certain case, seems to me also to be useful in other connections perhaps for the clarification of ideas. Where transport of oxygen by an enzyme is concerned one could call this sometimes for the purpose of closer characterization 'oxygen transportase'. An enzyme which transports water can in a similar way receive the name 'hydato-transportase'. Possibly the hydrolytic ferments are real hydato-transportases. Perhaps these enzymes work in the way that the water molecules accumulate on the complicated compounds, after which the altered intramolecular state of tension thus conditioned brings about a spontaneous breakdown of the molecule. Without being bound to this possibility, I allow myself to mention it here as one still discussible.

In conjunction with the older terminology, the substance which takes up the atom groups brought by the transportase may be called 'acceptor'. For the substance which gives up the atom groups concerned it would be advantageous to have a term. Possibly one could name it 'donator'. Although this designation sounds rather trivial, I have been able to find none better.[b]

Selection 64

As we have already seen, Warburg assumed that cellular oxidation was inseparably linked to the catalytic activity of iron present in cells. This supposition did not win universal acceptance and, after returning from war service, Warburg took up the theme once more in the early 1920s. He did this in conjuction with his respiration studies by employing cancerous tissue while improving the manometric method of gas analysis.[a]

Warburg's starting point was that cellular respiration was a cyclic reaction. Molecular oxygen entered into combination with bivalent iron to produce iron in a higher valency state. It was the latter which reacted with oxidizable organic material and in doing so returned to its former bivalent state.[b] Warburg's arguments in favour of his hypothesis were mainly based upon iron's being essential to life, its great reactivity and the specific effect of certain substances on iron. However, Warburg realized that he had not produced experimental support for his view and decided, therefore, to investigate artificial iron-containing systems purporting to show analogy to the reactions occurring in living cells.

Among the systems Warburg studied were the oxidation of amino acids by haemin charcoal,[c] the catalytic power of a mixture containing fructose and phosphate in the presence of iron, and the oxidation of cysteine and linolenic acid catalyzed by traces of iron. Warburg became satisfied that the behaviour of these systems, including the inhibition of their oxygen uptake by cyanide and urethane, pointed to the existence of an ubiquitous cellular iron-containing catalyst involved in the transport of oxygen in the cell, named *Atmungsferment* (respiratory ferment) by him.

As shown in the extracts below, Warburg pictured the catalytic effect of iron as due partly to a 'specific' chemical property and partly to an 'unspecific' physico-chemical property of the metal. He associated the first with iron readily undergoing alternate oxidation and reduction and the second with its ability to behave as a surface-active substance. As to the varying specific catalytic effects of iron on different substances, Warburg assumed that this depended on the way in which the iron atom was attached to particular groups of the compounds constituting the investigated artificial systems.

O. Warburg, *Biochem. Z.*, *152*, 479. *1924*

'Über Eisen, den sauerstoffübertragenden Bestandteil des Atmungsferments' (On iron, the oxygen-transferring constituent of the respiratory ferment)

pp. 487–8

When divalent iron reacts with oxygen, [or] iron of higher valency with the amino acids, the driving forces are specific chemical forces. Besides these in the oxygen transfer by means of haemin charcoal, the unspecific surface forces play an important role in that they bring the amino acids out of solution on to the iron. Substances[1] of the most different chemical properties, if they are only adsorbed by the charcoal, are in the state to displace the amino acids from the charcoal. Decisive for the displacing action of a substance is the strength with which the solid surface attracts it and this is proportional to the adsorption constant. The greater the adsorption constant so much the greater is the displacing activity, which shows that it is the unspecific surface forces which hold together the amino acids and catalyst.

The experience of physical chemistry teaches that through the action of unspecific surface forces chemical processes can be accelerated. Porcelain, quartz and other solid bodies accelerate the detonating gas reaction, hydrogen and oxygen become under the action of the surface forces capable of reaction, and we must assume the same in our case for the amino acids the structure of which, as one may imagine, is loosened through the surface forces.

Yet this loosening does not suffice to bring about a reaction between amino acids and molecular oxygen. Charcoals which contain neither iron nor carbon peroxide, even when they adsorb well, are throughout catalytically inactive.[2] A reaction occurs only then when chemical forces join the surface forces.

Thus arises that characteristic interplay of unspecific surface forces and specific chemical forces, as it is characteristic for the haemin charcoal as well as also for the living substances. Both systems react on the one hand in the manner of unspecific surface catalysis, on the other hand in the manner of specific metal catalysis. The specific anticatalyst is hydrocyanic acid, the unspecific anticatalysts are the narcotics.

pp. 488–9

If one has a broad enough concept of activation, then the iron concerned with the higher oxidation stages is 'activated' oxygen and the atoms of the loosened-up molecules are 'activated' carbon, hydrogen, sulphur, etc. All[3] molecules taking park in respiration are thus activated, the oxygen through chemical forces, the rest of the molecules through unspecific surface forces.

This result teaches that the earlier theories of respiration were partly correct partly false. The theories of Moritz Traube and Bach were correct in so far as they assumed the activation of oxygen, incorrect in so far as they overlooked the activation of the organic molecules. The theories of Pfeffer[4] and Wieland[5] were correct in so far as they assumed an activation of the organic molecules, incorrect in so far as they overlooked the activation of the oxygen.

The unsatisfactory nature of all earlier theories of respiration was their lack of definiteness. That a ferment activates the reacting substances is only another expression for the fact that it acts. Invertase activates cane sugar, zymase grape sugar. Not *whether* a ferment activates, but in what manner it activates, is the problem to be solved. If one satisfies oneself with the answer to the first question, there arises that [vicious] circle which explains respiration through activation and activation through the activating ferments.

pp. 493–4

If we survey the three models described: haemin charcoal and amino acids, fructose and phosphate, and cysteine and linolenic acid and compare the action of the iron, we recognize that the binding of the iron is of decisive influence on its action.

The iron bound to the nitrogen of the charcoal attacks neither sugar nor fatty acids, but only amino acids. The iron bound to fructose-phosphate attacks neither amino acids nor fatty acids, but only sugar and of these with appreciable speed only fructose, so that it would be easy with help of these artificial systems to separate fructose and galactose as far as with the help of fermenting yeast cells. Iron bound to the sulfhydryl group attacks neither amino acids nor sugar, but only fatty acids, and of these only the unsaturated.

This comparison teaches whereon the specific action of the respiration rests. Though it is always the same atom which takes up the oxygen and passes it further and always the same process – the valency change – which lies at the base of the oxidation, but *whether* this change occurs and *how quickly* it occurs, depends on the binding of the iron. With this the specific action is so far explained as is possible in the present state of physical chemistry. Why the iron in the one binding reacts quickly in the other slowly, we can explain as little as the rate of any one chemical process. The mechanism of chemical processes is unknown.

p. 494

If we ask ourselves in conclusion what the respiration ferment is, the answer obviously depends on how broadly we view the concept of respiration. If we confine ourselves to the process of oxidation the respiration ferment is the sum of all catalytically active iron compounds present in the cells. If we include in respiration processes in which, as in the reaction between fructose and phosphate, special affinities for iron yet arise then is also the phosphoric acid a constituent of the respiration ferment and with it all substances which through splittings or condensations create binding possibilities for the iron. Now whether we take a narrow or wide concept of respiration, iron as the oxygen-transferring constituent of the respiration ferment will always retain its central place, for the oxidation is that process by which respiration distinguishes itself from other ferment reactions, and in which the purpose of respiration, the gaining of energy, is fulfilled.

Selection 65

Not for the first time in the history of science two contradictory theoretical approaches, supported by experimental evidence, claimed to have produced valid scientific knowledge. This was the state of affairs regarding the Wieland-Thunberg 'hydrogen-activation' hypothesis and the Warburg 'oxygen-activation' hypothesis of the respiratory mechanism during the early 1920s.

As often happens in the history of science, the contradiction (either–or) was resolved by revealing the existence of both systems. And, similarly, not for the first time in the history of science, this was pointed out in two independent contributions appearing almost simultaneously (1924).

In 1923 the Swiss worker, Alfred Fleisch (1892–1973), while in Cambridge comparing the oxidation process in normal and sarcomatous tissue, investigated the mechanism of the aerobic oxidation of succinic acid. His experimental results led him to envisage the occurrence of double activation involving hydrogen and oxygen.[a] Although the experiments of Fleisch and his view as to the need for both activation processes were noted at the time, it would appear that their acceptance was mainly due to Albert von Szent-Györgyi (1893–1986).[b] Szent-Györgyi's paper, published somewhat later, consciously focused on the solution of the contradiction between the two approaches and, as shown in the extracts, he demonstrated ingeniously that they cannot be separated.

A. von Szent-Györgyi, *Biochem.Z.*, *150*, *195*. *1924*

'Über den Mechanismus der Succin- und Paraphenylendiaminoxydation. Ein Beitrag zur Theorie der Zellatmung' (On the mechanism of succinic and paraphenylenediamine oxidation. A contribution to the theory of cell respiration)

p. 198

This important question can, as I think, be decided by the following simple experiment. If one places muscle tissue in the respirometer with succinic acid, then this is oxidized and large oxygen uptake is shown. The oxidation process is now impaired by a suitable dose of cyanide. Suppose now that the reason for this stoppage is, as Warburg considers, the omission of oxygen activation and that the biological oxidation process consists, *besides* the oxygen activation, of a hydrogen activation not influenced by the poison cyanide. Then the oxidation can again be started if one replaces the oxygen activation with an artificial mechanism, and establishes a new link between the activated hydrogen and molecular oxygen. Such a link should be produced by the addition of methylene blue which is reduced by active hydrogen to leucomethylene blue; in presence of oxygen the latter is again promptly oxidized to methylene blue. If in this way through addition of methylene blue the oxidation is again set in motion, then this means that in spite of the presence of cyanide the hydrogen activation system was present with unchecked power of performance. If in spite of the hydrogen activation no oxidation takes place then this means that the molecular oxygen present (which now under the influence of the cyanide is no longer activated) is not in a state to oxidize the active hydrogen, thus cannot serve as hydrogen acceptor. That the oxidation now, in presence of methylene blue, runs well also without the active oxygen shows that the oxidation process comprizes surely, in part, hydrogen activation. That the process without methylene blue does not run further shows that the process comprizes, for the other part, oxygen activation and the oxidation of succinic acid takes place through the interaction of *both* processes, hydrogen and oxygen activation.

pp. 199–200

The result of such an experiment is shown in Figure 2. The ordinates give the pressure fluctuation in millimetres, 1 mm. nearly corresponds to 6 cmm. The abscissae give the time in minutes.

As one sees from the curve, the oxidation of the succinic acid is almost completely inhibited by the amount of cyanide used and after addition of the dye goes further nearly into its original range, without being troubled by the cyanide present. The experiment was naturally repeated several times. All experiments of this kind gave the same result. If the oxygen consumption at the beginning of the experiment was weaker, then after the methylene blue addition even an increase compared with the initial value could be observed.

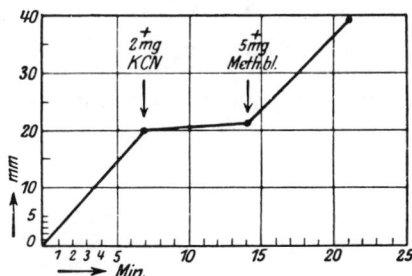

Figure 2 Cellular oxidation of succinic acid in the presence of potassium cyanide and methylene blue

Selection 66

In 1925 yet further impetus was given to the task of resolving the contradictory features of work on biological oxidation, work which was at times clouded by the bitter controversy between the chief protagonists of the hydrogen and oxygen activation theories.[a] It arose out of research on the respiration of parasitic insects and worms, which was carried out by the Polish-born biologist, David Keilin (1887–1963), at the Quick Laboratory (Molteno Institute for Research in Parasitology) in Cambridge from 1919. In 1923, with the aid of a small hand-held direct vision spectroscope, Keilin observed a four-banded spectrum belonging to a pigment present in the thoracic muscles of the adult fly, *Gasterophilus intestinalis*, the larvae of which exist parasitically in horses' stomachs.

The position of the absorption bands of the pigment resembled the spectrum described by MacMunn in his researches on myohaematin and histohaematin.[b] Keilin's investigations confirmed MacMunn's view that the pigment was a respiratory one and distinct from haemoglobins occurring in muscle and blood, and their derivatives.[c] They also revealed that it was a widely distributed respiratory pigment present not only in cells and tissues of animals but also in cells of higher plants and in bacteria and yeast. In fact, Keilin noticed the respiratory function of the pigment when, on passing a current of air through a yeast suspension, the absorption bands vanished, whereas on stopping it they again appeared.

Keilin attempted to extract the pigment in unaltered state, but failed. However, as shown below, on studying the absorption bands and other properties of the extracted products Keilin concluded that the pigment consists of three chemically related entities, resembling haemochrogen. This invalidated MacMunn's claim that the pigment is a simple haematin, and Keilin considered the terms myo- and histohaematin to be misnomers. Basing himself on the wide distribution of the intracellular respiratory pigment, he instead proposed the term 'cytochrome', without prejudice as to the future unravelling of its structure.

The striking demonstration of the reversible oxidation and reduction of the haematin was a crucial event in the twentieth-century history of biological oxidations. Cell respiration, during the period 1925–40, could profitably be interpreted in terms of a sequence of oxidizing-reducing catalyzed reactions ('respiratory chain').[d]

D. Keilin, *Proc.Roy.Soc. (B)*, **98**, **312. 1925**

'On cytochrome, a respiratory pigment common to animals, yeast and higher plants'.

pp. 314–15

Distribution of Cytochrome

In the course of my study on the respiration of parasitic insects and worms, I have found that the pigment myo- or histohæmatin not only exists, but has much wider distribution and importance than was ever anticipated even by MacMunn. Considering that this pigment is not confined to muscles and tissues, but exists also in unicellular organisms, and further, that there is no evidence that it is a simple hæmatin in the proper sense of the term, the names myo- and histohæmatin, given to it by MacMunn, are misleading. In fact, as we shall see later, there is ample evidence that this pigment is not a simple compound, but a complex formed of three distinct hæmochromogen compounds, the nature of which is not yet completely elucidated. I propose therefore to describe it under the name of *Cytochrome*, signifying merely 'cellular pigment,' pending the time when its composition shall have been properly determined. This name, which expresses also its intracellular nature, does not, however, relegate the pigment to any definite compound, an important consideration inasmuch as the properties of various compounds cannot hereafter be ascribed to it without good evidence . . .

The number and wide range of systematic distribution of the species which show this pigment clearly, and which have been enumerated either by MacMunn or myself, is so great that it may safely be concluded that cytochrome is one of the most widely distributed respiratory pigments. Moreover, cytochrome is not confined to animal cells alone. I have found it, and in great concentration, in cells of bacteria, those of ordinary bakers' yeast, and also in some of the cells of higher plants. To avoid all confusion in the terms which will be used in this paper it is important to mention beforehand that cytochrome (= myohæmatin = histohæmatin) is a pigment distinct both from blood hæmoglobin and from muscle-hæmoglobin (= myochrome of Mörner[a] = myoglobin of Günther[b]) or their derivatives. In many cells cytochrome may however, coexist with hæmoglobin.

pp. 315–17

General Characters of Absorption Spectrum of Reduced Cytochrome

Cytochrome in Animal Tissues. – The best material for the study of the absorption spectrum of cytochrome is provided by the thoracic muscles of the honey bee. Specimens of bees frozen at −7°C. are allowed rapidly to thaw. The head and abdomen are cut off, and by compressing the thorax laterally with the fingers the thoracic muscles are expelled in one mass through the anterior opening of the thorax. The muscles of 2 or 3 bees, compressed between a slide and coverslip and examined with the Zeiss microspectroscope, show clearly a very characteristic absorption spectrum (Figure 3) composed of four bands (*a, b, c, d*), the position of which can be determined only with the Hartridge-microspectroscope. For each band I have taken an average of 10 readings and although the pigment was examined *in situ*, the variations

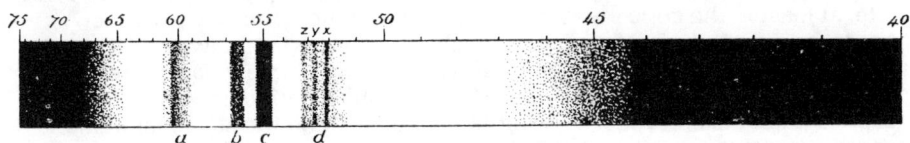

Figure 3 Absorption spectrum of cytochrome in thoracic muscles of a bee

between individual readings were only about 7 Ångström units. The position of maximum intensity of the bands in the bees is as follows: – a, 6046; b, 5665; c, 5502; d, 5210. The relative width and intensity of the bands, in other words, the general aspect of the spectrum; varies naturally with the thickness of the layer of tissue examined. In moderate concentration the relative width of the band is approximately as follows: – (in μμ): $a = 614\text{-}593$; $b = 567\text{-}561$; $c = 554\text{-}546$; $d = 531\text{-}513$. Figure 3 gives an idea of the absorption spectrum of cytochrome in a layer of muscle 0.65 mm. thick, examined with the Zeiss microspectroscope. It shows that band a is very asymmetrical, being darker near the border turned towards the red end of the spectrum; band b is symmetrical, but in bees is much lighter than the band c, the latter being the strongest band in the spectrum, band d which is faint and wide is also asymmetrical, being darker near its short wave end; it also shows a lighter space near the middle, giving to the whole band the appearance of being composed of at least two distinct bands (x and y) corresponding respectively to 5210 and 5280. In transparent muscles rich in cytochrome, a third very fine band (z) could be seen near the green end of the band d (532).

The absorption spectrum of cytochrome in other organisms differs very little from that of the honey-bee (Figure 4) . . .

	610	605	600	595	590	585	580	575	570	565	560	555	550	545	540	535	530	525	520
a Bee: wing muscles	6046									b 5665			c 5502						d 5210
Dyliscus: " "	6038									5664			5495						5203
Galleria : " "	6046									5657			5495						5200
Helix: radula "	6035									5650			5495						5200
Frog: heart muscle	6040									5660			5500						5205
Guinea pig: " "	6045									5662			5510						5205
a Yeast cells	6035									b 5645			c 5490						d 5190

Figure 4 Positions of the four absorption bands of cytochrome a, b, c and d in

pp. 319–20

Oxidized and Reduced Cytochrome

The absorption spectrum with four characteristic bands corresponds to the reduced state of cytochrome, while the spectrum of the pigment in its oxidized state, at least in the concentration found in the tissues, shows no distinct absorption bands, but only a very faint shading extending between 520–540 and 550–570 μμ. The oxidation and reduction of the pigment can be easily observed in yeast. If a shallow tube (30 mm. high) is half-filled with a suspension of bakers' yeast in water (20 per cent.), and the suspension then examined with the Zeiss microspectroscope, the four absorption bands may be clearly seen; but when the air is rapidly bubbled through the suspension the cytochrome becomes oxidized and the bands disappear. If the current of air is stopped the pigment becomes reduced and the four bands rapidly reappear.

A similar result can be obtained by shaking a 5 c.c. yeast-emulsion in a test-tube and examining it with the microspectroscope. When, instead of air, a current of N_2 is passed through the yeast emulsion, or when the latter is shaken with N_2, the cytochrome remains in a reduced state, showing all the time its characteristic four absorption bands. Similar results are obtained with the thoracic muscles of bees or the striated muscles of a guinea-pig . . .

The conditions which determine the oxidation and reduction of cytochrome are not equally influenced by the change of temperature. The oxidation depends upon the rate of diffusion of oxygen into the suspension of yeast on the tissue and is but little affected by temperature; its reduction is due to the chemical activity of the cells which contain the pigment, and therefore has a high temperature coefficient; e.g.: – (a) At the ordinary room temperature (18°–20°) the oxidized cytochrome of yeast usually becomes reduced by the activity of the cells in six to eight seconds. When the oxidized yeast is kept at 0° to −2°C. complete reduction takes place in 70 seconds to 2 minutes. (b) Again, in yeast emulsion kept in a thin layer in a Petri dish at −2° to −4°C. cytochrome usually remains oxidized. Similar results have been obtained with the thoracic muscles of blow-flies and those of bees. These tests show that a low temperature inhibits the reducing power of the tissue more easily than oxidation power of cytochrome.

pp. 320–1

Action of Narcotics on Cytochrome

When a drop of weak solution of KCN is added to the suspension of yeast, no matter how actively this suspension is shaken with air or pure O_2, the cytochrome remains completely reduced, and continues to show the characteristic bands, which do not differ in the slightest degree from the bands of an ordinary reduced cytochrome. The concentration of KCN which stops oxidation of cytochrome is about n/10,000, and a much lower concentration, such as n/100,000, inhibits to a great degree the oxidation power of the pigment.

Further and more important, when a drop of KCN is added to the suspension of yeast, kept at a low temperature and previously oxidized by a current of air, the cytochrome becomes immediately reduced, just as if KCN was acting as a powerful reducing agent. In fact, KCN does not act as a reducer, but inhibits the oxidation of cytochrome, while it does not inhibit other oxidation processes which may accompany the reduction of our pigment.

The action of sodium pyrophosphate is similar to that of KCN. It also inhibits the oxidation of cytochrome only, while it does not arrest the reduction of this pigment. Other substances, such as formaldehyde, ethyl alcohol, acetone, and ethyl urethane, act in a very different way. All these substances, even in concentrations which kill the cells of yeast, do not inhibit the oxidation of cytochrome. On the contrary, in such a concentration they completely stop the reduction of cytochrome, which then remains oxidized indefinitely.[1] In lower concentrations they delay the reduction of oxidized cytochrome, though they do not completely stop it.

If a suspension of yeast in ethyl urethane is shaken with air until the cytochrome becomes completely oxidized, and KCN solution is then added to this suspension, the cytochrome does not become reduced. It remains oxidized because the reducing action of the cells is inhibited, or even destroyed, by urethane, while KCN has no effect on oxidized cytochrome. When the reducing power of the cells is not completely destroyed by the action of urethane, on adding KCN to such suspension a gradual but slow reduction of cytochrome can be observed.

The facts given above show clearly that in relation to the oxidation process in cells in which cytochrome is involved all the inhibitors of oxidation can be separated into two distinct categories, the actions of which are fundamentally different. To one category belong KCN and sodium pyrophosphate; to the other such substances as alcohols, urethane, and aldehydes. The first category (A) inhibits the oxidation of cytochrome, the second (B) inhibits its reduction.

Diagrammatically this can be represented in the following way: –

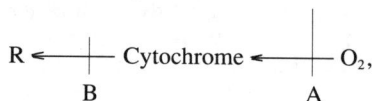

$$R \longleftarrow \!\!\!\mid\!\!\!- \text{Cytochrome} \longleftarrow \!\!\!\mid\!\!\!- O_2,$$
$$\qquad\;\; B \qquad\qquad\qquad\qquad A$$

R being the substances which reduce the oxidized cytochrome, B and A indicating the places of rupture in the oxidation system produced respectively by the substances of the corresponding category of inhibitors. This seems to indicate that, at least for oxidation systems similar to that of cytochrome, the problem of the action of narcotics needs further investigation and existing theories require careful revision.

pp. 325–30

Hæmochromogen Derivatives of Cytochrome

I. *Hæmochromogen in Water Extract of Yeast.* – Dried yeast powdered with sand and extracted with water gives an opalescent yellow fluid which shows two absorption bands. One band (α) is very intense, narrow, much resembling the band c of cytochrome and occupying almost the same position (= 5485); the other (β) is faint, situated in the region of the band d of cytochrome and seems to correspond to the portion x of that band; its position is approximately 521 $\mu\mu$. This pigment does not combine either with O_2 or with CO in neutral or acid solution; neither acid nor alkali shifts the position of its bands, but when KOH is added to the solution the pigment obtained can be oxidized with air and combines loosely with CO. The derivative thus obtained is therefore a hæmochromogen, which we will call hæmochromogen C. It is precipitated from the solution by alcohol, $HgCl_2$ and heat. The remaining yeast pulp still shows the presence of cytochrome, denoting that the extraction was only partial.

II. *Hæmochromogen in Water or KOH Extract of Acetone Yeast.* – Yeast was thoroughly shaken with acetone and allowed to stand for 48 hours, filtered, washed with fresh acetone and dried. The fine powder thus obtained, when wetted again, does not show any band, but on addition of $Na_2S_2O_4$ two bands appear, occupying the following positions: band α-5500, band β-5220. The main band (α) is asymmetrical, its darkest portion corresponding to 5490, in other words, to the band c of cytochrome, while the side turned towards the red end of the spectrum is fainter, and this gives the impression of a strong band c being fused with another band of lower intensity.

When acetone powder of yeast is treated with KOH and $Na_2S_2O_4$ is added,

610	605	600	595	590	585	580	575	570	565	560	555	550	545	540	535	530	525	520
6046 a	*Cytochrome Bee muscles*							5665 b				5502 c						5210 d
yeast, H₂O extract / Haemochromogen C									5585 α (C)								5215 β (C)	
Acetone yeast, H₂O extract / Haemoch. C + trace Haemoch. B									5500 α (C + trace B)									β
yeast extract in KOH / Haemoch. A + B + C							5760 α (A)				5520 α (B + C)						5240 β (A+B+C)	
yeast cells in KOH / Haemoch. A + B + C							5760 α (A)				5540 α (B + C)						5245 β (A+B+C)	
Bee muscles in KOH / Haemoch. A + B + C							5760 α(A)				5525 α (B + C)						5245 β(A+B+C)	
Haemochr in fat body Blow fly larva / probably B + C						5580 α												

Figure 5 Positions of the absorption bands in cytochrome and in its three haemochromogen derivatives

the same two bands appear, but the pigment now combines with O_2 and CO. This derivative is therefore also a hæmochromogen, though slightly different from the previous one.

The acetone solution separated from yeast is transparent and yellowish, and on spectroscopic examination in a tube of 10 cm. long shows faint bands resembling those of a porphyrin.

III. *Hæmochromogen in KOH Extract of Bees' Muscles and Yeast.* – In strong solution of KOH both bees' muscles and yeast rapidly change their colour; yeast becomes distinctly orange, while the muscles turn orange-red. Spectroscopic examination of bees' muscles reveals a marked change in the position of the bands of cytochrome; band *a*, which has become fainter, is moved towards the blue end of the spectrum and stops at 5760; bands *b* and *c* seem to fuse together, forming a wide and strong band with its axis at 5525; while band *d* has become slightly stronger, though it remains broad and lies at 5245.

When yeast is shaken with strong solution of KOH and allowed to stand, the clear solution becomes separated from the jelly-like mass of yeast cells. The latter, after 24 to 48 hours' soaking in KOH, shows three bands occupying the following positions: 5760/5540/5245. The yellow fluid covering the yeast has similar bands, the positions of which are 5760/5520/5240/. When this fluid is shaken with air it becomes oxidised and the absorption bands disappear. On adding $Na_2S_2O_4$, all the bands reappear. When a current of coal-gas or CO is passed through this solution the bands disappear, seeming to be replaced by very faint shading, corresponding approximately to 565 and 528. The three previous bands do not reappear, however, on adding a reducing substance. By passing through this solution a very strong current of air the CO becomes replaced by O_2 and on adding $Na_2S_2O_4$, the three bands reappear.

It is obvious, therefore, that the three-banded spectrum in KOH extract of yeast (or bees' muscles) belongs to a substance which has also the properties of a hæmochromogen compound, although it differs markedly from the two previously-described hæmochromogens.

Composition of Cytochrome

The question arises now as to how a pigment such as cytochrome can yield hæmochromogens so spectroscopically distinct, and as to what is the nature

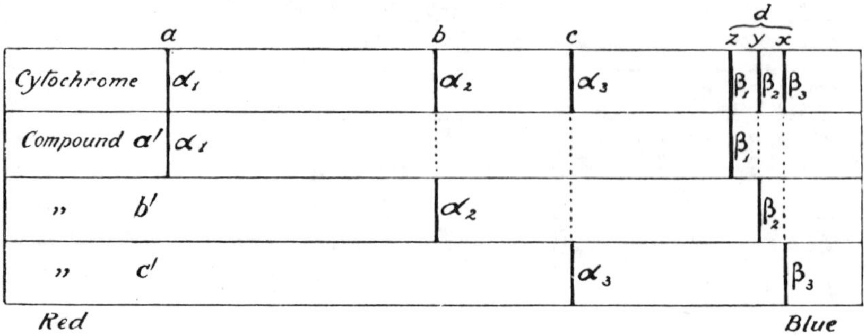

	a	b	c	d (z y x)
Cytochrome	α_1	α_2	α_3	β_1 β_2 β_3
Compound a'	α_1			β_1
" b'		α_2		β_2
" c'			α_3	β_3

Red   Blue

Figure 6 Diagram showing the three haemochromogen components of cytochrome

of the three-banded hæmochromogen obtained in the KOH extract. The only plausible answer to both these questions is the assumption that cytochrome is composed of three components (a', b', c'). Each of these components (a', b', c') resembles spectroscopically a hæmochromogen in showing two absorption bands (α and β) in the reduced form and very faint bands or none in the oxidized form. The bands a, b and c of cytochrome would correspond to the α-bands of these compounds, while the x, y and z of the band d would answer to their β-bands (Figure 6). These three components (a', b', c') may undergo some modifications which spectroscopically are manifested by the changes in the positions of their bands, and these modified components of the cytochrome in alkaline solution behave as hæmochromogens (A, B and C), each showing a two-banded spectrum in the reduced form (with a strong α-band) and combining both with O_2 and CO. We can see now that the water extract of ordinary yeast contained only the hæmochromogen C, while the extract of acetone yeast, in addition to hæmochromogen C, contained a very small quantity of hæmochromogen B, which shifted the middle of the α-band towards the red end of the spectrum and made this band asymmetrical.[2]

The KOH extract of yeast contains all the three hæmochromogens (A, B, and C). The band lying at 5760 is the α-band of hæmochromogen A,[3] and the wide strong band with axis at 5520 (or 5540 in KOH yeast pulp) corresponds to the two α-bands of mixed hæmochromogen compounds B and C fused together. As to the β-bands of these three hæmochromogens, they are fused into one broad band with the axis at 5240 (or 5245).

It is important to note that when CO-hæmochromogen in KOH extract is aerated in presence of a reducer, the α-band of hæmochromogen C appears first, occupying the position of the band c of cytochrome. This shows that in the KOH extract of yeast, just as in watery extracts of yeast or acetone-yeast, the position of the α-band of the c' component of cytochrome remains always the same. It also shows that in the KOH extract the component b' of cytochrome undergoes some change, giving rise to hæmochromogen B, the α-band of which, being nearer to that of C, mixes with it. We can also see that hæmochromogen B has greater affinity for CO than has hæmochromogen C.

The position of the α-band of hæmochromogen A at 576 $\mu\mu$ being far in the long-wave part of the spectrum has nothing exceptional in it; in fact, this band lies exactly in the middle between the initial and final positions of the axis of the α-hæmochromogen band of chlorocruorin, which according to Fox (1924)[c] moves during prepation from 584 μμ to 569 μμ.

The supposition that cytochrome is composed of three distinct hæmochromogen-like compounds (a', b', c') is supported by the following considerations: –

(1) As we have seen previously the band d of cytochrome is not simple but is composed of three narrow bands, x, y and z. The band d seems thus to contain the three β-bands of three hæmochromogens.

(2) Although the positions of the bands a, b, c and d of cytochrome are fairly uniform the relative intensity of the bands varies with the organism examined. Thus the band b, which in insects is much lighter than the band c is in yeast almost as strong as c, and in snails there is no difference between their intensity. The band a may also vary in intensity independently from the other bands. This variation in the relative intensity of the bands indicates that if the pigment is composed of three distinct compounds, the relative proportion of these compounds may vary to some extent.

(3) When the suspension of yeast is normally oxidized the four absorption bands disappear almost simultaneously. When, on the other hand, the suspension of yeast is treated with alcohol, urethane or formaldehyde, and then shaken with air, the band c disappears rapidly, while the band b seems to become even more intense and remains for some time together with the band a and at least a part of the band d. After a longer or shorter interval the band c may reappear or the other bands may disappear until the normal four-banded (reduced) or bandless (oxidized) cytochrome is reformed. In certain conditions, therefore, one part of the cytochrome may be reduced, while the other is oxidized.

(4) The thoracic muscles of bees spread on a slide and allowed to dry, show according to the spot examined, either all the four bands (a, b, c and d), or the bands a, c, d, or c and d, all the bands occupying their normal positions, or even no bands at all. But the combination a, c, d is most frequently seen. On slightly wetting the muscles and adding a reducing agent, all the four bands of normal cytochrome reappear. The result of this experiment shows that drying does not equally affect the rate of oxidation and reduction of the three components, and that here, contrary to experiment (3), the component b is the most rapidly oxidized.

Experiments (3) and (4) show that although under normal conditions in vivo the oxidation or reduction of all three components is synchronous, in abnormal conditions, the three compounds reveal a certain degree of independence.

(5) Finally, Hans Fisher [sic], in collaboration with Schneller and Hilger (1924).[d] has obtained coproporphyrin and Kämmerer's porphyrin from yeast. These two distinct porphyrins are probably derived from two of the hæmochromogens found in yeast.

Selection 67

In the meantime Warburg continued to probe deeply into the nature of the catalytic iron compound that he named *Atmungsferment*. What he did during the second part of the 1920s was to investigate the effect of light on the inhibition of the respiration of living cells caused by carbon monoxide.

The work was noteworthy for its integration of several theoretical and experimental approaches. They encompassed considerations of the processes of photochemistry in terms of the quantum theory and the combined utilization of such devices as the manometer, the photoelectric cell and the spectroscope.

From the shape of the curve obtained by plotting the effectiveness of light against its wavelength it was possible to deduce the resemblance between the respiratory ferment and haemins. For the recognition of the haemin-type nature of the respiratory ferment and the principles underlying this work Warburg was awarded the Nobel Prize in 1931.[a] By this time he proposed calling the respiratory ferment 'the oxygen-transferring ferment of respiration' and this was also the title of his Nobel Lecture from which the excerpts below are taken. In it Warburg summarized the results of the work done in his laboratory concerning the nature of the iron-containing respiratory catalyst, including the common origin of haemoglobin and chlorophyll. Pondering the evolutionary relations of the two pigments, Warburg thought that he had identified spectroscopically their common precursor in spirographis haemin, obtained from the blood pigment of the worm *Spirographis* which lives in the Adriatic Sea. While on reduction of spirographis haemin the derivative gave a spectrum similar to that of haemoglobin, on oxidation the spectrum was like that of chlorophyll.

O. Warburg in *Nobel Lectures. Physiology or Medicine 1922–1941* (Amsterdam-London-New York, 1965)

'The oxygen-transferring ferment of respiration'

p. 257

Not only cellular respiration but also simpler iron catalyses are reversibly inhibited by carbon monoxide and prussic acid. If one compares such iron catalyses and their inhibitions with cellular respiration and its inhibitions, then it appears that the catalyst of cellular respiration behaves like an iron compound in which the iron is bound to nitrogen. But it would never have been possible to reach a definite conclusion if Nature had not endowed the iron compounds of carbon monoxide with the remarkable property of becoming dissociated – with splitting off from the carbon monoxide – under the action of light.

If carbon monoxide is added to the oxygen in which living cells breathe, respiration ceases, as has already been mentioned, but if exposure to ultra-violet or visible light is administered, respiration recurs. By alternate illumination and darkness it is possible to cause respiration and cessation of respiration in living, breathing cells in mixtures of carbon monoxide and oxygen. In the dark, the iron of the oxygen-transferring ferment becomes bound to carbon monoxide, whereas in the light the carbon monoxide is split off from the iron which is, thus, liberated for oxygen transfer. This fact was discovered in 1926 in collaboration with Fritz Kubowitz.[a]

pp. 258–9

Photochemical dissociation of iron carbonyl compounds can be used to determine the absorption spectrum of a catalytic oxygen-transferring iron compound. One combines the catalyst in the dark with carbon monoxide, and so abolishes the oxygen-transferring power of the iron. If then this is exposed to monochromatic light of various wavelengths and of measured quantum intensity, and the effect of light W measured – the increase in the rate of catalysis – it is found that the effects of the light are proportional to the quanta absorbed.

The arrangement becomes very simple if the catalyst is present, as is usually the case, in infinitesimally low concentration in the exposed system. Then the thickness of the layers related to the amount of absorption of light can be considered to be infinitely thin, the number of quanta *absorbed* is proportional to the number of quanta *supplied by irradiation*, and the ratio of the absorption coefficients (β) of light is:

$$\frac{\beta_1}{\beta_2} = \frac{W_1}{W_2} \cdot \frac{i_2}{i_1} \tag{4}$$

Here, the effects of light W, i.e., the rate of increase of catalysis, and the incident quantum intensities i (both easily determinable figures) are on the right, while β, on the left, is the ratio (that is to be determined) of the coefficient of light β, so that the relative absorption spectrum of the catalyst, the position of the absorption bands and the intensity ratio of the bands can be estimated.

In collaboration with Erwin Negelein,[b] this principle was employed to measure the relative absorption spectrum of the oxygen-transferring respiratory ferment. The respiration of living cells was inhibited by carbon monoxide which was mixed with the oxygen. We then irradiated with monochromatic light of various wavelengths and of measured quantum intensity, and the increase of respiration measured together with the relative absorption spectrum – according to Eq. (4)....

p. 261

If the absorption coefficient is entered as a function of the wavelength, the absorption spectrum of the carbon-monoxide compound of the ferment is obtained, as shown in Figure 7. The principal absorption-band or γ-band lies in the blue, while to the right of this, lie the long-wave subsidiary bands α and β in the green and yellow, and, to the left of the principal band, lie the ultraviolet subsidiary bands δ and ε. This is the spectrum of a haemin compound, according to the position of the bands, the intensity state of the bands, and the absolute magnitude of the absorption coefficients.[c]

p. 266 [Figure 8 is from p. 260]

Still nearer the ferment in its spectrum, is a haemin occurring in Nature. This is spirographis haemin, which has been isolated from chlorocruorin, the blood pigment of the bristle-worm *Spirographis*, in collaboration with Negelein and Haas.[d] The bands of spirographis haemin, coupled to globin, are:

	Principal band	α-band
Carbon-monoxide compound of spirographis haemoglobin	434 μμ	594 μμ

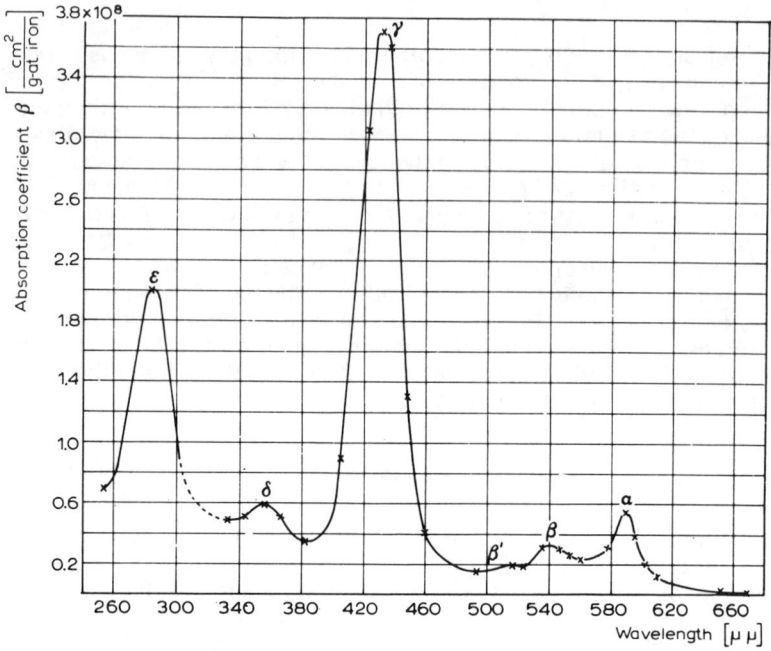

Figure 7 Absorption spectrum of the carbon monoxide
compound of the oxygen-transferring respiratory ferment

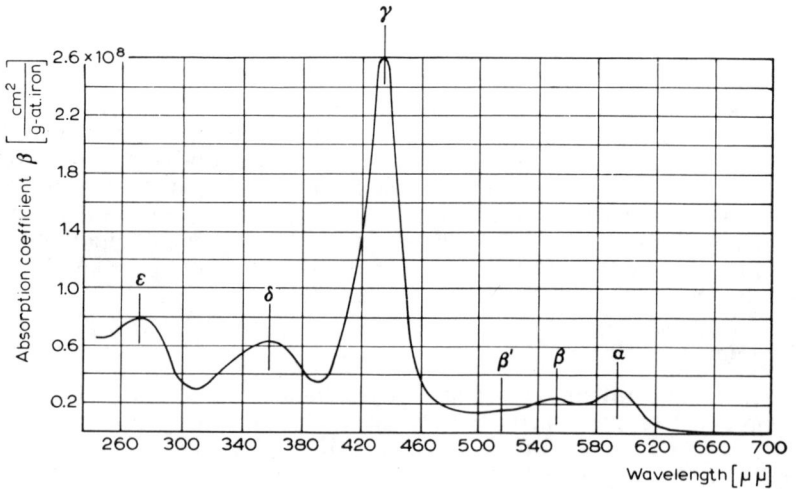

Figure 8 Absorption spectrum of the carbon monoxide
spirographis haemoglobin

pp. 267–8

If oxygen is passed through an aqueous solution of spirographis haemin, at ordinary temperature and under certain conditions, the haemin is oxidized. The previously mixed colour of the solution becomes green, and a band resembling that of chlorophyll appears in the red at 650 μμ. On the other hand, if hydrogen is passed through a solution of spirographis haemin at 37°, in the presence of palladium, the spirographis haemin undergoes reduction in the side-chain and a haemin resembling that of blood is formed. This is a genuine red haemin[1] which does not become mixed-coloured when acidified.

The unique intermediate status of the ferment-like haemins demonstrated by these simple experiments suggests the suspicion that blood pigment and leaf pigments have both arisen from the ferment – blood pigments by reduction, and leaf pigment by oxidation. For evidently, the ferment existed earlier than haemoglobin and chlorophyll.[e]

Selection 68

Around 1930 the significance of the experimental evidence concerning dehydrogenases, the oxygen-transferring ferment, and cytochrome was not in doubt. The discussion during the decade that followed centred rather on the interrelationship of these systems.

Considering Warburg's persistently negative view on the role of dehydrogenases it is of more than passing interest to note that it was the work performed in his laboratory at Berlin-Dahlem in the 1930s which profoundly contributed to a better understanding of their action. Warburg's antagonism had been based on the principle that the methylene blue technique widely employed in the study of dehydrogenases was unphysiological and, therefore, irrelevant for investigations on living tissues. Not until he had personally witnessed the apparent catalytic role of methylene blue in the respiratory process did he change his attitude. This happened in 1929 when he visited the laboratory of E.S.G. Barron (1898–1957) and G.A. Harrop, jun. (1890–1945) at the Johns Hopkins Medical School at Baltimore and watched experiments that these workers had described a year earlier.[a] If methylene blue and glucose were added to red blood corpuscles the rate of oxygen uptake increased and glucose was oxidized.

Admitting the import of these observations, Warburg used them as a point of departure, and he and his associates embarked on elucidating the nature of the oxidation system involved. A point to notice is that eventually the latter was studied under changed conditions. The whole red blood cells and glucose were replaced by cytolyzed red cells and Robison ester (glucose-6-phosphate) respectively. This procedure had the advantage that the biological fluid obtained could be subjected to ordinary chemical treatment. In addition, yeast extracts were used as experimental material.[b]

As the direct consequence of this endeavour particular elements of oxidation-reduction systems concerned in biological oxidations began to emerge. Interposed in the respiratory chain[c] between the activated hydrogen of the substrate and the cytochrome system these elements became associated with a group of complex substances known as 'pyridine nucleotides' and 'flavoproteins'. Their mode of action in respiring cells was dealt with by Warburg and his collaborators in a series of remarkable papers.[d]

The discovery of 'yellow enzyme' (*gelbes Ferment*), concerned with catalytic oxidation of glucose-6-phosphate in the presence of oxygen or methylene blue by red cells or yeast preparations, came first, chronologically (1932). Yellow enzyme was shown to be a conjugated protein of which the prosthetic group, a flavine phosphate or more specifically the isoalloxazine ring, turned out to be the hydrogen carrier.[e] The experimental evidence for this view, produced mainly by Theorell, is contained in the extract below.[f]

H. Theorell, *Biochem.Z.*, 275, 37. 1935

'Über die Wirkungsgruppe des gelben Ferments' (On the active group of the yellow enzyme)

In this journal[1] the author reported on the preparation of the pure yellow respiration enzyme and its reversible splitting into protein component and yellow active group.

If in the reversible experiment one replaces the yellow active group of the enzyme by lactoflavin – which I owe to the kindness of Herr P. Karrer – *no* yellow enzyme is formed. It follows from this that the active group of the enzyme is *not* the lactoflavin of Kuhn (although it is known that the active group and the lactoflavin have the same spectrum and both on exposure to light in alkaline solution give the same pigment, $C_{13}H_{12}N_4O_2$).[a]

If one mixes an aqueous solution of the pure yellow enzyme with three times the volume of methanol the yellow colouring matter goes into solution, whilst the protein component is for the most part precipitated (denatured). At the instigation of Herr Warburg I have investigated whether the yellow pigment so obtained contains phosphorus. When the amount of pigment was estimated photoelectrically and the phosphorus according to Briggs in the ashed solution, one atom of phosphorus to one molecule of pigment was found, for example:

> Preparation I 196γ pigment = 17γ P (1.06 mol. P/mol. pigment.)
> Preparation II 120γ pigment = 11γ P (1.1 mol. P/mol. pigment.)

The protein component on the other hand is phosphorus free.

With cataphoresis experiments (pH 7.2) it was further shown that the true active group of the enzyme moved *strongly towards the anode*, and indeed at the rate to be expected for a monophosphoric ester of this molecular size (v = 16 × 10^{-5} cm.^{2}/sec. volt). The lactoflavin showed at this pH no detectable movement.

The active group of the yellow enzyme is thus probably a phosphoric acid ester, a 'nucleotide' in which the purine base is replaced by Karrer's[2] dimethylalloxazine.

Selection 69

As we mentioned in the introduction to the previous selection, Warburg's investigations of the oxidation system in red cells led to the association of pyridine nucleotides with the oxidative mechanism. From 1931 Warburg, in collaboration with Walter Christian (1907–55), came to recognize that erythrocytes contained a factor acting as a co-enzyme (coferment) in the aerobic oxidation of the Robison ester (glucose-6-phosphate) to phosphohexonic (6-phosphogluconic) acid.

By 1935 Warburg and his collaborators, Christian and Alfred Griese (1918–43),[a] had come to assume that the co-enzyme preparations obtained from red blood cells comprized 1 molecule of adenine, 1 molecule of the amide of nicotinic acid, 3 phosphoric acid molecules and 2 pentose molecules. The purest co-enzyme preparations yielded about 14 per cent of the nicotinamide and, indeed, it was acknowledged that it was this pyridine component of the co-enzyme that constituted its active group. The activity of the co-enzyme was related to the dehydrogenated (oxidized) and hydrogenated (reduced) state of the nicotinate.[b] Nicotinic acid amide, like other pyridine compounds, has an absorption spectrum in the ultra-violet part, which changes markedly on reduction. Close study of these spectra threw considerable light on the activity of the ferments.

The excerpt below concerns the 'hydrogen-transporting coferment.' It indicates that Warburg and his group visualized cellular oxidation as 'reversible hydrogenation', linked to the pyridine ring of the 'coferment' and to the alloxazine ring of the yellow enzyme acting catalytically as hydrogen carriers.[c]

The name 'triphosphopyridine nucleotide' (later abbreviated to TPN) for this compound was introduced by Warburg and Christian in 1936. They differentiated it from the 'diphosphopyridine nucleotide' (DPN), the name given to yeast cozymase, known to be involved in alcoholic fermentation.[d] Warburg and Christian[e] found it to contain 1 molecule of pentose, 1 molecule of hexose (or 2 pentose molecules), 1 adenine molecule, 1 nicotinamide molecule and 2 phosphoric acid molecules. Thus the two pyridine nucleotides differed in the phosphoric acid content, cozymase containing 2 molecules and the co-enzyme from red blood cells 3 molecules. This result was confirmed in the Stockholm laboratory of v. Euler, who had studied cozymase for many years and believed it to be concerned in oxidation-reduction processes rather than in phosphorylation.[f]

O. Warburg, W. Christian and A. Griese, *Biochem.Z.*, *282*, 157. 1935

'Wasserstoffübertragendes Co-Ferment, seine Zusammensetzung und Wirkungsweise' (The hydrogen-transporting coferment, its composition and manner of working)

p. 161

With the hydrogenation of the coferment by platinum and hydrogen the pyridine ring of the coferment is hydrogenated (and no other part of the coferment) to the piperidine ring. At the same time the coferment loses its catalytic activity.

If one reduces the coferment first reversibly with hexosemonophosphoric acid, whereby it takes up one molecule of hydrogen, and then with platinum and hydrogen, now are no longer 3 but only 2 molecules of hydrogen taken up.

From this it follows that it is the pyridine ring of the coferment which with the reversible hydrogenation – that is to say with the activity – takes up the hydrogen. The reversible hydrogenation of the coferment is a partial hydrogenation of its pyridine ring.

If the pyridine ring is the hydrogen-transporting part of the coferment, we can state more exactly than formerly what happens if with our test reaction hexosemonophosphoric acid is oxidized to phosphohexonic acid: two nitrogen-containing rings inserted behind one another transport hydrogen from the substrate to molecular oxygen. Hydrogen, which the pyridine ring of the coferment removes from the substrate is further given to the alloxazine ring of the yellow ferment and from this to molecular oxygen.

Selection 70

By the mid 1930s the idea that catalytic hydrogen transport was of central importance to biological oxidation had become well established. From the previous examples we have seen something of the evidence for this view offered by the researches of Warburg and his group.

Further light was thrown on the subject by a significant body of experimental and theoretical work carried out by Albert (von) Szent-Györgyi and his associates at the University of Szeged in Hungary and published between 1934 and 1937.[a] They worked with suspensions of pigeon breast muscle in a saline medium that were obtained by mincing in a way that left the cellular mass appreciably intact. While respiring actively (high rate of oxygen consumption) it produced almost no lactic acid.

As noted before, the early history of the dehydrogenases was closely linked to the inquiries into the oxidation of succinic acid and fumaric acid by tissues. In a sense this was the source of Szent-Györgyi's idea for discovering more about the role of C_4-dicarboxylic acids in biological oxidation, a role which he pictured as a long-chain piecemeal combustion of hydrogen atoms.

The findings at Szeged showed that the falling rate of respiration of the muscle preparations could be restored firstly by catalytic amounts of succinate and fumarate and then also by malate and oxaloacetate. This pointed to their possible catalytic function in cellular respiration. Accordingly Szent-Györgyi proposed a scheme in which they were involved in transmitting hydrogen catalytically from the donator (for example carbohydrate) to what he called the Warburg-Keilin (WK) system composed of the *Atmungsferment* and cytochrome.

The excerpt below contains Szent-Györgyi's perhaps clearest attempt to formulate the arguments in support of his theory that C_4- dicarboxylic acids acted as catalysts rather than as metabolites in cell respiration.

A. von Szent-Györgyi in *Perspectives in Biochemistry*, eds. J. Needham and D.E. Green (Cambridge, 1937)

'Oxidation and fermentation'

pp. 169–70

From the very beginning of my biochemical studies my mind was bothered by the special position of the four-carbon atom dicarboxylic acids. I was taught that succinic acid was oxidized by most animal tissues at a very rapid rate to fumaric acid. Later I convinced myself that there is in fact no other substance oxidized by tissues as fast as succinate. Ogston & Green showed that the only substance cytochrome could act on was succinate.[a] It was also known that all tissues contained a very powerful enzyme, 'fumarase', which converts fumaric acid to malic, till the relative concentration of both is 1:3. In the same way it converts malic into fumaric. Later on I found that this enzyme is in fact one of the most powerful enzymes known. But what is its function? Nature is not extravagant, and yet neither succinic nor fumaric acid were regarded as among the most important metabolites. Also Thunberg showed that the isomer of fumaric acid, e.g. maleic acid, was a strong and specific poison of respiration.[b] I began to suspect that something must be wrong about the WK-Wieland theory. It might be true, but must be incomplete, and the C_4 dicarboxylic acid must play some very important catalytic role in respiration. So I investigated two things: (1) what happens to the respiration if we cut out the oxidation of succinic acid, and (2) what happens if we increase the minute quantity of fumarate normally present in the tissue. The possibility of inhibiting succinic oxidation in a specific way was opened up by J.H. Quastel, who showed that the oxidation of succinate can be poisoned fairly specifically by the C_3 dicarboxylic acid, malonic acid.[c]

The results were striking. Minute quantities of malonate poisoned respiration almost like cyanide. Fumaric acid strongly increased it. The rapidly declining respiration of tissues *in vitro* could be maintained constant for long periods by fumaric acid. As Baumann & Stare have shown in Keilin's laboratory, even a few γ of fumarate (γ = one millionth part of a gram) were active.[d]

It took several years of hard work to fit the contradictory observations into one theory. The theory is this: the C_4 dicarboxylic acids are a link in the respiratory chain between foodstuff and the WK system. Their function is to transfer the hydrogen of the foodstuff to cytochrome and to reduce by this hydrogen its trivalent iron again to the divalent form. Speaking more precisely, the cytochrome oxidizes off two hydrogen atoms from the succinic acid molecule. By the loss of two hydrogen atoms, the succinic acid is converted to fumaric acid. These two lost H atoms are replaced again by H coming from the foodstuff. The foodstuff, however, does not give its H immediately to fumaric acid. It gives its two H atoms to oxaloacetic acid, which is also a C_4 dicarboxylic acid. By taking up 2H oxaloacetic turns unto malic acid. Malic acid then gives its 2H to fumaric, and thus fumaric is converted to succinic acid. This can again be oxidized by cytochrome, while malic acid, after giving off its 2H becomes oxaloacetic, which can take up H from the foodstuff again, and so the play goes on, H being transmitted all the time from the foodstuff *via* oxaloacetic-malic-fumaric-succinic to the WK system.

The summary of the story is as follows:

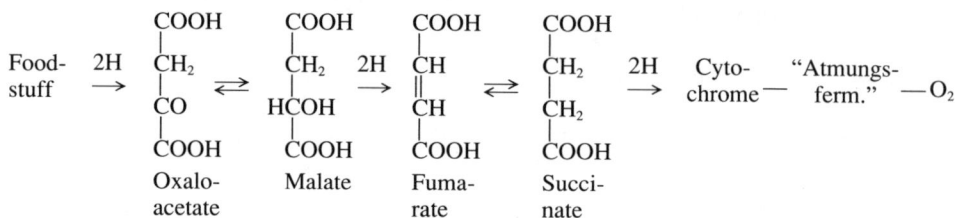

$$\text{Food-stuff} \xrightarrow{2H} \begin{array}{c} COOH \\ | \\ CH_2 \\ | \\ CO \\ | \\ COOH \\ \text{Oxalo-acetate} \end{array} \rightleftarrows \begin{array}{c} COOH \\ | \\ CH_2 \\ | \\ HCOH \\ | \\ COOH \\ \text{Malate} \end{array} \xrightarrow{2H} \begin{array}{c} COOH \\ | \\ CH \\ || \\ CH \\ | \\ COOH \\ \text{Fumarate} \end{array} \rightleftarrows \begin{array}{c} COOH \\ | \\ CH_2 \\ | \\ CH_2 \\ | \\ COOH \\ \text{Succinate} \end{array} \xrightarrow{2H} \text{Cyto-chrome} - \text{"Atmungsferm."} - O_2$$

Selection 71

If Szent-Györgyi had called attention to dicarboxylic acids, then in 1937 Sir Hans Adolf Krebs (1900–81), working in the Department of Pharmacology at Sheffield University, brought out the special place of a tricarboxylic acid, namely citric acid, in the aerobic oxidation of carbohydrates. Since the early 1930s, whilst still working in Germany, Krebs was struck by the absence of an adequate picture of the intermediate stages of oxidative breakdown of foodstuffs, especially of carbohydrate and fat. The weakness in this area became obvious when compared with the growth of knowledge resulting from studies of the anaerobic breakdown of sugar going on at the same time.[a] It was this that made Krebs continue to search into the pathway of sugar oxidation under aerobic conditions, after emigrating from Germany and settling in Britain in 1933.[b]

As in the case of other organic acids, the study of the oxidation of citric acid in living tissues goes back to the work of Torsten Thunberg.[c] It is of historical interest to note that in 1911 Battelli and Stern, in conjunction with their view about the non-enzymatic nature of main respiration, also found no ground for ascribing the combustion of citric acid to enzymatic activity.[d] Following his introduction of the vacuum tube technique, Thunberg regarded citric acid (in addition to succinic acid, fumaric acid and malic acid) as an intermediate component of metabolism, at least from 1920 onwards.[e]

As we have seen, the line taken by Szent-Györgyi was a different one. Instead of thinking in metabolic terms, he regarded the C_4- dicarboxylic acids (succinic, fumaric, malic and oxalacetic) as catalysts concerned in the aerobic oxidation of carbohydrates. Szent-Györgyi's work was seminal and Krebs became curious about whether the conclusions drawn from it would also apply to citric acid. He confirmed the catalytic effect of citric acid on the respiration of minced pigeon breast muscle early in 1937.

Then in February and March of the same year two publications appeared from the Institute of Physiological Chemistry at Tübingen which dealt with the breakdown of citric acid in living tissues. First, in a brief preliminary communication Carl Martius (b.1906) and Franz Knoop (1875–1946) suggested a metabolic pathway from citric acid to α-ketoglutaric acid and also pointed to the possibility of the reversible formation of citric acid.[f] In the second paper Martius offered experimental evidence favouring a postulated mechanism involving cis-aconitic acid and iso-citric acid in the conversion of citric acid to α-ketoglutaric acid. Following Knoop and Embden, the author also underlined the fact that α-ketoglutaric acid could be considered as a link in the formation of amino acids. Thus, according to Martius, citric acid – hitherto biochemists had hardly known what to make of it – was brought into the arena of protein synthesis as an important intermediate.[g]

At once recognizing the major importance of this work, Krebs began to understand more fully the significance of his own observations resulting from new experiments he proceeded to design. They included evidence concerning the rapid oxidation of citric acid, α-ketoglutaric acid and succinic acid and the formation of citric acid from oxaloacetic acid under anaerobic conditions. Moreover they showed that succinic acid could arise from oxaloacetic acid under aerobic conditions.[h]

As shown below, this led Krebs in collaboration with his assistant William Arthur Johnson (b.1913) to a novel theoretical treatment of the oxidation of carbohydrates based on a metabolic 'citric acid cycle'. By pointing to the oxidative nature of the steps from one metabolite to another operating the cycle, Krebs made it possible to disentangle the knot of the intertwined research threads originating in Szeged, Tübingen and Sheffield.

H.A. Krebs and W.A. Johnson, *Enzymologia*, *4*, 148. 1937

'The role of citric acid in intermediate metabolism in animal tissues'

pp. 149–50

II. Catalytic effect of citrate on respiration

If muscle tissue is minced and suspended in 6 volumes of phosphate buffer a high rate of respiration is observed initially, but after 20–40 minutes the rate begins to fall off. If citrate is added the rate of respiration is often increased and the falling off of respiration is always much retarded. This effect is brought about by small quantities of citrate, and comparing the extra respiration with the citrate added we find that the extra oxygen uptake is by far greater than can be accounted for by the complete oxidation of citrate. An example is the following experiment:

Table 2 Effect of citrate on respiration of minced pigeon breast muscle (Manometric experiment)

| | μl O_2 absorbed by 460 mg muscle (wet weight) suspended in 3 ccm phosphate saline | |
Time (min.)	No substrate added	0.15 ccm 0.02 M sodium citrate added
30	645	682
60	1055	1520
90	1132	1938
150	1187	2080

In this experiment the citrate caused an increased respiration of 893 µl, whilst 302 µl O_2 are calculated for the complete oxidation of the citrate added.

The magnitude of the effect of citrate shows considerable variations from experiment to experiment; the effect appears to be dependent on the amounts of citrate and other substrates preformed in the tissue. The effect is more pronounced if glycogen, or hexosediphosphate, or α-glycerophosphate are added to the muscle, and we presume therefore that the substrate the oxidation of which is catalysed by citrate, is a carbohydrate or a related substance . . .

The problem of the mechanism of this citrate catalysis can be approached in various ways . . .

We have chosen the investigation of the intermediate stages of the breakdown of citrate in the tissues. If all the stages are known the mechanism of the catalytic effect will be clear.

pp. 150–1

III. Rate of disappearance of citric acid in muscle

Since citric acid reacts catalytically in the tissue it is probable that it is removed by a primary reaction but regenerated by a subsequent reaction. In the balance sheet no citrate disappears and no intermediate products accumulate. The first object of the study of intermediates is therefore to find conditions under which citrate disappears in the balance sheet. We find that some poisons bring about this effect, for instance arsenite (Table 3) or malonate. If one of these two substances is present, very large amounts of citric acid disappear provided that oxygen is available. Obviously the poisons leave the breakdown of citric acid unaffected whilst they check the synthesis of citric acid.

IV. Conversion of citric acid into α-ketoglutaric acid

The oxidation of citric acid in the presence of arsenite or malonate is not complete. Only one or two molecules of oxygen are absorbed for each molecule of citric acid removed and the solution must therefore contain intermediate products of oxidation of citric acid.

Although it has long been known that citric acid is readily metabolised (see Östberg,[1] Sherman c.s.[2]) the pathway of the breakdown remained obscure until early 1937, when Martius and Knoop[3,4] working with citrico dehydrogenase from liver discovered that the oxidation of citric acid by methylene blue yields α-ketoglutaric acid. We are able to confirm Martius and

Table 3 Disappearance of citric acid in pigeon breast muscle in the presence of arsenite (3.10^{-3} mol.) (3 ccm muscle suspension containing 750 mg wet muscle were shaken for 40 min. at 40°)

µl citrate added	µl citrate found after 40 min.	µl citrate used	$Q_{citrate}$
1120	30	1090	−10,9
2210	972	1268	−12,7
4480	2790	1690	−16,9

Knoop's results with other tissues and with molecular oxygen as the oxidising agent.

pp. 151–5

V. Conversion of citric acid into succinic acid

In the presence of malonate the oxidation of citrate is checked at the stage of succinic acid as shown by the following experiment: 7.5 grammes (wet weight) minced pigeon muscle were suspended in 22.5 ccm phosphate buffer (0.1 M; ph = 7.4) and 3 ccm 0.2 M sodium citrate and 1 ccm 1 M malonate were added. The suspension was shaken for 40 min. in an atmosphere of oxygen, and then deproteinised by adding 34 ccm water, 2 ccm 50 per cent sulphuric acid and 2 ccm 15 per cent sodium tungstate. In the filtrate succinic and α-ketoglutaric acids were determined manometrically. 3 ccm contained 472 μl succinic acid and 80 μl α-ketoglutaric acid; α-ketoglutaric acid was also identified by the isolation of the 2,4-dinitrophenylhydrazone.

VI. Synthesis of citric acid in the presence of oxaloacetic acid

The new results of the citric acid breakdown, in conjunction with previous work on the oxidation of succinic acid in tissues may be summarised by the following series:

citric acid $\longrightarrow$ α-ketoglutaric acid $\longrightarrow$ succinic acid $\longrightarrow$ fumaric acid $\longrightarrow$ *l*-malic acid $\longrightarrow$ oxaloacetic acid $\longrightarrow$ pyruvic acid.

If it is true that the oxidation of citric acid is a stage in the catalytic action of citric acid then it follows that citric acid must be regenerated eventually from one of the products of oxidation. We are thus led to examine whether citric acid can be resynthesised from any of the intermediates of the citric acid breakdown.

Systematic experiments show that indeed large quantities of citric acid are formed if oxaloacetic acid is added to muscle anaerobically, whilst all the other intermediates, including pyruvic acid yield no citric acid under the same conditions. It is because the synthesis of citric acid from oxaloacetic acid does not require molecular oxygen and because citric acid is stable in the tissue anaerobically that it is possible to demonstrate the synthesis of citric acid in a simple experiment.

Minced pigeon breast muscle was suspended as usual in 3 volumes phosphate buffer and 3 ccm suspension were measured into a conical manometric flask the sidearm of which contained 0.3 ccm 1M oxaloacetate. In the centre chamber a stick of yellow phosphorus was placed and the gas space was filled with nitrogen. After the removal of oxygen the oxaloacetate was added to the tissue and the flask was shaken in the water bath for 20 mins. During this period about 1000 μl CO_2 were evolved. After the incubation, the suspension was quantitatively transferred into 25 ccm 6 per cent trichloracetic acid and the volume was made up to 50 ccm. Citric acid was determined in the filtrate and 0,0131 millimol (293 μl) citric acid were found. $Q_{citrate}$ is thus $\frac{293 \times 3}{150} = 5.86$. No citrate was present in the controls.

This experiment shows that muscle is capable of forming large quantities of citric acid if oxaloacetic acid is present and the question arises from which substance the two additional carbon atoms of the citric acid molecule are derived. Addition of various possible precursors such as acetate, or pyruvate, or of α-glycerophosphate had no effect on the rate of citric acid synthesis, but this negative result is no proof against the participation in the synthesis of one of these substances. Pyruvic acid and acetic acid arise rapidly from oxaloacetic acid and it may be that the tissue is already saturated with these substances if oxaloacetic acid alone has been added.

The fact that the catalytic effect of citrate is more pronounced if glycogen, or hexosemonophosphate, or α-glycerophosphate are present suggests that the substance condensing with oxaloacetate is derived from carbohydrate. We may term it provisionally as 'triose', leaving it open whether triose reacts as such or as a derivative for example as a phosphate ester, or pyruvic acid or acetic acid.

A synthesis of citric acid from a C_4-dicarboxylic acid and a second substance has often been discussed, especially with reference to the citric acid fermentation of moulds (see [5,6]), though it has not been shown before to occur in animal tissues.

Martius and Knoop[7] showed recently that citric acid is formed in vitro if oxaloacetate and pyruvate are treated with hydrogen peroxide in alkaline medium. This model reaction is an interesting analogy and it suggests that the synthesis of citric acid may be a comparatively simple reaction.[8]

VII. Role of citric acid in the intermediate metabolism

1. Citric acid cycle. The relevent facts concerning the intermediate metabolism of citric acid may now be summarised as follows:

   1. Citrate promotes catalytically the oxidations in muscle tissue, especially if carbohydrates have been added to the tissue.

   2. Similar catalytic effects are shown by succinate, fumarate, malate, oxaloacetate (Szent-Györgyi[9], Stare and Baumann[10])

   3. The oxidation of citrate in muscle passes through the following stages: citric acid → α-ketoglutaric acid → succinic acid → fumaric acid → l-malic acid → oxaloacetic acid.

   4. Oxaloacetic acid reacts with an unknown substance to form citric acid.

These facts suggest that citric acid acts as a catalyst in the oxidation of carbohydrate in the following manner:

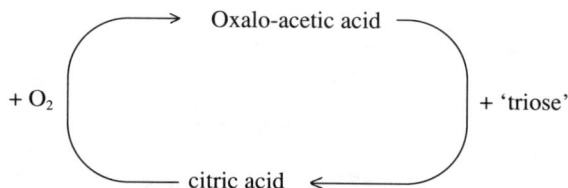

According to this scheme oxaloacetic acid condenses with 'triose' to form citric acid, and by oxidation of citric acid oxaloacetic acid is regenerated. The net effect of the 'citric acid cycle' is the complete oxidation of 'triose'.

The synthesis of citric acid from oxaloacetic acid as well as the oxidation of citric acid to oxaloacetic has been experimentally verified. The only hypothetical point in the scheme is the term 'triose', though we may consider it as certain that the substance condensing with oxaloacetic acid is related to carbohydrate.

The proposed scheme outlines a pathway for the oxidation of carbohydrate.

Many details must necessarily be left open at the present time, but a few points will be discussed in the following sections.

2. Origin of the C_4-dicarboxylic acid. According to the scheme succinic acid or a related compound is necessary as 'carrier' for the oxidation of carbohydrate and the question of the origin of succinic acid arises. We have shown previously[11] that succinic acid can be synthesised by animal tissues in small amounts if pyruvic acid is available. The physiological significance of the synthesis is now clear: it provides the carrier required for the oxidation of carbohydrate.

3. Further intermediate stages. (a) iso-Citric acid. Wagner-Jauregg and Rauen[12] and Martius and Knoop[3,4] have suggested that iso-citric acid is an intermediate in the oxidation of citric acid. We find that iso-citric acid is indeed readily oxidised in muscle, the rates of oxidation of citric acid and iso-citric acids being about the same.

(b) cis-Aconitic acid. cis-aconitic acid, discovered by Malachowski and Maslowski[13], was first discussed as an intermediate by Martius and Knoop[3] and Martius[4] showed that it yields readily citric acid with liver. We have examined the behaviour of cis-aconitic acid in muscle and other tissues and find that it is oxidised as readily as citric acid. The conversion of cis-aconitic acid into citric acid is also brought about by tissue extracts. One milligramme muscle tissue (dry weight) converts up to 0.1 mg cis-aconitic acid into citric acid per hour (40°; ph = 7.4).

Martius and Knoop[3,4] assume that the reaction cis-aconitic ⇌ citric acid is reversible and believe that it plays a role in the breakdown of citric acid. It cannot yet be said, however, whether the reaction is an intermediate step in the breakdown or in the synthesis of citric acid.

(c) Oxalo-succinic acid. The oxidation of iso-citric acid would be expected to yield in the first stage oxalo-succinic acid (Martius and Knoop). This β-ketonic acid is only known in the form of its esters, since the free acid is unstable in a pure state. In acid solution it is readily decarboxylated and yields α-ketoglutaric acid (Blaise and Gault[14]).

(d) Detailed citric acid cycle. The information available at present about the intermediate steps of the cycle may be summarised thus:

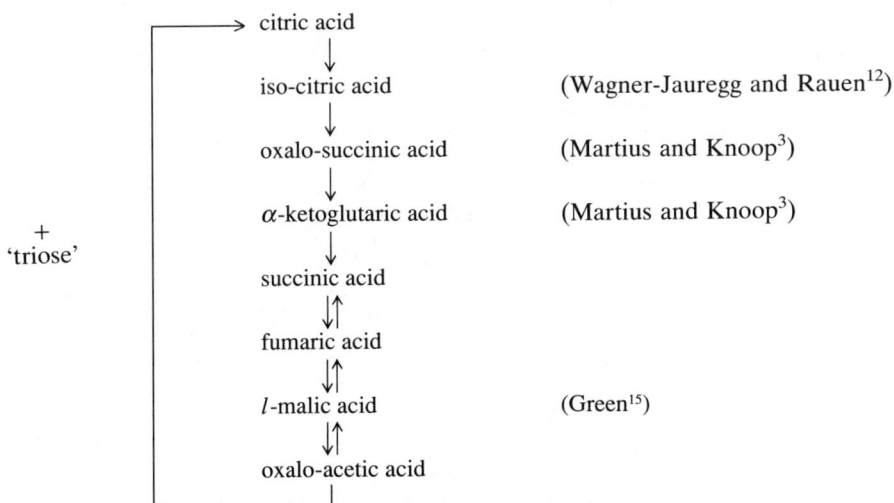

4. Reversible steps. Succinic acid arises according to our scheme by oxidative reactions from oxaloacetic acid, via citric and α-ketoglutaric acids. Anaerobic experiments, however, show succinic acid can also be formed by reduction from oxaloacetic acid (see also Szent-Györgyi). The reactions succinic acid → fumaric acid → *l*-malic acid → oxaloacetic acid are thus reversible under suitable conditions.

The outstanding problem in this connection is the question of the oxidative equivalent of the reduction. At least a partial answer may be given. The synthesis of citric acid as shown in section VI takes place anaerobically, although it is an oxidative process. A reductive process equivalent to the oxidation must therefore occur at the same time. The reduction of oxaloacetic acid to succinic acid is the only reduction of sufficient magnitude (see the next section) known so far to occur simultaneously with the citric acid synthesis and we assume therefore it is the equivalent for the synthesis of citric acid.

5. Effect of malonate. It follows from the preceding paragraph that succinic acid can arise from oxaloacetic acid in two different ways (a) oxidatively via citric acid and α-ketoglutaric acids (b) reductively via *l*-malic and fumaric acids. That two different ways and therefore two different enzymic systems bring about the conversion of oxaloacetic into succinic acid can be demonstrated with the aid of malonate. Malonate inhibits specifically the reaction succinic acid ⇌ fumaric acid. Aerobically it will therefore increase the yield of succinic acid from oxaloacetic acid since it prevents its secondary breakdown. Anaerobically, on the other hand, it will inhibit the formation of succinic acid, since in this case the succinic dehydrogenase is concerned with the formation of the succinic acid. The following experiment shows that the results are as expected. [see Table 4]

6. Citric acid cycle in other tissues. We have tested the principal points of the citric acid cycle in various other animal tissues and find that brain, testis, liver and kidney of the rat are capable of oxidising citric acid as well as synthesising it from oxaloacetic acid. Of these four tissues testis shows the highest rate of synthesis and this is of interest in view of the work of Thunberg's school on the occurrence of citric acid in spermatic fluid. 1 mg (dry weight) rat testis forms anaerobically up to 0.02 mg citric acid per hour if oxaloacetic acid is present.

Whilst the citric acid cycle thus seems to occur generally in animal tissues, it does not exist in yeast or in *B. coli*, for yeast and *B. coli* do not oxidise citric acid at an appreciable rate.

7. Quantitative significance of the citric acid cycle. Though the citric acid cycle may not be the only pathway through which carbohydrate is oxidised in

Table 4   Effect of malonate on the aerobic and anaerobic
conversion of oxaloacetic into
succinic acid (0.75 grammes wet muscle in 3 ccm phosphate buffer;
40° C; ph = 7,4)

Experimental conditions (final concentration of the substrates)	µl succinic acid formed in 40 min.
1. O_2; 0.1 M oxaloacetate;	1086
2. O_2; 0.1 M oxaloacetate; 0.06 M malonate	1410
3. N_2; 0.1 M oxaloacetate;	1270
4. N_2; 0.1 M oxaloacetate; 0.06 M malonate	834

animal tissues the quantitative data of the oxidation and resynthesis of citric acid indicate that it is the preferential pathway. The quantitative significance of the cycle depends on the rate of the slowest partial step, that is for our experimental conditions the synthesis of citric acid from oxaloacetic acid. According to the scheme one molecule of citric acid is synthesised in the course of the oxidation of one molecule of 'triose', and since the oxidation of triose requires 3 molecules O_2, the rate of citric acid synthesis should be one third of the rate of O_2 consumption if carbohydrate is oxidised through the citric acid cycle. We find for our conditions:

Rate of respiration $(Q_{O_2}) = -20$

Rate of citric acid synthesis $(Q_{citrate}) = +5, 8$

The observed rate of the citric acid synthesis as thus a little under the expected figure $(-6, 6)$, but it is very probable that the conditions suitable for the demonstration of the synthesis (absence of oxygen) are not the optimal conditions for the intermediate formation of citric acid, and that the rate of citric acid synthesis is higher under more physiological conditions. This is suggested by the experiments on the aerobic formation of succinic acid from oxaloacetic acid . . . $Q_{succinate}$, in the presence of malonate and oxaloacetate is $+14.1$, and if citrate is an intermediate stage the rate of citrate formation must be at least the same. But even the observed minimum figures of the rate of the synthesis justify the assumption that the citric acid cycle is the chief pathway of the oxidation of carbohydrate in pigeon muscle.

8. The work of Szent-Györgyi. Szent-Györgyi[9] who first pointed out the importance of the C_4-dicarboxylic acids in cellular respiration, came to the conclusion that respiration, in muscle, is oxidation of triose by oxaloacetic acid. In the light of our new experiments it becomes clear that Szent-Györgyi's view contained a correct conception, though the manner in which oxaloacetic acid reacts is somewhat different from what Szent-Györgyi visualised. The experimental results of Szent-Györgyi can be well explained by the citric acid cycle; we do not intend, however, to discuss this in full in this paper.

Selection 72

The elaboration and general acceptance of the citric acid cycle, including its wider significance as a link between the metabolism of primary foodstuffs (carbohydrates, fats and proteins), came after 1940. At the time of its publication the Krebs scheme was considered by most investigators (and to a certain extent by the author himself) to be an extension or variation of Szent-Györgyi's theory of the C_4-dicarboxylic acids accounting for their catalytic effect more fully. As pointed out previously, in the Szent-Györgyi scheme the C_4-dicarboxylic acids acted at catalytic carriers of hydrogen from the activated substrate molecules to the cytochrome system. In order to complete the historical picture during the decade 1930–40 let us turn our attention to this element of the oxidation pathway, by considering Keilin's investigations into the carrier action of cytochrome which followed his first paper introduced earlier (1925).

Experimenting with yeast cells and cell-free heart preparations during the late 1920s and the early 1930s, Keilin succeeded in isolating cytochrome c (cytochromes a and b attached to the insoluble material of the cells were found not to be extractable).[a] He also demonstrated that reduced cytochrome was re-oxidized by molecular oxygen in the presence of an enzyme he identified as indophenol oxidase.

Because of a resemblance in certain properties (e.g. inhibition by carbon monoxide in the dark, sensitivity to cyanide), Keilin suggested that indophenol oxidase and the *Atmungsferment* had to be treated as one. As for the reversible reduced and oxidized state of cytochrome, Keilin explained it by pointing out that cytochrome accepted hydrogen from the dehydrogenase-substrate component on the one hand, and donated hydrogen to the oxidase-oxygen component of the respiratory system of the cells, on the other.

Clearly, this interpretation of the respiratory mechanism of the cells acknowledged the participation of both activated hydrogen and activated oxygen in biological oxidation and, not surprisingly, was not looked upon with favour by Warburg. However, this is not the place to deal in more detail with the controversial issues raised by the differences in opinion between Keilin and Warburg about the mode of action and the function of cytochrome in cellular respiration up to 1934.[b]

Rather, the account given above is meant to provide a background to Keilin's return to the study of a cytochrome and the oxidase system by way of three papers and two notes written jointly with Edward Francis Hartree (b.1910) which appeared in the *Proceedings of the Royal Society* (B) and *Nature* between 1938 and 1940.[c] As stated in the introduction of the first paper, the literature on indophenol oxidase 'still contains a great bulk of inaccurate and controversial statements and faulty interpretations of observed facts. This was naturally responsible for the introduction of several unsound theories of the mechanism of intracellular respiration.'

Having ascertained that indophenol oxidase was solely concerned with catalytic oxidation of reduced cytochrome the authors changed the name of the enzyme to 'cytochrome oxidase', already proposed by M. Dixon (1899–1985) in 1929.[d] Intending to elucidate the nature of cytochrome oxidase they spectroscopically examined effects of respiratory inhibitors like CO, KCN, NaN_3, H_2S, NH_4OH, NaF and peroxides on cytochrome in heart muscle preparation reduced largely with sodium succinate. Painstaking analyses of spectroscopic data brought to light a new pigment a_3, closely connected with cytochrome a (see Figure 9), but differing from it in that it was autoxidizable and also capable of forming a CO-compound. Indirect evidence also suggested that it might contain copper as well as iron.

The considerations that led Keilin and Hartree to think of cytochrome a_3 as cytochrome oxidase and as a compound identical with Warburg's 'oxygen transferring respiratory ferment' are set out in the first excerpt below.

The second excerpt is from the last of the three papers by Keilin and Hartree referred to above. In the article the properties of the intracellular system catalyzing the aerobic oxidation of succinic acid were examined. It reflected the knowledge emerging during the 1930s that the required co-operation of the two portions of the respiratory mechanism, succinic dehydrogenase and the complete cytochrome system, depended on the mediation of another component. As indicated in the excerpt, they en-

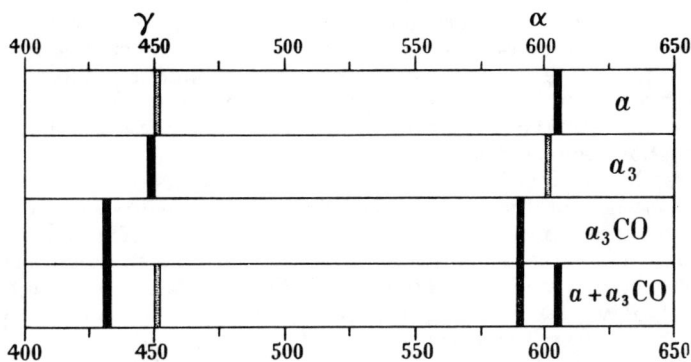

Figure 9 Diagram showing relationship between bands of cytochrome components a and a_3 and effect of carbon monoxide on a_3

visaged the possibility of a flavoprotein fulfilling this role by taking part in the hydrogen transfer (by virtue of undergoing hydrogenation and dehydrogenation).

D. Keilin and E.F. Hartree, *Proc.Roy.Soc.* (B), *127*, 167. 1939

'Cytochrome and cytochrome oxidase'

pp. 187–8

The study of heart-muscle preparations reveals the existence of a new haematin compound, a_3,[1] in addition to the components a, b and c of cytochrome. The existence of this compound was not previously recognized because in the oxidized state its absorption bands are invisible while in the reduced state they coincide with the corresponding bands of the component a. The evidence for its existence is obtained mainly from the study of the effects of certain respiratory inhibitors on the absorption spectrum of cytochrome. As these substances produce definite modification in the bands of component a it appears at first sight that they react directly with a. Careful analysis of these reactions reveals, however, that the absorption bands $a\alpha$ and $a\gamma$ do not belong to the component a only but to a mixture of a and a_3. The effects of the respiratory inhibitors on the appearance of the bands of cytochrome a is not due to their reaction with a but to the compounds they form with a_3.

One of the main properties of a_3 is its marked autoxidizability, and in this respect it differs from other components of cytochrome which are considered as non-autoxidizable haematin compounds. The term 'non-autoxidizable' should not, however, be taken in too strict a sense because cytochrome b, as we have seen, is to a great extent autoxidizable, although the rate of its oxidation by molecular oxygen is not as rapid as that of a_3 or of an ordinary haem or haemochromogen. The components a and c can also undergo a slow autoxidation even in presence of cyanide. Furthermore, a solution of pure cytochrome c cannot be preserved in the reduced state unless it is protected from oxygen. The autoxidation of a and c is, however, too slow to have any biological significance, while the autoxidation of b, although slower than that of a_3, may play a certain role in biological oxidation reactions.

It may be mentioned here that the property of 'non-autoxidizability' is very rare among haematin compounds.

The mere observation that an intracellular haematin compound is autooxidizable is, therefore, not sufficient to identify it with cytochrome oxidase. It must at the same time react with all the specific inhibitors and be influenced by all the factors which affect the oxidase reaction of the cell. It must also react with at least one of the non-autoxidizable components of cytochrome, of which c has already been found essential for the catalytic activity of the oxidase (Keilin and Hartree 1938a).[a]

So far, the component a_3 seems to be the only intracellular substance which answers most, although apparently not all, of these requirements.

In fact, the component a_3 is thermolabile and is affected by all treatment such as drying, freezing, acetone, alcohol, acids, alkali, etc., in the same way and to the same degree as is the oxidase activity of the preparation. It is autoxidizable and can be seen to undergo oxidations and reduction during the catalytic oxidation of metabolities. It forms two compounds with KCN: a divalent compound which is easily autoxidizable, and a trivalent compound which does not easily undergo reduction. In the trivalent state it combines with H_2S, NaN_3 and NH_2OH which, like KCN, stabilize it and prevent its reduction. In the divalent state it combines with CO forming a compound with bands occupying the same positions (590 and 432 mμ) as the corresponding bands in the photochemical absorption spectrum obtained by Warburg and his co-workers[b]. The component a_3 is therefore the only intracellular haematin compound which may be responsible for this photochemical absorption spectrum.

All this strongly supports the view of the identity of the component a_3 with cytochrome oxidase. This conclusion is, moreover, in agreement with the main results obtained by Warburg who has demonstrated that a haematin compound which with CO gives an absorption spectrum showing bands at about 590 and 432 mμ plays an essential role in cellular respiration. Component a_3 can therefore be identified with Warburg's respiratory or oxygen transporting enzyme.

D. Keilin and E.F. Hartree, *Proc.Roy.Soc. (B)*, *129*, 277. 1940

'Succinic dehydrogenase-cytochrome system of cells. Intracellular respiratory system catalysing aerobic oxidation of succinic acid'

pp. 293–4

The fact that preparations can be obtained in which the succinic system does not reduce c although it reduces methylene blue indicates that either the reduction of c takes place through an intermediary link destroyed in these preparations, or that the succinic system is accessible to methylene blue but not to c.

If there is a substance acting as a link between the succinic system and c it must be more fragile than a and a_3 of cytochrome. Such a link may be cytochrome *b* itself which is very labile and is always affected in some way in preparations not reacting with *c*. This does not exclude, however, the possibility of the existence of yet another labile link which may be either a haematin or a flavin compound. In fact, the intense absorption band which appears between *b* and *c* on treating the preparation with pyridine and reducer belongs probably to a pyridine haemochromogen of *b* mixed with that of another haematin compound the absorption spectrum of which is invisible in an untreated preparation.

On the other hand the tissue preparations show also two absorption bands (495 and 455 mμ) of a flavoprotein compound which on addition of succinic acid fade simultaneously with the appearance of the bands of reduced cytochrome (Keilin and Hartree 1939)[a], the disappearance of these bands may be either an apparent one and due to changes in the background during the reduction of cytochrome or it may be real and due to the reduction of the flavoprotein itself. In the latter case the flavoprotein would act as one of the components of the system oxidizing succinic acid and belonging either to succinic dehydrogenase itself or to one of the links uniting it with the cytochrome system.

Notes to Section III

Selection 46

Introduction:

a. On this subject see the reference to Holmes's very important study in Introduction to Selection 1, note f. Henry Guerlac's article in the *Dictionary of Scientific Biography* (New York, 1973), VIII, pp. 66–91, remains a convenient introduction to Lavoisier. Before and since then further work pertinent to the subject discussed has been published, e.g. R.E. Kohler, Jr, 'The origin of Lavoisier's first experiments on combustion', *Isis*, *63*, 349–55. 1972; M. Crosland, 'Lavoisier's theory of acidity', *Isis*, *64*, 306–25. 1973; R.E. Kohler, 'Lavoisier's rediscovery of the air from mercury, a re-interpretation', *Ambix*, *22*, 51–7. 1975; H. Guerlac, 'The chemical revolution: a word from Monsieur Fourcroy', *Ambix*, *23*, 1–4. 1976. See also Siegfried, 'Lavoisier's table'; Duncan, 'Lavoisier's list', note d. While this book goes to press it is impossible to do more than to point to C.E. Perrin's last paper (submitted shortly before he died), 'Document, text and myth: Lavoisier's crucial year revisited', *Brit.J.Hist.Sci.*, *22*, 3–25. 1989.

b. i.e. *De la fermentation des vins, et de la meilleure manière de faire l'eau-de-vie* (Lyon, 1770). Rozier's *Observations sur la physique, sur l'histoire naturelle et sur les arts*, appearing from 1771, became an important scientific journal. Cf. also Kohler, 'Lavoisier's first experiments', note a.

c. Cf. D. Keilin, *The History of Cell Respiration and Cytochrome* (Cambridge, 1966), p. 29.

d. The word 'oxygen' originated with Lavoisier in a paper read to the Academy in 1778 and published in 1780. Believing that the 'dephlogisticated' or 'eminently respirable air' was the common 'acidifying principle' of acids he named it *'principe oxygine'*. Cf. 'Considérations générales sur les acides et sur les principes dont ils sont composés', *Mém.Acad.Sci.*, 535–47. 1778 (1781). Carbonic acid' (*acide charbonneux*) was introduced by Lavoisier in 1781 (published in 1784) when he proved that 'fixed air' or 'chalky acid' was really a compound of charcoal (carbon) and oxygen. See 'Mémoire sur la formation de l'acide, nomme air fixe ou acide crayeux, que je désignerai désormais sous le nom d'acide du charbon', ibid., 448–67. 1781 (1784). In 1787 Guyton de Morveau (1736–1816) spoke of 'matter of fire' or 'matter of heat' as *calorique* in order to differentiate between the material cause and the sensation of heat to which it gave rise. See, de Morveau, Lavoisier, Bertholet [*sic*], de Fourcroy, *Méthode de nomenclature chimique* (Paris, 1787), pp. 30–1. Essentially weightless this subtle material, 'principle of heat', in combination with any body ('base') was supposed to produce 'elastic fluids' or 'gases', the term originally coined by van Helmont and adopted by Lavoisier. Accordingly 'oxygen gas' resulted from the union of the 'base' oxygen (that is the respirable portion of air) with 'caloric'. Lavoisier classed oxygen among 'principles' or 'elements', conceiving these as ultimate simple substances obtained by chemical decomposition. Revolutionary as the contribution of Lavoisier was to the development of chemistry, the break with the past was by no means absolute. Thus Lavoisier's list of thirty-six apparently known and also as yet unknown simple substances or elements in the *Traité élémentaire de Chimie*, with their old and new names, demonstrates clearly the link with earlier chemical thinking. Thus the traditional elemental entities: 'earth', 'water', 'air' and 'fire', accounting for transformations in the material world are discarded but not quite. The concept of 'fundamental' elements persists in Lavoisier's table of simple substances in the first group of all-embracing elements, five in number, characterized as 'belonging to all the kingdoms of nature, which may be considered as the elements of bodies': light, caloric, oxygen, azote (nitrogen), hydrogen. For a different focus on the background to Lavoisier's list of elements, see R. Siegfried, 'Lavoisier's table of simple substances: its origin and interpretation', *Ambix*, *29*, 29–48. 1982; also A.M. Duncan, 'The functions of affinity tables and Lavoisier's list of elements', *Ambix*, *17*, 28–42. 1970.

e.   Lavoisier and de La Place, 'Mémoire sur la chaleur', *Mém.Acad.Sci.*, 355–408. 1780 (1784).

Text (i):

a.   A misunderstanding cropped up, apparently, in the important study of C.A. Culotta, 'Respiration and the Lavoisier tradition: theory and modification, 1777–1850', when he took *ce viscère* to be the '*cavity of the intestines*', cf. *Trans.Am.Phil.Soc.*, N.S., *62*, (pt 3), 4. 1972.

b.   'Mémoire sur les changements que le sang eprouve dans les poumons et sur le mechanisme de la respiration'. Cf. Guerlac, *Dictionary*, p. 83.

c.   Calcined mercury (mercuric oxide) was known as *mercurius calcinatus per se* and *mercurius praecipitatus per se*. The first, the 'red calx of mercury', was formed on heating mercury in air. The second, the 'red precipitate of mercury', resulted from heating mercuric nitrate.

Text (ii):

a.   One of the older terms for oxygen employed by Lavoisier despite his repeated professions on 'the impossibility of separating the nomenclature of a science from the science itself'.

Selection 47

Introduction:

a.   See Introduction to Selection 31, Section II, note e.

b.   Selection 33, Section II.

c.   L. Spallanzani, *Mémoires sur la respiration*, translated into French from the unedited manuscript by J. Senebier (Geneva, 1803); L. Spallanzani, *Rapports de l'air avec les êtres organisés etc.* Taken from notebooks of observations by J. Senebier (Geneva, 1807), I–III.

d.   This draws on Teich, 'Circulation', see Introduction to Selection 31, Section II, note 3. Cf. also F. Duchesneau, 'Spallanzani et la physiologie de la respiration: révision théorique' in W. Bernardi and A. La Vergata (eds), *Lazzaro Spallanzani e la biologia del settecento* (Florence, 1982), pp. 45–65.

Text:

a.   Johann Kunckel (1630[3]–1702[3]). German chemist. His claim to have discovered phosphorus independently is in dispute. See M.B. Hall, *Dictionary of Scientific Biography* (New York, 1973), VII, pp. 524–6.

b.   Reference to the varying amount of carbonic acid given off by worms in contact with oxygen.

c.   See Introduction to Selection 46, note d.

Selection 48

Introduction:

a. H. Wollaston, 'On certain chemical effects of light', *Nicholson's Journal*, *8*, 293–7. 1804; W. Brande, 'Chemical experiments on guaiacum', *Phil.Mag.*, *25*, 105–12. 1806; see also 'Proceedings of Learned Societies', ibid., *23*, 269–70. 1806.
b. L.-A. Planche, 'Note sur la sophistication de la résine de jalap et sur les moyens de la reconnaître, etc.', *Bull.Pharm.*, *2*, 578–80. 1810.

Text:

1. The tincture of guaiacum, after having boiled, becomes coloured blue as before when one mixes it with milk which has not been heated; thus the heat acts only on the principle causing colouration.
a. See Wollaston, 'Chemical effects', Introduction, note a.
b. What is meant is the principle which produces blue colour and not the poisonous cyanogen (C_2N_2).

Selection 49

Introduction:

a. A.v. Humboldt, *Versuche über die gereizte Muskel- und Nervenfaser nebst Vermuthungen über den chemischen Prozess des Lebens in der Thier- und Pflanzenwelt* (Posen [now Poznan in Poland] and Berlin, 1797), II, p. 282f.
b. Georg Liebig's work followed from earlier unpublished observations (1843) by Emil Du Bois-Reymond (1818–76) – Johannes Müller's pupil and successor as Professor of Physiology in Berlin in 1858 – on production of carbonic acid in excised muscles. As with other themes in this volume, limited space imposes restraints on the tracing of the actual story of animal respiration from Lavoisier to G. Liebig. It was during this period that the interest in the physical aspects of respiration emerged (absorption and exchange of blood gases) in response to difficulties encountered in substantiating the Lavoisierian notion of respiration as slow combustion taking place in the lungs. The result was that the emphasis shifted from the lungs to the blood as the location of respiration. For more details consult the valuable and previously mentioned publication by C.A. Culotta, 'Respiration and the Lavoisier tradition: theory and modification, 1775–1850', *Trans.Am.Phil.Soc.N.S. (pt 3)*, *62*, 1–41. 1972.

Selection 50

Introduction:

a. Although there is a good deal of material on Schönbein, those aspects of his scientific practice and thought which continued to be influenced by *Naturphilosophie* have not received the attention they deserve. Cf. G.W.A. Kahlbaum and E. Schaer, *Christian Friedrich Schönbein, 1799–1868. Ein Blatt zur Geschichte des 19. Jahrhunderts* (Leipzig, 1899, 1901), I, II; J.R. Partington, *A History of Chemistry* (London, 1964), IV, p. 190f.

Selection 51

Introduction:

a. M. Traube, 'Zur Theorie der Gährungs- und Verwesungserscheinungen, wie der Fermentwirkungen überhaupt', *Ann.Phys.*, [4] *13*, 331–44. 1858. See also Selection 15, Section I.
b. See Selection 104, Section V.
c. Traube, 'Zur Theorie', 334.
d. Cf. Note by T. Thomson on indigo in *Ann.Phil.*, *15*, 466–7. 1820.

Text:

1. Traube, Theorie der Fermentwirkungen, p. 107. In so far as the chemical processes in the living organism mostly depend on ferment activities, an understanding of the chemistry of life without a correct theory of fermentation is altogether impossible.
a. Reference to *Theorie der Fermentwirkungen* (Berlin, 1858). See also Introduction to Selection 15, Section I, note a.

Selection 52

Introduction:

a. Professor of Physiology at Zurich (1868) and Königsberg (now Kaliningrad, RSFSR) from 1884 to 1913.
b. Introduction to Selection 81, Section IV.
c. W. Kühne, 'Untersuchungen über Bewegungen und Veränderungen der contractilen Substanzen IV', *Müllers Arch.*, *1*, 748–835. 1859; *Untersuchungen über das Protoplasma und Contractilität* (Leipzig, 1864).
d. Cf. L. Hermann, *Untersuchungen über den Stoffwechset der Muskeln, ausgehend vom Gaswechsel derselben* (Berlin, 1867). See also Selection 104, Section V; A. Fick and J. Wislicenus, 'Über die Entstehung der Muskelkraft', *Vierteljs.(Zurich)*, *10*, 317–48. 1865. An English version of the article called 'On the origin of muscular power' appeared in *Phil.Mag.* [4] *31*, 484–503. 1866.

Text:

a. In the German text the word used is '*krafterzeugende*' (force-generating). Cf. L. Hermann, *Grundriss der Physiologie des Menschen*, 5th edn (Berlin, 1874), p. 231.

Selection 53

Introduction:

a. T. Schwann, *Microscopical Researches into the Accordance in the Structure and Growth of Animals and Plants* (London, 1847), p. 200. See also Selection 163, Section VII.
b. See Introduction to Selection 170, Section VII.
c. It is not without interest to enquire into the beginnings of Ludwig's investigations of the blood gases. They go back to the very early period of his career when he taught and researched at Marburg (1842–9), where the eminent chemist R. Bunsen (1811–99), responding to the needs of the iron industry, was developing methods of gasometric analysis. 'To him [Bunsen]', wrote the notable English physiologist J.S. Burdon-Sanderson (1828–1905), 'there is reason to believe that he [Ludwig] was indebted for that technical knowledge of gas analysis which in later years was to bear such magnificent fruit'. See J.B.S., 'Obituary notices of Fellows deceased', *Proc.Roy.Soc.*, *59*, i-viii. 1896 (p. vii). Bunsen's classical account of methods for gas analysis simultaneously appeared in German and English: *Gasometrische Methoden* (Braunschweig, 1857); *Gasometry Comprising the Leading Physical and Chemical Properties of Gases* (London, 1857).
d. Apparently, the first perfusion experiments were performed by an investigator named Loebell in 1849. But the real starting point was the work by Alexander Schmidt (1831–94) in Leipzig under Ludwig in 1867. For information on the early phase of the subject, see K. Skutul, 'Über Durchströmungsapparate', *Pflügers Arch.*, *123*, 249–72. 1908.
e. See also the articles of Pflüger's collaborators: S. Wolffberg, 'Ueber die Athmung der Lunge', *Pflügers Arch.*, *6*, 23–43. 1872; G. Strassberg, 'Die Topographie der Gasspannungen im thierischen Organismus', ibid. 65–96; D. Finkler, 'II. Ueber den Einfluss der Strömungsgeschwindigkeit und Menge des Blutes auf die thierische Verbrennung', *Pflügers Arch.*, *10*, 368–71. 1875; E. Oertmann, 'Ueber den Stoffwechsel entbluteter Frösche', *Pflügers Arch.*, *15*, 381–98. 1877.

Selection 54

Introduction:

a. See Selection 162, Section VII.
b. See Selection 170, Section VII.
c. F. Hoppe-Seyler, 'Ueber die Prozesse de Gährungen und ihre Beziehung zum Leben des Organismus. Erste Abhandlung', *Pflügers Arch.*, *12*, 1–17. 1876.
d. In propounding his hypothesis of reductions and oxidations in connection with the processes of putrefaction, Hoppe-Seyler was largely influenced by the publications of Gottfried Wilhelm Osann (1797–1866) in Germany and Thomas Graham (1805–69) in Britain. It was out of their work in the 1850s and 1860s respectively that the terms 'nascent hydrogen' and 'active hydrogen' arose. It was suggested that the heightened chemical activity of this form of hydrogen was due to its existing in atomic condition. See J.S. Fruton, *Molecules and Life* (New York–London, 1972), pp. 316–17.

Selection 55

Introduction:

a. Selection 46.
b. Fourcroy, 'Expériences faites sur les matières animales', *Ann.Chim.*, *7*, 146–93. 1790. Fourcroy traced the observation of the presences of iron in blood to Nicolas Lemery (1645?–1715) and others. Cf. A.F. Fourcroy, *Système des connaissances chimiques, et de leurs applications aux phénomènes de la nature et de l'art* (Paris, 1801), IX, pp. 125–67.
c. W.C. Wells, 'Observations and experiments on the colour of blood', *Phil.Trans.*, *87*, 428. 1797.
d. L.-R. Lecanu, 'Études chimiques sur le sang humain', *Ann.Chim.*, *67*, 54–70. 1838.
e. J.J. Berzelius, *Lehrbuch der Chemie*, 4th edn (Dresden and Leipzig, 1840), IX, pp. 60–2.
f. F. Hoppe, 'Ueber die Einwirkung des Kohlenoxydgases auf das Hämatoglobulin Vorläufige Mittheilung', *Virchows Arch.*, *11*, 288–9. 1857; F. Hoppe-Seyler, 'Ueber das Verhalten des Blutfarbstoffes im Spectrum des Sonnenlichtes', *Virchows Arch.*, *23*, 446–9. 1862; F. Hoppe-Seyler, 'Ueber die chemischen und optischen Eigenschaften des Blutfarbstoffes Zweite Mittheilung', *Virchows Arch.*, *29*, 233–5. 1864. See also C.A. Culotta, 'On the colour of blood from Lavoisier to Hoppe-Seyler, 1777–1864: A theoretical dilemma', *episteme*, *4*, 219–33. 1970.
g. See Introduction to Selection 37, Section II.
h. Although MacMunn's observations attracted attention at the time, sixty years or so passed before their significance was more fully appreciated; see Selection 66. This was only partly due to Hoppe-Seyler whose views in this area carried great weight. To him myohaematin was not a pigment peculiar to pigeon breast muscle and functionally homologous to haemoglobin, as reported by MacMunn, but a decomposition product of haemoglobin. What apparently mattered more was the puzzling existence of the four-banded absorption spectrum in the reduced state of myohaematin (histohaematins). It made MacMunn's contemporaries doubt his claim that they were definite pigments, evolving into but not derived from haemoglobin. For an illuminating treatment of this question, consult D. Keilin, *The History of Cell Respiration and Cytochrome* (Cambridge, 1966), chap. 6.

Text (i):

1. These being all the animals which I have yet examined.
2. After suitable precautions had been taken to exclude the influence of the blood as fully described in the demonstration.
a. W. Kühne, 'Ueber den Farbstoff der Muskeln', *Virchows Arch.*, *33*, 79–94. 1865.

Text (ii):

a. C.A. MacMunn, 'Observations on the chromatology of Actiniae', *Phil.Trans.*, *176*, 641–63. 1885.
b. See Introduction to Selection 53, note e.

Selection 56

Introduction:

a. Selection 168, Section VII.
b. For an appraisal of Ehrlich's side-chain theory cf. E. Witebsky, 'Ehrlich's side-chain theory in the light of present immunology', *Ann.N.Y.Acad.Sci.*, *59*, 168–81. 1954–5. See also H. Bauer, 'Paul Ehrlich's influence on chemistry and biochemistry', ibid., 150–67; B. Schick, 'Ehrlich and problems of immunity', ibid., 182–9; J. Parascandola, 'The theoretical basis of Paul Ehrlich's chemotherapy', *J.Hist.Med.*, *36*, 19–43. 1981.
c. This was the beginning of the method of staining *in vivo*, which was influential before the introduction of radioactive isotopes. See N. Chandler Foot, 'Vital staining', *Ann.N.Y.Acad.Sci.*, *59*, 259–67. 1954–5.
d. Thesis that qualified the candidate, after it was passed, to offer courses of specialized lectures (*venia legendi*). The granting of the license to lecture gave the holder the right to the title of *Privatdozent* but not to a salary, which depended on getting a university post.

Text:

a. In specific detail the text of these translated excerpts varies from that in *The Collected Papers of Paul Ehrlich* (ed. F. Himmelweit with M. Marquandt, ed. direction Sir Henry Dale) (London and New York, 1956), I, pp. 440, 484.

Selection 57

Introduction:

a. Selections 50 and 54.
b. See a succession of Traube's articles on this topic in *Berichte der Deutschen chemischen Gesellschaft* (1882–6). In Traube's earlier work 'oxidation ferments' (*Oxydationsfermente*) appeared as 'decay ferments' (*Verwesungsfermente*). Cf. Selection 15, Section I.
c. Bach [Bakh] helped to develop modern biochemistry in Russia. Because of his revolutionary activities he was forced to leave the country and to live abroad until 1917. It seems that the Russian version of the paper in French from which the extracts have been selected was published in *Zhurnal Ruskogo fiziko-khimicheskogo obshchestva*, *29*, 375. 1891. Cf. A.N. Shamin, *Dictionary of Scientific Biography* (New York, 1970), I, pp. 360–3.
d. Prompted by Bach's article Carl Engler (1842–1925), in collaboration with W. Wild, dealt with a similar topic in 'Ueber die sogenannte 'Aktivierung' des Sauerstoffs und über Superoxydbildung', *Ber.chem.Ges.*, *30*, 1669–81. 1897. They examined the subject in purely chemical terms irrespective of biological questions.

Selection 58

Introduction:

a. T. Thunberg, 'Ein Mikrorespirometer. Ein neuer Respirationsapparat, um den respiratorischen Gasaustausch kleinerer Organe und Organismen zu bestimmen', *Skand.Arch.Physiol.*, *17*, 74–85. 1905; 'Studien über die Beeinflussung des Gasaustausches des überlebenden Froschmuskels durch verschiedene Stoffe. Erste Mitteilung', *Skand.Arch.Physiol.*, *22*, 406–29. 1909.
b. T. Thunberg, 'Studien über die Beeinflussung des Gasaustausches des überlebenden Froschmuskels durch verschiedene Stoffe. Zweite Mitteilung. Die Einwirkung von Oxalsäure, Malonsäure und Bernsteinsäure', *Skand.Arch.Physiol.*, *22*, 430–6. 1909.

Selection 59

Introduction:

a. F. Battelli and L. Stern, 'Die akzessorische Atmung in den Tiergeweben', *Biochem.Z.*, *21*, 487–509. 1909.

Text:

1. Battelli and Stern, Oxydation de l'acide succinique par les tissus animaux. Soc. de Biol. *69*, 301, 1910. The same, Influence de quelques facteurs sur l'oxydation de l'acide succinique par les tissus animaux. Ibid. *69*, 370, 1910.
2. F. Battelli and L. Stern, 'Die Oxydation der Citronen- Apfel- und Fumarsäure durch Tiergewebe', *Biochem.Z.*, *31*, 478–505. 1911.

Selection 60

Introduction:

a. H. Wieland, 'Über Hydrierung und Dehydrierung', *Ber.chem.Ges.*, *45*, 484–93. 1912.

Text:

1. Compare the important works of Palladin[b] and his pupils on the so-called respiration pigments e.g. B.H. *55*, 209 [1908]. Bio.Z. *18*, 151 [1909], *42*, 325 [1912], *44*, 317 [1912], *49*, 381 [1913].
2. O. Löw, B. *23*, 675 [1890]. Bach, Bio.Z. *33*, 288 [1911].
a. F. Schardinger, 'Ueber das Verhalten der Kuhmilch gegen Methylenblau und seine Verwendung zur Unterscheidung von ungekochter und gekochter Milch', *Z.Unters.Nahr.Genuss.*, *5*, 1113–21. 1902.
b. Vladimir Ivanovich Palladin (1859–1922) taught plant anatomy and physiology at the universities of Kharkov and St Petersburg (Petrograd, Leningrad). During his lifetime an influential figure for his studies of cellular respiration in plants.

Selection 61

Introduction:

a. Selection 20, Section I.
b. A. Harden and H. MacLean, 'The oxidation of isolated animal tissues', *J.Physiol.*, *43*, 34–45. 1911.
c. For an excellent sketch with critical evaluation of Warburg's life, work and personality, see Hans Krebs' (1900–81) obituary article (among the best of its *genre*) in *Biog.Mem.F.R.S.*, *18*, 629–99. 1972. Slightly expanded, the portrait was published in book form in German: H. Krebs, *Otto Warburg Zellphysiologe-Biochemiker-Mediziner 1883–1970*. Unter Mitwirkung von Dr Roswitha Schmid (Stuttgart, 1979). Translated (if this is the right description) by Krebs and Anne Martin, the study 'with a few minor changes and additions' appeared as *Otto Warburg Cell Physiologist Biochemist and Eccentric* (Oxford, 1981).
d. O. Warburg, 'Beobachtungen über die Oxydationsprozesse im Seeigelei', *Z.physiol.Chem.*, *57*, 1–16. 1908.
e. O. Warburg, 'Beiträge zur Physiologie der Zelle, insbesondere über die Oxydationsgeschwindigkeit in Zellen', *Erg.Physiol.*, *14*, 253–337. 1914.
f. O. Warburg, 'Über sauerstoffatmende Körnchen aus Leberzellen und über Sauerstoffatmung in Berkefeld-Filtraten wässriger Leberextrakte', *Pflügers Arch.*, *154*, 599–617. 1913.

    It may be asked whether there was any connection between Warburg's considerations of the structural aspects of cell respiration and the discussions on the existence and role of cell granules stemming from the work of Richard Altmann (1852–1901). It appears that in 1886 Altmann surmised that the granules he observed in a variety of cells might be involved in cellular respiration. After it was established that most cellular bodies described by Altmann were not artifacts, they were given the name 'mitochondria'. Knowledge concerning their oxidative properties began to emerge from the 1940s onwards and belongs to a period of the history of biochemistry outside the scope of this work. See A. Hughes, *A History of Cytology* (London and New York, 1959), p. 119f; D.E. Green, 'The cyclophorase system' in J.T. Edsall (ed.), *Enzymes and Enzyme Systems* (Cambridge, Mass., 1951), pp. 15–46.

g. For references to the development of the manometric method, see Introduction to Selection 41, Section II, note c.
h. See, for example, W. Spitzer, 'Die Bedeutung gewisser Nukleoproteide für die oxydative Leistung der Zelle', *Pflügers Arch.*, *67*, 615–56. 1897, which was cited by Warburg in the paper from which Selection 64 is taken. Spitzer suggested that the transfer of oxygen was mediated by organically-bound iron present in what he believed to be nucleoproteins.

Text:

1. Compare for example Dakin, Journ.biol.chemistry, Vol. 1, 17 (1906).
2. Hydrogen peroxide or iron peroxide.

Selection 62

Introduction:

a. H. Einbeck, 'Über das Vorkommen von Bernsteinsäure im Fleischextrakt und im frischen Fleische', *Z.Physiol.Chem.*, *87*, 145–58. 1913.

Selection 63

Text:

1. Reached the editorial office on 14 October 1916.
a. The year is given as printed on the title page of volume 35 of the journal. In the literature the article is referred to (even by Thunberg himself) as published in 1917. Cf. T. Thunberg, 'The hydrogen-activating enzymes of the cells', *Quart.Rev.Biol.*, *5*, 318–47. 1930.
b. The article contains a critical examination of Ehrlich's publication on the oxygen requirement of the organism (see Selection 56). Thunberg pointed out that two workers, H. Dreser and P. Ehrlich, simultaneously and unconnectedly introduced the methylene blue method into the study of oxidations and reductions occurring in the living body. Undoubtedly, Dreser had arrived at the methylene blue independently of Ehrlich. Cf. his 'Histochemisches zur Nierenphysiologie', *Z.Biol.*, *21*, 41–66. 1885. Prompted by this, Ehrlich published two papers with a critical discussion of some of Dreser's results. In a footnote to the first paper Ehrlich emphasized that he had known about methylene blue before he read Dreser's article and that he had lectured on it 18 December 1884. Cf. *Collected Papers Ehrlich, I*, pp. 497–508, Selection 56, note a.

Selection 64

Introduction:

a. O. Warburg, 'Versuche an überlebendem Carcinomgewebe (Methoden)', *Biochem.Z.*, *142*, 316–33. 1923; 'Verbesserte Methode zur Messung der Atmung und Glykolyse', ibid., *152*, 51–63. 1924.
b. Warburg might have drawn the idea of the cycle from Meyerhof's work on muscle chemistry in the early 1920s. Meyerhof visualized the breakdown of carbohydrate in muscle to lactic acid as the anaerobic phase and the synthesis of carbohydrate as the aerobic phase of 'a specific carbohydrate cycle' (*eines eigentümlichen Kohlenhydratkreislaufs*). Cf. O. Meyerhof, 'Die Energieumwandlung im Muskel, III. Kohlehydrat- und Milchsäureumsatz im Froschmuskel', *Pflügers Arch.*, *185*, 11–32. 1920; 'Die Energieumwandlungen im Muskel. IV. Mitteilung. Über die Milchsäurebildung in der zerschnittenen Muskulatur', ibid., *188*, 114–60. 1921. The idea of the cycle discussed by Meyerhof in these papers is referred to in O. Warburg, K. Posener and E. Negelein, 'Über den Stoffwechsel der Carcinomzelle', *Biochem.Z.*, *152*, 317. 1924. See also Selection 84, Section IV.
c. Warburg distinguished between blood charcoal, i.e. charcoal prepared by heating blood, and haemin charcoal prepared by heating so-called Teichmann's crystals (obtained by boiling blood with glacial acetic acid and sodium chloride). Warburg pointed out that

haemin charcoal was the active constituent of commercial blood charcoals. As the latter contained little of it they were comparatively weak catalysts. See footnote 1 on p. 485 of the paper from which the Selection has been taken. At that time, it seems, the suitability of Warburg's blood charcoal 'models' for the study of chemical aspects of a *physiological* process was not challenged, as later it would be, by Keilin. In his view after 'the incineration of haemin, treatment with acid and re-incineration, the charcoal that remained bore no resemblance to the original haemin and, as a catalyst, it could provide no more evidence for the intracellular function of haematin than could the somewhat more active charcoals obtained from commercial Bismarck brown or some other dyes'. Cf. D. Keilin, *The History of Cell Respiration and Cytochrome* (Cambridge, 1966), p. 137.

Text:

1. O. Warburg, this journal, *119*, 134., 1921.
2. O. Warburg und W. Brefeld, ibid., *145*, 461, 1924.
3. O. Warburg, ibid., *136*, 266, 1923.
4. W. Pfeffer, Beiträge zur Kenntnis der Oxydationsvorgänge in lebenden Zellen. Abhandl. der mathem.-physischen Klasse der Sächs. Ges. d. Wiss. *15*, No. 5. Leipzig, 1889.
5. H. Wieland, Oppenheimers Handb. d. Biochem., 2. Aufl., *2*, 252, 1923.

Selection 65

Introduction:

a. A. Fleisch, 'Some oxidation processes of normal and cancer tissue', *Biochem.J.*, *18*, 294–311. 1924.
b. Fleisch's picture of the mechanism was more elaborate. He was not quite sure about the presence of one or two enzymes (succinodehydrogenase, succinoxydase), although he concluded that 'it seems to be reasonable that only one enzyme should be present.' That the enzyme not only activates the substrate molecule but also acts as the initial hydrogen acceptor was not confirmed until 1951 (not covered in this volume). Cf. D. Keilin, *The History of Cell Respiration and Cytochrome* (Cambridge, 1966), pp. 126–7.

Selection 66

Introduction:

a. Due to limitations of space, other mechanisms of biological oxidation considered in the 1920s and the 1930s cannot be discussed. In particular they included the participation of hydrogen peroxide and organic peroxides (stemming from the peroxide theories of Bach and Engler mentioned already) and glutathione (discovered by F.G. Hopkins in 1921). See F.G. Hopkins, 'On an autooxidisable constituent of the cell', *Biochem.J.*, *15*, 286– 305. 1921; 'On current views concerning the mechanisms of biological oxidation', *Skand.Arch.Physiol.*, *49*, 33–59. 1926. (See Selection 175, Section VII); 'On glutathione: a reinvestigation', *J.Biol.Chem.*, *84*, 269–320. 1929. For a picture of these developments (up to the end of 1938), see C. Oppenheimer and K.G. Stern, *Biological Oxidation* (The Hague, 1939).
b. MacMunn gave the positions of three absorption bands only (cf. Selection 55).
c. Although Keilin discussed the work of MacMunn in his paper, it should be remembered that Keilin became familiar with it only after completing his own investigations. (Apparently, as a consequence of a remark by the eminent Cambridge physiologist, J. Barcroft (1872–1947), 'that an Irishman had reported something similar'). See E.C. Slater, 'Cytochrome', *TIBS*, *2*, 138–9. 1977.
d. See also Introduction to Selection 68, note c. Although Keilin's contribution to the development of knowledge of oxidation-reduction systems concerned in biological oxidations was recognized as outstanding, it did not earn him the Nobel Prize. It constitutes one of two notable omissions for salient work in the biochemical field, worthy of the award, during the period under review. The other overlooked achievement is Robin Hill's contribution to the understanding of the light reaction in photosynthesis (see Selection 44, Section II).

Text:

1. The oxidation of cytochrome in this case differs slightly from ordinary oxidation. In ordinary oxidation all the four bands fade away more or less simultaneously. In presence of urethane or formaldehyde band *c* disappears the first, while the other three bands remain and band *b* seems to be even intensified. On shaking the emulsion for a long time with air, all the bands disappear, band *b* being the last to go.

2. It is known now that different haemochromogens may exist showing different absorption spectra, and that some of them do not combine with CO in neutral solution, but will combine when pH is changed. For the detailed discussion of this and other problems concerning haemochromogen the reader is referred to the paper by Anson and Mirsky, which will shortly appear in the 'Journal of Physiology'.[e] The porphyrin found in acetone fluid was probably derived partly from haemochromogen *a'* and partly from *b'*.

3. The evidence as to the haemochromogen-nature of this compound is limited at present to the existence of two-banded absorption spectrum in the reduced state and to its O_2 and possibly CO compounds.

a. K.A.H. Mörner, 'Beobachtungen über den Muskelfarbstoff', *Nord.Med.Ark.*, *30* (Festband No. 2), 1–8. 1896. For obtaining a xerox-copy of this article thanks are due to N. Roll-Hansen.

b. H. Günther, 'Über den Muskelfarbstoff', *Virchows Arch.*, *230*, 146–178. 1921.

c. H. Munro Fox, 'On chlorocruorin. I.', *Proc.Cam.Phil.Soc.*, *Biol.Sci.*, *1*, 204–18. 1923–5.

d. H. Fischer and K. Schneller, 'Zur Kenntnis der natürlichen Porphyrine. VI. Mitteilung. Über die Verbreitung der Porphyrine in Organen. Nachweis eines Porphyrins in der Hefe'. *Z.physiol.Chem.*, *135*, 253–93. 1924; H. Fischer and J. Hilger, 'Zur Kenntnis der natürlichen Porphyrine. 8. Mitteilung. Über das Vorkommen von Uroporphyrin (als Kupfersalz, Turacin) in den Turakusvögeln und den Nachweis von Koproporphyrin in der Hefe', *Z.physiol. Chem.*, *138*, 49–67. 1924.

e. M.L. Anson and A.E. Mirsky, 'On haem in nature', *J.Physiol.*, *60*, 161–74. 1925.

Selection 67

Introduction:

a. Meanwhile the constitution of haemins had been established. They were known to be porphyrins containing four pyrrol rings linked by four methine groups. This was established by the extensive synthetical labours of Hans Fischer (1881–1945), who received the Nobel Prize for his achievement in 1930. See H. Fischer, 'On haemin and the relationship between haemin and chlorophyll', *Nobel Lectures Chemistry 1922–1941* (Amsterdam–London–New York, 1966), pp. 165–84.

Text:

1. According to its spectrum and the hydrochloric-acid number of its porphyrin, this haemin closely resembles mesohaemin, but has a free methine group in the β-position. For this reason spirographis haemin (C_{32}) contains two C atoms less than blood haemin (C_{34}). (Experiments in collaboration with E. Negelein.)

a. O. Warburg, 'Über die Wirkung des Kohlenoxyds auf den Stoffwechsel der Hefe', *Biochem.Z.*, *177*, 471–86. 1926. F. Kubowitz's collaboration in the production of this paper is not mentioned by Warburg. But he reports that during a visit to his laboratory at Dahlem (Berlin), A.V. Hill (1886–1977) alerted him to work by physiologists at Cambridge on the reduced affinity of haemoglobin for carbon monoxide under the influence of light. The work originated in observations of the diminishing stability of carboxyhaemoglobin in bright daylight, described by J.[S.] Haldane and J. Lorraine Smith, 'The oxygen tension of arterial blood', *J.Physiol.*, *20*, 497–520, 1896 (pp. 504–5). Hill's information to Warburg has attracted sufficient interest to have given rise to at least two accounts of it. According to the neurophysiologist, R.W. Gerard (1900–74), Hill talked about the work done in Cambridge at a dinner also attended by Warburg. The latter apparently 'soon excused himself from the party, and the next morning he sent in to *Naturwissenschaften* his classical note showing that light reverses the carbon monoxide inhibition of yeast respiration.' Cf. R.W. Gerard, 'The minute experiment and the large 'picture' in F.G. Worden, J.P. Swazey, G. Adelman (eds), *The Neurosciences: Path of discovery* (Cambridge, Mass., and London, 1975), p. 463. This somewhat colourful narrative is contradicted by Hans Krebs' substantiation of Warburg's statement. Krebs also provides further details of what took place. Krebs, who became involved in the ensuing research, recalls: 'I was present when A.V. Hill visited Warburg's laboratory. Warburg showed him the inhibition of yeast respiration by carbon monoxide and Hill then told him of the old work of Haldane and Smith on the light sensitivity of CO-haemoglobin. Within 24 hours Warburg tested the light sensitivity of the inhibition of yeast respiration and found the well known effect.' See M. Florkin, *A History of Biochemistry* (Amsterdam–Oxford–New York, 1975), pt. III, p. 210. Cf. also H. Krebs in collaboration with Roswitha Schmid, *Otto Warburg Cell Physiologist Biochemist and Excentric* (Oxford, 1981), pp. 26–7.

b. O. Warburg and E. Negelein, 'Über den Einfluss der Wellenlänge auf die Verteilung des Atmungsferments. (Absorptionsspektrum des Atmungsferments)', *Biochem.Z.*, *193*, 339– 46. 1928; O. Warburg and E. Negelein, 'Über die photochemische Dissoziation von Eisencarbonylverbindungen (Kohlenoxyd-Hämochromogen, Kohlenoxyd-Ferrocystein) und das photochemische Aquivalentgesetz', *Biochem.Z.*, *200*, 415–58. 1928.

c. O. Warburg and E. Negelein, 'Über die photochemische Dissoziation bei intermittierender Belichtung und das absolute Absorptionsspektrum des Atmungsferments', *Biochem.Z.*, *202*, 202–28. 1928; O. Warburg and E.Negelein, 'Absolutes Absorptionsspektrum des Atmungsferments', *Biochem.Z.*, *204*, 495–9. 1929; O. Warburg and E. Negelein, 'Über das Absorptionsspektrum des Atmungsferments', *Biochem.Z.*, *214*, 64–100. 1929.

d. O. Warburg, E. Negelein and E. Haas, 'Spirographishämin', *Biochem.Z.*, *227*, 171–83. 1930. See also O. Warburg and E. Negelein, 'Über das Hämin des sauerstoffübertragenden Ferments der Atmung, über einige künstliche Hämoglobine und über Spirographis-Porphyrin', *Biochem.Z.*, *244*, 9–32. 1932.

e. For work on reduction of spirographis haemin, see O. Warburg and E. Negelein, 'Notiz über Spirographishämin', *Biochem.Z.*, *244*, 239–42. 1932. A search for an original paper on its oxidation proved to be unsuccessful.

Selection 68

Introduction:

a. E.S. Barron and G.A. Harrop Jr, 'Studies on blood cell metabolism. II. The effect of methylene blue and other dyes upon the glycolysis and lactic acid formation of mammalian and avian erythrocytes', *J.Biol.Chem.*, *79*, 65–87. 1928. In view of Warburg's employment of blood-charcoal systems (Selection 64), his opposition to methylene blue technique is intriguing.

b. O. Warburg and W. Christian, 'Über Aktivierung der Robisonschen Hexose-Monophosphorsäure in roten Blutzellen und die Gewinnung aktivierender Fermentlösungen', *Biochem.Z.*, *242*, 206–27. 1932.

c. During the 1930s the term began to be used when discussing the oxidation-reduction sequence in relation to the hydrogen and oxygen ends of biological oxidation proceeding through 'carriers', progressively interpreted with the aid of the electronic theory of valency.

d. Because of the need to limit the size of this collection a greater number of selections from significant papers on this topic have had to be reluctantly omitted. For the included material no claim is made that it represents the wealth of work and issues enshrined in the contemporary publications concerning flavoproteins and pyridine nucleotides. Warburg's laboratory was particularly fruitful but important work also came from the laboratories of R. Kuhn (1900–67), P. Karrer (1889–1971) and Hans von Euler-Chelpin (1873–1964). That their work involved secrecy, so out of keeping with the accepted perception of the world of science, has been highlighted by Hugo Theorell (1903–82), one of Warburg's most prominent pupils, who worked in his laboratory at that time:

In December 1933 he [Warburg] showed me the first crystals of nicotinic acid amide as picrolonate. Nobody knew at that time what it was; that had to be found out. When I told Warburg I might go home to Stockholm for Christmas he hesitated, because there were living dangerous people like Hans von Euler and Karl Myrbäck who were on the same track. He finally agreed, but said: 'I am going to kill you if you say the word picrolinic acid in Stockholm.' That was easily promised.

Quoted by H. Krebs in collaboration with Roswitha Schmid, *Otto Warburg Cell Physiologist Biochemist Excentric* (Oxford, 1981), p. 32.

e.

Riboflavin
[6,7-dimethyl-9-(1′-D-ribityl)
isoalloxazine]

Riboflavin phosphate
(flavin mononucleotide)

Oxidized flavin   Reduced flavin

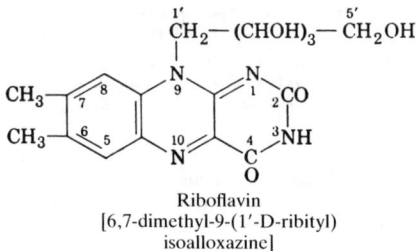

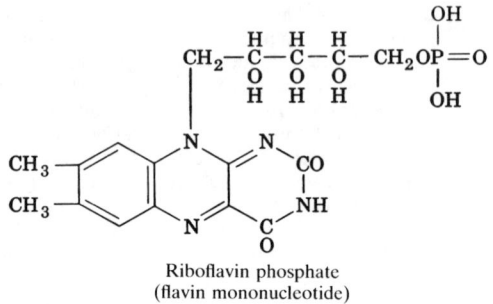

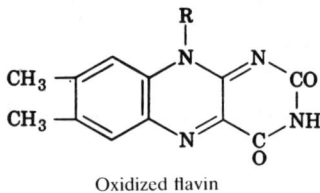

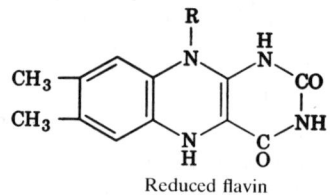

f.   See also H. Theorell, 'Das gelbe Oxydationsferment', *Biochem.Z.*, *278*, 263–90. 1935.

Text:

1.   H. Theorell, this journal, *272*, 155, 1934.
2.   P. Karrer, H. Salomon, K. Schöpp, E. Schlittler and H. Fritzsche, Helv. chim. Acta, XVII, 1010. 1934. [Cf. note ä.].
a.   Before Theorell published his findings it was widely thought that free lactoflavin constituted the active group of the yellow enzyme. Lactoflavin was obtained in the crystalline state by R. Kuhn, P. György and Th. Wagner-Jauregg. See 'Über Lactoflavin, den Farbstoff der Molke', *Ber.chem.Ges.*, *66*, 1034–9. 1933; also R. Kuhn, H. Rudy and Th. Wagner-Jauregg, 'Über Lactoflavin (Vitamin B₂)', *Ber.chem.Ges.*, *66*, 1950–6. 1933. The vitamin nature was pointed out by P. György, R. Kuhn and Th. Wagner-Jauregg in 'Das Vitamin B₂', *Naturwiss.*, *21*, 560–1. 1933. The relationship between flavins and vitamin B_2 was discussed by R. Kuhn, P. György and Th. Wagner-Jauregg in 'Über eine neue Klasse von Naturfarbstoffen (Vorläufige Mitteilung)', *Ber.chem.Ges.*, *66*, 317–20. 1933; 'Über Ovoflavin, den Farbstoff des Eiklars', *Ber.chem.Ges.*, *66*, 576–80. 1933. See also P. György, R. Kuhn and Th. Wagner-Jauregg, 'Darstellung von Vitamin B₂-Konzentraten', *Z.physiol.Chem.*, *223*, 27–35. 1934. On the first synthesis of the lactoflavin consult P. Karrer, K. Schöpp and F. Benz, 'Synthesen von Flavinen IV', *Helv.Chim. Acta*, *18*, 426–29. 1935; H. von Euler, P. Karrer, M. Malberg, K. Schöpp, F. Benz, B. Becker and P. Frei, 'Synthese des Lactoflavins (Vitamin B₂) und anderer Flavine', *Helv.Chim.Acta*, *18*, 522–35. 1935. An important contribution to the elucidation of the structure of lactoflavin came from Kurt Guenther Stern (1904–56) and Ensor Roslyn Holiday (*b*.1903) by methylation of alloxazine and preparation of, among others, 6,7-dimethyl-alloxazine, the compound identified by Karrer as the product of the photolysis of lactoflavin. See K.G. Stern and E.R. Holiday. 'Die Photo-flavine, eine Gruppe von Alloxazin – Derivaten', *Ber.chem.Ges.*, *67*, 1442–52. 1934.

Selection 69

Introduction:

a. O. Warburg, W. Christian and A. Griese, 'Wasserstoffübertragendes Co-Ferment, seine Zusammenfassung und Wirkungsweise', *Biochem.Z.*, *282*, 157–223. 1935. This paper contains references to previous work by Warburg and his collaborators in this area.

b

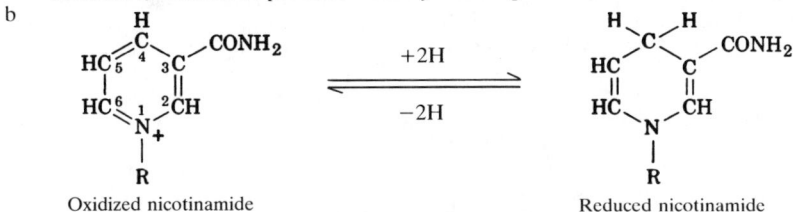

Oxidized nicotinamide

Reduced nicotinamide

The structure of the hydropyridine compound formed on reduction was not clarified until 1954. See J.S. Fruton and S. Simmonds, *General Biochemistry*, 2nd edn (New York – London, 1958) p.312.

c. Selection 68.

d. 'Cozymase' was the term given to Harden and Young's water soluble, heat resistant, dialyzable 'coferment' by H.v. Euler and K. Myrbäck, 'Gärungs-Co-Enzym (Co-Zymase) der Hefe. I.', *Z.physiol.Chem.*, *131*, 179–203. 1923. See also Selection 22, Section I.

e. O. Warburg and W. Christian, 'Pyridin, der Wasserstoffübertragender Bestandteil von Gärungsfermenten (Pyridin-Nucleotide)', *Biochem.Z.*, *287*, 291–328. 1936. Following the *Report of the Commission on Enzymes of the International Union of Biochemistry* (Oxford, 1961), the terms DPN and TPN were replaced by NAD (nicotinamide-adenine dinucleotide) and NADP (nicotinamide-adenine dinucleotide phosphate) respectively. In older literature cozymase was also known as co-enzyme I or codehydrogenase I (CoI) and Warburg's coferment was co-enzyme II or codehydrogenase II (CoII).

f. H.v. Euler, H. Albers and F. Schlenk, 'Chemische Untersuchungen an hochgereinigter Co-Zymase', *Z.physiol. Chem.*, *240*, 113–26. 1936.

Selection 70

Introduction:

a. For an outline of the problem that occupied Szent-Györgyi for many years and also for a summary of the work (including references to his own papers and those of his collaborators) see A.v. Szent-Györgyi, *Studies on Biological Oxidation and Some of Its Catalysts* (Budapest and Leipzig, 1937). Also published as *Acta Litterarum ac Scientiarum Reg. Univ. Hung. Francisco-Josephinae, Sectio Medicorum* (*Acta Med. Szeged*), *9* (pt 1), 1–98. 1937.

Text:

a. F.J. Ogston and D.E. Green, 'The mechanism of the reaction of substrates with molecular oxygen. I.II.', *Biochem.J.*, *29*, 1983–2004; 2005–12. 1935.

b. T. Thunberg, 'Zur Kenntnis des intermediären Stoffwechsels und der dabei wirksamen Enzyme', *Skand.Arch.Physiol.*, *40*, 1–91. 1920.

c. J.H. Quastel and M.D. Whetham, 'Dehydrogenations produced by resting bacteria. I.', *Biochem.J.*, *19*, 520–31. 1925; J.H. Quastel and W.R. Wooldridge, 'Some properties of the dehydrogenating enzymes of bacteria', *Biochem.J.*, *22*, 699–702. 1928. This investigation developed in connection with Quastel's work on the 'active centre' hypothesis. See also Selection 28, Section I.

d. F.J. Stare and C.A. Baumann, 'The effect of fumarate on respiration', *Proc.Roy.Soc.* (*B*), *121*, 338–57. 1936.

Selection 71

Introduction:

a.   Cf. Section IVb.
b.   Cf. 'For some time, from 1932, I had tested the oxidizability in various tissues (especially kidney, liver and muscle) of substances which, on the basis of knowledge of chemistry, might possibly be intermediates in the combustion of foodstuffs, and I had seen the ready oxidation of citrate. I, too, had tried to elucidate the chemical reactions of citric acid, but without success.' H. Krebs, *Reminiscences and Reflections* (Oxford, 1981), pp. 111–12.
c.   T. Thunberg, 'Studien über die Beinflussung des Gasaustausches des überlebenden Froschmuskels durch verschiedene Stoffe. Vierte Mitteilung', *Skand.Arch.Physiol.*, 24, 23–61. 1910 (pp. 54, 57). See also Selection 58.
d.   F. Battelli and L. Stern, 'Die Oxydation der Citronen-, Apfel-und Fumarsäure durch Tiergewebe', *Biochem.Z.*, 478–505. 1911.
e.   T. Thunberg, 'Zur Kenntnis des intermediären Stoffwechsels und der dabei wirksamen Enzyme', *Skand.Arch.Physiol.*, 40, 1–91. 1920.
f.   C. Martius and F. Knoop, 'Der physiologische Abbau der Citronensäure. Vorläufige Mitteilung', *Z.physiol.Chem.*, 246, I–II. 1937.
g.   C. Martius, 'Über den Abbau der Citronensäure', *Z.physiol.Chem.*, 247, 104–10. 1937. Also F. Knoop, 'Über den physiologischen Abbau der Säuren und die Synthese einer Aminosäure im Tierkörper', *Z.physiol.Chem.*, 67, 489–520. 1910; G.Embden and E. Schmitz, 'Über synthetische Bildung von Aminosäureen in der Leber', *Biochem.Z.*, 29, 423–8. 1910; see also Selections 122 and 123, Section V.
h.   In 1940 Krebs working with Leonard Victor Eggleston (1920–74) listed these three observations as forming the basis on which his original formulation had been built. See H.A. Krebs and L.V. Eggleston, 'The oxidation of pyruvate in pigeon breast muscle', *Biochem.J.*, 35, 442–59. 1940. Subsequent developments do not fall within the scope of this volume. For retrospective re-assessment of the events preceding and following the discovery of the cycle by Krebs see his Nobel Lecture (1953) 'The citric acid cycle' in *Nobel Lectures Physiology or Medicine* 1942–62 (Amsterdam–London–New York, 1964) pp. 399–410. In the second Verne R. Mason Memorial Lecture (University of Miami, 1969) Krebs discussed stimulatingly the historical, philosophical and motivational aspects of the work on the cycle. It has been published as 'The history of the tricarboxylic acid cycle' in *Persp.Biol.and Med.*, 14, 154–70. 1970. See also his article 'Errors, false trails and failures in research' in B. Kaminer (ed.), *Search and Discovery. A Tribute to Albert Szent-Györgyi* (New York–San Francisco–London, 1977), pp. 3–15 (9–10) and *Reminiscences*, chap. 9.

Text:

1.   Östberg, Skand. Arch. Phys. 62, 81 (1931).
2.   Sherman, Mendel, Vickery, Jl. of Biol. Chem. 113, 247, 265 (1936).
3.   Martius, Knoop, Zs. phys. Chem. 246, 1 (1937).
4.   Martius, Zs. phys. Chem. 247, 104 (1937).
5.   Bernhauer, Ergebn. Enzymf. 3, 185 (1934).
6.   Wieland, Sonderhoff, Ann. Chem. Pharm. Liebig 503, 61 (1933).
7.   Martius, Knoop, Zs. phys. Chem. 242, 1 (1936).
8.   [Here the reader is referred to] Claisen, Hori, Ber. Chem. Ges. 24, 120 (1891).
9.   Szent-Györgyi c.s., Biochem. Zs. 162, 399 (1925); Zs. phys. Chem. 224, 1 (1934), 236, 1 (1935), 244, 105 (1936), 247, 1 (1937).
10. Stare, Baumann, Proc. Roy. Soc. B 121, 338 (1936).
11. Krebs, Johnson, Biochem. Jl. 31, 645 (1937).
12. Wagner-Jauregg, Rauen, Zs. phys. Chem. 237, 227 (1935).
13. Malachowski, Maslowski, Ber. Chem. Ges. 61, 2524 (1928).
14. Blaise, Gault, C.R. 147, 198 (1908).
15. Green, Biochem. Jl. 30, 2095 (1936).

Selection 72

Introduction:

a. The reader is reminded that in the original paper (1925) the three haemochromogens were denoted as *a'*, *b'*, *c'*. See Selection 66.

b. Keilin's *The History of Cell Respiration and Cytochrome*, mentioned repeatedly earlier, while reflecting the author's point of view of the development of the subject is also valuable in that it contains a bibliography of his publications and also of works by others. It is necessary to point out that only the first eight chapters (up to 1926) reproduce the original manuscript found at the time of the author's death. With certain reservations the same applies to chap. 9 (1926–1933). The remaining five chapters are compiled, by Keilin's daughter, Joan, from unrevised drafts and published papers. For Warburg's view of the work on certain aspects of cellular respiration carried out by himself and his collaborators see his *Heavy Metal Prosthetic Groups and Enzyme Action* (translated by Alexander Lawson) (Oxford, 1949; published in German in Berlin in 1946 and again in Freiburg im Breisgau in 1949). For Keilin's critical review of the latter see his 'Metal catalysis in cellular metabolism', *Nature*, *165*, 4–5. 1950.

c. D. Keilin and E.F. Hartree, 'Cytochrome oxidase', *Proc.Roy.Soc. (B)*, *125*, 171–86. 1938; 'Cytochrome *a* and cytochrome oxidase', *Nature*, *141*, 870–1. 1938; 'Cytochrome and cytochrome oxidase', *Proc.Roy.Soc. (B)*, *127*, 167–91. 1939; 'Succinic dehydrogenase-cytochrome system of cells. Intracellular respiratory system catalysing aerobic oxidation of succinic acid', ibid., *129*, 277–306. 1940; 'Properties of cytochrome *c*', *Nature*, *145*, 934. 1940.

d. M. Dixon, 'Oxidation mechanisms in animal tissues', *Biol.Revs.*, *4*, 352–97. 1929.

Text (i):

1. The nature of the two components (a_1 and a_2) which have certain properties in common with a_3, and their relationship with the latter, will be discussed in a separate paper.[c]

a. D. Keilin and E.F. Hartree, 'Cytochrome oxidase', *Proc.Roy.Soc. (B)*, *125*, 171–86. 1938.

b. Selection 67.

c. There is no evidence that this actually happened. For more information one may consult chap. 12 in Keilin's *History*.

Text (ii):

a. Reference to paper from which the first excerpt in this selection is taken.

Section IV:
CARBOHYDRATES

a. From *c*. 1800 to *c*. 1860: analysis and identification. Synthesis in the animal body

Selection 73

We have seen that in 1789 Lavoisier estimated the proportion of constituents in 100 parts of sugar as follows: 8 parts of hydrogen, 64 parts of oxygen and 28 parts of carbon.[a] More accurate data were obtained more than two decades later by Joseph Louis Gay-Lussac (1778–1850) and Louis Jacques Thenard (1777–1857) by improving the combustion procedure already known to Lavoisier.

What they did was to carry out combustion of nineteen substances, derived from plants and animals that included sugar, with potassium chlorate [*muriate suroxigéné*], collect the gaseous products (oxygen, carbonic acid and possibly nitrogen) over mercury and measure them. But before that they determined in a blank experiment the amount of oxygen obtainable from the chlorate.

As shown below somewhat obscurely (no details are given), they evaluated the weight of carbon from the value of carbonic acid formed and absorbed in caustic potash. Moreover, they took it for granted that the remainder of oxygen produced water during combustion and on this basis they indirectly calculated the proportion of hydrogen in the substance from the amount of remaining oxygen. This approach contributed to the evolution of the belief that in sugars the proportion of hydrogen to oxygen corresponded to the proportion in which the two elements combined to form water and that they comprised a group conforming to hydrates of carbon.[b]

[G.L.] Gay-Lussac and [L.J.] Thenard, *Recherches physico-chimiques*, (Paris, 1811), II. (Physico-chemical researches)

'Méthode pour determiner la proportion des principes qui constituent les substances végétales et animales, et application de cette methode à l'analyse d'un grand nombre de ces substances' (Method for determining the proportion of the elements which constitute the vegetable and animal substances, and application of this method to the analysis of a great number of these substances)

pp. 265–6

It has been known for a long time that vegetable substances are formed of hydrogen, oxygen and carbon, and that animal substances contain nitrogen besides; but it is still not known what is the quantity of these substances contained in each of them. It is to this cause that one must attribute in great part the little progress which vegetable and animal chemistry have made up to the present. In fact it does not suffice to know the elements which constitute a body in order to have a conception of all the phenonena which can result from its contact with the others, it is still necessary for this to know the relation in which the elements are to one another.

pp. 267–8

To attain this aim our first idea, that which we soon gave up, has been to transform, with the help of oxygen, the vegetable and animal substances into water, carbonic acid and nitrogen. It was evident that if we could succeed in operating this transformation in such a way as to recover all the gases, this analysis became of very great exactitude and simplicity. Two obstacles blocked this way: one was complete burning of the hydrogen and carbon of these substances; the other was making the combustion in closed vessels.

One could only hope to surmount the first by means of metallic oxides which readily give up their oxygen or by means of potassium chlorate. Some trials soon made us give preference to this salt which was successful beyond our hopes.

pp. 285–6

So we have all the data necessary for knowing the proportion of the elements in the vegetable substance. We know how much we have burnt of the substance since we have the weight to half a milligram. We know how much oxygen has been necessary to transform it into water and carbonic acid since the quantity of this is given by the difference between that contained in the potassium chlorate and that which is contained in the gas. Finally we know how much carbonic acid has been formed and we calculate how much has been necessary to form water.

pp. 288–9

The sugar analyzed was sugar candy quite white and well crystallized . . .

Sugar used, total correction for weight applied	0.300
Potassium chlorate used, 7 fold the weight of sugar.	
Oxygen of potassium perchlorate	0.627
Oxyg. of carbonic acid produced	0.322
Oxyg. recovered	0.303
	0.625
Carbon of carbonic acid	0.1274
Then 100 parts of sugar are composed of:	
Carbon	42.47
Oxygen	50.63
Hydrogen	6.90
	100.00

Or expressed otherwise:

Carbon	42.47
Oxygen and hydrogen in the proportions necessary to make water	57.53
	100.00

Selection 74

An early consequence of the Continental Blockade during the Napoleonic Wars was the search for an alternative to cane sugar. It not only induced the starting of beet-sugar manufacture in many European countries but also fostered new knowledge of sugar chemistry. Among the scientists and technologists who turned their attention to devising a cane sugar substitute was Gottlieb Sigismund Constantin Kirchhof[f] (1764–1833), a German-born pharmacist working in Russia at St Petersburg (later Leningrad).[a] His success in obtaining in 1811 a sugary substance from starch treated with sulphuric acid was followed up by further investigations which shed considerable light on the chemical behaviour of sugars. Kirchhof published his results in journals which were not easily accessible, but his work was repeated by others, as shown in the extract below.

J.C.C. Schrader, *Schweiggers J.*, **4**, 108. 1812

'Ueber die neue von Kirchhof entdeckte Zuckergewinnung' (On the production of sugar newly discovered by Kirchhof)

pp. 108–9

Kirchhof prescribes taking for 100 parts of starch meal 1 part of sulphuric acid and 400 parts of water, and boiling for 36 hours. This is necessary with these proportions; but if one takes more sulphuric acid one can boil for a shorter time and still obtain a sweet product. With 5 and with 10 parts of sulphuric acid I used only 6 or 9 hours boiling. The product obtained was a wine-yellow syrup, equal in weight to the weight of the starch, which had a taste rather like roasted sugar, but still pleasant and certainly not to be compared with the usually unpleasant-tasting sugar syrup. Through careful evaporation and purification with charcoal powder this taste can be avoided and the product obtained whiter. This syrup coagulates after some days to a solid, which shows some crystallization in round granular groups. With complete drying this mass loses a further 17 per cent in weight and can then be ground to powder. The mass, like sugar, dissolves up in alcohol, a little starch remaining behind; but by cooling and evaporating such a solution, made by boiling in 85 per cent alcohol, I could still obtain no crystals. I set up some pounds of this syrup with a little yeast for fermentation, which however proceeded weakly and produced little spirit of wine.[a] Thus by its smaller fermentability and crystalizability this artificial sugar could be distinguished somewhat from solid cane sugar and be more like the mucilaginous sugar[b] although it has not the full sweetness of the latter.[c]

Selection 75

The industrial, agricultural and medical significance of carbohydrates greatly promoted an interest in their general chemistry and in their transformations in plant and animal life. Their study was effectively aided by the development of qualitative and quantitative methods of sugar analysis.

Among the most simple and useful ideas was that of employing yeast as a means for the identification and determination of fermentable sugars. Among the chemists who sought to work out a procedure along this line was the German chemist, Johann Wolfgang Döbereiner (1780–1849), and the extract below gives the news as it reached the British scientific community.

Extract of a Letter from Professor Wurzer, of Marburg, to Professor von Mons, *Phil.Mag.*, [4] *51*, 147. 1818

'Oxalic acid, fermentation, celestine,[a] &'

p.147

New experiments on fermentation have convinced M. Dobereiner that the smallest parcels of sugar concealed in any liquid may be discovered, and their affinity determined, by adding to such liquid some grains of yeast and inclosing the mixture in a vessel sealed with mercury. The fermentation, which at the temperature of from 15 to 20 degrees R. begins to manifest itself, and continues as long as any sugar remains, occasions the disengagement of a quantity of bubbles of carbonic acid gas, from which the quantity of sugar which the yeast has decomposed may easily be calculated. M.D. has found that five grains of sugar, dissolved in a half cubic inch or in a whole cubic inch of water, and put in contact with some grains of yeast, uniformly resolve themselves into 4.7 cubic inches of carbonic acid gas and 2.57 grains of alcohol.

Selection 76

The practical needs of sugar analysis led to the growing use of the polarimeter, 'perhaps the most significant analytical tool to influence biochemistry in the nineteenth century'.[a] The theoretical and practical foundations of polarimetry, resulting from the study of the optical activity of organic compounds, were largely laid by the French physicist, Jean-Baptiste Biot (1774–1862), during the second, third and fourth decades of the nineteenth century. Thus in 1816 when there were still doubts about the sameness of cane sugar and beet sugar Biot confirmed their identity by showing that both substances rotated equally the plane of polarization,[b] and by 1840 he designed an experiment for measuring optical rotation, using a quartz plate between polarizing and analysing mirrors.[c]

An improved form of the polarimeter was that of C. Ventzke (1797–1865) who replaced the mirrors by two Nicol prisms. Between the two prisms (the polarizer and analyzer) and rotating about the same axis there was a glass tube for holding the solution of the substance whose optical rotation was to be measured. The light source, placed in front of the polarizer, was an Argand burner, that is an oil burner having a cylindrical wick and providing a luminous light. The zero point corresponded to a position when the analyzer was set at extinction. The critical position was reached when the illumination of the field became red due to the extinction of the yellow component of the white light. The table of optical rotations of some sugars (see below) showed that fruit sugar was laevorotatory and grape sugar dextrorotatory.

[C.] Ventzke, *J.prakt.Chem.*, *1 (25)*, 65. 1842

'Ueber die verschiedenen Zuckerarten und verwandte Verbindungen, in Beziehung auf ihr optisches Verhalten und dessen praktische Anwendung' (On the different kinds of sugar and cognate compounds, in relation to their optical behaviour and the practical application of the latter)

p. 65

The management of the Schickler sugar factory in Berlin, zealously bent on safeguarding the continual advance of its industrial establishments, through its liberality has placed me in the position to undertake the necessary experiments: I hold myself in duty bound to recognize this with warmest thanks . . .

[The] beautiful discovery of Biot's must be equally welcome to chemists and technicians. For if one has good grounds for believing that bodies showing different polarization are *not* identical, then the chemist will certainly be induced to view more closely substances which till now he has considered as slight modifications of one another; and if he succeeds in characterizing them definitely and placing them separately, then also technical practice can benefit. Because when was there a time that a greater clarity in science did not, sooner or later, exert the most wholesome influence on the operation in factories?

p. 69

. . . the instrument is placed in the zero position. If one looks through the prism E^{1a} towards the flame of the lamp then in the case of the colourless fluid not producing circular-polarization, darkening occurs as for instance with pure water. In the opposite case [the field] appears bright and with the colour of the flame. Now one *turns* the front prism E^1 round *towards the right*. Colours appear in the following order:

Pale blue
dark blue
violet
purple
red
orange,

thus the *polarization* is undoubtedly *towards the right*. If the series of colours is shown on *turning E¹ towards the left*, then it is just as certain that the *circular-polarization* takes place *towards the left*. As soon as the red appears, which lies between the purple and orange, the degrees are read which the indicator gives, reckoned from the zero point.

pp. 72–3

In the following table are sugars, in the proper sense of the word, that is those which directly and indirectly make possible vinous fermentation. [They are] so arranged that those which mostly polarize towards the left lead off, those which give the largest deviation towards the right bring up the rear . . .

Table of the experiments
The length of the layer is in all cases 234 mm

Names of the aqueous solutions of the substances	Degree of polarization, 360 on the circle		Specific weight at 171/2° C	Percent of substances dissolved in liquid
	To left	To right		
No.				
1 Fruit sugar from grapes	35½	—	1,1056	—
2 Fruit sugar from honey	36	—	1,1056	—
3 Fruit sugar by the action of acids on cane sugar	35½	—	1,1056	—
4 Fruit sugar by fermentation of cane sugar	36	—	1,056	—
5 Syrup sugar	0	0	1,105	—
6 Milk sugar	—	43	1,102	25
7 Grape sugar of all kinds	—	46	1,095	25
8 Saccharate of sodium chloride	—	41	1,117	25
9 Cane sugar	—	56	1,1056	25
10 Dextrin sugar	—	92	1,1056	—
11 Dextrin	—	19	1,011	3,36
	...	...	...	...

Selection 77

So far we have concentrated on some pre-1850 developments of sugar analysis and identification, and indicated their connection with economic factors. Now we shall focus our attention on sugars as compounds of importance in the life of plants and animals. We shall be particularly concerned with the momentous work of Claude Bernard (1813–78) on the carbohydrate metabolism in the animal body. Although he was familiar with, and also interested in, diabetes Bernard's concern for the fate of sugar derived only in part from the study of this disease. When Bernard began his investigations he accepted the then generally held idea that plants were concerned with the formation of sugars and animals with their breakdown.[a]

When in the course of his researches he failed to locate the organ responsible for the destruction of the alimentary sugar he changed tack. He decided to look for its source in the animals, and located it in the liver in 1848. He injected known amounts of grape sugar into the right jugular vein and followed its path by taking blood samples from the carotid artery. Methodologically this was a pioneering contribution to the study of pathways of biochemical change.[b] Five years later Bernard elaborated on his discovery of the glycogenic function of the liver[c] and in 1857 he described the isolation of the sugar-forming substance (*matière glycogène*) from liver tissue. Because of its chemical and physiological closeness to the starch of the vegetable kingdom, following Bernard, glycogen became also known as 'animal starch'.

It should be noted that when the French researcher, André Sanson (1826–1902), announced shortly afterwards that he had come across glycogen in muscle, Bernard maintained that it could be found only in liver.[d]

Bernard's discovery that animals, like plants, were capable of synthesizing a starch-like substance brought a new perspective to the chemistry of life. For almost three decades it had been assumed that a chemical boundary divided the two kingdoms of plant and animal life. Bernard's work would eventually help to bring that boundary down.[e] It also reinforced aspects of Bernard's own ideas about the interrelationship between chemical change and physiological activity, ideas which he had apparently already been harbouring while he was engaged on investigations into digestion during the 1840s. He had come to believe that while the chemical change was not 'alive', the physiological activity was vital.[f]

As shown below, Bernard continued to argue on these lines after demonstrating glycogenic synthesis in animals. Formation of starch and glycogen was a vital process, common to plants and animals, only differing in the resulting product. But the transformation of starch and glycogen into sugar in the living organisms was a non-living fermentative, chemical, process. In view of Bernard's concern with establishing and characterizing 'general physiology' as an independent discipline,[g] his use of the term when discussing the newly recognized common feature of the chemistry of plant and animal life is of historical interest.

C. Bernard, *Comp.Rend.*, *44*, 578. 1857

'Sur le mécanisme physiologique de la formation du sucre dans le foie' (On the physiological mechanism of the formation of sugar in the liver)

pp. 582–3

After all the experiments which have been previously reported, it remains clearly established that the liver of dogs nourished exclusively on meat possesses a special property, not shared by any other organ of the body, of creating a sugar-forming substance (*matière glycogène*) quite analogous to vegetable starch and being able like the latter to change subsequently into sugar, passing through the intermediate stage of dextrin.

Without any doubt the study of the sugar-forming substance of the liver must not stop there. It is necessary to know exactly its elementary composition and its constitution; to know if this material changes completely into sugar and if, in this transformation, there are other products which arise and, in a word, to submit to a more profound study the parallelism so apparent, which is offered by the transformation into sugar of this sugar-forming substance of the liver with the transformation into sugar of vegetable starch. The cares of that study fall to the chemists. With regard to the present time, it suffices for me to have proved the existence of this special substance which always precedes the appearance of sugar in the liver [and thus] to have established a fact which is capable of powerfully illuminating the physiological mechanism of the formation of sugar in animals and of furnishing at the same time conclusions which are of interest to the highest degree for general physiology.

With regard to the physiological formation of sugar in animals it must necessarily be envisaged, as I say, not as a phenomenon of direct chemical decomposition of the blood components at the moment of the passage of the blood into the liver, but as a function made up of the succession and concatenation (enchaînement) of two essentially distinct actions.

The first entirely vital action, so called because of its performance does not take place outside the influence of life, consists in the creation of glycogen in the living hepatic tissue.

The second action, entirely chemical and capable of being carried out without vital influence, consists in the transformation of glycogen into sugar by the help of a ferment.

For the appearance of sugar in the liver, the coming together of these two sets of conditions is necessary. It is necessary that glycogen can be created by the vital activity of the organ; then that this substance may be put in contact with the ferment which ought to transform it into sugar.

Glycogen is formed like all the products of organic creation owing to the phenomena of slow circulation, which accompany the acts of nutrition. With regard to deciding whether, amongst the numerous blood vessels with which the liver is provided, there are some more specially charged with this nutritive circulation, whilst others would be more specially connected with the phenomena of chemical transformation of the sugar-forming substance – this is a physiological question which we have not taken up here for the moment.

As to the conclusions that we can actually deduce, from the point of view of general physiology, from the mechanism that we have indicated for the formation of sugar in the liver, it is impossible not to be struck by the similarity which exists in this connection between the glycogenic function of the liver and the production of sugar in certain acts of the vegetable organism. In a grain, for example, which produces sugar during germination, we have equally to consider two quite distinct series of phenomena: one primary [primitif], entirely vital, consists in the formation of starch under the influence of the vegetable life; the other consecutive, entirely chemical, able to take place outside the vital influence of the vegetable, is the transformation of starch into dextrin and into sugar by the action of the diastase. When a liver separated from the living animal continues still for a certain time to produce sugar, according to all the evidence the vital phenomenon of creation or of secretion of the sugar-forming substance has ceased; but the chemical phenomenon continues if the conditions of humidity and warmth necessary for its accomplishment are realized. In the same way in the grain separated from the plant, the vital phenomenon of secretion of starch has ceased with the vegetable life; but under the influence of favourable physicochemical conditions its transformation into dextrin and sugar, with the help of the diastase, can operate. Lastly, it is easy to see by these parallel observations that the formation of sugar in the liver of animals passes through three series of successive transformations quite analogous to that of the formation of starch, dextrin and sugar in the vegetable grain.

In summary, according to all the facts contained in this work, we can conclude that the question of the formation of sugar in animals has made important progress following on the isolation of the sugar-forming substance which constantly in the liver tissue pre-exists the sugar.

b. From *c*. 1880 to *c*. 1940: configuration. Anaerobic and aerobic breakdown, and phosphorylation

Selection 78

By 1860 carbohydrates occurring in nature were known to exist either as complex or simple sugars. Whereas the latter comprised only four recognized compounds – grape sugar (dextrose), fruit sugar (laevulose), galactose and sorbose – the former contained a number of them. This group included substances like starch, dextrins, cellulose and cane sugar but also milk sugar (lactose), malt sugar (maltose) and glycogen. Although the knowledge of the physical and chemical properties of carbohydrates (optical rotation, reducing reactions) underwent further extension during the next two decades the structure of even the simple sugars remained uncertain. They exhibited aldehydic or ketonic properties, they possessed hydroxyl groups and they were represented by the formula $C_6H_{12}O_6$. This was the situation in 1884 when Emil Fischer (1852–1919) began to study the behaviour of these substances towards phenylhydrazine, a compound he discovered in 1875.[a]

As shown below, in his first paper on the subject Fischer already suggested the usefulness of the aromatic base in identifying sugars. Fischer's discovery that in addition to hydrazones, phenylhydrazine yielded with sugars difficultly soluble but well crystalline yellow compounds which he called osazones[b] proved to be of great importance in the investigation of the configuration of sugars.[c]

E. Fischer, *Ber.chem.Ges.*, *17*, 579. 1884

'Verbindungen des Phenylhydrazins mit den Zuckerarten, I' (Compounds of phenylhydrazine with varieties of sugar, I)

p. 579

Up to the present time little is known of the nitrogen-containing derivatives of the sugars. Apart from glucosamine, whose connections with glucose are still not established with certainty, and from the complicated amorphous products which according to H. Schiff[1] are produced from glucose by means of ammonia and aniline, there remain only two rather better characterized compounds which Sachsse[2] obtained from milk sugar and aniline. More interesting is the behaviour of these bodies with phenylhydrazine. It appears that the base binds with all kinds of sugar which, like the aldehydes or ketonic alcohols, reduce alkaline copper solution. Those tested were dextrose, laevulose, galactose, cane sugar, milk sugar, sorbose and maltose which all yield hydrazine derivatives, whilst inositol and trehalose under the same conditions are unaffected by the base. The hydrazine compounds concerned are difficultly soluble in water and on this account easy to isolate. One will be able to use them in many cases for the recognition and separation of the individual kinds of sugar. The formation of the products proceeds in watery solution, but only at higher temperatures, best with heating on the water bath. The hydrazine is used as the hydrochloric acid salt with an excess of sodium acetate.

Selection 79

Continuing his researches on sugars, Emil Fischer turned his attention to glucosides and in 1893 reported a method for their synthesis based on the methylation of the hydroxyl groups of the alcohols and sugars. At this time, as shown below, he tentatively suggested that the glucosides had a formula containing a 5-atom ring.[a]

Among the very important consequences of the new formula was the recognition of the existence of two glucosidic stereoisomers. Fischer found, using the isomeric glucosides he had prepared, different behaviour with the enzymes invertin and emulsin – α-methylglucoside being hydrolyzed by invertin but not by emulsin while the β-methylglucoside behaved in the opposite way. A reference to this has already been made in connection with Fischer's contribution to the understanding of the nature of the combination of enzyme with substrate (1894).[b]

E. Fischer, *Ber.chem.Ges, 26*, 2400. 1893

'Ueber die Glucoside der Alkohole' (On the glucosides of the alcohols)

p. 2400

I have now in hydrochloric acid found a means to combine the kinds of sugar with the alcohols directly to products of glucoside nature. If one leads, with cooling, hydrochloric acid gas into a solution of grape sugar in methyl alcohol up to saturation then after a short time the mixture loses the power to reduce Fehling's solution, and then contains a beautifully crystalline product $C_6H_{11}O_6CH_3$, which thus originates from equal molecules of sugar and alcohol according to the equation

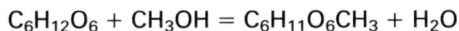

$$C_6H_{12}O_6 + CH_3OH = C_6H_{11}O_6CH_3 + H_2O$$

This reaction appears to be valid for all alcohols . . .

p. 2401

As for the alcohols the reaction is also in general valid for the glucoses . . .

Considering their constitution it seems to me suitable to use the general name glucosides for the new compounds just as Michael[a] has proposed for his synthesized phenol derivatives, and to distinguish the single products by addition of the alcohol, e.g. glycerolglucoside, lactic acid glucoside.

As however words like methylalcoholglucoside are too cumbersome, it may suffice here only to use the radicle in the name. I shall thus the compound of grapesugar with methylalcohol shortly name methylglucoside . . .

p. 2402

The knowledge of the new alcoholglucosides is of decisive meaning for the much discussed question, in what way are the glucosides and the more complicated carbohydrates constituted. As the behaviour with phenylhydrazine certainly shows, these compounds no longer contain the aldehyde group of the sugar. The latter must thus be fixed by the entering alkyl in a manner similar to that in the acetals. But as here the alteration is brought about by means of only one molecule of alcohol with liberation of water, in my opinion this is only to be explained on the assumption that one alcohol group of the sugar molecule itself takes part in the process through formation of an intramolecular ether group. As some characteristic reactions of the sugar, like the sensitivity towards alkalies and alkaline oxidizing agents or the transformation into osazones, are conditioned by the atomic group CO.CHOH, one could guess that the hydroxyl contained in them also takes part in glucoside formation. In order to test this assumption, I have investigated two easily obtainable ketoses which contain only that alcohol group, benzoylcarbinol[b] and benzoin.[c]

Whilst the first with alcoholic or aqueous hydrochloric acid undergoes a complex alteration, which for the question before us is without interest, the benzoin is very easily alkylated. However the products arising behave still as ketones and are quite differently constituted from the glucosides.

The atomic group CO.CHOH according to this appears not to suffice for the glucoside formation.

If however another hydroxyl of the sugar takes part, it is in all probability that which is found in the γ position to the aldehyde or ketone group. On this account I believe that methylglucoside must be given the structural formula

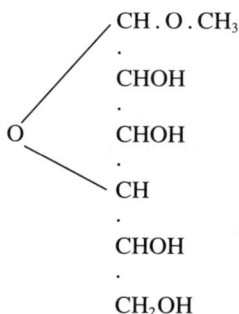

$$
\begin{array}{c}
\diagup \text{CH.O.CH}_3 \\
\cdot \\
\text{CHOH} \\
\cdot \\
\text{O} \qquad \text{CHOH} \\
\cdot \\
\diagdown \text{CH} \\
\cdot \\
\text{CHOH} \\
\cdot \\
\text{CH}_2\text{OH}
\end{array}
$$

p. 2404

The new glucoside formula, which can also obviously be carried over to the derivatives of the phenols, has a consequence truly worthy of consideration. It discloses the existence of two steroisomers, which arise from the same sugar; for through the glucoside formation itself the carbon atom of the original aldehyde group becomes asymmetric. Whether with the synthesis discussed such isomers arise simultaneously, I still cannot say with certainty, but I hold it for very probable . . .

A similar structure to that of the alcoholglucosides is also probably possessed by the more complicated carbohydrates.

Selection 80

Among the important consequences of the discovery of cell-free fermentation (1897)[a] was the developing knowledge of anaerobic breakdown of carbohydrates, linked with the gradual realization of parallel chemical reactions in yeast fermentation and muscle glycolysis. Undoubtedly, part of the story was the influence of the evolutionary outlook (derived from Darwin's work) reflected in the search for a chemical uniformity bearing on the variety of forms of life. For example, the contributions of the Czech worker, Julius Stoklasa (1857–1936), after the turn of the century, bear witness to his deliberate efforts in this area. What he sought was to demonstrate in cell-free extracts of higher plant and animal tissues the presence of zymase (or zymase-like enzymes) capable of producing alcoholic fermentation.[b]

Historically of greatest significance for the understanding of carbohydrate metabolism was the observation (1906) of the involvement of phosphate in alcoholic fermentation by Arthur Harden (1865–1940) and William John Young (1878–1942). Then their attention was drawn towards phosphates when they established that alcoholic fermentation depended on inorganic phosphate and on a fermenting complex contained in yeast juice.[c]

Harden and Young were not the first to observe the positive part phosphates played in fermentation. At the turn of the century, undertaking the isolation of the enzyme invertase and re-examining the work of Buchner, the Polish researcher A. Wróblewski also studied the influence of salts on zymase activity.[d] He established that the effect depended in general on the amount of the added neutral salts. Quantities of the order of 1 per cent as a rule weakened fermentation, amounts below this figure increased it. As for phosphates, Wróblewski found that they caused a significant speeding up of fermentation when they were added to yeast juice simultaneously with acids or alkalis. He assumed that the juice contained a relatively large amount of amphoteric phosphates and that their removal would affect fermentation in a negative way. Although not specifically employing the term, Wróblewski clearly recognized the rectifying role of phosphates in cell life. At the same time, the concept of 'buffer' (*tampon*) was being advanced by the French brewing scientists A. Fernbach (1860–1939) and L. Hubert.[e]

Whether Wróblewski's account of the regulatory operation of phosphates in living cells should be regarded as an independent contribution is difficult to say. Equally it is not easy to decide whether Harden and Young were completely ignorant of Wróblewski's findings.[f] What cannot be doubted was the latter's unconcern with the mechanism of the decomposition of glucose and, as shown below, the former's associating fermentation, in 1908, with phosphorylation of a cyclic nature and suggesting the equations for the chemical changes involved.

A. Harden and W.J. Young, *Proc.Roy.Soc*, *(B)*, *80*, 299. 1908

'The alcoholic ferment of yeast-juice. Part III. The function of phosphates in the fermentation of glucose by yeast-juice'

pp. 299–301

In a previous communication the authors have shown[1] that when a soluble phosphate is added to a fermenting mixture of glucose and yeast-juice the following phenomena are to be observed: (1) The rate of fermentation is at once greatly increased. (2) This acceleration lasts for a short time and the rate then falls off, and returns approximately to its orginal value. (3) During this period the extra amount of carbon dioxide evolved and alcohol produced are equivalent to the phosphate added. (4) The phosphate is converted into a form which is not precipitable by magnesia-mixture, and is then probably present as a salt of a hexosephosphoric acid.

(1) *Effect of the Addition of Phosphate on the Total Fermentation*

The addition of phosphate, however, does not simply produce this initial decomposition of an equivalent of glucose, but also, as a rule, a greater total fermentation, after allowance has been made for the amount decomposed during the initial period.

This is clearly shown by the results embodied in the following table.

In each experiment two or more portions of 25 c.c. of yeast-juice were taken, a solution of glucose alone, or one of glucose and phosphate, added, and the total volume made to 50 c.c. The solution of phosphate employed had a concentration of about 0.3 molar, and the concentration of glucose was 20 grammes per 100 c.c. in experiment 8, and 10 grammes per 100 c.c. in all the others. The fermentation was carried out at 25° in presence of toluene until the evolution of gas ceased. The numbers in the last column show the increase in the total fermentation produced during the period subsequent to the initial acceleration.

It will be seen that the increase which occurs after the initial period varies from about 10 per cent. of the original fermentation to as much as 150 per cent.

(2) *Recurrence of Phosphate*

The reason for this increase in the amount of sugar decomposed in the long period following the short initial period of acceleration appears to be that the phosphorus compound first formed, which is a hexosephosphate of the formula $C_6H_{10}O_4(PO_4R_2)_2$,[2] is slowly hydrolysed, probably by an enzyme, with the production of a phosphate and a hexose. The phosphate is thus slowly regenerated and then again undergoes the reaction, causing an increased fermentation in the same manner as when it was originally added.

Experiment	CO_2 evolved without phosphate	Cubic centimetres of phosphate solution added	CO_2 evolved in presence of phosphate	Increase due to phosphate	CO_2 equivalent to phosphate	Increase after initial period
	grammes	c.c.	grammes	gramme	gramme	gramme
1	0.484	10	0.717	0.233	0.132	0.101
2	1.280	5	1.584	0.304	0.066	0.238
3^a	0.422	10	0.634	0.212	0.132	0.080
3^b	0.422	20	0.748	0.326	0.264	0.062
4	0.440	10	1.258	0.818	0.132	0.686
5	0.405	5	0.515	0.110	0.070	0.040
6	0.603	5	0.735	0.132	0.066	0.066
7	0.438	5	0.593	0.155	0.057	0.098
8	1.016	15	1.632	0.616	0.198	0.418
	c.c.		c.c.	c.c.	c.c.	c.c.
9	369	10	629	260	63	197
10	337	10	569	232	56	176

This recurrence of phosphate is clearly shown by the following experiment. A known amount of phosphate was added to yeast-juice containing glucose, and the mixture incubated at 25° with toluene. At the close of the initial period a sample was removed, boiled and filtered, and the free and total phosphate present in it estimated, and this process was repeated at stated times. The results obtained are given below, the amounts of phosphate being expressed in grammes of $Mg_2P_2O_7$ per 10 c.c.

Experiment 11 – 215 c.c. of yeast-juice +20 grammes of glucose were made to 375 c.c. with a solution of potassium phosphate, the amount of the latter being equivalent to 0.133 gramme $Mg_2P_2O_7$ per 10 c.c. of the resulting liquid.

The slight increase in the total phosphate present is due to a corresponding degree of evaporation during the experiment. It will be seen that the free phosphate per 10 c.c. gradually increases from 0.021 to 0.226, so that 0.205 gramme is regenerated. Since the total phosphorus, expressed as phosphate in the original juice, was 0.266 − 0.133 = 0.133 gramme, it follows that at least 0.205 − 0.133 = 0.072 gramme of this has been derived from the hexose-phosphate produced during the initial period from the added phosphate.

Time in hours	Free phosphate as $Mg_2P_2O_7$ per 10 c.c.	Total phosphate as $Mg_2P_2O_7$ per 10 c.c.
	gramme	gramme
5.5	0.021	0.266
18.0	0.093	0.269
66.0	0.133	
138.0	0.175	
426.0	0.226	0.273

pp. 302–3

(3)   *Nature of the Chemical Change which occurs in the Fermentation of Glucose by Yeast-juice*

The cycle of changes which is undergone by a phosphate in the presence of yeast-juice and glucose appears from the foregoing to be as follows:

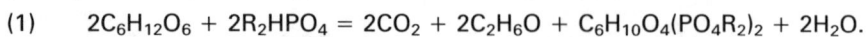

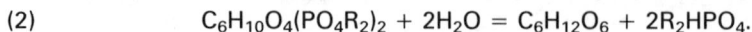

(1) $2C_6H_{12}O_6 + 2R_2HPO_4 = 2CO_2 + 2C_2H_6O + C_6H_{10}O_4(PO_4R_2)_2 + 2H_2O.$

(2) $C_6H_{10}O_4(PO_4R_2)_2 + 2H_2O = C_6H_{12}O_6 + 2R_2HPO_4.$

The first of these equations does not include the fermenting complex, without which, however, the change does not occur, and it is probable that both the glucose and the phosphate form an intermediate association with this complex, which then breaks down, giving rise to the substances on the right-hand side of the equation, and at the same time regenerating the fermenting complex.

Since free phosphate and a hexosephosphate are invariably present in the yeast-juice prepared by grinding yeast, it follows that at all events some portion of the fermentation is always due to the foregoing reactions. During the initial period of rapid fermentation, as long as free phosphate is still present, the greater part of the change is certainly due to this reaction. Whilst in the succeeding period of slower fermentation the constant production of free phosphate by the enzymatic hydrolysis of the hexosephosphate already formed, or by the action of proteoclastic enzymes on phosphoproteins, renders it equally certain that some portion of this greatly diminished fermentation must also be ascribed to the same reaction.

pp. 306–7

(4) *Influence of Concentration of Phosphate on the Course of the Fermentation*

When a phosphate is added to a fermenting mixture of glucose and yeast-juice, the effect varies both with the concentration of the phosphate and with the particular specimen of yeast-juice employed. With low concentrations of phosphate the acceleration produced is so transient that no accurate measurements of rate can be made. As soon as the amount of phosphate added is sufficiently large, it is found that the rate of evolution of carbon dioxide suddenly increases from 5 to 10 times, and then rapidly falls approximately to its original value.

As the concentration of phosphate is still further increased, it is first observed that the maximum velocity, which is still attained immediately on the addition of the phosphate, is maintained for a certain period before the fall commences, and then, as the increase in concentration of phosphate proceeds, that the maximum is only gradually attained after the addition, the period required for this increasing with the concentration of the phosphate. Moreover, with these higher concentrations the maximum rate attained is less than that reached with lower concentrations, and, further, the rate falls off more slowly. The concentration of phosphate which produces the highest rate, which may be termed the optimum concentration, varies very considerably with different specimens of yeast-juice.

Selection 81

Let us turn now to muscle glycolysis. Although the occurrence of lactic acid in muscle had been postulated as early as 1808[a] a considerable degree of uncertainty existed about its role in the activity of muscle. Beyond that it was recognized that lactic acid metabolism seemed to play a fundamental part not only in the biochemical transformations of carbohydrates, but also in the metabolic interrelationships of the latter with proteins and fats. This had indubitably been acknowledged by Walter Morley Fletcher (1873–1933) and Sir Frederick Gowland Hopkins (1861–1947) when they published, in 1907, the first reliable quantitative data on lactic acid in muscle, extracted by the employment of ice-cold ethanol, and given as percentages of anhydrous zinc lactate.

Their research was a follow-up of Fletcher's earlier work at the turn of the century on carbon dioxide production in excised surviving muscle which showed that the output of carbon dioxide could not be reconciled with the inogen theory of muscular activity propounded by Hermann. That is, that contraction is independent of oxygen supply.[b]

They wrote on the formation of lactic acid on anaerobic contraction of muscle and its removal on recovery in oxygen. Only lactic acid estimations were made, but the constancy of the *rigor mortis* lactic acid after several periods of stimulation and recovery suggested to them that the lactic acid might be restored in some way to a precursor state.

As mentioned, Fletcher and Hopkins generally shared the opinion regarding the pivotal role of lactic acid in the intermediate metabolism of living (animal) tissue. But this did not mean, as shown in the concluding remarks (not included in the excerpts), that they were prepared to apply the then widely held lactic acid theory of alcoholic fermentation to sugar breakdown in muscle. On this view, strongly promoted by J. Stoklasa and E. Buchner among others, lactic acid was an intermediate between sugar and alcohol in the case of muscle, a belief which Fletcher and Hopkins were unable to confirm.[c]

W.M. Fletcher and F.G. Hopkins, *J.Physiol.*, *35*, 247. 1907

'Lactic acid in amphibian muscle'.[1]

pp. 247–8

The observations to be described in this paper were undertaken in the hope of determining whether within a muscle itself means exist for an oxidative control of its own acid formation, or for the alteration or destruction of acid which has been formed, either there or by muscular activity elsewhere in the body. With this in view we examined in the first place the effect of an abundant supply of oxygen upon the development of lactic acid in a surviving excised muscle, and upon the stability of the acid within the muscle after its formation. For it has long been known that a surrounding atmosphere of oxygen has the effect of preserving the irritability of a resting excised muscle[2] and of delaying the course of fatigue when contractions have been performed,[3] and it has been shown that the oxygen atmosphere may indefinitely delay the onset of rigor mortis in a resting or fatigued muscle[4] Further we know that muscle entering the state of rigor as a result of fatigue, may actually be recalled to a flaccid resting condition by immersion in oxygen gas.[5]

Since an acid reaction of the muscle is, as most agree, a constant mark of the fatigued condition and a constant condition of the state of rigor, it is at once suggested that oxygen, when easily available, either restrains by some guidance of chemical events the yield of acid within the muscle, or is able to remove it after its production. And although the direct removal of lactic acid in the presence of oxygen by combustion or otherwise, does not take place under simple chemical conditions out of the body, yet an enquiry into the possibility of its occurrence within a muscle appeared very advisable in view of the facts that the resting survival yield of CO_2 by an excised muscle is at its minimum under anærobic conditions and is greatly increased in the presence of oxygen, and that the special yield of CO_2 due to contractions of the muscle is similarly increased in oxygen and may indeed be absent altogether in an oxygen-free atmosphere.[6]

Our earliest experiments gave decisive evidence that lactic acid within an excised surviving muscle is actually diminished in amount, or even wholly eliminated, after exposure of the muscle to abundant free oxygen.

A study of this removal of acid can only be based upon a knowledge of the rate of survival acid-production both in resting and in contracting muscle, under anærobic and ærobic conditions; and before dealing with the question of oxidative removal it will be necessary in the first place to give an account of the main facts of this acid production. For it is notorious that, quite apart from the question of oxidative removal of lactic acid – which has not previously, we think, been examined – there is hardly any important fact concerning the lactic acid formation in muscle which, advanced by one observer has not been contradicted by some other.

p. 249

We believe that the present confusion is not in chief, if at all, a result of the technical difficulties of lactic acid estimation, but that it is due to the difficulties inherent in the extractive treatment of an irritable muscle. For it is clear that in such a case no treatment for the extraction from muscle can be accepted which, acting itself as a stimulus, has among its effects an increase of the acid to be estimated . . .

As solvents of lactic acid, water and alcohol have been used for extraction, and in both cases the muscle must be reduced to small pieces, by cutting or grinding, in order to ensure complete extraction. Now it will be shown that chopping an irritable muscle is, as might be expected, an acid-producing stimulus; and it is obviously fallacious to consider that the extract of a muscle after such a treatment represents its previous condition. This applies in all cases where water is used for extraction; for water, whose virtue in this connection is that it is not itself a stimulant, by tolerating the maintenance of muscle irritability, allows time to pass after the chopping, during which, as we shall show, there occurs great augmentation of the acid yield.

p. 250

We hope to show that in our experiments we have avoided these dangers of alcohol stimulation – the magnitude of whose effects we never suspected before trial – by using ice-cold alcohol, which has no appreciable stimulant action, while it retains its killing and coagulative influence. Immersion in alcohol has been followed by immediate rapid and thorough grinding (with sand) of the muscle ice-cold, in ice-cold alcohol.

pp. 291–2

We have consistently found that the maximum lactic acid yield reached in heat-rigor by a given excised muscle is constant, not only for resting muscle and for fatigued, but is also constant whether, to a less or greater extent, the lactic acid yield has been diminished by exposure to oxygen. The disappearance of lactic acid, as such, from the muscle substance under the influence of oxygen, does not cause any reduction of the lactic acid maximum subsequently reached in heat-rigor.

p. 293

To put beyond a doubt the reality of the continued loss of lactic acid throughout long periods of activity in the presence of oxygen, and the constancy of the ultimate heat-rigor maximum, notwithstanding the successive losses, the process has been followed by successive estimations, during alternate intervals of production and loss. The results are expressed graphically in Figure 10.
    Four estimations of lactic acid due to heat-rigor are shown, two at the beginning, in the case of resting muscles, two at the 53rd hour, in the case of inexcitable muscles, which had gone through nine periods of severe stimulation alternated with periods of rest in an oxygen atmosphere. The enclosed areas represent time periods (drawn proportionate to abscissæ) of stimulation by strong interrupted shocks. × loss of excitability. Temperature 15°C. Continuous line shows course of acid loss as actually determined by estimation. Dotted line shows the presumed course of acid loss and gain during other alternate periods.

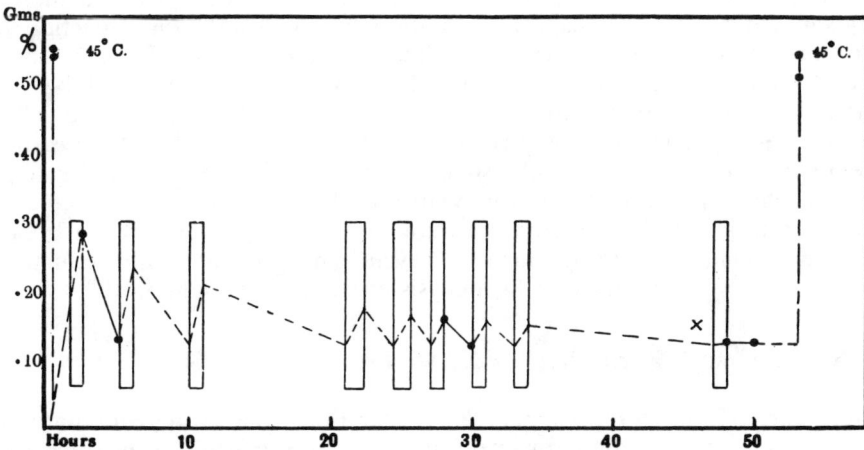

Figure 10 The relation of the heat-rigour lactic acid 'maximum' to the survival history of muscle

pp. 295–6

The chemical relationships of lactic acid, and the results of recent research, indicate that it is a substance which, in some sort, links proteid products with carbohydrates, and carbohydrates with fats. It is at least very possible that it is an intermediary substance in the processes of transformation of storage material in the body.[7] If such processes occur in metabolism, there is no reason why their progress should not come to light in the study of individual organs, and it is by no means fanciful to suppose that the appearance of lactic acid in muscle indicates a stage in the conversion of less readily available sources of energy into others which are more immediately available. Such processes of change would involve not only a breakdown which might continue anærobically, but also a subsequent synthesis, for which normal physiological conditions would probably be necessary. Without, however, venturing to insist upon any such details in the suggestion, we may at least indicate that, if during its disappearance in the oxygenative recovery of an excised muscle from fatigue, lactic acid does not undergo final oxidation, but shares rather in some constructive or reconstructive process, the circumstance would offer a ready explanation of the restoration of power to yield the normal acid maximum of the originally unfatigued muscle.

Selection 82

Concurrently with contributions noting the effect of phosphates, other work developed which ignored this influence or considered it to be of marginal importance in relation to anaerobic breakdown of carbohydrates. Thus Carl Neuberg (1877–1959), working in Berlin with several collaborators, believed that in sugar cleavage three-carbon compounds primarily participated and, therefore, research was to concentrate on the clarification of the nature and the sequence of reactions of these compounds which lead to the production of carbon dioxide and alcohol.

Among the possible intermediates great attention was given to methyl glyoxal and to its obvious transformation by the Cannizzaro reaction to the corresponding alcohol and acid, that is pyruvic acid. Then Neuberg and A. Hildesheimer observed in 1911 that pyruvic acid could be fermented by yeast and that this was a process catalyzed by an enzyme called 'carboxylase'.[a] The cleavage hypothesis depending on the three-carbon metabolites appeared to be plausible (despite the fact that methyl glyoxal was not fermentable), as shown below in the Neuberg-Kerb scheme of reactions in alcoholic fermentation. Notably, this hypothesis of the breakdown of sugar comprised stages of cyclic change and oxido-reduction, including enzyme catalysis, in its analysis of steps of fermentation.[b]

C. Neuberg and J. Kerb, *Biochem.Z.*, 58, 158. 1914

'Über zuckerfreie Hefegärungen. XIII. Zur Frage der Aldehydbildung bei der Gärung von Hexosen sowie bei der sog. Selbstgärung' (On sugar-free yeast fermentation. XIII. On the question of aldehyde formation in fermentation of hexoses as well as in so-called self-fermentation)

p. 158

It has long been known that with alcoholic fermentation small quantities of *acetaldehyde* appear. Opinions on the origin of these small quantities of aldehyde have changed in course of time.

Earlier a secondary formation through slight oxidation of the alcohol present was generally assumed. That such a manner of origin is in fact present appears from the work of Trillat[1] . . .

Only the finding of *pyruvic acid fermentation* authorizes the assumption of a direct origin of aldehyde during yeast fermentation and the possibility is discussed that such a partial step in fermentation could be a source of acetaldehyde.

Neuberg and Kerb[2] thus found acetaldehyde, as a result of theoretical studies, with self-fermentation or autolysis of yeast and yeast preparations.

p. 160

In order to make as few assumptions as possible, and to prejudice nothing concerning the *direct* fermentability of the trioses, one can consider in a preliminary way the following scheme.[3]

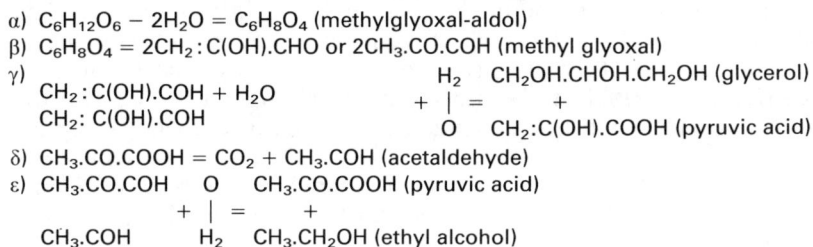

α) $C_6H_{12}O_6 - 2H_2O = C_6H_8O_4$ (methylglyoxal-aldol)

β) $C_6H_8O_4 = 2CH_2:C(OH).CHO$ or $2CH_3.CO.COH$ (methyl glyoxal)

γ)

$$CH_2:C(OH).COH + H_2O \quad\quad \begin{matrix} H_2 \\ + \\ O \end{matrix} \quad \begin{matrix} CH_2OH.CHOH.CH_2OH \text{ (glycerol)} \\ + \\ CH_2:C(OH).COOH \text{ (pyruvic acid)} \end{matrix}$$

CH_2:C(OH).COH

δ) $CH_3.CO.COOH = CO_2 + CH_3.COH$ (acetaldehyde)

ε) $CH_3.CO.COH \quad O \quad CH_3.CO.COOH$ (pyruvic acid)

$$+ \quad | \quad = \quad +$$

$CH_3.COH \quad\quad H_2 \quad CH_3.CH_2OH$ (ethyl alcohol)

One can let this formulation begin with the specially reactive modification (enol form?) of the sugar which perhaps arises with the breakdown of the known hexosediphosphoric acid ester. It can also start from the ordinary glucose and comes out finally to a repeated Cannizzaro reaction of the methyl glyoxal.[4] With the first reaction pyruvic acid appears and, with addition of water, *glycerol*. Then after the pyruvic acid is split by *carboxylase* into *carbon dioxide* and *acetaldehyde*, the Cannizzaro reaction takes place between two different aldehydes, acetaldehyde and methylglyoxal; in this way *alcohol* and *pyruvic acid* are formed. From the latter carboxylase produces always new carbon dioxide and acetaldehyde and so on. The *pyruvic acid* is always forming and constantly breaking down. It can never pile up, as little as can the *methyl glyoxal*. On the other hand it is clear that finally some *acetaldehyde* must remain over when all the methyl glyoxal is used up. According to this conception small amounts of acetaldehyde and glycerol would be *necessary* side-products.

Selection 83

Gustav Embden (1874–1933) was among those in the period who made important contributions in their researches on carbohydrate metabolism in the animal body by concentrating on lactic acid as the principal product in the degradation pathway. Embden worked with his pupils in Frankfurt-on-Main. Interested at first in the carbohydrate metabolism of the liver (a subject at the centre of attention since Bernard's discovery of its glycogenic function), Embden also investigated, by means of perfusion,[a] features of lactic acid formation in this organ. Familiar with the groundwork of Fletcher and Hopkins on lactic acid in amphibian muscle and influenced by Buchner's demonstration of cell-free fermentation, Embden wanted to investigate the possibility of cell-free lactic acid formation. By preparing muscle press-juice devoid of living cells he initiated developments which reinforced unitary ideas regarding knowledge of anaerobic metabolism of carbohydrates in yeast and in muscle.

In the first paper on the subject in 1912 Embden and his co-workers showed that in muscle press-juice from dogs lactic acid formation took place under certain conditions.[b] However, they could find no relationship between this lactic acid and glucose or glycogen (or inosite or d-l-alanine). On the assumption of rapid production of lactic acid during contraction they postulated a precursor of lactic acid in muscle, the so-called 'lactacidogen'. They assumed it to be a compound neutral or less acid in character than lactic acid itself. They also suggested that in some way it could be related to 'phosphocarnic acid', another supposedly energy-rich body reported in 1893 by M. Siegfried (1864–1920) to be present in muscle.[c] As it turned out, the existence of phosphocarnic acid and lactacidogen proved to be deceptive and both met the fate of the rejected inogen. Nevertheless, considerations of the way in which these illusory energy-rich compounds could decompose helped to reveal important aspects of the anaerobic breakdown of carbohydrates.

Thus influenced by early observations on phosphate excretion after strenuous muscle work (and also by phosphate participation in yeast fermentation), Embden and his colleagues two years later discovered the importance of phosphate[d] in lactic acid formation. In their view lactacidogen appeared to be a phosphorylated carbohydrate (see excerpt below).

In 1921 Meyerhof,[e] using chopped muscle, became convinced of the part played by phosphate in lactic acid formation. When the minced muscle was suspended in phosphate solution there was no falling off of lactic acid production as there is without the buffer; no other buffer tried had this effect.

G. Embden, E. Griesbach and E. Schmitz, *Z.physiol.Chem.*, *93*, 1. 1914

'Über Milchsäurebildung und Phosphorsäurebildung im Muskelpresssaft' (On lactic acid formation and phosphoric acid formation in muscle press-juice)

p. 3

In an investigation already mentioned we have referred to the possibility of a connection between lactacidogen and Siegfried's phosphocarnic acid. If as we wish to mention immediately here this guess is certainly not correct, yet it was the inducement for us to investigate whether perhaps the lactic acid formation in muscle press-juice is accompanied by phosphoric acid formation.

Now it has repeatedly been maintained in older and more recent times that strenuous muscle work leads to increased phosphoric acid excretion in the urine, and that also in the muscle *itself* phosphoric acid is set free.

Increased phosphoric acid excretion in the urine after strenuous work was possibly first observed by E.J. Engelmann.[1]

p. 30

As the total result of the experiments collected . . . we can accordingly indicate that, under the experimental conditions described with standing of the muscle press-juice from dog muscle for short periods, in the great majority of cases equimolecular amounts of lactic acid and phosphoric acid were formed.

p. 42

However we rather believe that by the observation that muscle press-juice under certain conditions forms equimolecular amounts of lactic acid and phosphoric acid, taken together with the fact that hexose phosphoric acid as the only one of all the substances investigated increases the extent of this lactic acid and phosphoric acid formation, it becomes extremely probable that also the lactocidogen is to be regarded as a carbohydrate phosphoric acid, or indeed contains a carbohydrate phosphoric acid complex in its molecule.

Selection 84

Of all the active tissues of the organism, muscle is the only one in which we can readily compare the chemical changes going on with the work done or the energy set free as heat. These possibilities were realized by Otto Meyerhof (1884–1951) who in 1919 and 1920 took up the question of the actual chemical happenings at different stages of an experiment such as that of Fletcher and Hopkins above, advancing the following points: that the respiratory quotient is unity so that lactic acid is actually burnt; that the oxygen uptake is one-third to one-quarter of that needed for complete oxidation of the lactic acid removed; that in the recovery period carbohydrate is indeed synthesized, in amount equivalent to the lactic acid not burnt; that the heat production in the recovery period is less than to be expected for the amount of lactic acid burnt – by an amount equivalent to the heat production during anaerobic contraction. In his first experiments on lactic acid oxidation (in the first paper quoted below) he believed that all the lactic acid was burnt; but as stated in his next paper he found he was mistaken in this.

The description quoted from the third paper shows the great care with which the experimental conditions were arranged for experiments in which lactic acid and carbohydrate changes, as well as oxygen consumption, were to be measured on one limb pair as in those quoted above.

O. Meyerhof, *Pflügers Arch.*, *175*, 88. 1919

'Zur Verbrennung der Milchsäure in der Erholungsperiode des Muskels' (On the combustion of lactic acid in the recovery period of the muscle)

p. 90

I have now made certain by a few direct experiments *that during the recovery period the respiratory quotient, that is the relation of the formed carbonic acid (not perhaps only that accidentally released) to the oxygen used, is exactly 1.0.* Thus not only is as much oxygen taken up, but also as much carbonic acid *produced*, as burning of the vanished lactic acid demands. This is a further support for the assumption *that indeed the lactic acid is completely burned in the recovery period.* It follows then with probability that at the same time, with strong binding of heat, a work of 'physical' restitution is performed the nature of which is completely unknown.

O. Meyerhof, *Pflügers Arch.*, *182*, 284. 1920

'Über die Energieumwandlungen im Muskel. II. Das Schicksal der Milchsäure in der Erholungsperiode des Muskels' (On the energy changes in muscle. II. The fate of the lactic acid in the recovery period of the muscle)

pp. 292–3

Where a lactic acid estimation was made after the course of the recovery period, a lactic acid content was obtained which was exactly equal to the resting value: 0.01–0.02% (only in one experiment where there was an irregularity in the estimation, and which is therefore without meaning, 0.04%). This result was after complete recovery so constant that I repeatedly no longer estimated the end value but, taking it from the other experiments, introduced it into the calculation . . . *Thus the lactic acid accumulated during stimulation completely vanished and at the same time an extra amount of oxygen is used, which in spite of all variations in the individual experiments, is always one third to one quarter as great as is needed for the burning of the lactic acid, that is 3 moles of O_2(96 g) to 3–4 moles of lactic acid (270–360 g).*

p. 312
Addendum while in press:Transformation of lactic acid back into glycogen during the recovery period.

Since the conclusion of the work the comparison (that was taken up) of the carbohydrate metabolism with formation and disappearance of lactic acid gave complete confirmation of the conclusions drawn in both pieces of work from the experimental data. Namely in the muscle were always estimated the glycogen as well as the total amount of the remaining carbohydrate which could be split to reducing sugars (all as glucose according to Bertrand).

1. When in the intact muscle lactic acid appears then the carbohydrate diminishes by exactly the same amount, and indeed as well on stimulation as in rest anaerobiosis. The alteration concerns quite predominantly the glycogen

content. The assumption that the lactic acid in rest anaerobiosis arises from the same source as with stimulation is thus confirmed. (Compare Part I, p. 268).[a]

2. If the lactic acid vanishes in the muscle in the recovery period, *then the carbohydrates increase to exactly the extent* calculated from the difference between the lactic acid disappearance and the oxygen consumption (recovery consumption + resting consumption). Again the change quite predominantly concerns the glycogen. *Thus a glycogen synthesis from lactic acid takes place.* The glycogen content at the end of the recovery period is the same as that before beginning of the stimulation minus the carbohydrate disappearance equivalent to the oxygen consumption.

3. In the resting respiration the glycogen vanishes quite or nearly (within the experimental error) corresponding to the oxygen consumption.

p. 316

When one measures the heat production in the recovery period in comparison to the oxygen, the former is clearly diminished in comparison with that calculated for the carbohydrate burnt. Whilst in the latter process for 1 ccm O_2 5 cal. must be formed, one finds for many hours during the recovery period only about 3.5 cal. If one works out the total deficit which originates during the complete recovery in relation to the calculated carbohydrate, one finds this is 0.6 to 1.1 cal. per 1 g. of muscle; this within the limits of accuracy of the experiments is equal to (or somewhat greater than) the heat produced in the contraction phase.

Thus the total heat formation connected with the recovery process amounts to about 1 cal. per g. of muscle, whilst in fatigue with corresponding lactic acid accumulation 0.8 cal. is produced. Hill's demonstration that with the single twitch the heat during the contraction and in the recovery phase each amounts to about 50% of the total heat[b] is confirmed for complete anaerobic fatigue and the subsequent restitution in oxygen; and at the same time also the connection required by theory is verified: heat of the anaerobic fatigue + heat of recovery in oxygen = total heat of the work in oxygen.

O.Meyerhof, *Pflügers Arch.*, *185*, **11. 1920**

'Die Energieumwandlungen im Muskel. III. Kohlenhydrat- und Milchsäureumsatz im Froschmuskel' (The energy changes in muscle. III. Carbohydrate and lactic acid metabolism in frog muscle)

pp. 21–3

The total result already given is *that in the recovery period carbohydrate synthesis from lactic acid takes place, which is equivalent to that part of the lactic acid not burnt; the latter is given by the difference between the lactic acid disappearing and the simultaneous oxygen consumption.*

The arrangement of the experiments set out from the consideration that the starting material in each experiment must have an exactly determined content in lactic acid and carbohydrate; and that the restitution in the respiration experiment must be accomplished at about the same rate as in the limbs used for performing chemically [*chemischen Aufarbeitung*]. I had earlier made certain that with indirect fatigue of the muscles, both with tetanus and with single twitches, the gastocnemii showed the same lactic acid content as the rest of the limb musculature. Therefore from each limb-pair one gastrocnemius was separated and used for respiration measurement at 14°. In this the total oxygen consumption was measured and calculated for 1 g. of muscle. Further, as discussed earlier, the value of the resting consumption was ascertained after expiration of the recovery period, and from the two values the amount of the recovery oxygen per unit weight of muscle. At the same time these muscles served as controls to show when the restitution had run its course and whether it was completely reversible, that is to say, whether the subsequent resting consumption fell within normal limits . . .

Directly after the fatigue period the limbs were cooled on ice, then the muscles on one side were prepared and worked up (except for the gastrocnemius), the intact limb was weighed exactly and hung in a wire frame under a bell-jar into which during the whole recovery period oxygen was slowly led. This served as a moist chamber. The very slight decrease in weight of the limb in the experimental time was exactly determined; two thirds of it was related to the muscle, one third to the bone. A stream of water ran over the bell-jar, its flow so regulated that the temperature under it remained the whole time at 14° (±1°), and thus agreed with the temperature of the oxygen measurements. When the respiration increase had run its course the limb from the moist chamber was immediately worked up.

p. 24

c) Balance calculation for 1 g muscle in milligrams of substance.

Lactic acid disappearance		Total oxygen consumption	Lactic acid burnt by O_2	Carbohydrate calculated	Synthesis measured
No.					
9	2.36	1.48	1.39	0.97	0.92
10	2.12	1.22	1.14	0.98	1.09
11	[2.8]	1.45	1.36	1.44	1.57
12	2.5	1.50	1.41	1.09	$(0.81)^1$
13	[2.8]	1.35	1.26	1.54	$(1.38)^1$
14	[2.8]	1.35	1.26	1.54	$(1.28)^{1a}$

Selection 85

For about fifteen years investigations on muscle glycolysis proceeded on the assumption that lactacidogen was the precursor of lactic acid. For more than a third of this time it was accepted that lactacidogen was hexosediphosphate. Then in 1927 Embden, in collaboration with Margarete Zimmermann, revised this opinion when they found that the hexose ester – to be known as Embden ester – which they obtained from fresh muscle tissue was a monophosphate.[a] As a result of this conclusion lactacidogen came to be identified with Embden ester. By this time it had become increasingly clear that working along these lines in the effort to understand the steps in sugar breakdown in alcoholic fermentation and glycolysis was not too satisfactory. There was, for example, the difficulty of explaining the lack of increase in free phosphoric acid corresponding to lactic acid formation which the lactacidogen hypothesis demanded.

New light was thrown on this problem in 1927 when Philip Eggleton (1903–54) and Grace Palmer Eggleton (1901–70), of University College London, demonstrated that the supposed inorganic phosphate of muscle was mainly an organic acid-labile phosphate which they called 'phosphagen'. It was not to be confused with lactacidogen as this was found to be acid resistant.[b] The occurrence of this organic phosphate in muscle was reported simultaneously and independently by Cyrus H. Fiske (1890–1978) and Yellapragrada Subbarow (1896–1948) of the Harvard Medical School.[c] Their identification of the compound as phosphocreatine (see below) was corroborated a year later by the Eggletons.[d]

C.H. Fiske and Y. Subbarow, *Science*, **65**, **401. 1927**

'The nature of the "inorganic phosphate" in voluntary muscle'

pp. 401–3

Some months ago we[1] described a colorimetric phosphate method, the special feature of which is the use of a very active agent (aminonaphtholsulfonic acid) for converting the phosphomolybdic acid to its blue reduction product. When we first made use of this method for the determination of inorganic phosphate in protein-free muscle filtrates, shortly after the details had been worked out, we found a marked delay in color production. The time required to reach a constant reading was about thirty minutes, whereas ordinarily the full color (relative to the standard) has developed within four minutes or less. This peculiar behavior appeared to indicate that muscle contains either some substance capable of retarding the color reaction or else a very unstable (presumably organic) compound which liberates o-phosphoric acid while the color is developing. While the course of the color development in muscle filtrates turned out to be quite different from anything which we had seen in testing out the method in the presence of known interfering substances, it is impossible to rely on this point as a means of distinguishing between the two alternatives. Ferric salts, for example, also behave in a way that is unique. A mixture of inorganic phosphate and ferric chloride certainly does not contain a highly unstable organic phosphorus compound, and the delayed color reaction found with muscle filtrates therefore does not constitute conclusive proof of the existence of such a substance in the muscles.

Further study of the course of color development nevertheless did bring out some interesting and suggestive points, notably the fact that the delay is hardly any more pronounced with 10 c.c. of muscle filtrate, for example, than with 5 c.c. or less. Every interfering substance which we investigated in the course of our work on the phosphate method, on the other hand, shows a much more marked effect when the phosphorus content of the sample is increased. Although these facts have been in our possession now for more than a year, we have until this time refrained from placing them on record, inasmuch as the phenomena observed could not with any certainty be ascribed to the presence of an organic compound of phosphoric acid until the compound had been isolated, or at least until the organic radicle had been identified. Both these things have now been done, although the isolated substance has not yet been obtained in the pure state, and the outcome appears likely to throw light on a field of biochemistry never before suspected of being in any way related to phosphoric acid.

Muscle filtrates from which all the inorganic phosphate has been removed (by precipitation with barium, silver, etc.), as well as material which has been still further purified, show the same delay in the production of the color. These facts, together with the knowledge that the delayed reaction really is associated with the hydrolysis of an organic compound of phosphoric acid, give real significance to the quantitative data which we have meanwhile been accumulating. Some of these data will now be presented before we proceed to a discussion of the nature of the substance.

The method which we have used for the determination of this unstable form of phosphorus (which we shall for the present designate as 'labile phosphorus') differs in no respect from our regular phosphate method, except that readings are taken at brief intervals (every minute, beginning with the third, until the unstable substance is about half hydrolyzed, and thereafter every few minutes until no further change occurs). The growth in color intensity for the first few minutes is practically linear, and the concentration of inorganic phosphate is found by extrapolating back to zero time. The sum of the inorganic and the 'labile' phosphorus is calculated from the final reading, and the amount of 'labile phosphorus' found by difference. From control analyses of solutions of the purified organic compound, with and without the addition of known amounts of inorganic phosphate, we believe that the results are accurate within 1 or 2 mg of phosphorus per 100 gm of muscle.

The principal results which we have obtained by this method of analysis are as follows: (1) The normal resting voluntary muscle of the cat shows generally about 60 to 75 mg of 'labile phosphorus' per 100 gm if removed with the greatest possible care and at once cooled to 0°C. or below. (The further preparation of the material for analysis consists simply in precipitating the protein with ice-cold trichloroacetic acid, filtering and immediately neutralizing the filtrate with sodium hydroxide.) The true inorganic phosphorus under these conditions is only 20 to 25 mg, instead of the 80 to 100 mg, shown by other methods of analysis. (2) After prolonged electrical stimulation, the 'labile phosphorus' usually falls to about 20 mg (in one instance to as little as 9 mg), while the inorganic phosphorus is correspondingly increased. (3) Stimulation with the blood supply shut off, in which case complete fatigue ensues, causes complete disappearance of the unstable compound, within the error of analysis. Merely shutting off the circulation for an equal length of time has little or no effect. (4) A period of rest after the muscle has been stimulated is accompanied by resynthesis of the organic compound. The maximum yield of 'labile phosphorus' so far observed under these conditions is 46 mg per 100 gm of muscle. The inorganic phosphate, however, falls to the normal level, so in all probability either some of the phosphate has been discharged into the circulation, or else the water content of the muscle has increased.

The compound under consideration is a derivative of creatine. On a small scale we have succeeded in separating it from all other phosphoric acid compounds. On a sufficiently large scale to yield material enough for a complete analysis, the separation is not so readily accomplished by our present methods, and the best products that we have so far procured contain

Mg per 100 gm muscle

	'Labile phosphorus'	Creatine in washed copper precipitate	Molecular ratio of creatine to phosphoric acid
Normal resting muscle	70	280	0.95
do	77	323	0.99
Normal resting muscle:			
(1) Fresh	65	269	0.98
(2) Stood few minutes after removal from body	55	222	0.96
(3) Stood for longer time	32	128	0.95
(4) Analysis of trichloro acetic acid filtrate (from fresh muscle) which had stood for several hours	0	2	
Muscle stimulated for 10 minutes with circulation intact	20	82	0.97
Muscle stimulated to fatigue with blood supply shut off, followed by 1 hour rest period	44	192	1.03

several per cent. of one or more other phosphorus compounds, since the full blue color can not be obtained unless the material is ashed. In spite of this contamination, the phosphate, creatine and base in two salts which have been prepared add up to about 85 per cent. of the weight of the entire substance, indicating the probable absence of a third component. The elementary analysis of amorphous products, unless they are of definitely established purity, proves very little. Our main evidence for the existence of 'phosphocreatine' in muscle is of a quite different nature. In the first place, we have found that the 'labile phosphorus' can be more or less completely separated by precipitation with a number of different reagents (six in all have been used to date), and that in each case there is precipitated with it one equivalent of creatine. One of these reagents (copper in slightly alkaline solution) has been applied to muscle filtrates prepared under various conditions, and the proportionality between creatine and 'labile phosphorus' has never failed. A few typical illustrations of our experience with copper precipitation are given in the accompanying table. They serve to show that creatine and the labile form of phosphate are precipitated together, in equivalent proportions, whether the muscle was fresh and in the resting state or whether it had been subjected to manipulations which alter the concentration of the 'labile phosphorus.' Thus, when the muscle is allowed to stand outside the body, the 'labile phosphorus' content progressively diminishes at a fairly rapid rate (it is entirely gone in less than twenty minutes), and at the same rate creatine is set free, as indicated by its failure to be thrown down by copper. If the trichloroacetic acid filtrate prepared from a sample of fresh muscle is kept (unneutralized) for several hours, the 'labile phosphorus' has completely disappeared, and the copper precipitate is then virtually free from creatine. The same parallelism holds for muscle which has been stimulated, showing that the liberation of inorganic phosphate during stimulation is associated with the conversion of the creatine to a form in which copper will not precipitate it. Finally, in case the muscle has been stimulated with the blood supply cut off – a procedure which, as stated, leads to the loss of all the 'labile phosphorus' – and then permitted to recover, the reappearance of 'labile phosphorus' is accompanied by the return of an equivalent quantity of creatine to the condition in which precipitation by copper does take place. Direct evidence for the synthesis of a creatine-phosphoric acid compound during recovery is thereby attained.

Quite aside from its obvious bearing on the mechanism of muscular contraction, the demonstration of 'phosphocreatine' in muscle should go far towards providing an explanation for a number of matters which in the past have been obscure. Among these may be mentioned the passage of administered creatine into muscle in spite of the large quantity already there, and the striking difference between resting (living) muscle on one hand and fatigued or dead muscle on the other in their capacity for retaining both creatine and phosphate, as shown by perfusion and dialysis experiments.

Selection 86

It will be recalled that Embden's researches into muscle chemistry were originally stimulated by his interest in the lactic acid formation in the liver. By the mid-1920s increased attention was given to the converse problem: the liver's capacity to form glycogen from lactic acid. Among those who not only realized clearly the significance of this question for the interdependence of carbohydrate metabolism in muscle and liver, but also set out to investigate it were Carl Ferdinand Cori (1896–1984) and Gerty Theresa Cori (1896–1957). It led the Prague-born researchers, settled in the USA since 1922, to formulate this problem of the carbohydrate metabolism of animals as follows:

Glucose derived from liver glycogen is convertible into muscle glycogen; it is, however, not definitely known whether lactic acid derived from muscle glycogen is convertible into liver glycogen. If this should prove to be the case, the glucose molecule would be capable of a complete cycle in the body; it could in turn be liver glycogen, blood sugar, muscle glycogen, blood lactic acid, and again liver glycogen.

This is from the introductory part of the article from which are taken the excerpts below. They recount how their experimental findings suggested the scheme known later as the 'Cori cycle'.

C.F. Cori and G.T. Cori, *J. Biol. Chem.*, *81*, 389. 1929

'Glycogen formation in the liver from d- and l-lactic acid'

p. 400

The main result of the present investigation is that d-lactic acid can be deposited as liver glycogen and that it is utilized several times faster in the rat than l-lactic acid. Also, the levo isomer is hardly able to form liver glycogen. This is another example of the discrimination of the body cells between two optical isomers.

pp. 400–2

Formation of liver glycogen from lactic acid is thus seen to establish an important connection between the metabolism of the muscle and that of the liver. Muscle glycogen becomes available as blood sugar through the intervention of the liver, and blood sugar in turn is converted into muscle glycogen. There exists therefore a complete cycle of the glucose molecule in the body, which is illustrated in the following diagram.

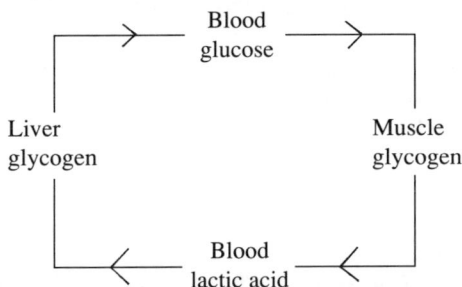

Epinephrine was found to accelerate this cycle in the direction of muscle glycogen to liver glycogen and to inhibit it in the direction of blood glucose to muscle glycogen; the result is an accumulation of sugar in the blood. Insulin, on the other hand, was found to accelerate the cycle in the direction of blood glucose to muscle glycogen, which leads to hypoglycemia and secondarily to a depletion of the glycogen stores of the liver. It will be investigated to what extent this cycle plays a role in the preservation of liver glycogen and hence of a normal blood sugar level during fasting. There is also a possibility that other hormones besides epinephrine and insulin influence this cycle.

Selection 87

Among the consequences of the attempts to isolate lactacidogen was the attention paid by the Embden school to another phosphorus-containing substance in muscle, which in 1914 was surmised to be of a nucleic acid type.[a] Then gradually in the later part of the 1920s information was gained indicating that the compound was adenylic acid,[b] indeed, the parent compound of inosinic acid which Liebig had discovered in meat extract (1847).[c] At first it was thought that the adenylic acid from muscle was identical with that already known and obtained from yeast nucleic acid. But his view changed when it was found that they were differently acted upon by a muscle enzyme obtained in a specific manner from bicarbonate extracts. The difference of yeast and muscle adenylic acid was adduced from the fact that only the latter was deaminated by the muscle enzyme.[d]

At about the same time the study of ammonia formation in muscle was taken up by Polish biochemists working in Lwow[e] under Jakob Karol Parnas

(1884–1949), who had for some years been interested in the ammonia content of blood. Recognizing the need for more substantial evidence that ammonia found in muscle was actually derived from adenylic acid, he re-investigated the question by an elaborate series of experiments employing small amounts of muscle. His results established that the traumatic and fatigue ammonia originated in the adenylic acid. As to resynthesis of adenylic acid from inosinic acid, it was accomplished under aerobic conditions, apparently through utilization of ammonia set free by oxidative deamination.

J.K. Parnas, *Biochem. Z.*, *206*, 16. 1929

'Über die Ammoniakbildung im Muskel und ihren Zusammenhang mit Funktion und Zustandsänderung. VI.' (On the ammonia formation in muscle and its connection with function and alteration of state. VI.)

p.37

Methods are given which permit of the estimation in small amounts of muscle of aminopurines and hydroxypurines, as well as of aminopurine nucleotides and hydroxypurine nucleotides in small amounts of muscle tissue.

By means of these methods the content in frog muscle of nucleotides of adenine and hypoxanthine as well as of adenine and hypoxanthine is estimated.

In fresh muscle which is worked up avoiding traumatic ammonia formation, the purine bases are contained for much of the greatest part as free nucleotides. Only about 4% of the aminopurine nitrogen can (if all the guanine is present as nucleic acid) be in the muscle a constituent of nucleic acid. Free aminopurine bases and nucleosides appear not to be present. Of the purine nitrogen in the fresh muscle of winter frogs, 82% is due to the adenine nucleus, 18% to the hypoxanthine nucleus; in summer the corresponding values found were 89% and 11%. Only about 1.4% of the purine-N occurred with the guanine nucleus.

Through mechanical destruction of the muscle, which calls forth the traumatic ammonia formation, the greatest part of the adenine nucleotide is changed into inosinic acid. A few minutes after the pulverizing of the muscle the muscle *brei* contains only 23% of the purine nitrogen in the adenine nucleus, on the other hand 77% in the hypoxanthine nucleus of the nucleotide . . .

When frog muscles are fatigued by stimulation of nerves under anaerobic conditions, the change of adenine into hypoxanthine is equivalent to the ammonia formation and appears to form the single source for this process.

When frog muscles work without fatigue under conditions of good oxygen supply, the change of adenine to hypoxanthine remains far behind the ammonia formation. This observation is explained in this sense that in the oxidative recovery processes, deamination of other substances brings about a resynthesis of adenine nucleotide from inosinic acid; and in this way a system is maintained which is over and over again renewed and conserved, readily available for anaerobic instantaneous splitting of ammonia.

Selection 88

At the beginning of the fourth decade of this century the long accepted view associating muscle contraction with lactic acid formation received a severe blow from Einar Lundsgaard (1899–1968). At first interested in the specific dynamic action of amino acids,[a] this Copenhagen scientist worked on muscle treated with iodine-substituted glycine. This led him to experiment with iodoacetate poisoned muscles when he discovered that in rigor they showed no lactic acid formation. Investigating also the phosphate metabolism he traced the energy source in normal muscle to phosphagen breakdown, while its resynthesis he related to energy derived from lactic acid production. But he also envisaged the feasibility of both processes playing the part of the combined source of energy that could be transferred to a third as yet unknown process. Time had to elapse before the relevance of this observation became clearer. Also it took time before Lundsgaard's findings on the role of oxidative breakdown in supplying energy for phosphagen synthesis were recognized as an early step in the development of the idea of oxidative phosphorylation.

The following excerpts are from two of Lundsgaard's papers which effectively heralded the end of the 'lactic acid era' in work on muscle metabolism.[b]

E. Lundsgaard, *Biochem.Z.*, *217*, 162. 1930

'Untersuchungen über Muskelkontraktionen ohne Milchsäurebildung' (Researches on muscle contraction without lactic acid formation)

pp. 162–3

Already two or three years ago I had the opportunity of observing, in experiments on rabbits, the symptoms of poisoning after intravenous injection of sodium monoiodoacetate . . . Intravenous injection of the acid, neutralized with sodium carbonate, in doses of 5 cg per kilogram, exerts a definitely lethal action. After the injection there follows a latent period of 5 to 20 minutes, according to the size of the dose, during which the animal behaves completely normally. Then the animal usually falls quite suddenly on its side, makes some lively, struggling movements with its extremities after which the whole musculature is overcome by a strongly pronounced rigor. The thorax remains fixed in the position of maximal inspiration and the animal dies from stoppage of respiration. The strong rigor continues. The animal remains, even to the most extreme parts of the hind feet, in a horizontal position, stiff as a piece of wood . . .

Characteristic for the alteration in the muscle caused by monoiodoacetic acid is the circumstance that this appears to be conditioned by the activity of the muscle. In deeply urethanised or curarised frogs no muscle alteration can be called forth. After section of the nerves of one hind foot the rigidity spreads over the whole animal up to the paralyzed limb. Now with subsequent stimulation of the nerve of the lamed leg, a short series of contractions is called forth, after which the extremity remains in maximal rigor. The maintenance of the latter is thus not conditioned by connection with the central nervous system. This dependence of the appearance of the special rigor form on the functioning of the poisoned muscle suggested that a closer study of these phenomena might have general interest for muscle physiology . . .

p. 168

Three experiments [on frogs] in which lactic acid estimations were undertaken yielded the following results in milligrams of lactic acid per cent.

	Normal			Poisoned		
Rest	17.9	27.1	40.00	14.9	14.0	25.0
Work	77.1	74.5	100.00	15.4	10.0	16.9

This result can scarcely be explained in any other way than that the poisoned muscles have performed work without lactic acid formation, work which in normal muscles called forth a lactic acid increase of 50 to 60 mg % . . .
   After this I have extended my experiments to the different phosphorus fractions of the muscle.

p. 169

Orthophosphate and phosphagen have been estimated according to Eggleton.[1] One experiment, in which these estimations are made, yields for example the following result. Phosphorus amounts are given in mg. P per cent.

		Lactic acid	Orthophosphate	Phosphagen
Normal	Rest	25	21	61
	Work	84	29	46
Poisoned	Rest	16	29	57
	Work	15	28	0

p. 172

It appears from my experiments that with stoppage of the lactic acid formation an amount of work, which in the normal muscle with maintained lactic acid formation calls forth only a very limited splitting of phosphagen, brings about complete breakdown of the latter. The most likely explanation of this observation must, in my opinion, be sought in the circumstance that lactic acid formation brings about a resynthesis of phosphagen. As can be seen from the experiment presented on p. 169 the phosphagen split in the normal muscle corresponds to 15 mg.% P. In the poisoned muscle, which has performed a corresponding amount of work, 57 mg.% of phosphagen P were split. The lactic acid formation yields the whole energy, while in normal muscle both have brought about a resynthesis of about 40 mg.% of phosphagen phosphorus. In the normal muscle during the work about 50 to 60 mg.% lactic acid were formed. It is now assumed that the deficit between the experimentally found and the theoretically ascertained lactic acid calorific quotient is to be referred to the participation of phosphagen splitting; so the calculation of the energy released through lactic acid formation must follow with the introduction of the theoretical calorific quotient, which makes only 280g-cal. This gives: $50-60 \times 280$ (= 15,500). The energy necessary for the resynthesis amounts to 40×360 (= 14,500).

E. Lundsgaard, *Biochem.Z.*, *227*, 51. 1930

'Weitere Untersuchungen über Muskelkontraktionen ohne Milchsäure-bildung' (Further researches on muscle contraction without lactic acid formation)

p. 81

I am of the opinion that for a process so complicated as muscle contraction it is simplest to assume that only energy of a quite definite kind is applicable, and that on this account at the moment one must reckon that the phosphagen splitting directly yields the energy of contraction. The role of the lactic acid formation, which is a more general anaerobic energy-yielding process demonstrable in almost all tissues must then be sought, according to my hypothesis, in this that it brings about a continuous resynthesis of the phosphagen so that only by way of the phosphagen the large store of carbohydrates is utilizable.

Nevertheless it must not be forgotten that the possibility remains that the coincidence of abolished lactic acid formation and increased phosphagen breakdown is caused by this: that the latter here on account of the disturbed lactic acid formation yields the whole energy, while in normal muscle both processes place their energy at the disposal of a third unknown reaction, which may be directly coupled with the course of the contraction.

p. 83

Muscles which were poisoned in a very careful manner performed in oxygen greater work with less decrease in phosphagen content than in nitrogen (oxidative resynthesis of phosphagen).

Selection 89

The vital role of phosphates in muscle metabolism, highlighted by the identification of Embden's lactacidogen as hexosemonophosphate and the discovery of phosphocreatine (phosphagen), was reinforced by the discovery of the wide distribution of pyrophosphate in many kinds of plant and animal cells. This was mainly due to Karl Lohmann (1898–1978),[a] working in the laboratory of Meyerhof, whose method of preparing KCl extracts from muscle (1926) contributed much to the understanding of intermediate stages in carbohydrate breakdown in muscle and yeast.[b] Almost thirty years later its impact on studies in this area was described as follows:

Up to that time experimental work had appeared to show that the architecture of the muscle was somehow necessary for the glycolytic change, and that where activity had appeared in extracts it was due to bacterial action ... It was shown how a simple cold KCl extract filtered through muslin contained all the enzymes necessary for a glycolysis from glycogen, and hexosephosphates, and that the extract would also work with glucose if a special activator known as hexokinase was also present. The point was that these changes occurred within 1 to 2 hours of making the extract, that is at a time when bacterial growth could not be influencing the results. Though the extracts were not free from solid matter, they were cell free.[c]

In 1929 Lohmann isolated from fresh muscle a compound which on neutral hydrolysis of its barium salt gave pyrophoshoric acid and adenylic acid.[d] At just the same time Fiske and Subbarow, working independently, isolated the crystalline calcium salt of a compound containing adenine, carbohydrate and the phosphate which Lohmann believed to be present in the muscle in the form of pyrophosphate.[e] The combination of adenylic acid and pyrophosphoric acid appeared to give rise to a compound important in the mechanism of muscular contraction, and acting not unlike a co-enzyme, known as adenylpyrophosphoric acid or adenosine triphosphoric acid.

This emerges from Meyerhof and Lohmann's examination of energetic changes playing part in the formation of phosphocreatine and the formation and decomposition of adenylpyrophosphoric acid. They showed that each of these substances liberated heat on hydrolysis, and this was taken as a guide to free energy liberation which was only measured much later.

O. Meyerhof and K. Lohmann, *Biochem.Z.*, 253, 431. 1932

'Über energetische Wechselbeziehungen zwischen dem Umsatz der Phosphorsäureester im Muskelextrakt' (On energetic correlations in the metabolism of phosphoric acid esters in muscle extract)

p. 460

We believe that, on the whole, our introductory thesis is well-grounded by the experiments described here: that the endothermic and not spontaneous synthesis of phosphocreatine can take place through a coupling of this process with the exothermic and spontaneous breakdown of adenylpyrophosphate whilst the resynthesis of the adenylpyrophosphate from adenylic acid and inorganic phosphate is brought about through the energy of lactic acid formation. A coupling of the synthesis may also be assumed here with the metabolism of the intermediate hexose phosphate, and so we may see in all the compounds concerned the common phosphoric acid group as the intrinsic carrier in the chemical coupling process. This would immediately make comprehensible how the adenylpyrophosphate acts as co-enzyme of lactic acid formation, in which process in some way the taking up of phosphoric acid by the hexose is connected with the release of phosphate on the part of the adenylpyrophosphate, or the giving up of phosphate on the part of the ester with the return of the pyrophosphate group to the adenylic acid. This idea would also make possible a more special interpretation for the fact that for the anaerobic splitting of the hexoses intermediate esterification with phosphoric acid is altogether necessary.

Certainly it is astonishing that in the structureless muscle extract such a fargoing coupling should exist between different processes as is here assumed, and that thereby such a considerable energy – and also power of work – should carry over through intermediate bodies from one process to a second.

p. 461

The heat of hydrolysis of phosphocreatine synthesized in muscle extract, as well as that of the pure preparation itself, amounts to 12,000 cal. per mole; the heat of hydrolysis of adenylpyrophosphate into inosinic acid, ammonia and o-phosphoric acid is 33,000 cal. per mole, of which 8,000 cal. are due to splitting off of ammonia and 25,000 to the splitting of pyrophosphate.

With the addition of adenylpyrophosphate to an inactive extract phosphagen synthesis is produced, which in certain circumstances is only explicable through splitting of adenylpyrophosphate, whilst it can be considerably increased by means of lactic acid formation. With addition of adenylic acid synthesis of adenylpyrophosphate appears, which however only takes place during lactic acid formation. The splitting of inosinepyrophosphoric acid acts similarly to that of the adenylpyrophosphoric acid . . . Up to now regarding purely enzymatic processes there is no analogy but, up to the present, chemical couplings and energy transformations associated with composite enzymatic processes such as lactic acid formation and alcoholic formation have not been explored at all.

Selection 90

So far the story of sugar metabolism has been presented as the story of processes occurring in the absence of air. This treatment inevitably reflects the historical importance occupied by the study of anaerobic transformations of carbohydrates and the emphasis placed on them for the understanding of alcoholic fermentation and muscle contraction. As to respiration, those who investigated the chemical aspects of muscle activity took into special consideration aerobic conditions. As previously noted, they considered them in relation to recovery processes. An important suggestion regarding the connection between the anaerobic breakdown of carbohydrates and the aerobic oxidation of lactic acid accompanied by glycogen synthesis was made by Meyerhof in 1925. It seemed to him that the connection was cyclic in nature and, indeed, that such cycles as this, known also as 'Meyerhof's cycle', probably characterized life.[a] This conception and the awareness of other evidence pointing to the significance of respiration in muscle activity (Parnas, Lundsgaard) influenced Vladimir Aleksandrovich Engelhardt (1894–1971) to investigate the part played by cellular respiration in phosphate metabolism. Employing avian red cells the Russian biochemist, working at the University of Kazan, demonstrated the mutual dependence of the anaerobic decomposition and the aerobic resynthesis of adenylpyrophosphoric acid.[b]

W.A. Engelhardt, *Biochem.Z.*, *252*, **343**. 1932

'Die Beziehungen zwischen Atmung und Pyrophosphatumsatz in Vogelery-throcyten' (The connection between respiration and pyrophosphate metabolism in bird erythrocytes)

p. 343

In a previous communication[1] experiments were reported which showed that in nucleus-containing bird erythrocytes the inhibition of respiration caused by cyanide, carbon monoxide or urethane is accompanied by a rapid increase in free phosphoric acid. The source of this phosphoric acid formation was found in the breakdown of the pyrophosphate fraction (adenylpyrophosphoric acid, henceforth simply called pyrophosphate). The lack of increase in phosphoric acid with maintained respiration was explained thus: that the respiration constantly reverses the pyrophosphate breakdown (which also takes place aerobically) by means of a resynthesis of the pyrophosphate coupled with the respiration.

By means of some experiments, where the poison (cyanide) inhibiting respiration was removed after a certain time by washing, it could be shown that in fact an aerobic resynthesis of the pyrophosphate broken down anaerobically can occur. But in these experiments the use of the poison inhibitory for the respiration brought in too many unphysiological and complicating circumstances; and now in experiments described below, the connection between respiration and pyrophosphate metabolism was to be investigated under conditions of pure anaerobiosis. It was further to be investigated not only how the anaerobiosis brings about breakdown, and how respiration brings about the resynthesis of the pyrophosphate, but there was further the need to ascertain now to what extent the breakdown of the pyrophosphate itself influences the respiration.

pp. 346–7

The results of the cyanide experiments were . . . completely confirmed by experiments with nitrogen anaerobiosis and now could be extended in different directions . . .

The anaerobic breakdown and the aerobic resynthesis of the pyrophosphate can be observed repeatedly in the same blood sample as appears in the experiment shown in Figure 11.

p. 353

On the grounds of certain considerations it was already assumed[2] in the first experiments that the pyrophosphate metabolism and the respiration in the nucleated red blood cells stood in a reciprocal relation to each other, that not only the respiration brings about a resynthesis of the anaerobically split

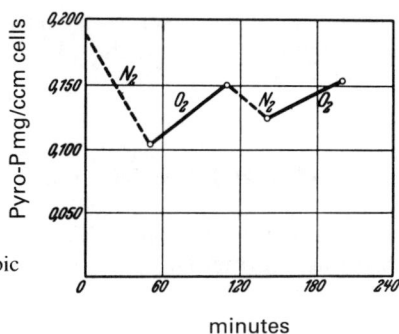

Figure 11 Repeated anaerobic breakdown and aerobic resynthesis of pyrophosphate

– – – – – – in nitrogen
————— in oxygen

pyrophosphate, but that the breakdown of the pyrophosphate itself stimulates the respiration.

Selection 91

Among those who in the period from the 1920s to the early 1930s looked at the stages in sugar breakdown in fermentation and in working muscle were Heinz Ohle[a] (1894–1959) and his collaborators,. They did much work on phosphorylated sugar, postulating that glucose-6-phosphoric acid → fructose-6-phosphoric acid → fructose-1,6-phosphoric acid. By oxidation and reduction experiments with various reagents the 6-carbon molecule was split into one molecule of dihydroxyacetone phosphoric acid and one molecule of glyceraldehyde phosphoric acid, these two finally losing their phosphate. Ohle remarks that by using a purely chemical procedure it is thus possible to develop a theory of glycolysis which is perfectly compatible with biological observation.

In 1932 Hermann Fischer (1888–1960) and Erich Baer (1901–75)[b] prepared dl-glyceraldehyde phosphoric acid which was shown in the next year by C.V. Smythe (b.1903) and W. Gerischer (b.1911)[c] to be fermented by yeast to the extent of 50 per cent (later work showed the d-isomer to be the reactive compound).

Then in 1933 Embden and his colleagues obtained further evidence for the formation of phosphorylated 3-carbon compounds in glycolysis.[d] They were investigating the difficultly hydrolyzable esters (believed to be hexose esters) which accumulated in muscle in the presence of certain anions (e.g. fluoride). On purification they got a compound which turned out to be the secondary barium salt of phosphoglyceric acid. Thus though fluoride hindered dephosphorylation, it did not prevent the splitting of the hexose. Phosphoglyceric acid added to fresh muscle mince ('brei') went very readily to pyruvic and phosphoric acids. The suggestion was that hexosediphosphoric acid yielded triosephosphoric acid which underwent a Cannizzaro dismutation to give phosphoglyceric and glycerophosphoric acids. From all the observations a scheme was proposed in which lactic acid formation is accompanied by triosephosphoric acid formation, the latter entering again into the cycle.

G. Embden, H.J. Deuticke and G. Kraft, *Klin.Wochensch.*, *12*, 213. 1933

'Über die intermediären Vorgänge bei der Glykolyse in der Muskulatur' (On the intermediate processes with glycolysis in muscle)

pp. 213–14

In researches from this Institute[a] for some time past it was shown that with muscle juice and muscle brei, by the action of certain salts, large amounts of inorganic phosphoric acid were caused to disappear. By addition of carbohydrate in the form of glycogen or starch this disappearance of phosphoric acid was enhanced, and in the experiments with muscle juice we succeeded in isolating a hexosediphosphoric acid which in its properties completely agreed with the Harden-Young ester formed on yeast fermentation.[b]

On a testing of these observations by Lohmann it was shown that the relations are in fact more complicated. That is Lohmann established that the mixture of phosphorylated products formed did not split off phosphoric acid under the action of hot acid, with the same ease as the Harden-Young ester, but is much more difficultly hydrolyzable. Therefore the major portion of the phosphoric acid ester formed in his experiments must consist of a substance different from the known form of hexosediphosphate. Isolation experiments led Lohmann to the view that with the difficultly hydrolyzable substance it was a question of a hexosediphosphate different from the Harden and Young ester.[c]

In the course of new experiments lately begun with quite other purpose on the effect of ions on synthesis of carbohydrate phosphoric acid esters, we obtained a large amount of a beautifully crystallizing barium salt, which could be identified as the secondary barium salt of a monophosphoric acid ester of l-glyceric acid.

The follow-up of this finding resulted, as we should like to believe, in throwing extensive light on the processes involved in glycolytic lactic formation in muscle, at least as far as glycolysis ensues as a consequence of intermediary phosphorylation.

It was first of all shown that under anaerobic conditions muscle brei forms phosphoglyceric acid from added hexosediphosphate still to a much greater extent than from starch; that thus in spite of the inhibition, in presence of certain ions, of the dephosphorylation of hexose esters, the breakdown of the 6-carbon chain into 2 fragments with 3 carbon atoms continues. We should like to believe that also in the analogous experiments of Lohmann the difficultly hydrolyzable ester consisted in good part of phosphoglyceric acid.

So far we have observed the formation of this substance under the action of sodium lactate, sodium fluoride and sodium oxalate.

Our conjecture that the newly found substance may be an intermediate product normally occurring on glycolysis in muscle could be supported by further experiments.

That is, it was first shown that fresh muscle brei splits added monophosphoglyceric acid with great ease into phosphoric acid and pyruvic acid. The latter substance could be isolated without use of a trapping agent by means of a method, certainly involving some losses, in a yield up to 80 per cent of the theoretically possible. Glyceric acid not esterified with phosphoric acid formed under the same conditions no detectable amount of pyruvic acid.

The glyceric acid, and the pyruvic acid split from its monophosphoric acid ester by the muscle, are more highly oxidized substances than the 6-carbon sugars and the lactic acid formed from these on glycolysis. These oxidation products arise under anaerobic conditions in very great amount and this oxidation process is naturally only possible with simultaneous occurrence of reduction processes.

On the basis of the experimental results reported so far we formed the view that the splitting of the 6-carbon chain in the middle that precedes the lactic acid formation, with the maintenance of phosphorylation, takes place in such a way that hereby from 1 molecule of hexosediphosphoric acid 2 molecules of triosemonophosphoric acid arise. For *fructose*diphosphoric acid this process would have to be formulated in the following way:[1]

Formula 1

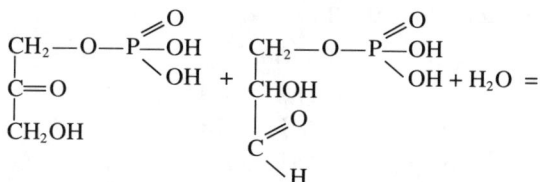

Dihydroxyacetonephosphoric
acid

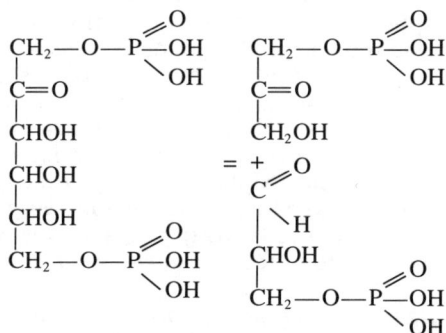

$$CH_2-O-P\underset{\diagdown OH}{\overset{\diagup O}{=}}OH$$

Fructosediphosphoric acid   Glyceraldehydephosphoric acid

By a dismutation process in the manner of the Cannizzaro reaction these two triosephosphoric acid molecules could be changed into 1 molecule of *glycerophosphoric* acid and 1 molecule of *phosphoglyceric* acid according to the following formula:

Formula 2

Dihydroxyacetonephosphoric acid   Glycerylaldehydephosphoric acid

$$\begin{array}{ccc}
= & \begin{array}{l} CH_2-O-P\overset{\displaystyle O}{\underset{OH}{<}}\!OH \\ CHOH \\ CH_2OH \end{array} & + \begin{array}{l} CH_2-O-P\overset{\displaystyle O}{\underset{OH}{<}}\!OH \\ CHOH \\ COOH \end{array}
\end{array}$$

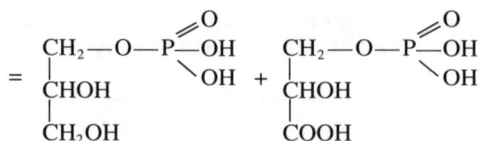

Glycerophosphoric acid   Phosphoglyceric acid

Of the two assumed dismutation products the glycerophosphoric acid, the abundance of which in the liver was recently shown by Fiske, has not yet been isolated.[2]

According to our findings just reported phosphyglyceric acid is changed into pyruvic acid with splitting off of phosphoric acid. We investigated now whether the pyruvic acid thus produced can be hydrogenated to lactic acid in this way: that the glycerolphosphoric acid possibly arising at the same time from the mentioned dismutation is dehydrogenated again to triosephosphoric acid.

Indeed, it could be shown that the slight lactic acid formation in muscle brei from *phosphoglyceric* acid alone and from *glycero*phosphoric acid alone is increased enormously by *simultaneous* addition of *both* substances. The participation of phosphoglyceric acid in this lactic acid formation corresponds to the demonstration that in the experiments with simultaneous addition of both substances the content of pyruvic acid is much smaller than in the experiments with phosphoglyceric acid alone; and the participation of glycerophosphoric acid is shown by the fact that the decrease in pyruvic acid content in the mixing experiments [*Mischversuchen*] is far less than corresponds to the increased lactic acid formation in these experiments . . .

The fact that the fermentative splitting of phosphoric acid from phosphoglyceric acid follows with such extraordinary ease in comparison with that from glycerolphosphoric acid seems to us in any case very noteworthy.

Our total experimental results so far lead to the following picture of the glycolytic lactic acid formation in so far as glycolysis takes place at all in conjunction with phosphorylation processes (see on this Bumm and Fehrenbach).[e]

First phase: Synthesis of hexosediphosphoric acid from 1 molecule of hexose and 2 molecules of phosphoric acid or from 1 molecule of hexosemonophosphoric acid[3] and 1 molecule of phosphoric acid.

Second phase: Breakdown of the hexosediphosphoric acid molecule into 2 molecules of triosephosphoric acid (see formula 1 above).

Third phase: Dismutation of 2 molecules of triosephosphoric acid into 1 molecule of glycerolphosphoric acid and 1 molecule of phosphoglyceric acid, thereby according to whether the hexosediphosphoric acid had a ketose or aldose character, α- or β- glycerolphosphoric acid could be formed (formula 2).

Fourth phase: Splitting of the phosphoglyceric acid into phosphoric acid and pyruvic acid according to formula 3 below:

Formula 3

$$\begin{array}{l} CH_2-O-P\overset{\displaystyle O}{\underset{OH}{<}}\!OH \\ CHOH \\ COOH \end{array} = \begin{array}{l} CH_3 \\ C=O \\ COOH \end{array} + H_3PO_4$$

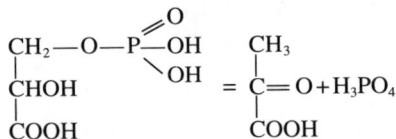

phosphoglyceric acid   pyruvic acid

In this formula finds expression that on the splitting glycerophosphoric acid [*Glycerinsäure*] does *not* appear as an intermediate, thus the splitting follows without taking up of water from outside.

Fifth phase: Reductive conversion of pyruvic acid into lactic acid at the expense of oxidative formation of triosephosphoric acid from glycerophosphoric acid according to formula 4:

Formula 4

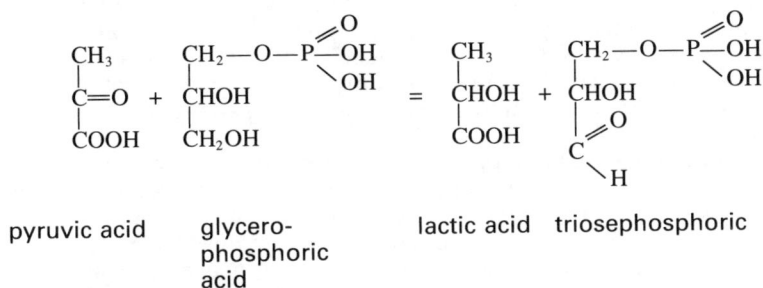

| pyruvic acid | glycero-phosphoric acid | | lactic acid | triosephosphoric |

The processes described as phases 3 to 5 are repeated on the triosephosphoric acid.

As can be seen in this picture of glycolysis methylglyoxal is missing; this substance as is well known is assumed by Neuberg as an intermediate product both in yeast fermentation and in glycolytic lactic acid formation in the animal body.[f]

Selection 92

The observation of the involvement of adenylpyrophosphoric acid in the synthesis of phosphocreatine by Meyerhof and Lohmann has already been noted.[a] These studies were carried a step further when Lohmann investigated the decomposition of phosphocreatine, the reverse reaction as it were. The supposition that this was a hydrolytic reaction caused by a single enzyme was not confirmed, after phosphocreatine was added to dialysed muscle extracts. What Lohmann did provide was evidence for adenylic acid playing a part in phosphocreatine breakdown. He attributed to it a co-enzyme function by observing that dialyzed muscle extracts to which adenylpyrophosphoric acid was added furthered the splitting of phosphocreatine. Moreover, he deduced that phosphocreatine breakdown in muscle contraction must be preceded by andenylpyrophosphoric acid breakdown. But he was cautious in not going beyond this position and not claiming that the formation and decomposition of adenylpyrophosphoric acid were in any direct way associated with the chemical and energetic changes of muscle contraction.

K. Lohmann, *Biochem.Z.*, **271**, **264.** **1934**

'Uber die enzymatische Aufspaltung der Kreatinphosphorsäure; zugleich ein Beitrag zum Chemismus der Muskelkontraktion' (On the enzymatic splitting of creatinephosphoric acid; at the same time a contribution to the chemistry of muscle contraction)

pp. 274–5

It is shown in the experimental part that for the splitting of creatinephosphoric acid a 'co-enzyme' ['*Co-Ferment*'] is needed. It seems surprising that the strongly exothermic splitting of creatinephosphoric acid (which is already in the cold hydrolyzed by dilute acid) in the muscle does not take place through a specific phosphatase, as was generally assumed until now. But that here an enzymatically conditioned reaction sequence involving adenylpyrophosphoric acid is present.

The heat formation of creatinephosphoric acid splitting in different estimation series with splitting through acids (this value was regarded as the least objectionable) was found to be 11,000 to 12,000 cal., with enzymatic 15,000 cal. per mole.[1] But according to new measurements also with enzymatic splitting after reactivation of inactivated extracts a value of 11,000 to 12,000 cal. per mole was given. From the work performed by monoiodoacetate-poisoned frog muscle by reckoning from the calorific quotient of the lactic acid formation, a mean value of 12,900 was obtained.[2] The dephosphorylation of one mole of adenylpyrophosphoric acid to one mole of adenylic acid and two moles of phosphoric acid gave on the other hand a heat formation[3] of 25,000 cal., that is to say, per mole of phosphoric acid split, 12,500 cal. On grounds of the measured heats of the equations (1) and (3)

Adenylpyrophosphoric acid = adenylic acid + 2 phosphoric acid + 25,000 cal. (1)
2 creatinephosphoric acid + adenylic acid = adenylpyrophosphoric acid + 2 creatine (2)
2 creatinephosphoric acid = 2 creatine + 2 phosphoric acid + 22,000 to 24,000 cal. (3)

it is thus seen that the splitting of creatinephosphoric acid with simultaneous esterification of the adenylic acid to adenylpyrophosphoric acid is, within the limits of accuracy of the measurements, a thermally neutral process.[a] Thus with the true breakdown of the creatinephosphoric acid no heat is set free (whether the reaction also runs without energy change must provisionally remain undecided, on account of the at present unknown free energy). The resynthesis of the adenylpyrophosphoric acid thus follows with practically 100% efficiency. A similar high efficiency was already observed for the anaerobic restitution of creatinephosphoric acid, which according to Meyerhof[4] takes place with 90% efficiency, according to Lundsgaard[5] with 100%.

If one transfers the results found with muscle extract to the intact muscle there appears, as shown *up to the present time*, the following series of chemical reactions during muscle contraction: 1. Breakdown of adenylpyrophosphoric acid, which 2. is reversed by the breakdown of creatinephosphoric acid. The resynthesis of creatinephosphoric acid follow. 3. partly anaerobically through the energy of lactic acid formation, partly aerobically through oxidative processes (usually carbohydrate combustion); these oxidative processes bring about moreover 4. (and indeed exclusively) the resynthesis of lactic acid to glycogen in the Pasteur-Meyerhof reaction.[b]

Before the breakdown of creatinephosphoric acid there thus occurs the
dephosphorylation of adenylpyrophosphoric acid; this is chronologically the
first exothermic reaction with the muscle twitch demonstrated until now. With
regard to the question in what chemical and energetic connection breakdown
and resynthesis of adenylpyrophosphoric acid stand to the fundamental
process of muscle twitch we must assume that probably a direct relation does
not exist.[c]

Selection 93

At this time, when increasing attention was focused on the part played by
adenylpyrophosphoric acid in glycolysis, there was uncertainty about its
chemical constitution. By 1935 Lohmann had established the formula of the
compound which is still accepted today. He based his proposal on the
reasoning given below.

K. Lohmann, *Biochem.Z.*, *282*, 120. 1935

'Konstitution der Adenylpyrophosphorsäure und Adenosindiphosphorsäure'
(Constitution of adenylpyrophosphoric acid and adenosine diphosphoric
acid)

pp. 120–3

Adenosine diphosphoric acid was prepared according to the directions in the
preceding communication.[1] After breakdown of the barium salt with the exact
amount of H_2SO_4, its caustic soda requirement and electrotitration curve were
investigated in the usual way.[2] (Adenosine diphosphoric acid breaks down on
short acid hydrolysis into one molecule each of adenine, pentosephosphoric
acid and inorganic phosphoric acid). According to this adenosine diphosphoric
acid itself contains *three*, the hydrolyzed compound however up to the weakly
alkaline region *four* acid valencies . . . The resulting curves are reproduced in
Figure 12.

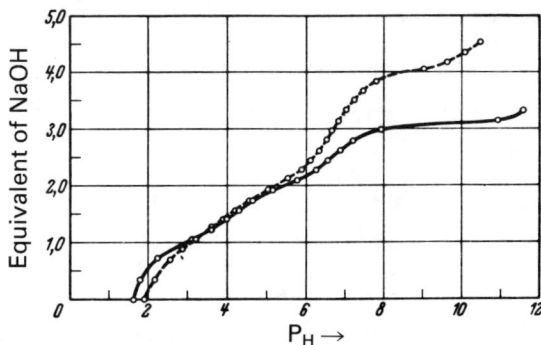

Figure 12 Electrotitration curve of adenosine diphosphoric acid with
sodium hydroxide before (o———o) and after hydrolysis (o‒‒‒‒‒o)

The course of the two curves shows that the first acid group each has about the strength of the first dissociation constant of phosphoric acid, the second group however both with the split and the unsplit compound is rather weaker and less sloping . . . The further course of the two curves of the adenosine diphosphoric as well as of the adenylpyrophosphoric acid now shows a fundamental difference: with the unsplit adenosinediphosphoric acid one finds still only *one* acid group of about the strength of the second dissociation constant of inorganic phosphoric acid, with the split compound *two* of these. Thus with the hydrolysis of adenosine diphosphoric acid a new acid group appears of the strength of the second dissociation constant of phosphoric acid, which in the unsplit compound is concealed.

This finding allows conclusions as well on the constitution of adenosineiphosphoric acid as of adenylpyrophosphoric acid (adenosine triphosphoric acid). Both substances are compounds of adenylic acid with one or two molecules of phosphoric acid.

$$
\begin{array}{l}
N{=}C\;.\;NH_2 \\
\;|\qquad| \\
HC\quad C{-}N \\
\;\|\qquad\qquad\diagdown CH\quad OH\quad OH\qquad\qquad\qquad\qquad OH \\
\;\|\qquad\quad\diagup\qquad\quad|\qquad\quad|\qquad\qquad\qquad\qquad\qquad| \\
N{-}C{-}N{-}CH{-}CH{-}CH{-}CH{-}CH_2{-}O{-}P{-}OH \\
\qquad\qquad\;\lfloor\;\;\;O\;\;\;\;\rfloor\qquad\qquad\qquad\qquad\quad\|\\
\qquad\qquad\qquad\qquad\qquad\qquad\qquad\qquad\qquad\qquad O
\end{array}
$$

Adenylic acid (adenosine monophosphoric acid).[3]

In both compounds the amino group in the adenine residue is free, since on treatment with nitrous acid it can easily be changed, without alteration of the P content, into the hydroxyl group. The adenine, which otherwise contains no esterifiable groups can therefore be excluded with regard to the binding of the rest of the phosphoric acid molecules.

What appears at hand next for the adenylpyrophosphoric acid is an esterification of the two hydroxyl groups of the pentose. With this the consideration must be fulfilled, however, that the adenylpyrophosphoric acid contains only four acid groups, and indeed three of the strength of the first dissociation constant of phosphoric acid and one of the strength of the second dissociation constant.[4] The hydrolyzed compound (1 molecule of adenine and pentosephosphoric acid each and 2 molecules of inorganic phosphate) contains, as to be expected, three acid groups of the strength of the first as well as of the second dissociation constant of phosphoric acid. Only the following formula could meet this demand:

$$
\begin{array}{l}
\quad OH\qquad OH \\
\quad\;|\qquad\quad| \\
O{=}P{-}O{-}P{-}O \\
\quad\;|\qquad\quad| \\
\quad\;O\qquad\;O\qquad\qquad\qquad\qquad OH \\
\quad\;|\qquad\quad|\qquad\qquad\qquad\qquad\quad| \\
{-}CH{-}CH{-}\!-\!CH{-}CH{-}CH_2{-}O{-}P{-}OH \\
\quad\;\lfloor\;\;\;O\;\;\;\rfloor\qquad\qquad\qquad\qquad\|\\
\qquad\qquad\qquad\qquad\qquad\qquad\qquad\qquad O
\end{array}
$$

Against this however is the fact that the two hydroxyl groups in question appear to be free, as the adenylpyrophosphoric acid (as well as the adenylic acid and the adenosinediphosphoric acid) can form a complex basic Cu-salt; and also that with the Böseken reaction with boric acid acidification results.

These two reactions however seem not completely convincing, as in a molecule with such a piling up of hydroxyl groups (near an amino group) the outcome of the reaction may not be unequivocal. The presence of only *three* acid groups in the adenosine diphosphoric acid and the appearance of the fourth group on hydrolysis shows now that its second molecule can *not* be esterified with a hydroxyl group of the pentose, as then no formula is possible to comply with this behaviour. For the adenosine diphosphoric acid then only the following constitution remains:

Adenosine diphosphoric acid

From this constitution for adenosine diphosphoric acid that of adenylpyrophosphoric acid results (both compounds can easily be converted enzymatically the one into the other by loss or gain of one molecule of phosphoric acid[5]):

Adenylpyrophosphoric acid (Adenosine triphosphoric acid)

Selection 94

As we have seen, glycerophosphoric acid, one of the intermediates in the glycolytic scheme proposed by Embden and his associates, had not yet been isolated. This was achieved by Meyerhof in collaboration with D. McEachern[a] who, just at the time when the paper of Embden, Deuticke and Kraft appeared, were working on pyruvic acid metabolism in muscle. Using sodium sulphite to trap pyruvic acid during hexosediphosphate break-down in muscle extract, and estimating both the pyruvate and lactate formed, they found that the phosphate set free was roughly equivalent only to these two acids; the reduced product must therefore still be present in phosphorylated form. They suggested and found α-glycerophosphate.

Following this up, Meyerhof and W. Kiessling (1901–58)[b] examined what happened when pyruvic acid and glycerophosphoric acid were added to muscle extract and found the results in accordance with the scheme of the Frankfurt school. But two years later they showed that it is not necessary to postulate continual dismutation of triosephosphoric acid to give glycero-phosphoric acid and phosphoglyceric acid. The oxidoreduction between pyruvic acid and triosephosphoric acid itself was found to be as rapid in dialyzed muscle extract as normal glycolysis and far more rapid than the oxidoreduction between pyruvic acid and glycerophosphoric acid. Comparing carbohydrate breakdown in muscle and yeast, they emphasized the correspondence of pyruvic acid and acetaldehyde and of lactic acid and alcohol.

O. Meyerhof and W. Kiessling, *Biochem.Z.*, *283*, 83. 1935–6

'Uber den Hauptweg der Milchsäurebildung in der Muskulatur' (On the principal way of lactic acid formation in muscle)

p. 113

Besides the only way considered so far for the enzymatic formation of lactic acid in muscle, where the α-glycerophosphoric acid arising through oxidoreduction reacts with the oxidized dismutation product, pyruvic acid, with ultimate formation of 2 moles of lactic acid, there is another way: there the pyruvic acid formed reacts with the primary esterification product of the hexoses. In this way 1 mole of pyruvic acid is reduced to lactic acid. This lactic acid production follows much faster than with the first-suggested way, and the individual partial reactions of this conversion go with greater speed than the total reaction. This is true also in such circumstances where α-glycerophosphoric acid can no longer enter into the reaction (as in extract of acetone powder). In this way the individual steps, which exactly correspond to those in fermentation, can be isolated partly by means of fluoride, partly by means of iodoacetic acid, whereby the pyruvic acid is the biological equivalent of the acetaldehyde, lactic acid the biological equivalent of the alcohol.

Selection 95

During the late 1920s and early 1930s opinions about the reactions involved in the phosphorylation of glycogen seem to have been uncertain. Considerable light had been thrown on this problem in 1935 when Parnas, together with T. Baranowski (*b*.1910), made the observation that in extracts thoroughly dialyzed at room temperature, glycogen could be phosphorylated, if Mg was added, to a difficultly-hydrolyzable ester by inorganic phosphate in the absence of adenosine triphosphoric acid. Further work performed by members of the Parnas laboratory, Pavel Ostern (1902–43) and J.A. Guthke, confirmed the suggestion that the ester formed was hexosemonophosphate. They also showed that further phosphorylation of this ester to hexosediphosphate did only take place at the expense of adenosine triphosphoric acid.

P. Ostern and J.A. Guthke, *Comp.Rend.Soc.Biol.*, *121*, 282. 1936

'Les transformations initiales de la glycogénolyse. La fonction de l'ester hexosemonophosphorique' (The initial transformations of glycogenolysis. The function of the hexosemonophosphoric ester)

p. 282

In a previously published work[1] Parnas and Baranowski have posed the problem of the first product of the binding of the phosphoric group to glycogen; we have taken up this question again.

Does there exist a transition between the hexosemonophosphoric ester produced by the muscle enzymes setting out from glycogen and free phosphate, and the hexosediphosphoric ester which can only be formed by transport of the phosphoric group from adenosinetriphosphoric acid? One may hope by changing the conditions to effect this transition in which Parnas and Baranowski were not successful.

We have first identified the ester formed in absence of adenosine triphosphoric acid in the course of the synthesis reported by Parnas and Baranowski . . . We have obtained . . . 3 gr. of pure substance, identified as the barium salt of the Robison-Embden ester by analysis, solubility and the curve of hydrolysis, identical to that established by Lohmann.

p. 283

Our experiments differ from those of Parnas and Baranowski only in one point: that the enzyme extracts used were more concentrated (equal weights of muscle and water).[2] If one adds for example the hexosemonophosphoric ester in the ratio of 2.5 mg. of phosphorus, and the adenosine triphosphoric acid in the ratio of 2.8 mg. of mobile phosphate, to 5 c.c. of enzyme extract poisoned with iodoacetate at the concentration of 1/400 mol., one finds after an hour that the adenosine triphosphoric acid has disappeared and that two thirds of the mobile phosphorus are contained [*renfermes*] in the hexosediphosphate. The results of experiments without iodoacetate poisoning are the same, although accumulation of the hexose hexosediphosphoric ester may be less pronounced; this is because this ester undergoes subsequent transformations.

The formation of the hexosediphosphoric ester indicated up to this point only by analytical experiments, has afterwards been proved by preparation of this substance.

p. 284

The hexosemonophosphoric ester is not according to this conception a product of 'stabilization',[a] but rather an activated product by comparison with the sugars; but the hexosediphosphoric ester may be considered as a product more 'active' in the course of glycolytic breakdown and this is why the latter does not accumulate in the muscle.

Selection 96

Also in 1936 the Coris identified a new hexosemonophosphate in frog muscle as the first phosphorylation product of glycogen – glucose-1-phosphoric acid, which was rapidly converted into the Embden ester. It became known as the 'Cori ester'. As C.F. Cori said later:

With the knowledge gained about the enzymatic reactions involved in the formation of glucose-6-phosphate . . . the idea that glucose-6-phosphate is a stabilization product of a reactive intermediate was given up.[a]

C.F. Cori and G.T. Cori, *Proc.Soc.Exp.Biol.Med.*, *84*, 702. 1936

'Mechanism of formation of hexosemonophosphate in muscle and isolation of a new phosphate ester'

p. 702

Experiments performed on intact frog muscle indicated that hexosemonophosphate, in contrast to hexosediphosphate, is formed by esterification with inorganic phosphate.[1] A further study of this problem was carried out on minced frog muscle which was almost completely inactivated (in regard to lactic acid formation) by 3 to 4 extractions with distilled water. When such muscle, which contains only 2 to 4 mg. % of orgainc, acid-soluble P, is incubated anaerobically for 3 hours in isotonic phosphate buffer, the organic P content rises from 8 to 13 mg. % due to the formation of hexosemonophosphate. Addition of small amounts of adenylpyrophosphoric or of adenylic acid greatly enhances the formation of hexosemonophosphate . . . The experiments indicate that hexosemonophosphate is formed from inorganic phosphate and that adenylic acid serves as the mediator of this reaction.

Observations after short periods of incubation showed that the first phosphorylation product is not hexose-6-phosphoric acid (Embden ester), but a new ester which is slowly converted to the 6-phosphoric acid under the conditions of these experiments.

p. 705

The first phosphorylation product proved to be a new ester which was isolated as the crystalline brucin salt and had the properties of glucose-1-phosphoric acid; when added to frog muscle extract it was converted in a few minutes to the Embden ester.

Selection 97

There were more ways of studying phosphate esterification in muscle than reported in the two previous selections. As we have seen, in 1935, Meyerhof and Kiessling drew attention to the oxidoreduction mechanism involving oxidation of triosephosphoric acid to phosphoglyceric acid and reduction of pyruvic acid to lactic acid.[a] Among researchers at the time who endeavoured to approach phosphate esterification from this point of view was Dorothy Moyle Needham (1896–1987) at Cambridge. In 1937, bearing on this, she postulated for muscle

an endothermic reaction coupled with an exothermic one – the formation of adenylpyrophosphate from adenylic acid and free phosphate, coupled with the oxido-reduction between pyruvic acid and triosephosphate:

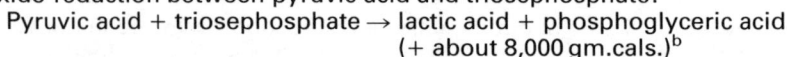

Pyruvic acid + triosephosphate → lactic acid + phosphoglyceric acid
(+ about 8,000 gm.cals.)[b]

At the time of her writing this passage there was no direct evidence for the postulated mechanism. But while in press, she and her colleague R.K. Pillai (1906–46) were able to demonstrate the existence of such a relationship and thus to make a distinguished contribution to the understanding of the anaerobic metabolism of carbohydrates.[c]

D.M. Needham and R.K. Pillai, *Nature*, *140*, 64. 1937

'Coupling of dismutations with esterification of phosphate in muscle'

pp. 64–5

The disappearance of free phosphate during glycogenolysis in muscle extract is a striking phenomenon. One mechanism by which this occurs has recently been elucidated by Parnas and his colleagues,[1] who showed that, even in long-dialysed extracts, phosphate and glycogen can react together giving hexosemonophosphate. This reaction is poisoned by $M/100$ phloridzin, but not by $M/400$ iodoacetic acid. We have observed a second method of phosphate esterification.

A certain dependence of esterification of carbohydrate upon dismutation has from time to time been observed, for example, by Meyerhof and Kiessling[2] in yeast and muscle extract, by Schäffner and Berl[3] in yeast extract, and by Dische[4] in red blood cells and hæmolysed blood. Dische mentions having found an increase in easily hydrolysable phosphate during this esterification, and suggested that the synthesis of adenylpyrophosphate might be an intermediate step. We now show that in the second method of phosphate esterification in muscle extract the dismutation of triosephosphate with pyruvic acid giving phosphoglyceric and lactic acids, is coupled with a synthesis of adenylpyrophosphate from adenylic acid and free phosphate.

As enzyme preparation we used extract of acetone muscle powder (rabbit) prepared according to Meyerhof's technique. In order to destroy adenylpyrophosphatase (the activity of which would have masked any adenylpyrophosphate synthesis) we kept the extract at 0 for five days before use; the extracts were also dialysed for four hours to remove any remaining traces of substrate. Typical results are given in Table 5. It will be observed that

Table 5 1 c.c. extract made up to final volume of 2.5 c.c.; with $M/100$ hexosediphosphate, $M/50$ pyruvate, $M/50$ fluoride, $M/30$ phosphate, M 200 adenylic acid, present in each sample. 0.4 mgm. co-zymase present in each. 30 min. incubation at 37° C. The figures give increases and decreases in mgm.

Exp.		Lactic acid	Phosphoglyceric P	Inorg. P.	Pyro P.
IV		+1.58	+0.57	−0.62	+0.66
V	No iodoacetate	+1.80		−0.78	+0.72
	+ iodoacetate	+0.32		0	0
XVIII	No phloridzin	+4.56		−0.51	+0.57
	+ phloridzin	+3.87		−0.45	+0.66
XIX	No arsenate	+4.26		−0.69	
	+ arsenate	+4.98		−0.18	

Iodoacetate $M/400$: phloridzin $M/100$: arsenate $M/75$.

(1) iodoacetate, which inhibits the dismutation, inhibits also the esterification, (2) phloridzin does not inhibit either dismutation or esterification, (3) arsenate inhibits the esterification without affecting dismutation.

The dismutation of α-glycerophosphate with pyruvate, and the dismutation of hexosediphosphate to give glycerolphosphate and phosphoglycerate can be coupled with the phosphorylation of adenylic acid.

As regards the mechanism of the coupling, it can be said (1) that we have tested our extracts for adenylpyrophosphatase activity and found it to be nil. It seems, therefore, that the coupled esterification cannot be dependent upon a reversed reaction catalysed by this enzyme. (2) We have examined the ratio of lactic acid formed to phosphorus esterified. The results indicate that, according to the conditions, a greater or less proportion of the dismutation can be coupled with adenylpyrophosphate synthesis, but that however favourable the conditions, probably not more than one atom of phosphorus can be esterified per molecule of lactic acid formed.

This coupled esterification of phosphate probably plays an important part during the anaerobic recovery period when creatinephosphate is resynthesized, free creatine and free phosphate disappearing. Making use only of the Parnas mechanism of esterification of glycogen by free phosphate, and of the mechanism whereby phosphate is transferred from phosphopyruvic acid to adenylic acid, it seems possible to account for the resynthesis of only $\frac{1}{2}$ mol. creatinephosphate per mol. lactic acid produced. Both Meyerhof[5] and Parnas[6] have pointed out that Lundsgaard[7] has actually observed a much greater synthesis than this – 2 mol. creatinephosphate per mol. lactic acid formed. By taking into account the coupled esterification, synthesis of $1\frac{1}{2}$ mol. of creatinephosphate per mol. lactic acid can be expected. This is a value much nearer Lundsgaard's observed results. Further, during anaerobic recovery there is very little heat output; this can only be explained if the energy of dismutation is not evolved. We here suggest that it is retained and used to make possible the endothermic synthesis of adenylpyrosphosphate.

These considerations will be dealt with in detail elsewhere.[a]

Selection 98

Although the observations reported by Lundsgaard (1930)[a] and Engelhardt (1932)[b] indicated a relationship between respiration and phosphorylation, the problem attracted little attention. This began to change when knowledge of aerobic metabolism progressed during the 1930s, especially under the influence of Szent-Györgyi's idea about the catalytic function of C_4-dicarboxylic acids of cell respiration.[c] The impact of this approach to phosphorylation is evident in the doctoral dissertation (submitted in 1938) by Hermann Moritz Kalckar (b.1909). Working in Copenhagen, where Lundsgaard's curiosity had shifted to the study of the active transport of glucose in the intestine and kidney, Kalckar investigated the stimulating influence of dehydrogenation of succinic to fumaric acid on phosphorylation in kidney.[d] Oxidative phosphorylation began to emerge as a new area of biochemical research.[e]

Among the early significant experiments conducted along this line were those of Vladimir Aleksandrovich Belitser (b.1906) and Elena T. Tsybakova from Moscow University. Using minced or homogenized pigeon-breast or rabbit-heart muscle with various substrates (pyruvate, citrate, ketoglutarate, succinate, fumarate, malate and lactate) they found respiration accompanied by synthesis of phosphocreatine. They also enquired into the efficiency of respiratory phosphorylation by measuring the ratio of the moles of synthesized phosphagen to moles of consumed oxygen. The found the coefficient of synthesis (called the 'P/O ratio') in the order of 5.

V.A. Belitser and E.T. Tsybakova, *Biokhimiya*, **4**, 516. 1939

'O mekhanizme fosforilirovaniya, sopryazhennogo s dykhaniem' (The mechanism of phosphorylation associated with respiration)[a]

pp. 516–17

Synthesis of adenosinetriphosphate and phosphagen (phosphocreatine) takes place in the muscle at the expense of the energy derived from either glycolysis or cell respiration.

From a number of indirect findings, however, it is apparent that some oxidizing processes may be linked with phosphorylation without having any direct connection with glycolysis. Braunshteyn and Severin[1,2] showed that oxidation of pyruvic acid, ketobutyric acid and glutamic acid, as well as of alanine, brings about a stabilization of adenosine triphosphate in nucleated erythrocytes. Grimlund[3] found that oxidation of lactic acid, pyruvic acid, and succinic acid increases the working capacity of a muscle in which glycolysis is obliterated. This was previously found to be true in the case of lactic acid, by Meyerhof and his coworkers,[4] who also stated that lactic acid oxidation brings about stabilization of phosphagen in muscles poisoned by iodoacetate. It can be presumed that stabilization of phosphorylated esters is the result of their resynthesis; then, on the basis of the investigations quoted above, a 'strictly respiratory' synthesis of these esters might well be inferred. This interpretation has received additional and more direct support. Kalckar[5] has described a 'respiratory' synthesis of phosphorylated hexoses in kidney cortex preparations, under conditions in which glycolysis is arrested. About the same time one of us[6] found that minced muscle tissue under aerobic conditions can sustain a synthesis of phosphagen from creatine and inorganic phosphate even after poisoning with monobromo-acetic acid.

Our goal in this investigation was to systematically study 'respiratory' synthesis of phosphorylated esters and to determine its mechanism. Some of the results in this area collected up to the summer of 1938 have already been published in a short communication.[7]

pp. 517–18

For the purpose of this study it was necessary to find a tissue which would permit the study of oxidative phosphorylation processes using selected exogenous-respiratory substrates. In order to observe a clear effect of exogenous substrates, it is necessary that the respiratory phosphorylation due to oxidation of preformed substrates be insignificantly small, under the experimental conditions chosen. We investigated the capability of muscle tissue from various animals to synthesize phosphagen after preliminary washing in water or saline. Various substances, particularly lactic acid, served as respiratory substrates. The majority of species of muscle investigated (rabbit, guinea pig, rat, and frog) were not able to synthesize phosphagen after subsequent mincing and washing. Only pigeon breast muscle and rabbit heart muscle, which evidently possess a more stable enzymic system of respiration, were still capable of respiratory synthesis of phosphagen after certain types of washing.

Muscle tissue freshly cut, chilled and cleared of connective and fat elements was minced on ice with scissors to the consistency of thin pulp and then washed with chilled 0.15 M phosphate solution of pH of 7.0−7.2 or in a mixture of 9 parts of 0.9% NaCl and one part of the phosphate solution (40 ml of solution for 5−7 gm. of pigeon muscle, or tissue from one rabbit heart). The tissue was left in the solution at 0°C for 20 minutes and was stirred several times. The liquid was then decanted, and the tissue dried quickly on filter paper and placed in a chilled Petri dish. It was divided into several portions, (usually 200 mg each) one or two of which were immediately fixed with chilled trichloracetic acid (end concentration 4%); the rest were placed in Warburg flasks containing phosphate buffer pH 7.5, 0.2 mg cozymase,[8] 0.14 mg magnesium in the form of $MgCl_2$ and 12 mg creatine; the substrates under investigation were added in amounts to give a concentration of 0.05 M or 0.025 M. In order to reach isotonicity KCl or NaCl were added. The total volume was 2 ml.

pp. 518–21

Considerable phosphagen synthesis occurred under aerobic conditions but only in the presence of pyruvic acid. Obviously the essential process governing the synthesis was oxidation of pyruvic acid . . .

We investigated a number of other respiratory substrates besides pyruvic acid. From all the collected data, it appears that any substrate which is more or less intensively oxidized in muscle tissue gives rise to phosphagen synthesis upon oxidation. Among these substrates are the following acids: citric, ketoglutaric [normal keto-pyrotartaric acid], succinic, fumaric, malic, lactic and pyruvic. Acetic acid does not cause any increased respiration or phosphagen synthesis. The activity of the various substrates is illustrated in Table 6.

Because the data in Table 6 were collected from a number of different experiments, showing a considerable variation in ability of tissue to synthesize phosphagen, it is necessary to refrain from a quantitative comparison of activity of the various substrates.

In experiments with unwashed frog muscle tissue, we repeatedly noted a direct relationship between increased respiration with the help of creatine − 'creatine effect' − and the quantity of synthesized phosphagen. In experiments with washed muscle tissue of warm-blooded animals we did not find such regularity. Increased respiration was insignificant (not more than 40%) and frequently completely absent.

pp. 526–7

In our previous experiments the ratio of moles of synthesized phosphagen to moles of consumed oxygen amounted to between 5 and 8 by allowing for the excess of consumed oxygen.[6] The applicability of such a calculation is

Table 6 Synthesis of phosphagen linked with oxidation of various substrates

Dates of experiment	Tissue	Substrate	Respiration in μl O_2 per 1 gm tissue for 30 minutes		Increase of phosphagen in mg of P_5O_2 per 1 gm tissue	
			without substrate	with substrate	without substrate	with substrate
1939						
April 1	Rabbit heart	Citric acid	263	399	4.00	7.25
April 19	"	Fumaric acid	95	386	0.20	5.62*
May 5	"	α-ketoglutaric acid	120	540	0.45	3.40*
April 7	"	Succinic acid	206	956	2.30	4.26
1938						
June 3	Pigeon muscle	Malic acid	280	420	0	1.84
May 16	"	Lactic acid	252	387	0	1.56
Oct. 26	"	Pyruvic acid	214	420	0	2.34
May 10	"	Acetic acid	170	153	0	0

*In the presence of 0.02 M NaF

debatable, however, since it was not proved that phosphagen synthesis is tied exclusively to the 'surplus' respiration, i.e. respiration resulting from added creatine. On the contrary, it is possible that 'basic respiration' (i.e., that portion of respiration that occurs in the absence of creatine) also participates in the synthesis. Our experiments with washed muscle, in which regular synthesis of phosphagen is frequently observed despite the predominant absence of excess respiration, point to this. If phosphagen synthesis in our earlier experiments is related not to 'excessive' but to ordinary respiration then the coefficient is 1−3 rather than 5−8.

Determination of the actual quantitative relationship between respiration and phosphagen synthesis is interesting not only for its own value but also in connection with the question of the mechanism of respiratory phosphorylation. Striving to obtain precise figures, we changed the experimental arrangement so as to decrease, as much as possible, basic respiration and to fix more precisely the moment of beginning and ending of synthesis.

Experiments were conducted at 0°C. A Warburg apparatus bath contained a mixture of water and snow, which was replenished at intervals throughout the experiment, with an excess of snow present at all times; 16 mg of dry creatine was placed in one of the side bulbs of the vessel. In the other bulb we put 0.2 ml of 40 per cent trichloracetic acid. The experimental specimen was unwashed minced frog muscle. The medium contained 0.037 M phosphate, pH 7.6; 0.15 mg Mg and 0.13 M KCl. The total volume was 2 ml, the tissue weight was 200 mg, and the gas was O_2.

After equilibration of the temperature (which required no less than half an hour) respiration was determined. Subsequently, creatine from the side bulbs was brought into the experimental vessel. Respiration measurement continued for no less than one hour. The trichloracetic acid from the second bulb was added, thus fixing the tissue in order to stop biological reactions. Phosphagen determination was done in the usual manner.

It should be noted that addition of the dry creatine from the side bulbs caused a small but definitely perceptible increase of gas pressure. This increased pressure, depending most likely on decreased solubility of O_2 under the influence of dissolving creatine, was measured in a separate vessel containing all ingredients except tissue, and was taken into account as a correction.

The coefficient:

$$\frac{\text{(synthesized phosphagen)}}{\text{(consumed oxygen)}},$$

in calculations involving the oxygen consumed in respiration in the period of synthesis, varied in different experiments between 3.8 and 4.3. If calculated on the basis of oxygen consumed in surplus, the coefficient amounted to 5.0–7.3.

It seems most reasonable to assume that the part of respiration directly resulting in synthesis is greater than the 'surplus', but smaller than the total respiration. Hence, the figure 5–7.3 are greater than the actual ones and the figures 3.8–4.3 are smaller. Until further refinement is attained, it is, therefore, necessary to consider possible variations of the coefficient in the range of 4 to 7.

Table 7   Coefficient of Synthesis (Experiment of March 23, 1939)

	Without incubation	Incubation with creatine	Incubation without creatine
Content of phosphagen {in mg P_2O_5	0.195	0.56	0.21
{in μM"	2.74	7.88	2.95
Consumption of oxygen {in μl O_2		27.5	12.5
{in μM "		1.23	0.55
'Excess' consumption of oxygen {in μl O_2		15.0	
{in μM "		0.67	

Coefficient of synthesis { for total respiration $\dfrac{7.88 - 2.74}{1.23} = 4.2$

{ for 'excess' respiration $\dfrac{7.88 - 2.95}{0.67} = 7.3$

Selection 99

In 1939 the biochemical world was startled by the discovery of V.A. Engelhardt and Militsa Nikolaevna Lyubimova (b.1898) that the protein myosin behaved as the enzyme adenosine triphosphatase (see below). What marks out this achievement in the history of biochemistry and physiology is the experimental identification of a direct relation between microstructure and chemical action. It is not without interest that their finding recalled some of the older ideas that myosin was an organic catalyst and also a means of transforming chemical into mechanical energy, though not in the form originally proposed.[a]

W.A. Engelhardt and M.N. Ljubimowa,[a] *Nature*, *144*, **668**. 1939

'Myosine and adenosinetriphosphatase'

pp. 668–9

Ordinary aqueous or potassium chloride extracts of muscle exhibit but a slight capacity to mineralize adenosinetriphosphate. Even this slight liberation of phosphate is mainly due, not to direct hydrolysis of adenosinetriphosphate, but to a process of secondary, indirect mineralization, accompanying the transfer of phosphate from the adenylic system to creatine, the corresponding enzymes (for which the name 'phosphopherases' is suggested) being readily soluble.

In contrast to this lack of adenosinetriphosphatase in the soluble fraction, a high adenosinetriphosphatase activity is associated with the water-insoluble proteins of muscle. This enzymatic activity is easily brought into solution by all the buffer and concentrated salt solutions usually employed for the extraction of myosine. On precipitation of myosine from such extracts, the adenosinetriphosphatase activity is always found in the myosine fraction, whichever mode of precipitation be used: dialysis, dilution, cautious acidification, salting out. On repeated reprecipitations of myosine, the activity per mgm. nitrogen attains a fairly constant level, unless denaturation of myosine takes place. Under the conditions of our experiments (optimal conditions have not been determined) the activity of myosine preparations ranged in different experiments from 350 to 600 microgram phosphorus liberated per mgm. nitrogen in 5 min. at 37°. Expressed as

$$Qp \left(= \frac{\mu gm.\ P/31 \times 22.4}{mgm.\ N \times 6.25 \times hour} \right)$$

this gives values of 500–850.

Acidification to *p*H below 4, which is known to bring about the denaturation of myosine,[1] rapidly destroys the adenosinetriphosphatase activity. Most remarkable is the extreme thermolability of the adenosinetriphosphatase of muscle: the enzymatic activity shown by myosine solutions is completely lost after 10 min. exposure to 37°. This corresponds with the well-known thermolability of myosine.[2] In respect of its high thermolability adenosinetriphosphatase resembles the protein of the yellow enzyme, which when separated from its prosthetic group is also rapidly inactivated at 38° (Theorell[3]). Evidently in the intact tissue of the warm-blooded animal (all experiments were performed on rabbit muscles), some conditions must exist which stabilize the myosine against the action of temperature. A marked stabilizing effect on the adenosinetriphosphatase activity seems to be produced by the adenylic nucleotide itself. As can be seen from the accompanying graph, in the presence of adenosinetriphosphate the liberation of phosphate proceeds at 37° over a considerable period (Curves I, Ia and Ib), whereas the same myosine solution warmed alone to 37° for 10–15 min. shows on subsequent addition of adenosinetriphosphate an insignificant or no mineralization whatever.

Crude buffer extracts accomplish a quantitative hydrolysis of the labile phosphate groups of adenosinetriphosphate; myosine, reprecipitated three times, liberates but 50 per cent of the theoretical amount of phosphorus (see

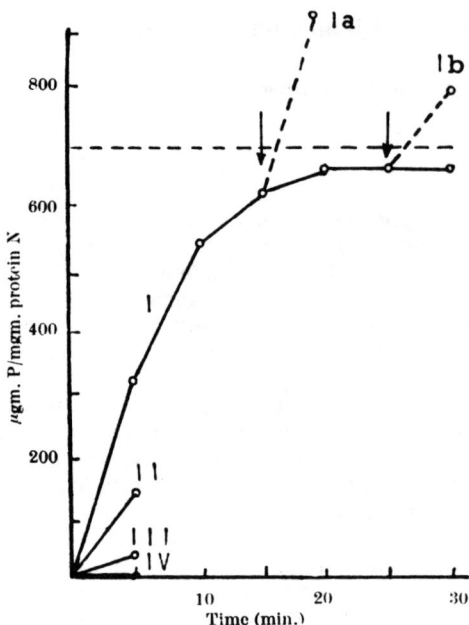

figure). It acts as true adenosine-*tri*-phosphatase and yields adenosinediphosphate, which is not further dephosphorylated and has been isolated in substance. This may serve as a convenient way of preparing adenosinediphosphate, instead of using crayfish muscle.[4] The adenosinediphosphatase is thus associated with the more soluble proteins, occupying an intermediate position between adenosinetriphosphatase and the most readily soluble phosphopherases.

Under no conditions tested could we obtain a separation of adenosinetriphosphatase from myosine. Either the activity was found in the myosine precipitate or else it was absent from the precipitates and from the remaining solution. This disappearance of the enzymatic activity we regard as the result of the start of denaturation of the very unstable myosine.

We are led to conclude that the adenosinetriphosphatase activity is to be ascribed to myosine or, at least, to a protein very closely related to and at present not distinguishable from myosine. Thus the mineralization of adenosinetriphosphate, often regarded as he primary exothermic reaction in muscle contraction, proceeds under the influence and with the direct participation of the protein considered to form the main basis of the contractile mechanism of the muscle fibre.

Selection 100

By 1940 a solid base appeared to have been established for the understanding of carbohydrate metabolism under anaerobic conditions. Harden and Young's observation some thirty-five years earlier that the rate of alcoholic fermentation related to the presence or absence of inorganic phosphate turned out to be indeed a key discovery.[a] It became the starting point for looking upon phosphorylation of carbohydrates as the ways and means of splitting them to obtain energy *and* to transfer it for biological purposes. Earlier passages have focused attention on the way in which knowledge increased about concordant intermediate steps in phosphorylitic cleavage of carbohydrates in yeast fermentation and muscle glycolysis, although little if any consideration was given to the underlying enzymic activity. What was actually emerging was that phosphorylation of carbohydrates could be seen as a metabolic cycle, ostensibly involving a sequence of reversible enzymatic reactions ('phosphate cycle'). The idea of cyclic phosphate transfer in muscle contraction was explicitly discussed by Dorothy Needham in 1937:[b]

There is little doubt that in appropriately prepared [muscle] extracts . . . reactions can take place with little if any formation or disappearance of free phosphate – the organic phosphate functioning in practically closed cycles.
    The presence in muscle extract of this beautifully co-ordinated series of enzymes – such that circulating phosphate can be kept always in the organic form, its transfer ensuring economical transfer also of energy – might tempt one to assume that, in the intact muscle itself, such cycles are used to the full.

Four years later, reviewing the field of intermediate carbohydrate metabolism, Meyerhof summarized current knowledge about the sequence of reactions in alcoholic fermentation and muscle glycolysis as follows:[c]

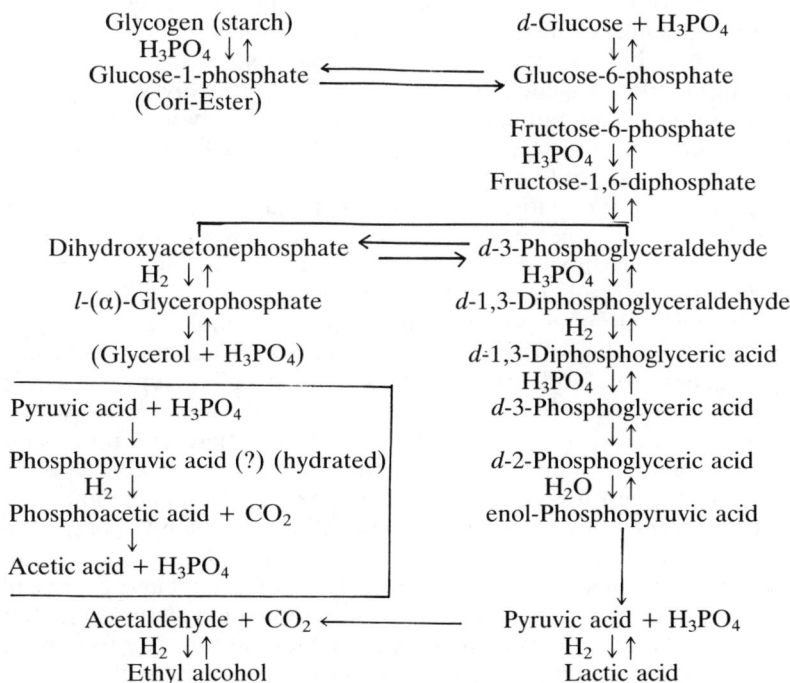

Glycogen (starch)   *d*-Glucose + H_3PO_4
  H_3PO_4 ↓↑   ↓↑
Glucose-1-phosphate ⇄⟶ Glucose-6-phosphate
(Cori-Ester)   ↓↑
                                        Fructose-6-phosphate
                                          H_3PO_4 ↓↑
                                        Fructose-1,6-diphosphate
                                                 ↓↑
Dihydroxyacetonephosphate ⇄⟶ *d*-3-Phosphoglyceraldehyde
  H_2 ↓↑ H_3PO_4 ↓↑
l-(α)-Glycerophosphate   *d*-1,3-Diphosphoglyceraldehyde
         ↓↑ H_2 ↓↑
(Glycerol + H_3PO_4)   *d*-1,3-Diphosphoglyceric acid
                                          H_3PO_4 ↓↑
Pyruvic acid + H_3PO_4 *d*-3-Phosphoglyceric acid
         ↓   ↓↑
Phosphopyruvic acid (?) (hydrated)   *d*-2-Phosphoglyceric acid
  H_2 ↓ H_2O ↓↑
Phosphoacetic acid + CO_2 enol-Phosphopyruvic acid
         ↓
Acetic acid + H_3PO_4 ↓

Acetaldehyde + CO_2 ⟵   Pyruvic acid + H_3PO_4
  H_2 ↓↑ H_2 ↓↑
Ethyl alcohol   Lactic acid

With this scheme in mind, it is not surprising that he who during the early 1920s postulated in muscle a closed cycle of interconversion of glycogen and lactic acid (anaerobic and spontaneous cleavage, oxidative and involuntary resynthesis) returned to it by stating:[d]

The original concept of a metabolic carbohydrate cycle involved the assumption that in the stationary state the quotient results from a continuous overlapping of anaerobic glycolysis and of oxidative resynthesis of the cleavage products – the endothermic resynthesis made possible by coupling with oxidation. Today it seems possible to refine this scheme and to modify it somewhat without rejecting the main argument. Indeed, in the past fifteen years a tremendous amount of material has been collected to prove that the general concept of these cycles in carbohydrate breakdown holds good, that every oxidative step is coupled with an involuntary phosphorylation, and that the several intermediate stages of the anaerobic breakdown can be reversed by means of the 'energy-rich phosphate bonds' . . . created in this way.

The concept of energy-rich phosphate bond, to which Meyerhof referred, was developed by Fritz Lipmann (1899–1986) in a review article in 1941, which attracted considerable attention. It originated in his studies of pyruvic acid oxidation by *Bacterium Delbrückii* when he demonstrated the participation of acetyl phosphate in this process (indicated in the insertion in the above diagram).[e] He suggested that this acetate was an 'active' intermediary, furnishing phosphate to the adenylic system, thus making the latter's role as a primary energy source for synthetic reactions carried out by cells more comprehensible.

Lipmann's purpose was to provide a deeper understanding of phosphate transfer of free energy changes (ΔF) in relation to the stepwise release of energy from metabolized carbohydrate. Lipmann suggested that 'phospho-organic compounds' (organic phosphates) could be roughly divided into two groups according to the amount of energy set free on their cleavage and, in turn, he related this to the type of phosphate linkage. Thus he assumed that on hydrolysis of phosphate esters ΔF amounted to about -3 kcal per mole and he designated the phosphate ester bond as 'energy-poor phosphate bond' ($-$ph). On the other hand substances containing linkages of the type P – O – P, N – P, carboxyl-P and enol-P, Lipmann classified as compounds with 'energy-rich phosphate bonds' ($\sim$ph) because on their hydrolysis the value for ΔF was estimated to be about -10 kcal per mole.

It is not without interest that, as shown below, in his attempt to make 'the metabolic generation and circulation of this peculiar type of chemical energy' more plausible, Lipmann resorted to comparison with the production of electrical current by a dynamo and its potentialities. Proceeding from an analogy between the chemical energy of the above 'high' phosphate groups and electrical energy, Lipmann imagined a 'phosphate current' playing 'in cell organization . . . a similar part as does electrical current in the life of human beings. It is a form of energy utilized for all-round purposes.' Like electricity, the potential energy present in energy-rich bonds was transmissible, and Lipmann pictured metabolic energy (or a large part of it) being made available in this way to living cells for mechanical and osmotic work and for biosynthetical activities.[f]

F. Lipmann, *Adv.Enzymol.*, *1*, 99. 1941

'Metabolic generation and utilization of phosphate bond energy'

p. 100

For a long time after its discovery by Harden and Young, phosphorylation of hexose in alcoholic fermentation was thought to be significant only as a means of modeling the hexose molecule to fit it for fermentative breakdown. However, as the outcome of intensive study of the intermediate reactions in fermentation and the relation between muscular action and metabolism, it later became evident that the primary phosphate ester bond of hexose changes metabolically into a new type of energy-rich phosphate bond.[1] In this bond large amounts of energy made available by the metabolic process accumulate. The recent recognition that in nature there occurs a widespread utilization of such phosphate bonds[2,3] as energy carriers, necessitates a still further revision of the earlier view concerning the biological significance of phosphate turn-over. During various metabolic processes phosphate is introduced into compounds not merely, or at least not solely, to facilitate their breakdown, but as a prospective carrier of energy.

pp. 101–2

The availability of the energy-rich phosphate bond (~ph) in absence of glycolytic or combustion energy and the ease and effectiveness with which glycolysis and combustion energy could be converted into ~ph, suggested[4] that the energy utilized in the mechanical set-up of muscle under all circumstances was derived from energy-rich phosphate bonds, supplied constantly by glycolytic or oxidative foodstuff disintegration. The manner in which this supply took place remained, however, entirely obscure.

The study of intermediate reactions in glycolysis and fermentation with tissue and yeast extracts furnished the first explanation of the chemistry of such energy transfers.

The understanding of the transfer mechanism in anaerobic glycolysis still left much unexplained as to how creatine-phosphate could be synthesized in purely aerobic metabolism, especially in the presence of iodoacetic acid.[5,6] The creatine in muscle must be considered as a natural trap or storehouse for ~ph. Every metabolic process utilizable for the rebuilding of creatine phosphate must generate energy-rich phosphate bonds. A partial explanation developed when it was found that keto acid oxidation, undoubtedly occurring to some extent in aerobic carbohydrate breakdown, can furnish energy-rich phosphate, which, when brought over to creatine, would reform creatine ~ph (Lipmann[7]). A more general study of purely oxidative phosphorylation, found to occur abundantly in extracts of kidney and liver, was initiated by the work of Kalckar[8] and is being continued in Cori's laboratory.[9] Here indications are found that present knowledge of the chemistry of generation and transfer of phosphate bonds is far from complete. More and more clearly it appears that in all cells a tendency exists to convert the major part of available oxidation-reduction energy into phosphate bond energy.

The metabolism of muscle is an almost unique case in nature of a straight-forward utilization of chemical energy. Here the need of organization into a uniform type is understandable. In all other cells the energy problem is much more complex. If, as in growth, foodstuff is transformed into protoplasm, the comparison of the free energies of starting material and final product frequently does not show appreciable difference, i.e., storage of energy may be insignificant. The extra 'energy of synthesis' needed here is used only in such a manner as to force chemical processes to go in desired directions.

pp. 121–3

Several reaction phases can be distinguished in the constantly occurring metabolic turn-over of phosphate: (1) introduction of inorganic phosphate into ester linkage, (2) generation of energy-rich phosphate bonds (~ph) by oxidation-reduction, (3) the taking over and distribution of ~ph by cell catalysts (e.g., adenylic acid), (4) utilization of ~ph and the regeneration of inorganic phosphate. The machine-like functioning of the revolving sequence of reactions appears from the schematic representation shown in Figure 13.

   For the undisturbed maintenance of this complicated series of reactions a well-balanced equilibrium is needed between the great number of enzymatic reactions involved. Oxido-reductive formation of ~ph and its removal by adenylic acid must follow each other in due course, in order to avoid obstruction of the smooth flow of reactions. A finely coordinated interplay between oxidation-reduction and phosphorylation-*de*-phosphorylation results. This explains why very often adenylic acid functions as apparent coenzyme of O/R reactions. In many if not all such cases the part played by the adenylic acid seems not related to O/R catalysis proper but to removal of obstructing phosphate groups. This was shown to be the case in the O/R reaction in the fermentation sequence (Warburg and Christian[10]):

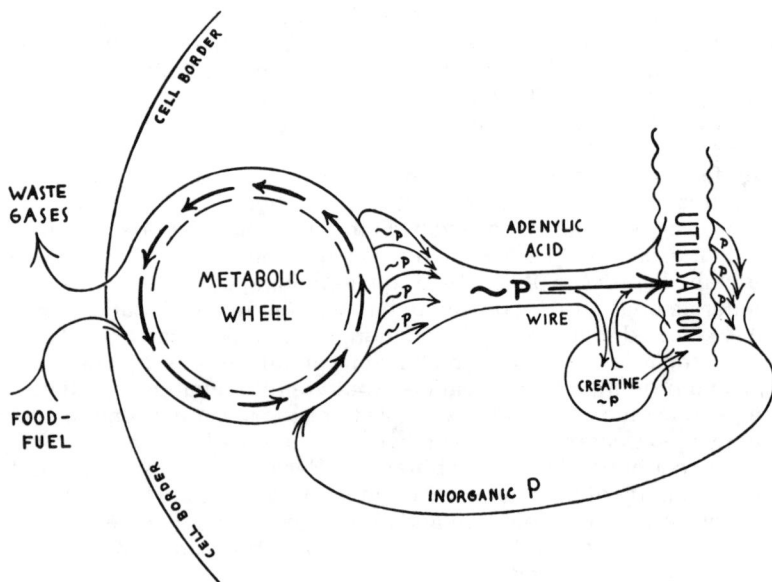

Figure 13 Schematic representation of the 'metabolic dynamo generating phosphate current'

$\longrightarrow$ pyridine + di-ph-glyceraldehyde $\rightleftarrows$

reduced pyridine + ph-glyceryl~ph   ph-glycerate $\longrightarrow$
+ $\longrightarrow$ +
ad-ph~ph   ad-ph~ph~ph

Ph-glyceryl~ph, if not dephosphorylated, would be accumulated which soon would cause the back reaction to take place to such an extent that a state of equilibrium would be established. Thus the forward reaction would come to a standstill and thereby the whole reaction chain would be broken. A similar case seems to be the oxidation of pyruvic acid where a phosphorylated oxidation product primarily is formed (Lipmann[11,3]) and adenylic acid is required for further oxidation (Banga, Ochoa, and Peters[12]). Disturbance of this finely balanced interplay is prone to lead to 'pathological' reactions in cell extracts which in many respects are certainly out of balance. Also in the body, such disturbances quite probably are the cause of pathological conditions. Recent work by Shorr, Barker, et al.,[13] is suggestive of the existence of a deficiency in phosphate turnover in diabetic tissues.

As an example of a reaction which is 'pathological' in certain respects, the Harden-Young fermentation in brewer's yeast extract might be cited,

$$2 \text{ hexose} + 2 \text{ phosphate} \longrightarrow \text{ alcohol} + CO_2 + \text{hexosediphosphate}.$$

Here, following and slightly expanding the interpretation of Harden,[14] it is assumed that the returning of inorganic phosphate into the cycle is out of function (see Figure 13). An 'over'-phosphorylation of hexose occurs possibly because of interruption of the normal utilization of ~ph, by way of which the inorganic phosphate is normally recovered. Through the erroneous course the ~ph takes, hexosediphosphate accumulates, and inorganic phosphate, needed in the cycle, eventually disappears. The result is an almost complete stop of fermentation when the inorganic phosphate, if present only in limited amounts, is irreversibly fixed.

In fact, the phosphate can only fulfill its function as a true catalyst in fermentation by entering and leaving the reaction cycle at the same rate. Nevertheless, in the preliminary phases, very often a more or less pronounced accumulation of phosphate ester does occur. Thus, in extracts of baker's yeast (Lipmann[15]) only in the beginning occurs extensive ester formation. The system never becomes entirely depleted of inorganic phosphate and fermentation continues with almost constant rate. Preliminary ester formation was likewise demonstrated at the onset of fermentation in living cells (Macfarlane[16], Wiggert and Werkman.[17] With slow fermenting fusarium lini, however, phosphorylation was not found in the beginning (Nord[18,19]). First in the later phases of this fermentation phosphorylation was observed. This was considered as indicating a non-phosphorylating fermentation mechanism.[a]

Notes to Section IV

Selection 73

Introduction:

a.   See Selection 1, Section I.
b.   The term 'carbohydrate' is due to C. Schmidt who introduced it in 1844. See his article 'Ueber Pflanzenschleim und Bassorin', *Ann.*, *51*, 29–62. 1844.

Selection 74

Introduction:

a.   See A.N. Shamin and A.I. Volodarsky, 'Kirchhof, Konstantin Sigizmundovich', *Dictionary of Scientific Biography* (New York, 1973), VII, pp. 378–9.

Text:

a.   Alcohol.
b.   Probably fruit sugar (fructose).
c.   There is no indication at this stage that Kirchhof identified the sugar with grape sugar (glucose), the isolation of which from grapes J.L. Proust (1754–1826) briefly reported in 'Sur le sucre de raisin', *J.Phys.*, *54*, 198. 1802. Returning to the subject four years later, Proust dealt with it more extensively and very interestingly. Among other recommendations, he urged governments to pay attention to grape sugar as a more readily available source of food than cane sugar. Cf. 'Mémoire sur le sucre de raisin', *J.Phys.*, *63*, 257–78. 1806; 'Suite . . . sur le sucre de raisin', *J.Phys.*, *63*, 341–62. 1806.

Selection 75

Text:

a.   Native strontium sulphate.

Selection 76

Introduction:

a.   A.I. Ihde, 'Chemical analysis and the growth of biochemistry', *Actes du XIe Congrès International d'Histoire des Sciences* (Warsaw-Cracow, 1965), IV, 143–7 (p. 143). See also R.E. Lyle and G.G. Lyle, 'A brief history of polarimetry', *J.Chem.Ed.*, *41*, 308–13. 1964.
b.   A.P. Crosland, 'Biot, Jean Baptiste', *Dictionary of Scientific Biography* (New York, 1970), II, p. 139.
c.   J.B. Biot, 'Sur la construction des appareils destinés à observer le pouvoir rotatoire des liquides', *Comp.Rend.*, *11*, 413–32. 1841.

Text:

a.   The analyzer of the polarimeter.

Selection 77

Introduction:

a.   Cf. Selections 162 and 164, Section VII.
b.   Thanks to historical scholarship going back close on twenty five years, we have a very much clearer picture about Bernard's work on the glycogenic function of the liver. See M.D. Grmek, 'The glycogenic function of the liver', *J.Hist.Biol.*, *1*, 141–54. 1968; F.L. Holmes, *Claude Bernard and Animal Chemistry* (Cambridge, Mass., 1974). For an earlier treatment of Bernard's work on the glycogenic function of the liver see two articles by F.G. Young, 'Claude Bernard and the glycogenic function of the liver', *Ann.Sci.*, *2*, 47–83. 1937; 'Claude Bernard and the discovery of glycogen. A century of retrospect', *Brit.Med.J.* (*1*), 1431–37. 1957.
c.   C. Bernard, *Nouvelle fonction du foie considéré comme organe producteur du matière sucré chez l'homme et les animaux* (Paris, 1853).
d.   A. Sanson, 'Note sur la formation physiologique du sucre dans l'economie animale', *Comp.Rend.*, *44*, 1323–5. 1857. Cf. 'La matière glycogène ou l'amidon animal se rencontre exclusivement dans le tissue du foie et aucun autre organe de l'économie n'en dénote le moindre trace' in C. Bernard, 'Remarques sur la formation de la matière glycogène du foie', *Comp.Rend.*, *44*, 1326. 1857.
e.   Selection 162, Section VII.

f. Holmes, *Claude Bernard*, pp. 406–7. For Bernard's work on the action of pancreatic juice on fats, see Selection 139, Section VI.
g. Selection 169, Section VII.

Selection 78

Introduction:

a. E. Fischer, 'Über aromatische Hydrazinverbindungen', *Ber.chem.Ges.*, *8*, 589–94. 1875.
b. E. Fischer, 'Verbindungen des Phenylhydrazins mit den Zuckerarten, II', *Ber.chem.Ges.*, *20*, 821–34. 1887.
c. C.S. Hudson, 'Emil Fischer's discovery of the configuration of glucose', A.I. Ihde and W.F. Kieffer (eds). *Selected Readings on the History of Chemistry* (Easton, 1965), pp. 262–6.

Text:

1. Liebigs Annal.d.Chem. *154*, 30 [1870].
2. Berichte d.d.chem.Gesellsch. *4*, 834 [1871].

Selection 79

Introduction:

a. Fischer's suggestion of a 5-atom ring in the sugars was taken up by Thomas Purdie (1843–1916) and James Irvine (1877–1952) in 1903 (see T. Purdie and J.C. Irvine, 'The alkylation of sugars', *J.Chem.Soc.*, *83*, 1021–37. 1903). Purdie had introduced a process whereby all the OH groups of glucosides could be methylated by means of methyl iodide and silver oxide, a method later improved by Walter Norman Haworth (1883–1950) during the 1920s. Use of this method showed clearly that the OH on the fourth carbon atom could not be taking part in the ring. The more stable form of glucose was found to have a 6-membered ring, while the more reactive form (known as γ-glucose) gave a 5-membered ring. See W.N. Haworth, *The Constitution of Sugars* (London, 1929). Also S.A. Barker and N. Baggett, 'Sugars', *TIBS*, *3*, 140–1. 1978.
b. Selection 19, Section I.

Text:

a. A. Michael, 'Sur la synthèse du phenolglucoside et de l'orthoformylglucoside ou helicine', *Comp.Rend.*, *89*, 355–8. 1879.
b. $C_6H_5CO. CH_2.OH$.
c. $C_6H_5. CH(OH). CO.C_6H_5$.

Selection 80

Introduction:

a. Selection 20, Section I.
b. J. Stoklasa, J. Jelinek, E. Vitek, 'Der anaerobe Stoffwechsel der höheren Pflanzen und seine Beziehung zur alkoholischen Gärung', *Hofmeisters Beitr.*, *3*, 460–509. 1903; J. Stoklasa and F. Czerny, 'Isolirung des die anaërobe Athmung der Zelle höher organisirten Pflanzen und Thiere bewirkenden Enzyms', *Ber.chem.Ges.*, *36*, 622–34. 1903; J. Stoklasa and F. Czerny, 'Beiträge zur Kenntniss der aus der Zelle höher organisisirter Thiere isolirten gährungserregenden Enzyme', *Ber.chem.Ges.*, *36*, 4058–69. 1903; J. Stoklasa, J. Jelinek, E. Šimáček, E. Vítek, 'Alkoholische Gährung im Thierorganismus und die Isolirung gährungserregender Enzyme aus Thiergeweben. I. Theil.', *Pflügers Arch.*, *101*, 311–39. 1904. Stoklasa's pursuits of this 'theoretical' theme grew out of his wide 'practical' agricultural research interests.
c. See Selection 22, Section I.
d. A. Wroblewski, 'Über den Buchner'schen Hefepressaft', *J.prakt.Chem.*, *64*, N.S. (*102*), 1–70. 1901. This is a shortened version of the Polish original which was published in the

Transactions of the Polish Academy of Sciences in Cracow. Wróblewski's preliminary communications on the subject had been appearing since 1898.

e.   A. Fernbach and L. Hubert, 'De l'influence des phosphates et de quelque autres matières minérales sur la diastase protéolytique du malt', *Comp.Rend.*, *131*, 293–5. 1900.

f.   Cf. also Korman, 'Discovery', and Kohler, 'Background', Introduction to Selection 22, Section I, note b.

Text:

1.   'Roy. Soc. Proc.', B, vol. 77, 1906, p. 405.
2.   Young, 'Chem.Soc.Proc.', 1907, vol. 65.

Selection 81

Introduction:

a.   In this year J.J. Berzelius (1779–1848) published the second volume of *Föreläsningar i djurkemien* (Lectures in Animal Chemistry) where he disclosed his detection (1807) of an acid in the flesh of freshly slaughtered animals, identified by him as lactic acid. The latter was discovered in 1780 in sour milk by his Swedish compatriot, C.W. Scheele (1742–86). Berzelius referred to his finding in communications published under 'Vermischte Notizen' (Miscellaneous Notes) in the *Journal für die Chemie, Physik und Mineralogie*, 7, 580–5. 1808 and 9, 585–9. 1810 (not seen). Lactic acid received a great deal of attention from J. Liebig (1803–73) and it was their respective claim of having pioneered the understanding of the chemistry and distribution of this organic compound that completely sealed the rift between the two. Cf. Selection 165, Section VII. See also J. Berzelius, *Jahres-Ber.*, *27*, 585–94. 1848; A.W. Hofmann (ed.), *Aus Justus Liebig's und Friedrich Wöhler's Briefwechsel in den Jahren 1829–1873* (Braunschweig, 1888), I, pp. 295–8; E. Hjelt (ed.), *Aus Jac.Berzelius' und Gustav Magnus' Briefwechsel in den Jahren 1828–1847* (Braunschweig, 1900), p. 185; O. Wallach (ed.), *Briefwechsel zwischen J. Berzelius und F. Wöhler* (Leipzig, 1901), p. 632. For information on Scheele's work on lactic acid (and other organic acids) consult F. Lieben, *Geschichte der physiologischen Chemie* (Leipzig and Vienna, 1935), p. 33.

b.   W.M. Fletcher, 'The survival respiration of muscle', *J.Physiol.*, *23*, 10–99. 1898–9; 'The influence of oxygen upon the survival respiration of muscle', *J.Physiol.*, *28*, 354–9. 1902; 'The relation of oxygen to the survival metabolism of muscle', *J.Physiol.*, *28*, 474–98. 1902. See also Selection 52, Section III.

c.   See Introduction to Selection 80, note b. For a contemporary review of the 'lactic acid theory of alcoholic fermentation' consult A. Harden, *Alcoholic Fermentation* (London, 1911), p. 90f.

Text:

1.   An account of the chief experimental results described in this paper was given to the Physiological Society, May 12, 1906.
2.   Liebig. *Arch.f.Anat.Phys.u.Wiss.Med.*, p. 393. 1850. (Humboldt's experiments in 1797 are given here).
3.   Ludwig & Schmidt. *Ludwig's Arbeiten* (Leipzig, 1869).
4.   Fletcher. This journal, xxviii. p. 474. 1902.
5.   Ibid. p. 480.
6.   Fletcher. This journal, xxviii. pp. 354 and 488. 1902.
7.   In connection with the subject matter of this and the preceding paragraph, compare J.B. Leathes, *Some Problems of Metabolism*,[a] 1906. Chapter iii. p. 50.
a.   The correct title: *Problems in Animal Metabolism*.

Selection 82

Introduction:

a.   C. Neuberg and A. Hildesheimer, 'Über zuckerfreie Hefegärungen.I.', *Biochem.Z.*, *31*, 170–6. 1911. This discovery led to the conception of what the Neuberg school termed 'sugar-free yeast fermentation' (*zuckerfreie Hefegärungen*). See also C. Neuberg and J. Kerb, 'Entsteht bei zuckerfreien Hefegärungen Äthylalkohol?', *Z.Gärungsphysiol.*, *1*, 114–23. 1912; also 'Über zuckerfreie Hefegärungen. XII. Über die Vorgänge bei der Hefegärung', *Biochem.Z.*, *53*, 406–19. 1913.

b. There was also the remarkable fact that the scheme was made use of in the industrial production of glycerol in Germany during the First World War. See J.S. Fruton, *Molecules and Life* (New York, 1972), p. 348.

Text:

1. [A. Trillat, *Compt. rend.de l'Acad. d. Sc.*, 23 March 1908; *Ann. d.l'Inst. Pasteur*, *22*, 1908; *24*, 1910].
2. C. Neuberg and J. Kerb, this journal, *43*, 494. 1912.
3. For further information see C. Neuberg, *Die Gärungsvorgänge und der Zuckerumsatz der Zelle*. Monogr. 1913, F. Fischer, Jena.
4. If for example glyceraldehyde is once certainly shown as an intermediate then its transition into methyl glyoxal, as well as a mutual Cannizzaro reaction between these two substances (which gives *glycerol* and *pyruvic acid*) fits in at once with the above scheme.

Selection 83

Introduction:

a. G. Embden and F. Kraus, 'Über Milchsäurebildung in der künstlich durchströmten Leber. I. Mitteilung', *Biochem.Z.*, *45*, 1–17. 1912. On Bernard's discovery of the glycogenic function of the liver, see Selection 77. On the beginnings of the perfusion technique, see Introduction to Selection 53, Section III, note a.
b. G. Embden, Fr. Kalberlah and H. Engel, 'Über Milchsäurebildung im Muskelpresssaft. I. Mitteilung', *Biochem.Z.*, *45*, 45–62. 1912.
c. F. Lieben, *Geschichte der physiologischen Chemie* (Leipzig and Vienna, 1935), p. 219.
d. It is interesting that during the First World War Embden sought to relate the emerging knowledge of the role of phosphate in muscular activity to problems of medical practice. He tested the physical performance of soldiers and miners and found it was improved by additions of phosphate to the diet. See E. Lehnartz, 'Gustav Embden', *Arbeitsphysiologie*, *7*, 481. 1933–4.
e. O. Meyerhof, 'Die Energieumwandlungen im Muskel. IV. Über die Milchsäurebildung in der geschnittenen Muskulatur', *Pflügers Arch.*, *188*, 117–160. 1921. (Essentially section II).

Text (i):

1. G.J. Engelmann, *Archiv.f.d.ges.Anatomie u. Physiologie*, 1871, p. 14.

Selection 84

Text (ii):

a. O. Meyerhof, 'Die Energieumwandlungen im Muskel. I. Über die Beziehungen der Milchsäure zur Wärmebildung und Arbeitsleistung des Muskels in der Anaerobiosis', *Pflügers Arch.*, *182*, 232–83. 1920.
b. A.V. Hill and W. Hartree, 'The four phases of heat-production of muscle', *J.Physiol.*, *54*, 84–128. 1920–1.

Text (iii):

1. Only glycogen.
a. In 1921 F.G. Hopkins welcomed Meyerhof's results: 'Calculated as for one gram of muscle the average of Meyerhof's results gives for the heat of contraction 0.75 cals; for that of recovery 1.0 cal.; and for the heat of the lactic acid (or carbohydrate) actually burnt during recovery 1.75 cals. If we calculate in round figures the calories corresponding to 1 gram of lactic acid either produced or removed we get the data in convenient form:

Heat of anaerobic contraction	Heat of recovery in oxygen	Total heat of of activity in oxygen	Heat of combustion of materials burnt
400 cals.	500 cals.	900 cals.	900 cals.

This is a balance sheet which any auditor would pass as satisfactory.' See F.G. Hopkins, 'The chemical dynamics of muscle', *Johns Hopkins Hosp. Bull.*, *32*, 365. 1921 (Lecture II of the Herter Series).

Selection 85

Introduction:

a. G. Embden and M. Zimmermann, 'Über die Chemie des Lactacidogens. V.', *Z.physiol.Chem.*, *167*, 114–36. 1927. Cf. also Selection 96, note a.
b. Ph. Eggleton and G.P. Eggleton, 'The inorganic phosphate and a labile form of organic phosphate in the gastrocnemius of the frog', *Biochem.J.*, *21*, (1), 190–5. 1927; 'The physiological significance of "phosphagen"', *J.Physiol.*, *63*, 155–61. 1927.
c. Also spelled: SubbaRow.
d. Ph. Eggleton and G.P. Eggleton, 'Further observations on phosphagen', *J.Physiol*, *65*, 15–24. 1928.

Text:

1. C.H. Fiske and Y. Subbarow, *J.Biol.Chem.*, vol. 66 (375) – 1925.

Selection 87

Introduction:

a. G. Embden and F. Laquer, 'Über die Chemie des Lactacidogens. I. Mitteilung. Isolierungsversuche', *Z.physiol.Chem.*, *93*, 94–123, 1914. Especially pp. 102, 118, 120.
b. G. Embden and M. Zimmermann, 'Über die Bedeutung der Adenylsäure für die Muskelfunktion. I. Mitteilung: Das Vorkommen von Adenylsäure in der Muskulatur', *Z.physiol.Chem.*, *167*, 137–40. 1927.
c. J. Liebig, 'Ueber die Bestandteile der Flüssigkeiten des Fleisches', *Ann.*, *62*, 257–369. 1847.
d. G. Schmidt, 'Über fermentative Desaminierung im Muskel', *Z.,physiol.Chem.*, *179*, 243–82. 1928. For excerpt translated from this paper see H.M. Kalckar, *Biological Phosphorylations Development of Concepts* (Englewood Cliffs, 1969), pp. 45–53.
e. Now Lvov in Ukrainian SSR.

Selection 88

Introduction:

a. Cf. Selection 118, Section V. note a.
b. See A. von Muralt, 'The development of muscle-chemistry, a lesson in neurophysiology' in D. Nachmansohn (ed.), *Metabolism and Function* (New York, 1950), p. 126.

Text (i):

1. P. Eggleton and G.P. Eggleton, *Biochem.Journal*, *21*, 190. 1927.

Selection 89

Introduction:

a. K. Lohmann, 'Über der Vorkommen und den Umsatz von Pyrophosphat in Zellen. I. Mitteilung: Nachweis und Isolierung des Pyrophosphats', *Biochem.Z.*, *202*, 466–93. 1928.

See further K. Lohmann, 'Über das Vorkommen und den Umsatz von Pyrophosphat in Zellen. II. Mitteilung. Die Menge der leicht hydrolysierbaren p-Verbindung in tierischen und pflanzlichen Zellen', *Biochem.Z.*, *203*, 164–71. 1928; idem, 'Über das Vorkommen und den Umsatz von Pyrophosphat in Zellen. III. Mitteilung: Das physiologische Verhalten des Pyrophosphats', *Biochem.Z.*, *203*, 172–207. 1928.
b.   O. Meyerhof, 'Über die enzymatische Milchsäurebildung im Muskelextrakt. I.', *Biochem.Z.*, *178*, 395–418. 1926.
c.   R.A. Peters, 'Otto Meyerhof', *Biog.Mem.F.R.S.*, *9*, 175–200. 1954 (p. 182).
d.   K. Lohmann, 'Über die Pyrophosphatfraktion im Muskel', *Naturwiss.*, *70*, 381–2. 1929.
e.   C.H. Fiske and Y. Subbarow, 'Phosphorus compounds of muscle and liver', *Science*, *70*, 381–2. 1929.

Selection 90

Introduction:

a.   O. Meyerhof, 'Über den Zusammenhang der Spaltungsvorgänge mit der Atmung in der Zelle', *Ber.chem.Ges.*, *58*, 991–1001. 1925. Cf. also Introduction to Selection 64, Section III, note b.
b.   See also Engelhardt's stimulating retrospective contribution on the subject: 'On the dual respiration', *Proceedings of the Conference on the Historical Development of Bioenergetics*, (sponsored by the American Academy of Arts and Sciences) (Boston, 1973), pp. 61–9. Engelhardt is here critical of Meyerhof though the latter influenced his own research. Thus the paper from which the previous selection was taken did provide Engelhardt with some clues for his own experimental work.

Text:

1.   W.A. Engelhardt, this journal, *227*, 16. 1930.
2.   W.A. Engelhardt, *Kasansk Mediz.J.*, *27*, 496. 1931. (In Russian.)

Selection 91

Introduction:

a.   H. Ohle, 'Die Chemie der Monosaccharide und der Glykolyse', *Erg.Physiol.*, *33*, 558–701. 1931. (See especially pp. 698–701).
b.   H.O.L. Fischer and E. Baer, 'Über die 3-Glycerinaldehydphosphorsäure', *Ber.chem.Ges.*, *65*, 337–45. 1932.
c.   C.V. Smythe and W. Gerischer, 'Über die Vergärung von Hexosemonophosphorsäure und 3-Glycerinaldehydphosphorsäure', *Biochem.Z.*, *260*, 414–16. 1933.
d.   On 3-carbon compound cleavage in alcoholic fermentation without involvement of phosphate, see Selection 82.

Text:

1.   In agreement with this view Embden and Jost[g] could show recently that under the conditions used in the present work sythetically prepared d-l-glyceraldehydephosphoric acid forms large amounts of lactic acid.
2.   If glucosediphosphoric acid is split according to the same scheme, then one of the two dismutation products would be phosphorylated not at the terminal C-atom but at a middle C-atom; thus, for example, instead of α-glycerophosphoric acid its β-form could arise.
3.   In the fresh muscle of animals investigated up to the present exclusively hemose*mono*phosphoric is found, according to Embden and Zimmermann; and therefore this was termed '*lactacidogen*' by the authors and named as the first *preformed* precursor of lactic acid. The change of hexosemonophosphoric acid into hexosediphosphoric acid follows, not only under the action of definite, specially added salts favouring synthesis, but, as recently shown, on simple grinding of muscle brei with large amounts of kieselguhr (Embden, Jost and Lehnartz).[h] The lack of hexose*di*phosphoric acid in fresh muscle is perhaps just to be explained by this, that it is much more easily broken down than the

hexosemonophosphoric acid. The idea that the breakdown of hexosemonophosphoric acid is preceded by its conversion into hexosediphosphate was already expressed by Meyerhof and his collaborators.[i]

a. Institute of Vegetative Physiology, University of Frankfurt on Main.
b. Selection 80.
c. K. Lohmann, 'Uber die Isolierung verschiedener natürlicher Phosphorsäureverbindungen und die Frage ihrer Einheitlichkeit', *Biochem.Z.*, *194*, 306–7. 1928.
d. C.H. Fiske and Y. Subbarow, 'Phosphorus compounds of muscle and liver', *Science, 70*, 381–2. 1929.
e. E. Bumm and K. Fehrenbach, 'Über verschiedene Wege des Zuckerabbaues im tierischen Organismus', *Z.physiol.Chem.*, *193*, 238–50. 1930; ibid., *195*, 101–12. 1931.
f. Selection 82. The idea of methylglyoxal as an intermediate between hexosediphosphate and lactic acid persisted for several years. It is mentioned as a current view in 1932. See P.A. Shaffer and A. Ronzoni, 'Carbohydrate metabolism', *Ann.Rev.Biochem.*, *1*, 247–66. 1932.
g. G. Embden and H. Jost, 'Über die Zwischenstufen der Glykolyse in der quergestreiften Muskulatur', *Z.physiol.Chem.*, *230*, 69–89. 1934.
h. G. Embden, H. Jost and M. Lehnartz, 'Über die Entstehung von Hexosediphosphorsäure bei der Herstellung von Muskelpresssaft', *Z.physiol.Chem.*, *189*, 261–72. 1930.
i. Consult O. Meyerhof, *Die chemischen Vorgänge im Muskel und ihr Zusammenhang mit Arbeitsleistung und Wärmebildung* (Berlin, 1930), pp. 155–9.

Selection 92

Introduction:

a. Selection 89.

Text:

1. Meyerhof and Lohmann, *Naturwiss*, *15*, 670, 768. 1927; Meyerhof and Suranyi, this journal, *191*, 125. 1927; Meyerhof and Lohmann, ibid., *196*, 49. 1928.
2. Lundsgaard, this journal, *227*, 58. 1930.
3. Meyerhof and Lohmann, ibid., *253*, 431. 1932.
4. Meyerhof and Schulz, ibid., *236*, 71. 1931.
5. Lundsgaard, ibid., *233*, 322, 1931.
a. H. Lehmann (1910–85) showed soon after that reaction 2 is freely reversible. See H. Lehmann, 'Über die Umesterung des Adenylsäuresystems mit Phosphagen', *Biochem.Z.*, *286*, 336–43. 1936.
b. A reference to the enzymatic process under aerobic conditions.
c. But see Selection 89.

Selection 93

Text:

1. K. Lohmann, this journal, *282*, 109. 1935.
2. O. Meyerhof and J. Suranyi, ibid., *178*, 427. 1926; O. Meyerhof and K. Lohmann, ibid., *196*, 49. 1928.
3. Concerning the depiction of the binding to N-atom 9 of the adeninepentose, the position of which is not yet certain, compare J.M. Gulland and E.R. Holiday, *Nature*, *132*, 782. 1933.
4. K. Lohmann, this journal, *254*, 381. 1932.
5. Compare previous communication.[a]
a. K. Lohmann, 'Über die Aufspaltung der Adenylpyrophosphorsäure und die Argininphosphorsäure in Krebsmuskulatur', *Biochem.Z.*, *282*, 109–19. 1935.

Selection 94

Introduction:

a.   O. Meyerhof and D. McEachern, 'Über anaerobe Bildung und Schwund von Brenztraubensäure in der Muskulatur', *Biochem.Z.*, *260*, 417–45. 1933.

b.   O. Meyerhof and W. Kiessling, 'Über das Auftreten und den Umsatz der α-Glyzerinphosphorsäure bei der enzymatischen Kohlenhydratspaltung', *Biochem.Z.*, *264*, 40–71. 1933.

Selection 95

Text:

1.   *C.R.de la Soc. de biol.*, 1935, t.*120*, p. 307.

2.   It must be noted that this difference can be very important, because the chances for the polymolecular reactions diminish considerably in inverse ratio of the concentration.

a.   This refers to Meyerhof's long-held idea of the known esters as stabilization products of active (labile) intermediates. According to this, the Embden ester in muscle and the Robison ester in yeast were formed as stabilization products of a monoester in *statu nascendi* (see Selection 98, note 4). The Embden and Robison esters are known to be very similar, comprizing about 80 per cent aldosephosphate and about 20 per cent ketosephosphate. Neuberg ester consists mainly of fructosephosphate, apart from a very small aldosephosphate component.

Selection 96

Introduction:

a.   C.F. Cori, 'Some highlights of the early period of bioenergetics', *Proceedings of the Conference on the Historical Development of Bioenergetics* (sponsored by the American) Academy of Arts and Sciences), (Boston, 1973), p. 162.

Text:

1.   Cori, G.T. and Cori, C.F., Summaries of Communications, XVth International Physiological Congress, p. 66, 1935.

Selection 97

Introduction:

a.   Selection 94.

b.   D.M. Needham, 'Chemical cycles in muscle contraction' in J. Needham and D.E. Green (eds), *Perspectives in Biochemistry* (Cambridge, 1937), pp. 201–14 (p. 204).

c.   The communication in *Nature* (here reprinted in full) appeared independently and very shortly after Meyerhof had shown that this phosphate uptake depends on the presence of the adenylic system. The presence of a phosphate acceptor was necessary for rapid response, and for this Meyerhof used creatine. See O. Meyerhof, 'Über die Synthese der Kreatinphosphorsäure im Muskel und die 'Reaktionsform' des Zuckers', *Naturwiss.*, *25*, 443–6. 1937.

Text:

1.   Parnas and Baranowski, *Comptes rend.Soc.Biol.*, *120*, 307 (1935). Ostern, Guthke and Tersakowec, *Z. physiol. Chem.*, *243*, 9 (1936). Parnas and Mochnacka, *Comptes rend. Soc. Biol.*, *123*, 1173 (1936).

2.   Meyerhof and Kiessling, *Biochem.Z.*, *281*, 449; *283*, 83 (1935).

3.   Schäffner and Berl, *Z. physiol.Chem.*, *238*, 111 (1936).

4.   Dische, *Naturwiss.*, *22*, 776 (1934); *24*, 462 (1936).

5. Meyerhof and Lehmann, *Naturwiss.*, *23*, 337 (1935).
6. Parnas and Ostern, *Bull. Soc. Chim. Biol.*, *18*, 1471 (1936).
7. Lundsgaard, *Biochem. Z.*, *233*, 322 (1931).
a. D.M. Needham and R.K. Pillai, 'The coupling of oxidoreductions and dismutations with esterification of phosphate in muscle', *Biochem.J.*, *31*, 1837–51. 1937.

Selection 98

Introduction:

a. Selection 88.
b. Selection 90.
c. Selection 70, Section III.
d. H. Kalckar, 'Phosphorylation in kidney cortex' in H.M. Kalckar, *Biological Phosphorylations Development of Concepts* (Englewood Cliffs, 1969), pp. 208–10. See also H.M. Kalckar, 'Phosphorylation in kidney tissue', *Enzymologia*, *2*, 47–52. 1937.
e. Cf. H.M. Kalckar, 'Origins of the concept of oxidative phosphorylation' in *Proceedings of the Conference on the Historical Development of Bioenergetics* (sponsored by the American Academy of Arts and Sciences), (Boston, 1973), pp. 171–89 (including discussion).

Text:

1. Braunshteyn, A.E.: 'Report to the XV International Physiological Congress in Leningrad' (1935).
2. Severin, V.A.: *Biokhimiya*, *2* (1937), 60.
3. Grimlund, K.: *Skand. Arch. Physiol.*, *73* (1936), 109.
4. Meyerhof, O. and Boyland E.: *Biochem.Z.*, *237* (1931), 406; Meyerhof, O., Gemmill, Ch., and Benatato, G.: *Biochem.Z.*, *258* (1933), 371.
5. Kalckar, H.: *Enzymologia*, *2* (1937), 47.
6. Belitser, V.A.: *Biokhimiya*, *2* (1937), 334; *3* (1938), 80.
7. Belitser, V.A.: *Byulleten' Experimental'noy Biologii i Meditsiny*, *7* (1939), 111.
8. Cozymase [DPN] was prepared from beer yeast by the Meyerhof and Ohlmeyer method[9] with some modifications by Ochoa. Our preparation, as judged by its activity in sampling with apoenzyme, contained about 60% of pure cozymase.
9. Meyerhof, O. and Ohlmeyer, P.: *Biochem.Z.*, *290* (1937), 334.
a. Translation taken from Kalckar, *Biological Phosphorylations*, pp. 211–27.

Selection 99

Introduction:

a. Selection 52, Section III.

Text:

1. v. Muralt, A. and Edsall, J.T., *J.Biol.Chem.*, *89*, 351 (1930).[b]
2. Bate Smith, E.C., *Proc.Roy.Soc.*, B, *124*, 136 (1937).
3. Theorell, H., *Biochem.Z.*, *272*, 155 (1934); *278*, 263 (1935).
4. Lohmann, K., *Biochem.Z.*, *282*, 109 (1935).[c]
a. Reproduced as given in *Nature*.
b. Selection 127, Section V.
c. Selection 93.

Selection 100

Introduction:

a. Selection 80.
b. D.M. Needham, 'Chemical cycles in muscle contraction' in J. Needham and D.E. Green (eds), *Perspectives in Biochemistry* (Cambridge, 1937), p. 201.

c.   O. Meyerhof, 'Intermediate carbohydrate metabolism' in *A Symposium on Respiratory Enzymes* (Madison, 1942), pp. 3–15 (p. 9).

d.   Meyerhof, *Symposium*, p. 6. Cf. also Selection 84 and O. Meyerhof, 'Über den Zusammenhang der Spaltungsvorgänge mit der Atmung in der Zelle', *Ber.chem.Ges.*, *58*, 991–1001. 1925.

e.   F. Lipmann, 'Role of phosphate in pyruvic acid dehydrogenation', *Nature*, *144*, 381. 1939; 'Phosphorylated oxidation product of pyruvic acid', *J.Biol.Chem.*, *134*, 463–4. 1940.

f.   The germ of Lipmann's ideas on the storage of energy and its utilization by the living cell can be traced to his work in Meyerhof's laboratory between 1927 and 1930. Cf. 'The Meyerhof period for me was equivalent to a biochemical adolescence and growing up. In the Freudian sense all that I did later was subconsciously mapped out there; it started to mature between 1930 and 1940 and was more elaborately realized from then on.' F. Lipmann, *Wanderings of a Biochemist* (New York, 1971), p. 18. In 1925, it is worth mentioning, Meyerhof compared the metabolic carbohydrate cycle to the function of a storage cell and correlated anaerobic glycolysis with discharging it and oxidative resynthesis with recharging it. Cf. Meyerhof, 'Atmung in der Zelle', *Ber.chem.Ges.*, *58*, 995. 1925.

Text:

1.   K. Lohmann and O. Meyerhof, *Biochem.Z.*, *273*, 60 (1934).
2.   E. Negelein and H. Brömel, Ibid., *303*, 132 (1939).
3.   F. Lipmann, *J.Biol.Chem.*, *134*, 463 (1940).
4.   E. Lundsgaard, *Biochem.Z.*, *233*, 322 (1931).
5.   E. Lundsgaard, Ibid., *227*, 51 (1930).
6.   A.V. Hill and J.L. Parkinson, *Proc.Roy.Soc.*, *B108*, 148 (1931). O. Meyerhof and E. Boyland, *Biochem.Z.*, *237*, 406 (1931). O. Meyerhof, Ch. L. Gemill, and G. Benetato, Ibid., *258*, 371 (1933).
7.   F. Lipmann, *Nature*, *143*, 281 (1939).
8.   H. Kalckar, *Biochem.J.*, *33*, 631 (1939).
9.   S.P. Colowick, M.S. Welch, and C.F. Cori, *J.Biol.Chem.*, *133*, 359 (1940).
10. O. Warburg and W. Christian, *Biochem.Z.*, *303*, 40 (1939).
11. F. Lipmann, *Cold Spring Harbor Symposia*, 7, 248 (1939).
12. I. Banga, S. Ochoa, and R.A. Peters, *Biochem.J.*, *33*, 1980 (1939).
13. E. Schorr, S.P. Barker, E. Cohen and M. Malam, *Am.J.Physiol.*, *129*, 463 (1940).
14. A. Harden, 'Alcoholic Fermentation', London, 1932.
15. F. Lipmann, *C.R. Carlsberg Lab.*, *22* (Volume jubilaire Sörensen), 317 (1937).
16. M.G. Macfarlane, *Biochem.J.*, *33*, 574 (1939).
17. W.P. Wiggert and C.H. Werkman, *Biochem.J.*, *32*, 101 (1938).
18. F.F. Nord, *Ergeb.der Enzymforsch.*, *8*, 149 (1939).
19. J.C. Wirth and F.F. Nord. *Science*, *92*, 15 (1940).

a.   When Lipmann's discussion of the energy-rich phosphate bond appeared there were not a few, biochemists and chemists, who found the concept puzzling. Recalling the time, Lipmann described the situation as follows:

> Some of the propositions made in that article must have been more novel than I realized. Much later, at a lively cocktail party, a younger colleague who had helped me with the English, confessed that when he was reading the manuscript he had not the slightest idea what I was talking about.
>
> My paper caused a stir when it appeared, although at first it was given what one might call a mixed reception. The talk of energy-rich bonds aroused forceful, sometimes virulent antagonism. Unwittingly, I had stepped into a hornet's nest by using 'bond energy' to express the potential energy derivable from a bond, brushing aside its accepted use for energy expended to form a bond. I was interested in the amount of free energy that could be derived from a bond and was groping for a definition of its capacity to carry potential chemical energy that could be used for synthesis. To amplify this I proposed alternatively the use of 'group potential', best expressed, probably, by the absolute value of the ΔF of hydrolysis.
>
> Many organic chemists were outraged by what they felt was a misuse of the term bond energy. This antagonism surprised me because I felt that organic synthesis as much as biosynthesis dealt largely with group activation and group transfer. I thought, therefore, that the new terminology I was forced to create should also have relevance to organic synthesis. It has made me happy recently to see the squiggle ($\sim$) I invented to denote energy-carrying bonds creep into borderline organic literature. But

the antagonism of the professional physichochemists [*sic*] still remains, and, as of this time, energy-rich phosphate bonds, when used, is apologetically surrounded by quotation marks to keep the distance from this nonprofessional label. The physical chemist remains aloof. He may be forced to accept the usage, but he usually refrains from referring to the dilettante who originated it.

Cf. Lipmann, *Wanderings*, p. 37. For an attempt to explain the role of phosphorus in this context from its position in the Periodic system, see G. Wald, 'Life in the second and third periods: or why phosphorus and sulfur for high-energy bonds?' in M. Kasha and B. Pullman, *Horizons in Biochemistry* (New York and London, 1962), pp. 127–42.

Section V: PROTEINS

a. From the 1770s to *c*. 1890: analysis. Nutritional and biological relationships

Selection 101

We have already touched upon the beginnings of the scientific study of proteins in mentioning Beccari's work on the component parts of wheat flour.[a] It was in 1728 that Iacopo Bartolomeo Beccari (1682–1776) of Bologna, who had been concerned with nutrition, reported the presence of two kinds of material in wheaten flour. One was a starch, a known component of flour. The other component was unknown, and, because it appeared to resemble glue, Beccari called it 'glutinous'. It was not until seventeen years later, following publication of Beccari's discovery,[b] that the news of the finding of an animal-like matter in a substance of vegetable origin began to circulate. This aroused great interest in the comparative chemistry of vegetable and animal products. Also in connection with the diet of herbivores the question was posed regarding the transformation of the vegetable matter into the animal one by way of a process termed 'animalization'.

After 1750 a good many analytical investigations employing the traditional dry distillation and the novel solvent extraction procedures[c] had been carried out in France, thanks to the influential work of Guillaume François Rouelle (1703–70) and his pupils. But it was his lesser known younger brother, Hilaire Martin Rouelle (1716–78), who began to pay particular attention to Beccari's glutinous matter and recognized that it was not confined to flour, but was distributed through the green parts of a plant. He called the substance 'vegeto-animal matter' (1773). He also noticed a similarity between it and the 'caseous portion' of milk.[d] In 1786 de la Metherie, reviewing the subject, considered the glutinous portion as the most complex constituent of the plant.[e]

By the end of the 1780s, in the light of Lavoisierian ideas about the ultimate nature of chemical composition, it became more widely acknowledged that the fundamental principles (elements) of vegetable and animal substances are carbon, hydrogen, oxygen and azote (nitrogen). Further, it was believed that their combination in different proportions leads to the formation of oils, acids, mucilages and other so-called 'immediate principles', that is, definite and constant chemical constituents of vegetable and animal matters.[f] As to the main chemical difference between them it was thought, following the work of Claude Louis Berthollet (1748–1822), that substances of animal origin contained a relatively greater proportion of nitrogen. To it was ascribed the production of a great deal of ammonia on dry distillation of animal substances. Nitrogen was also believed to be behind their tendency to putrefy.[g] Animalization was, then, the process by which certain immediate (also termed 'proximate') principles passing from plants into animals were chemically altered and acquired more nitrogen and appeared to be converted into animal matter.[h]

What was emerging from the study of the glutinous material occurring in substances produced by plants and animals was that it possessed certain chemical and physical properties in common with those of the egg white. It was from the Latin word for the egg white (*album ovi*)[i] that the generic name for these plant and animal chemical constituents derived: albuminous substances.

Very early on, coagulation by heat and acids and other treatments came to be considered as the characteristic reaction of albuminous substances. Coagulation began to be associated with the previously developed notions about the ultimate, 'mucilaginous', basis of the plant body and the ultimate, 'gelatinous', basis of the animal body. It was held that mucilaginous matter is the peculiar plant matter and gelatinous matter is the real animal matter, and that they resembled each other in their nutritive and regenerative functions.[j]

For an illuminating discussion of these questions at the end of the eighteenth century see the following selection from an article by one of the great French chemists of the day.

[A.F.] de Fourcroy, *Ann.Chim.*, *3*, 252. 1789

'Sur l'existence de la matière albumineuse dans les végétaux' (On the existence of albuminous matter in vegetables)

pp.252–5

The researches of modern chemists on plant substances offer to the philosopher who reflects on the progress of the sciences, two general results equally important; one is the discovery of many principles not known before in plants, the other is the analogy of several of these principles with the substances of another kingdom, and above all with those of the animal kingdom. The gluten found by Beccari in wheat flour was the first discovery which has, so to say, given birth to all the other successive discoveries; this scientist has rightly made us realize that the substance, already almost animalized, showed the closest connections with animal substances and made necessary a chemical comparison between the principles of the two kingdoms more exact than any which had been made so far. Rouelle, in speaking of plants which give ammonia in their analysis on heating, has named them animal plants, and has indeed determined in this way the intention, which he had, of comparing plant substances with those of the animal kingdom. The analogy found and recognized by a great number of chemists, between vegetable oils and animal fats, between emulsions and milk, between insipid mucilages and jellies, between plant sugar properly so-called and sugar of milk, could only direct all ideas and all efforts towards the comparative work of an analysis of these two classes of organized beings. Thus whilst the anatomist and physiologist investigated and demonstrated real relationships of structure and functions between the plants and the animals, at the same time a new kind of complex anatomy discovered an analogy in the form and uses of the organs of plants with those of animals, chemistry found on its side analogies in the nature of the substances which compose the bodies of the two classes of beings and confirmed in this manner their association in a single kingdom under the name of organic or organized kingdom, opposed to the dead brute and inorganic mass of the minerals.

M. Berthollet has just added again to this analogy between the principles composing plants and animals in finding, a few years ago, in several parts of the former nitrogen or the base of ammonia and nitric acid, etc.[a] which is fixed in great quantity in animal substances, and to which the latter owe, as this scientist has demonstrated, the property of producing ammonia on application of heat, and of passing promptly into putrefaction. Also the plant substances which contain nitrogen present absolutely the same characteristics; such are the gluten, the green fecula,[b] the cornaceous seeds. I have already collected in a particular chapter of the third part of my Elements of Chemistry[c] the most evident points of the analogies in nature and composition which the plants and animals present. It is to add a further touch to these analogies that I shall exhibit in this memoir a sequel of facts on the presence of albuminous matter in plants. I shall recall here first the chemical characteristics of animal albumen. A consistency often thick and flowing, insipid taste, solubility in cold water, precipitation by heat, solubility in alkalis and above all in ammonia, separating itself at the temperature of boiling water from all these liquids in which it is dissolved, passing into putrefaction without acidity, such are the properties which characterize the albuminous substance.

pp. 260–1

In working on the albuminous substance in all the plants which I have had occasion to analyze since the first discovery, I have recognized that all the plant substances, acids, and in particular the fruits, do not contain an atom of albumen, and that on the contrary one constantly finds jelly there. Such are the juices of orange, of lemon and of gooseberry, which give a great quantity of gelatinous matter, and in which one finds no trace of albuminous substance. On the other hand I have often seen albumen of the blood form with nitric, hydrochloric and acetic acids a kind of jelly soluble in water and coagulable by cooling. It would then be possible that the albumen which exists in all the plant substances which are young and deprived of acid is converted, by combination with the acids in proportion as these are formed by the increase in age and so on of the vegetation, into gelatinous substance.

M. Thouvenel[d] has already had the same ideas about the jelly of animal substances; according to that, could one not believe that all jelly is a combination of acid and albumen? Is it not because of this that all plant acids give no albumen, whilst those which contain albumen give no appearance of acid? Was it not the same with animal jelly? If this substance is not acid it has at least a great tendency to become so, and one knows that it suffices for it to absorb a small portion of oxygen to take on the character of acidity. In short, is it outside probability that the jelly differs from albumen only through a greater proportion of oxygen?

Albumen extracted from plants and dried gives on distillation ammonium carbonate, foetid red oil, hydrogen gas and carbonic acid; it leaves in the retort a light charcoal, difficult to incinerate, and of which I have not yet been able to obtain enough ashes to examine.

Selection 102

The action of sulphuric acid on plant materials attracted much attention in the early nineteenth century. As a result of such work, Kirchhof, as we have seen, succeeded in obtaining a sugary substance from starch treated with sulphuric acid in 1811.[a] Among those who extended this approach to animal materials early on was Henri Braconnot (1781–1855).

His experiments began with the study of gelatin prepared by boiling the skin, tendons, cartilage, etc. of animals. This substance, after treatment with concentrated sulphuric acid, was heated for some hours with water. The extract was saturated with chalk, filtered and evaporated; a syrup was obtained which, after standing for a month, deposited very sweet crystals. On distillation these gave a light, white sublimate and an ammoniacal product, showing that they contained nitrogen. This substance treated with nitric acid gave a new acid which he named 'nitro-saccharic acid'.

By extracting chopped muscle with water he obtained 'fibrin', believed to be the constituent of the muscle fibre. 30 g of this were treated with sulphuric acid, then chalk, in a similar manner to that used with gelatin. The filtered extract, on evaporation and treatment with potash, gave off ammonia. It was boiled several times with alcohol and from the total extract about 1 g of a peculiar white matter was deposited, which he provisionally name leucine (from the Greek *leukos* = white). Similar results were obtained with chopped wool.

The following contains Braconnot's conclusion summarizing his results in 1820.

H. Braconnot, *Ann.Chim.*, *13*, 113, 1820

'Sur la conversion des matières en nouvelles substances par le moyen de l'acide sulfurique' (On the conversion of animal subtances into new substances by means of sulphuric acid)

p. 125

It results from the principal facts contained in this Memoir:

1. That animal substances can be transformed into substances containing much less nitrogen by treatment with sulphuric acid;
2. That this transformation is brought about by a removal of hydrogen and nitrogen in the proportions necessary for ammonia formation, and probably by absorption of oxygen from the sulphuric acid;
3. That the gelatin can be thus converted into a species of very crystallizable sugar, *sui generis*, which probably does not exist in nature;[a]

4. That this sugar combines intimately with nitric acid apparently without decomposing it, even with the help of heat, and there results a particular crystalline acid that I have called *nitro-saccharic acid*;
5. That wool, and above all fibrin, treated with sulphuric acid, produce a particular white substance which I have called by the name *leucine*;[b]
6. That this substance, warmed with nitric acid, apparently does not decompose, and produces a crystallizable nitro-leucic acid;
7. Finally that other uncrystallizable and savoury [*sapides*] substances, analogous to certain vegetable principles, are also products of the reaction of sulphuric acid on the most insoluble animal substances.

Selection 103

It was the Swedish chemist, J.J. Berzelius (1779–1848), often referred to in this volume, who in the early years of the second decade of the nineteenth century considered fibrin, albumin and the colouring matter of the blood as modifications of one and the same substance and called them corporatively 'albuminous portions of the blood' (*les parties albumineuses du sang*).[a] The study of them took a new turn during the second half of the 1830s thanks to the researches of Gerardus Johannes Mulder (1802–80), who was advised by Berzelius with whom the Dutch worker kept closely in touch.

By that time the combustion analysis of vegetable and animal products devized by Liebig[b] proved its practical utility. Its underlying theory derived from the supposition that compounds of plant and animal origin are oxides of binary radicals consisting of carbon and hydrogen, or ternary radicals containing carbon, hydrogen and nitrogen. Essentially, this was an offshoot of the Lavoisierian concept assigning a central role to oxygen as well in relation to the composition of vegetable and animal substances. The view that they are oxides of compound radicals was clearly developed by Berzelius.[c]

Among the albuminous substances Mulder did study were those of blood (fibrin, albumin) and, of course, egg white (albumin). In order to obtain them he employed varied means of extraction and purification: heat coagulation, precipitation by acids, bases, heavy metal salts, alcohol, etc.[d] An important aspect of Mulder's work, guided by Berzelius, was the attention paid to the quantitative estimation of phosphorus and sulphur in the examined substances.

Mulder's analyses revealed in 1838 that the substances examined contained a constant core of carbon, hydrogen, nitrogen and oxygen, apparently in the same proportions $C_{400}H_{260}N_{100}O_{120}$, combined with 1 atom of phosphorus and 1 or 2 atoms of sulphur. Berzelius believed the presence of this organic oxide to be of the greatest importance but regarded the estimated amounts of oxygen as excessive and the assigned empirical formula (on the basis of equivalents) as improbable. It was in this connection that Berzelius, writing to Mulder in French, introduced the term *protéine* for the organic oxide presumed to be constantly present in albuminous substances, having a quarternary composition:[e]

The name protein that I propose to you for the organic oxide of fibrin and albumin I wish to derive from προτειοσ because it appears to be the primary [*primitive* in French] or principal substance of animal nutrition which the plants prepare for the herbivores, and which these then supply to the carnivores. To derive the name from the Greek word for the fibre would be less fitting, since the organic oxide is the basis also of the albumin and probably also of the coloring matter [of blood] as well as of others.

Berzelius's naming of the organic oxide was accepted by Mulder who began to use it in his publications. Following Mulder's work on 'protein-sulphuric acid', which seemed to support Berzelius's theory that the numbers of atoms in the organic oxide are too high by a factor of 10, the formula of 'protein' (Pr) began to be written $C_{40}H_{62}N_{10}O_{12}$ ($\overline{Pr}$ standing for 10 Pr).

The following extracts from a paper published in German in 1839 contain some of the findings on which Mulder based his notion that albuminous bodies contain the common 'protein' component in combination with phosphorus and sulphur in simple ratios. It also provides evidence for Berzelius's influence on Mulder concerning the plant origin of 'protein' and its fundamental role in animal nutrition.

G.J. Mulder, *J.prakt.Chem.*, *16*, 129. 1839

'Ueber die Zusammensetzung einiger thierischen Substanzen' (On the composition of some animal substances)

p. 129

I have been occupied for some time with the study of the most essential substances of the animal kingdom, the fibrin, the albumin and the gelatin. Since the publication of this work[a] I continued to study these substances. Berzelius communicated with me concerning the published results and gave me good advice for which I express my sincere thanks...

In my former experiments I did not consider the sulphur and phosphorus, which are found as integral parts of these bodies...

pp. 134–5

From the results quoted it can easily be concluded that in fibrin and in the egg albumin there is always 1 atom of sulphur to 1 atom of phosphorus, whilst with the serum albumin there is 1 atom of phosphorus to 2 atoms of sulphur.

We calculate now the atoms of the complex bodies according to the smallest atom number of the sulphur and phosphorus, thus we can describe the composition of these three bodies in the following way:[b]

	Fibrin and egg albumin		
Carbon	400[d]	30,574.80[e]	54.90[e]
Hydrogen	620	3,868.68	6.95
Nitrogen	100	8,851.80	15.89
Oxygen	120	12,000.00	21.55
Phosphorus	1	196.16	0.35
Sulphur	1	201.17	0.36
		55,692.61	100.00

Serum albumin

Carbon	400	30,574.80	54.70
Hydrogen	620	3,868.68	6.92
Nitrogen	100	8,851.80	15.84
Oxygen	120	12,000.00	21.47
Phosphorus	1	196.16	0.35
Sulphur	2	102.34	0.72
		55,893.78	100.00

These enormous numbers can be justified if it can be shown that SP binds with a quaternary organic body which has the composition $C_{400}H_{620}N_{100}O_{120}$, in order to form fibrin and the egg albumin, whilst SP binds with the same body to form the serum albumin.[f]

p. 138

The organic substance which is present in all constituents of the animal body, also as we shall soon see, in the plant kingdom, could be named *protein* from πρωτεῖοσ, *primarius*. Fibrin and the egg albumin thus have the formula P̄r + SP, the protein of the serum albumin P̄r + SP.

p. 140

It appears that animals draw their most essential nutrient ingredients directly from the plant kingdom. It is possible that the plant protein contains sulphur and phosphorus in a different relation from that in the animal protein which contains fibrin and so on, but the quaternary organic body is the protein itself.

The plant-feeding animals are thus, considered from this point of view, not different from the flesh-eating. Both are nourished by protein, by the same organic body, which plays a principal role in their economy. It remains still to know whether the starch and other substances, which are recognized as nutrient substances, in the animal body can be changed into protein.

The nourishing capacity of bread and other nutrients which contain protein is thus easily understood. They offer, unless the digestion works an alteration here, directly one of the most important parts of the animal body.

pp. 150–1

VIII. Proteinsulphuric acid . . .

This body thus has the following constitution:

	Found	Atoms		Calculated
Carbon	50.94	40	3,057.40	50.70
Hydrogen	6.93	62	386.86	6.41
Nitrogen	15.08	10	885.20	14.68
Oxygen	18.74	12	1,200.00	19.90
Sulphuric acid	8.34	1	501.17	8.31
			6,030.63	

Selection 104

During the late 1830s and the early 1840s chemists interested in the chemical composition and formation of plant and animal albuminous substances and also in their relationship, found the Mulder or, historically more appropriate perhaps, the Mulder–Berzelius 'protein' hypothesis attractive. It seemed to throw light on the similarities and differences of albumin, fibrin and casein believed to be present in plants as well as in animals, and assumed therefore to be their common basic constituents.

Thus in 1842 Liebig regarded them as isomeric compounds of the 'protein' radical with varying amounts of sulphur and phosphorus. While remaining uncertain whether 'protein' pre-existed in albuminous bodies Liebig was satisfied that it was obtained on their decomposition. As to the provenance of albuminous substances, Liebig also held that they were produced solely by plants and that, on ingestion by animals, the vital force built them up into living tissues of their bodies. The chemical processes involved, he assumed, had the character of oxidations and hydrations.

What Liebig wrote regarding their essentially identical chemical composition, which remained unaltered during the passage from the plant to the animal world, was widely accepted at that time. Liebig's view on the part played by albuminous substances in replacing the spent material of the tissues was to become equally important. Of particular significance was his opinion associating muscular action with their metabolism, together with the contention that the nitrogen appearing in the urine (as urea) could account for it.

These ideas put forward by Liebig in 1842 are contained in the following excerpts.

J. Liebig, *Animal Chemistry or Organic Chemistry in its Application to Physiology and Pathology* (Cambridge, Mass., 1842).

pp. 46–8

These three nitrogenized compounds, vegetable fibrine, albumen, and caseine, are the true nitrogenized constituents of the food of graminivorous animals; all other nitrogenized compounds, occurring in plants, are either rejected by animals, as in the case of the characteristic principles of poisonous and medicinal plants, or else they occur in the food in such very small proportion, that they cannot possibly contribute to the increase of mass in the animal body.

The chemical analysis of these three substances has led to the very interesting result that they contain the same organic elements, united in the same proportion by weight; and, what is still more remarkable, that they are identical in composition with the chief constituents of blood, animal fibrine, and albumen . . .

How beautifully and admirably simple, with the aid of these discoveries, appears the process of nutrition in animals, the formation of their organs, in which vitality chiefly resides! Those vegetables principles, which in animals are used to form blood, contain the chief constituents of blood, fibrine, and albumen, ready formed, as far as regards their composition. All plants, besides, contain a certain quantity of iron, which reappears in the coloring matter of the blood. Vegetable fibrine and animal fibrine, vegetable albumen and animal albumen, hardly differ, even in form; if these principles be wanting in the food, the nutrition of the animal is arrested; and when they are present, the graminivorous animal obtains in its food the very same principles on the presence of which the nutrition of the carnivora entirely depends.

pp. 102–3

As far, therefore, as our researches have gone, it may be laid down as a law, founded on experience, that vegetables produce, in their organism, compounds of proteine; and that out of these compounds of proteine the various tissues and parts of the animal body are developed by the vital force, with the aid of the oxygen of the atmosphere and of the elements of water.[1]

Now, although it cannot be demonstrated that proteine exists ready formed in these vegetable and animal products, and although the difference in their properties seems to indicate that their elements are not arranged in the same manner, yet the hypothesis of the preëxistence of proteine, as a point of departure in developing and comparing their properties, is exceedingly convenient. At all events, it is certain, that the elements of these compounds assume the same arrangement when acted on by potash at a high temperature.

All the organic nitrogenized constituents of the body, how different soever they may be in composition, are derived from proteine. They are formed from it, by the addition or subtraction of the elements of water or of oxygen, and by resolution into two or more compounds.

pp. 231–2

The act of waste of matter is called the change of matter; it occurs in consequence of the absorption of oxygen into the substance of living parts. This absorption of oxygen occurs only when the resistance which the vital force of living parts opposes to the chemical action of the oxygen is weaker than that chemical action: and this weaker resistance is determined by the abstraction of heat, or by the expenditure in mechanical motions of the available force of living parts . . .

The sum of the mechanical force produced in a given time is equal to the sum of force necessary, during the same time, to produce the voluntary and involuntary motions; that is, all the force which the heart, intestines, &c., require for their motions is lost to the voluntary motions.

The amount of azotized food necessary to restore the equilibrium between waste and supply is directly proportional to the amount of tissues metamorphosed.

The amount of living matter, which in the body loses the condition of life, is, in equal temperatures, directly proportional to the mechanical effects produced in a given time.

The amount of tissue metamorphosed in a given time may be measured by the quantity of nitrogen in the urine.

Selection 105

The supposition mentioned above regarding the animal organism's inability to produce albuminous substances and its dependence on the plant organism to supply them was in line with the chemical differentiation between the two organic kingdoms which had been advanced in 1829 by the German chemist, L. Gmelin (1788–1853) and restated in 1841 by the French investigators, J.B.A. Dumas (1800–84) and J.B. Boussingault (1802–87). Accordingly, the principal divide between living things was seen in the nature of the chemical reactions they promote: the plant organism being associated with the reduction and the animal organism with the oxidation process.[a]

A year later Dumas had occasion to reiterate this view in a joint paper with Auguste Cahours (1777–1859). It reflected clearly the recognition of the biological (nutritional) significance of nitrogen with which French chemists of the day were much concerned. Among other discoveries, they reported a small but definite difference in composition between two albuminous substances present in the blood corpuscles, globin and albumin; the former contained more nitrogen and less carbon than the latter. This was among the first factors to cast doubt on some of Mulder's analyses.[b]

These findings soon became the subject of a bitter controversy between Mulder and Liebig when the German chemist reversed his position and began to attack the 'protein' concept. Primarily the controversy arose from the work carried out in the Giessen laboratory on the sulphur content of albuminous substances. The principle of the method then used was to eliminate the sulphur in the substance by fusing with caustic potash and to assume that a sulphur-free 'protein' could then be obtained.

As shown in the selection below from the paper of a Russian student of Liebig's, N. Laskowski (1818–71),[c] published in 1846, this supposition could not be confirmed; and thus the reality of a sulphur-free protein, whether as a pre-existing constituent or as a decomposition product of albuminous substances, was called into question.[d]

N. Laskowski, *Ann.*, *58*, 129. 1846

'Ueber die Proteïntheorie' (On protein theory)

p. 164

What now is that body which according to the specified method of Herr Mulder one obtains under the name 'protein' and to which on paper one has attributed a constitution which it does not possess?

We have seen that, on treatment of albumin with caustic potash solution in the warm, products free from sulphur or poor in sulphur are formed; which, on saturation with acetic acid, precipitate with the substance not yet made sulphur-free, and of which not all can be removed by washing with water. The precipitate is thus always a mixture of different substances. The relative amounts of these substances must naturally be different according to the concentration of the alkali, the duration of its action, and especially the temperature at which the treatment takes place. For example if the treatment is made at 50°C, the precipitate (the protein material of Herr Mulder) consists for the most part of sulphur-containing substances showing the external properties of protein-like bodies and only a little of the sulphur-free products. On the other hand, after treatment at 70°C, the quantity of these latter products present in the precipitate is far more significant. Whether that sulphur-containing substances making the chief part of the so-called protein is plain albumin or a metameric body,[a] or something else – remains still to be determined.

pp. 165–6

It seems to us that, after all which has been remarked above about the theory of so-called protein compounds, one is necessarily led to the following conclusions:
1. *That the empirical formulae advanced by Herr Mulder for protein-like bodies are in no way acceptable, as they do not agree with the results of analysis;*
2. *that the substance prepared by Herr Mulder as protein is sulphur-containing, that is to say something different from that described by him under this name; and so far as our experience goes, the latter does not exist in the isolated state at all;*
3. *that as the conjecture of substance described by Herr Mulder as protein rests merely on the belief that it has been isolated sulphur-free – but the substance isolated by Herr Mulder contains sulphur and the one described by him is not isolatable – there is no ground for assuming the protein to be a hypothetical elementary substance [Grundstoff];*
4. *that therefore further upholding of Herr Mulder's assumed constitution of protein bodies would be something purely arbitrary.*

Selection 106

No doubt, the confusing practice of employing a variety of atomic weights and formulae when dealing with the same substance added to the difficulties of elucidating the chemical nature of substances such as albumin, fibrin and casein. The hitherto accepted supposition regarding their identical chemical composition and stability in the metabolic pathway came under attack.

By the mid-1840s it was becoming clear that no real insight could be gained into the constitution of serum albumin, fibrin and casein from the percentage amount of elements known to be present. Chemists were urged to adopt another procedure in order to unravel the composition of a complex organic substance. After splitting it into two or more simple products their composition and amount was to be compared with that of the compound intially taken. The importance of this approach was certainly brought out by Liebig in a paper in 1847 reflecting, among other things, his discovery of tyrosine as a decomposition product of casein. The paper also described further work on effective methods of preparing this substance carried out by his associate Friedrich Bopp (1824–49).[a]

In 1846, by fusing cheese with alkali, Liebig obtained, in addition to leucine and other products, a substance[b] which he named tyrosine in the 1847 paper referred to above. As leucine and tyrosine began to be considered the two characteristic breakdown substances for casein, Bopp started his experiments with the aim of discovering whether albumin and fibrin behave like casein when under the same conditions. He also searched for the most advantageous method of preparing these compounds and in the largest possible amounts.

Eventually, as shown below, Bopp found out that the best results were achieved by cleaving the albuminous substances not only by the already known means of sulphuric acid but also by the hitherto unknown aid of hydrochloric acid.

What emerges from Bopp's account is that at the time of publication leucine and tyrosine appeared to be the most important simple constituents of albuminous bodies whose complexity was not in doubt. Further, it hinted that in so far as albuminous bodies varied in properties, this was caused not by the difference in composition – they were supposed to be similarly composed – but possibly by the amounts in which the constituents occurred.

F. Bopp, *Ann.*, *69*, 16. 1849

'Einiges über Albumin, Casein und Fibrin' (Something on albumin, casein and fibrin)

pp. 20–1

As is well known, with the action of concentrated alkali on casein there appear, besides other products, especially two substances characteristic for this [compound], leucine and tyrosine. My first experiments had the aim of seeing whether albumin and fibrin under the same conditions behave like casein, and to find out possibly a method of preparing those two crystalline substances with certainty and in the largest possible amounts. In a series of some twenty experiments I subjected casein, albumin and fibrin in a comparative manner to a treatment as similar as possible as concerns the used amount of substance, indeed so that the single difference consisted in the duration of the operation. After some preliminary experiments it was noticeable that the action of the potash does not stop at the formation of those two substances but breaks them down further, at the same time in the comparative experiments no striking difference in the behaviour of albumin, fibrin and casein was to be noticed. It remained therefore still to find the most advantageous method of obtaining with certainty the largest amount of leucine and quite especially that of tyrosine, as anyhow the latter is formed in relatively small amount, and therefore the moment favourable for breaking off the operation is easily missed.

pp. 22–3

The obtaining of tyrosine rests on this, that it is only very little soluble in cold water, in hot water far more soluble ... The obtaining of leucine rests on this, that it is easily soluble in hot, less in cold water, a little in spirit of wine,[a] rather more in water comprizing free acetic acid, very easily in water containing free sulphuric acid or spirit of wine.

pp. 25–6

The great uncertainty especially in connection with the obtaining of tyrosine caused me to look around for another method which would provide greater certainty. The behaviour of leucine and tyrosine with hydrochloric acid in which indeed they are easily soluble, but not in the slightest degree broken down, and also with dilute sulphuric acid, with which acids they can be evaporated at a temperature not exceeding 100°, obliged me to investigate the *Action of hydrochloric acid and sulphuric acid on casein, albumin and fibrin*.

Besides the behaviour mentioned above of the two crystalline substances towards both mineral acids, it is also well known from experiments of Braconnot, that this chemist obtained leucine through treatment of muscle flesh, glue, wool and legumin (vegetable casein) with sulphuric acid[b] . . . It is well known that albumin, suffused [*übergossen*] with concentrated hydrochloric acid 4–5 times the amount of the dry substance and boiled for some time, dissolves with breakdown taking on a transitory beautiful violet colour. Casein and fibrin behave just the same. The violet colour goes over gradually into brown, but only with entry of air. It was now to be expected that in case the reaction went far enough to form leucine and tyrosine, one had to obtain the whole theoretical yield, as these acids do not, like potash, cause further breakdown. In fact I had the satisfaction of finding this confirmed by experiment. If one boils albumin or one of the other substances with 4–5 fold amounts of concentrated hydrochloric acid for 6–8 hours, then the breakdown has gone so far that, as described in the following, obviously the whole theoretical yields of leucine and tyrosine are present. If one boils with exclusion of air then the same happens with these two substances, but the violet colour remains, and goes into brown only gradually with entry of air.

p. 27

One removes the hydrochloric acid as much as possible, . . . one puts the mass into hot water and adds excess of milk of lime, boils some time in a metal container to remove the ammonia and filters as quickly as possible through a linen bag till quite clear. The lime dissolved in the fluid . . . is precipitated with little sulphuric acid and the excess of the latter by means of lead oxide in acetic acid [lead acetate]; the lead is removed by hydrogen sulphide and evaporation follows . . . With the evaporation one obtains a syrupy kind of mass in which soon crystals appear. One leaves it to stand for some days and then separates the crystallized leucine and tyrosine by means of 86 per cent spirit of wine from the syrup which dissolves therein. One then separates the two crystalline substances . . . through their so different solubility in water and use of some hydrate of lead oxide, and purifies them . . . by recrystallization and use of animal charcoal.

pp. 34–5

The leucine as well as the tyrosine are both extremely characteristic bodies, they are so easy to recognize already in their appearance, the crystal formation, the different solubility in water, spirit of wine, dilute acids and alkalis, that to me analysis for the purpose of *recognition* seemed quite unnecessary . . .

The multiformity of the products mentioned gives evidence on the composite nature of the atom[c] of the albuminous bodies investigated – these are as regards their composition completely similar, as it appears till the present time; casein, albumin and fibrin distinguish themselves not with reference to the quality of the breakdown products, whether there may be small differences in the quantity, this is at present not ascertainable; the known analyses seem to point to such differences.

Selection 107

It was in this period that the study of proteins also began to be affected by work bearing more immediately on physiological (digestion) problems.[a] By 1850, from the contributions of persons such as Louis Mialhe (1807–86) and Carl Gotthelf Lehmann (1812–63), as exemplified in the extracts below, the ideas emerged that under the action of pepsin and gastric juice respectively protein material decomposed into substances termed 'albuminoses' and 'peptones'.

Albuminoses were considered by Mialhe to parallel the compounds obtained from starchy material by diastase. Clearly, this theory originated in the belief that nutrition depended on the utilization of analogous simpler constituents of starchy and protein foods produced by fermentative (i.e. enzymic) activity.

[L.] Mialhe, *J. pharm.*, *10*, 161. 1846

'De la digestion et de l'assimilation des matières albuminoïdes' (On digestion and on the assimilation of albuminoid substances)

p. 165

All the albuminoid materials, without exception, are transformed by pepsin into an ultimate product, always showing the same chemical reactions although probably its centesimal composition differs a little according as it come from such or such albuminous compound, which I propose to designate under the name of *albuminose*, this name having the double advantage of recalling its origin and its physiological destination.

Albuminose is solid, white or yellowish white, offering a [faint] odour and a slight savour, but which recall a little however the odour and savour of meat, very soluble in water and completely insoluble in absolute alcohol.

Its aqueous solution is not precipitable by heat, nor by the bases, nor by the acids, nor finally by pepsin; it is on the contrary precipitated by a great number of metallic salts, such as those of lead, of mercury and of silver; chlorine precipitates it just as does tannin; even if this last reactant has received the addition of a certain amount of nitric acid.

Now, this product of transformation plays an immense role in the nutrition of animals, for it is to albuminous food what glucose is to starchy food, that is to say that alone it is apt to sustain the phenonomen of assimilation . . .

C.G. Lehmann, *Physiological Chemistry* (London, 1853), II[a]

p. 53

. . . we learn, from a positive experimental inquiry, what are the products which are developed during the process of digestion; and we ascertain that, by the action of natural or artificial gastric juice on protein-bodies or gelatigenous matters, there are formed thoroughly new substances, which although they coincide in their chemical composition and in many of their physical properties, with the substances from which they are derived, essentially differ from them, not only in their ready solubility (in water, and even in dilute alcohol), but in having now lost the faculty of forming insoluble combinations with most metallic salts. The formation of these substances, which we designate as *peptones*, depends solely on the action of the gastric juice, and occurs without the evolution or absorption of any gas, and without the production of any secondary substance.

pp. 55–6

Very little attention has hitherto been paid even to the best-known peptones; indeed, until Mialhe published his researches, positively nothing was known regarding their physical or chemical relations. This chemist erroneously regarded the soluble substances produced by digestion from the protein-bodies and from the gelatigenous tissues as perfectly identical. The following properties, which Mialhe attributes to his albuminose, are certainly correctly observed, and are common to most of the peptones; in the solid state the digested substances are white or of a pale yellow colour, possess little taste or odour, and dissolve readily in water and slightly in spirit, but not at all in absolute alcohol. The watery solutions of these substances are not precipitated by boiling, by acids, or by alkalies, but deposits are thrown down by metallic salts, by chlorine, and by tannic acid.

My own observations lead me to the belief that all the peptones are white, amorphous bodies, devoid of any odour, and having merely a mucous taste, soluble in every proportion in water, and insoluble in alcohol of 83%; their watery solutions redden litmus; they combine readily with bases – with alkalies as well as with earths – so as to form neutral salts, which are very soluble in water. The aqueous solutions of these salts are only precipitated by tannic acid, corrosive sublimate, and, if caustic ammonia has been previously added, by acetate of lead; all other metallic salts, even nitrate of silver and alum, produce no precipitate, and even basic acetate of lead only induces a slight turbidity, which disappears on the addition of an excess of the test. No precipitation or turbidity is produced by the addition of mineral or organic acids, either in a concentrated or in a very dilute state; even chromic acid fails to produce any appreciable effect. The ferrocyanide and ferridcyanide of potassium, when added to solutions acidified with acetic acid, occasion only a slight turbidity.

I have been unable to obtain the peptones perfectly free from mineral substances: I have, however, obtained them free from phosphates and hydrochlorates, so that their ash contained only alkaline carbonates or carbonate of lime, with small quantities of alkaline sulphates. With regard to the quantity of sulphur in the peptones, I found it to be constantly the same as that in the substances from which they were derived;

pp. 56–7

In my repeated analyses I have been unable to detect any differences between the quantities of nitrogen, carbon, and oxygen, contained in the peptone and in the substance from which it was derived, nor can I infer from my quantitative results, that the conversion of the protein-bodies into peptones is accompanied by an assimilation of water, as might have been supposed. The metamorphosis may be appropriately compared with that of starch into sugar, or even better perhaps, with that of cholic (Strecker's cholalic) acid into choloidic acid.[b]

Selection 108

Between 1840 and 1860 important developments took place in chemical science. They included efforts to elaborate systems of classifying organic compounds and to clarify the confusion over the use of various systems of atomic weights. However, Stanislao Cannizzaro's (1826–1920) suggestion in 1858 of returning to Avogadro's hypothesis (1811) and applying it to the determination of atomic and molecular weights was not taken up seriously until a good many years after the international meeting of chemists which was convened at Karlsruhe in 1860 to deal with the problem.[a]

Among those who played a major role in classifying organic compounds were the French chemists, Auguste Laurent (1808–53), and Charles Gerhardt (1816–56). In 1848, by applying the concept of homology which they introduced, Laurent and Gerhardt drew up a series corresponding to the general formual $C_nH_{2n+1}NO_2$ in which sugar of gelatine (glycocoll, glycine), leucine and sarcosine were included (C = 12, H = 1, N = 14, O = 16):[b]

CH_3NO_2	unknown
$C_2H_5NO_2$	sugar of gelatine
$C_3H_7NO_2$	sarcosine
$C_4H_9NO_2$	unknown
$C_5H_{11}NO_2$	unknown
$C_6H_{13}NO_2$	leucine

It should be added that they classified these nitrogenous compounds as alkaloids.

A compound which resembled glycocoll and leucine, especially in its capacity to combine with both acids and alkalis, was discovered unintentionally by Adolph Strecker in 1850. He obtained it after mixing a solution of acetaldehyde ammonia with prussic acid in the presence of hydrochloric acid and named it 'alanine'. Strecker then suggested that glycocoll, alanine and leucine are members of the following series (C = 6, H = 1, N = 14, O = 8):[c]

$C_4H_5NO_4$	glycocoll
$C_6H_7NO_4$	alanine
$C_8H_9NO_4$	unknown
$C_{10}H_{11}NO_4$	unknown
$C_{12}H_{13}NO_4$	leucine
etc.	

Strecker's unforeseen synthesis of alanine by way of cyanhydrin eventually became one of the methods of preparation of amino acids.[d] Mentioned in the chemical writings at least since 1845, the term 'amino acids' was not used in connection with glycocoll and leucine at the time.[e] In 1848, as we have seen, Laurent and Gerhardt treated them as alkaloids. It was a decade later that glycocoll, alanine and leucine were clearly listed as amino acids by Cahours, probably for the first time (C = 6, H = 1, N = 14, O = 8):[f]

$C_4H_4O_4$	$C_4H_3(NH_2)O_4$
acetic acid	glycocoll
$C_6H_6O_4$	$C_6H_5(NH_2)O_4$
propionic acid	alanine
$C_{12}H_{12}O_4$	$C_{12}H_{11}O_4(NH_2)O_4$
caproic acid	leucine

The disclosure of certain chemical relationships between glycocoll, alanine and leucine resulted from endeavours to bring order into the rapidly expanding chemistry of the carbon compounds. It certainly did not occur to systematizing chemists to study them as constituents or decomposition products of complex substances existing and playing a role in living bodies.[g] Perhaps this was because leucine and tyrosine were still regarded in the early 1860s as the two distinctive decomposition compounds obtained from proteins, as Liebig and his school had suggested a decade earlier. Furthermore, in line with the thinking emanating from Giessen, it was at that time commonly assumed that plants and animals contain a limited number of proteins of the same or very similar composition.

The situation began to change from the second half of the 1860s, primarily under the influence of the investigations of proteins from plant seeds carried out by Heinrich Ritthausen (1826–1912) who developed careful methods of extraction with water, weak alkaline and acid solutions, and dilute alcohol. On the basis of remarkable quantitative determinations of elements and considerations of solubility, Ritthausen was able to establish that plant seeds contain a larger number of diverse proteins than had been taken for granted. He also broke new ground by extending the qualitative and quantitative knowledge of amino acids in seeds, including its significance for the characterization of proteins.[h]

In 1866 Ritthausen showed that gluten, when boiled with sulphuric acid, yielded an unknown amino acid with the formula $C_5H_9NO_4$. It was given the name 'glutamic acid'.[i] In a similar manner he demonstrated during 1868–9 the unsuspected presence of the already known aspartic acid, while his co-worker, Ulrich Kreusler (1844–1921), discovered it in animal proteins (casein, egg white, yolk) in 1869.[j]

Then in 1871 they published jointly a paper of paramount historical importance from which the selection below is taken. Here the characterization of particular proteins in terms of the amounts of amino acids present is explored for the first time.

H. Ritthausen and U. Kreusler, *J.prakt.Chem.*, *3*, N.S. (*111*), 314. 1871

'Ueber die Verbreitung der Asparaginsäure und Glutaminsäure unter den Zersetzungsproducten der Proteïnstoffe' (On the distribution of aspartic acid and glutamic acid among the breakdown products of the proteins)

p. 314

a. *With plant proteins.* After we have investigated a greater number of these substances prepared from different seeds the following result appears: that if one boils them for a longer time in the familiar way with dilute sulphuric acid, besides tyrosine and leucine, *aspartic* ($C_4H_7NO_4$) and *glutamic* ($C_5H_9NO_4$) *acids always arise from them*; so that these are to be considered as *common breakdown products of all vegetable proteins*...

The amounts of each of the two acids which are formed from the individual proteins, on exactly similar treatment, diverge to some extent significantly from each other; so that we have cause specially to emphasize this difference which allows a conclusion (even if only a very general one) on the inner differences of the plant structure.

pp. 315–17

The following figures will to some extent illustrate this relationship; although neither aspartic nor glutamic acid can be estimated very quantitatively, the estimations performed can still serve as support.

Differences like those in these figures, especially for glutamic acid, (even if these still leave much to be desired in accuracy) can scarcely be seen as chance happenings; they depend on this, that the *protein bodies themselves*

Protein substances 100 parts by wt.gave:	Aspartic acid	Glutamic acid[1]
1. Mucedin[a,b]	Not estimated	25 p.c.
2. Maize fibrin[a,b]	1.4 p.c.	10 p.c.
3. Mixture of plant gelatin[b] mucedin and fibrin	1.1 p.c.	8.8 p.c.
4. Glutin-casein[a,b]	0.33 p.c.	5.3 p.c.
5. Conglutin[b] (lupins)	2.0 p.c.	3–5 p.c.
6. Legumin (from broad beans)[b]	3.5 p.c.	1.5 p.c.

from which they are derived, *must differ from one another*. The figures show further that, for the preparation of glutamic acid, wheat gluten is the most suitable material; which also, it must be added, yields the acid most beautifully crystallized and very pure. As conglutin can easily be prepared in large amount this also is very suitable for the same purpose, and gives besides at the same time considerable quantities of aspartic acid.

b. *With animal proteins.* From the experiments carried out by Kreusler, and already communicated,[2] it is shown that albumin (from eggs), casein and horny material boiled with sulphuric acid produce besides tyrosine and leucine also aspartic acid – though the horny material yields them only in small amount; *on the other hand no glutamic acid, which could not be found or detected in the decomposition fluid.* We are thus really entitled to draw the conclusion that glutamic acid does not originate in animal material but may be a characteristic breakdown product of plant proteins; nevertheless, in order to show this with complete safety, still further experiments with this and with larger quantities of material are necessary; it can, if its amount is small, easily be overlooked because then, like its copper salt, it can be prepared pure and crystallized only with difficulty.

According to the experiments under consideration, aspartic acid is indicated as a breakdown product *common* to *all* proteins and is thus to be placed together with tyrosine and leucine.

Selection 109

During the early 1870s notable support for Ritthausen's results came from the German-Bohemian phytochemist, Heinrich Hlasiwetz[a] (1825–75), collaborating with Josef Habermann (1841–1914). Their findings originated in the conjecture that metabolic links, based on close chemical, physico-chemical and physiological connection, existed between carbohydrates and proteins.

In their first paper[b] on proteins Hlasiwetz and Habermann justified this assumption by pointing out that carbohydrates as well as proteins included a number of isomeric and polymeric compounds. The resemblance appeared to extend also to external properties of certain compounds, as in the case of carbohydrate (plant) mucilages and protein (animal) mucilages. There were also reports indicating that when carbohydrates and proteins underwent decomposition by various means (fermentation, action of acids and alkalis), they yielded identical products (e.g. formic, oxalic, acetic acids) or closely related compounds (e.g. alcohols from carbohydrates and corresponding amides from proteins).

Hlasiwetz and Habermann decided to test the idea that carbohydrates and proteins have a common or related basis by making use of a method of oxidizing carbohydrates they had introduced previously. They found that sugars such as lactose and glucose were converted by bromine water into lactonic and gluconic acids respectively.[c] They thought that if by applying this method to proteins the same acids were produced, this would constitute sufficient proof that these two members of the essential groups of foodstuffs shared a common chemical arrangement.

Having failed to obtain the acids in question, Hlasiwetz and Habermann did not abandon the original idea but looked for other, as they said, smoother ways of decomposing proteins. They turned to the already known methods of breaking down proteins by hydrochloric and sulphuric acids in the belief that this action would be analogous to that which, under the same circumstances, occurred with the glucosides. Indeed, they maintained, nobody seemed to have looked for carbohydrates among the products of this decomposition.

However, as they noted in their second paper on the subject (from which the selection below is taken) this was not the only factor that made them study the most effective use of the acid breakdown of proteins. Another stimulus was the desire to elucidate whether glutamic acid was solely a product of the decomposition of vegetable proteins (until that time no one had succeeded in obtaining it from animal proteins).

By developing the method of acting on proteins with hydrochloric acid in the presence of stannous chloride and applying it to the decomposition of casein, they succeeded in producing glutamic acid, as well as aspartic acid, leucine and tyrosine. They drew attention to the source of ammonia eliminated in the breakdown, and also recognized that the properties of proteins depended on the differences in the amounts of amino acids. Last but not least, they had to abandon the supposition about the related nature of carbohydrates and proteins which had given impetus to their work in the first place.

H. Hlasiwetz and J. Habermann, *Ann.,* *169,* **150. 1873**

'Ueber die Proteïnstoffe; Zweite Abhandlung' (On the protein substances: second treatise)

p. 151

Casein

Already with our first experiments we had recognized in casein the type of protein with which one works most advantageously, because it undergoes change more easily and smoothly than the rest . . .

pp. 164–6

The facts presented are the result of eight experiments essentially run exactly in the same way, according to the method described [above]:
  Accordingly we consider as proved:
  1. That casein yields as breakdown products exclusively:
    a. Glutamic acid,
    b. Aspartic acid,
    c. Leucine,
    d. Tyrosine,
    e. Ammonia.
  2. It yields neither carbohydrate nor characteristic derivatives of the same.

Contrary to earlier conjectures carbohydrate cannot be concerned in its constitution.

3. It is highly probable that the ammonia which always appears is derived from primary compounds contained in the casein, which at the same time yield aspartic acid and glutamic acid.

This also would provide a natural explanation for the origin of the so-called 'loosely bound nitrogen' of the protein substances, on which repeated remarks have already been made[1] and the exact quantitative estimation of which O. Nasse[2] only recently undertook again.[a]

It is the nitrogen of that NH₂ group which comes from compounds like asparagine and glutamine in the form of ammonia, when aspartic acid and glutamic acid are formed.

Compounds of this kind, which on boiling with acid or alkali lose ammonia and yield these acids with uptake of water, must in general be considered pre-existent in casein and the protein substances.

Whether these are identical with the customary asparagine and the still to be prepared homologous glutamine, and whether the acids obtained are not products of a molecular re-arrangement and shifting, cannot be decided at present.

4. Glutamic acid is not characteristic exclusively of the plant protein substances as one could be tempted to assume according to the experiments of Kreusler (Journal für pract. Chemie *107*, 240), who could not obtain it out of animal material, but it is a constant and in amount a significant breakdown product of all the chief accepted forms of animal protein substances.[b] The maximum yield from casein we obtained was about 29 per cent.

We have also investigated the breakdown of protein substances according to our method with albumin, legumin and the plant albumin [*Pflanzeneiweiss*], and the qualitative result was the same as with casein.

We further endeavoured to find a quantitative expression for this manner of breakdown; but we have still not succeeded in separating and estimating the amounts of the single breakdown products with the desired exactitude.

Only this we can state on the grounds of numerous already collected data that the different protein modifications yield different amounts of these products, and it appears to us now already as more than a mere guess that the differences in the properties of the protein modifications are to be sought in a different proportion of the primary atom groups of which they are constituted.

Selection 110

Liebig's assumption (1842) concerning the direct relationship between muscular work and the amount of urinary nitrogen (as urea) has already been referred to.[a] During the following two decades it gave rise to experimental investigations which sought to substantiate it. Little by little, however, it began to be argued that the quantity of nitrogen in urine was not dependent on muscular work but corresponded to the amount of protein eaten. Space does not allow us to trace these developments in more detail.[b] What eventually emerged was the notion of two types of protein metabolism: one concerned with the fate of food protein and the other with that of tissue protein.

Before that, mention must be made of Carl (von) Voit (1831–1908) who from 1867 on repeatedly expressed the opinion that in order to understand the course of protein metabolism in animals it was necessary to differentiate between the protein dissolved and circulating in extracellular fluids (*circulirendes gelöstes Eiweiss*) and the tissue protein stored in cells, firmly bound, and frequently insoluble in water (*Organeiweiss*).[c]

Voit's assessment in 1881 of his own contribution to the elucidation of the problem of protein metabolism on the basis of the existence of two forms of protein, including the experimental evidence on which he drew, is summarized in the following excerpt.

C. von Voit, *Physiologie des allgemeinen Stoffwechsels und der Ernährung* (Leipzig, 1881)[a]

(Physiology of General Metabolism and Nutrition)

pp. 301–2

This protein dissolved in the juices [*Säfte*], which I have called 'circulating protein' was not discovered by me, for it was already long known that in the nutritional fluid [*Ernährungsflüssigkeit*] a protein solution flows through the organs; I have only brought it into a quite definite connection with the protein breakdown. I gave it this name not because it is broken down in the fluid stream [*Säftestrom*] or because the circulation is the cause of its breakdown, but in order to signify that it is dissolved in the nutritional fluid and through the intermediary or circulating fluid stream is brought to the cells, which provides the conditions of the breakdown. Thus I will not indicate by it a chemical difference, but first only a difference in the place in which it finds itself and then in its physiological connection to the breakdowns in the body. One and the same molecule of protein can at one definite moment be protein of the blood plasma, in the next protein of the nutritional fluid, in another protein of the lymph or also organ protein. According to the locality one gives the identical protein particles different names e.g. protein of the blood plasma, or of the lymph, or also circulating protein, if it finds itself dissolved in the intermediary fluid stream.

The results of my experiments determine me now to bestow on the protein of the nutritional fluid or the circulating protein an important role in the breakdown of protein. The dissolved protein of the juices to which that newly entering from the intestine is joined, is according to my experience more easily broken down than that firmly bound to the organized forms and consisting partly of water-insoluble organ protein. If this is successfully confirmed then the organized protein does not break down and the protein dissolved in the juices is used as substitute – it is the latter, as already had become probable through earlier considerations, which is destroyed under the influence of the cells.

In a hungry animal a considerable amount of protein is found piled up in the organs and yet of this only a small fraction is broken down daily, according to my estimations on a large dog not quite 1%. If on the other hand a certain amount of protein enters from the intestine, which amounts at most to 12% of the protein quantity to be found in the body during hunger, then the protein breakdown increases quite out of proportion and it becomes 15 times as great as in hunger. Thus the protein consumption is not at all proportional to the total protein amount in the body, but approximately to that quantity of protein coming out of the intestine; the newly brought in dissolved protein behaves quite differently, in connection with breakdown, from the protein piled up in far greater amount in the organs, whilst it either displaces organized protein or is itself very easily decomposed.

Selection 111

Although highly influential, Voit's supposition that dietary protein was metabolized in tissue fluids rather than in the tissues came under attack, notably by the physiologist, Eduard Pflüger, in 1893.[a] As shown elsewhere in this book, by uncovering the regulatory role of the cell in regard to oxygen consumption, he located the site of respiration in tissue cells. He also believed that ingested protein was 'dead' and became 'alive' through incorporation into living protoplasm, which was considered to be exceedingly unstable.[b]

Influenced by those beliefs, it appeared to Pflüger that it was the nutritional condition of the living cells of the animal themselves that controlled the breakdown of the protein entering them. In order to test this conjecture Pflüger designed a set of experiments that were carried out in his Bonn laboratory by B. Schöndorff.

They consisted of obtaining blood from starved dogs and conveying it through the hind limbs and liver either of dogs well nourished (on meat) or of starved dogs. Further, blood was taken from well fed dogs and circulated through the same organs of starving dogs. What was measured was the urea content of the blood before and after its passage.

It will be seen from the summary of the experimental evidence below that the amounts of urea found in the blood were held to express the protein breakdown in the muscle and liver cells, thereby favouring Pflüger's view.

B. Schöndorff, *Pflügers Arch.*, *54*, **420. 1893**

'In welcher Weise beeinflusst die Eiweissnahrung den Eiweissstoffwechsel der thierischen Zelle?' (In what manner does the protein diet influence the protein metabolism of the animal cell?)

p. 483

1. With the passing of the blood of a hungry animal through the organs and liver of a well nourished animal an increase in the urea content of the blood takes place.

2. With the passing of the blood of a hungry animal through the organs and liver of a starving animal no alteration in the urea content of the blood takes place.

3. With the passing of the blood of an animal richly nourished with protein through the organs and liver of a starving animal a diminution of the urea content of the blood takes place.

Thus:

I. *The size of the protein breakdown depends on the state of nutrition of the cell and not on the protein content of the 'intermediary fluid stream'.*

II. The *size* of the *urea content* of the blood depends on the *state of nutrition* of the animal; the same sinks with hunger to a *minimum* of *0.0348%* and rises in the stage of highest urea formation to a maximum of *0.1529%*.

III. The urea is *formed* in the *liver* from the *nitrogen-containing decomposition products*, probably *ammonium salts*, originating in the breakdown of protein in the organs.

b. From *c.* 1890 to *c.* 1940: nature and structure. Size and shape. Metabolism

The aforementioned clarification of the concept of the molecule (including the question of molecular size) and the development of ideas on carbon quadrivalency and molecular structure,[a] were two of the major steps which made it possible to obtain a more complete picture of the nature of organic compounds. In conjunction with the studies of decomposition and of synthesizing reactions, they provided the basis for new insights into the constitution of organic compounds and into their structural formulae, a task successfully completed by the late 1860s and early 1870s.

This approach, however, did not apply to protein substances. Thus in a textbook published in 1874 by one of the leading organic chemists working in Britain, Carl Schorlemmer (1834–92), he stated with respect to 'albuminoids or proteids':[b]

Their constitution is completely hidden in darkness; from the results of their ultimate analysis no formula can be calculated, on account of their high molecular weight, which, however, is not known, as they do not form definite compounds with other bodies, and are neither volatile nor crystalline.[c]

By 1881 in the revised and extended English version of a well-known German textbook the situation regarding the existence of proteins in crystalline form, their elemental composition, and their molecular size and structure is described as follows:

They are without exception amorphous, and all contain nitrogen in addition to carbon, hydrogen and oxygen, in most cases sulphur also. Accurate formulae have not yet been obtained for any of these bodies, but from the results obtained it is certain that their molecules are very large and of very complicated structure. In agreement with this they do not diffuse through membranes.

Here was a categorical statement regarding the amorphous existence of proteins. Yet a passage a few pages later reflects organic chemists' uncertainty about proteins' ability to form crystals.

The blood pigments, or haemoglobins, occupy a peculiar position amongst the protein substances . . . In a state of purity all haemoglobins appear to be crystalline, but do not all crystallize equally well nor with equal readiness.[d]

Indeed, from around the middle of the nineteenth century there was a steady accumulation of evidence which showed that proteins such as haemoglobins and seed globulins could crystallize.[e] Reports on crystalline proteins evoked controversies, especially as pure organic chemists were by and large neither prepared nor willing to recognize their significance at this time. It was from the interests of medical and biological scientists centred on the colloidal behaviour of proteins that crystallization by 'salting out' grew up as a preparative technique of protein chemistry.[f]

The paper by the Prague-born Franz Hofmeister (1850–1922) on the crystallization of ovalbumin, from which the selection below has been taken, is considered to be a landmark in the development of the subject. In view of the author's insistence that it was not a flukish discovery, a few remarks about its antecedents and conceptual significance will not be out of place.

Its immediate origins can be traced to 1885 when Hofmeister became Full Professor (*Ordinarius*) of Pharmacology at the German branch of the University of Prague. At that time he initiated research into resorption and the mode of action of drugs, believing that they related to the way proteins in the organism reacted with them. It was in this connection that the question of the true chemical individuality of proteins assumed both theoretical and practical importance and the method of their fractional precipitation by ammonium sulphate was developed. Further work on precipitation procedures by means of neutral salts followed, suggesting to Hofmeister that the salting-out of proteins depended on dehydration of proteins in solutions.[g] It was then that Hofmeister began to associate 'organization' of living systems with the natural interplay of the 'colloidal' and 'crystalloid' state based on the inhibition by and the withdrawal of water from proteins. He developed this theme of a physicochemical explanation of the machinery of life in a celebrated publication in 1901.[h]

F. Hofmeister, *Z.physiol.Chem.*, *14*, 165. 1890

'Ueber die Darstellung von krysallisirtem Eieralbumin und die Krysallisirbarkeit colloider Stoffe' (On the preparation of crystallized egg albumin and the crystallizability of colloidal substances)

pp. 165–6

Fresh egg white, which must be free of admixed yolk, is beaten with a good active eggwhisk to a fine foam, then left for 24 hours. The almost completely clear, thinly liquid egg white solution on the bottom of the vessel is poured off from the foam, and for the purpose of separating the globulin is treated with an equal volume[1] of cold-saturated, neutrally reacting ammonium sulphate solution. The resulting rather ample precipitate is filtered off and the completely clear, salt-containing fluid left on a large, flat dish with a level bottom to evaporation at room temperature.

After some days a more or less thick coat of a finely granular white, sometimes also yellowish or reddish coloured precipitate has deposited itself on the bottom, which on microscopic examination shows itself as made up of transparent, fairly large spheres or aggregates of spheres, diffracting light simply, without radiate or lamellate structure (*Globuliten*). The thin membrane which often covers the surface is also formed out of these same structural elements.

If one filters after an increase in the precipitate is not longer discernible, then one obtains almost the whole of the albumin in the form of a coarsely granular mass (pure white or a little coloured, completely soluble in water, separable from the mother liquor rather completely by pressure). For the purpose of further purification it is dissolved in half-saturated ammonium sulphate solution and is again left to gradual separation by means of free evaporation. This process is then repeated as long as the protein still separates in globulites.

p. 168

The protein body prepared appeared after closer examination to be identical with the hitherto purest egg albumin prepared by Starke[2] ... Now I can already communicate that the same procedure can also perform good service in the crystalline preparation of other protein substances and bodies related to the proteins.

pp. 168–9

After recent success in obtaining one of the typical 'colloids' in crystalline state it may be permitted to give space for some remarks on the reasons for the difficulties which oppose the crystallization of colloidal bodies ... To obtain the substances in crystalline state it is first of all necessary to isolate them and free them from admixtures.

pp. 170–1

If already from the beginning admixed impurities can impair the crystallizability of colloid substances, this still appears as a smaller difficulty than that which results from a very widespread characteristic of the colloids, namely from their tendency, on small external inducement, to go over into modifications which are insoluble but capable of swelling.

pp. 171–2

This tendency of colloidal bodies, for apparently quite unimportant causes, for example water loss on drying, or through the influence of chemically otherwise indifferent substances, the coagulating ferments,[a] to change into substances insoluble but capable of swelling appears to me for its characterization more important than the alleged crystallizability. So far as it really exists, it is not to be conceived as an inherent property of the colloid substances. That this tendency belongs particularly to the chemical substances which have an important part in the building up of the animal and plant tissues can be no accidental concurrence. The existence of substances which, themselves soluble, easily go over into insoluble compounds capable of swelling is a necessary preliminary condition for the building up of the cell, and with this of all organic life. Thus the artificial contrast, physically scarcely any longer tenable between 'colloid' and 'crystalloid' receives a physiological meaning. For just that property of the 'colloids', which makes them capable of building up the organism, their easy conversion into compounds insoluble but capable of swelling, is one of the principal obstacles which prevents attaining them in crystalline form.

Selection 113

Establishing that proteins combined an acidic and a basic nature was an integral part of the growth of knowledge about their solubility properties. The combination had been hinted at before by various researchers but the study of it received a great impetus from new physicochemical conceptions of solutions, including the theory of electrolytic dissociation which had been developed during the last quarter of the nineteenth century.[a]

Cardinal evidence for the view that proteins are amphoteric[b] electrolytes was obtained in 1899 by the many-sided Cambridge scientist, William Bate Hardy (1864–1934). Doubtful whether interpretations of structures seen in cells treated by fixatives were also applicable to their structural organization during life, Hardy decided to approach the problem through the study of colloidal systems.[c]

Among many other researchers he utilized the then novel method of observing the migration of protein particles in solution in an electrical field ('cataphoresis', later to be known as 'electrophoresis'). As shown below, Hardy discovered that the direction of their movement depended on whether the reaction of the fluid was acid or alkaline. Shortly afterwards he draw attention to the importance of the point, termed by him 'isoelectric', at which the acid-base relationship of the fluid was such that the colloidal system appeared to be free of any excess of positive or negative charge. Under these conditions the solubility of the suspended protein particles was at a minimum and consequently 'coagulation or precipitation' occurred.[d]

W.B. Hardy, *J.Physiol.*, *24*, **288. 1899**

'On the coagulation of proteid by electricity'.

p. 288

The material used in the experiments which form the subject-matter of this paper was the slightly opalescent fluid which is formed when white of egg is mixed with 8 or 9 times its volume of distilled water, filtered, and boiled. The proteid matter in solution is changed by the heat so that, according to Starke, the greater part presents all the characters of alkali albumen.[1]

p. 291

The [constant] current was supplied by storage cells at 105 volts. Platinum electrodes were used and the electromotive force was varied from 8, 13, to 105 volts in various experiments. The resistance was always very great. It was not thought necessary to measure it but the current passing was always less than 0.000001 ampère. The cell employed had the shape of a U-tube with vertical limbs. The bend of the U was narrowed in order to make it easier to withdraw the contents from the limbs without mixing . . .

The fluid has an alkaline reaction. The effect of the passage of a constant current is the formation of an opaque white coagulum about the anode. The density of the coagulum is dependent upon the electromotive force: with 105 volts it is yellowish with transmitted light and of a tough almost rubber-like consistency as though the particles had been impacted with considerable force.

pp. 294–5

The fluid has an acid reaction. The movements of the proteid particles and of the water are the reverse of those described. The particles now move with the positive stream from the anode to the cathode, the water moves with the negative stream . . .

Fluid is neutral. The coagulum which forms when dialysis is pushed far enough was broken up very thoroughly and suspended in distilled water. There is now so little movement of the particles under the influence of a current that it is difficult to detect, and what movement there is, is due to the fact that the material is not absolutely neutral. Thus, if the coagulum be formed by dialysing an acid fluid the particles when suspended in distilled water will move to the cathode.

p. 296

When the fluid is alkaline the proteid particles move with the negative stream and therefore they carry a negative charge, while the water moves with the positive stream and therefore carries a positive charge. When the fluid is acid, however, the proteid particles carry a positive charge and they therefore are positive to the water. When the fluid is neutral there is little or no difference of potential between the water and the particles – that is to say, the water and the particles form an electrically homogeneous mass.

The proteid particles therefore have this interesting property that their electrical characters are conferred upon them by the nature of the reaction, acid or alkaline, of the fluid. If the latter is alkaline the particles become electro-negative; and vice versâ.

p. 298

The experiments suggest certain ideas concerning the nature of the directive force which determines the aggregation and arrangement of the colloid particles in the formation of a hydrogel. Graham[a] attributes it to a quality of idio-attraction possessed by colloid molecules. Many colloidal solutions are however stable unless an electrolyte be added, and in this fact a clue to the nature of the force may perhaps be found.

In speaking of the precipitating action of crystalloids upon colloids, Hofmeister referred the property to the water-attracting power of the former, and the phrase has become current in physiological literature. It may I think be put aside once and for all. It is vague and meaningless in that it cannot be expressed in terms of any definite stoichiometrical quantity. Also any possible meaning it can bear must be shared by non-electrolytes, and these apparently possess no aggregating action upon colloid particles. The boiled egg-white solution, for instance, is freely miscible with alcohol.

Selection 114

Nine monoamino acids, three diamino acids and one sulphur-containing amino acid were obtained from proteins on hydrolysis and recognized as their constituents between 1820 and 1900. They included the following: glycine, leucine, tyrosine, serine, glutamic acid, aspartic acid, phenyl alanine, alanine, lysine, arginine, iodogorgoic acid, histidine and cystine. Usually hydrolysis with hydrochloric acid was employed, producing a mixture of amino acid hydrochlorides from which they or the copper, silver or other salts of the amino acids were isolated by fractional crystallization.

This was the situation when Emil Fischer (1852–1919) began to concern himself with protein chemistry in 1899. Improving on the previous work of Theodor Curtius (1857–1928),[a] he shortly afterwards pioneered the 'ester method' for the separation of the amino acids. It derived its name from the esterification of amino acids present in the hydrolysate with alcohol, followed by fractional distillation under reduced pressure of their esters.[b] When treated with boiling water or baryta water (barium hydroxide) the amino esters yielded the free amino acids.

Fischer's belief (see below) that the method (although not fully quantitative) could be of great service in extending knowledge of the amino acids occurring in proteins was vindicated.[c]

E. Fischer, *Ber.chem.Ges.*, *34*, **433**. 1901

'Ueber die Ester der Aminosäuren' (On the esters of the amino acids).[1]

pp. 434–5

The esters of the monoamino acids are, with the exception of the beautiful crystalline tyrosine derivative, fluids of alkaline reaction which distil completely under diminished pressure, and whose solubility in water decreases with increasing molecular weight. The derivatives of aspartic and glutamic acids dissolve conspicuously readily in pure water. The boiling points of the various esters differ considerably even at greatly diminished pressure, so that mixtures can be broken up by fractional distillation. These esters are also specially suited for isolation of the amino acids from complicated mixtures, and I have no doubt that they will be used in the future studies on the hydrolytic splitting of the proteins for recognition and purification of the amino acids; for the latter can very easily be regenerated from the esters by boiling with water or barium hydroxide and, moreover, the esters can be distinguished by their boiling points, by their different solubility in water or through the melting point of the most beautiful crystalline picrates . . .

In the esters the amino group is as active as in the ordinary amines; and as the esters also, unlike the free amines, are easily soluble in alcohol, ether and benzene it seems that they are specially suitable for the separation of numerous derivatives of the amino acids. I have convinced myself that they react energetically with acid anhydrides, acid chlorides, alkyl halogens, isocyanates, mustard oils, aldehydes, ketones, carbon disulphides, phosgene so that they can be substituted in all changes which are known for the simple primary amines.

Selection 115

By about 1900 researchers were beginning to ask basic questions about the structure of proteins. By a most remarkable coincidence, the discovery that protein structure consists of amino acids joining to form amide-like amino acid derivatives was made by two scientists, working quite independently of each other and communicating their conclusions simultaneously on the same day. It happened on 22 September 1902 at the 74th Assembly of the Society of German Naturalists and Physicians at Karlsbad (Karlovy Vary in Czechoslovakia), where the structure aspects of proteins were considered by Franz Hofmeister in the plenary session in the morning,[a] and by Emil Fischer in the chemical section in the afternoon.[b]

Their suggestion regarding the general principle underlying the structure of proteins pointed in the same direction. It depended on the formation of the CO-NH link by the amino group (NH_2) of one amino acid combining with the carboxyl group (COOH) of another. Nevertheless, they differed in their attitudes and concerns and, therefore, their individual share in the development of the 'Hofmeister-Fischer hypothesis of protein structure' will be considered separately.

In considering the various possibilities of combination between the two amino acids Hofmeister imagined that it occurred through a nitrogen atom. He described the characteristic structural unit of which proteins were built as —CO—NH—CH=, mainly on account of the liberation of amino-nitrogen and the disappearance of the biuret reaction during protein hydrolysis.

On reading the extensive review by Hofmeister (from which the selection below is taken) it emerges that he was also strongly influenced by biochemical (enzymological) considerations.

F. Hofmeister, *Erg.Physiol.*, *1*, 759. 1902

'Uber Bau und Gruppierung der Eiweisskörper' (On the structure and grouping of the protein bodies)

pp. 787–92

Of the decomposition products of typical proteins which, with as simple a structure as possible, still contain more than one nucleus, only arginine and leucine imide have a constitution exactly known to us. The manner in which the carbon nuclei[a] are united is seen in the following formulae:

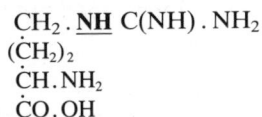

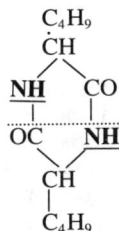

$$CH_2 . NH \ C(NH) . NH_2$$
$$(CH_2)_2$$
$$CH . NH_2$$
$$CO . OH$$

In both cases it is an imino group which joins the two carbon nuceli together, on the one hand the guanidine residue with ornithine, on the other two leucine complexes. The similarity of these ways of binding comes out even more clearly if only the carbons directly taking part are considered. Then we have the complexes:

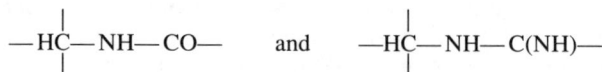

$$-\overset{\mid}{\underset{\mid}{HC}}-NH-CO- \quad\text{and}\quad -\overset{\mid}{\underset{\mid}{HC}}-NH-C(NH)-$$

In fact, as is shown below, a corresponding scheme for binding applies particularly to proteins.

... the protein molecule is mainly built up from amino acids, amongst which again the monoamino acids preponderate. A linking of the amino acids with one another by binding of carbon —C—C— would allow the giant protein molecule to appear as a single enormous branched carbon chain.[b] Its decomposability, but particularly fermentative breakdown to definite larger and smaller complexes, would be most difficult to comprehend. The breakdown of longer into shorter carbon chains by animal proteolytic ferments, even with the very active trypsin, has not been observed.[1]

Likewise an ether or ester type of linkage =C—C—C=, such as O. Nasse[2] was inclined to assume on grounds of the analogy of fermentative protein breakdown with the diastatic and fat-splitting action of some enzymes also can scarcely be considered, if one realises that the alcohol (OH) groups necessary for such a linkage have not been demonstrated in protein (apart from the hydroxyl of tyrosine), and thus are either lacking or may be only very sparsely present. Also an ester type of interaction of the carboxyl groups contained in the amino acid complex would (in contradiction to the observed facts) undoubtedly give a basic character for the whole molecule since, as Curtius[3] has first shown, in the amino acids after esterification, the acid-binding character of the NH_2 groups stands out strongly. On this ground also linkage in the manner of an acid anhydride is not to be considered.

Binary linking being presumed, the assumption of a binding in the manner ≡C—NC—C≡ appears much more probable, offering from the outset series of possibilities. If one neglects the sulphur, only sparingly present in protein, one may think of the following cases:

I	II		III
—CH$_2$—NH—CH$_2$	—CH$_2$—NH—CO—	or	—CH$_2$—NH—C(NH)—

With ternary binding still other assumptions could be made, for example

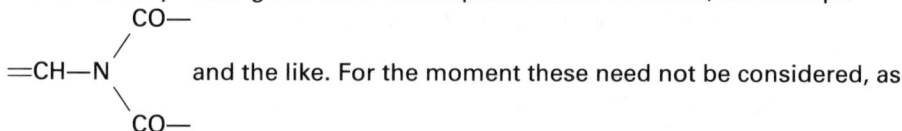

$$=CH-N\overset{\textstyle CO-}{\underset{\textstyle CO-}{}}$$

and the like. For the moment these need not be considered, as being unsuitable at the present time for fruitful discussion.

Against the more widespread distribution of a linkage according to Scheme I (which does in fact appear in individual nuclei, e.g. in the pyrrolidine nucleus) there is the circumstance that in this case (e.g. by analogous binding of two monoamino acids) the numerous carboxyl groups remaining free in the amino acid complex must give a strongly acid character to the whole molecule, which does not correspond to the facts.

Scheme III corresponds to the binding of the guanidine residue in arginine. As however the guanidine residue appears to occur in protein only in this single complex, this manner of binding in the building up of the protein molecule may be considered as of significance limited to the presence of arginine.

On the other hand there are numerous facts which speak for the very general distribution of a linkage according to Scheme II: —CH_2—NH—CO— (or —NH—CH_2—CO—NH—), which indeed is realized in leucine imide.

1. As we have just shown, only a small part of the total nitrogen of protein bodies is present in a form readily split off, which could therefore correspond to the binding method —CO—NH_2.[c] By far the greatest part of the nitrogen, up to over 90% (judging from the end-products), must be present in a form yielding on hydrolysis a small part as imine (from arginine), by far the greater part in the amino form. That indeed only a small amount of NH_2 groups are preformed in protein is shown in their behaviour with nitrous acid. According to O. Löw[4] and H. Schiff[5] only a relatively small part of the nitrogen is split off by this acid. As C. Paal[6] confirmed on the similarly constituted glutin peptone, substances behaving like nitrosamines are formed. Thus it is to be assumed that in the original protein molecule the NH_2 groups of the end products are preformed by NH groups.

2. The biuret reaction, which is considered specially characteristic for protein bodies, occurs according to Schiff,[7] with those substances which contain two CO—NH_2— complexes, or instead of these CS.NH_2— or C(NH).NH_2—; and in certain circumstances also with —CH_2.NH_2— complexes bound through their carbon atoms, or through intervention of a carbon atom (in a CH_2—, CH(OH)—, CO-group), or a nitrogen (in an NH— group).[d]

A formulation corresponding to this method of binding of the protein-contained α-amino acids, as well as to the arrangement favourable to the biuret reaction is conceivable in the simplest manner in the following way, where in the interest of simplicity the coupling of leucine and glutamic acid may serve as an example:

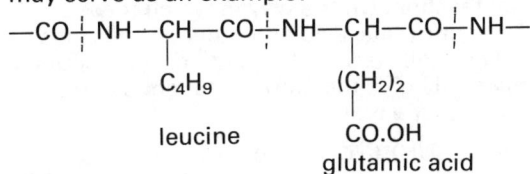

$$—CO—NH—CH—CO—NH—CH—CO—NH—$$

$$\text{leucine}\ C_4H_9 \qquad \text{(CH}_2)_2$$

$$\text{CO.OH}$$
$$\text{glutamic acid}$$

It is plain without further discussion that this manner of coupling with formation of the group —CH—NH—

CO—NH—

can repeat itself in the protein as often as an α-amino acid unites itself in the given manner with a second. Now are the glycine amide

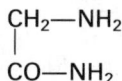

$$CH_2—NH_2$$
$$CO—NH_2$$

as well as sarcosine amide $CH_2—NH(CH_3)$

$$CO—NH_2$$

according to Schiff, aspartic acid diamide

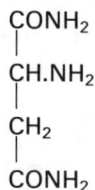

$$
\begin{array}{l}
\text{CONH}_2 \\
|\\
\text{CH.NH}_2 \\
|\\
\text{CH}_2 \\
|\\
\text{CONH}_2
\end{array}
$$

according to E. Fischer, bodies which give an intense biuret reaction. Thus they are relevant to the assumption that in the intact protein molecule (as well if not better in its near derivatives, the albumoses and peptones) the condition for the appearance of the biuret reaction is given by the presence of these $\text{CH}_2.\text{NH}_2$

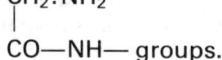

|

CO—NH— groups.

That nitrous acid without splitting the protein molecule causes this reaction to vanish is then immediately understandable through its known action on the NH_2– and NH– groups; there is a similar explanation for the effect of alkali, which, easily saponifying the CONH_2-groups, strikingly diminishes this reaction.[8]

3. Recent experience on the condensation of amino acids leads to similar conceptions about the linkage of carbon nuclei[9]. Schaal in 1871 by boiling aspartic chloride at 200° in a stream of carbonic acid obtained condensation products – Schiff later named them polyaspartic acids – which as Grimaux[10] found, gave the biuret reaction. Through fusion of asparagine with urea the latter first obtained artificial colloids which, besides other protein reactions also gave the biuret reaction splendidly. Starting from glycinethylester Curtius then obtained a series of condensation products amongst which one of basic nature gave the biuret reaction. However also through addition of glycocoll to hippuric acid Curtius[11] arrived at complicated bodies, of which one presented the same reaction. On the way in which these condensations result from the aminoacetic acid and other amino acids we are now sufficiently instructed by Emil Fischer. The simplest compounds of this kind, which of course only partly give the biuret reaction, e.g. the glycylglycine, are built according to the type:[12]

$$\text{NH}_2.(\text{CH}_2.\text{CO}.\text{NH})_n\text{CH}_2.\text{CO}.\text{OH}$$

Also the Curtius base has shown itself after closer research by M. Schwarzschild[13] to be an ester of a compound built just according to this type, $\text{NH}_2.(\text{CH}_2\text{CONH})_6\,\text{CH}_2.\text{CO}_2.\text{C}_2\text{H}_5$. All these facts teach that the amino acids are capable to an extraordinary degree of condensing *in vitro* in the specified manner to complexes of high molecular weight.

4. The coupling through formation of the –CH_2–NH–CO– group, which in these compounds plays so great a role – and the –CH_2–NH–C(NH)– group probably equivalent in this connection – is also very often met *in vivo*. One of the most familiar syntheses in the animal body, that of hippuric acid, $\text{C}_6\text{H}_5.\text{CO}.\text{NH}.\text{CH}_2\,\text{COOH}$ and its homologues follows this course; as well as the synthesis of the coupled bile acids[e] and the vital formation of the uramino acids through the attachment of the – $\text{CO}.\text{NH}_2$-residue to the NH_2-group of an amino acid and the attachment of acetyl to the same place.[f] This is, however, of significance inasmuch as it opens up an

understanding as to how the organism proceeds with the building up of adequate protein substances by means of the protein-breakdown material flowing out of the intestine [into the blood]. On the other hand, hippuric acid like the proteins is accessible to the hydrolytic action of acid and alkali as well as the ferments (histozyme[9] and bacterial enzymes). If in the former case hydrolytic splitting of the –CO–NH.CH$_2$– linkage takes place, so it becomes probable that also with the protein an analogous manner of binding occurs. Besides Schwarzschild[14] lately succeeded in showing that a synthetic amino acid derivative built up in this way (the Curtius base mentioned above) loses the biuret reaction on treatment with pure trypsin; this is the more remarkable as trypsin is not capable of splitting off ammonia from acid amides, e.g. acetamide or asparagine.

The type of condensation described here through formation of –CO–NH.CH$=$ groups may thus explain both the building up of protein substances in the organism, as well as their breakdown in the intestinal tract and in the tissues.

On the basis of these given facts one may therefore consider the proteins as for the most part *arising by condensation of* α-*amino acids, whereby the linkage through the group* –*CO–NH–CH*$=$ has to be regarded as the one regularly recurring [*die regelmässig wiederkehrende Verknüpfung*].

Selection 116

Emil Fischer believed that in order to make biology a more exact science it had to make use of the contributions offered by synthetic organic chemistry to the understanding of the constituents of biological systems.[a] He also felt that among his contemporaries he was supremely qualified to elucidate problems connected with the structure and synthesis of naturally occurring substances. It was these attitudes that induced him to work in turn on sugars,[b] purines and proteins from a chemical standpoint.

While developing the aforementioned ester method of separation of amino acids, in collaboration with Ernest Fourneau (1872–1949), Fischer found in 1901 that from the esters the bimolecular anhydrides of the amino acids (diketopiperazines, 'diacipiperazines') could be easily obtained. They started with the simplest member of the series, the glycine anhydride. On boiling it with concentrated hydrochloric acid they prepared the hydrochloride of an amino acid of the formula $C_4H_8H_2O_3$. They interpreted the process in the following way:

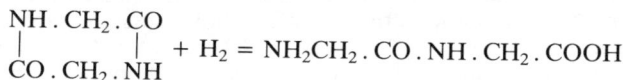

$$\begin{array}{l} \text{NH . CH}_2\text{ . CO} \\ | \qquad\qquad | \quad + \text{H}_2 = \text{NH}_2\text{CH}_2\text{ . CO . NH . CH}_2\text{ . COOH} \\ \text{CO . CH}_2\text{ . NH} \end{array}$$

Calling the group NH$_2$CH$_2$CO– glycyl, Fischer and Fourneau derived the name 'glycylglycine' for the compound comprizing two glycine molecules joined together by their unlike COOH and NH$_2$ ends. In addition to this product they obtained other compounds consisting of two or three amino acids linked in this way.[c]

As shown below, Fischer recognized the broad implications which stemmed from these findings regarding these amino acid residues combined together in what he called peptides. He clearly assumed that proteins contain, in the main,[d] this type of linkage between amino acids, which eventually became known as 'peptide linkage' or the 'peptide link'.

E. Fischer, *Chem.Ztg.*, *26*, 939. 1902

'Ueber die Hydrolyse der Proteïnstoffe' (On the hydrolysis of proteins)

p. 940

In conclusion the speaker considers the coupling of the amino acids in the protein molecule. The thought that here groups of acid amide nature play the chief role comes first, as indeed Hofmeister has assumed in his more general lecture held in the morning. The same conviction induced him 1½ years ago to investigate the synthetic chaining of the amino acids. The fruit of these studies was the finding of glycylglycine $NH_2.CH_2.CO.NH.CH_2.COOH$, of alanylalanine, leucylleucine, carbethoxy glycylglycylleucine etc.[a] In order now to supply the proof that similar groups are present in the protein molecule he investigated anew in collaboration with Dr. Bergell the hydrolysis of silk fibroin.[b] As was already known, the fibroin dissolves easily in cold strong hydrochloric acid, and with alcohol a product is then precipitated which Weyl has named sericoin.[c] But if one leaves the solution in the threefold amount of hydrochloric acid of specific weight 1.19 for 16 hours at room temperature, then there appears on treatment with a large amount of absolute alcohol only a little more precipitate, and the greatest part is now in the solution as the hydrochlorate of a peptone-like body. On evaporation under strongly reduced pressure and with a temperature not rising above 40° the salt remains as an amorphous mass. If it is treated in aqueous solution with silver carbonate and the filtrate is evaporated after exact precipitation of the dissolved silver, with hydrochloric acid, there results a colourless amorphous preparation which reacts alkaline, tastes bitter, gives the biuret and tyrosine colouration very strongly, is easily soluble in water, and in its whole character resembles the peptones. Specially interesting is its behaviour towards trypsin. If an aqueous solution is treated with ammonia and the fresh ferment the crystallization of nearly pure tyrosine begins after several hours, and after two days its splitting off is complete. By evaporation of the solution one now obtains a new product of peptone nature which is differentiated from the first especially by the absence of tyrosine. Limited activities of this kind by the ferments will doubtless play a great role in the further study of the protein group. This second peptone can be precipitated out of aqueous solution with alcohol and ether, and after washing with methyl alcohol forms a white non-hygroscopic powder which contains 40.6 per cent C, 6.6 per cent H and 18.4 per cent N, in aqueous solution rotates strongly to the left, and on total splitting with hydrochloric acid gives 40.1 per cent glycocoll and 28.5 per cent alanine. On 1−2 hours warming with 10 per cent baryta water at 90° it develops ammonia, loses its optical activity almost completely, and there arise now new products of which one belongs to the class of glycylglycine. That is the solution freed from baryta leaves on evaporation a syrup which, with longer standing, partly crystallizes in the dessicator. The mass [was] pressed from the syrup, but its further purification by re-solution encountered difficulties, and it was converted by treatment with β-naphthalene sulphonyl chloride and alkali into the naphthalene sulphonyl derivative which is difficultly soluble in cold water and crystallizes well. According to the analysis (found C 54.2, H 4.5, N 8.1) and the result of total hydrolysis, whereby glycocoll and alanine appear, the preparation is probably a β-naphthalene sulphonyl glycylalanine; in any case the product from which it arises belongs to the series of glycylglycine and analogous compounds. This observation, as is easily seen, is of general significance because it shows the possibility of obtaining crystalline products which stand between the peptones and amino acids. The speaker on this account makes the proposal, following the known division of the carbohydrates as disaccharides, trisaccharides, etc. to name the bodies of the type of glycylglycine dipeptides, and to designate anhydride-like combinations of a greater number of amino acids as tripeptides, tetrapeptides.[d]

Selection 117

We now turn to the state of knowledge of protein metabolism at the time when Hofmeister and Fischer had proposed the peptide hypothesis. To be more precise, we shall be concerned with the problem of the intestinal absorption of proteins.

It will be recalled that by 1850 the view was advanced that in the digestive tract proteins were broken down to albuminoses (later termed albumoses or proteoses) and peptones. By 1900 they were being distinguished from each other by certain physical and chemical properties. It was assumed that the molecular weight of proteoses and peptones, though lower than that of proteins, was still comparatively high. Both groups were considered to be incoagulable by heat but proteoses, like proteins, were supposed to be salted out whereas peptones were not. Also the greater ease with which the latter diffused through a semipermeable membrane was noted.

Peptones were regarded as the end products of protein gastro-intestinal digestion, but it was not clear what happened to them on absorption. In 1885 Hofmeister reported great proliferation of lymphocytes in the adenoid tissue of the mucous membrane of the small intestine of well nourished animals. He associated this with absorption and the building up of the peptone into protein within the lymphocytes of the intestine. Accordingly, it was in this assimilated form that peptone presumably entered the blood and was carried through the body.[a]

In 1890 Richard Neumeister (1857–1906) noted the disappearance of peptones from the blood diluted with peptone solution in which freshly excised pieces of intestinal mucous membrane were placed. But he remained uncertain as to whether, on vanishing, the peptone was built up again into protein or further broken down, as the presence of the amino acids leucine and tyrosine seemed to suggest.[b]

Neumeister's experiments were taken up by Otto Cohnheim (1873–1953), with the intention of isolating the protein formed synthetically by the cells of the intestinal wall. This he failed to achieve. Instead, as shown below, in a paper published in 1901 he demonstrated the breakdown of the peptone into crystalline decomposition products, effected not only by the mucous membrane itself but also by an extract of the membrane. He decided that a ferment (enzyme), differing from trypsin in its action and in other properties, was concerned; he named it 'erepsin'.

O. Cohnheim, *Z.physiol.Chem.*, *33*, 451. 1901

'Die Umwandlung des Eiweiss durch die Darmwand' (The transformation of protein by the intestinal wall)

p. 453

With the intestine artificially perfused with blood, the intestine lying in blood[a] and with use of the press-juice of the intestinal mucous membrane I observed indeed a disappearance of added albumoses and peptones, but on the other hand *never an increase in the protein*. Only the investigation of the fluid freed from protein and no longer containing peptone brought me on to the right track. *The peptone is not reconstituted*, but on the contrary is further broken down by the intestinal mucous membrane into crystalline split products.

As peptone I used the peptic digestion products of muscle.

pp. 454–5

With this peptone solution I have now first repeated Neumeister's experiments. 3–5 ccm. were added to diluted blood, small pieces of intestine from quite freshly killed dogs and cats were thrown into the fluid at body temperature and an oxygen stream was passed through. After 2 hours the protein of the blood was coagulated and filtered off through gauze, and the filtrate was tested for peptone by the biuret reaction; the test gave a negative result or showed a very weak violet colour, which undoubtedly originated in the protein residue or the albumoses arising from the protein coagulation; and which in any case was far from comparable with the extremely strong biuret reaction of a correspondingly diluted peptone solution. I then replaced the diluted blood with Ringer solution – containing 0.2 g $NaHCO_3$, 0.1 g $CaCl_2$, 0.07 g KCl and 8.5 g NaCl per litre – and found that also here the biuret reaction vanished. Admittedly, it seemed to me as to Neumeister that the action was less than with the use of blood. Nevertheless the third part of the small intestine of a cat or a small dog can in two hours with 5 ccm of the described peptone solution destroy the biuret reaction, thus transform about 0.6 g of peptone. The use of the common salt solution instead of blood affords the great advantage that one has only the small amount of protein in the fluid, that which in the course of the experiment passes out of the intestinal cell into solution.

pp. 456–8

The next question was now whether here a fermentative splitting of the peptone was concerned, or an action of the organized living intestinal wall. The latter could likewise result in a splitting; it could however also assimilate the peptone and thus eliminate it in another way. In order to decide this, in the following experiments I used not the intestine but its aqueous extract . . .

I now added to this extract of the intestine several cubic centimetres of my peptone solution, shook and coagulated immediately or after some minutes; thus I obtained in the filtrate a beautiful biuret reaction corresponding to the peptone amounts used. But if I left the peptone-treated extract some time in the warm, and then investigated it, the biuret reaction had disappeared; phosphotungstic acid however gave a rich precipitate of crystalline appearance, and the filtrate from the protein precipitate contained the total nitrogen of the peptone . . .

[The amounts of nitrogen][b] show undoubtedly that *the extract acts like the intestinal mucous membrane itself*, that the change in the peptone by the cells of the intestinal wall is not linked to their life, thus it is a fermentative process.

It was necessary to isolate this ferment. For this purpose I made use of the fractional salting out with ammonium sulphate, as it has been lately used by Jacoby[1] with such splendid results. The fraction 60 to 100 is quite inactive, on the other hand the ferment goes immediately into the fraction 0 to 60.[c]

Thus we have here a ferment which splits peptone, and the question was now whether I had to do with trypsin, which indeed could be present in the intestinal extracts of digesting animals. The limits of its precipitation with ammonium sulphate of course speak against this; for according to Jacoby[2] trypsin is precipitated only at 65% saturation. Certainly this limit is not quite sharp; I could occasionally find trypsin in the first fraction in small amounts.

On the other hand the ferment under consideration from the intestinal mucous membrane is *quite sharply distinguished from the trypsin of the pancreas*. Whilst it splits peptone quickly and abundantly, *it does not act at all on native protein bodies* . . .

Thus we have here without a doubt a special ferment which does not act on protein but really splits peptone. I propose to name this ferment *erepsin*, derived from ἐρείπω, I shatter. The erepsin would thus be linguistically a synonym of the trypsin.

Selection 118

By the beginning of the twentieth century the question of whether animals, like plants, were capable of producing proteins from simpler nitrogenous compounds remained unanswered.

In 1902 a notable contribution to this subject came from Otto Loewi (1873–1961), working in the Pharmacological Institute at Marburg. He attacked the problem by adopting the rather novel method of tissue autolysis.[a] More specifically, from autolyzed pancreas protein he obtained digestion products which did not show the biuret reaction. With these he carried out feeding experiments. The rest of the diet was made up of fat and carbohydrate. He showed that such non-protein material, as the only source of nitrogen in the food, could not only maintain 'nitrogen equilibrium' but also led to nitrogen deposition.[b] The results pointed to protein breakdown during digestion and also to protein resynthesis from a biuret product of digestion.

The paper is interesting for several reasons. By drawing on calorimetric data Loewi attempted to deal with the teleological objections of the well-known physiologist, Gustav (von) Bunge (1844–1920), who thought that the protein breakdown and resynthesis would be a waste of chemical potential energy and serve no purpose. Moreover, it indicates the growing awareness that animal organisms need certain constituents of food in very small amounts. Also Loewi detected that without carbohydrate in the diet, that is if it consisted solely of fat and the abiuret products of digestion, the state of nitrogenous equilibrium was not reached. However, this has not been included in the extract.

O. Loewi, *Arch.exp.Path.Pharm.*, *48*, 303. 1902

'Über Eiweisssynthese im Thierkörper' (On protein synthesis in the animal body)

p. 306

... I chose as food no homogenous substance, but rather a mixture of protein split products. The nitrogen supply must be present *only* in the form of non-protein bodies; this point is to be considered essential; yet a large series of very laborious researches on the nutritive value of certain bodies lose their value because *besides* them nitrogen in protein form was supplied; it is well known, as the experiments on the nutritive value of gelatin show, that very little of this suffices completely to prevent the breakdown. It was further necessary to consider that besides the nitrogen-containing substances as far as possible all the elements coming into consideration for the eventual synthesis should be sufficient. What we have to reckon with here is unknown. How necessary it is to reckon with this aspect appears for example from the experiments of Lunin[1] who with casein, fat, carbohydrate and salts did not succeed in maintaining the life of animals. We must assume that in our ordinary nourishment, besides protein, fat, carbohydrate and salts there are contained (if only in minimal amounts) substances which are irreplaceable in the normal course of life. One thinks of the significance, already known, of iodine or iron. So long as we do not know these 'minimum' substances, we must in all experiments on artificial nourishment reckon with the possibility of an uncontrollable error. If Lunin could not maintain the life of animals even with protein, it is really difficult to conceive why success should have been obtained with peptone or asparagine.

First I dared to hope that I might meet the requirements just described if I chose the total digestion product of a tissue which fed undigested to the organism maintained its stability. I chose the soluble products of pancreas self-digestion...

Everything now depended on proving that the fed material is protein-free. I dared to accept as evidence for this the lack of the biuret reaction.

p. 316

Just as my earlier experiments had shown success in bringing an animal into N equilibrium with biuret-free substances, the present experiment shows unexceptionally that biuret-free products can lead to considerable N deposit...

Thus it is finally resolved that the sum total of the biuret-free end-products can replace dietary protein; that is to say can substitute for all parts of the *body protein [Körpereiweisses] degraded during metabolism. Thus also the animal can build up protein and is not dependent on dietary protein, in so far as we wish to see in the biuret-giving group its characteristic features.*

The question is not herewith touched upon which groups may be lacking for the synthesis to become impossible.

pp. 328–9

After all this, it is highly probable to assume that the greatest part of the protein, perhaps the whole, normally is deeply split up before absorption and the split products . . . reach the circulation in a special linkage. For this all the support is indicated . . . in the results of the described research, according to which the split products can completely replace protein.

Bunge[2] maintains that a considerable formation of amino acids in the intestine is a waste of the chemical potential [*Spannkräfte*] which, with the splitting, is uselessly converted into living force; on the other hand it is to be emphasized as already Kutscher and Seeman[3] have adduced, that with plants such a process is proved, but above all also that we know nothing about the question whether and how much heat is consumed by the splitting . . . Rubner[a] to whom here I wish to express my best thanks has estimated the heat of combustion in the dry state of the pancreatic digestion solution used in the chief experiment . . .

The heat of combustion of 1 g. of substance in gram calories amounted to:

I	4,453
II	4,422
III	4,433
Mean	4,436

that is for ash-free substance
4,599 gram calories.

The value is about 10% lower than that determined by different investigators for flesh, but cannot be compared with this, as according to the kind and amount undetermined parts of the pancreas are contained in the undissolved residue of the digestion. Their heat value has remained outside consideration, but could raise the average importantly . . .

But even if in this way the 10 per cent by which we have found the analyzed part of the digestion products poorer in calories than flesh was not covered, if thus really with the protein breakdown some heat is lost, still the question remains always open whether the loss is 'purposeless waste' or is a great advantage for the organism.

Selection 119

Up to 1905 the approach to protein metabolism consisted of accepting one or other of the rival conceptions put forward by Voit and Pflüger and already described in this volume. It was in this year that the Swedish-born Otto Folin (1867–1934), who went to the United States in his teens, attempted to overcome their internal limitations. At the time, he was in charge of the first American biochemical laboratory attached to a hospital (McLean Hospital for the Insane at Waverley, Massachusetts).[a]

Originally desiring to learn whether knowledge of urine constituents could throw light on disorders of the mind, he carried out thorough investigations on the urine of healthy people. Struck by the appreciable variation in its composition, he related the varying amounts of urea (primarily conceived as originating in a series of hydrolytic splittings) to the nature of the diet, rich or poor in protein, and therefore assumed by him to represent 'exogenous metabolism'. The more constant quantities of creatinine and uric acid, unaffected by the amount of ingested protein, he associated with tissue breakdown and identified as endogenous metabolism.

All this led Folin to a critical appraisal of the theoretical and experimental basis of Voit's and Pflüger's opposing positions in an attempt to resolve the duality at one remove from both, as it were. Apart from examining the points on which the two protagonists disagreed, Folin interestingly and importantly also focused his attention on aspects of protein metabolism on which they were in accord. Urea may have derived from 'circulating protein' or from 'tissue protein', but both workers supposed that its rate of excretion was a measure of protein metabolism.[b] In the face of Folin's work this view could hardly be defended. Also it appeared to reinforce doubts, which began to be voiced at that time,[c] concerning the need for a high protein diet.

The basis of Folin's classification of protein metabolism together with his considerations regarding the physiological protein requirements are contained in the passage below.

O. Folin. *Am.J.Physiol.*, *13*, 117. 1905

'A theory of protein metabolism'

p. 118

In order that the reader may recall the extent of the changes in percentage composition of urine from normal persons shown by the experiments . . . a part . . . is here reproduced.

	July 13	July 20
Volume or urine	1170 c.c.	385 c.c.
Total nitrogen	16.8 gm.	3.60 gm.
Urea-nitrogen	14.70 gm. = 87.5%	2.20 gm. = 61.7%
Ammonia-nitrogen	0.49 gm. = 3.0%	0.42 gm. = 11.3%
Uric acid-nitrogen	0.18 gm. = 1.1%	0.09 gm. = 2.5%
Kreatinin-nitrogen	0.58 gm. = 3.6%	0.60 gm. = 17.2%
Undetermined nitrogen	0.85 gm. = 4.9%	0.27 gm. = 7.3%
Total SO_3	3.64 gm.	0.76 gm.
Inorganic SO_3	3.27 gm. = 90.0%	0.46 gm. = 60.5%
Ethereal SO_3	0.19 gm. = 5.2%	0.10 gm. = 13.2%
Neutral SO_3	0.18 gm. = 4.3%	0.20 gm. = 26.3%

pp. 122–3

We have seen . . . that the composition of urine, representing 15 gm. of nitrogen, or about 95 gm. of protein, differs very widely from the composition of urine representing only 3 gm. or 4 gm. of nitrogen, and that there is a gradual and regular transition from the one to the other. To explain such changes in the composition of the urine on the basis of protein katabolism, we are forced, it seems to me, to assume that that katabolism is not all of one kind. There must be at least two kinds. Moreover, from the nature of the changes in the distribution of the urinary constituents, it can be affirmed, I think, that the two forms of protein katabolism are essentially independent and quite different. One kind is extremely variable in quantity, the other tends to remain constant. The one kind yields chiefly urea and inorganic sulphates, no kreatinin, and probably no neutral sulphur. The other, the constant katabolism, is largely represented by kreatinin and neutral sulphur, and to a less extent by uric acid and ethereal sulphates. The more the total katabolism is reduced, the more prominent become these representatives of the constant katabolism, the less prominent become the two chief representatives of the variable katabolism.

The fact that the urea and inorganic sulphates represent chiefly the variable katabolism does of course not preclude the possibility that they also represent to some extent the constant katabolism; but I have reason to believe that it is possible to plan feeding experiments which will yield urines containing very much smaller per cents of these two constituents than I have yet obtained. We know from the experiments of Sivén[a] that it is possible to reduce the total protein katabolism still more, and I am confident that in such cases the per cent of urea-nitrogen will sink still lower, and that the nitrogen of the other constituents, particularly of the kreatinin, will again show a corresponding increase.

If there are two distinct forms of protein metabolism represented by two different sets of waste products, it becomes an exceedingly interesting and important problem to determine, if possible, the nature and significance of each. The fact that the kreatinin elimination is not diminished when practically no protein is furnished with the food, and that the elimination of some of the other constituents is only a little reduced under such conditions, shows why a certain amount of protein must be furnished with the food if nitrogen equilibrium is to be maintained. It is clear that the metabolic processes resulting in the end products which tend to be constant in quantity appear to be indispensable for the continuation of life; or, to be more definite, those metabolic processes probably constitute an essential part of the activity which distinguishes living cells from dead ones. I would therefore call the protein metabolism which tends to be constant, *tissue* metabolism or *endogenous* metabolism, and the other, the variable protein metabolism, I would call the *exogenous* or intermediate metabolism.[1]

pp. 128–9

According to the views here presented . . . only a small amount of protein, namely, that necessary for the endogenous metabolism, is needed. The greater part of the protein furnished with standard diets like Voit's,[b] *i.e.* that part representing the exogenous metabolism, is not needed, or, to be more specific, its nitrogen is not needed. The organism has developed special facilities for getting rid of such excess of nitrogen so as to get the use of the carbonaceous part of the protein containing it. The first step in this process is the decomposition of protein in the digestive tract into proteoses, amido acids, ammonia, and possibly urea.[2] The hydrolytic decompositions are carried further in the mucous membrane of the intestines, and are completed in the liver, each splitting being such as to further the formation of urea.

In these special hydrolytic decompositions, the result of which is to remove the unnecessary nitrogen, we have an explanation of why and how the animal organism tends to maintain nitrogen equilibrium even when excessive amounts of protein are furnished with the food. This excess of protein is not stored up in the organism, as such, because the actual need of nitrogen is so small that an excess is always furnished with the food, except, of course, in carefully planned experiments. The ordinary food of the average man contains more nitrogen than the organism can use, and increasing the nitrogen still further will therefore necessarily only lead to an immediate increase in the elimination of urea, and does not increase the protein katabolism involved in the kreatinin formation any more than does an increased supply of fats and carbohydrates....

The normal human organism can be made at almost any time to store up fats and carbohydrates. The katabolism of these products consists chiefly of oxidation, a decomposition which sets free large quantities of heat which can be converted into mechanical energy useful to the organism. The hydrolytic removal of nitrogen from the protein involves by comparison a very small transformation of energy and yields a non-nitrogenous rest of great fuel value. This non-nitrogenous rest derived from protein may partly be directly transported to the different tissues, and thus at once supply oxidative material where needed, but in all probability is partly converted into fats, or at least into carbohydrates, and then becomes subject to the laws governing the katabolism of these two groups of food products.

Selection 120

The growing recognition, during the early years of the twentieth century, of the importance of substances in minimal quantities to maintain life has been referred to previously – in connection with Loewi's investigations. In 1906, due to Frederic Gowland Hopkins's (1861–1947) new insights into the influence of minimal *qualitative* variations in the dietaries were offered.

By that time feeding experiments with gelatin led to the general assumption that its inadequacy to maintain life had something to do with the absence of the indol-containing amino acids, tyrosine and tryptophane. In addition, Hopkins was much impressed by contemporary developments in the study of 'chemical messengers' or 'hormones', especially by the information that adrenaline was an aromatic nitrogenous substance.[a] Hopkins conceived that adrenaline was produced by the cells of the suprarenal gland from an aromatic precursor, set free during the digestion of the protein of the food in the alimentary canal.

From all this Hopkins drew the conclusion that it was necessary to differentiate between the minimal amount of protein needed for tissue repair and that required for total maintenance. Considerations such as these led him, in collaboration with Edith G. Willcock, to feeding experiments on a diet of zein (known to be deficient in tryptophane and lysine), carbohydrate and fat.[b]

The experiments yielded evidence, as the extract below shows, that young mice failed to maintain growth on the 'zein diet', but they survived longer if it was supplemented with tryptophane. Nothing comparable happened when tyrosine was added. It was also suggested that the tryptophane was utilized as the normal precursor of some specific 'hormone' or other substance essential to the processes of the body.

E.G. Willcock and F.G. Hopkins, *J.Physiol.*, *35*, 88. 1906–7

'The importance of individual amino-acids in metabolism. Observations on the effect of adding tryptophane to a dietary in which zein is the sole nitrogenous constituent'

p. 98

... the following diagram ... shows the effect of adding tryptophane to the zein diet compared with that of adding tyrosine. The animals on zein only are not included.

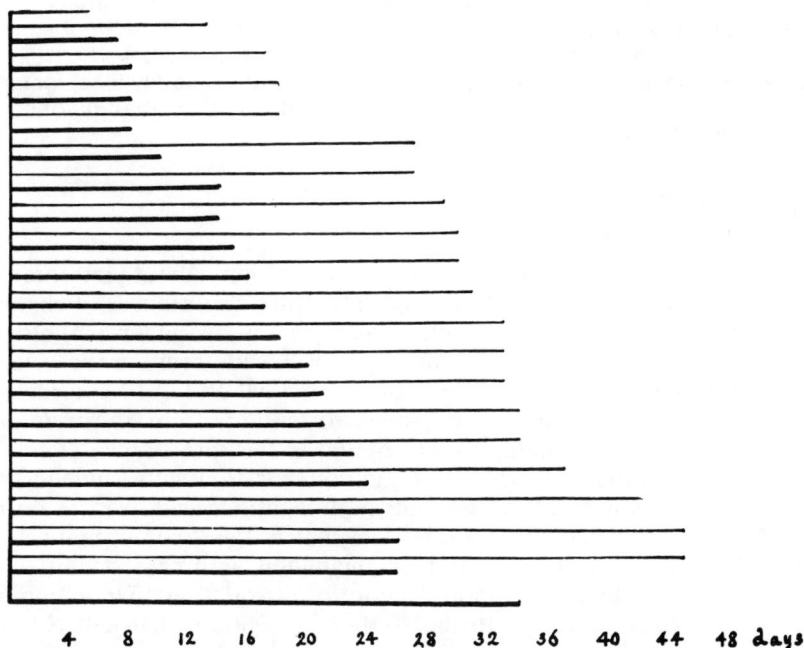

Diagram constructed from the results of Exps. V, VI, and VII. The thick lines show the survival periods (in days) of twenty-one individual mice upon the zein diet with tyrosine added. The thin lines show the same for nineteen mice upon the zein diet with tryptophane added.

pp. 100–1

Zein, it is clear, has no power whatever of maintaining growth in the young animal; loss of weight begins the moment it forms the sole nitrogenous supply. The addition of the missing tryptophane group to the zein has, it is also clear, no power to convert such loss into equilibrium or gain; a fact possibly due to other deficiences in the zein molecule, such as the absence of lysine, or the lack of some other amino-acid not yet observed. There was no close relationship in our experiments between the loss of weight and the length of survival period. In many individual cases the mice upon tryptophane lost a considerably larger percentage of their weight before death than, on the average, did those without it. Such differences may be largely due to differences in the nutritional condition of individuals at the outset, but the results appear to show that death was not determined by a critical percentage loss. On the other hand the figures show that, on the average, the loss of body weight was slower with tryptophane than without it. But this result might well be expected, even if the tryptophane administered undergoes utilization without directly contributing to tissue formation or structural maintenance. If it serves as a basis for the elaboration of a substance absolutely necessary for life – something, for instance, of an importance equal to that of adrenaline – then, in starvation, or when it is absent from the diet, a supply is likely to be maintained from the tissue-proteids; the demand for it would become one of the factors determining tissue breakdown. In the case of young animals which directly benefit from the addition of a protein constituent, otherwise absent from their diet, to the extent of a well nigh doubled life, and marked improvement in general condition, but at the same time steadily lose, instead of gaining, weight, the utilization of the constituent would seem to be of some direct and specific nature.

Selection 121

An important approach to obtaining information about intermediary metabolism had been the study of abnormal substances ('chemical malformations'), such as homogentisic acid,[a] in urine under pathological conditions. Normally metabolized, homogentisic acid occurs in the urine of patients suffering from alkaptonuria. Between 1890 and 1910 considerable interest was focused on its possible formation from tyrosine and phenylalanine, following findings that pointed in this direction.[b]

In 1904 Otto Neubauer (1874–1957), in collaboration with Wilhelm Falta (1875–1950), accepted the view that homogentisic acid was a normal intermediate product of the metabolism which the two aromatic acids underwent.[c] The work in this field amalgamated with that on fatty acid metabolism, on the basis of Knoop's theory of β-oxidation.[d] Upon applying the method of liver perfusion to the study of the fate of leucine, tyrosine, phenylalanine and homogentisic acid it was found that acetone was formed, as in the metabolism of fatty acids.[e]

It became clear that in the formation of homogentisic acid by alkaptonuric patients, fed on phenylalanine and tyrosine, the reaction involved removal of a $CHNH_2$ group. (When tryptophane was administered homogentisic acid was not produced.) Neubauer investigated which possible products of their metabolism could go on to become homogentisic acid. That led him, as shown below, to his conclusion in 1909 on the steps in metabolic degradation of these compounds and to the suggestion that formation of CO from $CHNH_2$, i.e. an α-ketonic acid, can be a normal process for amino acids (oxidative deamination).

O. Neubauer, *Arch.klin.Med.*, **25**, **211**. 1909

'Über den Abbau der Aminosäuren im gesunden und kranken Organismus' (On the degradation of the amino acids in the healthy and diseased organism)

pp. 252–4

... one can form a rather clear conception of the course of protein degradation in the healthy (and the diseased) organism, which can be expressed in the following theses.

1. The protein substances in the body first undergo a hydrolytic splitting, whereby amino acids and diamino acids arise.
2. The *amino acids* are then converted by oxidative deamination into the corresponding keto acids.
3. The (non-aromatic) *keto acids* go further, by splitting off CO_2 and oxidation, into the fatty acid poorer by one carbon atom, the further oxidation of which is subject to the known laws valid for the combustion of fatty acids.
4. Of the aromatic amino acids *tyrosine*, like the rest of the amino acids, is first changed into the corresponding keto acid, which is oxidized to the corresponding quinol and further rearranged as hydroquinone pyruvic acid; this goes then, like the other keto acids, with splitting off of CO_2 and oxidation, into the fatty acid poorer by one C-atom, that is to say homogentisic acid. The further breakdown includes the benzene ring and leads to its disruption; thereby acetone bodies[a] appear, which are oxidized in the normal way to CO_2 and H_2O.
5. The *phenylalanine* goes either by the way of phenylpyruvic acid or by tyrosine into p-hydroxyphenylpyruvic acid, which then, as considered under 4, is burnt further.
6. *Tryptophane* is burnt by another, still unknown way, which does not lead over homogentisic acid.
7. With *alkaptonuria* the normal degradation of tyrosine and phenylalanine (see 4 and 5) is hindered, so that it remains stationary at the stage of homogentisic acid; the degradation of the rest of the amino acids, including tryptophane, is undisturbed.

Refer to 2 and 3:

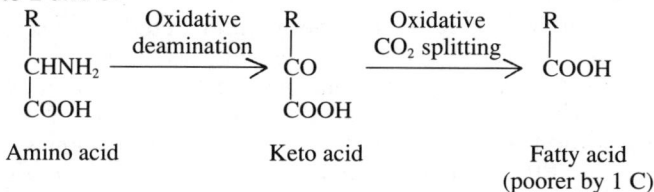

$$\underset{\text{Amino acid}}{\overset{\text{R}}{\underset{|}{\overset{|}{\text{CHNH}_2}}}\atop\overset{|}{\text{COOH}}} \xrightarrow[\text{deamination}]{\text{Oxidative}} \underset{\text{Keto acid}}{\overset{\text{R}}{\underset{|}{\overset{|}{\text{CO}}}}\atop\overset{|}{\text{COOH}}} \xrightarrow[\text{CO}_2\text{ splitting}]{\text{Oxidative}} \underset{\substack{\text{Fatty acid}\\ \text{(poorer by 1 C)}}}{\overset{\text{R}}{\underset{|}{\text{COOH}}}}$$

Refer to 4 and 5:

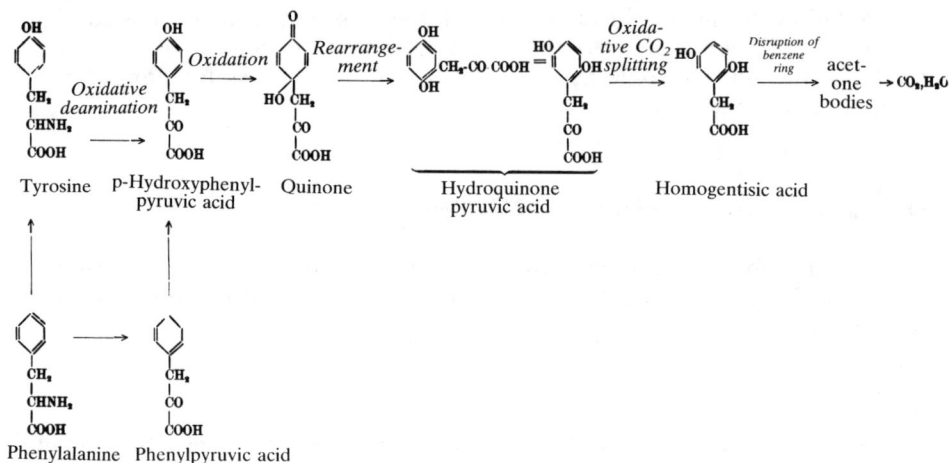

Tyrosine · p-Hydroxyphenyl-pyruvic acid · Quinone · Hydroquinone pyruvic acid · Homogentisic acid

Phenylalanine   Phenylpyruvic acid

Selection 122

Corresponding to the work of Neubauer on the oxidative breakdown of amino acids, dealt with in the previous selection, the possibility of the *in vivo* synthesis of amino acids was contemplated at the same time. By the end of the first decade of the twentieth century accumulated clinical and physiological evidence supported the assumption that carbohydrates were converted into fats, and proteins and fats into carbohydrates.[a] Out of this the belief was materializing the possibility that a close relationship existed between the metabolism of carbohydrates, fats and proteins. This suggested that chemical intermediates between the foodstuffs and their oxidation end-products might be found.

Bearing this in mind Franz Knoop (1875–1946) went on to consider whether in any one stage of animal metabolism a non-nitrogenous decomposition product could bind nitrogen and become involved in protein metabolism. He turned to the feeding of dogs with an α-ketonic acid since the reactivity of ketonic acids with different nitrogenous compounds was well recognized. He rejected the use of a compound that could possibly result in the formation of a physiological amino acid.

Instead, he made use of the non-physiological β-benzylpyruvic acid, $C_6H_5CH_2CH_2CO.COOH$, by giving the acid and the sodium salt of the acid to dogs *per os* and subcutaneously. Each time he found no difficulty in isolating the crystalline acetyl product of the corresponding dextrorotatory amino acid, β-benzlalanine.[b]

In the paper (published in 1910) Knoop interpreted this experimental result as pointing to the possibility of biosynthesis of particular protein breakdown products out of ammonia and substances that were not previously linked with the physiology of protein metabolism.

Knoop's elaboration of this view is contained in the following extracts. At that time, it should be noted, chemically minded physiologists and nutritionists paid great attention to the question of the relation of carbohydrates and fats (the non-nitrogenous components of the diet) to the body requirements for protein.

F. Knoop, *Z.physiol.Chem.*, *67*, 489. 1910

'Über den physiologischen abbau der Säuren und die synthese einer Aminosäure im Tierkörper' (On the physiological degradation of the acids and the synthesis of an amino acid in the animal body)

pp. 499–500

The question now arises: has the behaviour found here of an α-ketonic acid a physiological interest, are there present in the body-ketonic acids which in this way can take part in protein metabolism? This question must be answered in the affirmative. First the ketonic acids from the protein bodies themselves. An equilibrium must exist between them, the ammonia split off and the amino acids; and if the ketonic acid is not further changed immediately after its appearance, it can again find conditions which allow it to go through the same process in the opposite direction. Thus the organism could undertake a rearrangement of the stereomeric forms and, as the synthesis goes symmetrically, transform an unusable l-modification into a dextrorotatory. Normal fatty acids cannot come into consideration here; my own experiments, in which no α-oxidation products could be shown, are against such participation. At best the formation of aspartic acid (which is both an α- and a β-amino acid) could be considered from succinic acid. However, there are in the intermediary metabolism other substances which could be quoted, and which arise not from fat but from carbohydrate.

p. 501

If we see in the animal body fatty acids formed physiologically from carbohydrates, then there is no hesitation in the assumption of the reduction processes necessary for this purpose. The question about the possibility of an addition of NH_2-groups finds its principal solution in the experiments quoted.

In the physiological sense the question of the nitrogen minimum is dependent on this. If split products of other foodstuffs may again unite with the split protein nitrogen and make it newly useful before it terminates [*verfällt*] in urea synthesis, then will the artificial addition of such substances probably reduce the protein minimum. Perhaps thus is explained a part of the protein-sparing action of the nitrogen-free foodstuffs, which just for the carbohydrates is indicated as greater than for the fats.[1]

Selection 123

Another worker who concerned himself at that time with the formation of amino acids in the animal body from nitrogen-free material and ammonia was Gustav Embden (1874–1933). He made use of the perfusion technique but his first attempts to obtain alanine from ammonium lactate on perfusion of the liver did not lead to a positive result. Embden attributed this to the limitations of the chemical method.

It was the aforementioned work of Knoop that induced Embden to take up these experiments again by adding pyruvic acid (or a derivative) and lactic acid in the form of ammonium salts to the perfusing blood. The perfusion was broken off after 60 to 90 minutes and blood and liver were worked up for the suspected amino acids. As shown below, the results pointed strongly to the participation of carbohydrate in the synthesis of protein.

G. Embden and E. Schmitz,[1] *Biochem.Z.*, *29*, 423. 1911

'Über synthetische Bildung von Aminosäuren in der Leber' (On synthetic formation of amino acids in the liver)

pp. 425–7

Our experiments were performed on the glycogen-poor liver. Perfusion experiments on the glycogen-poor liver, in which no substance was added in the blood, showed us that under these conditions indeed apparently a certain increase in the amino acids of the blood appeared, yet by means of the methods we used a definite amino acid was not isolated.

On the other hand we succeeded, after perfusion with about 5 g of the p-oxyphenylpyruvic acid, in winning nearly 0.5 g of the dinaphthalinsulphotyrosine compound, after perfusion with addition of 5.7 g γ-phenylpyruvic acid about 0.5 g of the uramino acid of the phenylalanine.

In a series of experiments with addition of pyruvic acid alanine was formed in substantial quantities and was isolated as naphthalinsulphoalanine.

We wish still specially to emphasize that the tyrosine and the alanine appear in optically active, and indeed the natural form, which shows with certainty that the amino acids isolated by us are not to be conceived as chemical laboratory-products. On the optical behaviour of the phenylalanine we can meanwhile make no statement . . .

Some further experiments now appear of special interest, these were performed in connection with the previous work of *Fräulein* Fellner.[a] In these experiments it could be shown that, just like pyruvic acid, also lactic acid added to perfusing blood forms alanine, though to all appearance in smaller amount than pyruvic acid. This quite agrees with the statements of Knoop on the aromatically substituted keto acid and hydroxy acid.[b]

The transformation of lactic acid into alanine shows that carbohydrates, especially glycogen and grape sugar, can be changed into alanine just as, according to researches published by us, lactic acid appears as breakdown product of the carbohydrates named on the perfusion of the liver.[c]

We mention here, concerning the possibility of an alanine formation from lactic acid and thus from carbohydrate, that Knoop also calls attention to his own and our earlier experiments.

If by the alanine formation from lactic acid the conversion of carbohydrate into an amino acid is already to be taken as proved, then this proof is successful in a still more direct manner on the perfusion of the glycogen-rich liver with addition of small amounts of ammonium chloride in the perfusing blood.

Also in these experiments naphtalinsulphoalanine was obtained in not inconsiderable amount.

There is success thus, with simple perfusion for 1½ hours of the glycogen-containing liver, in transforming a part of the glycogen through lactic acid and probably pyruvic acid into alanine.[2]

Selection 124

In tracing some of the main lines of work on proteins we shall turn our attention next to developments arising from the study of colloidal systems. By 1910 it became widely accepted that the term 'colloid' should be applied to a state of matter rather than to a kind of matter. Opinions differed, however, on the theoretical approach to colloidal systems. The question was whether the basic principles of physical chemistry were relevant to the understanding of the properties and behaviour of protein solutions.

Most investigators inclined to the view vigorously advanced by, for example, the German worker, Wolfgang Ostwald (1883–1943), or the American Wilder Dwight Bancroft (1867–1953), that protein solutions could not be regarded as molecularly dispersed systems. Therefore it was not to be expected that quantitative relations, derived from true solutions and involving such properties as their freezing and boiling points and osmotic pressure, applied to them.

But there was another contending view which maintained that the nature of protein solutions could be better comprehended by concentrating on points of resemblance between them and true solutions rather than on the difference. This view emerged particularly strongly after the publication in 1917 of systematic investigations of proteins which had been carried out in the Carlsberg Laboratory under Sørensen, the originator of the pH concept.[a]

The short extract below is from a long paper dealing with carefully conducted osmotic pressure measurements of egg albumin solutions. In particular Sørensen explored the dependence of the osmotic pressure on the ammonium sulphate concentration, on the hydrogen ion concentration and on the protein concentration. On the basis of osmotic pressure measurements he calculated the molecular weight of egg albumin which came to about 34,000.

This result seemed to confirm assumptions which had been gaining ground since 1900 about the high molecular weights of proteins. Indeed it exceeded by a factor of 7 the molecular weight of the largest molecule prepared in the laboratory shortly before that time.[b]

S.P.L. Sørensen, *Comp.Rend.(Carlsberg),* **12, 262. 1917**

'Studies on proteins. V. On the osmotic pressure of egg-albumin solutions'[a]

pp. 313–15

The form in which we have above comprised the result of a measurement of osmotic pressure simply represents the values found by direct measurement. In dealing with the results of a considerable number of experiments, however, it is necessary, for the sake of comparison, to re-calculate the values in such a manner as to make the results illustrative of the dependence of the osmotic pressure on the albumin concentration.

The simplest way of expressing albumin concentration is that commonly adopted, in which the concentration, c, is reckoned as proportionate to the amount of protein nitrogen contained in volume unit of the solution, the osmotic pressure P being reckoned as proportionate to c, so that P will be equal to $R.T.c$. It is thus taken for granted – just as we have done, for simplicity's sake, in our considerations regarding the 'Donnan counter-pressure'[b] – that van't Hoff's law holds good of the solutions considered. It is to be remarked, however, that this law applies solely to very weak solutions, where the volume of the dissolved substance is quite minimal as compared with that of the dissolvent;[c] but in the case of stronger solutions it is not certain that the law will hold true, and we have therefore thought necessary to take into account the volume of the egg-hydrate.

... the just-mentioned reason has induced us to state the albumin concentration, in general, not in proportion to the volume of the solution, but to the weight or volume of the dispersion medium.

... by the ammonium sulphate content of the dispersion medium, S, we understand the weight, in g., of the ammonium sulphate per 100 g. of water in the dispersion medium – and by the egg-hydrate concentration, E, the number of milligram-equivalents of protein nitrogen per 100 g. water in the dispersion medium. In such cases – as for instance in determinations of the magnitude of osmotic pressure – where it is convenient to reckon with the egg-hydrate concentration, ε, per volume unit of dispersion medium, we shall have ε equal to E/V_s, where V_s means the volume at 18°, in c.c., of the weight of dispersion medium containing 100 g. water ...

If now we provisionally suppose the magnitude of the osmotic pressure to be proportionate to the concentration ε, as defined above, and if we suppose that each albumin molecule or albumin particle acting as a molecule contains n nitrogen atoms, then we shall have

$$P = \frac{R.T.\varepsilon}{n} = \pi.\varepsilon$$

and, hence,

$$\pi = \frac{P}{\varepsilon} = \frac{P.V_s}{E},$$

where $\pi \left(= \dfrac{R.T}{n} \right)$ signifies the osmotic pressure per milligram-equivalent protein-nitrogen per cubic centimeter of dispersion medium. The pressure being measured in cm. of water, and the volume unit used being the cubic centimeter, we get, at the temperature 18°:

$$R.T = \frac{p_o \cdot v_o}{T_o} T_{18} = \frac{22.412 \cdot 76 \cdot 13.596 \cdot 1 \cdot 291}{273} = 24685 \text{ cm.}$$

p. 356

... we are as yet unable to give a definite answer to the question as to the value acquired by π under experimental circumstances where neither the albumin condensation nor the combined effect of albumin salt dissociation and 'Donnan counter-pressure' come into play, so that the osmotic pressure may be reckoned as proportionate to the concentration of non-condensed egg-albumin. On the other hand, it is evident from those results that the value of π must under such circumstances be comprised between 60 and 70 cm. water-pressure. We are thus enabled to judge of the magnitude of the non-condensed albumin particles, inasmuch as π is equal to $\frac{RT}{n}$, where RT is equal to 24685 cm. water-pressure, and n means the number of nitrogen atoms in each albumin particle ... As will easily be seen, *the calculation gives the result that, supposing π equal to 65 cm., n is about 380, so that the weight of an albumin particle – or molecule – in the anhydrous state will be approximately 380 · 14.01 · 6.45 = about 34000.*

Selection 125

Around 1920 much importance was attached to the work of Jacques Loeb (1825–1924) on the physical chemistry of proteins. He was widely known for his thoroughgoing physicochemical reductionist attitude to life phenomena which he set forth in the influential book *The Mechanistic Conception of Life* (1912).

An interesting application of this approach was his theoretical and experimental treatment of the colloidal behaviour of proteins. He regarded osmotic pressure, swelling and a certain type of viscosity as the specifically colloidal properties of proteins. Working with gelatin, and paying careful attention to the hydrogen ion concentration, he concluded that these properties were influenced only by the valency of the ion and not by its chemical nature ('Hofmeister ion series').[a] Accepting that proteins are amphoteric electrolytes and recognizing the general validity of Donnan's theory of membrane equilibria,[b] Loeb considered that the colloidal behaviour of proteins ultimately depended on membrane equilibria.

The following recapitulation of Loeb's standpoint is taken from the second edition of his work (first published in 1922), in which he presented his theoretical views and experimental results in a systematic manner.

J. Loeb, *Proteins and the Theory of Colloidal Behaviour*, 2nd edn (New York and London, 1924)

pp. 21–2

If we now recapitulate the history of this subject, we notice that in the colloidal literature proteins were assumed to combine not stoichiometrically but by adsorption. Because proteins are amphoteric electrolytes and their salts are strongly hydrolyzed, it is necessary to measure the hydrogen ion concentration of protein solutions with the hydrogen electrode before any conclusions can be drawn concerning the nature of the combination of proteins. If this advice is followed, it is found that proteins react stoichiometrically with acids and alkalies, and that there is no difference between the chemistry of proteins and that of crystalloids.

It was further argued that solutions of genuine proteins in water are always diphasic systems in which the particles of protein are kept in solution on account of electrical double layers, due to an alleged preferential adsorption of ions by particles. It was shown that certain proteins are kept in solution by the same forces which determine the solution of crystalloids, e.g., the amino-acids from which the proteins are built up. The forces which keep such genuine proteins in solution do not differ from the forces which keep crystalloids, like amino-acids, in solution.

Finally, the belief in the reality of the Hofmeister ion series lent further support to the belief in the adsorption theory; yet it is possible to show that these series for swelling, osmotic pressure, and viscosity, were also based on a methodical error, namely, the failure to measure the hydrogen ion concentration of the solutions. It is, therefore, obvious that the so-called colloid chemistry of proteins is a system of errors based on inadequate and antiquated methods of experimentation.

The question then arises: Are there any peculiarities in the behavior of proteins which are not found in crystalloids? To this the answer must be given that such peculiarities exist and that they are found in the influence of electrolytes on the osmotic pressure of protein solutions, on the viscosity of certain (but not all) protein solutions, and on the swelling of gels. These peculiarities are as follows:

1. The addition of little acid or alkali to originally isoelectric protein increases the osmotic pressure and viscosity of the protein solution and the swelling of protein gels, until a certain limit is reached; after this the addition of more acid has a depressing effect on these properties.
2. The addition of neutral salts has only a depressing effect on these properties.
3. The depressing effects of electrolytes increase with the valency of that ion which bears a charge opposite to that of the protein ion.
4. Only the valency but not the chemical nature of the crystalloidal ions influences the above-mentioned properties of proteins (except where the electrolyte has secondary effects which influence these properties in an indirect way).

The writer undertook measurements of a new property which had been overlooked in the colloidal literature, namely, the membrane potentials of protein solutions, i.e., the potential differences between solutions of protein salts contained in a collodion bag and outside aqueous solutions free from protein, at the point of osmotic equilibrium. These measurements led to the result that membrane potentials are influenced by electrolytes in a similar way as osmotic pressure, viscosity, or swelling. Since the membrane potentials could be correlated mathematically and quantitatively with Donnan's theory of membrane equilibria, the possibility arose that membrane equilibria might account also for the similar influence of electrolytes on the other three properties; and this was found to be correct.

Selection 126

We have seen that a molecular weight of 34,000 for egg albumin was suggested by Sørensen in 1917, on the basis of osmotic pressure measurements. The molecular weight obtained by the osmotic pressure method gave no indication whether the protein solution contained protein molecules (particles) of a uniform size or a mixture of different sizes and not all workers accepted proteins as chemical individuals. In 1926 Dorothy Jordan Lloyd, a noted protein chemist, described their position as follows:

> They consider that the groupings in the protein are not held in association by the ordinary valency bonds of chemical combination. On this theory of protein constitution the disintegration of a protein into its units is not a chemical degradation but a physical separation, and the calculation of molecular weight for proteins is a proceeding of somewhat dubious value.[a]

About that time further support for the idea that proteins are substances of relatively high molecular weight began to come from the Laboratory of Physical Chemistry at Uppsala. Working there, The(odor) Svedberg (1884–1971), developed two ultracentrifugal methods for the determination of the molecular weights of proteins, also with regard to their stability within a given pH range. Either the sedimentation rate in an ultracentrifuge cell rotating at great speeds was measured, or the solution was allowed to reach an equilibrium and the concentration measured at distances from the axis of rotation, with the ultracentrifuge rotating at moderate speeds. The measurements related to evaluated microphotometric records obtained from photographs of changes that took place in the ultracentrifuge cell during the run. A line distinctly separating the solvent from the sedimented protein suggested that the latter was homogeneous.

The extracts below demonstrate Svedberg's approach, based on the calculation of the molecular weight of proteins from ultracentrifugation data, to the molecular individuality of proteins. Further they reveal Svedberg's attempt at a classification of proteins which hinged upon the concept that their molecular weight was a multiple of a unit molecular weight believed to be 34,500. Last but not least, in discussing the meaning of these apparent number relationships Svedberg sought to connect them with a general structure of proteins, associated with the arrangement of their components.

T. Svedberg, *Koll.Z.*, *57*, 10. 1930

'Ultrazentrifugale Dispersitätsbestimmungen an Eiweisslösungen' (Measurements of ultracentrifugal dispersity in protein solutions)

pp. 11–12

A centrifugal field can be used in two ways for the estimation of particle size or molecular weight. On the one hand, one can measure *sedimentation rate*, on the other hand the equilibrium state – the so-called *sedimentation equilibrium* – which is reached after longer centrifuging.

If the particles of the colloid are of equal size and we designate the molar particle weight (= molecular weight) as M, the partial specific volume as V, the density of the solution as ρ, the molar frictional coefficient as f (f_s or f_D), the diffusion coefficient by D and the sedimentation in unit field as s = [(dx/dt)/$\omega^2 x$], where x means the distance from the rotation axis, t the time and ω the angular rate, so for sedimentation rate measurements the following representations are valid:

$$M = \frac{f_s s}{1 - V\rho} \tag{1}$$

$$f_D = \frac{RT}{D} \tag{2}$$

and for the case where $f_s = f_D$

$$M = \frac{RT\ s}{D(1 - V\rho)} \tag{3}$$

The sedimentation in unit fields we will call the sedimentation constant. It is, like the diffusion constant, at a given temperature and in a given solvent, a natural constant characteristic of the substance under investigation.

With sedimentation equilibrium measurements the following equation is valid:

$$M = \frac{2\,RT\ln(c_2/c_1)}{(1 - V\rho)\,\omega^2(x_2^2 - x_1^2)} \tag{4}$$

Both formulae 3 and 4 for the calculation of the molecular weight are derived on the assumption that the osmotic rules of dilute solutions hold good. The equations contain no hypothetical values, but only well-defined measurable quantities. *The derivations which have led to equations 3 and 4 do not use Stokes's law, and their validity on this account is independent of the form of the molecule or particle.*[a] Further conditions for validity entail that no electric forces should disturb the sedimentation or diffusion. If a dissociable substance is being used one must centrifuge either in the electrolyte-free state at the isoelectric point or in the presence of a non-sedimenting electrolyte of sufficiently high concentration.

In the derivation of equation 3 the assumption was made that the molar frictional coefficient f_s, which appears on sedimenting, is identical with the molar frictional coefficient of diffusion f_D. As our experience has taught us, this is not always the case. A diminution of diffusion through forces acting between the particles, weak gel formation and so on will, with the use of the diffusion estimation, give a value for f_D which is much greater than the effective value f_s of this coefficient. With the diffusion the particles must be in a state in which their reciprocal positions can change without influencing them, in order that one may obtain a correct value for the frictional coefficient; with the sedimentation this free movement need not be present. In the latter case indeed one measures only the shifting of the particle swarm as a whole in the centrifugal field. On this account cases occur where the sediment rate can give only the sedimentation constant, not the molecular weight. These diffusion anomalies often disappear with increasing dilution of the solution; we have however learnt to recognize cases where these persist even at concentrations of a few hundreths per cent.

The finding of these diffusion anomalies throws new light on certain earlier diffusion estimations in the realm of highly dispersed organic colloids. Many discrepancies must be explained by the fact that the diffusion constant of a highly dispersed organic colloid (e.g. a protein or carbohydrate) is very dependent on the concentration and also on small differences in experimental conditions. The estimation of the molecular weight by diffusion estimations can on this account lead to large errors . . .

It is very remarkable that the estimation of the sedimentation equilibrium (equation 4) yields normal values for the molecular weight also in cases where the diffusion is diminished. This holds as well for protein as for cellulose solutions. The establishment of the equilibrium needs indeed more time than in the case of normal diffusion . . . The practical consequence now is that the sedimentation equilibrium method is often more reliable for estimation of the molecular weight than the sedimentation velocity method. The latter is indeed indispensable for study of homogeneity of the degree of dispersion, for the ascertaining of the degree of symmetry of the particles (or molecules) as well as for the estimation of the range of pH and temperature stability of a protein.

The advantages of the sedimentation rate method are the following. The sedimentation constant can be measured in 2−6 hours. An equilibrium measurement demands 2−6 days. Apart from the considerable gain in time there is a principal advantage: with short experiments one avoids the often very disturbing effect upon the protein through acid or alkaline reaction or higher temperatures. For the case of a mixture of particles or molecules of different sizes, the measurement of sedimentation rate permits a much sharper analysis of the dispersity than does the equilibrium method.

One obtains the molar frictional coefficient f by combining an estimation of sedimentation constant s by means of the rate method with an estimation of the molecular weight M obtained by means of the equilibrium method. It is

$$f = \frac{M (1 - V\rho)}{s} \tag{5}$$

For a spherical particle one has also

$$f_o = 6\pi\eta N \left(\frac{3MV}{4\pi N}\right)^{1/3} \tag{6}$$

The relation f/f_o must thus equal unity for spherical particles. Deviation from the value of unity is thus a measure of the deviation of the particle from spherical symmetry.

pp. 19–21

Up to the present time 28 of the proteins described in the literature have been investigated by the method of centrifugation. Of these 14 have turned out to be monodisperse: namely ovalbumin, Bence-Jones protein, haemoglobin, serum albumin, serum globulin, amandin, edestin, excelsin, legumin, C-phycocyanin, R-phycocyanin, H-haemocyanin, L-haemocyanin. The other 14 are polydisperse and in general very unstable: namely euglobulin, fibrinogen, gelatin, gliadin, globin, glutenin, histone, casein, lactalbumin, legumelin, leucosin, muscle globulin, ovoglobulin, pseudoglobulin. These products ought

Table 8 Monodisperse protein bodies.

Nomenclature	Source	Molecular weight	Sedimentation constant = S 20⁰	Molar frictional coefficient = f 20⁰	disymmetry number = f/f_0	Radius of molecule = r in $\mu\mu$[b]
Group 1						
Class 1						
Bence-Jones protein[1]	urine	$35{,}000 = 1.01 \times 34{,}500$	3.55×10^{-13}	2.48×10^{16}	1.00	2.18
Egg albumin[2,3]	hen's egg	$34{,}500 = 1.00 \times 34{,}500$	3.32×10^{-13}	2.63×10^{16}	1.06	2.17
Class 2						
Haemoglobin[4,5]	horse blood	$68{,}000 = 1.97 \times 34{,}500$	4.37×10^{-13}	3.89×10^{16}	1.25	molecule not spherical
Serum albumin[6]	horse blood	$67{,}500 = 1.96 \times 34{,}500$	4.21×10^{-13}	4.01×10^{16}	1.29	molecule not spherical
Class 3						
Serum globulin[6]	horse blood	$103{,}800 = 3.03 \times 34{,}500$	5.66×10^{-13}	4.57×10^{16}	1.28	molecule not spherical
Class 4						
Amandin[7]	almond	$208{,}000 = 6.04 \times 34{,}500$	11.4×10^{-13}	4.53×10^{16}	1.03	3.94
Edestin[8]	hemp seed	$208{,}000 = 6.04 \times 34{,}500$	12.8×10^{-13} [15]	4.16×10^{16} [15]	0.93	3.94
Excelsin[7]	Brazil nut	$212{,}000 = 6.15 \times 34{,}500$	11.8×10^{-13}	4.63×10^{16}	1.02	3.96
Legumin[9]	vetch	$208{,}000 = 6.04 \times 34{,}500$	11.5×10^{-13}	4.60×10^{16}	1.02	3.96
C-Phycocyanin[10]	Cyanophyceae (algae)	$208{,}000 = 6.04 \times 34{,}500$	11.2×10^{-13}	4.74×10^{16}	1.05	3.94
R-Phycocyanin[11,10]	Rhodophyceae (algae)	$206{,}000 = 5.97 \times 34{,}500$	11.1×10^{-13}	4.51×10^{16}	1.00	3.95
R-Phycoerythrin[11,10]	Rhodophyceae (algae)	$209{,}000 = 6.06 \times 34{,}500$	11.5×10^{-13}	4.61×10^{16}	1.02	3.95
Group 2						
H-Haemocyanin[12,13]	snail blood	$5{,}000{,}000$	98×10^{-13}	13.5×10^{16}	1.05	12.0
L-Haemocyanin[14]	king-crab blood	$2{,}000{,}000$	35.7×10^{-13}	14.9×10^{16}	1.56	molecule not spherical

to be considered as partly broken down, originally monodisperse proteins. Such a conception is supported by the circumstance that also the monodisperse proteins become polydisperse on insufficiently careful isolation. Similarly the polydisperse proteins globin, euglobulin and pseudoglobulin originate from monodisperse proteins through chemical interference. Probably later it will be possible under proper conditions to prepare in the monodisperse state the proteins which are now found in the polydisperse state. In Table 8 are the observed molecular weights, sedimentation constants, molar friction coefficients, dissymmetry constants f/f_0 and molecular radii of the monodisperse proteins in the isoelectric state. On consideration of the table one sees at once that these proteins, in relation to their molecular weight, can be divided into two principal groups: I. those with molecular weights between 35,000 and 210,000, and II. those with molecular weights of millions. The latter principal group includes the haemocyanins. In the first principal group four different classes can be distinguished. The proteins of the first class have a molecular weight of 34,500, those of the second a molecular weight about twice 34,500, those of the third class a molecular weight of about three times 34,500, and those of the fourth class a molecular weight of about six times 34,500. Inside each class the rest of the constants – sedimentation constants, friction coefficients, dissymmetry constants – are nearly the same. The proteins of the first and fourth classes have almost spherical particles with radii of about 2.2 or 4.0,[b] the particles of the proteins of the second and third classes diverge considerably from the spherical form.

These completely unexpected regularities are hard to explain. One cannot assume that the different proteins are simply aggregates of primary protein of molecular weight 34,500; for the great differences of the elementary composition and the hydrolysis products compel the conclusion that the amino acids and other building blocks enter into the different proteins in different relations. According to these determinations the simple aggregation hypothesis is excluded. It appears much more likely that one must assume that the molecules of the proteins are formed according to a common building plan which is responsible for the fact that the molecular weight values fall so regularly. This hypothesis is thus to be conceived as a modified aggregation hypothesis. Submolecules of weight about 34,500 enter into all protein molecules, the different smaller building blocks (amino acids, etc.) can change with different limits.

Selection 127

Apart from the centrifugal method, there were other physical approaches around 1930 that helped to throw light on the dispersion and shape of proteins in solution. In this connection it is appropriate to refer to investigations on double refraction of flow exhibited by muscle globulin preparations, carried out by the Swiss–American pair of scientists, Alexander L. von Muralt (*b*. 1903) and John Tileston Edsall (*b*. 1902) at the Department of Physical Chemistry at Harvard Medical School.[a] It should be pointed out that originally the work was aimed at contributing to the theoretical understanding of muscle contraction.[b]

Experimentally the arrangement consisted of placing the muscle globulin solution between two concentric cylinders of which the external one rotated at a chosen rate. The annular space was viewed in plane-polarized light through a nicol in a direction parallel to the axis of concentric cylinders. As long as the cylinders were at rest, the field of vision was dark. On setting the external cylinder in motion, the annular space lit up and a black cross ('cross of isocline') appeared. Ascribed to shearing forces arising between the adjoining concentric layers in the fluid, the change was associated with a definite orientation of previously randomly oriented asymmetrical particles (molecules) present in the fluid. The two situations are schematically represented in Figure 14, which also shows the planes of vibration of the light transmitted by polarizer (P) and analyser (A).

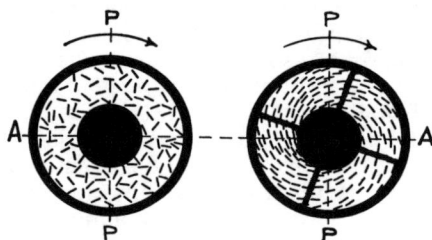

Figure 14 Orientation of muscle globulin particles in the annular space between two concentric cylinders, one of which rotated, in polarized light

At first what was measured was the 'angle of isocline' ψ. As stated by the authors, this was the English rendering of the German term *Kreuzwinkel* denoting the larger of the two angles which the optic axis of the oriented particles forms with the radius drawn to the particle. Theoretically angles of isocline between 45° and 90° might be expected. In Figure 15 three angles of isocline are shown: $\psi = 90°$, 45°, and 78°.

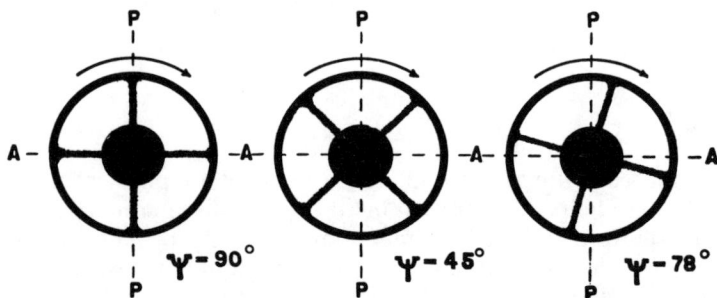

Figure 15 'Angles of isoclines' for sols of vanadium pentoxide and solution of muscle globulin

Muralt and Edsall found the angle of isocline to be constant (77.5–78.5°) for intermediate and high concentrations of muscle globulin solutions at low temperature (2–4°C). They interpreted this result (in the first paper) as indicating that the muscle particles were of uniform size and shape.[c]

Then they proceeded to measure the size of birefringence by making use of a Sénarmont compensator consisting of a mica plate producing a phase difference of a quarter wave-length for light within the range 540 to 550 micromillimetres (accurate within 1%). Without entering into technical detail, let it be said that the amount of double refraction produced by the flowing solution was determined by the angle ($\Delta°$) through which the analyzer had to be rotated (the two nicols being crossed) in order to extinguish the plane polarized light emerging from the quarter wave-length plate.

Muralt and Edsall established a relation between double refraction ($\Delta°$) and angular velocity in revolutions per minute (Ω), reflected in characteristic curves, which varied with protein concentrations (see Figure 16).

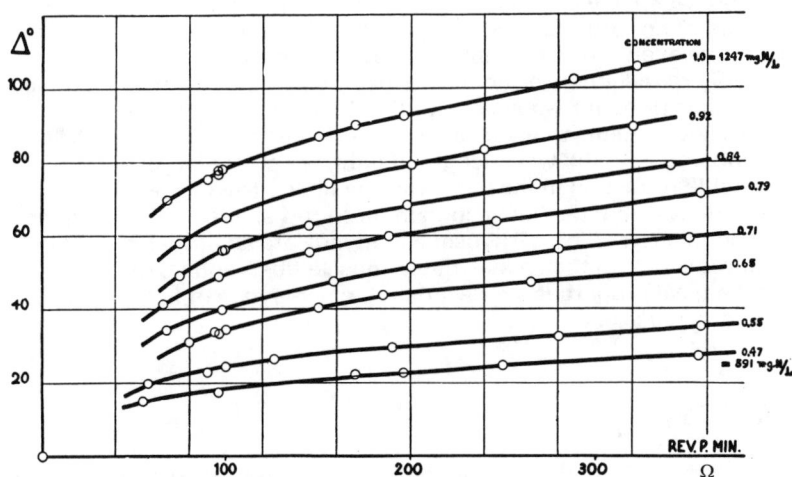

Figure 16 Characteristic curves for various muscle globulin concentrations showing relation between double refraction and the rate of rotating cylinder

As shown in the extract from the (second) paper below, Muralt and Edsall explained their observations by associating the double refraction of flow with the presence of elongated particles in myosin (later recognized as actomyosin). They tentatively located it in the anisotropic disc (A band) of the cross-striated muscle fibre.

A.L. von Muralt and J.T. Edsall, *J.Biol.Chem.*, *89*, 351. 1930

'Studies in the physical chemistry of muscle globulin. IV. The anisotropy of myosin and double refraction of flow'.

pp. 385–6

Double refraction (Δ) and angular velocity (Ω) are found to be related by characteristic curves, which vary with protein concentration. These Δ-Ω curves are reproducible, for any given preparation, over a protracted period of time. There is no suggestion of the ageing effect observed with many doubly refractive colloids, such as the V_2O_5 sol. The general form of the Δ-Ω curves is the same for all preparations examined.

These facts suggest that the double refraction is intimately related to the chemical nature of the protein solution. It is found, in accordance with this view, that typical denaturing agents produce rapid and complete destruction of the double refraction of flow. It would appear that only the undenatured protein is responsible for the double refraction.

The double refraction, is ascribed primarily to the orientation of anisotropic protein particles, due to the shearing stresses which arise during flow. In future it may be possible to relate the double refraction quantitatively to the size, shape, and optical properties of the protein particle. Neither experiment nor theory is yet sufficiently advanced for the accomplishment of this task . . .

The evidence appears to us to favor the view that the anisotropie protein particles are of uniform size and shape (monodisperse). The preparations obtained, however, are not completely free of other proteins which do not show double refraction. It is suggested that the name myosin be restricted to the anisotropic protein responsible for the double refraction of flow . . .

Stuebel's work points to the existence of oriented rod-shaped particles, small compared with the wave-length of light, which are largely responsible for the double refraction in the intact muscle fiber.[a] The properties of myosin solutions suggest that they may contain these rod-shaped particles. In any case the double refraction of myosin would indicate its probable location in the anisotropic disc of the cross-striated muscle fiber, and its general physicochemical properties suggest that it may play a part in the functional activity of the muscle.

Selection 128

In the 1930s another physical method concerned with the size, shape and structure of the protein molecule began to produce results that attracted wide interest. It was X-ray analysis, an outgrowth of the methodology for examining crystals developed by Max von Laue (1879–1960) at the University of Munich in 1912, in collaboration with Walther Friedrich (1883–1968) and Paul Knipping (1883–1935). Only a year later, it would appear, the first attempt to obtain X-ray photographs of fibrous materials, such as silk and wool, bamboo and hemp, were made by Shoji Nishikawa (1884–1952) at the University of Tokyo.

Not until the early 1920s, however, was X-ray analysis applied in a systematic manner to the study of natural and artificial cellulose fibres and also to the study of animal fibres with a protein basis. At this stage the chief centre of this line of research was the Institute of Fibre Chemistry in Berlin-Dahlem headed by Reginald Oliver Herzog (1878–1935).

In the following decade it was the pioneering pursuits of two research groups in Britain that proved the value of obtaining information on proteins by means of X-rays. The groups were headed by John Desmond Bernal (1901–71) and William Thomas Astbury (1898–1961) respectively. Having served, as it were, their apprenticeship with (Sir) William Henry Bragg

(1862–1942) at the Davy-Faraday Laboratory (Royal Institution) in London, where they learned the principles of X-ray diffraction, Bernal moved to Cambridge (1927) and Astbury to Leeds (1928).[a]

Bernal turned first to X-ray photography of amino acids and of sterols. These studies evolved from his early and abiding interest in the nature of the unity and diversity of non-living and living matter, which blended with the research work on many aspects of the chemistry of life which was being carried out at the Cambridge Biochemical Laboratory under Hopkins's stimulating direction.[b]

By the very nature of his appointment to a lectureship in textile physics Astbury concentrated on X-ray investigations of textile fibres by considering the relationship between their molecular structure and their elastic properties.[c] Soon after taking up the study of wool in the unstretched ('normal') and stretched state Astbury interpreted the X-ray diffraction pattern as supporting the view that the protein keratin, present in the wool, contained polypeptide chains. Whereas in the unstretched state the polypeptide chain was visualized as being folded into loops (α-keratin), the stretched state was associated with the elongated polypeptide chain (β-keratin). Astbury extended this approach to crystalline proteins such as pepsin, which in the dried state appeared to give an X-ray diagram similar to that shown by fibrous proteins.

It is against this background that the uncertainty regarding the interpretation of X-ray photographs of wet pepsin crystals must be seen. The photographs were taken and discussed by Bernal and his collaborator Dorothy Crowfoot (*b*. 1910), as she then was, in 1934. The division of proteins into two types with respect to shape, the globular and the fibrous, may be effectively taken to date from their communication and that of Astbury and his co-worker to *Nature* (see below).

At Cambridge, however, the researchers were by no means convinced that the globular proteins were built up of polypeptide chains. At Leeds, on the other hand, the workers mooted the possibility of the globular type converting into the fibrous one through extension of polypeptide chains.[d]

J.D. Bernal and D. Crowfoot, *Nature*, *133*, 794. 1934

'X-ray photographs of crystalline pepsin'

pp. 794–5

Four weeks ago, Dr G. Millikan[a] brought us some crystals of pepsin prepared by Dr Philpot[b] in the laboratory of Prof. The Svedberg, Uppsala. They are in the form of perfect hexagonal bipyramids up to 2 mm. in length, of axial ratio $c/a = 2.3 \pm 0.1$. When examined in their mother liquor, they appear moderately birefringent and positively uniaxial, showing a good interference figure. On exposure to air, however, the birefringence rapidly diminishes. X-ray photographs taken of the crystals in the usual way showed nothing but a vague blackening. This indicates complete alteration of the crystal and explains why previous workers have obtained negative results with proteins, so far as crystalline pattern is concerned.[1] W.T. Astbury has, however, shown that the altered pepsin is a protein of the chain type like myosin or keratin giving an amorphous or fibre pattern.

It was clearly necessary to avoid alteration of the crystals, and this was effected by drawing them with their mother liquor and without exposure to air into thin capillary tubes of Lindemann glass. The first photograph taken in this way showed that we were dealing with an unaltered crystal. From oscillation photographs with copper K α-radiation, the dimensions of the unit cell were found to be $a = 67$ A., $c = 154$ A., correct to about 5 per cent. This is a minimum value as the spots on the c row lines are too close for accurate measurement and the c axial length is derived from the axial ratio. The dimensions of the cell may still be multiples of this. Using the density measured on fresh material[2] as 1.32 (our measurements gave 1.28), the cell molecular weight is 478,000, which is twelve times 40,000, almost exactly Svedberg's value arrived at by sedimentation in the ultracentrifuge. This agreement may however be quite fortuitous as we have found that the crystals contain about 50 per cent of water removable at room temperature. But this would still lead to a large molecular weight, with possibly fewer molecules in the unit cell.

Not only do these measurements confirm such large molecular weights but they also give considerable information as to the nature of the protein molecules and will certainly give much more when the analysis is pushed further. From the intensity of the spots near the centre, we can infer that the protein molecules are relatively dense globular bodies, perhaps joined together by valency bridges, but in any event separated by relatively large spaces which contain water. From the intensity of the more distant spots, it can be inferred that the arrangement of atoms inside the protein molecule is also of a perfectly definite kind, although without the periodicities characterising the fibrous proteins. The observations are compatible with oblate spheroidal molecules of diameters about 25 A. and 35 A., arranged in hexagonal nets, which are related to each other by a hexagonal screw-axis. With this model we may imagine degeneration to take place by the linking up of amino acid residues in such molecules to form chains as in the ring-chain polymerisation of polyoxy methylenes. Peptide chains in the ordinary sense may exist only in the more highly condensed or fibrous proteins, while the molecules of the primary soluble proteins may have their constituent parts grouped more symmetrically around a prosthetic nucleus.

At this stage, such ideas are merely speculative, but now that a crystalline protein has been made to give X-ray photographs, it is clear that we have the means of checking them and, by examining the structure of all crystalline proteins, arriving at far more detailed conclusions about protein structure than previous physical or chemical methods have been able to give.

W.T. Astbury and R. Lomas, *Nature*, *133*, 795. **1934**

['On the configurational paradox in pepsin: globularity and chain system']

p. 795

It is now some time since we first took X-ray powder photographs of crystalline pepsin kindly sent by Prof. J.H. Northrop, but no really satisfactory interpretation of these photographs presented itself because they show features which we have learnt recently to associate with the fibrous proteins:[1] even single crystals, so far as we could judge with the minute crystals available, appeared to give results similar to those produced by many crystals in random orientation. The two chief rings have spacings of about 11.5 A. and 4.6 A. at ordinary humidity, corresponding to the 'side-chain spacing' and the 'backbone spacing', respectively, of an extended polypeptide.[1]

It was difficult, of course, to reconcile such findings with external morphology and the Law of Rational Indices, but the photographs of Bernal and Miss Crowfoot, taken before the degeneration which we now see the crystals must have undergone on drying, clear up this long-standing problem at once. Furthermore, their photographs tend to confirm the suggestion[2] that the numbers 2, 3, 4, and 6 occurring in Svedberg's multiple particle weights are fundamentally of *crystallographic* significance, even though their conclusions to date appear to be against the chain mechanism proposed for the building-up of the various crystallographic groups.[2]

We are left now with the paradox that the pepsin molecule is both globular[3] and also a real, or potential, polypeptide chain system, and the immediate question is whether the chains are formed by metamorphosis and linking-up of the globular molecules, or whether the initial unit is the chain itself, which is afterwards folded in some neat manner which is merely an elaboration of the intra-molecular folding that has been observed in the keratin transformation.[1] What is either an exceedingly valuable clue or else only a fantastic coincidence is found in the fibre photograph of feather keratin,[4] a study of which will be published shortly; for if, as Bernal thinks, the pepsin molecules are piled, perhaps in a screw, along the hexad axis, their length in this direction is 140.6, that is, about $23\frac{1}{2}$A., which is almost exactly the strongest period along the fibre-axis of feather keratin, a period which is again repeated probably six (or a multiple of six) times before the fundamental period is completed! The innermost equatorial spot of the feather photograph also corresponds to a side-spacing of about 33 A. (though this is probably not the maximum side-spacing), which again is in simple relation to the side dimensions of the pepsin unit cell. As just said, these resemblances may be only accidental, but we cannot afford to overlook anything in such a difficult field, and it is not impossible that we have here an indication of how very long, *but periodic*, polypeptide chains can arise by the degeneration and linking-up of originally globular molecules.

Selection 129

Before considering further work on the structure of proteins in the 1930s let us take note of investigations into protein and amino acid metabolism which revived again during this period.

Ever since 1842, when Liebig had argued that tissue metabolism in the animal body was represented by the amount of urinary nitrogen measured as urea, the latter's part in protein metabolism had been the subject of continuous interest. Between 1870 and 1900 it was gradually recognized that urea was formed by the mammalian liver but physiologists, physiological chemists (biochemists) and the pathologists were for a long time uncertain about the mechanism of its synthesis.

A good many favoured the view that urea was formed in the liver through dehydration of ammonium carbonate which was thought to have resulted from the reaction between ammonia (detached from amino acids) and carbon dioxide in the blood. There were others who, for instance, inclined towards a 'biological' version of Wöhler's preparation of urea from ammonium cyanate.[a]

The synthesis of urea by the mammalian liver was still a problem awaiting solution in 1931, when it was taken up by Sir Hans Adolf Krebs (1900–81), aided by Kurt Henseleit (1907–76) who was just completing his medical studies. In that year, it should be explained, Krebs obtained a position at the University of Freiburg im Breisgau which enabled him to develop biochemical studies in relation to clinical problems. Previously Krebs had spent four years (1926–30) in the laboratory of Otto Warburg at Berlin-Dahlem, where he became familiar with the tissue-slice technique employed in researches on processes of disintegration (glycolysis, respiration). It occurred to him that this method, quantitatively more accurate and also simpler than the organ perfusion technique, was more generally applicable and therefore could be extended to investigations of biosynthetic processes such as the formation of urea in the liver. In developing the procedure Krebs paid great attention to establishing an accurate method for determining urea. He was also anxious to provide an appropriate medium for the tissue slices.

The amount of urea formed in the liver was measured by the urease method modified to suit the Warburg apparatus. A significant contribution was the introduction of a physiological saline, the ionic composition of which was close to that of mammalian blood serum.

Employing mainly slices of rat liver, Krebs found among other things that the addition of ammonium salts and certain amino acids affected variously the rate of urea formation. However, in the case of ornithine (α, δ – diaminovaleric acid) and ammonium chloride the rate of synthesis, as well as the amount of urea produced, was comparatively high pointing to a catalyzed reaction. How Krebs concluded that the formation of urea from carbon dioxide and ammonia in the liver takes place in a chemically unpredictable manner[b] is shown below. He interpreted it as a cyclic sequence of steps involving the conversion or ornithine into citrulline (α - amino – δ – ureidovaleric acid), followed by the production of arginine from citrulline with the regeneration of ornithine under the influence of liver arginase.[c]

H.A. Krebs and K. Henseleit, *Z.physiol.Chem.*, *210*, 33. 1932

'Untersuchungen über die Harnstoffbildung im Tierkörper' (Researches on urea formation in the animal body)

p. 49

The liver procures ornithine in the organism from food proteins, which without exception contain preformed ornithine. In hunger the ornithine supply sinks and one can trace the fall in urea formation to lack of ornithine, since ornithine addition abolishes the difference in urea formation conditioned by the state of the nourishment.

If one inquires about the cause of the ornithine action, one would first think that the ornithine may be the mother substance (*Muttersubstanz*) of the urea. Yet this is improbable because ornithine in absence of ammonia scarcely forms urea . . .[1]

That ornithine cannot be the source of the urea is shown with certainty by the fact that during the urea formation – measured in the amino-nitrogen content of the solution – ornithine does not essentially fall . . .

If the ornithine, whilst it acts, does not fall in concentration on balance, then it should be possible to obtain *large* effects with *small* amounts of ornithine. This is indeed the case . . .

pp. 51–2

[Accordingly] the action of ornithine in the living cell resembles catalytic action . . .

If one wishes to understand the action of ornithine one must assume that it takes part as an intermediate substance in the urea synthesis, that the synthesis does not run according to the simple equation

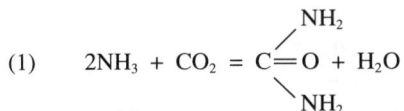

$$(1) \qquad 2NH_3 + CO_2 = C{\overset{\displaystyle NH_2}{\underset{\displaystyle NH_2}{=}}}O + H_2O$$

but in several partial reactions, with primary formation of an intermediate product which must be produced from ornithine, ammonia and carbonic acid. It must be required of the intermediate substance sought that it in the living cell breaks down with at least the same rate as reaction (1), splits off urea and forms ornithine again. Because [bringing the equation] into balance we find that ornithine is not used up. A substance which possesses the required properties for the intermediate has long been known: it is arginine. Arginine can be pictured as formed from ornithine, ammonia and carbonic acid and is, as required, split with great speed in the liver to ornithine and urea . . . We therefore draw the conclusion that arginine is the sought intermediate of urea synthesis in the liver, and that the primary reaction of the synthesis from ammonia and carbonic acid is:

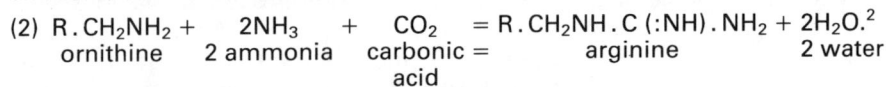

$$(2)\ R.CH_2NH_2 + \quad 2NH_3 \quad + \quad CO_2 \quad = R.CH_2NH.C\,(:NH).NH_2 + 2H_2O.^{[2]}$$

ornithine	2 ammonia	carbonic = acid	arginine	2 water

The second reaction of the urea synthesis is the splitting of arginine by the action of arginase, known since the work of Kossel and Dakin:[3]

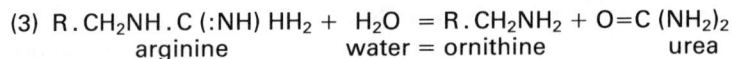

$$(3)\ R.CH_2NH.C\,(:NH)\,HH_2 + \ H_2O\ = R.CH_2NH_2 + O{=}C\,(NH_2)_2$$

arginine	water = ornithine	urea

If one adds the two equations (2) and (3) one obtains the balance equation (1).

pp. 53–4

Already Clementi,[4] and after him Hunter and Dauphinée[5] saw, when they investigated the distribution of arginase in the animal kingdom, that the ferment is present in the liver only of such animals as excrete urea as the end product of protein metabolism . . .

The single substance with which we have found similar properties as with ornithine is . . . citrulline. The experiments with citrulline are of special interest because they explain more clearly the chemical mechanism of reaction (2), the synthesis of arginine out of ornithine, ammonia and carbonic acid.

Citrullin was isolated in 1914 by the Japanese Y. Koga and S. Odake[6] from the water-melon (Citrulus vulgaris). Its chemical constitution was explained by M. Wada in the year 1930 and confirmed by synthesis.[7] It is an α-amino-δ-ureido- valeric acid:

$$HOOC . CHNH_2 . CH_2 . CH_2 . CH_2NH . CO . NH_2$$

Citrulline can be considered as formed from ornithine, carbonic acid and ammonia with loss of water. The addition of a further molecule of ammonia to citrulline with loss of a further molecule of water would lead to arginine.

This connection of citrulline with ornithine and arginine brings up the question whether the building up of arginine from ornithine, ammonia and carbonic acid goes through citrulline according to equations

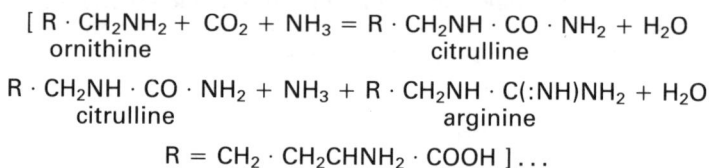

$$[R \cdot CH_2NH_2 + CO_2 + NH_3 = R \cdot CH_2NH \cdot CO \cdot NH_2 + H_2O$$
ornithine   citrulline

$$R \cdot CH_2NH \cdot CO \cdot NH_2 + NH_3 + R \cdot CH_2NH \cdot C(:NH)NH_2 + H_2O$$
citrulline   arginine

$$R = CH_2 \cdot CH_2CHNH_2 \cdot COOH] . . .$$

then it would be required:

1. that in presence of citrulline + ammonia in the liver urea should be formed at least as quickly as in presence of ornithine + ammonia + carbonic acid;
2. that the urea nitrogen formed in presence of citrulline + ammonia arises half from the citrulline, the other half from the ammonia.
   Both conditions were fulfilled [experimentally] . . .

Selection 130

Among the notable features of biochemical history of the first four decades of this century was the enduring belief in the dichotomy of 'endogenous' and 'exogenous' protein metabolism. It took thirty years before Folin's hard and fast division of protein metabolism was being called into question. It was in 1935 that the published results of Henry Borsook (b. 1907) and Geoffrey Lorrimer Keighley (b. 1901) on the rate of nitrogen and sulphur excretion cast real doubts on its validity. It should be noted that their original aim was to estimate the stimulating effect (which they failed to establish) of the 'specific dynamic action'[a] of protein (amino acids) on the endogenous metabolism.

Instead, as shown below, Borsook and Keighley interpreted their experimental results as pointing to the existence of what they termed 'continuing' metabolism of nitrogen.

Accordingly they accepted that the animal organism was capable of storing nitrogen. But they argued that this capacity could not be connected with amino acids since these were involved in unceasing synthetic processes producing polypeptides and protein. They presumed that the fixation of nitrogen occurred within the latter.

Although departing from the historically evolved distinction between two kinds of nitrogenous metabolism (Voit, Pflüger, Folin), the concept of continuing nitrogen metabolism still retained it in the sense that it was thought to include both food and tissue in almost equal measure.

H. Borsook and C.L. Keighley, *Proc.Roy.Soc.*, *(B)*, *118*, **488**. **1935**

'The 'continuing' metabolism of nitrogen in animals'

p. 489

The experiment was carried out as follows. Men were maintained on a constant diet until the urinary nitrogen and sulphur excretion were constant. The regime was then interrupted by a twenty-four-hour period when the original diet was changed for one in which the nitrogen was supplied with little or no sulphur. After equilibrium was re-established with the original diet, there was a second twenty-four-hour interruption by a period of nitrogen starvation but adequate caloric intake. The sulphur excretion on the first experimental day is a measure of the extent of the continuing metabolism on that day, and, compared with the day of nitrogen starvation, indicates the extent to which the continuing metabolism is normally in operation.

The data on the nitrogen excretion on the days succeeding the nitrogen starvation and experimental days also contribute to a decision on this question. The lag in attaining a nitrogen balance in passing from one level of nitrogen intake to another, and the conformance of the data to a first-order reaction indicate that the rate of the continuing metabolism is proportional to the extent of the available store of metabolizable nitrogen. This store is greater at a higher than at a lower level of protein intake. If the nitrogen katabolized is derived mainly from the protein ingested on that day, the effect of a day of protein starvation should be noticeable only on that day. The resumption of the diet on the following day should be accompanied by nearly the same nitrogen excretion as on the day preceding the day of starvation. On the other hand, if there is a large continuing metabolism the effects of such an interruption will appear not only on the day of the experiment but also, on the resumption of the original diet, in the quota of the continuing metabolism of the succeeding days.

p. 495

In a man in nitrogen equilibrium at a urinary level of 10–11 gm daily, 50 to 60 per cent of this nitrogen is derived from the continuing metabolism. The remainder is the exogenous quota, i.e., derived from the nitrogen ingested on that day. When physiologically incomplete nitrogen is ingested it may be expected that this quota will be a little larger than with a complete protein, because the extent of resynthesis (*see* below) into labile nitrogen will be somewhat less. As a result, there will be a small negative nitrogen balance.

This negative balance should be less than the figure for the daily endogenous metabolism. When large quantities of single amino acids are taken the negative nitrogen balance may be augmented by the escape of a considerable quantity of the amino acid into the urine unmetabolized. For example, in the glycine experiments . . . the free amino nitrogen in the urine in subject 1 was 830 mg, and in subject 2,973 and 749 mg, whereas the normal figures were of the order of magnitude of 200 mg daily.

This conclusion regarding the extent of the continuing metabolism rests on the assumption that the daily quantitative changes in sulphur metabolism are not widely divergent from that of the nitrogen in the mixture of amino acids undergoing metabolism. The warrant for this assumption is that there is a lag in the excretion of sulphur as well as of nitrogen in passing from one level of protein intake to another.

pp. 512–13

We are therefore led to a conception of normal protein metabolism which differs in an important respect from the generally accepted theory first enunciated by Folin. According to Folin's theory, anabolic processes, such as the synthesis of protein, are restricted in the adult organism in equilibrium to not more than the replacement of the 'wear and tear' or endogenous quota. This quota corresponds to not more than the minimum excretion of nitrogen on a diet containing little or no nitrogen, but otherwise adequate. When more than this minimal quantity of protein nitrogen is ingested the excess over the endogenous requirement is considered to be stored in the tissues as amino acids. These are quickly metabolized, and the nitrogen appears in the urine chiefly as urea. According to our view, the abundant data in the literature on nitrogen storage, the lag in attaining equilibrium in passing from one level of nitrogen intake to another, observations . . . and the data on the constancy of the free amino nitrogen of the tissues, indicate that the anabolic processes continually in operation must be more extensive than postulated in the current theory of protein metabolism, and in men may normally amount to 50 per cent or more of the nitrogen intake.

It follows from this that the extent of the continuing metabolism in different animals, under comparable conditions of nitrogen intake, carbohydrate metabolism, and work will be a function of the extent of the synthetic processes normally in operation. This need not be the same in all animals, and may be a characteristic of the species. The amount of nitrogen which can be stored therefore, and *pari passu* the extent of the continuing metabolism, need not be the same in all animals.

The continuing metabolism in the normal dietary state in man is quantitatively more important than the endogenous metabolism postulated by Folin. If the metabolism which leads to the urinary creatinine be excluded it is an open question whether the remainder of the endogenous metabolism yielding urea, ammonia, and part or possibly all the uric acid has any physiological reality. We approach, therefore, the point of view of Mitchell, Nevens, and Kendall (1922)[a] that under normal dietary conditions there may be no breakdown of tissue. But the hypothesis proposed here differs from that of Mitchell, Nevens, and Kendall in the same respect as it does from Folin's original theory in postulating much more extensive synthetic processes continually in operation.

Selection 131

Earlier in the century, we have noted, a good deal of attention centred on the problems of human (animal) nutrition. This led to inquiries into the ways in which proteins, carbohydrates and fats interrelated and replaced each other in nutrition. The accompanying interest in their 'normal' and 'erroneous' intermediate metabolism generated in its turn work which pointed to the existence of general metabolic mechanisms, including chemical convertibility of metabolites. In this connection we have already mentioned the findings by Neubauer (1909) on the oxidative deamination of amino acids to keto acids and Knoop's method (1910) of producing them *in vivo* from keto acids.[a] Since this work, little more had been added to knowledge of the mechanism of deamination and of the transference of amino groups from amino acids to other molecules during 1900–30.

Later the face of protein metabolism studies began to change once more. Evidence in favour of oxidative deamination of amino acids was obtained in 1935 by Krebs, who worked with kidney and liver slices. He found two deaminating enzyme systems differing in the way they attacked the naturally occurring *l*-series and the non-natural *d*-series of amino acids and also in the way they behaved on extraction of the ground tissue.[b]

As to what became known as 'transamination', one of the earliest hints relating to it is contained in a paper published by Dorothy M. Needham (1896–1987) in 1930.[c] What she suggested was that an amino group could enter into combination with some reactive carbohydrate residue and be retained as a new amino acid. Her suggestion was based on her findings that glutamic or aspartic acid, added to minced muscle under anaerobic conditions, produced succinic, malic and fumaric acids without formation of ammonia or change in the soluble nitrogen fractions. There was no change in the latter, either, under aerobic conditions, when the two amino acids used more oxygen (and the aspartic acid disappeared).

As shown below, the Soviet biochemists Alexander Evseyevich Braunstein (*b*. 1902) and Maria Grigorievna Kritzmann (1904–71) in 1937 provided evidence for enzyme catalyzed transamination (*Umaminierung*) in muscle tissue.

A.E. Braunstein and M.G. Kritzmann, *Nature*, *140*, 503. 1937

'Formation and breakdown of amino-acids [*sic*] by intermolecular transfer of the amino group'

pp. 503–4

We have previously reported[1] the discovery in muscle of a highly active metabolic mechanism, by the action of which the amino group and two hydrogen atoms of glutamic acid are transferred to pyruvic acid (added or of metabolic origin) with the formation of alanine (and ketoglutaric acid). This is the key to the puzzling fact that glutamic acid is transformed into succinic acid by muscle tissue without the formation of either ammonia or amide nitrogen (D. Moyle-Needham).

Further work, to be published in detail elsewhere[a], showed this reaction to be a reversible one. From alanine and α-ketoglutaric acid muscle tissue rapidly forms glutamic and pyruvic acids, equilibrium mixtures of similar composition resulting in both the direct and the reversed reaction.

The presence of the enzyme system responsible for this process of *Umaminierung* can be readily demonstrated not only in muscle, but likewise in other organs (heart, brain, liver, kidney), irrespective of their capacity or inability to metabolize amino-acids by oxidative deamination or reductive amination respectively. In organs containing Krebs's aminodehydrogenase the intermolecular transfer of the amino group competes with oxidative deamination and, owing to its greater velocity, completely inhibits the latter process in the presence of an excess of pyruvic acid. Only with nucleated erythrocytes and malignant tissues no evidence of *Umaminierung* has been obtained up to the present.

α-Ketoacids other than pyruvic, for example, α-ketobutyric, α-ketocaproic, oxaloacetic or phenylpyruvic acid, may equally serve as acceptors for the amino group of glutamic acid. On the other hand, all α-amino acids readily give up their amino groups to α-ketoglutaric acid in the presence of muscle tissue; the formation of glutamic acid has been demonstrated with sixteen different natural and racemic amino-acids including such as glycine or histidine, known to be difficultly accessible to oxidative deamination by surviving tissues. The rate of *Umaminierung* is astonishingly high; thus, no less than 6.5 mgm. glutamic acid is formed in two minutes from glycine and ketoglutaric acid by one gram of muscle tissue. This would correspond to a value of $Q_{Umam} = 152$. In our opinion, the mechanism of this biocatalytic reaction is the same as that discussed by Herbst[2] in recent studies on a somewhat similar model reaction, occurring when amino-acids are heated with ketoacids in aqueous solution.

Aspartic acid behaves in the same way as glutamic acid (Kariagina, this laboratory), and oxaloacetic acid – as ketoglutaric acid, in the function of amino nitrogen donators or acceptors respectively. The 'trapping' of oxaloacetic acid by the acceptance of amino groups, interfering with the action of Szent-Györgyi's 'fumarate system', offers an adequate explanation for the inhibitory effect of amino-acids on the respiration of muscle and other non-deaminizing tissues, repeatedly observed by different authors. An attractive hypothesis on the origin of diabetic ketosis was recently put forward by Koranyi and Szent-Györgyi,[3] who consider this condition as a result of damage to the fumarate system, possibly brought about by *Umaminierung* or a similar mechanism.

It deserves special attention that no transfer of the amino group occurs unless either the amino-acid or the ketoacid is a dicarboxylic one. No instance of a direct amino nitrogen transfer between two monocarboxylic acids has yet been observed, but some evidence has been obtained showing that the reaction can be brought about by the catalytic action of dicarboxylic amino- or ketoacids, added in small amounts and functioning as intermediary amino nitrogen carriers. The above data indicate that the dicarboxylic amino- and ketoacids probably play an important and specific part in the intermediary nitrogenous metabolism, bearing some analogy to the catalytic function of the C_4-dicarboxylic acids in tissue respiration.

Selection 132

We already know that in 1907 Willcock and Hopkins reported the beneficial effect of tryptophane additions to zein food.[a] Following this there was a drive from workers such as Osborne and Mendel[b] to establish

whether an amino-acid is essential or not, in the sense that it cannot be manufactured in the animal organism and must accordingly be supplied in the intake if nutrition adequate for the existing condition of the individual is to be maintained.[c]

In the paper from which this quotation is taken the Americans demonstrated the need for a supply of lysine in the diet of rats, in addition to confirming the observations of Willcock and Hopkins on tryptophane. Furthermore, while obtaining evidence for the indispensability of tryptophane for maintenance and lysine for growth they also found that lysine could not replace tryptophane in maintenance. They suggested the nutritional value of proteins had to be viewed not only in relation to their containing certain amino acids as indispensable components, but also from the point of view of whether the latter were essential to growth and maintenance respectively.

Two years later Hopkins, in a paper under his name and that of Ackroyd, described experiments made during 1914 and 1915 which explored the roles of arginine and histidine as precursors of the purine ring. On removal of both diamino acids from the diet, nutritive failure and fall in allantoin excretion followed. On restoration of one or the other in sufficient amount maintenance and even growth were observed, with no marked fall in allantoin excretion. It was suggested 'that this is because each one of them can, in metabolism, be converted into the other'.[d]

It was not until after William Cumming Rose (1887–1985), based at the University of Illinois (Urbana), became involved during the mid-1920s that the study of the significance of the individual amino acids as dietary components entered a new phase. The actual impetus that set him working was when he[e]

saw a lowly rat, on his way to oblivion because of histidine starvation, suddenly and dramatically change his course and begin to grow after consuming a tiny bit of the missing acid. Nothing in metabolism could be more spectacular or arouse more reverence for the marvelous chemistry that takes place in living things.

What Rose did (in collaboration with G.J. Cox) was to confirm the essential nature of histidine only, while failing to establish the interchangeable relationship between arginine and histidine.[f] He then developed an experimental approach that brought a hitherto sporadically employed procedure into general use. The protein rations in the diet were replaced by known mixtures of purified amino acids. This was seen to be a better way of obtaining information about the indispensability of certain amino acids than feeding experiments which used purified proteins or protein hydrolysates from which one or more amino acids have been removed. In the first case the protein was unsuitable to be the dietary source of nitrogen if it was deficient in only one amino acid. In the second case the difficulty lay in ensuring the complete removal of particular amino acids. Hence contamination by their residual traces could not be excluded and uncertainty in evaluating the experimental data prevailed.

The extract below is taken from Rose's article containing a critical review of literature and a summary of experimental work bearing upon the indispensability of amino acids. By 1938 ten amino acids were believed to be indispensable.

W.C. Rose, *Physiol.Revs.*, *18*, 109. 1938

'The nutritive significance of the amino acids'

pp. 129–31

The final classification of the amino acids with respect to their growth effects is summarized in table 9.

As will be observed, only ten are indispensable. In arriving at the above classification, the usual procedure was to omit one or two from the food at a time. It seemed not improbable, therefore, that a diet carrying only the ten essentials might fail to support growth inasmuch as the organism would be called upon to synthesize twelve cleavage products of proteins *simultaneously*. As a matter of fact, feeding trials with this simplified diet yielded results which exceeded our expectations.[1]

The animals gained in weight just as rapidly as when all of the protein components were supplied preformed. This finding is all the more remarkable in view of the fact that the food contained only 11.2 per cent of active amino acids.

Summary. The use of diets containing mixtures of highly purified amino acids in place of proteins has not only led to the discovery and identification of the new growth essential, threonine, but has served as a relatively simple and thoroughly trustworthy method of determining the nutritive significance of the individual amino acids. By this procedure, seven amino acids have been added to the list of indispensable dietary components. These, with tryptophane, lysine, and histidine, are the only cleavage products of proteins which are necessary for the normal growth of the rat. Twelve amino acids have been shown to be dispensable. The omission of any or all of these from the ration exerts no inhibitory action upon the growth processes.

THE PHYSIOLOGICAL AVAILABILITY OF THE OPTICAL ISOMERS OF THE NATURAL AMINO ACIDS. Of the ten indispensable amino acids listed in table 9, five are either difficult to prepare from their natural sources, or may be obtained more readily in quantity by synthetic methods. These are valine, isoleucine, phenylalanine, lysine, and methionine. Furthermore, natural, *l*-leucine is ordinarily contaminated with isoleucine and methionine, and cannot be used in nutrition studies where exact information concerning the composition of the diet is required. It is generally customary, if a synthetic product is employed, to double the amount in order to insure the presence of the active isomer at the desired level. Thus, it becomes a matter of considerable practical importance to know whether the antipodes of the natural amino acids can be utilized for

Table 9 Final classification of the amino acids with respect to their growth effects

Indispensable	Dispensable
Lysine	Glycine
Tryptophane	Alanine
Histidine	Serine
Phenylalanine	Norleucine
Leucine	Aspartic acid
Isoleucine	Glutamic acid
Threonine	Hydroxyglutamic acid
Methionine	Proline
Valine	Hydroxyproline
*Arginine	Citrulline
	Tyrosine
	Cystine

* Arginine can be synthesized by the animal organism, but not at a sufficiently rapid rate to meet the demands of normal growth.

growth purposes. If they are available physiologically, the use of increased amounts of the racemic modifications is unnecessary. This is not a negligible consideration in view of the cost of such materials. But aside from the practical aspects of the question, the behavior in the organism of compounds differing from each other only in spatial configuration is of unusual theoretical interest. The literature records numerous illustrations of optically isomeric compounds which are unlike in their physiological or pharmacological action, or in the ease with which they are oxidized in the body. But the possibility of replacing indispensable amino acids by their enantiomorphs has not been extensively investigated.

pp. 132–3

During the past two years, a number of experiments of this nature have been completed in this laboratory. As is our practice, the diet contained mixtures of amino acids in place of proteins. This method has the advantage of excluding the presence of all traces of the compounds under investigation, and to this extent is less likely to lead to erroneous conclusions . . .

The above facts are summarized in table 10. Of the antipodes of the indispensable amino acids only that of *d*-arginine remains to be tested.

No satisfactory explanation is at hand to account for the fact that both optically isomeric forms of certain amino acids are utilized for growth, while only the natural modifications of others are available. A number of investigators have suggested that when an inversion takes place in the animal body the unnatural isomer undergoes oxidative deamination, and the resulting α-keto acid is asymmetrically transformed into the natural acid. That α-keto acids can be converted into the corresponding α-amino acids has been thoroughly established. Therefore, if the postulated mechanism of inversion is correct, the failure of certain unnatural amino acids to support growth must be due to the fact that they are not readily deaminized. In this connection it is interesting to note that α-keto-γ-methylvaleric acid[2] and α-keto-β-methylvaleric

Table 10   Nutritive value of the optical isomers of the
indispensable amino acids

Only natural isomer promotes growth	Both isomers promote growth
Valine	Tryptophane
Leucine	*Histidine
Isoleucine	Phenylalanine
Lysine	Methionine
Threonine	

The unnatural form of arginine has not been tested.
d-Histidine appears to be somewhat less efficient than
l-histidine.

acid[3] induce excellent growth in animals deprived of l-leucine and d-isoleucine, respectively. We are not convinced, however, that the above theory will suffice to explain all of the facts concerning the behavior of the unnatural enantiomorphs. In any event, interesting problems for future investigation are suggested by the experimental data already obtained.

Summary. Of the ten indispensable amino acids, l-tryptophane, l-histidine, l-phenylalanine, and l-methionine can be replaced for growth purposes by their antipodes. On the contrary, only the natural forms of valine, leucine, isoleucine lysine, and threonine are available for the uses of the growing organism. The possible substitution of l-arginine for d- arginine has not been investigated.

Selection 133

Though the work by Borsook and Keighley (to which we have previously referred) might have challenged Folin's conception of distinct 'endogenous' and 'exogenous' sources of urinary nitrogen and might even have created doubts about it, most biochemical and physiological writers stood by it during the 1930s. Their picture of protein metabolism remained essentially a static one. Clearly there was a gap between the general acceptance of the dynamic approach to intermediate processes of metabolism in terms of 'simple substances undergoing comprehensible reactions', advocated by Hopkins in the renowned Presidential Address to the Physiology Section of the British Association in 1913,[a] and the specific adherence to two separate sources of urinary nitrogen (dietary and tissue protein), still prevalent a quarter of a century later.

Underlying this contradiction was the fact that there was no direct way of 'seeing' the paths of the metabolities involved in the sequences of reactions. Of particular historical interest in this respect are the neglected observations of the Italian worker Cesari Bertagnini (1827–1857) on the elimination of nitro-benzoic acid as nitro-hippuric acid in the 1850s.[b] Significantly, Hopkins touched upon this work in his 1913 Address when he was dealing with the way knowledge of biochemical synthesis developed. Perhaps overstating Bertagnini's actual intention Hopkins interpreted it as follows:

This investigator wished to earmark, as it were, the benzoic acid administered to the animal, in order to make sure that it was the same molecule which reappeared in combination. He so marked it with a nitro-group, giving nitro-benzoic acid and observing the excretion of nitro-hippuric acid. Later on he continued this interesting line of research by giving other substituted benzoic acids, and showed that in each case a corresponding substituted hippuric acid was formed.[c]

It was with the development of the isotope trace technique (itself an outgrowth of the work on the structure of the atom) that Hopkins's insight into the 'dynamic side of biochemistry' was powerfully amplified. It appears that the earliest biological application of the isotope technique was due to George Hevesy (1885–1966) in 1923. He investigated the transport of lead in *Vicia Faba* (horse bean), the roots of which were submerged in a solution containing the radioactive lead isotope thorium B.[d]

After the discovery of deuterium by Harold Clayton Urey (1893–1981)[e] in 1932, steps were taken to determine the possibility of using it in biological research. When the isotope of nitrogen, N^{15}, became available it became possible to synthesize amino acids from isotopic ammonia to study their fate *in vivo* under (nearly) physiological conditions. Initially this was thanks to the work of Rudolph Schoenheimer (1898–1941) and his two collaborators, David Rittenberg (1906–70) and Sarah Ratner (*b.* 1903).[f]

By 1939, in the light of their isotope experiments, there were cogent reasons for abandoning Folin's theory. Instead, as shown below, a dynamic picture of protein metabolism emerged. It was visualized that the nitrogen eliminated in the course of metabolic transformations derived from a 'reservoir' of amino-acids of mixed origin, drawn from dietary and tissue proteins.

R. Schoenheimer, S. Ratner and D. Rittenberg, *J.Biol.Chem.*, *130*, 703. 1939

'Studies in protein metabolism. X. The metabolic activity of body proteins investigated with $l(-)$-leucine containing two isotopes'

pp. 729–31

According to the prevalent view of the fate of dietary protein, distinction is generally made between exogenous and endogenous sources of urinary nitrogen . . .
It is scarcely possible to reconcile our findings with any theory which requires a distinction between these two types of nitrogen. It has been shown that nitrogenous groupings of tissue proteins are constantly involved in chemical reactions; peptide linkages open, the amino acids liberated mix with others of the same species of whatever source, diet, or tissue. This mixture of amino acid molecules, while in the free state, takes part in a variety of chemical reactions: some reenter directly into vacant positions left open by the rupture of peptide linkages; others transfer their nitrogen to deaminated molecules to form new amino acids. These in turn continuously enter the same chemical cycles which render the source of the nitrogen

indistinguishable. Some body constituents like glutamic and aspartic acids and some proteins like those of liver, serum, and other organs are more actively involved than others in this general metabolic mixing process. The excreted nitrogen may be considered as a part of the metabolic pool originating from interaction of dietary nitrogen with the relatively large quantities of reactive tissue nitrogen.

SUMMARY

1. Four adult, non-growing rats were kept on an ordinary stock diet containing casein to which was added, during the last 3 days, an amount of isotopic leucine corresponding to 23 mg. of nitrogen per day. The leucine had been synthesized in such a manner that the carbon chain was marked by heavy hydrogen and the amino group by heavy nitrogen. The racemic material was resolved and only the 'natural' isomer, *l*(−)-leucine, was used. The animals maintained a constant weight during the entire experiment.

2. The material was excellently absorbed, but less than one-third of the marked nitrogen appeared in the urine; the bulk (57 per cent) had replaced nitrogen of the body proteins and only a small amount (8 per cent) was present in the non-protein fraction. The proteins of blood and various organs revealed different activities in regard to the 'acceptance' of leucine nitrogen. The isotope concentration in the nitrogen of the serum was highest, that of the internal organs was slightly less, while that of muscle and skin was relatively low. Nevertheless, owing to their large size, the last two tissues had the greatest share in the uptake of dietary nitrogen.

3. From the proteins of liver, intestinal wall, and the rest of the body a total of twenty-two amino acid samples was isolated. Their isotope content showed that various chemical processes take place continuously in the proteins of normal animals.

4. The presence of deuterium and N^{15} in the leucine from the proteins indicated extensive replacement of amino acids by the same species derived from the diet. The process requires continuous opening and closing of peptide linkages. At least 32 per cent of the dietary leucine had found its way into the proteins by this reaction, leading to a replacement of 24 per cent of liver leucine and 7 per cent of carcass leucine. As the process of removal and introduction must also lead to the reincorporation of leucine from tissue proteins, the extent of the actual replacement must have been considerably higher.

5. α-Nitrogen atoms of amino acids have been transferred to other amino acids, leading to a distribution of isotopic nitrogen among all amino acids except lysine. Apart from leucine, glutamic and aspartic acids had the highest isotope content. The transfer process also involves a temporary opening of peptide linkages. Amide nitrogen is also continuously replaced.

6. The presence of isotope in the amidine group of arginine from protein suggests that this amino acid, even though it exists in protein linkage, is available for urea formation by the Krebs cycle.

7. Ornithine isolated from arginine contained isotope, indicating that the former like most other amino acids was involved in a chemical process (synthesis and destruction or nitrogen shift).

8. Only one-third of the isotopic nitrogen in the body proteins had entered directly with the carbon chain of the ingested leucine; the remaining

two-thirds was deposited as a result of continuous nitrogen transfers from one amino acid to others.

9. Deuterium and N^{15} determinations of leucine from body proteins indicate that this indispensable amino acid, like histidine, not only yields nitrogen but also accepts it from other amino acids; at least one-third of the isotope in the leucine was replaced by ordinary nitrogen in the latter process.

10. These results, in conjunction with earlier findings, demonstrate that these continuous chemical processes occur extensively in the tissue proteins, even under conditions of nitrogen balance. The dietary nitrogen enters the pool of these spontaneous reactions which tend to the ultimate replacement of protein nitrogen. The urinary nitrogen represents a sample of the metabolic mixture originating from the chemical interaction of the dietary nitrogen with the relatively large amounts of reactive tissue nitrogen.

Selection 134

We now return to investigations of protein structure or architecture (as it also used to be termed) during the second half of the 1930s. First we shall concern ourselves with the then much debated cyclol hypothesis of protein structure by Dorothy M. Wrinch (1896–1976). As shown below, she pictured the possibility of building not only one-dimensional but also two-dimensional and three-dimensional protein molecules from 'cyclised polypeptides' or 'cyclols' based on hexagons (formed from two amino acid residues).

There were considerable doubts about the cyclol hypothesis even in its brief heyday during 1936–9. In retrospect Dorothy Hodgkin describes the attitude of Bernal and herself as follows:

She was a friend of ours, and as a mathematician was very anxious to become acquainted with biology and biochemistry and to contribute mathematical ideas to the problems on which we were working ... We were friends of hers, and had helped to develop her theories, but we did not believe in them, and that was our trouble.[a]

Yet it is essential to view Dorothy Wrinch's ideas just as much in their personal as in their historical context. She was an active member of the forward looking interdisciplinary group of scientists that included Bernal and J. Needham, which in 1935 proposed the establishment of an Institute for Mathematico-Physico-Chemical Morphology in Cambridge. The proposal grew out of the conviction, among others, that X-ray analysis (especially of fibres) had a role to play 'in bridging the gulf between morphology and biochemistry'.[b] Whilst the proposal came to grief, the hypothesis of protein structure suggested by Dorothy Wrinch in 1936 created a great stir. In a well-known textbook, published in 1949, it was stated: 'No theory of protein structure has stimulated as much thought or experimentation as has the cyclol theory.'[c]

It is against this background that the following appraisal of Wrinch's cyclol hypothesis by a protein chemist of high standing, made in 1937, has to be seen:

The cyclol theory of the protein molecule, which is derived directly from the polypeptide theory, has its origin in the growing realization that the geometry of a molecule is equally as important as a quantitative knowledge of its constitution.

This modern trend in chemical thought is well illustrated by the history of the evolution of modern views on the protein molecule, but it can be seen also in any branch of chemistry where large molecules are concerned. It certainly arose from the study of substances of biological origin, and a realization of the significance of form may perhaps be regarded as the contribution of biology to chemical theory.

The cyclol theory of protein structure leads to the view that all protein molecules have a dorsiventral structure. Moreover, it leads to the view that on the one side (which may be taken as the back) they have a common pattern. The significance of this is considerable. The paradox has long been realized that though life is intimately connected with the colloidal state yet biochemical reactions appear to be intimately connected with small molecules. Again, though every species yields different proteins, yet these small active molecules seem common to large numbers of species.[d]

Last but not least, while the hypothesis was though 'to be too much at variance with experimental facts to find wide acceptance', it was still considered as 'the one coherent hypothesis' as late as 1945.[e]

D.M. Wrinch, *Nature*, *137*, 411. 1936

'The pattern of proteins'

pp. 411–12

Any theory as to the structure of the molecule of simple native protein must take account of a number of facts, including the following:

1. The molecules are largely, if not entirely, made up of amino acid residues. They contain —NH—CO linkages, but in general few —NH$_2$ groups not belonging to side chains, and in some cases possibly none.
2. There is a general uniformity among proteins of widely different chemical constitution which suggests a simple general plan in the arrangement of the amino acid residues, characteristic of proteins in general. Protein crystals possess high, general trigonal, symmetry.
3. Many native proteins are 'globular' in form.
4. A number of proteins[1] of widely different chemical constitution, though isodisperse in solution for a certain range of values of *p*H, split up into molecules of submultiple molecular weights in a sufficiently alkaline medium.

The facts cited suggest that native protein may contain closed, as opposed to open, polypeptides, that the polypeptides, open or closed, are in a folded state, and that the type of folding must be such as to imply the possibility of regular and orderly arrangements of hundreds of residues.

An examination of the geometrical nature of polypeptide chains shows that, *since all amino acids known to occur in proteins are α-derivatives*, they may be folded in hexagonal arrays. Closed polypeptide chains consisting of 2, 6, 18,

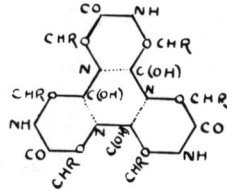

Figure 17 The 'cyclol 6' molecule

42, 66, 90, 114, 138, 162 . . . (18 + 24n) . . . residues form a series with threefold central symmetry. A companion series consisting of 10, 26, 42, 58, 74, 90, 106, 122 . . . (10 + 16n) . . . residues have twofold central symmetry. There is also a series with sixfold central symmetry: others with no central symmetry. Open polypeptides can also be hexagonally folded. The number of free —NH, groups, in so far as these indicate an open polypeptide, can be made as small as we please, even zero if we so desire. The hexagonal folding of polypeptide chains, open or closed, evidently allows the construction of molecules containing even hundreds of amino acid residues in orderly array, and provides a characteristic pattern, which in its simplicity and uniformity agrees with many facts of protein chemistry.

The stability of these folded polypeptide chains cannot be attributed to electrostatic attractions between the various CO . NH groups, for the appropriate distance between carbon and nitrogen atoms in these circumstances[2] lies between 2.8A. and 4.2A., whereas the distance in our case is at most 1.54A. By using the transformation* suggested by Frank in 1933 at a lecture given by W.T. Astbury to the Oxford Junior Scientific Society,

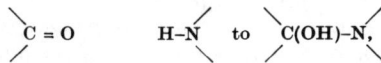

$$\diagdown{C} = O \qquad H\text{--}N\diagup \quad \text{to} \quad \diagdown{C(OH)\text{--}N,}\diagup$$

which has already proved useful in the structure of α-keratin,[3] the situation is at once cleared up and we obtain (Figure 17) the molecule 'cyclol 6' (the closed polypeptide with six residues), 'cyclol 18', 'cyclol 42' (Figure 18) and so on, and similarly open 'cyclised' polypeptides (Figure 19).

Hexagonal packing of polypeptides suggests a new *three dimensional* unit,

—CHR—C(OH)—N$\diagup$, which may be used to build three-dimensional

molecules of a variety of types. These are now being investigated in detail. At

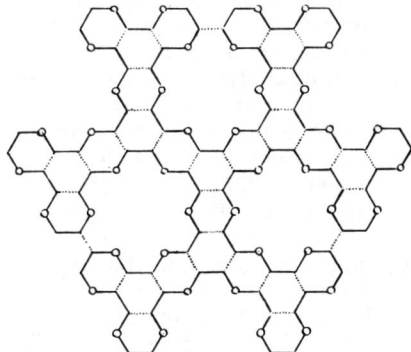

Figure 18 A 'cyclol 42' molecule

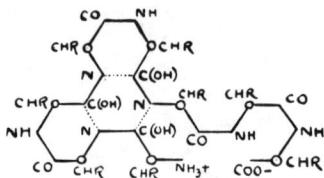

Figure 19 'Cyclised' polypeptides

the moment we direct attention only to single cyclised polypeptides forming hexagons lying approximately in one plane. The cyclol layer molecule is a fabric the thickness of which is one amino acid residue. *Since all naturally occurring amino acids are of lævo type*[4] this fabric is dorsiventral, having a front surface from which the side chains emerge, and a back surface free from side chains. Both front and back carry trios of hydroxyls normal to the surface in alternating hexagonal arrays. Such a layer molecule and its polymers, formed also by the same transformation, can cover an area of any shape and extent. It offers suggestions as to the structure of the solid protein film when it is one amino acid residue thick. In its most compact form, the cyclol layer molecule gives an area per residue of about $9.9 \, A.^2$ Less dense layers can be built, for example, from polymers of cyclol 18 and of cyclol 66 respectively, where the corresponding areas per residue are $13.2 \, A.^2$ and $16.2 \, A.^2$ respectively. The figures for unimolecular films of gliadin, glutenin, egg albumin, zein, serum albumin, serum globulin range from $1.724 \times 10^{7-}$ gm./ cm.^{2} for serum globulin to 1.111×10^{-7} gm./cm.^{2} for serum albumin[5,6,7]. With an average residue weight of 120, these densities give an area per residue ranging from $11.48 \, A.^1$ to $18.82 \, A.^2$. On the basis of the proposed hexagonal packing of polypeptides I therefore suggest that the upper limit of density of which a protein film is capable without buckling, provided that it is only one amino acid residue thick, is one residue per $9.9 \, A.;^2$ further, that a higher density implies that the film, though it may still be unimolecular, is more than one amino acid residue thick.

Cyclol layers may also be used to build molecules and molecular aggregates with extension in three dimensions, since they may be linked front to front by means of the side chains, in particular, by cystine bridges, and back to back by means of hydroxyls.[8] The single-layer cyclol is a fabric capable of covering a two-dimensional area of any shape and extent; a three-dimensional array can then be built, layer upon layer, the linkage being alternately by means of side chains and hydroxyls. The idea that native proteins consist largely, if not entirely, of cyclised polypeptides therefore implies that some native proteins, including those of 'globular' type, may have a layer structure.

Linkage by means of hydroxyls recalls the structures of graphitic oxide and montmorillonite, etc. Such a structure suggests a considerable capacity for hydration, an outstanding characteristic of many proteins. Further, since alternate layers are held together by means of hydroxyls, and contiguous molecules may also be held together in the same way, a protein molecular aggregate will, on this theory, necessarily be sensitive to changes in the acidity of the medium; in particular, a sufficiently high *p*H will cause such an aggregate to dissociate into single-layer units or into two-layer units joined by cystine bridges or side chains in covalent linkages. Svedberg's results, according to which a number of different native proteins break up into smaller molecules with sub-multiple molecular weights,[1] here find a simple interpretation. The particular sub-multiples which occur may be regarded as affording evidence as to the type of symmetry possessed by the layers out of which the molecular aggregates are built.

The hypothesis that native proteins consist essentially of cyclised polypeptides thus takes account of the facts mentioned in (1), (2), (3), (4) above. Further, it derives support from the case of α-keratin, for with Astbury's 'pseudo-diketopiperazine' structure[3] the polypeptides may be regarded as partially cyclised since they are cyclised at regular intervals, one out of every three (CO . NH) groups being involved. It is also suggestive in relation to a variety of other facts belonging to organic chemistry, X-ray analysis, enzyme chemistry and cytology. I cite the following:

(1) The rhythm of 18 in the distribution of amino acids in gelatin found by Bergmann,[9] and the suggestion of Astbury[10] that in gelatin 'the effective length of an amino acid residue is only about 2.8 A.'.
(2) The low molecular weight not exceeding 1,000 found by Svedberg[11] for the bulk of the material from which lactalbumin is formed.
(3) Secretin,[12] a protein with molecular weight of about 5,000, containing no open polypeptide chains.
(4) The nuclear membrane, which, consisting of proteins and lipoids, plays an important part in mitosis on account of its variable permeability.
(5) Bergmann's findings[13] with respect to dipeptidase; these suggest that the dipeptide substrate, upon which this enzyme acts, has a hexagonal configuration.

Finally, the deduction from the hypothesis of cyclised polypeptides, that native proteins may consist of dorsiventral layers, with the side-chains issuing from one side only, suggests that immunological reactions are concerned only with surfaces carrying side-chains. Hence, such reactions depend both on the particular nature and on the arrangement of the amino acids.

Full details of the work, which is to be regarded as offering for consideration, a simple *working hypothesis*, for which no finality is claimed, will be published in due course.

Selection 135

Developing her hypothesis on topological lines Dorothy Wrinch conceived of three-dimensional models of protein molecules constructed from truncated tetrahedra. She thought that they were formed by polymerization of 'cyclol 6'. She visualized a cyclol 'fabric' or 'network' lying in the plane surfaces of the tetrahedron with the sidechains either located within the network or projecting 'outside' or 'inside' it. She further suggested that the first and smallest tetrahedron accommodated 72 amino acids, the second 288 and that a theoretical series of model protein molecules could be set up, containing $72 n^2$ amino acid residues. Calculations of molecular weights of proteins on this basis gave values corresponding to those obtained by ultracentrifugation.[a]

---

* The application of this transformation to these molecules was suggested to me by J.D. Bernal.

That protein molecules appeared to carry 288 amino acid residues (or multiples of this number) emerged also from the work produced by Max Bergmann (1886–1944) and Carl Niemann (1908–1964) at that time. Thus they reported that the molecules of cattle globin and cattle fibrin each contained $576 = 2^6 \times 3^2$ amino acid residues and that the molecule of egg albumin comprized $288 = 2^5 \times 3^2$ amino acid residues. In general, it seemed as if the various protein molecules comprized $2^n \times 3^m$ amino acid residues, where n and m were positive whole numbers. Also they discovered that on multiplying the number of amino acid residues present in 1 molecule of a given protein by the average weight of the residues, the minimum molecular weights of the proteins were obtained. Further, they submitted that there were grounds for believing in a distribution of the various amino acids at regular intervals ('periodicity', 'frequency') along the potential polypeptide chain.[b]

The following extract from the paper on the structure of silk fibroin gives an indication of Bergmann and Niemann's approach to protein structure. It should be noted that the treatment in their papers is 'somewhat condensed and difficult to follow', as already noted by their contemporaries.[c]

M. Bergmann and C. Niemann, *J.Biol.Chem.*, *122*, 577. 1937–8

'On the structure of silk fibroin'

pp. 578–82

The percentages of glycine, alanine, tyrosine, and arginine which were obtained as the result of our analysis of silk fibroin are given in Column 1 of Table 11, together with the earlier values for histidine and lysine which were

Table 11   Ratio of Amino Acids in Silk Fibroin after Hydrolysis

| Amino acid | Weight (1) | Mol. wt. (2) | Gm. molecule per 100 gm. protein | | Ratio (5) | Fraction of total residues (frequency) (6) |
			Found (3)	Calculated* (4)		
	per cent					
Glycine	43.8	75	0.584_0	0.584_0	1296	2
Alanine	26.4	89	0.296_6	0.292_0	648	4
Tyrosine	13.2	181	0.072_9	0.073_0	162	16
Arginine	0.95	174	0.005_5	0.005_4	12	216
Lysine	$0.25^\dagger$	146	0.001_7	0.001_8	4	648
Histidine	$0.07^\dagger$	155	0.0004_5	0.0004_5	1	2592

* Base $= 0.584_0$ (gm. molecule of glycine).
[†] Vickery and Block[1].

determined by Vickery and Block.[1] The percentage values were recalculated on a gm. molecular basis and are presented in this form in Column 3 of Table 11. The gm. molecular ratios of the six amino acids are listed in Column 5 of Table 11. The mean molecular weight of the amino acids formed by the complete hydrolysis of fibroin was estimated to be 102, which in turn leads to the value of 84 for the average residue weight. From this latter value, it was calculated that 100 gm. of silk fibroin must, on complete hydrolysis, produce approximately 1.190 gm. molecules of the hypothetical average amino acid.[a] When one compares this value with the values for the individual amino acids given in Column 3 of Table 11, it is apparent that the amino acids glycine, alanine, tyrosine, arginine, lysine, and histidine account for $\frac{1}{2}$, $\frac{1}{4}$, $\frac{1}{16}$, $\frac{1}{216}$, $\frac{1}{648}$, and $\frac{1}{2592}$ of the total number of amino acids produced by the complete hydrolysis of fibroin.[b] From the values given in Columns 5 and 6 of Table 11 it follows that silk fibroin must contain $2592 = 2^5 \times 3^4$ amino acid residues or a whole number multiple thereof, and on multiplying 2592 by the average residue weight, i.e. 84, it is evident that the minimum molecular weight of silk fibroin is approximately 217,700. When the value of 3.5 Å. is accepted as the distance between two adjacent peptide bonds[2] and when it is assumed that the peptide chain is situated along the fiber axis, it follows that the length of the minimum molecule is 9.072 Å. It should be pointed out here that the percentage value of histidine which was employed in the above calculation was accepted with considerable hesitation because of the extraordinarily low histidine content of fibroin. Because of this latter fact, we have not attempted a new estimation of this amino acid, and it appears that this point cannot be settled until a more satisfactory method for the determination of histidine is devised. When all consideration of the histidine content of fibroin was omitted, it was found that the minimum number of amino acid residues in the fibroin molecule was changed to $1,292 = 2^4 \times 3^4$ and the minimum molecular weight to approximately 108,500. However, it is unlikely that these latter values are correct, since if this were the case, the above histidine value would have to be in error by 100 per cent. Therefore, although a final commitment is not possible at this time, we favor the value of $2,592 = 2^5 \times 3^4$ as the number of amino acid residues in the fibroin molecule, which in turn leads to the molecular weight of approximately 217,700.

The six hydrolytic products of silk fibroin the stoichiometrical relationships of which have been determined account for approximately 85 per cent of all of the amino acid residues originally present in the intact protein. Thus, with this high percentage of cleavage products accounted for, it was possible to arrange them in a sequence which offered a rather complete structural characterization of the fibroin molecule. A fragment of this structure is given below.

—G—A—G—T—G—A—G—Ar—G—A—G—X—G—A—G—X—
          [G—A—G—T—G—A—G—X—G—A—G—X—G—A—G—X]$_{12}$
G—A—G—T—G—A—G—X—G—A—G—X—G—A—G—Ar—
          [G—A—G—T—G—A—G—X—G—A—G—X—G—A—G—X]$_{13}$

(The symbols G, A, T, and Ar refer respectively to the residues of glycine, alanine, tyrosine and arginine. The symbol X refers to the residues of the various other amino acids.)

The above fragmentary formula containing 432 amino acid residues is subject to a series of variations in which the frequencies of the glycine, alanine, tyrosine, and arginine residues are not altered. That is, one can withdraw one or more amino acid residues from the beginning of the chain and transfer them, in their entirety and without altering the sequence of the amino acid residues in the transferred segment, to the end of the chain. In this manner, one can construct 431 isomeric variations of the above structural fragment. One of these 432 isomers, when arranged in a linear sequence with five additional identical fragments, leads to the correct structure of the complete fibroin molecule containing 2,592 amino acid residues. It should be emphasized again that in the above structural variations the sequence of the amino acid residues within the peptide chain is not altered; the only difference being that the beginning of the peptide chain is at a different position from that given in the fragment formulated above. As soon as it is experimentally possible to ascertain how far distant the first arginine residue is from either end of the peptide chain of the intact fibroin molecule, then a decisive selection of one of the 432 isomers mentioned above can be made.

Selection 136

From the outset the Bergmann-Niemann periodicity (frequency) hypothesis found itself the centre of interest and controversy among protein chemists. In a way the proposition that there was an order in which amino acids were linked to form proteins represented a continuation of the line first taken by Mulder and Berzelius in the 1830s and developed forcefully (in more than one sense) by Svedberg a century later. Thus by 1937 Svedberg suggested that the molecular weights of the proteins fall into multiples of 17,600, instead of 34,500 as given previously. Again he interpreted this as indicating an underlying common principle in the architecture of proteins.[a]

Among those who criticized the frequency hypothesis of Bergmann and Niemann was Albert Neuberger (b. 1908), on the grounds that the average error in the determination of individual amino acids in a protein may be fairly taken to be 5–6 per cent. He pointed out that on this basis there was a probability of up to 80 per cent that a purely random distribution would give values apparently in accord with the formula of the frequency hypothesis. However, he concluded that though it was not convincingly established with respect to the complex globular proteins, 'in the case of the simple fibrous proteins it rests on a more secure basis'.[b] Similar doubts and criticisms were expressed by Norman Wingate Pirie (b. 1907). Nevertheless, he looked upon it as an 'interesting generalization'.[c] Indeed, it stimulated the application of more accurate procedures in amino acid analysis after 1940 (partition chromatography, electrophoresis, ionic exchange).

One of the consequences of the frequency hypothesis was the evident degree of restriction on the number of isomers, in comparison with the case of the random distribution of amino acid residues. Doubtful about this limitation 'on the astronomical possibilities of protein isomerism permitted by the Hofmeister-Fischer hypothesis', A.H. Gordon (b. 1916) A.J.P. Martin (b. 1910) and R.M.L. Synge (b. 1914) came up with a funda-mental recommendation. They stated: 'It is clear that the only direct evidence in favour of a hypothesis concerning the order of amino-acid residues in a peptide chain can come from the isolation and identification of fragments of this chain containing at least two amino-acid residues.' They prepared lower peptides by partial hydrolysis of wool, edestin and gelatin. Using electrodialysis in a diaphragm cell they were able to isolate a large proportion of the basic amino acid residues as dipeptides. This approach pointed to a more complicated structure than the one derived on the basis of the frequency hypothesis (e.g. silk). None the less, Gordon, Martin and Synge made it equally clear that their work 'does not by any means form a rigid refutation of the Bergmann–Niemann hypothesis'.[d]

Around 1940 even those protein chemists most sceptical in their attitude to the Bergmann–Niemann hypothesis did not disown it completely. Because of the limitations in amino acid analysis of the day the evidence against it was not clear-cut. This also emerges from A.C. Chibnall's (1894–1988) authoritative review of knowledge of the structure of proteins in the light of data obtained by chemical and X-ray analysis which he gave when he delivered the Royal Society's Bakerian Lecture in 1942 from which the extract below is taken.

Almost a quarter of a century later he wrote about it as follows:

To me, in retrospect, this lecture marks the close of an era ushered in at the turn of the century by Emil Fischer and Kossel,[e] in which the emphasis was on good old-fashioned gravimetric analysis. Chromatography in those days was still in the exploratory stage and no one would have predicted that the progress in its application would be so rapid that the older procedures were to be ousted within a very few years.[f]

Indeed, the study of the chemical structure of proteins, at once continuous with and separate from the endeavours of the past, was entering a new phase, lying outside the scope of this volume.[g]

A.C. Chibnall, *Proc.Roy.Soc.*, *(B)*, *131*, 136. 1942

'Amino-acid analysis and the structure of proteins'

pp. 157–9

My examination of the relevant evidence adducible from amino-acid analysis, titration curves and free amino groups leads to the conclusion that the molecules of edestin, β-lactoglobulin, egg albumin and insulin are systems of peptide chains containing six, nine, about four, and eighteen components respectively. Evidence given ... suggests that the haemoglobin molecule likewise contains sixteen component chains. Let me discuss the bearing that these findings have on the validity of the Bergmann–Niemann hypothesis, for we have seen that the straight evidence for this has been somewhat conflicting. In the case of edestin the analytical data ... suggest that in the smallest possible submolecule the total number of residues and also the number of residues of thirteen amino-acid and acid-amide species are severally factors of 432 ($2^4 \times 3^3$), and that the molecular weight of the submolecule is 50,000. There is also a body of independent evidence which gives support to this submolecular weight, sufficient I think to constitute a strong presumption that it is a true verdict; it might be held with some justice therefore that the molecule of this particular protein, in which all the constituent peptide chains are of like composition, is one which lends support to the Bergmann–Niemann hypothesis.

In making this assertion, however, I should be guilty of omission if I did not emphasize that evidence on two very important points is entirely lacking. In the first place no account has been taken of the 224 residues which constitute the remainder of the edestin submolecule. These represent the hydrolysis products for which methods of analysis hitherto employed do not possess the required degree of accuracy, and it is gratifying to note that several fundamentally new lines of approach to overcome these difficulties are being explored at the present time (Bergmann & Niemann 1938; Ussing 1939; Rittenberg & Foster 1940; Martin & Synge 1941a,b). In the second place an evaluation of the amino-acids given on strong acid hydrolysis, however complete we may be able to make it eventually, cannot provide any direct evidence for the regular frequency of recurrence which the Bergmann–Niemann hypothesis implies; as Gordon, Martin & Synge (1941) have pointed out, such evidence can be obtained only by the isolation and identification of partial hydrolysis products of chain length sufficient to contain the necessary minimum number of residues, and their researches on these lines will be followed with great interest by protein chemists. Failure to obtain direct confirmatory evidence nevertheless will not necessarily imply that all the Bergmann–Niemann generalizations are false, for certain of the residues may be incorporated in the peptide chain in such a way that they exhibit more than one periodic interval of recurrence, overlapping of which would result in two or more near positions in the chain being occupied by the same residue species at more distant intervals. This may well be true with respect to the large number of arginine and dicarboxylic acid residues in edestin, whereby the close proximity of ionizable groups might create a relatively few foci of strong charge which could play some part in the hydrolysis attending edestan formation (vide Bailey 1942).

It is easy to see that under such conditions it would be difficult to prove regularities in structure by chemical analysis, and the problem becomes even more complicated if these conditions apply to proteins like β-lactoglobulin, haemoglobin and insulin in which the molecule is a system of peptide chains of varied composition, or to proteins known to be polydisperse. Clearly the analyst, whose activities at present are restricted to the simpler hydrolysis products, can contribute but little on his own, and there is need for more co-operation with the physical chemists and crystallographers, who are able to investigate the properties and structure of larger units and of the intact protein molecule itself. The need for co-operation moreover is mutual, but progress may be slow for the art of the analyst is an exacting one. Meanwhile, I think that those interested in proteins would be wise to regard the Bergmann–Niemann hypothesis as still tentative and in any case as applicable only to the component peptide chains of the molecule, for much of the evidence hitherto brought forward to support it has been based on inadequate experimental data and has demonstrated nothing more than the hypnotic power of numerology.

Notes to Section V

Selection 101

Introduction:

a. See Introduction to Selection 2, Section I.
b. Cf. E.F. Beach, 'Beccari of Bologna. The discoverer of vegetable protein', *J.Hist.Med.*, *16*, 354–73. 1961. See also 'Sur le froment', *Collection académique . . . de la partie étrangère, contenant les Mémoires de l'Académie des sciences de l'Institut de Bologne, 10*, 1–5. 1773. The original version 'De frumento' is in *De Bononiensi scientiarium et artium Instituto atque Academia commentarii, 2*, (pt 1). 1745 (not seen).
c. F.L. Holmes, 'Elementary analysis and the origins of physiological chemistry', *Isis, 54*, 50–81. 1963; 'Analysis of fire and solvent extractions: the metamorphosis of a tradition', *Isis, 62*, 129–48. 1970.
d. 'Éloge de M. Rouelle le jeune', *Observ.phys.*, *16*, 165–74. 1780 (1789); also Rouelle, 'Éxperiences', *J.med.chir.pharm.*, *39*, 250–266. 1773. 'Observations sur les fécules ou parties vertes des plantes et sur la matiere [*sic*] glutineuse ou végéto animale', *J.med.chir.pharm.*, *40*, 59–67. 1773 (p. 65). For providing a xerox-copy of this article thanks are due to N.W. Pirie.
e. de la Metherie, 'Discours préliminaire contentant un précis des nouvelles decouvertes (1)', *Observ.phys.*, *28*, 28–36. 1786. Jean-Claude de Lametherie (1743–1817) edited *Observations sur la physique* from 1785. His annual reviews provide an historical background to the state of science at the time.
f. de Fourcroy, *Elements of Natural History and Chemistry*, 3rd edn (London, 1790), III, p. 296. According to Lavoisier plant materials were composed of carbon, hydrogen, oxygen, nitrogen, phosphorus and sulphur, see A.-L. Lavoisier, *Elements of Chemistry* (Edinburgh, 1790), pp. 123, 145. But he was not consistent, cf.: '. . . of vegetables which contain azote, such as the cruciferous plants, and of those containing phosphorus . . .' (p. 126).
g. Berthollet, 'Recherches sur la nature des substances animales, et sur leurs rapports avec les substance végétales', *Mém.Acad.Sci.*, 120–5. [1780] (1784); 'Sur l'analyse animale comparée à l'analyse végétale', *Observ.phys.*, *28*, 272–5. 1786. An English version of this article is in de Fourcroy, *Elements of Natural History and Chemistry*, 2nd edn (London, 1788), IV, pp. 283–7.
h. de Fourcroy, *Elements*, 3rd edn, III, p. 297.
i. Cf. '*cum ovi albo*' in Pliny, *Natural History* (The Loeb Classical Library, Cambridge, Mass.; London, 1963), VIII, p. 48 (book 28, 18, §66).
j. [P.J. Macquer], *Dictionaire de chimie* (Paris, 1776), II, pp. 131–2.

Text:

a. Ordinarily in Lavoisier's system the term 'base' (or 'radical') meant the non-oxygen component of oxides. In the particular case of ammonia, known to be composed only of nitrogen and hydrogen, the former, as the principal element of the compound, was regarded as the 'base'. Cf. also Introduction to Selection 1, Section I.
b. Regarding the ambiguity of the term 'fecula' see Selection 6, Section I, note a.
c. de Fourcroy, *Elements*, 3rd edn, pp. 293–7.
d. Pierre Thouvenel (1747–1815). According to the Soviet historian A.N. Shamin, Thouvenel was general inspector of mineral resources in France before the Revolution and under Napoleon. Shamin lists two books by Thouvenel bearing on the subject under review: *De corpore nutritivo et nutritione tentamen chymico-medicum* (Piscenis, 1770); *Mémoire chimique et médicinale sur la mechanisme et les produits de la sanguification* (St Petersbourg, 1777). Cf. A.N. Shamin, *Istoriya khimii belka* (Moscow, 1977), pp. 27–8, 32.

Selection 102

Introduction:

a.   See Selection 74, Section IV.

Text:

a.   Braconnot named it *sucre de gélatine*, rendered into German as *Leimzucker*. Subsequently
     it was established that the substance was not sugar at all and Eben Norton Horsford
     (1818–93), an American pupil of Liebig, termed it 'glycocoll' in 1846. Horsford pointed
     out that glycocoll possessed simultaneously the properties of an acid, a base and a salt. It
     was renamed 'glycine' by Berzelius in 1847. See E.N. Horsford, 'Ueber Glycocoll
     (Leimzucker) und einige seiner Zersetzungsproducte', *Ann.*, *60*, 1–57. 1846; J. Berzelius,
     'Leimzucker. Glycin', *Jahres-Ber.*, *27*, 652–65. 1847. Both terms were still in use in the
     1930s.
b.   References are found in the literature stating that the *acide caséique* and the *oxide caséeux*
     obtained (1819) by Joseph Louis Proust (1754–1826) from cheese were possibly impure
     leucine. See J.L. Proust, 'Sur le principe qui assaisonne les fromages', *Ann.Chim.*, *10*,
     29–49. 1819. Cf. F. Lieben, *Geschichte der physiologischen Chemie* (Leipzig and Vienna,
     1935), p. 339. In most historical accounts it is stated that the history of the discovery of
     amino acids begins with the work of Braconnot on glycine and leucine. While this is
     technically true it should be remembered that neither the term 'amino acid' nor their
     significance for the differentiation of albuminous substances (proteins) was contemplated
     at the time.

Selection 103

Introduction:

a.   J. Berzelius, 'Sur la composition des fluides animaux', *Ann.Chim.*, *88*, 26–72. 1813 (p.
     71).
b.   J. Liebig, 'Ueber einen Apparat zur Analyse organischer Körper, und über die Zusam-
     mensetzung einiger organischer Substanzen', *Ann.Phys.*, *21*, 1–43. 1831. On the
     development of organic analysis M. Dennstedt's account is still valuable. Cf. his 'Die
     Entwickelung der organischen Elementaranalyse', *Sammlung Chemischer und chemisch-
     technischer Vorträge*, *4*, No. 1. 1899. F. Szabadváry surveys the area in *History of
     Analytical Chemistry*, translated by G. Svehla (Oxford, 1966), pp. 284–308. For a critique
     of positivistic assumptions by historians of organic analysis, see E. Hickel, 'Die organische
     Elementaranalyse im Schnittpunkt von Theorie und Praxis der frühen Biochemie',
     Pharmazie in unserer Zeit, *8*, 1–10. 1979.
c.   Cf. J.J. Berzelius, *Lehrbuch der Chemie*, 4th edn (Dresden and Leipzig, 1840), IX, p. 5f.
d.   E. Glas, 'The protein theory of G.J. Mulder (1802–1880)', *Janus*, *62*, 289–308. 1975.
e.   *Brefväxling mellan Berzelius och G.J. Mulder* (1834–47), *Correspondence entre Berzelius
     et G.J. Mulder* (1834–47) in *Jac. Berzelius Bref – Jac. Berzelius Lettres*, ed. H.G.
     Söderbaum (Uppsala, 1916), V, pp. 71–80, 104–9. (p. 109).

Text:

a.   On Mulder's published work in Dutch journals see H.B. Vickery, 'The origin of the word
     protein', *Yale J.Biol. and Med.*, *22*, 387–93. 1950; also Glas, 'The protein theory of G.J.
     Mulder'.
b.   The concept of molecular weight was unknown at that time. Empirical formulae were
     determined from analysis but uncertainties existed with respect to the ratio of atoms of the
     elements in the compounds concerned. Mulder's calculations were based on Berzelius'
     atomic weights using $O = 100.000$ as base, $H = 6.2398$, $C = 76.437$, $N = 88.08$, $S =
     201.165$, $P = 196.155$.
c.   This column contains percentage amount of each element.
d.   The round numbers of atoms of C, H, N, O related to the following calculated amounts:

carbon   = 30,678.43
hydrogen = 3,883.70
nitrogen = 8,879.42
oxygen   = 12,042.26

e. Amounts calculated from the round numbers of atoms of C, H, N, O.
f. Barred sulphur ($\bar{S}$) represented, in effect, double atoms of sulphur.

Selection 104

Text:

1. The experiment of Tiedemann and Gmelin, who found it impossible to sustain the life of geese by means of boiled white of egg,[a] may be easily explained, when we reflect that a graminivorous animal, especially when deprived of free motion, cannot obtain, from the transformation of waste of the tissues alone, enough of carbon for the respiratory process. 2 lbs. of albumen contain only 3½ oz. of carbon, of which, among the last products of transformation, a fourth part is given off in the form of uric acid. -L.
a. F. Tiedemann and L. Gmelin, *Die Verdauung nach Versuchen* (Heidelberg and Leipzig, 1827), II, pp. 197–8. On the significance of this two-volume collaborative effort in the development of biochemistry, see Introduction to Selection 159, Section VII.

Selection 105

Introduction:

a. See Selections 162 and 164, Section VII.
b. Dumas and Cahours, 'Sur les matières azotées neutres de l'organisation', *Ann.Chim.*, [3] *16*, 385–448. 1842. In the concluding passage of this paper the authors stressed the social value of their research. It is interesting to note who they thought could primarily benefit from their experiments when they stated: '. . . we have sought to establish on a sure basis the rules to follow in the calculation of the diet of the soldier, the worker or the prisoner, as also the rules which must direct administrators in establishments dedicated to public charity'.
c. More appropriately: Lyaskovsky (Nikolay Erastovich).
d. It was shortly afterwards that Joseph Redtenbacher (1810–70), then professor in the University of Prague, demonstrated the unreliability of this method for removing sulphur from organic compounds. In analysing taurine he failed to split off its sulphur. Cf. J. Redtenbacher, 'Ueber die Zusammensetzung des Taurins', *Ann.*, *57*, 179–84. 1846; also F. Lieben, *Geschichte der physiologischen Chemie* (Leipzig and Vienna, 1935), pp. 325, 360–1. In the main, the dispute between Liebig and Mulder has been treated from the point of view of the superiority or inferiority of the analytical procedures adopted in their laboratories. That there was more to the disagreement between the two protagonists has been emphasized by E. Glas, 'The Liebig-Mulder controversy. On the methodology of physiological chemistry', *Janus*, *63*, 27–66. 1976. The author suggests that the controversy reflected the increasing polarization of two approaches to the chemistry of life: the quantitative ('statistical') view of Liebig and the qualitative ('dynamical') view of Mulder. For further treatment of the subject see Glas's *Chemistry and Physiology in their Historical and Philosophical Relations* (Delft, 1979).

Text:

a. The concept of molecular formula was not then known and the combination of elements in a given compound was referred to as its 'atom'. To account for varying relationships in compounds of identical composition with different properties, Berzelius introduced the terms isomerism, metamerism and polymerism in the early 1830s. See 'Körper von gleicher Zusammensetzung und verschiedenen Eigenschaften', *Jahres-Ber.*, *11*, 44–8.

1832; 'Isomerie, Unterscheidung von damit analogen Verhältnissen', *Jahres-Ber.*, *12*, 63–5. 1833.

Selection 106

Introduction:

a.   J. Liebig, 'Ueber die Bestandtheile der Flüssigkeiten des Fleisches', *Ann.*, *62*, 257–369. 1847. It is in this long and renowned paper that Liebig introduced the names 'tyrosine', 'sarcosine' and 'inosinic acid' derived from the Greek: *tyros* (cheese), *sarkos* (meat) and *is,inos* (muscle). Liebig mentions that it was not actually he who had come up with the terms but a certain von Ritgen (1811–99) (architect and since 1853 professor of *Kunstwissenschaft* at Giessen).
b.   [J. Liebig], 'Baldriansäure und ein neuer Körper aus Käsestoff', *Ann.*, *57*, 127–9. 1846. Liebig found that the substance, although easily soluble in alkali, combined with acid.

Text:

a.   Alcohol.
b.   Selection 102.
c.   That is, molecule.

Selection 107

Introduction:

a.   Cf. Selections 7 and 10, Section I. On this subject one may consult F.L. Holmes, *Claude Bernard and Animal Chemistry* (Cambridge, Mass., 1974), p. 290f.

Text (ii):

a.   Translation of *Lehrbuch der physiologischen Chemie*, 2nd edn (Leipzig, 1850).
b.   According to the German chemist, Adolph Strecker (1822–71), the reaction could be designated as follows ($C = 6$, $H = 1$, $O = 8$):

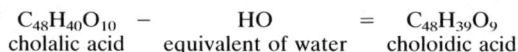

$$C_{48}H_{40}O_{10} \quad - \qquad HO \qquad = \quad C_{48}H_{39}O_9$$
cholalic acid   equivalent of water   choloidic acid

See A. Strecker, 'Untersuchung der Ochsengalle', *Ann.*, *65*, 1–37. 1848; 'Vorläufige Notiz über die Spaltung der Cholsäure in Glycocoll und stickstofffreie Säuren', ibid., 130–2. 1848.

Selection 108

Introduction:

a.   S. Cannizzaro, *Sketch of a Course of Chemical Philosophy*, (Edinburgh, 1910). The original Italian version appeared in *Il Nuovo Cimento*, *7*, 321–66. 1858. Though often discussed, the literature on the Karlsruhe Congress is meagre. A short informative account of the background and events of the Congress is contained in Clara de Milt's 'The Congress at Karlsruhe' in A.H. Ihde, W.F. Kieffer (eds), *Selected Readings in the History of Science* (Easton, 1965), pp. 37–40. See also her 'Carl Weltzien and the Congress at Karlsruhe', *Chymia*, *1*, 153–69. 1948.
b.   A. Laurent and Ch. Gerhardt, 'Sur la composition de l'aposépédine ou leucine (oxyde caséique de Proust)', *Comp.Rend.*, *27*, 256–8. 1848.
c.   A. Strecker, 'Ueber die künstliche Bildung der Milchsäure und einen neuen, dem Glycocoll homologen Körper', *Ann.*, *75*, 27–45. 1850. Strecker was seeking, in reality, to obtain lactic acid in the same manner as mandelic acid was known to have been produced by the action of prussic acid and hydrochloric acid on oil of bitter almonds (benzaldehyde). By making alanine react with nitrous acid Strecker prepared lactic acid. Thus he reached his original aim as he put it 'in a roundabout way', noting that alanine was a derivative (*Modification*) of lactic acid.

d.   The course of the Strecker synthesis was mostly interpreted as follows:

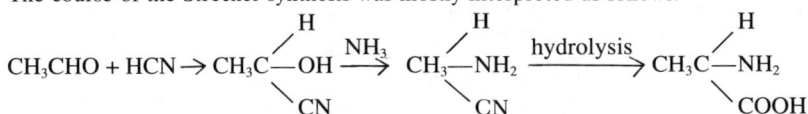

$$CH_3CHO + HCN \rightarrow CH_3C\overset{\displaystyle H}{\underset{\displaystyle CN}{\diagup}} OH \overset{NH_3}{\longrightarrow} CH_3\overset{\displaystyle H}{\underset{\displaystyle CN}{\diagup}} NH_2 \overset{hydrolysis}{\longrightarrow} CH_3C\overset{\displaystyle H}{\underset{\displaystyle COOH}{\diagup}} NH_2$$

It is not without interest that Strecker wondered whether the synthesis could be made into a general method for the preparation of higher homologues of alanine by employing the corresponding members of the nitrile series (written as $C_{2n}H_{2n-1}N$) of which prussic acid was the first. However, on reflection Strecker expressed doubts about the possibility of success on these lines because of the great differences in chemical behaviour between prussic acid and the higher homologues of the series of alkyl cyanides, designated by him as the nitriles of the 'so-called volatile fatty acids'.

e.   Cf. 'Comme on le voit, l'acide hippurique ressemble par ces reactions aux acides amides' in 'Extrait d'une lettre de M. Dessaignes à M. Dumas' published as 'Nouvelles recherches sur l'acide hippurique, l'acide benzoïque et le sucre de gelatine', *Comp.Rend.*, *21*, 1224–7. 1845, (p. 1226).

f.   A. Cahours, 'Recherches sur les acides amides', *Ann.Chim.*, *53*, 322–59. 1858.

g.   That was part of a process which had been continuing since about 1840 and which separated the erstwhile unified chemical science into three divisions: the study of the chemical properties of the elements (inorganic chemistry); the study of the chemical properties of the carbon compounds (organic chemistry); and the study of the chemistry of plant, animal and human life (plant, animal and human physiological chemistry).

h.   Ritthausen's work was rooted in a lifetime spent in agricultural teaching and research. He recognized that a better knowledge of the composition of proteins is necessary for the advance of the physiological side of animal nutrition. This comes out clearly in his pioneering book *Die Eiweisskörper der Getreidearten, Hülsenfrüchte und Ölsamen Beiträge zur Physiologie der Samen der Culturgewächse, der Nahrungs- und Futtermittel* (Bonn, 1872).

i.   H. Ritthausen, 'Ueber die Glutaminsäure', *J.prakt.Chem.*, *99*, 454–62. 1866. See also Werther, 'Ueber Glutaminsäure', *J.prakt.Chem.*, *99*, 6–7. 1866. The formula $C_{10}H_9NO_8$ is also used (C = 6, H = 1, N = 14, O = 8). Cf. also 'Ueber die Glutansäure, das Zersetzungsproduct der Glutaminsäure durch salpetrige Säure', *J.prakt.Chem.*, *103*, 239–42. 1868.

j.   H. Ritthausen, 'Ueber die Zersetzungsproducte de Legumins und des Proteïnkörpers der Lupinen und Mandeln beim Kochen mit Schwefelsäure', *J.prakt.Chem.*, *103*, 233–8. 1868; 'Asparaginsäure und Glutaminsäure, Zersetzungsproducte des Legumins und Conglutins beim Kochen mit Schwefelsäure', *J.prakt.Chem.*, *106*, 445–6. 1869; 'Asparaginsäure und Glutaminsäure, Zersetzungsproducte des Legumins und Conglutins beim Kochen mit Schwefelsäure', *J.prakt.Chem.*, *107*, 218–40. 1869; U. Kreusler, 'Asparaginsäure als Zersetzungsproduct thierischer Proteïnstoffe etc.', *J.prakt.Chem.*, *107*, 240–5. 1869.

Text:

1.   The values for glutamic acid for 1-5 give the amount which crystallizes directly from the reaction fluid and is purified by solution in water etc.; so that the loss on recrystallization and the amounts of impregnating aspartic acid (which could not be determined with certainty) remained uncalculated. For 6 on the other hand the acid is precipitated with the aspartic acid and is separated from the latter. The aspartic acid, if its amount was small, was weighed not as the crystalline free acid but as copper salt in the air-dried state; and thus the amount was calculated according to the composition ascertained by Ritthausen.

2.   This journal, *107*, 140–245.

a.   A.C. Chibnall in his Silliman Lectures (originally published in 1939) *Protein Metabolism in the Plant* (New Haven and London, 1964), p. 13 gives the following:

> Mucedin = Wheat gliadin
> Maize fibrin = Maize glutelin
> Gluten-casein = Wheat glutelin.

b.   Cf. also the statement by H.B. Vickery in 'The contribution of the analytical chemist to protein chemistry', *Ann.N.Y.Acad.Sci.*, *47*, 63–94. 1946 (p. 66):

The differentiation and characterization of these protein materials was by no means easy. Ritthausen, in fact, failed to make entirely clear the grounds upon which he relied. He depended upon solubility primarily, but he used the terms albumin (*Eiweiss*), plant casein, or plant gelatin as group designations. Under the heading plant casein, he placed legumin and conglutin, as well as the less-clearly differentiated gluten casein, which is now called glutenin. In justice to Ritthausen, it is only fair to state that, with respect to this last product, we have today only given another name to a protein system that is still scarcely at all understood. Plant gelatin was a group designation employed for the alcohol-soluble proteins of cereal seeds, gliadin, mucedin and gluten-fibrin, which were thought to be present in different proportions in the various cereal grains. We define gliadin today as the system of alcohol-soluble protein components of wheat seeds, and have dropped Ritthausen's other terms in favour of specific designations (such as zein for the alcohol-soluble protein of maize) or, occasionally, of Osborne's[c] general term, prolamins, for proteins of this type.

c.   Thomas Burr Osborne (1859–1929) worked in the Connecticut Agricultural Station and contributed significantly to the development of protein chemistry. Cf. also Introduction to Selection 132.

Selection 109

Introduction:

a.   Phytochemistry was cultivated in Austria and Bohemia a good deal. Thus it is possible to draw a link between the work of one of the founders of modern phytochemistry, Friedrich Rochleder (1819–74), his pupil Hlasiwetz, Zdenko Hans Skraup (1850–1910) who developed a general method for the preparation of quinolines, and Fritz Pregl (1869–1930) who refined microchemical analysis.

b.   H. Hlasiwetz and J. Habermann, 'Ueber die Proteïnstoffe. Erste Abhandlung', *Ann.*, *159*, 304–33. 1871.

c.   H. Hlasiwetz and J. Habermann, 'Zur Kenntnis einiger Zuckerarten (Glucose, Rohrzucker, Levulose, Sorbin, Phloroglucin)', *Ann.*, *155*, 120–44. 1870.

Text:

1.   Erlenmeyer und Schöffer, Journal für pract. Chemie 1860, 357; Theile, Chemisches Centralblatt, 1867, 385; Wanklyn, Pharm. Journal and Transact. *1*, 66; Hüfner, Chemisches Centralblatt 1872, 152.
2.   Pflüger's Archiv für Physiologie *6*, 589; *7*, 139.
a.   Cf. also Selection 115.
b.   Cf. Selection 108.

Selection 110

Introduction:

a.   Selection 104.
b.   For an incisive treatment see F.L. Holmes, 'Introduction' to the facsimile of the Cambridge (Mass.) edition of Liebig's (1842) *Animal Chemistry or Organic Chemistry in its Application to Physiology and Pathology* (New York and London, 1964).
c.   Originally Voit called the protein located in extracellular fluids 'reserve protein' (*Vorrathseiweiss*). Cf. 'Der Eiweissumsatz bei Ernährung mit reinem Fleisch', *Z.Biol.*, *3*, 1–85. 1867. The term 'circulating protein' occurs first in his article 'Ueber den Eiweissumsatz bei Zufuhr von Eiweiss und Fett, und über die Bedeutung des Fettes für die Ernährung', *Z.Biol.*, *5*, 329–68. 1869.

Selection 111

Introduction:

a.   E. Pflüger, 'Ueber einige Gesetze des Eiweissstoffwechsels (mit besonderer Berücksichtigung der Lehre vom sogenannten 'circulirenden Eiweiss')', *Pflügers Arch.*, *54*, 333–419. 1893.
b.   See Selection 53, Section III and Selection 168, Section VII.

Selection 112

Introduction:

a. Due mainly to contributions by Archibald Scott Couper (1829–96), Friedrich August Kekulé (1829–96) and Alexander Mikhailovich Butlerov (1828–86) between 1858 and 1861. For material on this subject see articles reprinted from *Journal of Chemical Education* in A.J. Ihde, W.F. Kieffer (eds), *Selected Readings in the History of Chemistry* (Easton, 1965) by H.C. Brown, 'Foundations of the structure theory' (pp. 229–35); O.T. Benfey, 'Kekulé – Couper Centennial Introduction' (pp. 236–7); E.N. Hiebert, 'The experimental basis of Kekulé's valence theory' (pp. 237–44); H.M. Leicester, 'Contributions of Butlerov to the development of structural theory' (pp. 257–61); A.J. Rocke, 'Kekulé, Butlerov, and the historiography of chemical structure', *Brit.J.Hist.Sci.*, *14*, 27–57. 1981.

b. There was neither a recognized nomenclature nor classification of proteins at that time.

c. C. Schorlemmer, *A Manual of the Chemistry of the Carbon Compounds; or, Organic Chemistry* (London, 1874), p. 491.

d. J. Wislicenus, *Adolph Strecker's Short Textbook of Organic Chemistry*, translated and edited, with extensive additions, by W.R. Hodgkinson and A.J. Greenaway (London, 1881), pp. 752, 764.

e. Cf. J.T. Edsall, 'Proteins as macromolecules: an essay on the development of the macromolecule concept and some of its vicissitudes', *Arch.Biochem.Biophys.*, *Suppl.1*, 12–20. 1962; J.T. Edsall, 'The development of the physical chemistry of proteins, 1898–1940', *Ann.N.Y.Acad.Sci.*, *325*, 53–74. 1979.

f. Apparently in 1853 the renowned German pathologist, Rudolf Virchow (1821–1902), was the first to investigate in a systematic manner precipitation of proteins by salts ('salting out') such as magnesium sulphate, sodium sulphate, potassium sulphate, alum (aluminium potassium sulphate), calcium chloride and sodium chloride. He noted that by and large the capacity to coagulate proteins depended on the solubility of salts. He also became convinced that coagulation had to be interpreted as a dehydration process. Cf. R. Virchow, 'Ueber ein eigenthümliches Verhalten albuminöser Flüssigkeiten bei Zusatz von Salzen', *Virchows Arch.*, *6*, 572–80. 1853.

    Conscious of and interested in physico-chemical aspects of pathological processes (at the time it was dropsy), Virchow became alerted to the precipitation of albuminous substances from their solutions by inorganic salts: his interest was aroused by references to the employment of magnesium sulphate in Ch. Robin and F. Verdeil's *Traité de chimie anatomique et physiologique normale et pathologique etc.* (Paris, 1853), III, pp. 298–304, 454. Robin and Verdeil were stimulated to turn to magnesium sulphate by Claude Bernard's work on pancreatic juice during the 1840s. Bernard found that a material, not unlike albumin and emulsin, was precipitated from the juice by means of magnesium sulphate. See his 'Mémoire sur le pancréas et sur le role due suc pancréatique dans les phénomènes digestifs particulièrement dans la digestion des matières grasses neutres', *Comp.Rend.*, *Suppl.1*, 379–563. 1856 (p. 439). It is also of historical interest to note that at the time the terms 'precipitation' and 'coagulation' were interchangeable. By 1900 there was a tendency to distinguish between 'chemical' precipitation and 'physical' coagulation, though not consistently.

    Virchow's article is also interesting because he made use of concepts termed '*Colloidmassen*' and '*Colloidsubstanzen*'. Believed to be jelly-like precipitations, they manifested themselves in fluids and organs under pathological conditions. The introduction of the term 'colloid' is generally associated with Thomas Graham's (1805–69) paper 'Liquid diffusion applied to analysis', *Phil.Trans.*, *151*, 183–224. 1861. While there is little doubt that this marks the beginning of the scientific study of the colloidal state, its medical background is less well appreciated. Before and around 1850 the term 'colloid' had been applied to describe 'viscid jelly-like elements' found in certain types of cancerous cells. Cf. *The Cyclopaedia of Anatomy and Physiology*, ed. R.B. Todd (London, 1852), IV, p. 137.

g. Cf. G. Kauder, 'Zur Kenntnis der Eiweisskörper des Blutserums', *Arch.exp.Path.Pharm.*, *20*, 411–25. 1885; J. Pohl, 'Ein neues Verfahren zur Bestimmung des Globulins im Harn und in serösen Flüssigkeiten', ibid., *20*, 426–38. 1885; S. Lewith, 'Zur Lehre von der Wirkung der Salze. Erste Mittheilung. Das Verhalten der Eiweisskörper des Blutserums gegen Salze', ibid., *24*, 1–16. 1888; F. Hofmeister, 'Zur Lehre von der Wirkung der Salze. Zweite Mittheilung. Ueber Regelmässigkeiten in der eiewessfällenden Wirkung der Salze und ihre Beziehung zum physiologischen Verhalten derselben', ibid., *24*, 247–60. 1888; 'Zur Lehre von der Wirkung der Salze. Dritte Mittheilung. Ueber die wasserentziehende Wirkung der Salze', ibid., *25*, 1–30. 1889. These studies of the effects of the neutral salts laid the groundwork for what became known as the 'Hofmeister lyotropic ion series'. By taking into account the electrovalence of anions and cations (and also, in the case of the latter, their atomic weight), they could apparently be arranged into two separate series of anions and cations, according to their influence on a variety of properties of proteins such as the salting out or swelling. By 1940 it had become widely accepted that the precipitating effect decreased along the series of anions: citrate > tartrate > SO_4 > acetate > Cl > NO_3 > Br > I > SCN. The same was thought to apply, though to a lesser degree, to the series of cations: Ba > Sr > Ca > K > Na > Li. See H.B. Bull, *Physical Biochemistry* (New York, 1943), p. 84.

h.   See Selection 172, Section VII.

Text:

1.   Compare Kauder, Arch. f. exp. Path. u. Pharm., vol. 20, p. 412 and Pohl, ibid., p. 426.
2.   Jahresbericht für Thierchemie for 1881, p. 19.
a.   Rennet and thrombin.

Selection 113

Introduction:

a.   Among chemists who gave particular attention to these matters were: Francois-Marie Raoult (1830–1901), Professor in Grenoble; Svante Arrhenius (1852–1927), who became Director of the Nobel Institute for Physical Chemistry at Stockholm; Wilhelm Ostwald (1853–1932), Professor in Riga and Leipzig; Jacobus Henricus van't Hoff (1852–1911), active in Amsterdam and in Berlin.
b.   The term 'amphoteric' goes back to 1800 when it was employed by the Professor of Chemistry and Botany at the University of Budapest, Jacob Joseph Winterl (1732–1809). By this expression he denoted certain oxides that behaved as acidic or basic oxides. See J.R. Partington, *A History of Chemistry* (London, 1962), II. pp. 599–600. We have mentioned Horsford's observation in 1846 when he pointed out that glycocoll could be regarded as an acid as well as a base or a salt (Selection 102, note a). A valuable contribution to the elucidation of the amphoteric character of glycocoll – without using the term – came from the noted A. Strecker whose synthesis of alanine starting with cyanhydrin has already been referred to (Introduction to Selection 108). In 1868 Strecker indicated that the basic amino group and the acidic carboxylic group were capable of forming what became known as an 'internal' salt. Cf. A. Strecker, 'Ueber das Lecithin', *Ann.*, *148*, 77–90. 1868 (pp. 87–88).
c.   (Sir) William Bate Hardy contributed to the understanding of colloids, films and surfaces including lubricants. He wrote on theoretical aspects of biology and was also concerned with applied biology (food and marine research). When Biological Secretary of the Royal Society he was the guiding spirit behind its Food (War) Committee (1915–19). For a brief account of Hardy's life and work see *Obit.Not.F.R.S.*, *1*, 327–33. 1932–5. His articles were made available as *Collected Scientific Papers* (Cambridge, 1936).
d.   See W.B. Hardy, 'On the mechanism of gelatin in reversible colloidal systems', *Proc.Roy.Soc.*, *66*, 95–125. 1899–1900, (p. 112). It may be noted that the concept of pH was unknown at the time; Sørensen formulated it a decade later. See Selection 23, Section I.

Text:

1.   J. Starke, Ref. in *Malys Berichte*, xxvii, p. 19. 1897.
a.   See Introduction to Selection 112, note b.

Selection 114

Introduction:

a. For Curtius' appreciation of his own contributions and of Fischer's work in this area see his '34. Verkettung von Amidosäuren', *J.prakt.Chem.*, *70* N.S. (*178*), 57–262. 1904.
b. Fischer's development of distillation *in vacuo* interacted with the contemporary advances in the electrical field. In particular they were the vacuum technique stimulated by the needs of the electric bulb industry and the adaptability of the electric motor (in driving the pump). Much insight into this, including Fischer's links with the well-known electrical firm *Siemens and Halske* and also information on the price of the apparatus, is given in E. Fischer and C. Harries, 'Ueber Vacuumdestillation', *Ber.chem.Ges.*, *35*, 2158–63. 1902.
c. Fischer himself separated and identified valine (1901–6), proline (1901–4) and oxyproline, now generally called hydroxyproline (1902). See E. Fischer, 'Ueber die Hydrolyse des Caseins durch Salzsäure', *Z.physiol.Chem.*, *3*, 151–76. 1901, 'Spaltung der α-Aminoisovaleriansäure in die optisch activen Componenten', *Ber.chem.Ges.*, *39*, 2320–8. 1906; E. Fischer, 'Notizen', *Z.physiol.Chem.*, *35*, 227–30. 1902; E. Fischer and U. Suzuki, 'Synthese von Polypeptiden. III. Derivate der α-Pyrollidincarbonsäure', *Ber.chem.Ges.*, *37*, 2842–8. 1904. E. Fischer and A. Skita, 'Ueber das Fibroin und den Leim der Seide', *Z.physiol.Chem.*, *35*, 221–6. 1902. E. Fischer, 'Ueber eine neue Aminosäure aus Leim', *Ber.chem.Ges.*, *35*, 2660–5. 1902. The name 'oxyproline' is due to H. Leuchs, 'Synthese von Oxy-pyrollidin-carbonsäuren (Oxy-prolinen), *Ber.chem.Ges.*, *38*, 1937–43. 1905.

Text:

1. Presented to the Berlin Academy [of Sciences] on 29 November 1900. See Sitzungsberichte *1900*, 1062.

Selection 115

Introduction:

a. Cf. *Chem.Ztg.*, *26*, 919. 1902.
b. See *Chem.Ztg.*, *26*, 939. 1902 and Selection 116.

Text:

1. So far as such fermentative breakdown is known at all, it concerns however the otherwise well reported specific case of oxidation, where a terminal carboxyl splits off in the form of CO_2.
2. Nasse, O., Über die Wirkung der Fermente. Rostocker Zeitung, 15 Dec. 1894.
3. Curtius, Th. and Göbel, Fr., Über Glykokolläther. Journ. f. prakt. Chemie, *37*, 150.
4. Loew, O., Über das Eiweiss und die Oxydation desselben. Journ. f. prakt. Chemie, *31*, 129.
5. Schiff, H., Über Desamidoalbumin, Ber. d. deutsch. chem. Ges. *29*, 1354.
6. Paal, Über die Peptonsalze des Glutins. Ber. d. deutsch. chem. Ges. *25*, 1235.
7. Schiff, H., Biuretreaktionen. Ber. d. deutsch. chem. Ges. *29*, 298; Über Polyaspartsäuren. Ibid., *30*, III. 2449 and Liebigs Annalen, *299*, 236 and *319*, 300.
8. In this way also is explained the apparent contradiction that the biuret reaction of the protein is already abolished by nitrous acid with only a small evolution of nitrogen, although the appearance of split products giving an exquisite biuret reaction indicates wide occurrence of the group concerned in the reaction. In fact, it appears according to Schiff that the presence of NH_2 end-groups is necessary: it is easy to see that by hydrolysis of the CONH . CH. groups NH_2-groups must always be formed again from the protein.
9. Schaal, Ed., Über einige Derivate der Aspartsäure. Liebigs Annal. *157*, 24.
10. Grimaux, Ed., Sur des colloides azotes. Bull. soc. chim., 1882, *38*, 64.
11. Curtius, Th., Über einige neue der Hippursäure analog konstituierte synthetisch dargestellte Amidosäuren. Journ. f. prakt. Chem. *26*, 145.
12. The compound $C_2H_5 . CO_2 . NH . CH_2 . CO . NH . CH_2 . CO . NH_2$ gives the biuret reaction as does an analogous compound, which contains a leucine residue.

13. From still unpublished experiments.
14. A communication to appear shortly will go further into this. We will only mention that a specially purified trypsin was used, completely free from substances giving the biuret reaction as well as from steapsin.[h]
a.   In Hofmeister's usage 'carbon nuclei' represented presumed but inexactly known structural units of which proteins were built.
b.   See Selection 168, Section VII.
c.   This refers to developments originating in the work of Otto Nasse (1873–1903) on the estimation of ammonia produced during alkaline hydrolysis of proteins in the early 1870s. The observation of a difference in the rates at which ammonia was produced (rapid at first, slower afterwards) led him to the concept of 'loosely bound ammonia'. Taking into account the relationship of asparagine to amides and amino acids (*Amidoaminosuccinsäure*), Nasse assumed that the 'loosely bound ammonia' was formed by the acid amide group present in the protein molecule. Cf. O. Nasse, 'Studien über die Eiweisskörper', *Pflügers Arch.*, 6, 589–616. 1872; 7, 139–55. 1873.
d.   Ostensibly, Ferdinand Rose (1809–61), in the paper 'Ueber die Verbindungen des Eiweiss mit Metalloxyden', *Ann.Phys.*, 28 (*104*), 132–42. 1833, was the first to describe the violet colour reaction given by proteins with copper sulphate in alkaline solution but he did not attempt to explain it. The term 'biuret' was proposed in 1847 by Gustav Wiedemann (1826–99) for the substance, in view of its composition, obtained by him on heating urea or urea nitrate. See 'Neues Zersetzungsproduct des Harnstoffs', *J.prakt.Chem.*, 3 N.S. (*42*), 255–6. 1847. Following Hugo Schiff (1834–1915), who began to publish on this topic in 1896, the biuret reaction became connected with substances containing at least two –CO . NH– groups. See 'Ueber Biuret und Biuretreaktion', *Ann.*, 299, 236–66. 1896.
e.   That is, glycocholic acid $C_{23}H_{36}(OH)_3CO . NH > CH_2COOH$ (cholic acid coupled with glycine) and taurocholic $C_{23}H_{36}(OH)_3CO . NH . CH_2CH_2SO_2OH$ (cholic acid in combination with taurine).
f.   That is, uramino acid $R . CH(NH . CO . NH_2)COOH$.
g.   A fat-digesting enzyme preparation obtained from pancreas.
h.   Enzymic extract of animal tissues capable of hydrolyzing hippuric acid into benzoic acid and glycine.

Selection 116

Introduction:

a.   E. Fischer, 'Synthetical chemistry in its relation to biology', *J.Chem.Soc.*, *91*, 1749–65. 1907.
b.   See Selections 78 and 79, Section IV.
c.   E. Fischer and E. Fourneau, 'Ueber einige Derivate des Glykocolls', *Ber.chem.Ges.*, *34*, 2868–84. 1901.
d.   Four years later, in a lecture given to the *Deutsche chemische Gesellschaft*, Fischer emphasized that, apart from the amide formation, it was highly probable that piperazine and hydroxyls of oxyamino acids also contributed to protein structure. See E. Fischer, 'Untersuchungen über Aminosäuren, Polypeptide und Proteïne', *Ber.chem.Ges.*, *39*, 530–610, 1906 (pp. 607–8).

Text:

a.   See Fischer and Fourneau, Introduction to this Selection, note c.
b.   E. Fischer and F. Bergell, 'Ueber die β-Naphtalinsulfoderivate der Aminosäuren', *Ber.chem.Ges.*, *35*, 3779–87. 1902.
c.   Th. Weyl, 'Zur Kenntniss der Seide. I.', *Ber.chem.Ges.*, *21*, 1407–10. 1888.
d.   In 1906 Fischer stated that in coining the term 'peptide' he also desired to make use of the word 'peptone'. Apparently, from the beginning he had expected to find a marked relationship between the artificial peptides and the natural peptones. All subsequent developments reinforced this viewpoint, 'in other words that peptones constitute a mixture of polypeptides hitherto incapable of being separated'. See E. Fischer, 'Untersuchungen über Aminosäuren, Polypeptide und Proteïne', *Ber.chem.Ges.*, *39*, 551. 1906.

Selection 117

Introduction:

a. F. Hofmeister, 'Untersuchungen über Resorption und Assimilation der Nährstoffe. Erste Mittheilung', *Arch.exp.Path.Pharm.*, *19*, 1–33. 1885.
b. R. Neumeister, 'Zur Physiologie der Eiweissresorption und zur Lehre von den Peptonen', *Z.Biol.*, *27*, 309–73. 1890.

Text:

1. M. Jacoby, Z. f. physiol. Chem., vol. XXX, pp. 135 and 166. 1900. – The same, Schmiedeberg's Arch., vol. 46, p. 28. 1901. – The same, Hofmeister's Beiträge zur chemischen Physiologie u. Pathologie. vol. I, p. 51. 1901.
2. M. Jacoby, Schmiedebergs Arch., 46, 28. 1901.
a. Cf. Introduction to Selection 53, Section III, note d.
b. Determined by the nitrogen method developed by Johan Kjeldahl (1849–1900) at the Carlsberg Laboratory, in connection with studies of protein metabolism during barley germination. The idea behind this was to oxidize the organic material with concentrated sulphuric acid and permanganate and convert the nitrogen into ammonium sulphate. On distilling the digest with caustic soda the ammonia formed was collected in standard acid and estimated volumetrically. Its description first appeared in print in 1883. See J. Kjeldahl, 'Neue Methode zur Bestimmung des Stickstoffes in organischen Körpern', *Z.anal.Chem.*, *22*, 366–82. 1883. For an English summary see 'New method for the determination of nitrogen', *Chem. News*, *48*, 101–2. 1883. It may be useful to add that the quantitative method in use for the determination of nitrogen was originally due to J.B. Dumas. See 'Lettre de M. Dumas à M. Gay-Lussac, sur les Procédés de l'analyse organique', *Ann.Chim.*, *47*, 198–215. 1831. For works on the history of organic analysis see Introduction to Selection 103, note b.
c. Reference to per cent saturation with ammonium sulphate.

Selection 118

Introduction:

a. Apparently first clearly described by E. Salkowski (1844–1923) who observed in yeast (1889) and in animal organs (1890) the process which he called 'autodigestion'. In 1903 M. Jacoby (*b.* 1872) introduced the term 'autolysis', which came into use. Cf. F. Lieben, *Geschichte der physiologischen Chemie* (Leipzig and Vienna, 1935), pp. 385–6. See also A. Neuberger, 'History of our understanding of proteolysis', *TIBS*, *6*, 139–41. 1981.
b. An animal was said to have reached the condition of 'nitrogen equilibrium' when it was shown that the nitrogen content of the ingested food could be accounted for by the amount of excreted nitrogen (essentially as urea). Originating in Voit's investigations (cf. Selection 110), the concept of nitrogen equilibrium was to remain for many years central to enquiries about foods and the processes of their utilization in the animal body.

Text:

1. Lunin, Zeitsch. f. physiol. Chem. vol. V. p. 31. 1881.
2. Bunge, Lehrb. der physiol. Chem. p. 116. 1887.
3. Kutscher and Seemann, *Zeitschr.f.physiol.Chem.*, vol. XXXIV, p. 528. 1902.
a. Max Rubner (1854–1932), the leading German nutritionist of the day. While concerned with physiological and social aspects of nutrition, Rubner became convinced that its progress depended on precise data of the caloric content of foods. It is scarcely necessary to remind the reader that this approach reflected the view that the human (animal) body was effectively a steam engine and food its fuel, supplying the necessary energy for its running. The German word for the latter is *Betrieb* which, on combining with the expression *Stoffwechsel* (metabolism) became *Betriebsstoffwechsel*. It was no accident that this term, a combination of designations embodying industrial operations and chemical transformations in living bodies, was widely used by Rubner in his writings. He introduced concepts such as 'isodynamic values' and 'specific dynamic action' of foodstuffs and

undoubtedly they carried a lot of weight. The first term related to the idea that proteins, fats and carbohydrates could replace each other in nutrition in proportion to their caloric values. The second term was associated with observations that purported to demonstrate the production of characteristic amounts of extra heat, on ingestion of each class of foodstuffs. See M. Rubner, 'Die Vertretungswerthe der hauptsächlichsten organischen Nahrungsstoffe im Thierkörper', *Z.Biol.*, *19*, 313–496. 1883; 'Calorimetrische Untersuchungen', ibid., *21*, 250–334; 337–410. 1885; *Die Gesetze des Energieverbrauchs bei der Ernährung* (Leipzig and Vienna, 1902), pp. 327–55.

Selection 119

Introduction:

a.   J.S. Fruton, 'Folin, Otto', *Dictionary of Scientific Biography* (New York, 1972), V, 53.
b.   But this should be seen against the background of prevailing uncertainty regarding the nature of various metabolic end-products found in mammalian urine, on ingestion of proteins. Thus in 1885 Max Rubner, the eminent pupil of Voit, wrote that it was highly probable that substances other than urea were produced. In this manner the possibility of creatinin formation was not to be ignored. See M. Rubner, 'Calorimetrische Untersuchungen', *Z.Biol.*, *21*, 277. 1885.
c.   R.H. Chittenden, *Physiological Economy in Nutrition* (London, 1904), pp. 127–30.

Text:

1.   These names are chosen because they have already been frequently used as indicating the origin of special metabolism products.
2.   Kossel and Dakin: Zeitschrift für physiologische Chemie, 1904, xlii, p. 181.
a.   V.O. Sivén, 'Ueber das Stickstoffgleichgewicht beim erwachsenen Mensch', *Skand.Arch.Physiol.*, *10*, 91–148. 1900; 'Zur Kenntniss des Stoffwechsels beim erwachsenen Menschen, mit besonderer Berücksichtigung des Eiweissbedarfs', ibid., *11*, 308–32. 1901.
b.   118 grams of protein, 500 grams of carbohydrate and 56 grams of fat. See C. von Voit, 'Physiologie des allgemeinen Stoffwechsels und der Ernährung' in L. Hermann, *Handbuch der Physiologie* (Leipzig, 1881), VI/1, pp. 519–25.

Selection 120

Introduction:

a.   For a review of work occurring at the time, see S. Vincent, 'The ductless glands', *Science Progress*, *3*, 410–24. 1908–9. The term 'hormone' was introduced by E.H. Starling in 1905. See 'The Croonian Lectures on the chemical correlation of the functions of the body', *Lancet* (II), 339–41, 423–25, 501–3, 579–83. 1905 (p. 340).
b.   The diet consisted of the zein mixed with two and a half times its weight of carbohydrate (two parts of starch and one part of cane-sugar) and fat (varied in the form of butter, etherial extract of cheese, bacon fat, olive oil, occasional addition of a drop of cod liver oil). Lecithin was added to offset the phosphorus deficiency in zein. The ash of dog biscuits and oats served as the source of salts. Charcoal (macerated filter paper at first) was also added to the mixture.

Selection 121

Introduction:

a.

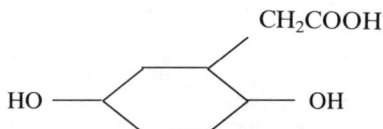

b. M. Wolkow and E. Baumann, 'Ueber das Wesen der Alkaptonurie', *Z.physiol.Chem.*, *15*, 228–85. 1891; W. Falta and L. Langstein, 'Das Entstehen von Homogentisinsäure aus Phenylalanin', ibid., *37*, 513–17. 1903. A by-product of the work on this theme was, effectively, the beginning of biochemical genetics, as set forth in A.E. Garrod's renowned *Inborn Errors of Metabolism* (London, 1909).

c. O. Neubauer and W. Falta, 'Ueber das Schicksal einiger aromatischer Säuren bei der Alkaptonurie', *Z.physiol.Chem.*, *42*, 81–101. 1904.

d. See Selection 145, Section VI.

e. G. Embden, H. Salomon and Fr. Schmidt, 'Über Acetonbildung in der Leber, Zweite Mitteilung: Quellen des Acetons', *Hofmeisters Beitr.*, *8*, 129–55. 1906.

Text:

a. Acetone (ketone) bodies – term used for the following compounds: acetone, acetoacetic acid and β-hydroxybutyric acid.

Selection 122

Introduction:

a. For information on the subject during this period see J.B. Leathes, *Problems in Animal Metabolism* (London, 1906).

b. Knoop's suggestion was that the acetyl group originated in pyruvic acid. He based his suggestion on reports showing the use of pyruvic acid (or its homologue) in the preparation of acetyl amino acids. See A.W.K. de Jong, 'L'action de l'acide pyruvic sur son sel d'ammonium', *Rec.Trav.Chim.*, *19*, 259–310. 1900; E. Erlenmeyer jun. and J. Kunlin, 'Ueber die Synthese des Phenylacetylphenylalanins', *Ann.*, *307*, 163–70. 1899; E. Erlenmeyer jun., 'Ueber die Einwirkung von Ammoniak an ein Gemisch zweier α-Oxonsäuren', *Ber.chem.Ges.*, *36*, 2525–6. 1903.

Text:

1. I should like in this connection to refer to a new work of E.P. Cathcart who has established that, in hunger, N-free carbohydrate diet without addition of fat diminishes the urine nitrogen; N-free fat diet with carbohydrate increases it. He summarizes: The hypothesis is put forward that *the carbohydrates are absolutely essential for endocellular synthetic processes in connection with protein metabolism.* (Journal of physiology, vol. XXXIX, p. 4 [Oct. 1909].) These physiological results perhaps find their explanation in the hypotheses developed here.

Selection 123

Text:

1. A lecture held on 1 November 1910 in the Scientific Association at the Municipal Hospital of Frankfurt-on-Main.

2. We should like in this place to meet the obvious objection that in the experiments described the observed amino acids are set free through a beginning autolysis of the liver. Disregarding that in the blank experiments on the glycogen-poor liver only an inconsiderable formation of amino acids took place, in experiments with addition of definite keto- and oxyacids always only *that* amino acid was isolated, which was theoretically to be expected. Despite the fact that the isolation procedure applied down to the smallest detail, for example to leucine and phenylalanine, was the same.

a. Cf. H. Fellner, 'Über synthetische Bildung von Aminosäuren in der Leber. IV. Mitteilung. Bildung von Alanin und Glykogen', *Biochem.Z.*, *38*, 414–20. 1912.

b. Selection 122.

c. For reference to this work see G. Embden and Fr. Kraus, 'Über Milchsäurebildung in der künstlich durchströmten Leber. I. Mitteilung'. *Biochem.Z.*, *45*, 1–17. 1912. Cf. also Selection 83, Section IV.

Selection 124

Introduction:

a.   Selection 23, Section I.
b.   In the light of his synthetical work on compounds with relatively high molecular weight, Emil Fischer assigned to them a maximal value of about 5,000. Higher findings raised doubts about the homogeneity of the proteins concerned. Cf. E. Fischer, 'Synthese von Depsiden, Flechtenstoffen und Gerbstoffen', *Ber.chem.Ges.*, *46*, 3253–89. 1913.

Text:

a.   In collaboration with I.A. Christiansen, Margrethe Høyrup, S. Goldschmidt and S. Palitzsch.
b.   This is a reference to the work of Frederick George Donnan (1870–1956), who published a seminal paper on membrane equilibria in 1911. Considering the effect of colloids on the distribution of diffusible ions, he established their unequal concentrations on the two sides of the membrane. He pointed out that this produced an osmotic increment additional to the osmotic pressure of the non-diffusible (colloidal) ion alone. Cf. F.G. Donnan, 'Theorie der Membrangleichgewichte und Membranpotentiale bei Vorhandensein von nicht dialysierenden Elektrolyten. Ein Beitrag zur physikalisch-chemischen Physiologie', *Z.Elektrochem.*, *17*, 572–81. 1911.
c.   The equation $P = c.R.T.$ was developed by J.H. van't Hoff by drawing an analogy between gaseous pressure and osmotic pressure of relatively dilute solutions. See his 'Die Rolle des osmotischen Druckes in der Analogie zwischen Lösungen und Gasen', *Z.physik.Chem.*, *1*, 481–508. 1887.

Selection 125

Introduction:

a.   Cf. Selection 112.
b.   See Selection 124, note b.

Selection 126

Introduction:

a.   D. Jordan Lloyd, *Chemistry of the Proteins and its Economic Applications*, Introduction by Sir Frederick Gowland Hopkins (London, 1926), pp. 32–3.

Text:

1.   T. Svedberg and B. Sjögren, Journ. Amer. Chem. Soc. *51*, 3594 (1929).
2.   T. Svedberg and J.B. Nichols, Journ. Amer. Chem. Soc., *48*, 3081 (1926).
3.   J.B. Nichols, unpublished research.
4.   T. Svedberg and R. Fåhraeus, Journ. Amer. Chem. Soc., *48*, 430 (1926).
5.   T. Svedberg and J.B. Nichols, Journ. Amer. Chem. Soc., *49*, 2920 (1927).
6.   T. Svedberg and B. Sjögren, Journ. Amer. Chem. Soc., *50*, 3318 (1928).
7.   T. Svedberg and B. Sjögren, Journ. Amer. Chem. Soc., *52*, 279 (1930).
8.   T. Svedberg and A.J. Stamm, Journ. Amer. Chem. Soc., *51*, 2170 (1929).
9.   B. Sjögren, unpublished research.
10.   T. Svedberg and T. Katsurai, Journ. Amer. Chem. Soc., *51*, 3573 (1929).
11.   T. Svedberg and N.B. Lewis, Journ. Amer. Chem. Soc., *50*, 525 (1928).
12.   T. Svedberg and E. Chirnoaga, Journ. Amer. Chem. Soc., *50*, 1399 (1928).
13.   T. Svedberg and F.F. Heyroth, Journ. Amer. Chem. Soc., *51*, 550 (1929).
14.   T. Svedberg and F.F. Heyroth, Journ. Amer. Chem. Soc., *51*, 539 (1929).
15.   The sedimentation constant and the frictional coefficient for edestin are uncertain on account of the high salt concentration of the solutions investigated.
a.   Stoke's law is an empirical one. It states that $f = 6\pi\eta r$, where f is the molar frictional coefficient, η is the viscosity of the solvent, and r is the radius of the particle.
b. $1\mu\mu = 1$ millimicron $= 10^{-9}$ m. The term formerly used for nanometer (nm).

Selection 127

Introduction:

a. The main aim of the Department of Physical Chemistry (established in 1920) was the study of proteins. It was headed by Edwin Joseph Cohn (1892–1953) whose researches concentrated on elucidating the nature of their solubility at that time. Among other things, he showed that the linear equation $\log s = \alpha m + \beta$ described quantitatively the salting out of proteins. In this widely employed equation s constituted the solubility of the protein, m the molecular concentration of the salt, α the proportionality factor, and β another constant. See E.J. Cohn, 'The physical chemistry of the proteins', *Physiol.Revs.*, *5*, 349–437. 1925 (pp. 413–14). Cf. also J.T. Edsall, 'Edwin J. Cohn and the physical chemistry of proteins', *TIBS*, 6, 336–7. 1981.

b. The idea that biological materials consist of crystalline, birefringent particles ('molecules', 'micelles') was conceived of and developed by Carl Wilhelm (von) Nägeli (1817–91) from the late 1850s onwards. For a careful historical treatment of knowledge stemming from Nägeli's inquiries into the submicroscopical structure of biological substances, including birefringence of flow (apparently first observed by Clerk Maxwell in 1847), see J.S. Wilkie, 'Nägeli's work on the fine structure of living matter', *Ann.Sci.*, *16*, 11–40, 171–207, 209–39. 1960; ibid., *17*, 27–62, 1961. For an absorbing personal account of the work presented in this selection see J.T. Edsall, 'Double refraction of "myosin" in flowing solutions', *TIBS*, *5*, 228–30. 1980. It took two years to complete before it was published in two extensive papers.

c. A.L. von Muralt and J.T. Edsall, 'Studies in the physical chemistry of muscle globulin. III. The anistropy of myosin and the angle of isocline', *J.Biol.Chem*, *89*, 315–50. 1930.

Text:

a. H. Stübel, 'Die Ursache der Doppelbrechung der quergestreiften Muskelfaser', *Pflügers Arch.*, *201*, 629–45. 1923; H. Stuebel and T.-Y. Liang, 'Untersuchungen ueber die Veraenderungen der Doppelbrechung bei verschiedenen Starrezustaenden des Muskels', *Ch.J.Physiol.*, *2*, 139–50. 1928.

Selection 128

Introduction:

a. For material on the early X-ray crystallography and its development see the substantial commemorative volume, P.P. Ewald (ed.), *Fifty Years of X-Ray Diffraction* (Utrecht, 1962).

b. By contributing materially to the elucidation of the carbon skeleton of sterols in 1932, Bernal demonstrated X-ray crystallography's potential in clarifying the structure of complex organic substances. See J.D. Bernal, 'Carbon skeleton of the sterols', *J.Soc.Chem.Ind.*, *51*, 466. 1932 and Selection 151, Section VI. Much later Bernal stated that while the selection of sterols for X-ray structure analysis was not accidental, the establishment of the structure was:

At the same time in close connection with the Biochemical Laboratory of Professor Hopkins, work was started, first on amino acids and then on sterols. There, owing to what was effectively a happy chance of being able to discover, by X-rays in the first place, the correct carbon skeleton of the sterols, Bernal was able to unify the structure of these important bodies which were then of particular interest in connection with vitamins and sex hormones.

See J.D. Bernal, chapter 17 in Ewald (ed.), *Fifty Years of X-Ray Diffraction*, p. 378. Bernals life and work is examined by Dorothy M.C. Crowfoot in *Biog.Mem.F.R.S.*, *26*, 17–84. 1980. For a shorter appreciation see C.P. Snow, 'J.D. Bernal, A personal portrait' in M. Goldsmith and A. Mackay (eds), *The Science of Science Society in Technological Age* (London–Toronto, 1964), pp. 19–29 (produced to commemorate the 25th anniversary of the publication of Bernal's *Social Function of Science* (1939)). M. Goldsmith's *Sage A Life of J.D. Bernal* (London, 1980), weak on analysis and strong on gossip, is hardly the 'masterly biography' (blurb) it claims to be.

c. W.T. Astbury's *Fundamentals of Fibre Structure* (London, 1933) provides valuable information regarding the state of X-ray fibre research around 1930, with a perceptive appreciation of its industrial significance. Astbury's scientific activities are dealt with interestingly by Bernal in *Biog.Mem.F.R.S.*, *9*, 1–35. 1963.

d. Looking back, Dorothy Crowfoot Hodgkin presents an appraisal of the attitudes and opinions of researchers involved in this work and of the reactions of other scientists to it in 'Crystallographic measurements and the structure of protein molecules as they are', *Ann.N.Y.Acad.Sci.*, *325*, 121–45. 1979. See also 'Discussion of the paper', ibid., 145–8. Edited by P.R. Srinisivan, J.S. Fruton and J.T. Edsall, this particular volume of *Annals of the New York Academy of Sciences*, with the title *The Origins of Modern Biochemistry: A retrospect on proteins*, comprizes in collected form other noteworthy papers pertinent to the history of proteins. They were given to a conference held by the New York Academy of Sciences between 31 May and 2 June 1978. The volume was critically reviewed by Pnina Abir-Am, *Brit.J.Hist.Sci.*, *15*, 301–5. 1982. For the rejoinder see N.W. Pirie, 'Letter to the Editor', *Brit.J.Hist.Sci.*, *16*, 273–4. 1983.

Text (i):

1. G.L. Clark and K.E. Korrigan (*Phys.Rev.*, (ii), *40*, 639; 1932) describe long spacings found from crystalline insulin, but no details have been published.
2. J.H. Northrop, *J.Gen.Physiol.*, *13*, 739; 1930.
a. Glenn Allan Millikan (1906–47) – American physiologist.
b. John St. L. Philpot (1907–) – British biochemist.

Text (ii):

1. W.T. Astbury, *Trans. Faraday Soc.*, *29*, 193; 1933. W.T. Astbury and A. Street, *Phil. Trans. Roy. Soc.*, *A*, 230, 75; 1931. W.T. Astbury and H.J. Woods, NATURE, *126*, 913, Dec. 13, 1930. *Phil. Trans. Roy. Soc.*, *A*, 232, 333; 1933. W.T. Astbury and W.R. Atkin, NATURE, *132*, 348, Sept. 2, 1933.
2. W.T. Astbury and H.J. Woods, NATURE, *127*, 663, May 2, 1931.
3. J.St. L. Philpot and Inga-Britta Eriksson-Quensel, NATURE, *132*, 932, Dec. 16, 1933.
4. W.T. Astbury and T.C. Marwick, NATURE, 130, 309, Aug. 27, 1932.

Selection 129

Introduction:

a. Selection 161, Section VII.
b. Cf. 'The facts as revealed have just that degree of unexpectedness – if I may use the phrase – which was to be expected in a biochemical phenomenon. I often find myself compelled to assert that though biochemical events are, of course, limited by chemical possibilities, they are not safely to be predicted by chemical probabilities, even when these are strong.' From Sir F. Gowland Hopkins' comments on the published results by Krebs and Henseleit in his Anniversary Address as President of the Royal Society on 30 November 1932, *Proc.Roy.Soc. (B)*, *122*, 175. 1932.
c. There is little doubt that the discovery of the ornithine cycle in the liver was the beginning of Krebs' concern with the general significance of cycles in living matter. It appears, for the first time, in the article 'Cyclic processes in living matter', *Enzymologia*, *12*, 88–100. 1946–8; and for the last time, shortly before his death in J.E. Baldwin and H. Krebs, 'The evolution of metabolic cycles', *Nature*, *221*, 381–2. 1981. Historians of biochemistry, in particular, have cause to appreciate Krebs' insistence on analysis of the process that led to the discovery of the urea cycle. For a knowledgeable attempt to reconstruct its history based on the study of original notebooks and also in the light of their retrospective evaluation by Krebs, see F.L. Holmes, 'Hans Krebs and the discovery of the ornithine cycle', *Fed. Proc.*, *39*, 216–25. 1980.

Text:

1. Felix and Röthler, this journal, *143*, 134 (1925).
2. $R = CH_2 . CH_2 . CHNH_2 . COOH.$
3. This journal, *41*, 321 (1904); *42*, 181 (1904).

4.   A. Clementi quoted after Hunter and Dauphinée.
5.   A. Hunter and J.A. Dauphinée, Proc.Roy.Soc.Subl. -B. *97*, 227 (1925).
6.   Y. Koga and S. Odake, J. Tokyo Chem.Soc. *35*, 519 (1914), quoted after Wada.
7.   M. Wada, Biochem.Z. *224*, 420 (1930).

Selection 130

Introduction:

a.   See Selection 118, note a.

Text:

a.   H.H. Mitchell, W.B. Nevens and F.E. Kendall, 'The relation between the endogenous catabolism and the non-protein constituents of the tissues', *J.Biol.Chem.*, *52*, 417–37. 1922.

Selection 131

Introduction:

a.   Selections 121 and 122.
b.   H.A. Krebs, 'Metabolism of amino-acids. III. Deamination of amino-acids', *Biochem.J.*, *29*, 1620–44. 1935. Regarding stereoisomers the *d*- and *l*- designations were commonly used at that time.
c.   D.M. Needham, 'A quantitative study of succinic acid in muscle. III. Glutamic and aspartic acids as precursors', *Biochem.J.*, *24*, 208–27. 1930.

Text:

1.   *Bull.Biol. et Med.Exper., Moscou*, *3*, 230 (1937). *Biochimia*, 2, 242. (1937).
2.   Herbst and Engel, *J.Biol.Chem.*, *107*, 505 (1934). Herbst, *J.Amer.Chem.Soc.*, *58*, 2239 (1936).
3.   Koranyi und Szent-Györgyi, *Deutsch. Med. Wochenschr.*, No. 27 (1937).
a.   A.E. Braunstein and M.G. Kritzmann, 'Überden Ab – und Aufbau von Aminosäuren durch Umaminierung,' *Enzymologia*, 2, 129–46. 1937.

Selection 132

Introduction:

a.   Selection 120.
b.   Lafayette Benedict Mendel (1872–1935), Professor of Physiological Chemistry at Yale. On Osborne see Selection 108, note c.
c.   T.B. Osborne and L.B. Mendel (with the co-operation of E.L. Ferry and A.J. Wakeman), 'Amino-acids in nutrition and growth', *J.Biol.Chem.*, *17*, 325–49. 1914 (p. 327).
d.   H. Ackroyd and F.G. Hopkins, 'Feeding experiments with deficiencies in the amino-acid supply: arginine and histidine as possible precursors of purines', *Biochem.J.*, *10*, 551–76. 1916 (p. 575). Harold Ackroyd (1877–1917), who worked at the Institute for the Study of Animal Nutrition at Cambridge, did not see the paper in published form – he was killed in action. (He was a recipient of the Victoria Cross, the highest British military award for bravery.)
e.   W.C. Rose, 'How did it happen?', *Ann.N.Y.Acad.Sci.*, *325*, 229–34. 1979 (p. 234). About this volume of *Annals of the New York Academy of Sciences*, see Introduction to Selection 128, note d.
f.   W.C. Rose and C.J. Cox, 'The relation of arginine and histidine to growth', *J.Biol.Chem.*, *61*, 747–73. 1924; 'Further experiments on the alleged interchangeability of arginine and histidine in metabolism', ibid., *68*, 217–23. 1926.

Text:

1.   Meyer, C.E. and W.C. Rose. Unpublished data.
2.   Behrens, O.K. and W.C. Rose. Unpublished data.
3.   Hancock, E. and W.C. Rose. Unpublished data.

Selection 133

Introduction:

a. F.G. Hopkins, 'The dynamic side of biochemistry', *Rep.Brit.Ass.*, 652–68. 1923 (p. 653). See also Selection 174, Section VII.
b. Cf. C. Bertagnini, 'Note sur la formation de l'acide nitro-hippurique dans l'économie animale', *Comp.Rend.*, *31*, 490–1. 1850.
c. Hopkins, *Rep.Brit.Ass.*, 655. 1913.
d. G. Hevesy, 'The absorption and translocation of lead by plants. A contribution to the application of the method of radioactive indicators in the investigation of the change of substance in plants', *Biochem.J.*, *17*, 439–45. 1923. See also, by the same author, 'Historical sketch of the biological application of tracer elements', *Cold Spr.Harb.Symp.*, *13*, 129–50. 1948.
e. H.C. Urey, 'Deuterium and its compounds in relation to biology', *Cold Spr.Harb.Symp.*, *2*, 47–56. 1934.
f. Cf. R. Schoenheimer, *The Dynamic State of Body Constituents* (Cambridge, Mass., 1942); see also S. Ratner, 'The dynamic state of body proteins', *Ann.N.Y.Acad.Sci.*, *325*, 189–209. 1979.

Selection 134

Introduction:

a. D. Hodgkin, 'Discussion of the paper' by N.W. Pirie, 'Purification and crystallization of proteins', *Ann.N.Y.Acad.Sci.*, *325*, 33. 1979. See also J.T. Edsall's 'Dorothy Wrinch, the cyclol hypothesis, and the X-ray crystallographers' in D. Bearman and J.T. Edsall (eds), *Archival Sources for the History of Biochemistry and Molecular Biology. A Reference Guide and Report* (Ann Arbour, 1980), pp. 6–8. There is a thought-provoking biographical sketch of Wrinch Prima G. Abir-Am in P.G. Abir-Am and D. Outram (eds.), with a foreword by M.W. Rossiter, *Uneasy Careers and Intimate Lives, Women in Science, 1789–1979* (Rutgers University Press, 1987) pp. 239–80.
b. See Selection 180, Section VII. The story of the Institute is examined in P. Abir-Am, 'The discourse of physical power and biological knowledge in the 1930s: a reappraisal of the Rockefeller Foundation's 'policy' in molecular biology', *Social Studies of Science*, *12*, 341–82. 1982.
c. R.A. Gortner, jun. and W.A. Gortner (eds), *Outlines of Biochemistry*, 3rd edn (New York–London, 1949), p. 368.
d. D. Jordan Lloyd, 'Recent developments in our knowledge of the protein molecule' in J. Needham and D.E. Green (eds), *Perspectives in Biochemistry* (Cambridge, 1937), pp. 32–3.
e. J.B. Bateman in R. Höber, *Physical Chemistry of Cells and Tissues* (London, 1945), pp. 140–1. Since then alkaloids from ergot have been found to contain cyclol structures. See R.L.M. Synge, '75 years of the Fischer-Hofmeister theory of protein structure' in H. Zahn and K. Ziegler (eds), *Deutsches Wollforschungsinstitut an der Technischen Hochschule Aachen e.V.: 25 jähriges Jubiläum am 1. und 2. April 1977. Fachvorträge und Ergänzungen der Festschrift* (Aachen, 1977), pp. 17–25.

Text:

1. Svedberg, *Science*, *79*, 327 (1934).
2. International Tables for the Determination of Crystal Structure.
3. Astbury and Woods, *Phil. Trans. Roy. Soc.*, *232*, 333 (1933).
4. Jordan Lloyd, *Biol.Rev.*, 7, 256 (1932).
5. Gorter, *J.Gen.Phys.*, *18*, 421 (1935); *Amer.J. Diseases of Children*, *47*, 945 (1934).
6. Gorton and van Ormondt, *Biochem.J.*, *29*, 48 (1935).
7. Schulman and Rideal, *Biochem.J.*, *27*, 1581 (1933).
8. Bernal and Megaw, *Proc.Roy.Soc.*, *A*, *151*, 384 (1935).
9. Bergmann, *J.Biol.Chem.*, *110*, 471 (1935).
10. Astbury, Cold Spring Harbor Symposia on Quantitative Biology, 2, 15 (1934).
11. Sjogren and Svedberg, J. Amer. Chem. Soc., *52*, 3650 (1930).
12. Hammersten et al., *Biochem.Z.*, *264*, 272 and 275 (1933).
13. Bergmann et al., *J. Biol. Chem.*, *109*, 325 (1935).

Selection 135

Introduction:

a. D.M. Wrinch, 'Cyclol hypothesis and 'globular' proteins', *Proc.Roy.Soc.* (*A*), *161*, 505–24. 1937.

b. M. Bergmann and C. Niemann, 'On blood fibrin. A contribution to the problem of protein structure', *J.Biol.Chem.*, *115*, 77–85. 1936. Idem, 'On the structure of proteins: cattle hemoglobin, cattle fibrin and gelatin', ibid., *118*, 301–14. 1937. Bergmann and Niemann were not alone in contemplating this type of structure. In 1934, for example, in a seminal paper Astbury, discussing the X-ray photographs of collagen and gelatin, suggested that 'the amino-acid residues are somehow grouped in sets of three'. Drawing his evidence from recent chemical analyses of gelatin, Astbury stated: 'The residues of the two chief acids, glycine and oxyproline, account, respectively, for about one third and one ninth of the total number of residues, that is to say, every third residue could be a glycine residue and every ninth an oxyproline residue.' W.T. Astbury, 'X-ray studies of protein structure', *Cold Spr.Harb.Symp.*, *2*, 15–27. 1934 (p. 22).

c. See T.W.J. Taylor, 'The chemistry of the proteins and related substances', *Ann.Rep.Prog.Chem.*, *34*, 302–26. 1937 (p. 316). Cf. also discussion of C. Nieman's paper 'Chemistry of protein structure', *Cold Spr.Harb.Symp.*, *6*, 58–66. 1938.

Text:

1. Vickery, H.B. and Block, R.J., *J.Biol.Chem.*, *95*, 105 (1931).
2. Astbury, W.T., *Trans.Faraday Soc.*, *29*, 193 (1933).

a. It should be borne in mind that the Bergmann and Niemann scheme of calculation derived essentially from the stimulus given by Astbury in his already mentioned paper of 1934, where he introduced the concept of the 'mean (average) weight of an amino-acid residue'. Astbury's procedure may be summarized as follows. Let R be the residue weight and M the molecular weight of the amino acid, then R = M-18. Taking into account known chemical analyses and denoting the weight of the amino acid from 100 g of gelatin as W. Astbury calculated that the latter contained $\frac{W}{M}$ = 0.875 + 0.169 = 1.044 g residues of the average amino acid residue. From this he obtained $\frac{100}{1.044}$ = 96 as the mean weight of an amino acid residue in gelatin.

   In fact the value 1.044 included the total obtained from known analytical data (0.875) and the total related to amino acids not accounted for by analysis and assumed to be monamino acids (0.169 = $\frac{2.36}{14}$). The figure was deduced indirectly from consideration of the total nitrogen content of the amino acids found (15.64) and that of the protein molecule (18.00): 18 – 15.64 = 2.36. See Astbury, *Cold Spr.Harb.Symp.*, *2*, 21–2. 1934.

b. Obtained by dividing the values in column 3 by 1,190. Thus $\frac{0.584}{1.190}$ = 0.49 = $\frac{1}{2}$; $\frac{0.2966}{1.190}$ = 0.249 = $\frac{1}{4}$ and so on.

Selection 136

Introduction:

a. T. Svedberg, 'The ultra-centrifuge and the study of high-molecular compounds', *Nature*, *139*, 1051–62. 1937. See also summary of this: 'The ultra-centrifuge in biochemistry', ibid., 1046. 1937. For an earlier account see Selection 126.

b. A. Neuberger, 'Chemical criticism of the cyclol and frequency hypothesis of protein structure', *Proc.Roy.Soc.* (*B*), *127*, 25–6. 1939.

c. N.W. Pirie, 'Amino-acid analysis and protein structure', *Ann.Rep.Prog.Chem.*, *36*, 351–3. 1939 (p. 351).

d. A.H. Gordon, A.J.P. Martin and R.M.L. Synge, 'A study of the partial acid hydrolysis of some proteins, with special reference to the mode of linkage of the basic amino-acids', *Biochem.J.*, *35*, 1369–87. 1941.

e.   Albrecht Kossel (1843–1927) contributed to the development of nucleoprotein chemistry. Cf. also Selection 173, Section VII.

f.   A.C. Chibnall, 'The road to Cambridge', *Ann.Rev.Biochem.*, *35*, 1–22. 1966 (p. 13). See also R.C. Olby's 'Essay review' of *The Origin of Modern Biochemistry: A retrospect on proteins* (cf. Introduction to Selection 128, note d, *Hist.Phil.Life Sci.*, *4*, 159–68. 1982 (p. 161)).

g.   This would have to take note, for example, of the 'industrial' connection in the study of the fundamental aspects of the subject. The work of Martin and Synge and their collaborators was developed at Leeds in a laboratory serving industrial interests (Wool Industries Research Association).

Text:

References

Bailey, K. (1942) *Biochem. J.*, *36*, 140.
Bergmann, M. and Niemann, C. (1938) *J.Biol.Chem.*, *12*, 577.
Gordon, A.H., Martin, A.J.P. and Synge, R.L.M. (1941) *Biochem.J.*, *35*, 1369.
Martin, A.J.P. and Synge R.L.M. (1941 a,b) *Biochem.J.*, *35*, 91, 1358.
Rittenberg, D. and Foster, G.L. (1940) *J.Biol.Chem.*, *133*, 737.
Ussing, H.N. (1939) *Nature, Lond.*, *144*, 977.

Section VI: LIPIDS

a. From the 1780s to c.1890: analysis. Saponification and absorption. Lipase.

Selection 137

Together with carbohydrates and proteins, lipids have been recognized and studied as the third major group of substances fundamentally involved in biochemical events. Regarding their chemical properties they certainly form a more heterogeneous group of compounds than the carbohydrates and proteins. Among lipids substances have been progressively included, such as oils and fats, waxes, phosphatides, cerebrosides and products containing sterols. This contributed a good deal of confusion in their assignation and nomenclature.[a]

In 1783[b] the great Swedish chemist, Carl Wilhelm Scheele (1742–86), published a paper in the *New Transactions* of the Swedish Academy of Sciences pointing out that oils and fats contain a substance with a sweetish taste. Without actually naming it, Scheele obtained glycerol by treating oils and fats with 'calx of lead' (lead oxide). It is because Scheele recognized that this particular compound was a constant constituent of naturally occurring oily and fatty substances that it is fitting to begin this section with this discovery.

As shown below, Scheele interpreted his results in phlogistic terms. According to Georg Ernst Stahl (1659–1734), who elaborated the doctrine of phlogiston, most substances of whatever origin contained phlogiston, some more, some less, which they released on burning. It fully permeated vegetable and animal fats but no parts of vegetables and animals were without it. The exception was if they contained adventitious water. As to mineral bodies, phlogiston was not, or hardly, present in water, common salt, pure vitriol and light-coloured sand. But coal and pitch contained a good deal and so did sulphur. The same applied to combustible metals.

On calcination (heating, burning), combustible metals gave off their phlogiston and changed into products called 'calces'. If the combustion of metals involved 'dephlogistication', the reduction of calces was interpreted as 'phlogistication'. Stahl's suggestion that reduction of ores to metals, and their combustion, could be envisaged as reversible processes, as it were, represented a major theoretical advance. But Stahl went beyond that in suggesting that phlogiston not only embodied the subtle matter of combustibility but also constituted *the* common material principle of and, indeed, the *link* between the vegetable, animal and mineral kingdoms. In fact, he conceived of a global circulation of phlogiston uniting inanimate and animate nature.

Directly, unhindered, indeed instantly, phlogiston was transferred from vegetables and animals to minerals and metals buried in the ground. Thence, in turn, phlogiston passed back to plants, to animals deriving their nourishment from plants, and to flesh-eaters. It was in this connection that Stahl offered a remarkable observation regarding chemical operations occurring in living organisms. He wrote in 1718 that growing plants were continuously building up fats from accumulated phlogiston, and also decomposing them. 'Fatsynthesis' was carried out only by plants and could not be achieved artificially. Neither did it happen in animals whose fats were ultimately of vegetable origin.[c]

C.W. Scheele, *The Collected Papers* (London, 1931),[a] p. 255

'Experiments relating to the peculiar saccharine principle in expressed oils and fats'[b]

pp. 255–6

It is probably still unknown that all fats and expressed oils are endowed by nature with a sweet substance that is distinguished by its special behaviour and properties from the generally known saccharine principles which the vegetable kingdom produces. This sweet substance shows itself when these oils are boiled with calx of lead and water; for there is then produced a hard mass which is known in apothecaries' shops under the name of *emplastrum simplex*.[c] This plaster is a species of soap which, although it cannot be dissolved in water, is nevertheless soluble in part in strong spirit of wine[d] by aid of trituration.

The actual method to obtain this sweet substance is this: one part of finely pulverised calx of lead, or litharge, is boiled with two parts of fresh olive oil and one part of water, during continuous stirring with a spatula, until all the calx of lead is dissolved by the oil; then one part of water is again added and it is boiled anew for a couple of minutes, after which the boiler is taken from the fire and set aside. When the plaster has cooled, the water which lies above it and which contains the sweet substance, is poured off: this water is filtered and is boiled down until the residue becomes as thick as syrup.

pp. 257–8

Hereupon I examined the behaviour of milk fat, or butter, with calx of lead. The calx dissolved very rapidly, the butter became hard like plaster, and here also I obtained the same sweet substance. I now boiled fresh hog's lard with this calx and the result was the same.

Concerning the nature and properties of this sweet substance, I examined it, 1st, by the method of crystallisation: I evaporated it to a proper consistence and although it stood several months in a cold place, still no crystals were formed.

2. I poured a little of this thick material upon a tea saucer and placed it upon red-hot charcoal, when the vapour took fire and the substance burned with a blue sulphur flame and left a light charcoal behind.

3. Its behaviour in the method by distillation is remarkable. Pure water goes over first, whereupon the boiling slackens and although the fire is increased no more will go over; but when the retort becomes almost red-hot, the syrup enters anew into ebullition and the receiver becomes filled with vapour which afterwards collects again in drops. It is thick like syrup, and retains its sweet taste, although it is mixed with some empyreumatic matter, like spirit of tartar.[e] When the heat is at last increased till the bottom of the retort is red-hot, the receiver becomes filled with a brown vapour which, after it has condensed, yields a black oil with some black liquid of a bitter taste and very penetrating smell. The residuum in the retort was a light shining charcoal which did not contain any traces of lead.

4. I mixed one part of this syrup with four parts of water and placed it in a luke-warm position where it remained for several months; but the mixture had not undergone any fermentation or any other change during this time; neither was tincture of litmus changed by it.

5. To ascertain what the acid of nitre would do to it I mixed this syrup with three times as much of this acid and distilled it again therefrom; and, as the acid became much phlogisticated by this method, I abstracted the same quantity several times, yet so that each time pure acid of nitre[f] was taken; when eventually, after the ninth distillation, what was left behind in the retort completely thickened to a considerable quantity of crystals which behaved like common acid of sugar.[g]

6. I mixed this sweet substance into an alkaline tincture (*tinct. tartari*)[h] whereby no visible change occurred; but if, on the contrary, such a tincture is mixed with sugar syrup, or honey, the sugar attracts the dissolved alkali from the spirit of wine and falls with it to the bottom as a stiff mucilage.

Hence it is seen from this that all fatty oils contain a sweet substance which differs from sugar and honey in these respects: (1)That it cannot be brought to crystallisation. (2) That the sweet substance can not only withstand much stronger heat before it is destroyed, but also that it passes over into the receiver in part unchanged, with retention of its sweetness. (3) That it cannot enter into any fermentation, and (4) that it mixes with spirituous alkaline solutions. All these special differences appear to arise from a great quantity of phlogistic material which this sweet substance from fat contains, as is proved by the great amount of acid of nitre that is required for its dephlogistication before the acid of sugar shows itself; because much less acid of nitre is required for ordinary sugar.

Selection 138

If anyone deserves the title 'father of the chemistry of fats', it is Michel Eugène Chevreul (1786–1889).[a] Seeking to comprehend the chemical basis of soap-making he began to investigate it in 1811. On this subject he published a series of nine papers entitled 'Recherches chimiques sur plusieurs corps gras, et particulièrement sur combinaisons avec les alkalis' in *Annales de chimie* and *Annales de chimie et de physique* between 1813 and 1817.

In 1823 when his *Recherches chimiques sur les corps gras d'origine animale* appeared, containing an expanded summary of this earlier work, it was widely acknowledged as a major achievement of scientific chemistry. What he did was to extend the laws of constant and multiple proportions to the study of fatty substances, and in doing so he converted it into the then most advanced branch of the chemistry of substances of biological origin. Primarily undertaken to understand the existing practice of soap-making, Chevreul's results became a substantial factor in the development of the manufacture of soaps and candles on a large scale.[b]

Before Chevreul's findings it was held that the chemical process occurring in soap-making was merely the combination of fats with alkaline substances such as potash and soda. Moreover, under the impact of the oxidation theory, it was assumed that the presence of oxygen was required.

Chevreul established that saponification involved the decomposition of fats into an acidic component (fatty acid) and Scheele's sweet principle, which he named *glycérine*. He also demonstrated that the production of soaps could take place *in vacuo*. In the light of his studies Chevreul came to regard soaps as salts of fatty acids and fats as a group of ether-like compounds[c] composed of fatty acids and glycerine. Chevreul found that in certain fatty substances glycerine was replaced by other compounds such as 'ethal' (cetyl alcohol) in spermaceti. As to the crystalline compound obtained from human biliary calculi, considered at the time to be a fatty body, Chevreul showed that it could not saponify. In 1816 he named it *cholestérine*, derived from the Greek words for bile (*chole*) and solid (*stereos*).

The following selection is from the English version of Chevreul's first paper describing the isolation of the fatty acid, which he first called 'margarine'. Later renamed 'margaric acid', it eventually proved to be an equimolecular mixture of stearic and palmitic acids.

[M.E.] Chevreul, *Phil.Mag.*, *44*, 193. 1814

'Chemical inquiries into the nature of several fatty substances, and particularly on their combinations with the alkalis. Of a new substance, called *margarine*, obtained from the soap made from the fat of pork and potash'

p. 194

When we immerse soap made of pork grease and potash in a large quantity of water, one part is dissolved, while another is precipitated in the form of several brilliant pellets, which I shall call mother of pearl substance.

p. 195

The fatty substance, separated from the salifiable bases, was dissolved in boiling alcohol: on cooling, it was obtained crystallized and very pure, and in this state is was examined. As it has not been hitherto described, it ought to be distinguished from other substances by a peculiar name: consequently I purpose to call it *margarine*, from the Greek word signifying pearl, because

SECTION VI: LIPIDS   383

one of its characters is to have the appearance of mother of pearl, which it communicates to several of the combinations which it forms with the salifiable bases.

p. 197

I now come to speak of one of the most remarkable combinations of margarine, being that which it forms with potash. It truly characterizes this substance, and leads me to examine some points of the chemical doctrines respecting acidity.

p. 205

. . . turnsole appears to be adopted by all chemists to detect acidity.[a] . . . we shall be compelled to rank margarine among the [acids]; since it reddens turnsole, takes up potash from the carbonic acid, and since its combinations with this base have the greatest analogy with the salts. If it be objected that its composition removes it too far from the series of the acids, we may cite a single example, that of sulphuretted hydrogen, which evidently possesses the characters of acidity, as M. Berthollet has proved. All the chemists, in regarding this body as an acid, have, I believe, established that in the chemical system the analogy of properties has been consulted in preference to that of composition.[b]

Selection 139

So far the discussion has focused on the chemical rather than the biochemical aspects of fats. A significant breakthrough in this area came from Claude Bernard's (1813–78) studies of the digestion of fats in the late 1840s. In the course of this work Bernard established that pancreatic juice played an essential part in the emulsification of fats, including their splitting.

Bernard's studies of the action of pancreatic juice on fats yielding fatty acids and glycerine originated in his previous digestion researches which largely involved saliva and gastric juice. Indeed, in 1845, in collaboration with the chemist, Charles Louis Barreswil (1817–70), he arrived at the conclusion that while the digestive action of saliva, gastric juice and pancreatic juice was, in general, due to a common organic principle present in these fluids, their specific activity was related to the nature of the conditions under which it took place (acidity, alkalinity).

In opposition to this view, Louis Mialhe (1807–86) suggested in 1846 that digestion of each of the three major types of food was brought about by a specific ferment. He held that diastase, specifically, was involved in the digestive breakdown of starchy foods and pepsin in that of protein ('albuminoid') foods, converting them into 'albuminoses'.[a] But there was no known ferment that was or could be associated with the digestion of fats at that time.

It is worth pointing out at this stage that Bernard's investigations into the digestive function on pancreatic juice were greatly facilitated early in 1848 by his superior technique in producing a pancreatic fistula. It enabled him to obtain samples of natural pancreatic juice, without unduly distressing the animal. It also led Bernard, in co-operation with Barreswil and another chemist, Fréderic Margueritte, to the discovery of its fat-splitting property in the spring of 1848.

It is appropriate to direct attention to the ideas about the nature of pancreas that were current when Bernard began to be preoccupied with the effect of the pancreatic juice on fats. As he pointed out in the introductory remarks of the paper from which the excerpt below is taken, for a long time anatomists had considered the pancreas to be an abdominal salivary gland. There could be no doubt, Bernard continued, that this comparison was wrong and was far from revealing the specific function on the pancreatic juice in relation to the digestion of fats. So, effectively reversing his former standpoint on the existence of a common organic principle in saliva, gastric juice and pancreatic juice, Bernard moved closer to the position of Mialhe.[b]

Cl. Bernard, *Ann.Chim.*, 25, 474. 1849

'Recherches sur les usages du suc pancréatique dans la digestion' (Researches on the uses of the pancreatic juice in digestion)

pp. 479–80

I have said, in beginning, that the pancreatic juice was designed, to the exclusion of all the other intestinal fluids, to modify in a special manner the neutral fatty matters which can be encountered in the food. Nothing is so easy to demonstrate as making the pancreatic juice act directly in a test-tube on the fatty matters.

When one mixes at the temperature of 38 to 40 degrees pancreatic juice with oil, butter or fat, one sees that the fatty matter finds itself instantly emulsified in the most complete manner by the pancreatic juice. There results a liquid whitish and creamy similar to chyle. On examining more closely the characteristics of this emulsion, it soon becomes evident that under the influence of the pancreatic juice the fatty matter has not simply divided up and emulsified, but that it has been moreover chemically modified. In fact, at the moment of the mixing of the neutral fatty matter with the alkaline pancreatic juice, the whole possesses a very clear alkaline reaction, whilst that soon after the liquid acquires a reaction very manifestly acid. In the laboratory of M. Pelouze,[a] with Messieurs Barreswil and Margueritte, we have examined these products, and we have easily recognized that the fat had been decomposed into fatty acid and glycerine. When one chooses butter to bring about the emulsion with the pancreatic juice, soon the butyric acid is recognizable by its characteristic odour.

From the foregoing it results clearly *that the pancreatic juice possesses, outside the animal, the property of emulsifying instantaneously and in a complete manner the neutral fatty matters and of decomposing them then into fatty acid and glycerine.*

The pancreatic juice alone enjoys this property, to the exclusion of all the other fluids of the [animal] economy I have tried comparatively on the neutral fatty matters, the action of the bile, of the saliva, of the gastric juice, of the blood serum, of the cerebrospinal fluid, and none of these fluids has emulsified or modified the fat.

All the preceding results have been reproduced a great number of times and the experiments are so clear and so simple to repeat that everyone will be able to verify them with ease.

But here is the place to recall the essential distinction which we have established between the normal pancreatic juice and the pancreatic juice diseased or altered. In fact this instantaneous emulsion of the neutral fatty matters and their decomposition into glycerine and fatty acid are effected only by the normal pancreatic juice, that is to say by the pancreatic juice viscous, alkaline, coagulating in mass by means of heat and acids. If, on the contrary, one mixes by shaking with oil or fat some of the pancreatic juice diseased or altered, that is to say some pancreatic juice still alkaline but turned aqueous, non-viscous, and coagulating no more with heat nor acids, its action on the fatty matters is nil, and soon a separation occurs between the inactive pancreatic juice and the unmodified fat.

Selection 140

We have seen that Bernard was assisted by chemists in his studies of the fat-splitting action of the pancreatic juice. Subsequent illumination of the chemical side of this physiological phenomenon came from the work of Marcellin Berthelot (1827–1907), who relinquished medicine for chemistry.

The following excerpts are from his article, published in 1854. It constitutes something of a landmark, both in terms of the chemical knowledge of fats and also in terms of correlating it to the elucidation of the action of the pancreatic juice on them.

Taken from the article's résumé and conclusion, the excerpts first show that Berthelot confirmed Chevreul's view on the composition of fats. Further, he demonstrated their preparation by heating fatty acids with glycerine. By recognizing similarities and divergencies between the properties of alcohol and glycerine Berthelot provided the groundwork for the study of polyalcohols as well.

Berthelot also followed up Bernard's findings concerning the role of pancreatic juice in the digestion of fats. As shown in a further excerpt below, Berthelot was able to corroborate them by proving that artificially prepared monobutyrin was decomposed almost completely into butyric acid and glycerine under the influence of pancreatic juice.

M. Berthelot, *Ann.Chim.*, [3] *41*, 216. 1854

'Mémoire sur les combinaisons de la glycérine avec les acides et sur la synthèse des principes immédiats des graisses des animaux' (On the compounds of glycerine with the acids and on the synthesis[a] of the proximate principles[b] of the fats of animals)

p. 307

§1. *On the formation of the artificial glyceric compounds*

1. All these bodies are obtained by the direct union of their two proximate principles: acid and glycerine. This union is accomplished under the influence of a prolonged contact in closed vessels, with the aid of a more or less raised temperature. The elements of water separate themselves simultaneously.

   A great number are produced already at the ordinary temperature, but in very small quantity.

   At 200 degrees, in presence of an excess of glycerine, one generally obtained the bodies of the first series: monostearin, monobenzoycin.

   Those of the third series: tristearin, triolein, a series identical with the natural fat bodies; these are prepared by making react on those of the first or better of the second series a great excess of acid at a temperature of 240 or 260 degrees.

   As for the bodies of the second series: distearin, diolein, diacetin; certain of these result from the action exercised at 200 degrees by the acid in excess of the glycerine; . . . I have likewise prepared some by making the glycerine react at 200 degrees on the natural fat bodies belonging to the third series.
2. The fat bodies are likewise produced by making at 100 degrees another [*auxiliaire*] acid, sulphuric, hydrochloric, phosphoric, tartaric, react on the mixture of glycerine and fatty acid.

pp. 308–9

§2. *On the decomposition of the artificial glyceric compounds . . .*

1. Treated by the alkalis, potash, baryta,[c] lead oxide all the glyceric compounds slowly reproduce at 100 degrees the original (*primitif*) acid and the glycerine, with fixation of the elements of water[d] . . .
2. Treated by concentrated hydrochloric acid at 200 degrees, the glyceric compounds split into fatty acids and glycerine . . .

pp. 313–14

§3. *On the physical properties of the artificial glyceric compounds*

1. The physical properties of these bodies present certain general characteristics . . .
2. The melting-point of the neutral fatty bodies is always less elevated than that of the acids from which they are derived, a property already noted with regard to the natural compounds.[1]

pp. 315–19

§4. *On the constitution of the glyceric compounds*

1. The compounds, which I have studied are neutral; treated by potash they change into neutral salts with equivalent production of a body identical for all, the glycerine. It is this same body which had produced, in uniting itself directly to the acid, the original compound, and it remains capable of

regenerating it in the same manner. When the glycerine unites with an acid, water is eliminated, and the properties of the acid become latent; when it separates, water attaches itself to the elements of the compound, and the properties of the acid reappear.

These same phenomena of decomposition manifest themselves in the most varied circumstances, often under the mildest influences.

These conditions, these phenomena, these products are precisely the same which accompany the decomposition of the natural fat bodies, as forty years ago the works of M. Chevreul have shown.

2. ... Moreover, direct and reciprocal reactions establish the same behaviour [*l'équivalence*] of glycerine and alcohol towards the acids ...

3. Nevertheless, if glycerine resembles alcohol regarding the nature of the combinations to which the acids give rise, the formula of these same compounds, the existence of several neutral compounds between glycerine and the same acid establish between glycerine and alcohol a profound difference. In fact, whilst alcohol forms only one single series of neutral combinations with the acids, glycerine produces three distinct series ...

4. ... glycerine presents towards alcohol precisely the same relation as phosphoric acid towards nitric acid. In fact, whilst nitric acid produces only a single series of neutral salts, phosphoric acid gives rise to three distinct series of neutral salts, the ordinary phosphates, the pyrophosphates and the metaphosphates. These three series of salts, decomposed by strong acids in presence of water, reproduce the same single phosphoric acid.

In the same way, whilst alcohol produces only a single series of neutral ethers, glycerine gives rise to three distinct series of neutral combinations. These three series, by their total decomposition in presence of water, reproduce one and the same body, glycerine.

pp. 273–4

Action of the pancreatic juice on the monobutyrin and on the neutral fatty bodies

In accordance with the wish of [Bernard] I have sought to isolate the products (acid and glycerine) of the decomposition of the following neutral fatty bodies: monobutyrin and pork fat,[2] this decomposition being brought about under the influence of the pancreatic juice.

This is how I have worked:

I. To about 20 grams of pancreatic juice fresh and of good quality, I have added some decigrams of monobutyrin and have maintained the whole at a gentle heat during twenty-four hours. At the end of this time, the liquid had become a milky white, and exhaled a very strong odour of butyric acid.

I have increased its volume of water and shaken three times with ether to dissolve the butyrin not decomposed and the butyric acid. A fourth treatment has extracted only traces of fatty matter, a fifth has furnished of it no more at all. I have thus obtained: (A) an etheric solution of the fat bodies; (B) an aqueous liquid rid of the fat bodies, but able to contain glycerine.

A. The ether has been evaporated on the water-bath. To the residue which it has left, I have added a little water, and as this residue gave an acid reaction, I have saturated it exactly by means of a titrated solution of baryta. The baryta used corresponded to 0.106 grams of free butyric acid. I have shaken immediately and repeatedly with ether to dissolve the butyrin, until a last treatment on evaporation gave no longer any residue. I have thus obtained an etheric liquid (a) and aqueous liquid (b).

(a) The etheric liquid evaporated has furnished only some centigrams of *butyrin*. This body had then been almost entirely decomposed by the action of the pancreatic juice.

(b) The aqueous liquid evaporated on a stove has provided crystallized butyrate of baryta. This salt corresponded precisely to the *free butyric acid* produced by the action of the pancreatic juice on butyrin.

B. The aqueous liquid from which I had separated the fat bodies should contain the glycerine corresponding to the butyric acid. I have filtered the liquid, and I have evaporated it to dryness on the water-bath, in presence of an excess of lead oxide. I have dissolved one time only the residue by means of absolutely cold alcohol. I have thus obtained an alcoholic liquor (c) and an insoluble residue . . .

(c) The alcoholic liquor has been diluted with water and hydrogen sulphide was added, which precipitated a little lead oxide dissolved in this liquor. I have evaporated on the water-bath the filtered liquid and have obtained in notable quantity a syrup with a taste first sugary, then slightly saline, insoluble in the ether and deliquescent. These characteristics, combined with the dissolving of the lead oxide and the origin of the product agrees with the presence of *glycerine*.

Selection 141

Liebig's *Animal Chemistry* (1842), which has been repeatedly mentioned here before, provided fresh impetus to research into the general metabolism of fats, just as it had done with other areas of metabolic study.

In *Animal Chemistry* Liebig suggested a connection between the formation of fat in the body of the animal and the decomposition of the non-nitrogenous, carbohydrate, constituents of food. Moreover, he associated the depositing of fat in the animal body with a deficiency in the supply of oxygen during respiration.[a]

At first opposed, this conception gradually received wide approval but remained under active review during the next two decades. Then in 1865 the Munich physiologist, Carl (von) Voit (1831–1908), re-opened the question of the source of the fat stored up in animals.[b] Working with Max (von) Pettenkofer (1818–1901),[c] he had found that, on feeding dogs with meat, only part of the carbon passed into the urine and faeces or was got rid of in respiration. The rest, it was suggested, was utilized for the production of animal fat. In other words, the animal body could make its fat from nitrogenous constituents of the diet.[d]

It was recognized as a problem and immediately again taken up by John Bennet Lawes (1814–99) and Joseph Henry Gilbert (1817–1901) at Rothamsted. The results of their investigations, summarized in the table and the extract below, were published in 1866.[e]

First Lawes and Gilbert looked into the comparative fattening qualities of oxen, sheep and pigs. They found that in proportion to the original weight of the animal and to the food consumed within a given time the pig was the most suitable animal for the study of the origin of the fat in the food.

The latter consisted of bean-meal, lentil-meal, maize-meal, barley-meal and bran, given as mixtures or separately. In experiment 4 only maize-meal and in experiment 5 only barley-meal (considered as *the* fattening food of the pig) was given. 'The result was', wrote Lawes and Gilbert,

that in both these experiments the proportion of non-nitrogenous to nitrogenous substance in the food was very nearly, though rather higher than, the average in that which is recognized as the most appropriate fattening food of the animal.

From these two experiments it certainly appeared that about 40 per cent of the carbon of the produced fat, not accounted for either by the carbon of the fatty or nitrogenous constituents in the food, was supplied by carbohydrates. None the less, the possibility of fat being derived from protein in the food was not excluded.

J.B. Lawes and J.H. Gilbert, *Phil.Mag.*, [4] *32*, 439. 1866

'On the sources of the fat of the animal body'

p. 446

Relation of the total Fat in the Increase to the ready-formed fatty matter in the Food, and of the Carbon in the Fat produced within the body to that in the nitrogenous substance consumed, in experiments with Fattening Pigs.

pp. 449–50

It is hardly necessary to point out that, according to the mode of illustration we have adopted, the figures show not only the utmost proportion of the carbon of the stored-up fat which could possible have had its source in the nitrogenous substance of the food, but even notably more than could possibly have been so derived. Thus, to say nothing of other considerations, it has been assumed for simplicity of illustration, and granted for the sake of argument, that the whole of the ready-formed fatty matter of the food contributed to the fat stored up, that the whole of the nitrogenous substance of the food not stored up as increase would be perfectly digested and become available for the purposes of the system, and that in the breaking up of the nitrogenous substance for the formation of fat no other carbon-compounds than fat and urea would be produced. It is obvious, however, that these assumptions are in part improbable, and in part quite inadmissible, and that the tendency of each of them is to show too large a proportion of the produced fat to have been possibly derived from the nitrogenous constituents of the food.

The amount of fat necessarily derived from other sources than the nitrogenous constituents of the food must therefore be greater than our mode of estimate can indicate; and it is obvious, from the figures given in the Table, that the less the excess of nitrogenous substance in the food, the greater was the proportion of produced fat which must necessarily have had its source in the carbo-hydrates of the food, and that, at any rate in those cases in which the proportion of non-nitrogenous to nitrogenous constituents supplied was the

Experiments	1	2	3	4	5	6	7	8	9
Conditions, and actual results of experiment.									
Number of animals	1	3	3	3	3	3	3	3	3
Duration of experiment (weeks)	10	8	8	8	8	10	10	10	10
Non-nitrogenous substance to one nitrogenous substance in food	3.6	3.3	2.0	6.6	6.0	4.1	4.1	4.7	3.9
Original live-weight (lbs.)	103	429	440	431	448	286	285	281	292
Final live-weight (lbs.)	191	685	743	652	739	533	533	553	604
Increase in live-weight (lbs.)	88	256	303	221	291	247	248	272	312
Increase on 100 original weight	85.4	59.7	68.9	51.3	64.9	86.4	87.0	96.8	106.8
Calculated for 100 increase in live-weight.									
Fat { Stored up increase	63.1	73.9	69.6	79.0	71.2	64.1	63.9	62.0	59.9
Fat { Ready-formed in food	15.6	20.4	11.2	26.3	12.4	7.9	7.9	7.3	6.6
Fat { Not ready-formed (produced)	47.5	53.5	58.4	52.7	58.8	56.2	56.0	54.7	53.3
Nitrogenous-substance { Consumed in food	100.0	107.0	138.0	57.0	64.0	81.0	81.0	74.0	82.0
Nitrogenous-substance { Stored up in increase	7.8	6.1	6.7	5.3	6.5	7.5	7.6	8.0	8.2
Nitrogenous-substance { Not stored up (available for fat, & c.)	92.2	100.9	131.3	51.7	57.5	73.5	73.4	66.0	73.8
Carbon { In 'produced' fat	36.6	41.2	45.0	40.6	45.3	43.3	43.1	42.1	41.0
Carbon { In 'available' nitrogenous substance *minus* urea	44.0	48.1	62.6	24.7	27.4	35.1	35.0	31.5	35.2
Carbon { Difference	+7.4	+6.9	+17.6	15.9	17.9	8.2	8.1	10.6	5.8
Calculated for 100 carbon in estimated 'produced' fat.									
Carbon { In 'available' nitrogenous subtance *minus* urea	120.2	116.7	139.1	60.8	60.5	81.1	81.2	74.8	85.9
Carbon { Not available from nitrogenous substance	...	...	...	39.2	39.5	18.9	18.3	25.2	14.1

more nearly that occurring in the admittedly most appropriate fattening food of the animal, the proportion of the fat which must necessarily have been derived from the carbo-hydrates was very large, even allowing all that was possible to have been produced from the nitrogenous substance of the food.

That, nevertheless, fat may be produced in the animal body at the expense of nitrogenous substance, in greater or less degree according to the character of the animal and of the food, not only chemical and physiological considerations, but direct experimental evidence would lead us to conclude.

Selection 142

Bernard's aforementioned discovery that pancreatic juice decomposed and emulsified fats was one thing, the way fat absorption took place was another. The researches into the absorption of fatty acids reported by Immanuel Munk (1852–1903) in 1880 are of particular historical importance in this context.

Up to then it was widely held that although decomposition of ingested fats occurred in the intestine, it was slight. Nevertheless, it was thought to be sufficient for the formation of soaps with the alkali present in the intestinal juice and bile. The effect of the soaps was to emulsify the remaining fat in order to facilitate its passage to the intestinal villi and to the lacteals, ultimately reaching the thoracic duct as chyle.

As shown below, Munk envisaged that fatty acids, produced in the small intestine from fat by a pancreatic ferment with a lipolytic action, played a part in fat metabolism. Apparently they were not only absorbed but, following absorption, underwent resynthesis to fat. While admitting this possibility, Munk stressed that the breaking down of the intestine fat was not necessarily fat's universal metabolic course.

Munk's feeding of fatty acids, instead of fats, to dogs was part of the then widely displayed interest in learning about the replacement of proteins by carbohydrates and fats in nutrition ('protein-sparing'). The adopted procedure was to conduct feeding experiments on animals kept in 'nitrogen equilibrium', that is in a situation where the nitrogen intake equalled the amount excreted.[a]

Munk found that dogs maintained under these conditions did not have their metabolic balance upset when fed on fatty acids. Like fats, the fatty acids were also to be looked upon as protein-sparers.

I. Munk, *Virchows Arch.*, *80*, 10. 1880

'Zur Kenntnis der Bedeutung des Fettes und seiner Componenten für den Stoffwechsel' (Contribution to the knowledge of the significance of the fat and its components for the metabolism)

p. 19

After a longer preliminary feeding it was possible to bring [the] dog with only 600 g meat and 100 g fat into N-equilibrium and also nearly into body weight equilibrium. After this state of N-equilibrium was sufficiently established (based on the determination of N-excretion for 5 days), *during 21 days* 100 g of preparable [*darstellbare*] fatty acids were fed instead of the fat. The N-excretion through the urine, as well as the body weight, were determined on all of 21 days, the estimation of the nitrogen content of the faeces happened only for the first and last 5 days. Finally for 5 days again, as in the preliminary period, fat was given. The particular values necessary for the judging of the metabolic relations are recorded in the accompanying table.

pp. 20–1

Big domestic dog, 600 g meat, 350–400 cc water.

Date	Diet	Urine in cc	N in urine	N in faeces	Body weight in kg
			[g]	[g]	
2. Jan 1879	Always 100g	573	19,992		31.08
3. — —	fat	578	20,216		30.95
4. — —		554	20,328	2.11[1]	30.8
5. — —		456	18,872		30.75
6. — —		487	20,888		30.89
7. — —		484	19,88		30.84
8. — —		503	20,02		30.81
9. — —		499	19,768	2.8[2]	30.82
10. — —		566	20,272		30.76
11. — —		607	20,468		30.7
12. — —		552	21,448		30.65
13. — —	Fatty acids	507	20,888		30.64
14. — —	always from	484	19,264		30.65
15. — —	100g fat	481	18,676		30.42
16. — —		469	19,096		30.4
17. — —		455	19,432		30.27
18. — —		439	18,704		30.4
19. — —		451	18,592		30.42
20. — —		460	19,068		30.59
21. — —		538	18,508		30.44
22. — —		437	18,956		30.57
23. — —		446	19,516		30.65
24. — —		470	18,872		30.69
25. — —		481	18,396	2.19[3]	30.79
26. — —		497 (?)	19,404		30.74
27. — —		495	18,676		30.85
28. —		514 (?)	19,32		30.88
29. —		559	21,028		30.84
30. —		633	22,148	2.003[4]	30.79
31. —	Always 100g	567	21,728		30.69
1. Feb —	fat	594	21,952		30.51

On the basis of this feeding series lasting several weeks, at any rate this much can be asserted that the protein breakdown in the body of the carnivore remains unaltered no matter if fat or only the equivalent amount of fatty acid is fed. If we consider further that also the body weight of the animal maintained on a fatty acid diet for three weeks does not essentially differ from the body weight retained during the preliminary period, then the experimental series leads to [this] remarkable conclusion. *A dog which, fed upon a diet of meat and fat, finds itself in N and body weight equilibrium, persists in the equilibrium also if through 21 days, instead of the fat, only the fatty acids contained in the latter are given; thus the fatty acids as protein sparers have the same significance as the fat.*

p. 33

Further, the preceding experiments lead to the noteworthy result that, *after introduction of pure fatty acids, a considerable increase in the fat content of the chyle can be observed,* 9–10 times the normal and even under the most unfavourable experimental conditions . . . where a part of the fed fatty acids are vomited . . . After feeding of pure fatty acids the relatively high content of the chyle in fat and its much smaller content in fatty acids can indeed not be otherwise explained than that *the fatty acids are not only reabsorbed, but on the way from the intestinal cavity to the thoracic duct are submitted to a conversion to fat, to a synthesis.*

p. 37

The experiment has established that the fatty acids in doses up to 100 g, are utilized, that is resorbed in the intestine to the same extent as the corresponding amount of neutral fat. Further, it has been shown that also as protein sparers they have the same value as the fats. Finally, no small portion of them, at any rate before their entry into the blood stream, becomes fat by the synthetic way. Thus would scarcely anything be adduced against it if one assumed that also normally the fat in the intestinal canal through the pancreatic and putrefaction ferment[a] breaks down into fatty acids and glycerine and these decomposition products, after their entry into the resorption path, are again regenerated to fat. If, as in this case besides the fatty acid at the same time glycerine is found in the resorption path, it must follow that one can assume the synthesis to fat in still greater extent. Yet I should like to emphasize that even if according to my experiments nothing would stand in the way of this hypothesis, yet out of the proof of the possibility of such a process of fat resorption, still no binding conclusion follows that normally also each fat particle in fact undergoes this splitting in the intestine.

Selection 143

In tracing the advances in the biochemistry of lipids it is appropriate to refer now to the original chemical investigations of brain by Johann Ludwig Wilhelm Thudichum (1829–1901) undertaken between 1874 and 1884.[a] German-born Thudichum came to London to practice medicine in 1854. He became known for his contributions to otolaryngology and for his treatment of gallstones. He paid attention to urine and discovered and named its chief pigment, urochrome. Regarding other biological pigments he detected 'luteines' (carotenoids) and prepared 'cruentine' (an iron-free haematoporphyrin). His wide-ranging interests included the making of wines and beers, and cookery, reflecting the historically close relation between biochemistry and drinks and foods.

Above all he cleared the ground for the systematic study of the chemistry of the brain which was regarded by him as 'the most diversified chemical laboratory of the animal economy' involving 'complicated chemical structures and processes'. He firmly believed that 'the explanation of the mental phenomena, and of their aberrations under the influence of disease' rested on their knowledge. As to his own researches in this area, Thudichum took them very seriously emphasizing that 'all further developments in chemical neurology must start from them as a basis'. If this pronouncement appears to shown an immodest sense of one's own achievements, it can be traced to the extremely adverse opinions of them, expressed in Germany and Britain by scientists of distinction, such as F. Hoppe-Seyler (1825–95) and A. Gamgee (1841–1909). Be that as it may, Thudichum has emerged as a major figure in establishing what he termed 'chemical biology' during the second half of the last century.

Thudichum advocated greater utilization of chemistry in medical practice and was deeply concerned with the public aspects of health. Indeed, it was under the auspices of the then Medical Department of the Privy Council and subsequently of the Local Government Board that Thudichum carried out his systematic investigations on brain chemistry.[b] 'The matter of the work is therefore public property,' stated Thudichum, 'and under these circumstances I think it my duty specially to avow my responsibility for its contents.' Along with the already quoted observations he wrote this in the preface of the monograph from which the excerpts below are taken.[c]

They indicate the principles of Thudichum's historically influential classification of fat-like bodies occurring in the brain. He distinguished between three groups, depending on whether the bodies contained nitrogen and phosphorus, nitrogen only, or were devoid of both nitrogen and phosphorus.

For the group comprizing nitrogen and phosphorus Thudichum used the term 'phosphatides'. Here the classification was further determined by the $N:P$ ratio. Thus Thudichum perceived in lecithin and kephalin (which he discovered) the $N:P$ ratio as $1:1$ and in sphingomyelin (also obtained by him) as $2:1$. A further suggested difference was that the latter did not contain glycerol.

The group of nitrogenous compounds lacking phosphorus was termed 'cerebrosides' by Thudichum. It included substances such as phrenosin and kerasin which he discovered and studied. Their cleavage products pointed to the presence of a sugar, fatty acid and alkaloids (i.e. base-like compounds). He named the sugar 'cerebrose' having the formula $C_6H_{12}O_6$;[d] he characterized the fatty acid as an isomer of stearic acid $C_{18}H_{36}O_2$ and named it 'neurostearic acid';[e] he first prepared and named the 'alkaloid' sphingosin, suggesting the formula $C_{17}H_{35}NO_2$.[f]

No excerpt will be given concerning the group free from nitrogen and phosphorus. However it should be noted that among these compounds present in the brain Thudichum discovered, as he put it, 'the variety of lactic acid known as *lactic acid from flesh, or sarkolactic acid*'.[g]

J.L.W. Thudichum, *A Treatise on the Chemical Constitution of the Brain* **(London, 1884)**

pp. 3–5

The great quantity of these matters occurring in the brain forms three groups; the members of one contain at least five, sometimes six, elements, amongst which is phosphorus; hence they may be termed PHOSPHORISED BODIES. The members of the second group contain four elements, amongst them nitrogen, but no phosphorus, and therefore are termed NITROGENISED BODIES. The members of the third group contain only three elements, carbon, hydrogen, and oxigen, present also in the other two groups, but neither phosphorus nor nitrogen, and may be termed OXIGENISED BODIES.

The group of the PHOSPHORISED BODIES contain the phosphorus in the form of phosphoric acid combined with from two to five organic compound radicles. As the earliest known body of this group, *lecithin*, yielded its phosphoric acid mainly in combination with glycerol, as glycerophosphoric acid, it was supposed that the phosphorised bodies, of which a number were theoretically admitted to exist, were constituted like the fats, by combination of compound organic radicles with the radicle of glycerol, in other words, that they were ethers of the alcohol glycerol, and contained the phosphoric acid as an inserted, and not as a fundamental radicle. But as we now know at least one phosphorised principle from the brain which does not contain any glycerol, and does therefore not yield, on chemolysis, any glycerophosphoric acid, but phosphoric acid merely without any attached organic radicle, we thereby obtain a new insight into the chemical constitution of the phosphorised substances altogether, and are under the necessity of subjecting their theory to a revision. According to the result of this revision the phosphorised substances are not glycerides at all, as commonly defined, and have nothing in common with fats considered as glycerides, except that some of them contain certain fatty acids also present in fats, while they differ in physical and chemical properties widely from fats. In accordance with this new knowledge, I have termed the phosphorised substances *phosphatides*, that is to say, substances which are similar to (but not by any means identical with) phosphates, on the assumption that their basal or principal joining radicle is that of phosphoric acid.

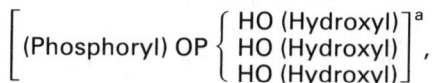

$$
\left[\text{(Phosphoryl) OP} \left\{ \begin{array}{l} \text{HO (Hydroxyl)} \\ \text{HO (Hydroxyl)} \\ \text{HO (Hydroxyl)} \end{array} \right. \right]^{a},
$$

and that in this acid one, two, or three molecles of hydroxyl may be replaced by radicles of alcohols, acids, or bases, and that to a molecle [*sic*] formed by three such substitutions there may yet be attached, either by substitution of an element in a radicle itself already substituted (side-chain), or by addition with elimination of water from the added radicle, a fourth radicle . . . The bodies . . . contain the phosphorised radicle once, and may therefore be termed *monophosphatides*; but there are present in the brain and other protoplastic centres, bodies which contain the phosphorised radicle twice, and which may therefore be described as *diphosphatides;* . . . I shall at once pass to a short consideration of the mononitrogenised monophosphatides, of which lecithin is the earliest known, and was, before the institution of my researches, the only one of which any closer knowledge existed.

In this subgroup nitrogen is to phosphorus in the proportion of one atom to one atom, a relation to be expressed by the formula $N:P = 1:1$. Of the educts of the brain four species with all their varieties belong to this subgroup, namely, the *lecithins, kephalins paramyelins*, and *myelins*; a product also may be alluded to, obtained from a dinitrogenised educt by the loss of a nitrogenised radicle, namely, *sphingomyelic acid*. This latter contains no glycerol; the lecithins, and kephalins, and paramyelins probably contain glycerol, and yield glycerophosphoric acid; of the myelins, the constitution in this particular respect has yet to be ascertained.

pp. 10–11

I now come to the subgroup of *dinitrogenised monophosphatides*, of which I had given a preliminary notice in my researches of 1874 under the name of *amidomyelin* and *apomyelin*, but of which the former has been isolated only lately, while the latter has been more fully investigated, and classified with a body isolated and investigated under the name of *sphingomyelin*...

Amidomyelin is analogous to sphingomyelin and apomyelin in this, that it contains two atoms of nitrogen upon one atom of phosphorus; ...

pp. 16–18

The NITROGENISED NONPHOSPHORISED substances of the brain imitate in many respects, but with little intensity, the properties of the phosphorised ... This group includes six great subgroups, ...

The frist subgroup is that of the *cerebrosides*, or bodies which contain a peculiar sugar, *cerebrose*, in which different radicles of acids and alkaloids (it is not known whether of alcohols also, in some cases) are inserted. Thus, of *phrenosin*, $C_{41}H_{79}NO_8$, the definition and formula of constitution are the following:

Neurostearo-Sphingoso-Cerebroside.

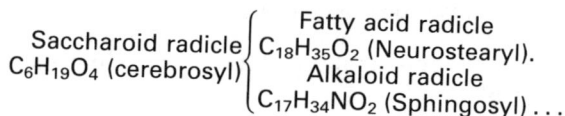

$$\text{Saccharoid radicle}\left\{\begin{array}{l} \text{Fatty acid radicle} \\ C_{18}H_{35}O_2 \text{ (Neurostearyl).} \\ \text{Alkaloid radicle} \\ C_{17}H_{34}NO_2 \text{ (Sphingosyl)} \dots \end{array}\right.$$
$$C_6H_{19}O_4 \text{ (cerebrosyl)}$$

The second cerebroside, and accompanying phrenosin, is *kerasin*, ... Kerasin may, like phrenosin, comprise a number of analogously constituted bodies ...

Selection 144

At the close of the period dealt with in this subsection, only the case of fermentative or enzymic fat splitting in the intestinal contents was known. As we have seen, it was believed that the intestinal fat-digesting ferment originated from the pancreas.

In 1896, however, this situation changed when the French writer, Maurice Hanriot (1854–1933), published a paper regarding the presence of a ferment with lipolytic action in the blood and serum which he designated 'lipase'.

The following selection is from this paper which is remarkable principally for two reasons. First, for the suggestion that lipase was the enzyme generally responsible for the breakdown of fats throughout the animal and vegetable kingdoms. Second, for the description of an early attempt at quantitative testing for enzyme activity. This was achieved by the determination of the point of neutralization of the fatty acid (liberated on account of lipase action) by alkali.

M. Hanriot, *Comp.Rend.*, *123*, 753. 1896

'Sur un nouveau ferment du sang' (On a new ferment of the blood)

pp. 753–5

The fats are not, so to speak, attacked by sodium carbonate at the body temperature; hence it was not possible that the feeble alkalinity of the blood plasma sufficed to saponify them; I therefore have searched to see if the blood should not contain a ferment capable of realizing this attack.

The natural fats lend themselves badly to this study; by their insolubility by that of the fatty acids which result from their splitting, they are scarcely wetted by the blood which affects them only very slowly. So I addressed myself first to the fatty acid ethers properly so-called (*éthers à acides gras proprement dits*),[a] and more particularly to an ether little soluble in water but easily emulsified, the *monobutyrin*, discovered by M. Berthelot, who has reported its easy saponification by pancreatic juice.[b] With the help of this reagent one can easily follow the course of the saponification when it takes place, by titrating with sodium carbonate the butyric acid set free.

First I have established that the blood serum saponifies monobutyrin easily and very actively, when the solution is neutral or slightly alkaline; but this slows considerably if one does not take care to neutralize as the acid is set free. Further in equal intervals of time, and using equal quantities of butyrin, the acidity increases uniformly with the quantity of serum employed, which allows one, up to a certain point, to compare between them the activeness of the different sera and, as a result, their richness in ferment. In determinations of this kind, one must take count only of the results obtained in the first hours, for the action slows as the result of a phenomenon to which I shall return.

In order to establish well that it is a case, in these experiments, of a true diastatic fermentation,[c] I have worked aseptically in a manner to avoid the influence of the presumed ferments. However, the acidity does not appear in the control test-tubes containing, in one butyrin alone at the same dilution, in the other serum alone. Finally, in a last experiment, I have brought serum to 90° in order to destroy the diastase and I have established that it could no longer acidify the solution of butyrin.

The following figures give the number of drops of a solution of sodium carbonate, 5 g per litre, necessary to neutralize 10 c.c. of a solution of monobutyrin [containing] 25 [parts] in 10,000. They show thus the influence of the proportion of the ferment.

I have likewise verified that the oils and natural fats are saponified by serum, but here the action is slower and cannot be followed as simply; one is obliged to isolate the fatty acid formed; . . .

Time in min.	Serum alone	Butyrin alone	Butyrin and 0.5 c.c.	Butyrin and 1 c.c.	Butyrin and 1 c.c.	Butyrin and serum heated
20	0	0	4	9	9	0
40	0	0	8	17	17	0
60	0	0	12	24	24	1
80	0	0	16	32	31	1
170	1	1	27	50	49	1
305	1	2	39	66	66	2

I have further ascertained that air does not affect the phenomenon which goes on equally well in the absence of oxygen.

This ferment, for which I propose the name *lipase*, is very stable; it persists in the serum for a very long time; at the end of eight days it has appeared to me as active as at the beginning of the experiment.

The presence of lipase wherever there is a fatty reserve to utilize . . . shows that the catabolic phenomena seem, in animals and plants, to be enacted like those of digestion by the intermediary of soluble ferments.

b. From *c*. 1900 to *c*. 1940: fatty acid metabolism, lipase action and cholesterol

Selection 145

As already mentioned, in the early 1820s the chemical understanding of the fats surpassed that of the carbohydrates and proteins. Although it is probably true to say that the attention subsequently paid to fats by biochemists did not overshadow that received by proteins and carbohydrates, the study of oxidative degradation of fatty acids was to figure significantly in efforts to establish metabolic links between the three major groups of foodstuffs. By the beginning of this century it was accepted that oxidative degradation of fatty acids belonged to the most widespread type of chemical reactions occurring in the animal body. Further, physiological chemists were alive to the notion that lower fatty acids represented common intermediates derived from the degradation of fats, proteins and carbohydrates.

This attitude found its expression in the work of Franz Knoop (1875–1946) on the oxidation of fatty acids at the β-position in the animal organism. He began his investigations at the Institute of Physiological Chemistry at Strassburg (Strasbourg) and continued them at the Medical Department of the Chemical Institute at Freiburg im Breisgau. Arriving there in the autumn of 1903, Knoop intended to make the results of his research the subject of his qualifying thesis for lecturing at a university (*Habilitation*). But there was some reluctance to allow this, for the topic, more comprehensively treated for the first time in *Der Abbau aromatischer Fettsäuren im Tierkörper* (1904), was received with certain misgivings, both by chemists and physiologists. Whereas the former found Knoop's approach not sufficiently chemical, the latter considered it to be excessively so.[a]

Breaking from the traditional drawing up of balance sheets of the intake and output of substances, including heat values, Knoop stressed the need to concentrate on the study of the intermediate chemical changes for an understanding of the metabolic relations in the animal. His method was to study the oxidative breakdown of fatty acids by 'labelling'[b] them by means of the benzene group which was not affected by oxidation in the organism. The idea behind this was that the group would attach itself to breakdown products of these fatty acids and in this way it would be possible to throw light on the stages of their oxidation.

Among the workers who had been giving attention to the fate of aromatic acids undergoing oxidation in the animal body and who clearly stimulated Knoop were Ernst Salkowski (1844–1923) and Heinrich Salkowski (1846–1929). They were concerned with a number of homologues of benzoic acid which appeared on putrefaction of protein and were known to occur in the intestine under some conditions. They concluded that acids homologous to benzoic acid were broken down to benzoic acid if the side-chain contained more than two carbon atoms or was weakened in its stability by substitution of an atom of H by OH or of two H atoms by O.[c]

However, this 'rule of E. and H. Salkowski' was known to be contradicted by the often observed fact that mandelic acid, $C_6H_5CHOH\,COOH$, passed unaltered through the animal body. This, together with the failure of phenylacetic acid, $C_6H_5CH_2COOH$, to undergo oxidation in the body, became the starting point for Knoop's study which appeared in 1905 and is presented in the excerpts below.

It pointed to the degradation of fatty acids by loss of two carbon atoms by oxidation in the β-position. The indication was that an acid with an even number of carbon atoms in the chain (e.g. phenylbutyric acid) produced phenylacetic acid and an acid with an odd number of carbon atoms (e.g. phenylpropionic acid) formed benzoic acid. Excreted in the urine (in combination with glycine, NH_2CH_2COOH) as phenaceturic and hippuric acid respectively, they were to provide information about the intermediate stages of fatty acid oxidation in the animal body:

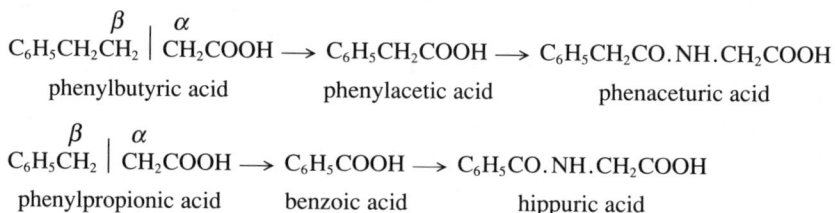

$$\overset{\beta}{C_6H_5CH_2CH_2} \mid \overset{\alpha}{CH_2COOH} \longrightarrow C_6H_5CH_2COOH \longrightarrow C_6H_5CH_2CO.NH.CH_2COOH$$

phenylbutyric acid   phenylacetic acid   phenaceturic acid

$$\overset{\beta}{C_6H_5CH_2} \mid \overset{\alpha}{CH_2COOH} \longrightarrow C_6H_5COOH \longrightarrow C_6H_5CO.NH.CH_2COOH$$

phenylpropionic acid   benzoic acid   hippuric acid

F. Knoop, *Hofmeisters Beitr.*, **6**, 150. 1905

'Der Abbau aromatischer Fettsäuren im Tierkörper' (The degradation of aromatic fatty acids in the animal body)

p. 153

It was first necessary to ascertain whether the resistance of phenylacetic acid and its derivatives is a matter of single occurrence, or whether it is the single example of a general rule to be more closely substantiated. In any case it was not to be assumed that the phenylpropionic acid, if it is burnt to benzoic acid, first would become phenylacetic acid; since in this case phenylacetic acid or its conjugate [phenaceturic acid] would have to appear in the urine; so at present no other assumption remains than that the oxidation with phenylpropionic acid takes place not on the α- but on the β-carbon atom. Whether such an oxidation on the side chain in the β-position is a rule or not was to be ascertained by means of comparative experiments with higher homologues of phenylpropionic acid or its derivatives.

p. 154

In order to obtain comparable results I have carried out all experiments (with the exception of that with ethyl benzene) at least once on the same dog, as well as a number of parallel experiments on other dogs, without coming upon any exceptions.

In this way I found by checking the older results:

I. In normal urine a vanishingly small amount of hippuric acid.

By feeding of:

II. 2 g of phenylpropionic acid (Kahlbaum):[a] hippuric acid and no phenaceturic acid.

III. 2 g of mandelic acid (inactive, Kahlbaum): unaltered mandelic acid.

IV. Phenylacetic acid (Kahlbaum): phenaceturic acid, no increase in hippuric acid.

V. Ethyl benzene (Merck):[a] hippuric acid, no phenaceturic acid.

pp. 155–6

2 g of the [phenylbutyric] acid were fed twice. Both times I obtained from the urine an acid of the typical crystal form of phenaceturic acid which, like the latter, melted at 142°, and after splitting with dilute sulphuric acid in ether gave phenylacetic acid of melting point 76°. Hippuric acid was not found.

So it appeared in fact that an oxidation in the β-position had taken place here, a result for which it was of importance to obtain further proof. Next at hand was to continue in the series and to investigate phenylcaproic acid – it should go by way of phenylbutyric acid to phenaceturic acid. Unfortunately however this acid has not been prepared. The intermediate *phenylvaleric acid* could be obtained; it must according to the same principle go through phenylpropionic acid to hippuric acid, if also here β-oxidation occurred. I have prepared it according to the instructions of Fitting and Hoffman[1] . . .

1.5 g of the acid obtained was fed and out of the urine exclusively hippuric acid was obtained, in amount almost 0.5 g.

Fed	Excreted	Observed alteration[2]
$C_6H_5.COOH$	$C_6H_5.COOH$	Unchanged
$C_6H_5.CH_2.COOH$	$C_6H_5.CH_2.COOH$	"
$C_6H_5.CH(OH).COOH$	$C_6H_5.CH(OH).COOH$	"
$C_6H_5.CH(NH_2).COOH$		" (deaminated)
$C_6H_5.CH_2.CH_2.COOH$		
$C_6H_5.CH(OH).CH_2.COOH$	$C_6H_5.COOH$	Oxidized at the
$C_6H_5.CO.CH_2.COOH$		β-carbon atom
$C_6H_5.CH = CH.COOH$		
$C_6H_5.CH_2.CH(NH_2).COOH$		
$C_6H_5.CH_2.CH(OH).COOH$	0	Apparently totally
$C_6H_5.CH_2.CO.COOH$		oxidized
$C_6H_5.CH = C(NH_2)_2.COOH$		
$C_6H_5.CH_2.CH_2.CH_2.COOH$		Oxidized at the
$C_6H_5.CO.CH_2.CH_2.COOH$	$C_6H_5.CH_2.COOH$	β-carbon atom
$C_6H_5.CH = CH-CH_2.COOH$		
$C_6H_5.CH_2.CH_2.CH_2.CH_2.COOH$	$C_6H_5.COOH$	Oxidized at the δ-carbon atom
$C_6H_5.CH.CH_2.CH_2.CO$ with O–COOH ring	Unchanged	0
$C_6H_5.CH.CH.CH_2.CO$ with O ring		

p. 160

In the following [table above] I put together the pertinent experiments up to this time, so far as they concern aromatic acids not substituted in the nucleus and with simple side chains . . .

Selection 146

The idea of β-oxidation was welcomed by certain workers in the field, for example Gustav Embden (1874–1933) and his collaborators, interested in acetone formation in the body. Perfusion experiments with isobutyric, isovaleric and isobutyl acetic acids led them to the opinion that the aliphatic fatty acids were broken down by the splitting off of two carbon atoms from the carboxyl end.[a]

As shown below, they put this conclusion to experimental proof, using the higher normal homologues of butyric acid up to decoic acid.[b] They interpreted their results as meaning that with the acids containing an even number of carbon atoms the three carbon atoms farthest removed from the carboxyl end of the fatty acid, and of each intermediate acid, gave acetone.

G. Embden and A. Marx, *Hofmeisters Beitr.*, *11*, 318. 1908

'Über Acetonbildung in der Leber. Dritte Mitteilung' (On acetone formation in the liver. Third communication)

pp. 318–19

In our experiments we started out from the following consideration: if upon degradation of the normal higher homologues with an even number of C atoms 2 C atoms were always split off, then out of these substances there must again arise compounds (probably acids) with an even number of C atoms.

If these assumptions were correct, β-hydroxybutyric acid, acetoacetic acid and acetone[a] must be formed from normal decoic acid, via octoic and caproic acids, butyric acid, whilst with the corresponding acids containing an uneven number of C atoms, this should not be the same . . .

The results of these experiments are seen in Table 12.

p. 320

From columns 3 and 4 it is seen that the regularity [*Gesetzmässigkeit*] we expected in fact exists. That is to say that only the acids with an even number of C atoms cause an increase of acetone formation in the liver, whilst with the acids of uneven C-atom number the acetone amounts formed are not greater than those found earlier by Embden and Kalberlah[1] on perfusion of the liver with normal blood.

The idea that the degradation of fatty acids – at least of those with unbranched chains – follows the earlier assumed course gains exceedingly in probability through these experiments.

Table 12

1 No.	2 Substance added to the perfusing blood	3 Acetone formed per litre	4 Acetone formed in the whole amount	5 Mean acetone formation per litre in 2 parallel experiments
		mg	mg	mg
1	2 g normal butyric acid	148	237 ⎫	
2	2 g "	108	173 ⎭	128
3	2 g normal valeric acid	19	30 ⎫	
4	2 g "	21	34 ⎭	20
5	2 g normal caproic acid	89	142 ⎫	
6	2 g "	111	178 ⎭	100
7	2 g normal heptylic acid	9	14 ⎫	
8	2 g "	15	24 ⎭	12
9	2 g normal octoic acid	60	96 ⎫	
10	2 g "	60	96 ⎭	60
11	2 g normal nonoic acid	9	15 ⎫	
12	2 g "	219	46 ⎭	19
13	2 g normal decoic acid	71	114 ⎫	
14	1.5 g "	46	74 ⎭	58

Not only the manner of origin of hydroxybutyric acid, acetoacetic acid and acetone is brought nearer to understanding through the results obtained by us, but also the formation of lower fatty acids from higher through degradation at the carboxyl end is made very probable through them; even if the direct proof for this manner of formation of the lower fatty acids – through their isolation after perfusion with the higher – still remains to be ascertained.

This manner of formation of the lower fatty acids would be in agreement with the fact that in the animal organism and in nature altogether are found quite preponderantly those with an even number of carbon atoms; Knoop[2] has already drawn attention to this on the grounds of his feeding experiments with aromatically substituted fatty acids.

It is seen from column 5 of the Table 12 that the amount of acetone formed from the normal fatty acids with even number of C atoms falls from 128 mg with butyric acid to 58 mg with decoic acid. This is quite understandable. In the first place, the process of acetone formation is so much the more complicated the longer the carbon chain is, the more fatty acids as intermediates thus have to be formed; and secondly, from each acetone-forming fatty acid only the three carbon atoms farthest removed from the carboxyl end pass into acetone. Equal amounts of acetone therefore cannot arise from equal but from equimolecular weights of the different acetone-forming fatty acids in the organism.

Selection 147

Let us return to Knoop's table containing the summary of his findings after he administered to dogs the sodium salts of phenyl derivatives of straight chain fatty acids. Regarding phenylvaleric acid, it will be noticed that Knoop implied that it was oxidized at the δ-carbon atom.[a]

Further study of the oxidation of this fatty acid was undertaken by Henry Drysdale Dakin (1880–1952), who suggested that in reality no δ-oxidation occurred. He based his position on identifying phenyl-β-oxypropionic acid, cinnamoylglycocoll and acetophenone in the urine of animals that had received injections of sodium phenylvalerate in doses of about 0.8 g per kilo. These substances were known to be intermediate products of catabolism of phenylpropionic acid. It was possible, Dakin maintained, that phenylvaleric acid via phenyl-β-oxyvaleric acid underwent β-oxidation yielding phenyl-propionic acid. By another β-oxidation the latter was converted into benzoic acid.

In Dakin's view, as shown below, it was probable that repeated β-oxidation was the reaction characteristic for the normal catabolism of straight chain saturated fatty acids.

H.D. Dakin, *J.Biol.Chem.*, **6, 221. 1909**

'The mode of oxidation in the animal organism of phenyl derivatives of fatty acids. Part V. Studies on the fate of phenylvaleric acid and its derivatives'

pp. 223–5

With the assumption that phenylpropionic acid is formed from phenyl-β-oxyvaleric acid, with the possible intermediary formation of phenylpropionylacetic acid, the remaining steps in the catabolism of phenylvaleric acid will be identical with those of phenylpropionic acid. The whole series of changes may be represented as follows:

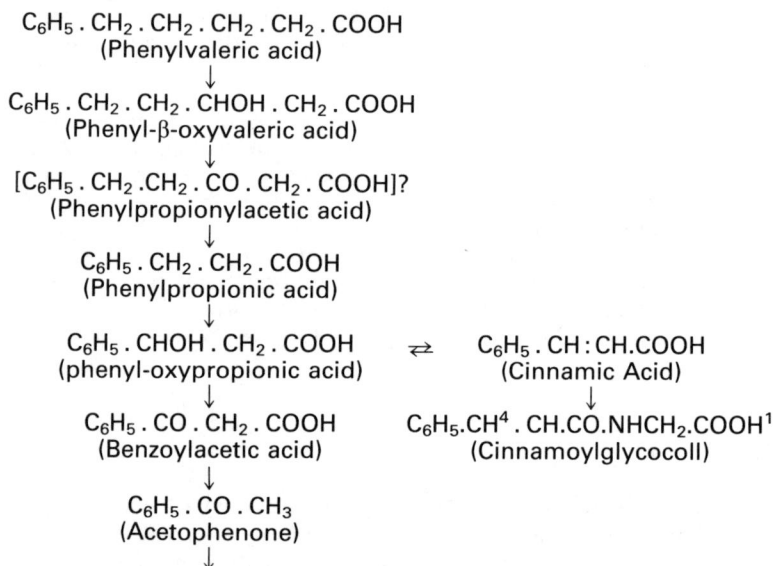

$C_6H_5 . CH_2 . CH_2 . CH_2 . CH_2 . COOH$
(Phenylvaleric acid)
↓
$C_6H_5 . CH_2 . CH_2 . CHOH . CH_2 . COOH$
(Phenyl-β-oxyvaleric acid)
↓
$[C_6H_5 . CH_2 .CH_2 . CO . CH_2 . COOH]$?
(Phenylpropionylacetic acid)
↓
$C_6H_5 . CH_2 . CH_2 . COOH$
(Phenylpropionic acid)
↓

$C_6H_5 . CHOH . CH_2 . COOH$ ⇄ $C_6H_5 . CH:CH.COOH$
(phenyl-oxypropionic acid)   (Cinnamic Acid)
↓   ↓
$C_6H_5 . CO . CH_2 . COOH$ $C_6H_5.CH^4 . CH.CO.NHCH_2.COOH$[1]
(Benzoylacetic acid)   (Cinnamoylglycocoll)
↓
$C_6H_5 . CO . CH_3$
(Acetophenone)
↓

C₆H₅COOH

C_6H_5COOH

(Benzoic acid)

↓

$C_6H_5 . CO . NH . CH_2 . COOH$

(Hippuric acid)

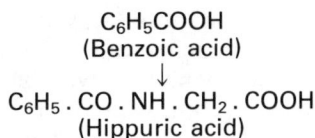

In order to determine whether this mode of oxidation of the side chain of phenylvaleric acid by which the four carbon groups are removed in two pairs, constituted a general type of reaction it was decided to investigate the fate of a number of derivatives of phenylvaleric acid. The substances examined were as follows:

Phenyl-α-β-pentenic acid	$C_6H_5.CH_2.CH_2.CH : CH.COOH$
Phenyl-β-γ-pentenic acid	$C_6H_5.CH_2.CH : CH.CH_2.COOH$
Cinnamylidene acetic acid	$C_6H_5.CH : CH.CH : CH.COOH$
Phenyl-γ-oxyvaleric acid	$C_6H_5.CH_2.CHOH.CH_2.CH_2.COOH$
Cinnamylidenemalonic acid	$C_6H_5.CH : CH.CH : C(COOH)_2$

Of these substances the first three were oxidized to benzoic acid and excreted as hippuric acid in the urine. In the case of each of these substances evidence was obtained that the oxidation took place in such a fashion that the five-carbon atom side-chain was converted primarily into a three-carbon atom side-chain and the latter again oxidized with a further loss of two carbon groups. Acetophenone and phenyl-β-oxypropionic acid, and probably cinnamoylglycocoll, were detected in the urine in each case. *The mode of oxidation of these three substances is completely analogous to that of phenylvaleric acid and phenyl-β-oxyvaleric acid.*

The two remaining substances, phenyl-γ-valeric acid and cinnamylidenemalonic acid, were scarcely attacked when administered to cats, for the greater part was excreted unchanged. The resistance of phenyl-γ-valeric acid to oxidation in the body is analogous to the similar behavior of phenyl-γ-oxybutyric acid. It appears that γ-oxy-acids are commonly oxidized in the body with difficulty and appear to be converted into lactones and excreted in the urine. The lactones as a class are much more resistant to oxidation *in vitro* than are the oxy-acids from which they are derived through loss of a molecule of water.

To sum up: evidence has been obtained that five acids of the type Ph. C.C.C.C.C.COOH undergo oxidation in the body in such a way that four carbon atoms are removed from the side chain in *two pairs*. In every case benzoic acid was the end product.

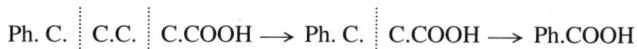

Ph. C. ┆ C.C. ┆ C.COOH → Ph. C. ┆ C.COOH → Ph.COOH

This type of oxidation may be termed successive β-oxidation, and I see no reason to suppose that it is not a general biochemical reaction.

Selection 148

In the meantime work appeared criticizing the theory of β-oxidation. Among those who came to object to it was Ernst Friedmann (1877–1956), working at the Charité Hospital in Berlin. His attention was drawn to the study of Adam Loeb, who reported regular and marked increases in acetoacetic formation, on perfusion through the surviving liver with blood to which acetic acid was added.[a] As these results contradicted his own,[b] Friedmann proceeded to ascertain the cause for the divergency, and traced it to the presence or absence of glycogen in the liver. The experiments indicated that the formation of acetoacetic acid was inhibited by the presence of glycogen.

As shown below, Friedmann interpreted the experiments as indicating the possibility that acetylation played a part in carbohydrate metabolism. Moreover, Friedmann believed that his results did not support the occurrence of the β-oxidation of fatty acids.

E. Friedmann, *Biochem.Z.*, **55**, 436. 1913

'Zur Kenntnis des Abbaues der Karbonsäuren im Tierkörper. XVII. Mitteilung. Über die Bildung von Acetessigsäure aus Essigsäure bei der Leberdurchblutung' (Concerning the knowledge of the degradation of carboxylic acids in the animal body. XVII. Communication. On the formation of acetoacetic acid from acetic acid during the perfusion of the liver with blood)

pp. 441–2

The fact that only in livers poor in glycogen, in contrast to livers rich in glycogen, a formation of acetoacetic acid after acetic acid addition takes place, could perhaps be explained thus that in the latter case the acetic acid predominantly finds use in the acetylation of carbohydrates or of substances which stand in physiological connection with these. This conjecture makes the assumption that the glycogen-rich livers are also capable of causing the disappearance of acetic acid added in the blood stream, a question with which we are engaged in experimental testing.

For the chemical processes, which control the physiological breakdown of the fatty acids, the observation that acetic acid considerably increases the range of acetoacetic acid formation in the surviving liver is of decisive meaning. As it is generally assumed that the normal, saturated fatty acids are broken down by acid splitting through the β-ketonic acids to acids poorer by two carbon atoms:

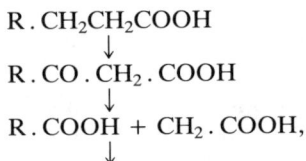

$$R \, . \, CH_2CH_2COOH$$
$$\downarrow$$
$$R \, . \, CO \, . \, CH_2 \, . \, COOH$$
$$\downarrow$$
$$R \, . \, COOH + CH_2 \, . \, COOH,$$
$$\downarrow$$

thus according to this conception all normal fatty acids should on their physiological breakdown yield acetic acid. As now acetic acid considerably raises the acetoacetic acid formation in the surviving liver it would be expected that all normal fatty acids would also form acetoacetic acid. In fact acetoacetic acid arises only from the normal fatty acids with an even number of carbon atoms, not from the normal fatty acids with an uneven number of carbon atoms. It therefore follows from this *that the coupled breakdown of the normal saturated fatty acids does not proceed through acetic acid splitting.* From this conclusion it further follows that the assumption *that the normal saturated fatty acids are broken down by acid splitting through the β-ketonic acids to fatty acids poorer by two carbon atoms is not justified.*

Selection 149

In our treatment of fatty acid metabolism so far, the part played by enzymic action has been absent. It will be useful, therefore, to look at developments following Hanriot's already mentioned discovery of the fat-splitting enzyme, lipase (1896).[a]

Shortly afterwards it occurred to the American worker, Arthur Salamon Loevenhart (1878–1929), that the enzyme could also be involved in fat synthesis. This idea then led to one of the earliest experiments on the reversibility of enzyme action, carried out by Loevenhart himself, in association with Joseph Hoeing Kastle (1864–1916). In their paper published in 1900 they showed that lipase was indeed capable of a reversible action by effecting a synthesis of ethyl butyrate from butyric acid and alcohol.[b] Also Hanriot independently, at the time of this work, reported the synthesis of monobutyrin from butyric acid and glycerine by the means of lipase.[c] Following Loevenhart's efforts to throw more light on the role of lipase in fat metabolism,[d] further evidence of its synthetic capability began to accumulate.

The selection below contains results on the synthetic activity of lipase reported in 1914 by the Czech biochemist, Antonín Hamsík [Hamsik] (1878–1963), who was among the first to investigate systematically enzymic hydrolysis and synthesis of fat.[e]

A. Hamsik, *Z.physiol.Chem.*, *90*, 489. 1914

'Zur synthetisierenden Wirkung der Endolipasen' (On the synthetic action of the endolipases)

pp. 490–1

In the research communicated in the following the organ ferments were investigated not on their fat-splitting, but on their synthetic capability ...

As the fat-splitting power of the organ ferments (with exception of the pancreas) is insignificant, their synthetic activity could be expected to be still weaker, because the synthetic action of the ferments always remains behind their splitting action. Negative results thus do not exclude the presence of small amounts of lipase, and indeed so much the less, as also the method used is not sufficiently sensitive; the positive results on the other hand point to an already rather greater lipase content of the organs concerned.

With the organs investigated positive results were obtained – apart from the pancreas and the intestinal mucosa respectively, which served as controls – only with the liver of the ox and the lung of the dog. The synthetizing action of these organs fell strongly behind that of the pancreas and could only be established with longer lasting experiments. The synthetizing action of the intestinal mucosa of the pig is indeed also much weaker than that of the pancreas, however still much stronger than that of the ox liver and of the dog lung. The sequence runs: pancreas, intestinal mucosa of the pig, liver of the ox, intestinal mucosa of the horse, lung of the dog; . . .

pp. 491–4

Experimental methodology: fresh organs were cut up into small pieces, treated with alcohol and ether, ground and sieved; the resulting powdered organs were used directly in the experiments. A number of small flasks was filled with a measured amount of oleic acid, glycerol (occasionally amyl alcohol) and sometimes n/10 soda solution; finally the organ powder was added. Then the acidity of the mixture was immediately estimated in a series of small flasks; the others were placed in thermostats (T = 37° during the day) and were only after the lapse of a definite time tested for their acidity . . .

Two lots of experiments were set up, with and without toluene addition. As the organ material through the treatment with alcohol was sterilized and the mixture richly contained glycerol, bacterial action was not to be feared. Without toluene therefore was one experimental series set up because for example occasionally the synthetizing action of the clear, active pancreas-glycerol-extract with greater toluene addition does not appear. As example should here be adduced: 3 small flasks were filled with 2 c.c. oleic acid; 1. besides the first flask contained 10 c.c. glycerol, 0.1 g pancreas powder from the pig and 5 c.c. toluene, 2. the second: 10 c.c. glycerol extract (1 g powdered pancreas extracted with 100 c.c of glycerol followed by clear filtration and 5 c.c. toluene, whilst 3. the third flask was prepared like the second, but without toluene. The initial acidity corresponded to 64.3 c.c n/10 KOH; after the flasks had stood 10 days in thermostats the following figures were obtained on titration: 1. 19.5 c.c., 2. 64.00 c.c., 3. 20.3 c.c n/10 KOH . . .

The results of the experiments are put together in Table [13]. The experiments with pancreas powder and those with amyl alcohol (instead of glycerol) serve for the control of the experimental arrangement. The diminution in acidity is greatest with use of pancreas (No. 3, 4, 15–17), with use of intestinal mucosa of the pig is it already smaller (No. 5); then follows the series: the liver of the ox (No. 42–27) and the intestinal mucosa of the horse (No. 51). The smallest however still undoubtedly a diminution was attained by the use of the lung of the dog (No. 25–28). The diminution in acidity obtained with the rest of the organ powders lies already within the limits of experimental error.

Selection 150

It will be remembered that *cholestérine* was the term chosen by Chevreul to denote the substance obtained from gallstones (1816).[a] Among the compounds associated with fats, cholesterol (as it became known in English) proved to be among the most difficult to deal with.

Thus not before 1888 did the molecular formula $C_{27}H_{46}O$, established by Friedrich Reinitzer come to seem probable.[b] Uncertainty about the constitution of cholesterol persisted much longer. Among researchers who brought particular insights to the constitutional formula of cholesterol was Adolf Windaus (1876–1959). In 1919, reviewing the current state of

Table 13

No.	Organ and amount of powdered organ	Oleic acid c.c.	Glycerine c.c.	Amyl alcohol c.c.	n/10-Soda-solution c.c.	Toluene c.c.	instantly	after 3 days (dog: 10 days)	after 3 weeks	after 4 weeks	Acidity diminution in c.c. n/10 KOH
	1. Without powdered organ										
1	—	2	10	—	—	—	64.1	—	63.5	63.1	0.6, 1.0
2	—	2	—	10	—	—	64.2	—	58.1	57.8	6.1, 6.4
	2. Powdered organ from the pig										
3	Pancreas 0.1	1	5	—	—	5	32.6	19.3	—	—	13.3
4	Pancreas 0.1	2	10	—	0.5	—	64.3	43.6	—	19.5	44.8
5	Intestinal mucosa 0.3	2	10	—	0.5	—	64.5	—	—	26.5	20.9, 38.0
6	Liver 1	2	10	—	—	—	64.6	—	65.2	—	0
7	Liver 1	2	10	—	—	2.5	64.6	—	63.4	—	1.2
8	Kidney 1	2	—	10	—	—	64.6	—	56.9	—	7.7
9	Kidney 1	2	10	—	—	—	64.6	—	64.6	—	0
10	Spleen 1	2	—	10	—	—	64.6	—	58.2	—	6.4
11	Spleen 1	2	10	—	—	—	64.6	—	63.0	—	1.6
12	Thymus 1	2	—	10	—	—	64.6	—	57.4	—	7.2
13	Thymus 1	2	10	—	—	—	64.6	—	62.0	—	2.6
14	Thymus 1	2	—	10	—	—	64.6	—	57.3	—	7.3
	3. Powdered organ from the dog										
15	Pancreas 0.2	1	5	—	—	5	32.5	14.0	—	—	18.5
16	Pancreas 0.1	2	10	—	0.5	—	64.3	—	—	21.1	43.2
17	Pancreas 0.1	2	—	10	0.5	—	64.3	—	—	6.5	57.8
18	Intestinal mucosa 0.5	2	10	—	0.5	—	64.3	—	—	62.4	1.9
19	Intestinal mucosa 1	2	10	—	—	—	64.3	—	61.9	—	2.4
20	Intestinal mucosa 1	2	10	—	—	2.5	64.3	—	63.9	—	0.4
21	Gastric mucosa 0.5	2	—	10	0.5	—	64.3	—	—	55.4	8.9
22	Gastric mucosa 0.5	2	10	—	0.5	—	64.3	—	—	61.1	3.2
23	Lung 0.5	2	—	10	0.5	—	64.3	—	—	58.6	5.7
24	Lung 0.5	1	5	—	—	5	32.3	31.4	—	—	0.9
25	Lung 0.5	2	10	—	0.5	—	64.3	—	—	59.8	4.5

Table 13 (Cont'd)

No.	Organ and amount of powdered organ		Oleic acid c.c.	Glycerine c.c.	Amyl alcohol c.c.	n/10-Soda-solution c.c.	Toluene c.c.	Acidity in c.c. n/10 KOH				Acidity diminution in c.c. n/10 KOH
								instantly	after 10 days	after 3 weeks	after 4 weeks	
	3. Powdered organ from the dog											
26	Lung	1	2	10	—	—	—	64.3	—	56.4	—	7.9
27	.	1	2	10	—	0.5	2.5	64.3	—	56.2	—	8.1
28	.	0.5	2	—	10	—	5	64.3	—	—	41.9	22.4
29	Liver	0.5	1	5	—	0.5	—	32.5	32.4	—	—	0.1
30	.	0.5	2	10	—	—	2.5	64.3	—	—	62.4	1.9
31	.	1	2	10	—	—	—	64.4	—	62.9	—	1.5
32	.	1	2	10	—	—	5	64.4	—	62.9	—	1.5
33	.	1	2	—	10	0.5	—	64.4	—	58.9	—	5.5
34	Kidney	0.5	1	5	—	—	2.5	32.5	32.2	—	—	0.3
35	.	0.5	2	10	—	—	—	64.4	—	—	61.4	3.0
36	.	1	2	10	—	0.5	5	64.4	—	60.8	—	3.6
37	.	1	2	10	—	—	—	64.4	—	62.8	—	1.6
38	Spleen	0.5	2	—	10	—	2.5	64.4	—	—	56.3	8.1
39	.	0.5	1	5	—	0.5	—	32.5	32.1	—	—	0.4
40	.	0.5	2	10	—	—	5	64.5	—	—	60.1	4.4
41	.	1	2	10	—	—	—	64.5	—	63.8	—	0.7
	4. Powdered organ from the ox								after 3 days			
42	Liver	0.5	1	5	—	—	—	32.5	31.4	—	—	1.1
43	.	0.5	2	10	—	0.5	—	64.5	—	—	28.8	35.7
44	.	0.5	5	2.5	—	0.25	2.5	157.6	—	—	120.0	37.6
45	.	1	2	10	10	—	—	64.5	—	48.4	—	16.1
46	.	1	2	10	—	—	2.5	64.5	—	58.0	—	6.5
47	Intestinal mucosa	0.5	2	10	—	0.5	—	64.5	—	—	18.3	46.2
48	Intestinal mucosa	1	2	10	—	—	—	64.5	—	60.0	—	4.5
49	.	1	2	10	—	—	2.5	64.5	—	62.2	—	2.3
50	.	1	2	—	10	—	—	64.5	—	57.4	—	7.1
	5. Powdered organ from the horse											
51	Intestinal mucosa	1	2	10	—	—	—	64.5	—	40.2	—	24.3

knowledge, he pinpointed three obstacles to clear identification of the carbon skeleton of cholesterol. First was the great number of carbon atoms (27) directly bound together in the molecule. Secondly, cholesterol (like many other plant and animal products) appeared to contain complicated hydrogenated ring systems, which were not then accessible to synthesis and therefore scarcely studied systematically. Thirdly, with the employment of harsh degradation methods cholesterol yielded decomposition products mutually hindering their crystallization, definite separation and purification.

Nevertheless, as shown below, Windaus put forward a constitutional formula of cholesterol. It was based on previous work which included the following points: the presence of one hydroxyl group that was part of a secondary alcoholic group;[c] the presences of a double linkage;[d] the relationship of cholesterol to the hydrocarbon cholestane $C_{27}H_{48}$, allowing the inference that cholesterol possessed four hydrogenated carbon rings in the molecule. (Cholestane contains eight hydrogen atoms less than the corresponding saturated hydrocarbon $C_{27}H_{56}$). Albeit tentative, the formula was found to be fruitful because it tied together much of the available data for the first time.

A. Windaus, *Nachr.Ges.Wiss.(Göttingen)*, **237.** 1919

'Die Konstitution des Cholesterins' (The constitution of cholesterol)

p. 254

On the ground of the existing experimental material the formula of cholesterol may be solved as follows:

$$(CH_3)_2 C \overset{H}{-} C \overset{H_2}{-} C \overset{H_2}{-} C_{13}H_{24}$$

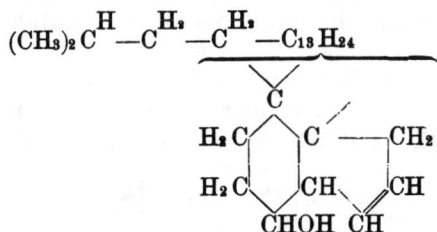

If one considers that proof has been brought for an octyl residue the formula can be written as follows:

$$(CH_3)_2 C \overset{H}{-} C \overset{H_2}{-} C \overset{H_2}{-} C \overset{H_2}{-} C \overset{H}{-} CH_3$$

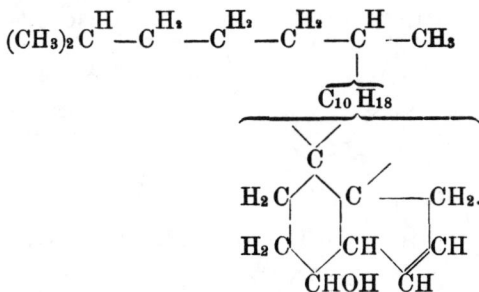

In the residue with 10 carbon atoms two saturated rings must be present ... Details determined with certainty are lacking at present, but structural formulae nevertheless can be derived which give a nearly correct picture of the large cholesterol molecule. One of the possible constitutional formulae is, for example, the following:

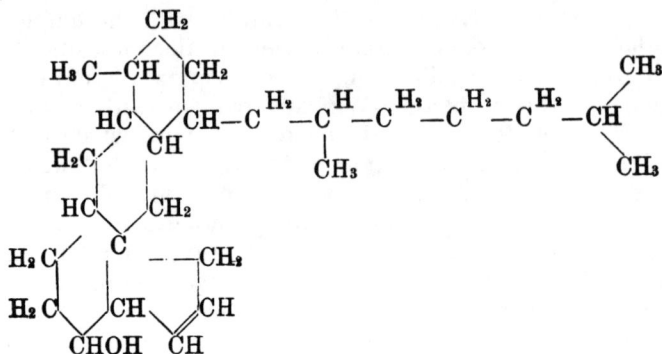

Selection 151

Since 1919 the recognition of the structural relation of cholesterol to the bile acids had been an important factor in gaining further understanding of its constitution. In that year it was shown in Windaus' laboratory that, by oxidation with chromium trioxide, pseudo-cholestane (a stereoisomer of cholestane) was converted to acetone and cholanic acid, $C_{24}H_{40}O_2$.[a] Three years previously (1916) the latter was proved, in the laboratory of Heinrich Wieland (1877–1957) at Munich, to be the parent substance of the bile acids.[b] It was also there, a decade later (1926), that the reverse process was demonstrated, resulting in the synthesis of the pseudo-cholestane from cholanic acid.[c]

By the late 1920s the bile acids and cholesterol were thought to contain a nucleus made up of four fused rings, as represented in the following formulae, devised by Wieland and Windaus respectively.[d]

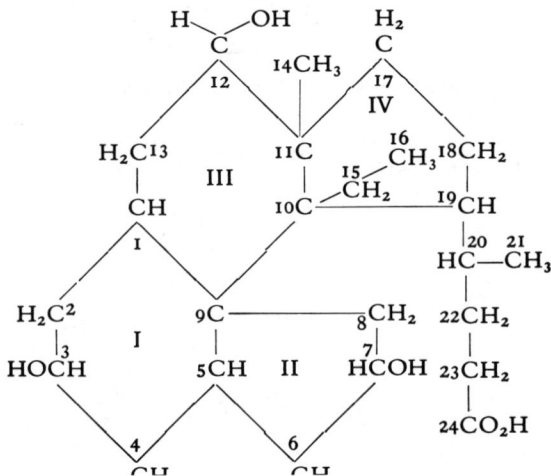

$$CH_2$$

$$H_2C \quad\quad CH_3$$
$$C\text{————}CH_2$$

$$HC \quad\quad CH_2\text{—}CH_3 \quad CH_2$$
$$C\text{————}CH\text{—}CH\text{—}CH_2\text{—}CH_2\text{—}CH_2\text{—}CH$$

$$CH_3 \quad\quad CH_3$$

$$H_2C \quad C\text{————}CH_2 \quad\quad CH_3$$

$$H_2C \quad CH \quad CH$$

$$HCOH \quad CH$$

Cholesterol

In the meantime another centre for studies of cholesterol and the bile acids was developing at Kiel, under Otto Diels (1876–1954). The work there was an outgrowth of Diels' interest in the relationship between cholesterol and cholestane earlier in the century (1908) and it has been noted in the introduction to the previous selection.

Between 1925 and 1929 Diels and his associates established that dehydrogenation of cholesterol with palladium charcoal and selenium led to the production of chrysene, $C_{18}H_{12}$, and two other hydrocarbons, $C_{18}H_{16}$ and $C_{25}H_{24}$.[e]

As attested by Windaus, the significance of these results for the study of the cholesterol and bile acids structures was not fully appreciated at first.[f] But interest in Diels' work mounted during the early 1930s when it became apparent that the four fused rings did not compose the nucleus of these substances, as envisaged by Windaus and Wieland. Of growing importance was the hydrocarbon, $C_{18}H_{16}$ ('Diels' hydrocarbon'), the constitution of which was not clear. (Chrysene was known to contain four benzenoid rings.)

The uncertainty arose, in part, from X-ray studies on calciferol, the crystalline preparation of vitamin D, undertaken by John Desmond Bernal (1901–71) at Birkbeck College (London). By 1930 he was concerned to extend X-ray crystallography to the study of structures of biologically important compounds.[g] Calciferol, known to be related to cholesterol and the bile acids, was apparently the compound Bernal investigated first. From the very beginning his X-ray measurements led him to distrust the formula ascribed to calciferol by Windaus and Wieland.[h]

Calciferol was also investigated chemically at that time by Otto Rosenheim (1871–1955) and Harold King (1887–1956) at the National Institute for Medical Research in London. Sharing Bernal's doubts about the structures developed by Wieland and Windaus, they turned to the work of the group at Kiel.

As shown in the first extract, Rosenheim and King drew on the information provided by Diels and his associates and proposed to replace the nucleus in the formulae of Wieland and Windaus by a chrysene nucleus. The second extract contains Bernal's endorsement of the new skeleton.

How did Windaus and Wieland respond? To Windaus the Rosenheim and King's formulation was 'immediately congenial though it was not at once clear to me how the new formula could be brought in line with Wieland's results.'[i] As to Wieland, he and his collaborator Elisabeth Dane regarded the proposal of Rosenheim and King 'as a very valuable stimulation' but did not agree with them that ring IV should become six-membered – it was to remain a five-membered structure.[j]

As shown in the extracts from the third paper below, Rosenheim and King recognized Wieland's modification of ring IV. As to Diels' hydrocarbon, $C_{18}H_{16}$, they suggested that it was a methylcyclopentenophenanthrene.

O. Rosenheim and H. King, *J.Soc.Chem.Ind.*, *51*, 464. 1932

'The ring-system of sterols and bile acids'

p. 465

On the basis of the available experimental evidence, which for lack of space cannot be reviewed here, we suggest that the ring-system of the sterols and bile acids is that of chrysene, and consists of four six-membered rings as shown in (VI). Chrysene is obtained in good yield by dehydrogenation of cholic acid with selenium, and the same reaction[1] leads from both cholesterol ($C_{27}H_{46}O$) and ergosterol ($C_{28}H_{44}O$) to the same partially hydrogenated hydrocarbons: $C_{18}H_{16}$ and $C_{25}H_{24}$ ($C_{26}H_{24}$?). The analytical results for the latter are in better agreement with the formula $C_{26}H_{24}$ that with $C_{25}H_{24}$.

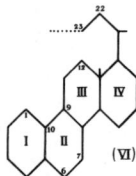

From (VI) the constitutional formulæ of these hydrocarbons may be arrived at, i.e., $C_{26}H_{24}$ by ring closure and loss of CH_3 groups, and $C_{18}H_{16}$ by splitting off of the side-chain and CH_3. These formulæ are in excellent agreement with the established experimental facts (formation of a mono-ketone and of a dinitro-derivative), thus affording additional evidence for the view that the chrysene skeleton exists preformed in the molecule.

J.D. Bernal, *J.Soc.Chem.Ind.*, *51*, 466. 1932

'Carbon skeleton of the sterols'

p. 466

X-Ray examination of sterols[1] pointed clearly to the fact that the older accepted formulæ for these compounds could not be made to fit into the crystallographic cells. The new formula, proposed by Rosenheim and King (see above), however, is not open to this objection, and although it cannot be demonstrated by X-ray methods as the correct formula, it serves to explain most of the general properties of the crystals whose structures have been studied.

Of three formulæ for the ergosterol ring system I and II are the earlier and

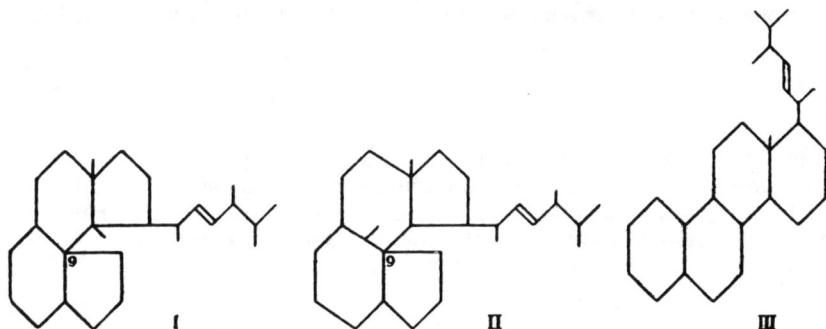

more recent accepted forms and III is Rosenheim and King's formula. If models are made to scale of these formulae and their dimensions measured with due allowance for the space occupied by hydrogens, the following results are obtained:

Observed	$7.2 \times 5.0 \times 17{-}20$Å.
III	$7.5 \times 4.5 \times 20$
I	$8.5 \times 7.0 \times 18$
II	$11.0 \times 7.5 \times 15$

The impossibility of I and even more so of II is apparent; this is due to the three rings meeting on C_9 which cannot be in plane configuration. The new formula, on the other hand, gives very reasonable agreement, the greater length being accounted for by tilt of the molecules. Optical evidence is also in favour of III.

The same considerations hold for the related structures of cholesterol itself and suprasterol II where the cell dimensions are essentially the same, although the symmetry is in one case triclinic and in the other orthorhombic.

O. Rosenheim and H. King, *J.Soc.Chem.Ind.*, **52**, 299. 1933

'The ring-system of sterols and bile acids. Part III. Observations on the structure of the Diels' hydrocarbon $C_{18}H_{16}$'

p. 299

The nature of ring IV in the sterol and bile acid formula, which we proposed a year ago, still remains undecided. The available experimental evidence, which has recently been critically reviewed by Ruzicka and Thomann,[1] is inconclusive and allows the formulation of ring IV as either a cyclopentane or cyclohexane ring.

A knowledge of the constitution of the hydrocarbon $C_{18}H_{16}$, obtained by Se-dehydrogenation from cholesterol,[2] ergosterol,[3] and cholic acid[1] is of decisive importance in this respect. Owing to the difficulties connected with its isolation and purification however, this hydrocarbon is obtainable in such small amounts only that an elucidation of its structure by direct chemical methods has hitherto not been feasible. We had therefore recourse to physical methods, and by a comparative study of the ultra-violet absorption spectra of various chrysene- and phenanthrene-derivatives arrived at the conclusion that the hydrocarbon $C_{18}H_{16}$ is a substituted phenanthrene (this JOURNAL, 1933, p. 10).

p. 300

Since the spectroscopic evidence affords, however, conclusive proof that the hydrocarbon $C_{18}H_{16}$ is a phenanthrene derivative, we consider, taking in account the other evidence, that its constitution may be represented as that of a methyl-1 : 2-cyclo-pentenophenanthrene, i.e., a substituted hydrindene with a secondary methyl group in the cyclopenteno-ring.

p. 301

We suggest that when the side-chain at C_{17} is removed from cholesterol (ergosterol or cholic acid) by selenium, the methyl group at C_{13} takes up the adjacent vacant position on C_{17}, thus preserving the saturation of the five-membered ring IV. The hydrocarbon $C_{18}H_{16}$ on this view has the constitution 3-methyl-6 : 7 : α : β-naphthohydrindene (II).

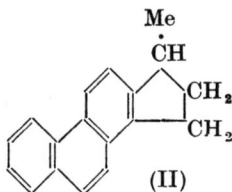

(II)

Selection 152

The major division of biochemically important substances into proteins, carbohydrates and fats originated in an appreciation of their place in human nutrition.[a] The interest was centred in these nutrients in efforts to establish the desirable qualitative and quantitative composition of a diet, including their relations and interchangeability. It was during the First World War that the practical aspects of these problems became of central concern to the governments and peoples of the warring countries. Thus the official Allied view, based on top scientific advice, was that the 'average man' needed 70 g of protein and 70 g of fat per day as the necessary minimum.[b]

Nevertheless, in 1920, Osborne and Mendel, who had long been concerned with identifying 'indispensable' or 'essential' substances in diets,[c] admitted that there was 'no available information respecting the actual requirement of the healthy mammal for fat'. The reason why this question had not been answered they found in 'the experimental difficulties heretofore inherent in its isolation'.[d] In fact, this was no more than a reiteration of what they had already stated eight years previously, in the face of the laborious nature of the work in which they were engaged: the complete removal of fats from food materials in which they are normally found.[e]

In about 1926 George Oswald Burr (*b*.1896) from the University of Minnesota (Minneapolis) took up the question of the requirement of animals for fat by devising new diets of high purity which were extremely low in fat. Between 1929 and 1932 he and his associates reported pathological phenomena which had not been previously noticed in rats put on fat-free diets. Symptoms included scaliness of the skin, kidney degeneration, water imbalance and sexual disorders. The American workers characterized them as a fat deficiency disease arising out of the absence of certain unsaturated acids from the diet.

The following excerpts from the third paper of the series contain the results of the studies that led to the conception of dietary 'essential fatty acids' which must be supplied in food if normal metabolic processes are to continue, as the animal (human) body itself cannot produce them.

G.O. Burr, M.M. Burr and E.S. Miller, *J.Biol.Chem.*, 97, 1. 1932

'On the fatty acids essential in nutrition. III'

p. 1

After it was demonstrated[1] that a deficiency disease was caused by the lack of fatty acids in the diet, a study of the well known natural fatty acids was undertaken. In the second paper of this series[2] it was shown that none of the saturated fatty acids occurring in hydrogenated coconut oil was effective in curing the disease and promoting renewed growth of the animal. Pure methyl linolate is highly effective in curing sick animals and all oils which contain appreciable amounts of this acid are likewise good.

Since butter (30 per cent oleic acid) gave very poor results at the high level of 300 mg. daily (3 per cent of the diet) it was postulated that oleic acid is entirely negative and the small effects due to butter were due to traces of linoleic acid. But a sample of commercial methyl oleate gave good results and the value of oleic acid was left uncertain. Other common fatty acids which are now being studied are linolenic, arachidonic, and eleostearic. It is the object of this paper to report the results of some studies of these acids.

p. 2

Rats are weaned when 21 days old and put on the low fat diet. They must weigh over 36 gm. on weaning day. The weight curves reach a plateau when the rats are about 150 gm. in weight and when it has been established that they have reached their maximum weight and are actually declining slightly they are used as cures. Positive results are marked by a clearing of the skin, improvement of hair coat, and renewed growth both in length and weight. Increase in weight is used as the quantitative measure of the effectiveness of an oil or fatty acid.

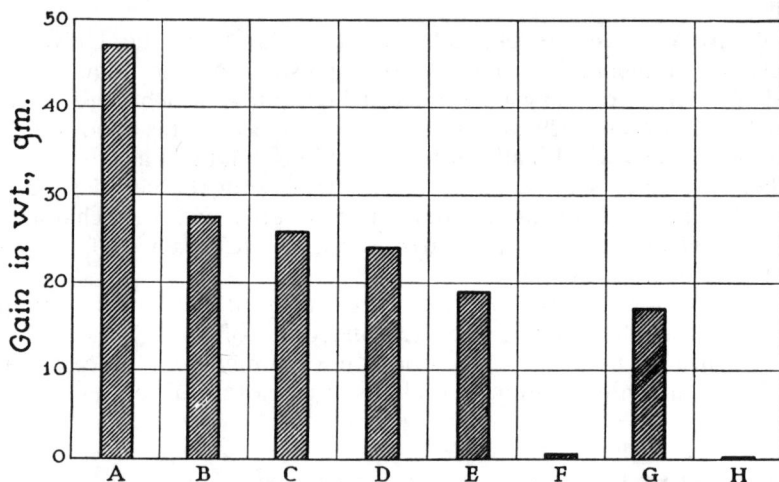

CHART V. A summary of the results given in the previous charts. The columns give the average maximum gains in weight of groups receiving the supplements over a period of 40 days. Column A, $7\frac{1}{2}$ drops of methyl linolate daily; Column B, 3 drops of methyl linolate daily; Column C, 3 drops of methyl linolenate daily; Column D, 3 drops of methyl linolate plus methyl linolenate (1 : 1 mixture) daily; Column E, 3 drops of methyl linolate plus methyl linolenate (1 : 1 mixture) plus 10 per cent methyl arachidonate daily; Column F, 5 drops of methyl oleate (from butter and from olive oil) daily; Column G, 5 drops of tung oil (90 per cent eleostearic acid) daily; Column H, 5 drops of methyl-α-eleostearate daily.

pp. 7–8

A general summation of the comparisons of oleic, linoleic, linolenic, and α-eleostearic acids is given in Chart V. In all cases where considerable growth took place the skin cleared and the rats were generally improved. A better muscle tone is always noticeable after a rat has been cured.

By this work oleic acid has been definitely grouped with the saturated acids as ineffective in the curing of rats subnormal because of the lack of fat. This substantiates the arguments put forth in the second paper of this series[2] that it is possible for animals to synthesize from carbohydrates large amounts of fat and still suffer from a fat deficiency ... it seems, clear that warm blooded animals synthesize only the saturated acids and oleic acid and that they are dependent upon the food supply for linoleic and linolenic acids. One of these two acids must be ingested by the rat if it is to survive and our findings indicate that they are interchangeable in the tissues . . .

The comparison of whole tung oil with methyl-α-eleostearate is interesting. Since the α-eleostearic acid does not have any curative effect it is evident that there is an acid in tung oil in small amounts which causes the renewed growth. Similar effects were seen when 15 drops of butter were fed daily to rats[2]. Since pure oleic acid and the saturated fatty acids are ineffective, small amounts of undetermined acids are assumed to be present. These acids are probably linoleic or linolenic.

A mixture of linoleic and linolenic esters is of no more value than either of the esters alone (Chart V, Column D). This is interesting since tissues normally have a mixture of the two. When methyl arachidonate was added as 10 per cent of the mixture the animals uniformly showed less response (Chart V, Column E). The reason for this is not at all clear. Lard contains appreciable amounts of arachidonic acid and it is one of the best curative fats. Liver and liver fat are rich sources of arachidonic acid. Both have been used by us as preventives for the fat deficiency and have proved highly effective. Since there is no reason to attribute toxic effects to small amounts of arachidonic acid it seems probable that some of the purified arachidonic acid which we have fed has been altered in the process of preparation.

Selection 153

We now return to studies of oxidation of fatty acids during the period 1930–40 which arose from doubts about whether fatty acids underwent degradation only by β-oxidation in the animal and human body. In this context, an active research programme was carried out by Pieter Eduard Verkade (1891–1979) and his co-workers at the Dutch Commercial University at Rotterdam.

The starting point was their discovery of considerable quantities of undecane dicarboxylic acid, $HOOC(CH_2)_{11}COOH$, in urine after administration of triundecylin (the triglyceride of undecylic acid, $C_{11}H_{22}O_2$). Their interpretation was that the terminal methyl group of the component acid of triundecylin was oxidized to a carboxyl group:

$$CH_3(CH_2)_9COOH \longrightarrow HOOC(CH_2)_9COOH$$

They termed the oxidation of the terminal CH_3 group to COOH (leading to a long-chain dicarboxylic acid) ω-oxidation.[a]

This was followed by attacking the question

whether the administration of chemically pure simple triglycerides of other higher and lower, odd and even saturated fatty acids leads similarly to the excretion of corresponding dicarboxylic acids and if so, to what extent.[b]

Regarding the ω-oxidation of the component fatty acid, their results suggested no difference between triglycerides derived from acids with odd and even numbers of carbon atoms. The largest effect was found with triundecylin and tricaprin, derived from C_{11} and C_{10} acids respectively; the effect diminished on descending the series and none was given by the higher triglycerides.

Continuing their work along these lines, the Dutch investigators pointed to the probability that both modes of oxidation took place. They envisaged that initially ω-oxidation produced a dicarboxylic acid which underwent consecutive β-oxidations to lower dicarboxylic acids. They introduced the term 'bilateral β-oxidation' for this type of successive removal of pairs of carbon atoms at one or both ends of the molecule, in order to distinguish it from the 'unilateral β-oxidation' which they associated with simple β-oxidation of fatty acids proposed by Knoop.

The general scheme of decomposition containing the ω- and β-oxidations was pictured in a later paper as follows:

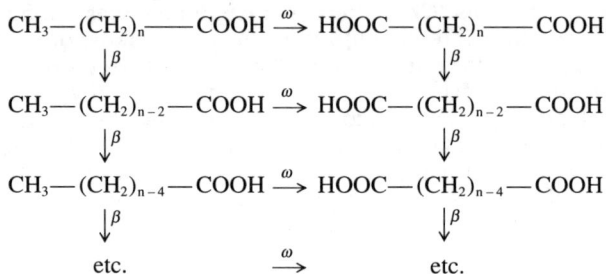

$$CH_3-(CH_2)_n-COOH \xrightarrow{\omega} HOOC-(CH_2)_n-COOH$$
$$\downarrow \beta \qquad\qquad\qquad\qquad \downarrow \beta$$
$$CH_3-(CH_2)_{n-2}-COOH \xrightarrow{\omega} HOOC-(CH_2)_{n-2}-COOH$$
$$\downarrow \beta \qquad\qquad\qquad\qquad \downarrow \beta$$
$$CH_3-(CH_2)_{n-4}-COOH \xrightarrow{\omega} HOOC-(CH_2)_{n-4}-COOH$$
$$\downarrow \beta \qquad\qquad\qquad\qquad \downarrow \beta$$
$$\text{etc.} \qquad\qquad \xrightarrow{\omega} \qquad\qquad \text{etc.}$$

It is from this paper that their conclusion, not unreservedly accepted by other workers in the field, is given in the extract below.

P.E. Verkade and J. van der Lee, *Z.physiol.Chem.*, *227*, 213. 1934

'Untersuchunge über den Fettstoffwechsel. IV. Zweitseitige β-Oxydation der durch ω-Oxydation gesättigter Fettsäuren entstandenen Dicarbonsäuren' (Researches on fat metabolism. IV. Bilateral β-oxidation of the dicarboxylic acids produced by ω-oxidation of saturated fatty acids)

p. 218

By ω-oxidation of the fatty acids a corresponding dicarboxylic acid is formed which then, passing through lower dicarboxylic acids, arising from β-oxidation, is further broken down. It is indeed highly interesting that also with this mode of degradation the rule of β-oxidation found by Knoop[1] again predominates. One could be inclined to see in the ω-oxidation only an auxiliary measure of the organism whereby the fatty acid molecule receives a second point of attack for the β-oxidation, namely a second terminal carboxyl group; the possibility for the appearance of a process which we will name *bilateral β-oxidation* is thus created.

Selection 154

An early challenge to the overall validity of the β-oxidation theory, as formulated by Knoop, came from W.H. Hurtley, author of an extensive study of the state of diabetics (1915–16).[a] Under diabetic conditions, it was known that ketone bodies were excreted in the urine, due to inordinate fat metabolism. As their genesis was attributed to butyric acid, Hurtley was puzzled that the body of diabetics did not smell of butyric acid, although their breath smelled of acetone. He also found it perplexing that neither the fat from the tissue nor the lipaemic blood of diabetics, who died in coma, smelled of butyric acid. In the light of the theory of β-oxidation, it was expected that the chief fatty acids of the food (palmitic, stearic, oleic) produced intermediate lower fatty acids of the series C_nH_{2n+1}.COOH that included the fourth (butyric acid) to the twelfth member (lauric acid). Since they were not detected in the human body, Hurtley suggested that probably the high fatty acids were not merely attacked at the β-carbon atom but also at other carbon atoms along the whole length of the chain. The transformation was pictured as follows:

$$-CH_2CH_2.CH_2COOH \longrightarrow -CO.CH_2CO.CH_2COOH$$

It was to take some twenty years before Maurice Jowett and Juda Hirsch Quastel (1899–1987) supplied evidence that appeared to favour this type of oxidation of fatty acids, which they called 'multiple alternate oxidation'.

The point at issue was that the study of the metabolism of fatty acids had been hampered by the difficulty in obtaining quantitative results and comparisons. The change was to come from the application of manometry and the tissue slice technique (as developed by Warburg) to problems of fatty acid metabolism by Quastel, in collaboration with A.H.M. Wheatley in 1933. They found that fatty acids were oxidized at considerable rates by slices of liver *in vitro*, giving acetoacetic acid as one of the oxidation products.[b]

Jowett and Quastel's hypothesis of multiple alternate oxidation of fatty acids derived from their paying further attention to the quantitative yields of acetoacetic acid. They found the yields were greater than could be expected on the basis of the β-oxidation theory.

Their discussion of their results in relation to the inadequacy of the β-oxidation theory which led them to put forward the conception of oxidation at alternate carbon atoms along the chain is presented below.

M. Jowett and J.H. Quastel, *Biochem.J.*, *29*, 2159. 1935

'Studies in fat metabolism. II. The oxidation of normal saturated fatty acids in the presence of liver slices'

pp. 2177–9

We have found, in accordance with Embden,[a] that the rate of production of acetoacetic acid from fatty acids varies in an alternate manner, the even-numbered acids producing much more acetoacetic acid than their odd-numbered neighbours. A new result is that, with guinea-pig liver at least, the odd-numbered fatty acids – apart from formic and propionic acids – also give rise to small but significant amounts of acetoacetic acid.

It is therefore likely that β-oxidation, if it occurs at all, is not the only type of oxidation which takes place with the odd-numbered acids.

The fatty acids produce β-hydroxybutyric acid as well as acetoacetic acid. The hydroxy-acid is probably produced by reduction of acetoacetic acid, for this has been shown to be true in the case of crotonic and butyric acids . . . [b] The production of total ketones is therefore in all probability the best measure of the gross acetoacetic acid production.

The increase in respiration of guinea-pig liver brought about by odd-numbered acids is greater than that due to their even-numbered neighbours, which suggests that the odd-numbered acids are more completely oxidised. This more complete oxidation would be expected if odd-numbered acids yield a three-carbon acid which is readily oxidised by the liver, in contrast to acetoacetic acid, which is little oxidised.

The yields of total ketone bodies obtained from the even-numbered acids of 4 to 10 carbon atoms with guinea-pig liver are of the order of one molecule of ketone bodies per molecule of fatty acid.

So far, then, we have a number of facts which are in general accord with the β-oxidation theory, although there is a suggestion that the odd-numbered acids possess also another mode of oxidation which may not be quantitatively of great importance. The facts next to be considered may also be interpreted on the β-oxidation theory, if we add to this an assumption regarding the nature of the product formed from the two carbon atoms lost in each stage of oxidation. We have, in fact, found that fixed acid is produced during the oxidation of the higher fatty acids by guinea-pig liver. The acid (or acids) has not yet been identified. If it should be acetic acid, or another 2-carbon acid, our results regarding acid production are in qualitative agreement with the predictions of the β-oxidation theory. These results have already been discussed.

The suggestion that acetic acid is a product of fatty acid oxidation leads to the query whether acetic acid might not be the source of the acetoacetic acid formed by the higher odd-numbered fatty acids. Since these acids form acetoacetic acid at a higher rate than does acetic acid with guinea-pig liver, it is probable that the explanation does not suffice, and that β-oxidation will not explain the acetoacetic acid production of odd-numbered acids.

Considering more quantitatively the yields of ketone bodies obtained from the even-numbered fatty acids with guinea-pig liver, we see that butyric acid gives a lower yield than do the higher fatty acids. Here, again, we may ask whether the higher yield obtained from the higher acids may not be due to their yielding acetic acid, which may be a source of additional ketone bodies. From our data, we consider that this possible explanation of the higher yields is inadequate. Now, according to the β-oxidation theory, butyric acid may give a higher yield of ketone bodies than do the higher fatty acids but cannot give a lower yield. Taking into account acetic acid as a possible intermediary, we find the position little altered. We, therefore, consider that the facts regarding yields are incompatible with the β-oxidation theory.

A more striking discrepancy still is found in the effect of benzoate on the rates of acetoacetic acid and total ketone body production from the even-numbered fatty acids. The rates are much more strongly inhibited in the case of butyrate than with the higher fatty acids. In the absence of an inhibitor the rates of acetoacetic acid production from the 4-, 6- and 8-carbon acids are approximately equal. In the presence of benzoate the 6-, 8- and 10-carbon acids yield acetoacetic acid and total ketones at definitely higher rates than does butyric acid. Since the production from acetic acid is low, and is as strongly inhibited by benzoate as that from butyric acid, it does not avail to suggest that acetic acid is an intermediary and is causing an apparent discrepancy with the β-oxidation theory. We must therefore conclude that the even-numbered fatty acids are not broken down through butyric acid as an intermediary. Nor, according to our data, can hexanoic acid be an intermediary for the higher acids.

We are therefore forced to abandon the β-oxidation theory. In attempting to formulate a new hypothesis to replace it, we will first point out that at the present stage such a hypothesis must differ from the β-oxidation theory more in the mechanism of oxidation which it proposes than in the nature of the products which it supposes to be formed.

We propose the view that fatty acids undergo, at a common enzyme an oxidation throughout the fatty chain, alternate carbon atoms being affected. The oxidised product then breaks down. We may term this a theory of 'multiple alternate oxidation'[1]

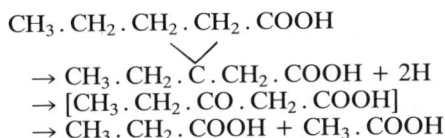

At the present time, the detailed application of this theory is necessarily somewhat speculative. The oxidised intermediary substances, which may exist only in combination with the enzyme, may be supposed to be capable of breaking down in a number of ways.

The theory is illustrated below by the supposition that an even-numbered acid, octanoic acid, undergoes oxidation at its β-, σ- and γ-carbon atoms before the chain breaks, and that an odd-numbered acid, valeric acid, may undergo either β-oxidation or oxidation at the α- and γ-carbon atoms.

(1) Even-numbered acid:

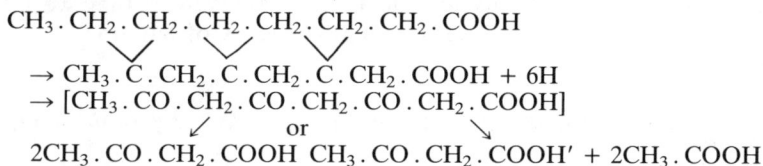

$$CH_3.CH_2.CH_2.CH_2.CH_2.CH_2.CH_2.COOH$$

$$\rightarrow CH_3.\overset{.}{C}.CH_2.\overset{.}{C}.CH_2.\overset{.}{C}.CH_2.COOH + 6H$$
$$\rightarrow [CH_3.CO.CH_2.CO.CH_2.CO.CH_2.COOH]$$
$$\text{or}$$
$$2CH_3.CO.CH_2.COOH \quad CH_3.CO.CH_2.COOH' + 2CH_3.COOH$$

(2) Odd-numbered acid:

(a) β-oxidation

$$CH_3.CH_2.CH_2.CH_2.COOH$$

$$\rightarrow CH_3.CH_2.\overset{.}{C}.CH_2.COOH + 2H$$
$$\rightarrow [CH_3.CH_2.CO.CH_2.COOH]$$
$$\rightarrow CH_3.CH_2.COOH + CH_3.COOH$$

(b) αγ-oxidation

$$CH_3.CH_2.CH_2.CH_2.COOH$$

$$\rightarrow CH_3.\overset{.}{C}.CH_2.\overset{.}{C}.COOH + 4H$$
$$\rightarrow [CH_3.CO.CH_2.CO.COOH]$$
$$\downarrow \text{ or } \searrow$$
$$CH_3.CO.CH_2.COOH \qquad CH_3.COOH$$
$$+ CO_2 \text{ (or } H.COOH) \qquad + CH_3.CO.COOH$$

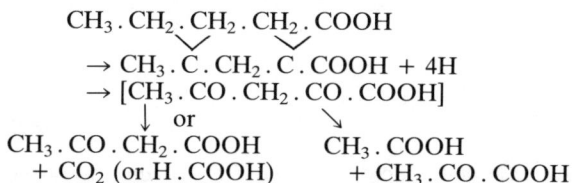

The detailed working out of our hypothesis is still, as just pointed out, a matter of speculation. We cannot dismiss the possibility that even-numbered fatty acids may undergo oxidation at their α-, γ-, ε-... carbon atoms, of the possibility that ω-oxidation may occur with the higher fatty acids to some considerable extent.

We can do little more now than present the hypothesis that oxidation proceeds along the whole chain of a fatty acid molecule at alternate carbon atoms, with the ultimate formation of acetoacetic acid and other acid products. Oxidation at the β-, δ-... carbon atoms is considered to be the major process in the oxidation of fatty acids containing four or more carbon atoms.

Selection 155

During the years around 1940 interest continued in the production of ketone bodies and their relationship to fat metabolism. Acetoacetic acid was considered to be the primary ketone body and it was to the question of its metabolic origin that considerable attention was devoted. It followed on from the already mentioned liver perfusion experiments of A. Loeb and E. Friedmann earlier this century (1912, 1913), which had connected the formation of acetoacetic acid with acetic acid.[a]

The excerpts below are taken from a study which was specifically 'designed to determine whether or not acetic acid is ketogenic in the intact organism'. On feeding acetic acid to a phlorrhizinized[b] dog or to fasting rats, an increase in the excretion of ketone bodies in the urine was observed by Eaton Macleod MacKay (1900–73) and his collaborators. Not due to changes in the renal threshold, the results were interpreted as evidence for the conversion of acetic acid to acetoacetic acid. Moreover, they were taken to support a modified view of the β-oxidation theory of Knoop.

E.M. MacKay, R.H. Barnes, H.O. Carne and A.N. Wick, *J.Biol.Chem.*, *135*, 157. 1940

'Ketogenic activity of acetic acid'

p. 157

When acetic acid is fed to fasting rats (Table 14), there is always an increase in the ketonuria. This is true whether the acetic acid is fed as such or partially or entirely neutralized. Blood acetone body determinations indicate that this increased urinary excretion is not a result of changes in the renal threshold but results from an increased production of acetone bodies.

p. 159

p. 160

This demonstration that acetic acid may give rise to acetone bodies in the intact mammalian organism has an important bearing upon the current conception of the method by which fatty acids are oxidized in the organism under circumstances such that acetone bodies are being formed. Acetone bodies are produced in the liver and carried by the blood stream to the muscles and other tissues for final oxidation. The old conception of

Table 14 Ketogenic Activity of Acetic Acid in Fasting Rats

Experiment No.	No. of rats*	Average body weight (gm.)	Average body surface (sq. dm.)	Solution fed in doses† of 1 cc. per sq. dm. body surface	Urine N per sq. dm. body surface per day (mg.)	Urine acetone bodies per sq. dm. body surface per day‡				Blood acetone bodies at end of			
						1st day (mg.)	2nd day (mg.)	3rd day (mg.)	4th day (mg.)	1st day (mg. per cent)	2nd day (mg. per cent)	3rd day (mg. per cent)	4th day (mg. per cent)
1	6	173	3.5	1.0 M Na HCO₃	30.5	2.7	5.0	3.0	2.7				
	6	173	3.5	1.0 ' , ' + 0.5 M acetic acid	31.6	4.9	28.2	8.5	5.0				
2	3	199	3.9	0.75 ' , '	15.7	0.6	16.0	35.9	35.5				
	8	198	3.9	0.75 ' , ' + 0.75 M acetic acid	16.4	17.4	50.6	54.4	51.2				
3	6	183	3.7	0.75 ' , '	18.3	7.1	12.3	24.9	25.5				
	6	185	3.7	0.75 ' , ' + 1.0 M acetic acid	24.9	50.9	72.1	61.6	66.8				
4	4	151	3.2	0.75 ' , '	21.5	0.8	2.2	2.5					
	4	147	3.2	0.75 ' , ' + 0.75 M acetic acid	26.7	2.9	27.2	28.1					
5	3	201	3.9	0.25 M Na NaCl	19.0	1.3	25.3	17.6	15.1		50		67
	3	198	3.9	0.25 ' , ' + 0.25 M acetic acid	19.5	3.5	30.4	31.0	24.3		51		60
6	4	261	4.6	0.25 ' '	16.9	15.3	18.9	13.6	10.8	41	47	40	45
	4	264	4.7	0.25 ' ' + 0.25 M acetic acid	19.9	15.9	20.2	30.9	20.6	45	46	49	43
7	6	237	4.3	Water	15.3	12.1	22.7	16.2	5.1		58	61	45
	6	237	4.3	0.25 M acetic acid	18.7	22.2	24.4	20.9	15.1		61	72	59
	6	237	4.3	0.25 M Na HCO₃ + 0.25 M acetic acid	15.3	75.2	51.0	39.2	20.4		66	68	59

* Male rats were used in Experiments 1, 6, and 7. Female rats comprised the other experiments.

† Two doses of the solutions listed were fed each day except in Experiments 5 and 6 in which three and four doses respectively were administered daily.

‡ The rats in Experiments 1 and 4 were fasted directly from the stock diet, while all of the rest received a low protein diet for 4 to 23 days prior to fasting.

β oxidation with the production of only 1 molecule of acetone body from 1 molecule of even the long chain fatty acids is no longer tenable. On this basis an impossible amount of fat would have to be burned to furnish the quantity of acetone bodies which is known to be utilized during a ketosis . . .

p. 161

With the evidence which has been presented in the experimental portion of this paper that acetic acid is an acetone body former the classical β oxidation hypothesis may be modified so that it may explain all of the findings which have been cited against it.

When a fatty acid is oxidized by β oxidation, 2 carbon atoms are dropped from the chain together. What form this cleavage takes is unknown but it is usually pictured as productive of acetic acid. If this were true and the acetic acid formed acetone bodies which we know it may do, there would be little difficulty in retaining the β oxidation hypothesis in the light of our present knowledge. All of the carbon atoms in a fatty acid molecule with an even number of carbon atoms could yield acetone bodies (1 molecule of C_{16} → 4 molecules of acetoacetic acid; 2 molecules of C_{18} → 9 molecules of acetoacetic acid; etc.) which would account for the high acetone body production which made the mechanism of β oxidation appear impossible.[1,2,3]

Selection 156

We have already considered the use of isotopes in the study of intermediary reactions concerning, in particular, photosynthesis[a] and proteins.[b] Knowledge gained in this way unmistakably pointed to an essentially dynamic state of chemical components of living systems, involved in continuously occurring synthetic and degradation reactions.

What is perhaps less appreciated is that when Rudolf Schoenheimer (1898–1941) and his associates developed the first metabolic experiments with isotopes in the mid-1930s, they concerned fats.[c] As a result of this work a great deal of information was obtained about fat depots of animals, undergoing continuous turnover of glycerides. This despite the tendency of animals to reproduce depots of fat with properties characteristic of the species.

The paper by De Witt Stetten (*b*.1909) and Schoenheimer from which the extracts below are taken dealt with the question of interconversion of dietary fatty acids in the animal body. It was considered to be one of the mechanisms responsible for the constant composition of the depot fat.

The experiments indicated the animal's ability to extend or reduce the chain length of fatty acids rapidly. It is not without interest that the degradation of palmitic acid to myristic acid $C_{14}H_{28}O_2$, and possibly lauric acid, $C_{12}H_{24}O_2$, was interpreted by the authors as a confirmation of β-oxidation.

De W. Stetten, jun. and R. Schoenheimer, *J.Biol. Chem.*, *133*, 329. 1940

'The conversion of palmitic acid into stearic acid and palmitoleic acids in rats'

pp. 331–2

In the present experiment the test substance was deuteropalmitic acid containing 22.4 atom per cent deuterium. It was administered as the ethyl ester in a concentration of only 0.56 per cent of the diet. In order to insure an adequate supply of a variety of normal fatty acids, 6.4 per cent butter was also incorporated in the diet. From the quantity and deuterium content of the test substance, and from the amount of normal, non-isotopic palmitic acid known to occur in butter, it may be calculated that the mean deuterium content of the mixed dietary palmitic acid was 5.7 atom per cent.

Three growing male rats were used as test animals, growing rats being employed in preference to adult animals because of their ability to absorb and deposit fat rapidly. The diet was offered to them *ad libitum* for 8 days, and the animals were then killed. The lipids isolated from the livers and from the remainder of the carcasses were fractionated by methods that have been previously described from this laboratory.[1,2]

p. 332

Of the deuterium fed as palmitic acid at least 92 per cent was found to have been absorbed and 44 per cent was recovered in the fatty acids of the depot fat...

In addition to the presence of deuterium in the depot fat, relatively high deuterium concentrations were found in the fatty acids of the liver. Both the saturated and unsaturated fatty acids of this organ were found to have deuterium concentrations about 2.5 times as great as the corresponding fractions of the depot fat...

The total fatty acids of the carcass, containing 0.50 atom per cent deuterium, were fractionated, and several individual fatty acids were isolated (Table 15).

pp. 332–3

As will be seen in Table 15 the fatty acid of the carcasses that was richest in deuterium was palmitic acid... If the deuterium content of the palmitic acid of the carcass be compared with that of the dietary palmitic acid, it will be found that 1.38/5.7 = 0.24, or about *24 per cent of all the palmitic acid in the carcasses was derived from the direct deposition of dietary palmitic acid*...

In addition to its occurrence in significant concentration in the carcass palmitic acid, deuterium was also found (Table 15) in several other fatty acids...

Table 15 Deuterium Content of Fatty Acids Isolated from Carcass Fat of Rats Fed Deuteropalmitic Acid
The mean deuterium content of dietary palmitic acid was 5.7 atom per cent.

Fatty acid	Deuterium
	atom per cent
Palmitic	1.38
Stearic	0.53
Mixture of myristic and lauric	0.32
Palmitoleic	0.36
Oleic	0.06
Linoleic	0.02

pp. 333–4

The sample of stearic acid isolated from the depot contained 0.53 atom per cent deuterium, or about 38 per cent of that found in the carcass palmitic acid. This finding eliminates the possibility that the palmitic acid was first degraded to small fragments, such as acetaldehyde, and these subsequently recondensed to form an 18-carbon chain. In such a series of reactions, the bulk of the deuterium would have been lost, first, during the oxidative degradation, second, by exchange of deuterium with the normal hydrogen of the body water, and third, by dilution with the normal hydrogen required to reduce the double bonds that would result from aldol condensations. The high deuterium content of the stearic acid must be taken as evidence for the direct elongation of the 16-carbon skeleton of palmitic acid to the 18-carbon chain of stearic acid, by the addition of 2 carbon atoms ... Part of the dietary palmitic acid was degraded to acids of lower molecular weight, chiefly myristic and possibly lauric acids. The deuterium content of this fraction (Table 15) is 23 per cent of that found in the carcass palmitic acid.

   This finding is analogous to that described in an earlier experiment,[3] in which, after the ingestion of large quantities of deuterostearic acid, deuteropalmitic acid was recovered from the carcass. It was taken as direct experimental confirmation of the theory of Knoop[4] and Dakin[5] that fatty acids are degraded by a process of one-sided β oxidation.

p. 335

The data presented in Table 15 are compatible with the following series of reactions.

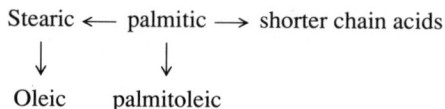

$$\text{Stearic} \longleftarrow \text{palmitic} \longrightarrow \text{shorter chain acids}$$
$$\downarrow \qquad\qquad \downarrow$$
$$\text{Oleic} \qquad \text{palmitoleic}$$

Or, when these results are combined with those already obtained from previous isotope studies,[1–3] the reactions may be formulated

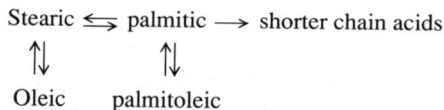

$$\text{Stearic} \leftrightharpoons \text{palmitic} \longrightarrow \text{shorter chain acids}$$
$$\updownarrow \qquad\qquad \updownarrow$$
$$\text{Oleic} \qquad \text{palmitoleic}$$

   In consideration of the transformations that the labeled palmitic acid underwent after ingestion, it should again be emphasized that each of the fatty acids into which the test substance was transformed was liberally supplied in the butter of the diet. In other words, *the rat can and does continuously convert palmitic acid into several other fatty acids, even though these products are available from other sources.*

Notes to Section VI

Selection 137

Introduction:

a. In 1940 – the essentially terminal date for the period covered by this *Documentary History* – the situation was characterized by a leading worker on fats as follows:

Unanimity has not yet been reached in the terminology to be adopted in classifying the various types of naturally occurring compounds in which higher fatty acids are present. These types are broadly as follows:

(I)   *Compounds containing only carbon, hydrogen, and oxygen:*
    (A) Esters of higher fatty acids with glycerol (triglycerides).
    (B) Esters of higher fatty acids with alcohols other than glycerol (higher aliphatic alcohols, sterols, etc.).

(II) *Compounds containing other elements in addition to carbon, hydrogen and oxygen:*
    (C) Esters of glycerol with fatty acids and also phosphoric acid coupled with a nitrogen base.
    (D) Compounds of fatty acids with a carbohydrate and containing nitrogen.
    (E) A few fatty acid derivatives also containing either nitrogen or sulphur.

Even a collective title for the whole group is not generally settled. Following I. Smedley-Maclean, most British workers refer to the whole group as *lipoids*, the Germans employing the corresponding word (*Lipoide*); but American biochemists have adopted Bloor's generic term *lipids*.

The sub-groups also share several alternative titles, the latter being roughly tabulated as follows:

	British	*German*	*American*
Group Title	Lipoids	Lipoide	Lipids
Sub-group I			Simple lipids
Type A	Fats	Fette	Fats
Type B	Waxes	Wachse	Waxes
Sub-group II	Lipins		{ Compound / Complex } lipids
Type C	Phosphatides	Phosphatide	Phospholipids
Type D	Cerebrosides	Cerebroside	Cerebrosides

(see T.P. Hilditch, *The Chemical Constitution of Natural Fats* (London, 1940), pp. 2–3).

Ida Smedley-MacLean (1877–1944) was a prominent researcher in fats who had evolved a classification widely adopted in Britain in the 1930s. See H.B. Bull, *The Biochemistry of the Lipids* (New York, 1937), p. 3. Cf. also W.R. Bloor, 'Biochemistry of the Fats', *Chem.Revs.*, 2, 243–300. 1926.

The term 'lipoid' was apparently first employed by V. Kletzinsky (1826–82) in his *Compendium der Biochemie* (Vienna, 1858), p. 120. (On Kletzinsky see Introduction to Selection 170, Section VII, note d.) Following the work of Ernest Overton (1865–1933) and Hans Horst Meyer (1853–1939) at the turn of the century, it came into use with the lipoid theory of cell permeability. The suggestion was that the penetration of substances across the cell surface was governed by their low or high solubility in the 'lipoid' layer. See O. Rosenheim, 'Proposals for the nomenclature of the lipoids', *Biochem.J.*, 4, 331–6. 1909 (p. 331).

b.   The date also mentioned in literature is 1779.

c.   See Teich, 'Circulation', Introduction to Selection 1, Section I, footnote f.

Text:

a.   Translated from the Swedish and German originals by L. Dobbin.

b.   Published in *Kongl. Vetenskaps Academiens Nya Handligar*, 4, 324–9. 1783, under the title 'Roen betraeffande ett saerskildt Socker – Aemne uti exprimerade Oljor och Fettmor'.

c.   Lead oleate.

d.   Ethyl alcohol.

e.   Substance obtained on dry distillation of crude tartar.

f.   Nitric acid.

g.   Oxalic acid.

h.   Alcoholic solution of caustic potash.

Selection 138

Introduction:

a.   On Chevreul and his work see A.B. Costa, *Michel Eugene Chevreul Pioneer of Organic Chemistry* (Madison, 1962).

b.   F.N. Gibbs, 'The history of the manufacture of soap', *Ann.Sci.*, 4, 169–90. 1939 (pp. 184–7).

c. The term 'ether' was applied to compounds of alcohol and acids since 1787. See de Morveau, Lavoisier, Bertholet [*sic*] and de Fourcroy, *Méthode de nomenclature chimique* (Paris, 1787), p. 73. It competed for a long time with the term 'ester', introduced in 1848. Cf. L. Gmelin, *Handbuch der Chemie* (Heidelberg, 1848), IV, p. 182. See also J.R. Partington, *A History of Chemistry* (London, 1964), IV, p. 342.

Text:

a. *Tournesol* – French designation for litmus.
b. In 1787 Claude Louis Berthollet (1748–1822) showed that prussic acid consisted only of hydrogen, carbon and nitrogen. Cf. 'Mémoire sur l'acide prussique', *Mém.Acad.Sci.*, 148–62. 1787 (1789). This cast doubt upon the validity of Lavoisier's concept of acids always containing oxygen.

Selection 139

Introduction:

a. See Selection 107, Section V.
b. This introduction draws on F.L. Holmes's *Claude Bernard and Animal Chemistry* (Cambridge, Mass., 1974), especially chapters X–XVII.

Text:

a. Théofile Jules Pelouze (1807–67), well-known French chemist.

Selection 140

Text:

1. I have observed that glycerine frozen at −40 degrees becomes almost solid and similar to gum.
2. This body does not contain a volatile acid.
a. For the probably first use of the term 'synthesis' for the artificial preparation of organic compounds, see Introduction to Selection 161, Section VII, note d.
b. As previously pointed out, the constitution of plant material had been presented from the standpoint of 'proximate (immediate) principles' since the end of the eighteenth century (see Selection 3, Section I, note d). Eventually, the term applied to well-defined groups of chemical compounds in naturally occurring substances of biological origin, such as sugars, organic acids, alcohols, etc. Obviously Berthelot became alive to the possibility of making them the starting point for synthetic reactions.
c. Barium oxide.
d. Reflecting the prevalent uncertainty regarding equivalents, atoms and molecules, Berthelot wrote the formula of water as HO. Cf. also Introduction to Selection 108, Section V.

Selection 141

Introduction:

a. J. Liebig, *Animal Chemistry or Organic Chemistry in its Application to Physiology and Pathology* (Cambridge, Mass., 1842) pp. 79f.
b. K. Voit, 'Ueber Fettbildung im Thierkörper und Mästung', *Landw. Versuchs-St.*, 8, 23–8. 1866.
c. Pettenkofer was Professor of Hygiene at the University of Munich. He designed a respiration aparatus – a chamber large enough to hold an individual – fitted with two pipes for allowing air to enter it and for passing it out. From the known rate of flow of air through the chamber and from measured amounts of the carbon dioxide and water of the expired air that were absorbed Pettenkofer and Voit obtained respiratory and metabolic data, regarded as the most accurate at the time. Cf. Pettenkofer, 'Ueber den Respirations- und

Perspirationsapparat im physiologischen Institute zu Müchen', *J.prakt.Chem.*, *82*, 40–50. 1861; idem, 'Ueber die Respiration', *Ann.*, *2 Supp.*, 1–52. 1862–3.

d. M. Pettenkofer and C. Voit, 'Untersuchungen über die Respiration', *Ann., 2 Supp.*, 52–70. 1862–3 (p. 57); idem, 'Ueber die Producte der Respiration des Hundes bei Fleischnahrung und über die Gleichung der Einnahmen und Ausgaben des Körpers dabei', ibid., 361–77. 1862–3 (pp. 373–4).

e. For preceding work referred to by Lawes and Gilbert see M.J. Dumas and M.J.B. Boussingault, *The Chemical and Physiological Balance of Organic Nature: An Essay*, 3rd edn (London, 1844), pp. 116f; Dumas and Milne Edwards, 'Note sur la production de la cire des abeilles', *Comp.Rend.*, *17*, 531–45. 1843 (including discussion); Persoz, 'Expériences sur l'engrais des oies', *Ann.Chim.*, [3] *14*, 408–19. 1845; Boussingault, 'Sur le développement de la graisse pendant l'alimentation des animaux' ibid., 14, 419–82. 1845.; 'Expériences statiques sur la digestion', ibid., *18*, 444–78. 1846; J.B. Lawes and J.H. Gilbert, 'On the composition of foods, in relation to respiration and the feeding of animals', *Rep.Brit.Ass.*, 323–53. 1853. On Boussingault's researches inquiring into the formation of fat in animals consult F.L. Holmes, *Claude Bernard and Animal Chemistry* (Cambridge, Mass., 1974), chap. V.

Selection 142

Introduction:

a. Cf. Selection 118, Section V.

Text:

1. Dry weight of faeces, 41.8 g containing 5.05 per cent N.
2. Dry weight of faeces, 56.3 g containing 4.97 per cent N.
3. Dry weight of faeces, 43.4 g containing 5.05 per cent N.
4. Dry weight of faeces, 46.9 g containing 4.27 per cent N.

a. This reference to fermentative lipolytic action has to be seen in conjunction with the description of the effect of the 'pancreatic fat-splitting ferment', also conceived as the 'intestinal putrefaction ferment', on lecithin in 1877. The resulting products were specified as glycerophosphoric acid, neurin (a base) and fatty acids. It was thought that these breakdown products were at least partly absorbed, probably as salts, for after lecithin-rich nourishment the phosphoric acid excretion increased in urine and faeces, but no trace of lecithin or glycerophosphoric acid was to be found. Cf. A. Bókay, 'Ueber die Verdaulichkeit des Nucleins und Lecithins', *Z.physiol.Chem.*, *1*, 157–64. 1877.

Selection 143

Introduction:

a. For biographical reference, see D.L. Drabkin, *Thudichum Chemist of the Brain* (Philadelphia, 1958). For a brief but perceptive appreciation of Thudichum's scientific work, see K. Dixon, 'Some biochemical signposts in the progress of neurology' in J. Needham (ed.), *The Chemistry of Life* (Cambridge, 1970), pp. 85–91.

b. Thudichum started as a government-aided scientist in 1865 and this association lasted until 1882. During this time the *Annual Reports of the Medical Officer of the Privy Coucil* became the immediate repositories of his scientific and medical researches.

c. J.L.W. Thudichum, *A Treatise on the Chemical Constitution of the Brain* (London, 1884), pp. viii–ix.

d. By 1890 the sugar was identified as galactose. Cf. H. Thierfelder 'Über die Identität des Gehirnzuckers mit Galactose', *Z.physiol.Chem.*, *14*, 209–16. 1890. See also H.T. Brown and G.H. Morris, 'Note on the identity of cerebrose and galactose', *J.Chem.Soc.*, *57*, 57–9. 1890.

e. Since 1928 it has been accepted that the constituent acid of phrenosin is hydroxylignoceric acid $C_{24}H_{48}O_3 = CH_3(CH_2)_{21}CHOH.COOH$. See E. Klenk, 'Über die Oxysäuren der Cerebroside des Gehirns (7. Mitteilung über Cerebroside)', *Z.physiol.Chem.*, *174*, 214–

32, 1928; idem, 'Über die Cerebronsäure. (9. Mitteilung über Cerebroside)', ibid., *179*, 312–19. 1928.

f. This formula was shown to require modification: $C_{18}H_{37}NO_2$. See E. Klenk, 'Über Sphingosin. (10. Mitteilung über Cerebroside.)', *Z.physiol.Chem.*, *185*, 169–82. 1929.

g. Thudichum, *Treatise*, p. 205.

Text:

a. Square brackets in original.

Selection 144

Text:

a. See Introduction to Selection 138, note c.
b. See Selection 140.
c. In contemporary French usage soluble ferments (enzymes) were classed as *diastases*. Hence '*diastatique*' meant enzymic.

Selection 145

Introduction:

a. F. Knoop, *Oxidationen im Tierkörper. Ein Bild von den Hauptwegen physiologischer Verbrennung* (Stuttgart, 1931), p. 3. Published in H. Grossmann (ed.), *Sammlung chemischer und chemisch-technischer Vorträge*, N.S. no. 9. Cf. also K. Th[omas], 'Franz Knoop zum Gedächtnis', *Z.physiol.Chem.*, *283*, 1–8. 1948; Introduction to Selection 56, Section III, note d.
b. Cf. Introduction to Selection 133, Section V.
c. E. Salkowski and H. Salkowski, 'Ueber das Verhalten der aus dem Eiweiss durch Fäulniss entstehenden aromatischen Säuren im Thierkörper', *Z.physiol.Chem.*, *7*, 161–77. 1882–3.

Text:

1. *Ann.f.Chem.*, *283*, 314.
2. A possible conjugation is not considered here.
a. Firm producing fine chemicals.

Selection 146

Introduction:

a. G. Embden, H. Salomon and Fr. Schmidt, 'Ueber Acetonbildung in der Leber. Zweite Mitteilung: Quellen des Acetons', *Hofmeisters Beitr.*, *8*, 129–55. 1906.
b. The names are given as reported in contemporary literature: heptylic acid, octoic acid, nonoic acid, decoic acid.

Text:

1. G. Embden and Fr. Kalberlah, Über Acetonbildung in der Leber. This journal 7, 121 (1906).
2. Fr. Knoop, Der Abbau aromatischer Fettsäuren im Tierkörper. This journal *6*, 150.
a. Acetone (ketone) bodies – see Selection 121, Section V, note a.

Selection 147

Introduction:

a. See Selection 145.

Text:

1. The mode of formation of cinnamoylglycocoll is not clear. A discussion of this question is contained in the preceding paper (p. 204);[a] it probably is not derived from the direct coupling of glycocoll and cinnamic acid as represented in the diagram which is intended merely to show the structural relations of the substance.

a. H.D. Dakin, 'The mode of oxidation in the animal organism of phenyl derivatives of fatty acids. Part IV.', *J.Biol.Chem.*, *6*, 203–19. 1909.

Selection 148

Introduction:

a. A. Loeb, 'Über das Verhalten der Essigsäure bei künstlicher Durchblutung der Leber', *Biochem.Z.*, *47*, 118–26. 1912.
b. E. Friedmann, 'Zur Kenntnis des Abbaues der Karbonsäuren im Tierkörper. Achte Mitteilung. Über das Verhalten der α, β-ungesättigten Säuren bei der Leberdurchblutung', *Hofmeisters Beitr.*, *11*, 371–75. 1908.

Selection 149

Introduction:

a. Selection 144.
b. J.H. Kastle and A.S. Loevenhart, 'Concerning lipase, the fat-splitting enzyme, and the reversibility of its action', *Am.Chem.J.*, *24*, 491–525. 1900.
c. Hanriot, 'Sur la reversibilité des actions diastasiques', *Comp.Rend.Soc.Biol.*, *53*, 70–2. 1901; 'Sur le mécanisme des actions diastatiques', *Comp.Rend.*, *132*, 212–15. 1901.
d. A.S. Loevenhart, 'On the relation of lipase to fat metabolism – lipogenesis', *Am.J.Physiol.*, *6*, 331–50. 1902. Lovenhart's choice of the term 'lipogenesis' was made to emphasize its correspondence to the term 'glycogenesis' which had come into use since the mid-1850s in connection with Claude Bernard's work on the glycogenic function of the liver. See Selection 77, Section IV.
e. A. Hamsik, 'Reversible Wirkung der Darmlipase', *Z.physiol.Chem.*, *59*, 1–12. 1909; 'Ueber den Einfluss der Galle auf die durch die Pankreas- und Darmlipase bewirkte Fettsynthese', ibid., *65*, 232–45. 1910; 'Zur Kenntnis der Pankreaslipase', ibid., *71*, 238–51. 1911.

Selection 150

Introduction:

a. Selection 138.
b. F. Reinitzer, 'Beiträge zur Kenntniss des Cholesterins', *Sitzber.(Wien)*, *97* (I. Abth.), 167–87. 1888.
c. M. Berthelot, 'Sur plusieurs alcools nouveaux', *Ann.Chim.*, [3] *56*, 51–98. 1859 (pp. 54–69); O. Diels and E. Abderhalden, 'Zur Kenntniss des Cholesterins. (II. Mittheilung)', *Ber.chem.Ges.*, *37*, 3092–103. 1904.
d. J. Wislicenus and W. Moldenhauer, 'Ueber das Cholesterindibromür', *Ann.*, *146*, 175–80. 1868.
e. O. Diels and K. Linn, 'Zur Kenntniss des Cholesterins (VI. Mittheilung)', *Ber.chem.Ges.*, *41*, 544–50. 1908; J. Mauthner, 'Neue Beiträge zur Kenntniss des Cholesterins (IV. Mittheilung)', *Monatsh.Chem.*, *39*, 635–47. 1909.

Selection 151

Introduction:

a. A. Windaus and K. Neukirchen, 'Die Umwandlung des Cholesterins in Cholansäure (28. Mitteilung über Cholesterin)', *Ber.chem.Ges.*, *52*, 1915–19. 1919.

b. H. Wieland and H. Sorge, 'Untersuchungen über die Gallensäuren. III. Mitteilung. Die strukturellen Beziehungen zwischen Cholsäure und Desoxycholsäure', *Z.physiol.Chem.*, *98*, 59–64. 1916–17.

c. H. Wieland and R. Jacobi, 'Die Synthese des Pseudo-cholestans aus Cholandsäure', *Ber.chem.Ges.*, *59*, 2064–7. 1926.

d. H.O. Wieland, 'The chemistry of the bile acids', *Nobel Lectures Chemistry 1922–1941* (Amsterdam, 1961), p. 96; A. Windaus, 'Constitution of sterols and their connection with substances occurring in nature', ibid., p. 106.

e. O. Diels and W. Gädke, 'Über die Dehydrierung des Cholesterins (Vorläufige Mitteilung.)', *Ber.chem.Ges.*, *58*, 1231–3. 1925; idem, 'Über die Bildung von Chrysen bei der Dehydrierung des Cholesterins', ibid., *60*, 140–7. 1927; O. Diels and A. Karstens, 'Über Dehydrierung mit Selen. III. Mitteilung. Dehydrierung des Ergosterins, der Cholsäure und des Schellacks', *Ann.*, *478*, 129–37. 1930.

f. A. Windaus, 'Über die Konstitution des Cholesterins und der Gallensäuren', *Z.physiol.Chem.*, *213*, 147–87. 1932 (p. 172).

g. Cf. J.D. Bernal, 'The place of X-ray crystallography in the development of modern science', *Radiology*, *15*, 1–12. 1930 (pp. 9–10).

h. D.M.C. Hodgkin, 'John Desmond Bernal', *Biog.Mem.F.R.S.*, *26*, 17–84. 1980 (p. 37).

i. A. Windaus, 'Konstitution des Cholesterins', *Z.physiol.Chem.*, *213*, 173. 1932.

j. H. Wieland and E. Dane, 'Untersuchungen über die Konstitution der Gallensäuren. XXXIX. Mitteilung. Zur Kenntnis der 12-Oxy-cholansäure', *Z.physiol.Chem.*, *210*, 268–81. 1932 (p. 275).

Text (i):

1. Lieb.Ann., 1927, *459*, 1; Lieb.Ann., 1930, *478*, 129.

Text (ii):

1. J.D. Bernal, Nature, Feb. 20 and May 14, 1932.

Text (iii):

1. Helv.Chim.Acta, 1933, *16*, 216.
2. *Lieb.Ann.*, 1927, *459*, 1.
3. *Lieb.Ann.*, 1930, *478*, 129.

Selection 152

Introduction:

a. See Selection 160, Section VII.

b. E.H. Starling, 'The food supply of Germany during the war', *J.Roy.Statist.Soc.*, *83*, 225–45. 1920 (pp. 252–3). Let us recall that Voit's recommendations were: 118g of protein and 56g of fat. See Selection 119, Section V, note b.

c. See Introduction to Selection 132, Section V.

d. T.B. Osborne and L.B. Mendel, 'Growth on diets poor in true fats', *J.Biol.Chem.*, *45*, 145–52. 1920–1 (p. 145).

e. T.B. Osborne and L.B. Mendel (with the cooperation of E.L. Ferry), 'Feeding experiments with fat-free food mixtures', *J.Biol.Chem.*, *12*, 81–9. 1912.

Text:

1. Burr, G.O., and Burr, M.M., *J. Biol.Chem.*, *82*, 345 (1929).
2. Burr, G.O., and Burr, M.M., *J. Biol.Chem.*, *86*, 587 (1930).

Selection 153

Introduction:

a. P.E. Verkade, M. Elzas, J. van der Lee, H.H. de Wolff, A. Verkade-Sandbergen and D. van der Sande, 'Untersuchungen über den Fettstoffwechsel. I.', *Z.physiol.Chem.*, *215*, 225–57. 1933.

b.   P.E. Verkade and J. van der Lee, 'Researches on fat metabolism. II.', *Biochem.J.*, *28*, 31–40. 1934 (p. 31).

Text:

1.   Hab.-Schrift Freiburg i.B. 1904; Hofmeisters Beitr. *6*, 150 (1905).

Selection 154

Introduction:

a.   W.H. Hurtley, 'The four carbon acids of diabetic urine', *Quart.J.Med.*, *9*, 301–408. 1915–16 (pp. 373–6).
b.   J.H. Quastel and A.H.M. Wheatley, 'Oxidation of fatty acids in the liver', *Biochem.J.*, *27*, 1753–62. 1933.

Text:

1.   A somewhat similar view was put forward by Hurtley (1916).
a.   G. Embden and H. Engel, 'Über Acetessigsäurebildung in der Leber', *Hofmeisters Beitrg.*, *11*, 323–47. 1908.
b.   M. Jowett and J.H. Quastel, 'Studies in fat metabolism. I. The oxidation of butyric, crotonic and β-hydroxybutyric acids in presence of guinea-pig liver slices', *Biochem.J.*, *29*, 2143–58. 1935.

Selection 155

Introduction:

a.   See Selection 148.
b.   In metabolic studies of fats and carbohydrates the glucoside phlorrhizin (phloridzin, $C_{21}H_{24}O_{10}$) was employed since it induced, on injection, temporary diabetes in animals.

Text:

1.   Barnes, R.H. and Drury, D.R., *Proc.Soc.Exp.Biol. and Med.*, *36*, 350 (1937).
2.   Butts, J.S., Cutler, C.H., Hallman, L. and Deuel, H.J. Jr, *J.Biol.Chem.*, *109*, 597 (1935).
3.   Blixenkrone-Møller, J., *Z.physiol.Chem.*, *252*, 169 (1938).

Selection 156

Introduction:

a.   Selection 45, Section II.
b.   Selection 133, Section V.
c.   Cf. R. Schoenheimer and D. Rittenberg, 'Deuterium as an indicator in the study of intermediary metabolism. III. The role of the fat tissues', *J.Biol.Chem.*, *111*, 175–81. 1935.

Text:

1.   Schoenheimer, R., and Rittenberg, D., *J.Biol.Chem.*, 113, 505 (1936).
2.   Rittenberg, D. and Schoenheimer, R., *J.Biol.Chem.*, *117*, 485 (1937).
3.   Schoenheimer, R. and Rittenberg, D., *J.Biol.Chem.*, *120*, 155 (1937).
4.   Knoop, F., *Beitr.chem.Physiol.u.Path.*, *6*, 150 (1904).
5.   Dakin, H.D., *J.Biol.Chem.*, *4*, 77, 227, 419 (1908).

Section VII:
CONCEPTUAL AND DISCIPLINARY ISSUES

a. From the 1770s to *c*. 1880: vital force. Chemistry of plant and animal life. Cell theory and evolutionary theory. Chemistry, physiology and biochemistry

Selection 157

One way to unravel the knot of conceptual threads is to try to trace them to the perennial and basic problems of biology and philosophy which have intertwined and interacted over the centuries. Those fundamental questions have concerned the priority of the immaterial and the material, and the principle which underlies the working of the living organism as a whole. Viewed in this way, we can see that scientists have, broadly speaking, been guided either by vitalist idealist views on the one hand, or by mechanist materialist notions on the other. From the standpoint of vitalist idealism (unlike that of mechanical materialism), the material originates, ultimately, in the immaterial and the living organism is not to be regarded merely as a mechanism or an assembly of physicochemical mechanisms.[a]

In practice, however, most scientists did not identify in such a clear-cut way with one trend or the other. Instead, they found it compatible to incorporate elements from both into their outlook on the origins and nature of life. The contradictions that underlay their study sustained vitalist–mechanist controversies on these issues and in turn gave rise to questions and answers which proved to be stimulating to scientific investigations into the material basis of life.

A good deal of the debate turned on the question of how to see the 'vital force' which, during the last quarter of the eighteenth century, emerged as the principle supporting the phenomenon of life. Was it an immaterial principle entering matter from the outside, or was it an entity intrinsic to matter?

Friedrich Casimir Medicus (1736–1808) has been credited with being the first to introduce and discuss the notion of the vital force[b] in a lecture entitled *Von der Lebenskraft* (On the vital force) on 5 November 1774.[c] Medicus put forward his theory in order to 'reconcile' two different approaches to the relation between mind and matter, soul and body, organism and mechanism, which were in vogue. One trend derived from the teachings of Georg Ernst Stahl (1659–1734), who looked upon the immaterial 'soul' as the agency that animated non-living material systems by organizing, regulating and directing them.[d] As Medicus saw it, Stahl exaggerated the commanding control of the soul over *all* manifestations of life. But Medicus was not happy about the other tendency either. He was critical of its bias towards an explanation of the phenomenon of life which flowed from physics and mechanics and virtually ignored the question of their ultimate immaterial source. Medicus considered Friedrich Hoffmann (1660–1742),[e] Hermann Boerhaave (1668–1788)[f] and Albrecht (von) Haller (1708–77)[g] as the leading exponents of this approach.

Medicus started from the following propositions: the constituent components of the animal body were 'organized matter and the soul'; whereas animals had a sensory-sentient (*sinnlich- empfindende*) soul, human beings possessed a rational soul; organized matter was complex but, whatever its complexity, in the last analysis it remained inert matter bereft of life; on the other hand, the soul created by God was a simple immaterial substance, immortal and conscious of itself.

The question brought out by Medicus was this: since the human soul was concerned with conscious activities, i.e. thinking and willing, it could not be accountable for unconscious (vegetative) functions relating to digestion, secretions, blood formation and circulation, the heart-beat, nutrition and other essential vital processes (*Geschaefte des Lebens*). Their existence was well known from experience but nothing was known of their mode of action; neither the application of physical nor of mechanical laws was sufficient to elucidate it. Accordingly, Medicus asserted that there had to be another simple immaterial entity, also owing its existence to God, that controlled living activities. He variously termed it 'animating force' (*belebende Kraft*), 'vital force' (*Lebenskraft*) or 'vital spirit' (*Lebensgeist*).

Medicus was aware of this implied relatedness of the soul and the vital force and, therefore, sought to distinguish between them in a threefold way. First, Medicus pointed out, the soul tired and this state of the soul found its expression in the tiredness of the body to which it belonged. Vital force did not tire but, if it did, life ceased for ever. Secondly, Medicus specified, human souls differed. At the same time, however, and in accordance with personal desires, they were capable of individual development and perfection. This did not apply to the vital force activating the vegetative functions continuously, steadily and independently of any volition. For example, no exercise of will could bring an at times much needed rest to a human heart. Thirdly, Medicus therefore concluded, what distinguished the soul from the vital force was the presence of the will in the manifestations of the former and its absence from the operations of the latter.

With the help of the vital force concept Medicus attempted to link, but also to demarcate, all forms of life. Implanted by God at the inception of plant, animal and human life, the vital force was the source of its activities, but it affected them quantitatively and qualitatively in a different manner. Therefore Medicus saw in the vital force the operating influence that produced not only the characteristics of each plant and animal species but also those of an individual of any species. It could be added that he made the proposition that the vital force of humans and animals was located in the brain and in the spinal cord, and that of plants in the wood (xylem).

Because of the rather simple structure of plants, Medicus suggested, the range within which the vital force could function in them was limited. It had greater scope in animals, which were more completely structured than plants. But there was no hard and fast line of demarcation between the plant and animal kingdoms, as demonstrated by the fact that some organisms were classed either with plants or animals. As for man, although like all living beings he also possessed a vital force sustaining vegetative operations, there was no blurring of the boundary between him, on the one hand, and the plants and animals, on the other. The capability of thinking, judging and discovering which stemmed from the human soul raised man above all other creatures.

There is virtually nothing in literature on the immediate effect of Medicus's elaboration of the vital force concept. Twenty years later, however, it was taken for granted, though not necessarily in the idealist form propounded by Medicus. Thus in the opening article, *Von der Lebenskraft*, of the first issue of the *Archiv für die Physiologie* (1796), the vital force was treated from the materialist standpoint. Its author was the editor of the newly established first journal in Germany devoted to physiology, Johann Christian Reil (1759–1813), then Professor of Clinical Medicine at Halle. The article was preceded by an open letter addressed to his university colleagues, the physicist and chemist, F.A.C. Gren (1760–98), and the philosopher, L.H. Jakob (1759–1827). Emphasizing in it the need for the clarification of scientific concepts, Reil attacked the idealist tendencies in physiology (p. 4):

We look for the cause of animal phenomena in a transcendental substrate, in a soul, in a general world spirit, in a vital force, which we imagine as something incorporeal and in this way we are hampered and led astray.

In the long article from which the excerpts below are taken, he developed the main theoretical proposition that the phenomena displayed by living bodies were materially conditioned. In reading the passages on the vital force one is struck by the reluctance with which Reil uses this term. This is shown in numerous explanatory footnotes which accompany the otherwise clear text.

J.C. Reil, *Arch.Physiol.*, *1*, 8. 1796

'Von der Lebenskraft' (On the vital force)

p. 14

That we do not find in inanimate nature phenomena belonging to those of the animate world depends on the specific kind of the organic material which is not found in inanimate nature. Can we, indeed, on this account deny to all others special properties, which a certain kind of matter lacks? Must we, therefore, deduce the magnetic property of iron from something other than matter because we observe no magnetic phenomena in tin, in stone and in wood?

p. 23

The material of animate nature is markedly different from the material of inanimate nature. The vegetable and animal matter have a certain similarity and constituents which are common to both. Therefore we perceive also in the animal and plant phenomena a certain specific and unmistakable similarity. Therefore we include, and indeed rightly, animals and plants under the common name of *organic beings* and separate them from inanimate nature. But although the composition and form of the animal and vegetable matter has similarity it is still not the same.[1] Therefore the plants and also the animals display each its specific manifestations, by which they distinguish themselves.

 Why are the phenomena of animal bodies so necessarily bound to a certain composition [*Mischung*] and form of matter? If we moisten, dry, stretch, relax, compress the animal bodies, in short alter the physical condition of the matter, then immediately is also the tone [*Stimmung*] of the vital force altered. A change of the matter causes a change in all its forces and we have no means, as many a physician may well think, which affect only the vital forces and others which affect only the inanimate forces. Why also do stones, Vaucanson's[a] automats and Kempelen's[b] chessplayer not live, if nothing further belongs to life than that one implants a soul or a vital spirit into inanimate matter? Why has a man never begot a marrow, why has a donkey never predicted anything and why has the oak never moved its branches of its own free will, as the animal its limbs?

 The composition of the animal matter extending from the simplest elements to the most perfect organs is highly specific. We find everywhere different elements [*Grundstoffe*], a different proportion of their combination and several orders of simple and composite components.[c] Already through mere senses we perceive that each organ has its own composition, and that the same organ always has the same composition. How characteristic is the composition of matter in muscle flesh, in nerve medulla, cell tissue,[d] viscera, bones? How do they differ from one another? A nerve which works as a nerve has its own and never another matter. Why this stability in the composition of matter? Why does one always find in the nerve tubes nerve medulla, never jelly or anything else?

pp. 25–6

The organic matter is indeed as much specific in the organic realm and never is to be met in inanimate nature. But the very beginnings of the same lie certainly already totally on hand in the womb of the inanimate nature.

pp. 28–9

One has always cherished the opinion that in nature a certain subtle principle comprises the proximate cause of its phenomena . . . especially was this subtle principle assumed to be the basic source of the phenomena in the organic nature.[e]

pp. 40–3

Yet I wish finally to warn against an error, namely that one must not believe that these subtle *substances are alone the force or at least the substrate or organic beings*. The foundation of life lies in matter as a whole, in the composition and form of all that which is visible and invisible. The subtle matter can as little produce life by itself as can the coarse matter alone. All that is must be there, if from it the final result: *life*, is to emerge ...

In the way the composition and mixture of the animal substance is specific and perfect, its *form* and *development* is just as specific and perfect. It is a wondrously artistic construction, within the animal body, in its principle simple and in its connections most varied that is so superior to the structure of the inanimate nature and the works of art. Not only the whole body or its coarse components, but even its smallest parts are machines; everything, down to the smallest fibre, dissolves itself into nothing but purposefully formed bodies. The whole body consists of several large components; each component again [consists] of muscles, vessels, nerves; the muscle again of skin, fibres, vessels. What ingenious and composite mechanics! How many stages and arrangements of the same! How far has anatomy reached in its analysis and yet has not exhausted it. How imperfect on the other hand is the mechanics of works of art and that of inanimate nature! *Here is only the whole a machine*[f] and the parts of the whole are natural bodies without purposeful development. To the regular mechanism of the animal body there also belongs both the *coarse* and the *fine tissues* of the fibres, the *articulations* of the large parts, the proportion of the *size* of the parts to one another, the *number* of the same etc. Through the union of these countless organs, which by different stages combine together into a whole machine there are equally composite forces communicated to it. Through this arrangement it is also capable of manifold manifestations, which in inanimate nature are not possible.

To the formation of the substance of animate beings we have given a specific name *organization* on account of its excellent perfectness. *Organ and organization is thus formation and structure of animate bodies ...*

It would be advantageous for theoretical and practical medicine if we could analyse the different *kinds of degrees of organization*, if we could reduce their most complex tissues to their most simple elements and if we were able to follow them from the original most elemental organ to the most complex animal organs. We would then be able, more happily, to analyse many phenomena and to reduce them more accurately to their principles. Simple organs which are formed out of homogeneous material must indeed produce the same phenomena; in contrast the composite organs are the embodiment of the forces of the simpler organs behaving as the simpler organs behave, from which they are built up.

pp. 45–50

All phenomena in the corporeal world are results of a definite form and composition of matter ...

The relation of the phenomena to the properties of matter through which they are generated I name force. Force can thus be considered general, special, and individual in the way that the relations of the effects to the causes, and the phenomena to the properties of matter, may be thought about. Force is thus something inseparable from matter, a property of the same through which it brings forth phenomena.

Force is a subjective concept, the form in which we think of the relation between cause and effect. If it were possible for us to think at the same time clearly of each body as it is, of the nature of all its elements and their combination, and of their composition and their form, then we would not need the concept of force, which gives occasion for quite a number of erroneous conclusions. In the case of phenomena not related to the senses, for instance the ability to think, we are inclined to look for the force in the substrate of the senses or in a metaphysical substrate, and we tend to invest such a substrate if we cannot demonstrate it. With material things we often try to perceive the cause of their manifestations in still other things than those which we have before us and perceive with the senses. We put in the muscle additionally an irritable substance, in the nerve additionally nerve spirit and regard these things as the characteristic substrate of the force or as the final principle of animal phenomena. We are inclined to think of the force as something different from matter and to regard matter as the vehicle of the force, although its manifestations are inseparable from it and are results of its properties. Matter is nothing other than a force, its accidents[9] are its effects, its being is action, and its specific being is to act in a specific way. Alkali and acid combine to give a neutral salt [*Mittelsalz*], because this is the property of these things which cannot be separated from them. Besides the alkali and the acid there is no third thing which brings about this combination. Common salt grows into cubic crystals because it is common salt, which as a specific kind of matter is wont to grow in this way.

The forces, or the partnership which finds place between the phenomena and the properties of matter, we define subjectively according to the more or less general nature of the phenomena. According to this definition of the concept force, the word[2]

1) *Physical force* indicates: the most general manifestations of matter and its relation to more general properties that we meet both in inanimate and animate nature.[3]

2) *Vital force*[4] indicates the relation of more individualized phenomena to a special kind of matter which we encounter only in living nature with plants and animals. The most general attribute of this particular kind of matter is a special kind of crystallization. Besides, we cannot offer any genetic [*genetische*] definition of this force as long as chemistry has not made known to us more exactly the elements of organic matter and their properties. Until then it will be impossible to set a definite boundary to the different realms of nature. Are the phenomena of life effects the sum total of properties, which are met with in the animal matter, or is there a single matter on which alone depend the phenomena of life, as the expansibility of all natural bodies depends on caloric [*Wärmestoff*]?[h] The latter view appears not probable to me, because we never meet in nature a simple matter which has life, but find it always only in the known combination with visible substances, because life expresses itself through so very different phenomena; and finally because through alteration of the visible matter, or by addition of different substances, electricity, heat, oxygen, opium, etc. we can now raise, now lower the vital force.

3) *Vegetative force* and its product *plant life*, is a property of a special matter of which plants consist.

4) *Animal force* indicates the property of a more individualized matter which we meet in the animal kingdom, characterized by a specific phenomenon, namely *muscular movement*. One can divide the animal force according to the special modifications of matter in muscle and nerves again into *sensitive force* and *motive force*, although this division is not logically correct. We can think of the forces as individualized as often as matter may be individualized. Only one must imagine in this way nothing further than another chemically combined [*gemischte*] and formed animal matter, which thus gives rise also to other phenomena.

5) Finally remains the *faculty of reasoning* which is peculiar only to man.

If we go in this way from the most general phenomena of bodies, by degrees, to their special manifestations: then we come at last to phenomena which can only find place in an individual body, indeed only in a single part of the same; in the brain, in the nerves, and so on.

The forces of the human body are thus properties of its matter, and its special forces are results of its specific matter. The phenomena of matter are as different as are its properties and the relation between phenomena and properties as manifold as the properties of matter are manifold. However much these relations can be thought of as varied, the concept of force is equally as varied.

pp. 52–4

The physical, chemical and mechanical forces of animal bodies, it is said, are subordinated to the vital force, at the same time bound by it and are only freed from this subordination by the death of the animal and again set in authority.[5] But such authority and subordination actually cannot be imagined in nature, but everything works in it according to eternal and immutable laws. Our subjective concepts, which we carry over into nature, not seldom deceive the mind of stupid human beings and hand them, instead of reality, a plaything. In nature one finds no separation of the forces, no universality, authority or subordination, but the bodies are concrete and bring forth their phenomena according to their matter. In the muscle fibre matter, as it is, produces all its phenomena; its matter is heavy, it coheres, it is extensible, elastic, slippery, soft, it has special chemical properties, it is sensitive to stimulus, it contracts on stimulation. These different phenomena of the muscle fibre we must not seek, perhaps, in special principles of the same, but their matter, such as it is, is heavy, irritable, etc., and contains the cause of all its phenomena. A matter which is heavy remains, so long as it is this matter, unaltered in its heaviness and no subordination can suppress the expressions of the heaviness. Suppose also it was bound also with another matter which had just as much absolute lightness as it possessed of heaviness, then our senses would indeed be deceived but thereby the effect of heavy matter would not be nullified.

Also it is said of the chemical laws of affinity that they are subordinated to the vital force and rescinded by it. But no law can be rescinded in nature, so long as the conditions, under which it operates, last. If the conditions alter then the law is rescinded, not in nature but in our mind. The animal bodies consist of a specific matter, thus have also their own laws of affinity, as also the bodies of inanimate nature have their own. Putrefaction is brought up as an example and it is contended that it is a natural law of animal substances that they putrefy. The law, however, is subordinated to life. But putrefaction is only a property of the inanimate and not of the animate animal matter. With loss of life some or other component of animal matter, in presence of which no putrefaction could occur, will be eliminated and removed. After the elimination of this substance the properties of the rest of the matter are altered, including its chemical laws. Thus also no dead flesh putrefies so long as one adds alcohol; no barley ferments as long as one of its components, gluten, is not removed through germination.[6]

Selection 158

From what has been shown previously it is fairly evident that Reil's view of the vital force was anything but mysterious. He believed that various manifestations of life would in due course be explained in terms of chemistry and that it would be possible to dispense with the essentially vague notion of vital force. This, however, did not happen, largely because of the considerable religious influence on scientific theoretical thought.

While practically occupied in the laboratory with the chemical side of the biological phenomena, scientists did not refer to religion and virtually conducted themselves as mechanical materialists. But the situation changed when it came to settling whether knowledge of the chemical operations underlying the manifestations and the maintenance of life was in itself sufficient to comprehend its peculiarity. It was in that connection that scientists stopped behaving as mechanical materialists. By accepting the notion of God-created life sustained by an ultimately unfathomable vital force, they effectively adopted the position of vitalist idealism.

During the first half of the nineteenth century, chemists concerned with plant, animal and human chemistry were, as a rule, imbued with this ambivalence but few voiced it in their writings as unambiguously as the eminently influential Jöns Jacob Berzelius (1779–1848). This may be exemplified by the passages below, taken from Berzelius's report on the situation of animal chemistry which he presented to the Swedish Academy of Science on 13 August 1810. It was translated into English by G. Brunnmark, the chaplain to the Swedish legation in London, and published in 1818.[a]

At the time of writing this report, Berzelius was a relatively young man and it may well be asked whether he eventually changed ideologically and the answer is 'no'. Indeed, shortly before his death Berzelius was at pains to emphasize that life was something foreign to matter and something unfathomable to man. It did not arise spontaneously from matter. What could be presupposed was that life was implanted into matter by an immaterial force. Once this occurred, life produced conditions for the operations of natural forces which could not be separated from matter. But how this happened in the first place was a riddle that would probably never be solved. What was open to investigation were the natural conditions of life and here, although far from comprehending them, the scientists succeeded in some small measure in revealing their underlying chemical basis.[b]

In trying to evaluate Berzelius's general viewpoint on the nature of life it is possible to characterize it as idealist and vitalist with respect to the origins of life, which he believed were derived from an immaterial force. But he was a materialist when it came to the composition of plant and animal substance and their changes in living bodies: these he dealt with in chemical and physical terms without invoking a chemical vital force. An historically important example of this is Berzelius's development of the concept of 'catalytic force', which has been noted earlier.[c]

The simultaneous existence of Berzelius's two apparently conflicting attitudes to the complexity of life has been discussed by the historians of science, Bent Søren Jørgensen[d] and Timothy O. Lipman.[e] It led Jørgensen to the suggestion that Berzelius was not a convinced or genuine vitalist and was a 'physiological' rather than a 'chemical' vitalist. As Lipman points out, this proposition has the disadvantage of not recognizing that a scientist like Berzelius 'could believe in the simultaneous operation of both a vital force and the natural forces'. But the last part of Lipman's statement urging historians of science 'to examine the physical, chemical aspects of the scientists' work and thought, and *then leave it there* (our italics) is disputable. This positivist position effectively precludes an analysis of the extra-scientific relations of scientific thinking which, with Berzelius, clearly included religion.

I.J. Berzelius, *A View of the Progress and Present State of Animal Chemistry* (London, 1818)

(*Öfversigt af djurkemiens framsteg och närvarande tillstånd*)
(Stockholm, 1812)

pp. 4–8

The constituent parts of the animal body are altogether the same as those found in unorganised matter, and they return to their original unorganic [sic] state by degrees, partly during the progress of life, partly when the body, after death, undergoes its final change. And independently of this, there exists processes between the unorganic constituent, or elementary particles within the animal body, which have sometimes not the least resemblance to those we see in unorganised matter. We may consider the whole animal Body as an instrument, which, from the nourishment it receives, collects materials for continual chemical processes, and of which the chief object is its own support. But, with all the knowledge we possess of the forms of the body, considered as an instrument, and of the mixture and mutual bearings of the rudiments to one another, yet the cause of most of the phenomena within the Animal Body lies so deeply hidden from our view, that it certainly will never be found. We call this hidden cause *vital power*; and like many others, who before us have in vain directed their deluded attention to this point, we make us of a *word* to which we can affix no idea. This *power to live* belongs not to the constituent parts of our bodies, nor does it belong to them as an instrument, neither is it a simple power; but the result of the mutual operation of the instruments and rudiments on one another – a result, which varies as the operations vary, and which often, from small changes and obstructions, ceases altogether. When our elementary books inform us, that the vital power in one place produces from the blood the fibres of the muscle; in another, a bone; in a third, the medulla of the brain; and in another again, certain humours, which are destined to be carried off; we know after this explanation as little as we knew before.

This unknown cause of the phenomena of life is principally lodged in a certain part of the Animal Body, viz. in the nervous system, the very operation of which it constitutes. The brain and the nerves determine altogether the chemical processes which occur within the body: and although it cannot be denied that the exercise of their functions tends to produce chemical effects; yet we are constrained to confess, that the chemical operations therein are so far beyond our reach, that they entirely escape all our observations. Our deepest chemical researches, and the finest discoveries of later times give us no information on this subject. Nothing of what Chemistry has taught us hitherto, has the smallest analogy to the operation of the nervous system, or affords us the least hint toward a knowledge of its occult nature. And the chain of our experience must *always* end in something inconceivable; unfortunately, this *inconceivable something* acts the principal part in Animal Chemistry, and enters so into every process – even the most minute, that the highest knowledge which we can attain, is the knowledge of the nature of the productions, whilst we for ever are excluded from the possibility of explaining how they are produced. Permit me here to show by an example the embarrassment of the student of Animal Chemistry on all occasions, when the inconceivable nervous system exercises its operations: It is well known that blood, which is always formed from the food of the animal, is the raw material out of which the body recruits and re-produces its parts; and that this blood, which is everywhere of the same nature, is conveyed through the arteries to the different parts of the body. From this blood the kidneys form urine; the glands near the ear and under the tongue, saliva; those in the breasts of women, milk; and so forth: all which are humours of the most different nature. The most acute anatomical investigation has proved, beyond all doubt, that the vessels in these parts, while they extend themselves, proceed in an uninterrupted course, without communicating with any others – that no foreign humours which could affect the blood, have access to them; and that consequently the blood is not exposed to the influence of any mixed chemical agency. But what is it that here effects the chemical process, which, from the same particles of the blood, forms those of saliva, milk and urine? It cannot be form and flexure of the vessels, since that can only cause a greater or less delay; and that this alone cannot determine the formation of the secreted matter, common Chemistry will shew. Consequently, there remains only the influence of the nerves, which enter into these parts, and which determine as well the nature of the secreted matter as its quantity; but until our experiments on unorganised matter shall have furnished us with a chemical phenomenon, which has any analogy with the operations of the nerves on these occasions, we shall never be able to discover the laws of those operations, nor explain the intimate nature of these processes. And if the knowledge of the transformation of the blood into other humours, which knowledge does in itself bear an analogy to chemical phenomena in general, is so deeply hidden from our view, how shall we attempt to explain the renewal of the solid animal parts, whereby the body is supported during the constant exchange of its elements? But still more astonishing are the operations of the brain. How amazing, that our thought, even in its sublimest flight, and when in the most hidden recesses of nature, should depend on a previous chemical process, which, if in the least disturbed as to its correctness, would distract this very thought, change it into madness, or make it cease altogether; and yet this is an incontrovertible truth. But is it not probable, that human understanding, which is capable of so much cultivation, which has calculated the laws of motion for distant worlds, and explored in so many instances the beauty and wonders of surrounding nature, and even attained a degree of perfection, the summit of which is concentrated in GOD, may one day explore itself and its nature? I am convinced it will not.

Selection 159

As we have seen in preceding sections, nutrition and digestion proved to be singularly important areas in the expanding knowledge of the chemistry of life. During 1820–30, overall, the most notable contribution to the study of digestion emerged from the fruitful collaboration of the physiologist, Friedrich Tiedemann (1781–1861), and the chemist, Leopold Gmelin (1788–1853). Originally presented as a piece of work for a prize of the French Academy in 1823, the results of their investigations were clearly set forth in their two-volume work, *Die Verdauung nach Versuchen*, published simultaneously in German and French in 1826 and 1827.

Above all they turned to the problem of identifying the properties of the digestive juices, clearly accepting that their study had to include more than one kind of animal 'because these juices could perhaps exhibit differences not only with the animals of different classes, but even with carnivores and herbivores of one class, and thus with one-sided knowledge of the same we would have drawn false conclusions'.[a] This conviction led them to study digestion in mammals, birds, reptiles, amphibians and fish. In their effort to follow, in chemical terms, the breakdown of ingested food, they made use of specific identification tests that had been developed not long before. Thus the behaviour of starch towards iodine[b] and the capacity of sugar to ferment indicated their presence in the alimentary canal. They also made use of the reported conversion of starch into sugar in the presence of gluten (and malt).[c]

Great as were the particular achievements of Tiedemann and Gmelin in elaborating experimental procedures and in advancing, thereby, the knowledge of digestion, there is no scope here to deal with them.[d] It is rather to the general aspects of this work that we must turn now. In penetrating to the chemical levels underlying the manifestations of life went the question of how to regard their true nature. Was digestion a purely chemical process or was it a vital one? The answer that Tiedemann and Gmelin provided is shown below. They differentiated between the chemical side, on the one hand, and the biological side of the digestive process controlled by a vital force, on the other, and subordinated the chemical to the biological. In the sense that the chemical transformations in the stomach depended on its being an organ of a living, and not a lifeless animal. It was perhaps this attitude that led them to view 'artificial digestion' as an intrinsically limited method of studying digestion. They hardly made use of it and did not develop it.[e]

F. Tiedemann and L. Gmelin, *Die Verdauung nach Versuchen* (Heidelberg and Leipzig, 1826), I.

pp. 333–4

With the solution, which takes place by means of the fluids of the stomach, there appears to be connected with several foodstuffs a special breakdown at the same time. This is shown in the case of starch which on dissolving has lost its property of blueing iodine and is changed into sugar and gum.[a] Something similar may also take place with other materials. To such conversions perhaps not only the free acids of the stomach fluid contribute, but also the salivary-like matter and osmazôme-like matter[b] contained in it, since a similar action of gluten on starch is known.

Into the dissolving action of butyric acid on the simple foodstuffs the chemists have still not inquired. As it is present in the gastric juice of the horse and in the rennet-fluid of ruminant animals, one may conjecture that it acts perhaps likewise dissolving the foods.

As the gastric juice (secreted by the living stomach from the arterial blood as the result of a stimulus), in virtue of its content of water, acetic acid, hydrochloric acid and different salts, has the chemical property of dissolving the simple nutrients, then it is clear that it must have this action also on the composite foods which, as chemistry teaches, are only a different combination of the simple nutrients.

Regarding the digestibility of foods, they are according to this theory, the more easily digestible and the digestion happens the more quickly, the more they are easily soluble in the gastric juice, on account of their composition. Digestible most easily and in the shortest time are such foodstuffs which contain substances already themselves easily soluble in warm water, such as foods primarily composed of sugar, mucilage, fluid albumen and gelatin. More difficultly digestible are those which come from foods which must be dissolved by the help of acids, such as those containing much gluten, coagulated albumen, fibrin and casein. Finally, indigestible are such materials which be dissolved at all by the juice, like the husks of cereals, the very hard plant or wood fibres, the shells of some leguminous plants, the pips and stones of fruits, hairs, feathers and so on.

pp. 335–6

The absolute digestibility of foods thus depends on their properties and composition. We must however distinguish also a relative digestibility, which is conditioned by the nature and the dissolving property of the digestive juices of the different animals. In general, the digestive juice of the carnivores appears to be less active than that of the herbivores, and therefore the former can indeed digest animal substances and easily soluble vegetable substances, but they are not capable of digesting coarse vegetable foods, like raw herbs, grass and straw. These on the other hand are digested by animals that feed on plants, which mostly have a more complicated digestive apparatus. Plant-eating animals can also digest animal substances, and it is sufficiently known that horses and ruminants can be trained to take such substances.

Even though gastric juice, in virtue of its chemical composition, is the dissolving agent of both the simple and the composite foods, and its action on food is a chemical one, digestion is still a vital process, an event conditioned by the life of the animal. This is in so far as only the living stomach through its vital forces is able to secrete the dissolving agent, the gastric juice. Should the stomach secrete this fluid, then it must be nourished and have its specific form and composition, through which it is made capable of performing and displaying its activity.

Selection 160

At the time when Tiedemann and Gmelin published their work on digestion, the nature of gastric juice was regarded as highly uncertain. They themselves observed that the chemists and physiologists were more divided in their opinions regarding the properties of gastric juice than they were about any other animal fluid. Indeed, while confirming Prout's findings (1824) that the gastric juice of animals contained hydrochloric acid,[a] Tiedemann and Gmelin opposed his contention that it was the only acid present.[b]

Not long ago the English physician, William Prout (1785–1850), was primarily known as the person who proposed the hypothesis that atomic weights of elements were integral multiples of the atomic weight of hydrogen. His scientific contributions on the medical and scientific side, including the investigations into the composition and analysis of substances derived from plants and animals, were hardly mentioned. Even less appreciated was Prout's belief that the exact determination of the proportions of carbon, hydrogen, oxygen and nitrogen in the compounds concerned would help to unravel the mathematical basis of the general laws of nature.[c]

Prout early recognized that the knowledge of composition of substances originating in organisms ('organized bodies') had an essential part to play in the understanding of physiological and pathological processes. On developing this approach in England, Prout (who was a most skilled analyst) resembled, say, J.B. Dumas (1800–84) in France or J.Liebig (1803–73) in Germany.

While engaged on the systematic analytical enquiry into the chemical composition of vegetable and animal products and concerned also with their digestion and assimilation, Prout gradually realized that the products could be classified as constituents of food. This, in effect, led him to inaugurate the historical division of foodstuffs into three classes: carbohydrate ('saccharine'), fat ('oily') and protein ('albuminous') in 1827.

Prout's examination of the nature of the transition from the crystalline to the organic state is also noteworthy. Prout termed the transformation process 'merorganization' and believed that it was brought about by water or small amounts of extraneous admixtures present. Prout saw it as a natural change and found it unnecessary to invoke the direct influence of the vital force, although he held firmly to the view of an ultimately immaterial source of things. Prout averred that the product obtained through merorganization had, as a rule, the same chemical composition as the crystalline substance. Thus without naming it, Prout tackled the phenomenon later known as isomerism.[d]

These questions are discussed by Prout in the following extracts from the 1827 paper.

W. Prout, *Phil.Trans.*, 355. 1827

'On the ultimate composition of simple alimentary substances; with some preliminary remarks on the analysis of organised bodies in general'

pp. 356–7

Organic chemistry[a] is confessedly one of the most difficult departments of the science; and though much has been done, and more attempted on the subject, it is yet in a very imperfect and unsatisfactory state; and it must be frankly admitted that Physiology and Pathology have derived less advantage from this most promising and really powerful of the auxiliary sciences, than might have been expected. To explain this perhaps would not be difficult; but as the explanation would be misplaced here, I shall merely observe, that dissatisfied with the old modes of inquiry, I determined to attempt a different one, and keeping in view the notions I had originally formed respecting chemical combinations, proposed to myself to investigate the modes in which the three or four elementary substances entering into the composition of organized bodies are associated, so as to constitute the infinite variety occurring in nature.

With these views my first object was to determine the exact composition of the most simple and best defined organic compounds, such as sugar, and the vegetable acids, a point that had been several times before attempted, but, as it appeared to me, without complete success. About the same time also albumen[b] and other animal products, as urea, lithic acid,[c] &c. were examined with similar views. The subject of digestion, however, had for a long time occupied my particular attention: and by degrees I had come to the conclusion, that the principal alimentary matters employed by man, and the more perfect animals, might be reduced to three great classes, namely, the *saccharine*, the *oily*, and the *albuminous*: hence, it was determined to investigate these in the first place, and their exact composition being ascertained, to inquire afterwards into the changes induced in them by the action of the stomach and other organs during the subsequent processes of assimilation.

pp. 374–5

It has been known from the very infancy of chemistry, that all organised bodies, besides the elements of which they are essentially composed, contain minute quantities of different foreign bodies, such as the earthy and alkaline salts, iron, &c. These have been usually considered as mere mechanical mixtures accidentally present; but I can by no means subscribe to this opinion. Indeed, much attention to this subject for many years past has satisfied me that they perform the most important functions; in short, that organization cannot take place without them. This point will be more fully investigated hereafter: at present it is sufficient merely to observe, that many of those remarkable changes which crystallized bodies undergo on becoming organized, are more apparent than real; that is to say, their chemical composition frequently remains essentially the same; and the only points of difference that can be traced, is the presence of a little more or less of water, or the intimate mixture of a minute portion of some foreign fixed body. There is no term at present employed which expresses this condition of bodies, and hence, to avoid circumlocution, I have provisionally adopted the term[1] *merorganized*, (μέξοσ *pars* vel *partim*) meaning to imply by it that bodies on passing into this state become partly, or to a certain extent, organized. Thus starch I consider as *merorganized* sugar, the two substances having, as we shall see presently, the same essential composition, but the starch differing from the sugar by containing minute portions of other matters, which we may presume, prevent its constituent particles from arranging themselves in the crystalline form, and thus cause it to assume totally different sensible properties.[2]

Selection 161

When Friedrich Wöhler (1880–82) published his paper on the preparation of urea from cyanate and ammonia in 1828 (the English version is given in full below),[a] scientific contemporaries recognized it at once as being of great import. They regarded it as 'remarkable', 'unexpected', 'brilliant', 'extraordinary', 'epoch-making', etc., because a substance normally formed in a living animal body was prepared by laboratory means.[b] A century later Wöhler's achievement was regularly coupled with the beginning of the end of chemical vitalism.

Interest in this aspect of the discovery which led to obtaining a physiological product by a chemical operation in a test tube has revived since the historian of science, Douglas McKie, returned to it in 1944.[c] He asserted then that it was a legend that Wöhler's preparation of urea from ammonium cyanate rang the death-knell of vitalism in organic chemistry. His main thesis was that it was highly arguable whether one could talk about a synthesis, as scientific historians usually did, and that it would be better to talk about transformation of the ammonium cyanate into urea. McKie defined synthesis as 'the compounding of a substance from the elements that compose it' and pointed out that the cyanate as then prepared originated from organic matter. McKie considered the first synthesis to be the synthesis of acetic acid achieved by Kolbe in 1845.[d] He further maintained that neither Wöhler himself, nor Berzelius, nor the prominent French organic chemist, Gerhardt,[e] saw Wöhler's preparation of urea as a refutation of vitalism. McKie concluded that 'Vitalism in organic chemistry was rejected not by a sudden and dramatic synthesis – for science does not advance and Nature does not reveal herself in that way – but by steady accumulation of contradictory facts.'[f]

Much of what McKie wrote was true, yet his evaluation of Wöhler's *transformation* of ammonium cyanate into urea was inadequate. In fact, the preparation of this substance revealed one of those contradictory facts which McKie thought – and rightly so – helped to banish vitalism from organic chemistry. As the same time it forced vitalists to rethink and readjust their approach to what science had learned about the physicochemical levels of life.

Which contradiction had been revealed by Wöhler's preparation of urea from ammonium cyanate? It is the one which Wöhler describes in his letter to Berzelius (22 February 1828) as follows 'and I must tell you that I can prepare urea without requiring kidneys or an animal, either man or dog.'[g] Wöhler was puzzled by this because he produced a substance artificially which was otherwise produced only by living creatures. He did not think that he drove vitalism out of organic chemistry because he was not particularly concerned with this philosophical question.[h] It may well be that this, in fact, helped to increase his mental uncertainty as to how to interpret his experiment. Thus in the same letter to Berzelius he noted:

This artificial formation of urea, can one regard it as an example of the formation of an organic substance from inorganic materials? It is conspicuous that one must have for the production of cyanic acid (and also of ammonia) always initially after all, an organic substance, and a *Naturphilosoph* would say that the vital aspect [*das Organische*] has not yet disappeared from either the animal carbon or the cyanic compounds derived therefrom and an organic body, therefore, may always be produced from it.[i]

It would be entirely wrong to think that Wöhler or Berzelius did not realize the significance of the artificial production of urea. From Berzelius's answer to Wöhler (7 March 1828) it would appear that the Swedish chemist took Wöhler's communication rather lightly, but in fact he said: 'It is quite an important and nice discovery which Hr. Doktor effected and I was indescribably pleased to hear of it.'[j] Then he continued to discuss peculiarities of the reaction from a purely chemical point of view. He did the same in his textbook on animal chemistry which appeared in German in 1831 where he stated very clearly: 'Wöhler made the remarkable discovery that urea can be compounded *artificially*.' (our italics – M.T.)[k]

Undoubtedly, the artificial method of producing a substance which was normally regarded as biological in origin created a puzzle which scientists like Berzelius had to face. He could not deny the fact, and he did not want – or was not prepared – to face its implications. Therefore he suggested that urea was on the borderline between organic and inorganic substances, and that may have been the reason why it could be produced artificially. This view was taken up by others who also were unable to solve the puzzle. Thus J. Müller (1801–58), the famous German physiologist, wrote in 1835 in the second edition of his well-known *Handbuch der Physiologie des Menschen*: 'The way that elements combine in organic bodies is peculiar and conditioned by peculiar forces. Though chemistry can dissolve organic compounds it cannot create them.'[l] After rejecting claims by various authors to have produced artificial organic substances, Müller admitted only the validity of Wöhler's discoveries, and then continued: 'However, urea is placed at the extreme border of organic substances and is more of an excretion than a component of the animal body. Perhaps urea is not at all a compound with the characteristic properties of organic products.'[m]

For a long time Wöhler's achievement remained an isolated example of its kind. Yet it is an important landmark in the history of biochemistry. The contradiction which it raised could be solved only by the development of both its chemical and biological aspects. In fact, the previously unified science of chemistry began to show definite signs of splitting into inorganic and organic chemistry: organic chemistry which initially occupied itself with the chemical properties of substances of vegetable and animal origin became in the end the chemistry of carbon compounds. That involved a process of working out a theoretical basis which no longer needed to fall back on the vital force concept.[n] A different situation was obtained with plant and animal physiology which studied the changes these substances were undergoing in living bodies. There the vital force controversy profoundly influenced explanations of the nature of these changes.

F. Wöhler, *Ann.Phys.*, *12 (88)*, 253. 1828

'Ueber künstliche Bildung des Harnstoffs' (On the artificial production of urea)

pp. 253–6

In a brief earlier communication, printed in Volume III of these Annals,[a] I stated that by the action of cyanogen on aqueous ammonia, besides several other products, there are formed oxalic acid and a crystallizable white substance which is certainly not cyanate of ammonia, but which one nevertheless always obtains when one attempts to combine cyanic acid with ammonia for instance by so-called double decomposition. The fact that in the union of these substances they appear to change their nature, and give rise to a new body, drew my attention anew to this subject, and research gave the unexpected result that by the combination of cyanic acid with ammonia, urea is formed, a fact that is the more noteworthy inasmuch as it furnishes an example of the artificial production of an organic, indeed, a so-called animal substance, from inorganic materials.

I have already stated that the above-mentioned white crystalline substance is best obtained by the decomposition of cyanate of silver with sal ammoniac solution or of cyanate of lead by aqueous ammonia. In the latter way I prepared for myself the not unimportant amounts employed in this research. I obtained it in colorless, clear crystals often more than an inch long in the form of slender four-sided, dull-pointed prisms.

With caustic potash or chalk this substance evolved no trace of ammonia; with acids it showed none of the breakdown phenomena of cyanic acid salts, namely, evolution of carbonic acid and cyanic acid; neither would it precipitate lead and silver salts as genuine cyanic acid salts do; it could, therefore, contain neither cyanic acid nor ammonia as such. Since I found that by the last-named method of preparation no other product was formed and that the lead oxide was separated in a pure form, I imagined that an organic substance might arise by the union of cyanic acid with ammonia, possibly a substance like a vegetable salifiable base [an alkaloid]. I therefore made some experiments from this point of view on the behavior of the crystalline substance with acids. It was, however, indifferent to them, nitric acid excepted; this, when added to a concentrated solution of the substance, produced at once a precipitate of glistening scales. After these had been purified by several recrystallizations, they showed a very acid character, and I was already inclined to take the compound for a peculiar acid, when I found that after neutralization with bases it gave salts of nitric acid, from which the crystallizable substance could be extracted again with alcohol, with all the characteristics it had before the addition of nitric acid. This similarity to urea in behavior induced me to carry out comparative experiments with completely pure urea isolated from urine, from which it was plainly apparent that urea and this crystalline substance, or cyanate of ammonia, if one can so call it, are completely identical compounds.

I will describe the properties of this artificial urea no further, since it coincides perfectly with that of urea from urine, according to the accounts of Proust,[b] Prout[c] and others, to be found in their writings, and I will mention only the fact, not specified by them, that both natural and artificial urea, on distillation evolve first large amounts of carbonate of ammonia, and then give off to a remarkable extent the sharp, acetic acid-like odor of cyanic acid, exactly as I found in the distillation of cyanate of mercury or uric acid, and especially of the mercury salt of uric acid. In the distillation of urea, another white, apparently distinct substance also appears, with the examination of which I am still occupied.

But if the combination of cyanic acid and ammonia actually gives merely urea, it must have exactly the composition allotted to cyanate of ammonia by calculation from my formula for the cyanates; and this is, in fact, the case if one atom of water is added to cyanate of ammonia, as all ammonium salts contain water, and if Prout's analysis of urea is taken as the most correct.

According to him[1] urea consists of:

		Atoms
Nitrogen	46.650	4
Carbon	19.975	2
Hydrogen	6.670	8
Oxygen	26.650	2
	99.875	
	[sic]	

But cyanate of ammonia would consist of 56.92 cyanic acid, 28.14 ammonia, and 14.74 water, which for the separate elements gives:

		Atoms
Nitrogen	46.78	4
Carbon	20.19	2
Hydrogen	6.59	8
Oxygen	26.24	2
	99.80[2]	

One would have been able to reckon beforehand that cyanate of ammonia with I atom of water has the same composition as urea, without having discovered by experiment the formation of urea from cyanic acid and ammonia. By the combustion of cyanic acid with copper oxide one obtains 2 volumes of carbonic acid and 1 volume of nitrogen, but by the combustion of cyanate of ammonia one must obtain equal volumes of these gases, which proportion also holds for urea, as Prout actually found.

I refrain from all the considerations which so naturally offer themselves, particularly those bearing upon the composition relations of organic substances, upon the like elementary and quantitative composition of compounds of very different properties, as for example fulminic acid and cyanic acid, a liquid hydrocarbon and olefiant gas.[d] The deduction of a general law awaits further experiment on several similar cases.

Selection 162

As already noted, Tiedemann and Gmelin were confronted with the problem of reconciling chemistry and life when it came to explaining the mechanism of digestion. Notwithstanding the dilemma (or because of it), Leopold Gmelin (1788–1853) shortly afterwards pondered the chemical operations underlying plant and animal life. Trying to demarcate the chemical boundary between the two kingdoms he identified it with two opposite processes, desoxidation in plants and oxidation in animals. Gmelin associated desoxidation with the building up of 'high level' organic compounds in plants and oxidation with their degradation to a 'lower level'. Accordingly, the higher organic compounds contained relatively less oxygen and more carbon and hydrogen than the lower ones.

This idea of distinguishing between the chemical activities of plants and animals was advanced by Gmelin shortly after Wöhler had announced that he had succeeded in preparing artificial urea. Although the conception expressed in Gmelin's paper endured in various forms for some fifty years, it has received little attention from historians,[a] compared with Wöhler's celebrated article. In fact, Wöhler's preparation of urea (the news of which reached Gmelin before his own paper went to press) was of great interest to him for its theoretical implications, in connection with his own deliberations on the chemical transformation of organic compounds, both in living nature and in the test tube. Wöhler's work was an important influence but Gmelin also took into account other investigators' results, such as those made in connection with the studies of the breakdown of starch and gelatine by means of sulphuric acid (Kirchhof,[b] Braconnot[c]).

Gmelin believed that the difference between inorganic and organic compounds was tied up with the strength of the chemical forces which held together the elements that composed them. In inorganic compounds relatively strong chemical forces obtained that it was due to them that elements had the tendency to form simple inorganic compounds, spontaneously in nature or artificially in the laboratory (where higher temperature, electricity or other promoting factors come into play). In organic compounds, on the other hand, relatively weak chemical forces operated and this was the reason, Gmelin argued, why it was difficult but not impossible to prepare them artificially on the side, as it were.

For Gmelin organic compounds were primarily products of plant life that were derived from the conversion of carbonic acid, water and air, with the assistance of sunlight. For the reaction to take place, Gmelin believed, a very small amount of a previously formed organic compound was also required. Ultimately the system was under the control of the vital force.

Gmelin stressed that the various plant products did not result from one single act, but from linked desoxidative actions, involving successive conversions of lower organic compounds to higher ones that were accompanied by the evolution of excess oxygen. Gmelin envisaged a variety of 'constructive' transformations in the living plants. Among them he regarded the route starting with plant acids and proceeding through sugar, gum, starch to wood fibre as probable, though not completely established.

Gmelin also examined the artificial transformation of organic compounds and compared it with that effected by the living activity of plants. The main difference, he found, was that organic compounds were commonly brought to a lower state of composition when artificially transformed. In this respect many of the transformations of organic compounds in animals appeared to Gmelin to have been achieved in the same manner as those produced artificially.

The artificial transformation of organic compounds was effected either by the use of oxidants or water. In reactions in which an oxidating agent was involved, Gmelin interpreted, the supplied oxygen removed the carbon and hydrogen from the organic compound. The addition of water meant that hydrogen and oxygen were transferred to the organic compound. According to Gmelin, both types of reactions were oxidative and led to lower organic end-products. But he also allowed for the possibility of a simultaneous transformation of organic compounds into lower organic or even inorganic compounds, on the one hand, and into higher organic compounds, on the other. He found evidence for this, for example, in the conversion of sugar into carbonic acid and alcohol during fermentation.[d]

The questions raised by Gmelin in two papers given to the Heidelberg Society for the Study of Nature and Medicine on the 23 January 1827 and the 31 May 1828 were summarized by him as follows.

L. Gmelin, *Z.Physiol.*, *3*, 172. 1829

'Ueber die chemische Umwandlung der organische Verbindungen' (On the chemical transformation of organic compounds)

pp. 193–4

The facts and considerations collected together here lead to these conclusions.

1) The organic compounds occupy different levels of organic composition.
2) In the living plants, with the participation of light, out of the inorganic compounds present, especially out of water and carbonic acid, accompanied by development of oxygen gas, first lower organic compounds are produced and these through further acts of the vegetation whereby always more oxygen is developed, are constantly transformed into higher ones.
3) The less oxygen an organic compound contains, and the more carbon and hydrogen, the higher it stands.

4) Whilst in the plants, by means of the desoxidation process occurring there, always higher compounds are produced, this is reversed in the animals. To the extent that here a continuous oxidation process is given, and equally through artificial action by the addition of oxygen or water, [organic compounds] for the most part are transformed again into lower organic or into inorganic compounds. There can be, however, exceptions to this, in so far as with some artificial transformations an organic compound appears to break down on the one hand into a higher, on the other hand into a very low organic or into an inorganic compound, and the same may also occur in the animal body. As now some of the lowest organic compounds, like oxalic acid, urea and so on can be prepared from inorganic compounds, the possibility, at least out of hand, cannot be disputed that through suitable decomposition of the artificially formed lower organic compounds also the higher ones could be produced.

Selection 163

The central position of the idea of metabolism in the development of biochemistry cannot be doubted. Yet there is still little appreciation of the background leading to the formulation of this concept by Theodor Schwann (1810–82) in his classic book on the cell. It was the culmination of a series of brilliant investigations, achieved within an unbelievably short time (1835–9), embracing the examination of the tension-length relationship in muscle;[a] the study of digestion, including the discovery of pepsin;[b] the experimental enquiry into the nature of putrefaction and the demonstration of the improbability of spontaneous generation;[c] the elucidation of the connection between the growth of 'sugar fungi' (yeast globules) and the production of alcohol during fermentation.[d] By putting together his results on digestion, on alcoholic fermentation, and on the cell, Schwann assumed that the study of fermentation held the key to the understanding of cell-life. It was in this connection that he introduced the idea of metabolic change in the cell.

Schwann published the first results of his findings, dealing with the analogies in the structure and growth of plants and animals, in three articles in 1838.[e] They were then incorporated into the first and second sections of the book *Mikroskopische Untersuchungen über die Uebereinstimmung in der Struktur und dem Wachsthum der Thiere und Pflanzen* which were published in two parts in 1838 and contained Schwann's cell theory. In addition to this Schwann produced also a theory of cells forming the third section of the book, which bears 1839 as the publishing date.[f]

There had been others before Schwann who groped towards a cell theory but none of them had gone further and formulated a theory of cells.[g] It is the combination of both with makes Schwann's work a landmark in the history of morphology and biochemistry. This comes out clearly, for instance, when the achievements of Schwann and Purkyně[h] are compared. Purkyně perhaps more than anybody else before Schwann came nearest to the formulation of a comparative theory of microscopic structures in plants and animals together with a rudimentary theory of their formation. But he recognized the merit of Schwann's attempt to work out a generalization in a field where others (including himself) had tried before and had not succeeded.[i]

The usual interpretation of Schwann's contribution to the establishment of the cell theory is that he introduced the notion of the cell as the basic structural unit of all plant and animal tissues. Schwann himself saw his achievement in a different light. He stressed that the main feature of *his* cell theory consisted in the recognition of the formation of cells to be the *common* principle in the development of plants and animals.[j]

As earlier and contemporary microscopists had done, Schwann accepted the cell to be the most widespread elementary part (*Elementarteil*) of plants but not of animals. The elementary parts of animals, according to Schwann, were diverse; apart from cells, tubes and globules, they consisted mostly of fibres. In fact, the aim of his famous treatise was to show that whatever the function of the elementary parts, their origin was the same and they related to the whole. Because of this, the elementary parts of organisms ceased to be formations existing side by side, without any mutual relations. The cell became just as much a morphological as a physiological concept.

Then deeply convinced that teleological explanations of living processes should be eschewed as much as possible, Schwann outlined a theory of cells. In the particular case of living activity he was concerned with, that is, the formation of cells, he perceived at that time, at any rate, no difficulty in visualizing the natural steps from chemistry to morphogenesis. 'As the elementary materials of organic nature,' he wrote, 'are not different from those of the inorganic kingdom, the source of the organic phenomena can only reside in another combination of these materials, whether it be in a peculiar mode of union of the elementary atoms to form atoms of the second order, or in the arrangement of these conglomerate molecules when forming either the separate morphological elementary parts of organisms, or an entire organism.'[k]

Schwann linked the formation of cells directly to cell-life and in doing so he differentiated between the *plastic* (morphological) and *metabolic* processes. He associated the morphological features of cell formation, for instance the building up of fibres, with presumably physical interaction at molecular level, whereas the metabolic phenomena he conceived of as chemical changes involving either the constituents inside the cell, or the exchange of materials between the cells and its surrounding medium. Schwann considered the structureless substance in the interior of the cell and the intercellular substance surrounding the cell to be essentially the same. Nevertheless, he distinguished between the cell contents and the external *cytoblastema*, and because of this, he ascribed to the cell membrane a major active role in the transport of materials into and out of the cell. It is interesting that he questioned whether the processes in the passage across the membrane were not electrochemical, similar to those taking place in a galvanic pile. Finally, he discussed the possibility that the formation of the elementary parts of organisms and their growth could involve crystallization of the structureless substance, capable at the same time of imbibition.

Although scientific progress undermined Schwann's particular view on cytogenesis, which he owed to Matthias Jakob Schleiden (1804–81),[l] he was proved right in visualizing the chemistry, morphogenesis and physiology of the cell as interrelated aspects of the same process. This appears clearly from the way he linked up his studies on fermentation with his work on the development of cells.

After 1839 there was a perceptible shift in the scientific output of Schwann, who, although comparatively young, did not keep up his previous exertions. It has now become more clear that this was primarily due to the confrontation of the scientific interpretation of the phenomena of life and of his religious beliefs. Schwann began his great active phase of scientific activity assuming, as his pioneering experiments in muscular physiology showed, that biological processes could be completely reduced to physics and chemistry. Perhaps the discovery that alcoholic fermentation was due to a vital activity contributed to Schwann's subsequent decision to examine in more detail the relations of scientific knowledge of living phenomena and the teachings of the Catholic Church, which he accepted. Beginning with the cell, he intended to explore all fundamental aspects of life, including the brain and the nature of consciousness, culminating in enquiry into the position of man as a biological and moral being, but he never finished it. For many years he used to enter his thoughts into a Notebook or he wrote them up into short sections, but he refrained from publishing practically any of it. Judging from the account given by Marcel Florkin, who had seen the handwritten material, there is little doubt that while Schwann abandoned his erstwhile materialist outlook, his endeavour to integrate scientific findings within the framework of Christian beliefs constituted the other side of the coin.[m]

T. Schwann, *Microscopical Researches into the Accordance in the Structure and Growth of Animals and Plants* (London, 1847)

pp. 190–3

We set out, therefore, with the supposition that an organized body is not produced by a fundamental power which is guided in its operation by a definite idea, but is developed, according to blind laws of necessity, by powers which, like those of inorganic nature, are established by the very existence of matter. As the elementary materials of organic nature are not different from those of the inorganic kingdom, the source of the organic phenomena can only reside in another combination of these materials, whether it be in a peculiar mode of union of the elementary atoms to form atoms of the second order, or in the arrangement of these conglomerate molecules when forming either the separate morphological elementary parts of organisms, or an entire organism. We have here to do with the latter question solely, whether the cause of organic phenomena lies in the whole organism, or in its separate elementary parts. If this question can be answered, a further inquiry still remains as to whether the organism or its elementary parts possess this power through the peculiar mode of combination of the conglomerate molecules, or through the mode in which the elementary atoms are united into conglomerate molecules.

We may, then, form the two following ideas of the cause of organic phenomena, such as growth, &c. First, that the cause resides in the totality of the organism. By the combination of the molecules into a systematic whole, such as the organism is in every stage of its development, a power is engendered, which enables such an organism to take up fresh material from without, and appropriate it either to the formation of new elementary parts, or to the growth of those already present. Here, therefore, the cause of the growth of the elementary parts resides in the totality of the organism. The other mode of explanation is, that growth does not ensue from a power resident in the entire organism, but that each separate elementary part is possessed of an independent power, an independent life, so to speak; in other words, the molecules in each separate elementary part are so combined as to set free a power by which it is capable of attracting new molecules, and so increasing, and the whole organism subsists only by means of the reciprocal[1] action of the single elementary parts. So that here the single elementary parts only exert an active influence on nutrition, and totality of the organism may indeed be a condition, but is not in this view a cause.

In order to determine which of these two views is the correct one, we must summon to our aid the results of the previous investigation. We have seen that all organized bodies are composed of essentially similar parts, namely, of cells; that these cells are formed and grow in accordance with essentially similar laws; and, therefore, that these processes must, in every instance, be produced by the same powers. Now, if we find that some of these elementary parts,not differing from the others, are capable of separating themselves from the organism, and pursuing an independent growth, we may thence conclude that each of the other elementary parts, each cell, is already possessed of power to take up fresh molecules and grow; and that, therefore, every elementary part possesses a power of its own, an independent life, by means of which it would be enabled to develop itself independently, if the relations which it bore to external parts were but similar to those in which it stands in the organism. The ova of animals afford us examples of such independent cells, growing apart from the organism. It may, indeed, be said of the ova of higher animals, that after impregnation the ovum is essentially different from the other cells of the organism; that by impregnation there is a something conveyed to the ovum, which is more to it than an external condition for vitality, more than nutrient matter; and that it might thereby have first received its peculiar vitality, and therefore that nothing can be inferred from it with respect to the other cells. But this fails in application to those classes which consist only of female individuals, as well as with the spores of the lower plants; and, besides, in the inferior plants any given cell may be separated from the plant, and then grow alone. So that here are whole plants consisting of cells, which can be positively proved to have independent vitality. Now, as all cells grow according to the same laws, and consequently the cause of growth cannot in one case lie in the cell, and in another in the whole organism; and since it may be further proved that some cells, which do not differ from the rest in their mode of growth, are developed independently, we must ascribe to all cells an independent vitality, that is, such combinations of molecules as occur in any single cell, are capable of setting free the power by which it is enabled to take up fresh molecules. The cause of nutrition and growth resides not in the organism as a whole, but in the separate elementary parts – the cells. The failure of growth in the case of any particular cell, when separated from an organized body, is as slight an objection to this theory, as it is an objection against the independent vitality of a bee, that it cannot continue long in existence after being separated from its swarm. The manifestation of the power which resides in the cell depends upon conditions to which it is subject only when in connexien with the whole (organism).

The question, then, as to the fundamental power of organized bodies resolves itself into that of the fundamental powers of the individual cells. We must now consider the general phenomena attending the formation of cells, in order to discover what powers may be presumed to exist in the cells to explain them. These phenomena may be arranged in two natural groups: first, those which relate to the combination of the molecules to form a cell, and which may be denominated the *plastic* phenomena of the cells; secondly, those which result from chemical changes either in the component particles of the cell itself, or in the surrounding cytoblastema, and which may be called *metabolic* phenomena (τὸ μεταβολικὸν, implying that which is liable to occasion or to suffer change).

pp. 197–201

The cytoblastema, in which the cells are formed, contains the elements of the materials of which the cell is composed, but in other combinations: it is not a mere solution of cell-material, but it contains only certain organic substances in solution. The cells, therefore, not only attract materials from out of the cytoblastema, but they must have the faculty of producing chemical changes in its constituent particles. Besides which, all the parts of the cell itself may be chemically altered during the process of its vegation. The unknown cause of all these phenomena, which we comprise under the term metabolic phenomena of the cells, we will denominate the *metabolic power.*[a]

The next point which can be proved is, that this power is an attribute of the cells themselves, and that the cytoblastema is passive under it. We may mention vinous fermentation[2] as an instance of this. A decoction of malt will remain for a long time unchanged; but as soon as some yeast is added to it, which consists partly of entire fungi and partly of a number of single cells, the chemical change immediately ensues. Here the decoction of malt is the cytoblastema; the cells clearly exhibit activity, the cytoblastema, in this instance even a boiled fluid, being quite passive during the change. The same occurs when any simple cells, as the spores of the lower plants, are sown in boiled substances.

In the cells themselves again, it appears to be the solid parts, the cell-membrane and the nucleus, which produce the change. The contents of the cell undergo similar and even more various changes than the external cytoblastema, and it is at least probable that these changes originate with the solid parts composing the cells, especially the cell-membrane, because the secondary deposits are formed on the inner surface of the cell-membrane, and other precipitates are generally formed in the first instance around the nucleus. It may therefore, on the whole, be said that the solid component particles of the cells possess the power of chemically altering the substances in contact with them.

The substances which result from the transformation of the contents of the cell are different from those which are produced by change in the external cytoblastema. What is the cause of this difference, if the metamorphosing power of the cell-membrane be limited to its immediate neighbourhood merely? Might we not much rather expect that converted substances would be found without distinction on the inner as on the outer surface of the cell-membrane? It might be said that the cell-membrane converts the substance in contact with it without distinction, and that the variety in the products of this conversion depends only upon a difference between the convertible substance contained in the cell and the external cytoblastema. But the question then arises, as to how it happens that the contents of the cell differ from the external cytoblastema. If it be true that the cell-membrane, which at first closely surrounds the nucleus, expands in the course of its growth, so as to leave an interspace between it and the cell, and that the contents of the cell consist of fluid which has entered this space merely by imbibition, they cannot differ essentially from the external cytoblastema. I think therefore that, in order to explain the distinction between the cell-contents and the external cytoblastema, we must ascribe to the cell-membrane not only the power in general of chemically altering the substances which it is either in contact with, or has imbibed, but also of so separating them that certain substances appear on its inner, and others on its outer surface. The secretion of substances already present in the blood, as, for instance, of urea, by the cells with which the urinary tubes are lined, cannot be explained without such a faculty of the cells. There is, however, nothing so very hazardous in it, since it is a fact that different substances are separated in the decompositions produced by the galvanic pile. It might perhaps be conjectured from this peculiarity of the metabolic phenomena in the cells, that a particular position of the axes of the atoms composing the cell-membrane is essential for the production of these appearances.

Chemical changes occur, however, not only in the cytoblastema and the cell-contents, but also in the solid parts of which the cells are composed, particularly the cell-membrane. Without wishing to assert that there is any intimate connexion between the metabolic power of the cells and galvanism, I may yet, for the sake of making the representation of the process more clear, remark that the chemical changes produced by a galvanic pile are accompanied by corresponding changes in the pile itself.

The more obscure the cause of the metabolic phenomena in the cells is, the more accurately we must mark the circumstances and phenomena under which they occur. One condition to them is a certain temperature, which has a maximum and a minimum. The phenomena are not produced in a temperature below 0° or above 80° R.;[b] boiling heat destroys this faculty of the cells permanently; but the most favorable temperature is one between 10° and 32° R.[c] Heat is evolved by the process itself.

Oxygen, or carbonic acid, in a gasceous form or lightly confined, is essentially necessary to the metabolic phenomena of the cells. The oxygen disappears and carbonic acid is formed, or *vice versa*, carbonic acid disappears, and oxygen is formed. The universality of respiration is based entirely upon this fundamental condition to the metabolic phenomena of the cells. It is so important that, as we shall see further on, even the principal varieties of form in organized bodies are occasioned by this peculiarity of the metabolic process in the cells.

Each cell is not capable of producing chemical changes in every organic substance contained in solution, but only in particular ones. The fungi of fermentation, for instance, effect no changes in any other solutions than sugar; and the spores of certain plants do not become developed in all substances. In the same manner it is probable that each cell in the animal body converts only particular constituents of the blood.

The metabolic power of the cells is arrested not only by powerful chemical actions, such as destroy organic substances in general, but also by matters which chemically are less uncongenial; for instance, concentrated solutions of neutral salts. Other substances, as arsenic, do so in less quantity. The metabolic phenomena may be altered in quality by other substances, both organic and inorganic, and a change of this kind may result even from mechanical impressions on the cells.

Such are the most essential characteristics of the fundamental powers of the cells, so far as they can as yet be deduced from the phenomena. And now, in order to comprehend distinctly in what the peculiarity of the formative process of a cell, and therefore in what the peculiarity of the essential phenomenon in the formation of organized bodies consists, we will compare this process with a phenomenon of inorganic nature as nearly as possible similar to it. Disregarding all that is specially peculiar to the formation of cells, in order to find a more general definition in which it may be included with a process occurring in inorganic nature, we may view it as a process in which a solid body of definite and regular shape is formed in a fluid at the expense of a substance held in solution by that fluid. The process of crystallization in inorganic nature comes also within this definition, and is, therefore, the nearest analogue to the formation of cells.

p. 215

The view then that organisms are nothing but the form under which substances capable of imbibition crystallize, appears to be compatible with the most important phenomena of organic life, and may be so far admitted, that it is a possible hypothesis, or attempt towards an explanation of these phenomena. It involves very much that is uncertain and paradoxical, but I have developed it in detail, because it may serve as a guide for new investigations. For even if no relation between crystallization and the growth of organisms be admitted in principle, this view has the advantage of affording a distinct representation of the organic processes; an indispensable requisite for the institution of new inquiries in a systematic manner, or for testing by the discovery of new facts a mode of explanation which harmonizes with phenomena already known.

Selection 164

It was a considerable time before Schwann's idea concerning the uniformity of chemical processes on the cellular level gained much attention. It can be fairly said that it did not really come into its own before the establishment of intracellular respiration (1875).[a]

As for relating it to fermentation, we find that the majority of chemists (influenced by Berzelius and especially by Liebig) completely rejected the suggestion that fermentation had anything to do with life. On the contrary, Liebig defended for many years the view that fermentation was purely a chemical process closely related to the putrefaction of those organic bodies which in fact had ceased to live.[b] Under these circumstances it is not really surprising that Schwann's suggestion was not taken up, and that the study of fermentation at first developed independently of the cell theory.

Behind this delay were conflicting impulses affecting the interaction between chemistry and physiology. That interaction would eventually lead to the formation of physiological chemistry as an essential branch of physiology (c. 1840–c. 1880).

On the one hand, there were men like Liebig and Dumas who pursued chemical studies of plant and animal life because they were convinced that these furnished the key to physiology. Apart from its physiological significance, they considered their work to be of practical value in other areas (agriculture, medicine, industry). However, in their efforts the cell theory played no part – they proceeded experimentally and theoretically straight from test tube findings to physiological processes.

On the other hand, there were a few, often connected with the medical profession, such as Rudolf Virchow (1821–1902), who were alive to the significance of the cell theory for physiology but were not successful in applying it productively. In the end he, the author of the famed *Die Cellularpathologie in ihrer Begründung auf physiologische und pathologische Gewebelehre* (1858), found the anatomical rather than the physiological approach more rewarding.[c]

Moreover, the study of the chemical sides of plant and animal life remained very much under the influence of ideas that stressed an underlying difference between them. They found their expression in Gmelin's characterization (already noted) of the chemical activities in plants as desoxidation and of those occurring in animals as oxidation.[d] Such an approach did not preclude thinking about a connection between both sets of phenomena. Indeed, such thinking would seem to have been derived from concepts which replaced the cycle derived from the doctrine of phlogiston with a cycle incorporating the oxygen theory.[e]

The cyclic pathway was conceived of as depending on the atmosphere providing the link between the plants and animals whereby the former solely performed desoxidative reactions and the latter oxidative reactions. Accordingly it was in the nature of plants, aided by the light from the sun, to remove the carbonic acid breathed out by animals and renew the oxygen in the atmosphere.

The passage below illustrates the case for the cycle operated by contrary but complementary activities of plants (reduction) and of animals (combustion). It was specifically presented in a lecture by Dumas (on concluding a course on organic chemistry at the École de Médicine) which first appeared under his and Boussingault's[f] names in 1841.[g] Because, as he put it, 'our modes of viewing every subject that interests us, has engendered a community of opinion between us, in which it would be difficult for each to specify what belongs to him in particular'. In the appended historical notes they made explicit their concern for the priority of their own contribution and also for that of French workers in general, to the understanding of the chemistry of life.

The extracts are from the English version of the third edition of the lecture published in 1844.

J. Dumas and J.B. Boussingault, *The Chemical and Physiological Balance of Organic Nature: An essay* **(London, 1844) [unpaginated]**

PROGRAMME OF THE DISCOURSE

AN ANIMAL	A VEGETABLE
is	is
AN APPARATUS OF COMBUSTION;	AN APPARATUS OF REDUCTION;
Possesses the faculty of Locomotion;	Is fixed;
Burns Carbon, Hydrogen, Ammonium;	Reduces Carbon, Hydrogen, Ammonium;
Exhales Carbonic Acid, Water, Oxide of Ammonium, Azote;[a]	Fixes Carbonic Acid, Water, Oxide of Ammonium, Asote;
Consumes Oxygen, Neutral Azotised matters, Fatty matters. Amylaceous matters,[b] sugars, gums;[c]	Produces Oxygen, Neutral Azotised matters, Fatty matters, Amylaceous matters, sugars, gums;
Produces Heat, Electricity;	Absorbs Heat, Abstracts Electricity;
Restores its elements to the air or to the earth;	Derives its elements from the air or from the earth;
Transforms organised matters into mineral matters.	Transforms mineral matters into organic matters.

p. 2

Vegetables, animals, man, contain matter in their composition. Whence comes it? What part does it play in their tissues and in the fluids which bathe them? What becomes of it when death breaks the chain by which its various parts and forms were so closely conjoined?

pp. 3–8

These words will bring to your minds with what amazement we discovered together, that of all the elements of modern chemistry, organic nature made use of but three or four; that of those vegetable and animal substances which are now multiplied almost to infinity, general physiology requires no more than some ten or twelve species; and that all the phenomena of life, so complex in appearance, may be referred in their essence to a single general formula, so simple, that in a few words every thing seems stated, every thing having been recalled to mind, every thing foreseen.

Have we not, in fact, found, by a multitude of results, that an animal, in a chemical point of view, constitutes a true apparatus of combustion, by which carbonaceous matters, burnt incessantly, are returned to the atmosphere in the shape of carbonic acid; in which hydrogen, burnt incessantly, is returned as water; whence, in fine, free azote is ceaselessly exhaled in the breath, and, in the state of oxide of ammonium, is thrown off in the urine?

From the animal kingdom, therefore, as a whole, carbonic acid, watery vapour, and azote or oxide of ammonium, are continually escaping, – simple substances, and few in number, the formation of which is intimately connected with the history of the atmosphere itself.

Have we not, on the other hand, found that vegetables, in their natural and healthy state, decompose carbonic acid incessantly, fixing the carbon, and setting free the oxygen; that they decompose water, seizing on its hydrogen, and disengaging its oxygen as before; lastly, that they either abstract azote directly from the air, or take it indirectly from oxide of ammonium, or nitric acid; thus acting, in every particular, inversely or in opposition to animals? If the animal kingdom constitute an immense apparatus of combustion, the vegetable kingdom, in its turn, constitutes an immense apparatus of reduction, where carbonic acid decompounded leaves its carbon, water its hydrogen, and oxide of ammonium and nitric acid their ammonium or their azote.

If animals incessantly produce carbonic acid, water, azote, and oxide of ammonium, vegetables consequently consume, without cease, oxide of ammonium, azote, water, and carbonic acid. What the one gives to the atmosphere, that the other takes from it; so that, surveying these facts from the loftiest point of view, and in connexion with the physics of the globe, it would be imperative on us to say that, in so far as their truly organic elements are concerned, plants and animals are the OFFSPRING OF THE AIR; that they are but condensed or consolidated air; and that, to form a true and accurate idea of the constitution of the atmosphere at the epochs which preceded the birth of organised beings, it would be necessary to restore to it, by calculation, the whole of the carbonic acid and azote, the elements of which were appropriated by vegetables and animals when they appeared.

Vegetables and animals, therefore, come from the atmosphere, and return to it again; they are true dependants of the air.

Vegetables, then, assume from the atmosphere the elements which animals exhale into it; viz. carbon, hydrogen, and azote, or rather carbonic acid, water, and ammonia.

But how do animals procure the elements which they give to the atmosphere? Let us inquire particularly into this point. Now it is impossible to contemplate, without admiration, the sublime simplicity of the laws of nature here, as every where! Animals always derive their elements primarily from vegetables.

We have found, in fact, by results beyond the reach of question, that animals do not create any of the truly organic substances, but that they consume or destroy them; that vegetables, on the contrary, habitually create these substances, and that they destroy but few, and this only for particular and determinate ends.

It is, in the vegetable kingdom therefore, that the great elaboratory of organic life is found; it is there that both vegetable and animal substances are compounded: and they are all alike formed at the cost of the atmosphere.

From vegetables these substances pass ready-formed into the bodies of herbivorous animals, which destroy one portion of them, and store up another in their tissues.

From herbivorous animals they pass ready-formed into the bodies of carnivorous animals, which destroy or lay them up, according to their wants.

Finally, during the life of these animals, or after their death, the organic substances in question return to the atmosphere from whence they originally came, in proportion as they are destroyed.

Thus is the mysterious circle of organic life upon the surface of the globe completed and maintained! The air contains or engenders the oxidised substances required, – carbonic acid, water, nitric acid, and ammonia. Vegetables, true-reducing apparatus, seize upon the radicals of these, carbon, hydrogen, azote, ammonium; and with them, they fashion all the variety of organic or organisable matters which they supply to animals. Animals, again, true apparatuses of combustion, reproduce from them carbonic acid, water, oxide of ammonium, and azotic or nitric acid, which return to the air to reproduce the same phenomena to the end of time.

And if, to this picture, already so striking by its simplicity and grandeur, we add the indubitable part performed by the solar light, which is alone possessed of power to bring into play this immense, this unparalleled apparatus, constituted by the vegetable kingdom, in which the oxidized products of the atmosphere are subjected to reduction . . .

Selection 165

Chemistry at the time of Lavoisier was conceived as a unified branch of natural science. From the purely chemical point of view Berzelius, developing the concepts of Lavoisier, saw the difference between inorganic and organic substances in that the former were oxides of simple radicals whereas the latter were oxides of compound radicals. With plants the compound radicals consisted in general of carbon and hydrogen and with animals of carbon, hydrogen and nitrogen. The radicals were believed to retain their identity in chemical changes. In that sense a compound radical in organic chemistry took the place of an element in inorganic chemistry.

Berzelius was also a strong advocate of applying electrochemical concepts both to inorganic and organic compounds. Since elements were considered to be either electropositive or electronegative, the same could be thought about radicals. Metals, hydrogen or alcohol radicals were electropositive whereas chlorine, benzoyl or other acid radicals were considered to be electronegative. By 1838 it became increasingly obvious that it was impossible to cover the facts of organic chemistry on the basis of an electrochemical theory of radicals. One of the facts concerned was that chlorine, generally accepted to be an electronegative element, was able to replace in some organic compounds the electropositive hydrogen. This became the basis for the development of the substitution-theory initiated by Dumas.[a]

From the study of the correspondence between Berzelius and Liebig it is possible to get an intimate view of some of the problems which gave rise to the separation of organic from inorganic chemistry. From a convinced advocate of Berzelius's views who was a resolute opponent of splitting chemistry Liebig became an opponent of Berzelius without, however, having been able to solve the difficulties.

The discussions between Berzelius and Liebig produced no results. The older chemist did not budge and the younger not really knowing the answer became very unsettled. The letter of Liebig to Berzelius of 26 April 1840 contains the following sentences:

I confess here in advance, this is the expression of an insurmountable loathing and revulsion against what is at present happening in chemistry. The quarrel about the substitution theory has been pushed to extreme lengths . . . I am becoming completely dispassionate, cooler and more sensible . . . I have asked myself earnestly of what use are those arguments. Neither medicine nor physiology, nor industry, can gain useful applications from them. Everything we have is too young and new that we may hope to develop laws which would last longer than a month . . . For four months I have been devoting myself to quite another aspect of the science. I have been studying organic chemistry in relation to its laws which in their present stage have a bearing on agriculture and physiology.[b]

Liebig was referring to his studies which brought organic chemistry into close relation to plant and animal physiology.

Physiology – the science of life – underwent profound changes on the basis of the cell theory and chemistry since the 1840s. Liebig had taken hardly any notice of the cell theory for physiology but was fully aware of the importance of chemistry. As already stated, Liebig had begun to enquire more deeply into the usefulness of chemistry for agriculture, medicine and industry, when he became unable to see his way in the theoretical discussion on organic chemistry. Not only physiology but also chemistry went through great changes at that time. The separation of inorganic and organic chemistry, so deplored by Berzelius and not understood by Liebig, was temporary. Following the Congress at Karlsruhe (1860) where the confusion over atoms, molecules and equivalents was highlighted and the proposal made to solve it on the basis of Avogadro's hypothesis, both were able to unite. By the 1860s the character of organic chemistry was changing too. Instead of dealing with chemical substances and processes in plants and animals it became primarily concerned with the properties of carbon compounds as such.[c]

Physiological chemistry in reality inherited the problems which were previously investigated by organic chemistry. The significance of Liebig in this field does not lie so much in his scientific achievements because he was not infrequently in error, but in that he understood the necessity of linking chemistry, particularly organic chemistry, to physiology, and that he fought hard to put plant and animal physiological chemistry directly into the service of agriculture, medicine and industry. This he expressed very clearly in his famous book *Organic Chemistry in its Applications to Agriculture and Physiology*, first published in 1840 and written at the request of the British Association for the Advancement of Science. In the dedication to this body one reads:

I have endeavoured to develop, in a manner correspondent to the present state of science, the fundamental principles in chemistry in general, and the laws of organic chemistry in particular, in their applications to agriculture and physiology; to the causes of fermentation, decay, and putrefaction; to the vinous and acetous fermentations and nitrification . . . Perfect agriculture is the true foundation of all trade and industry – it is the foundation of the riches of states. But a rational system of agriculture cannot be formed without the application of scientific principles; for such a system must be based on an exact acquaintance with the means of nutrition of vegetables, and with the influence of soils and action of manure upon them. This knowledge we must seek from chemistry, which teaches the mode of investigating the composition and of studying the characters of the different substances from which plants derive their nourishment.[d]

The companion book, *Animal Chemistry in its Application to Physiology and Pathology*, appeared two years later. The brief dedication to Berzelius in the German original was left out of the English version.[e] Instead, Liebig again addressed the British Association for the Advancement of Science, stating that the work constituted the second part of his report on the state of organic chemistry. He further mentioned his intention of following it up with a third part, dealing with the food of man and animals, within two years.[f]

Despite the dedication which he did formally and publicly acknowledge, Berzelius reacted to *Animal Chemistry* negatively. Indeed following its publication, the rift that had been developing between the two chemists (accentuated by differences in age and temperament) entered into its final phase. For some time Berzelius and Liebig had not seen eye to eye about electrochemistry and the concept of catalysis;[g] now they parted company on account of their different approach to the chemical study of physiological processes.

As Berzelius perceived it, the tendency in contemporary physiological chemistry to ignore the anatomical aspects of physiology led to chemical speculations. He clearly saw them exemplified in the work of Liebig and Dumas. Berzelius patently deplored the fact that speculations such as these became the source of priority disputes between the German and French chemists over something which in his view amounted to an 'easy kind of physiological chemistry created at the writing desk'. Consequently there was the danger that a mere possibility and probability would be conceived as truth. This, Berzelius believed, was to the detriment of real solutions to the problems physiological chemistry was seeking to solve.[h]

Reviewing *Animal Chemistry*, Berzelius concentrated on areas where he thought that Liebig went astray by failing to distinguish between a secured and a hypothetical body of scientific knowledge.[i] First on the list was the question of the source of animal heat. Berzelius objected to what he believed was axiomatically asserted by Liebig, that it results from chemical transformations, that is, from the combination of oxygen circulating in the blood with the constituents of food. This was a possibility but not the only one: additional mechanical, electrical and nervous mechanisms were not to be excluded. Secondly, Berzelius took exception to the manner in which Liebig represented the course of the chemical reactions taking place in the animal body in terms of addition of hydrogen or oxygen or both and substraction of carbonic acid, ammonia or water. Convinced that the whole exercise was premature, Berzelius predicted that all Liebig's equations would be invalidated within a few years. Thirdly, Berzelius dealt with Liebig's view on animal respiration and especially with his emphasis on the share of iron in it, due to the presence of the metal in the blood corpuscles. Berzelius found this idea of Liebigs on the role of iron in respiration plausible, indeed probable. But it was a hypothesis in need of experimental corroboration. If only Liebig had supplied it, concluded Berzelius, this portion of *Animal Chemistry* would have constituted its 'most beautiful pearl'.

Not surprisingly, Berzelius's observations aroused Liebig's wrath and he gave vent to it in a long article from which the extracts below are taken.

Also not unexpectedly, both Berzelius and Liebig proved to be right as well as wrong on this or that particular or general issue, whether their position is examined in a contemporary light or in retrospect. It is beyond the scope of this volume to examine this part of the story in more detail.

However, what is appropriate is to focus briefly on the methodological side of the dispute, following the publication of *Animal Chemistry*. To begin with, Berzelius's scientific methodology cannot be divorced from his world outlook, strongly beholden to religion, which made him believe in a mysterious vital force governing the organism as a whole. What were fathomable were the physical, chemical, anatomical and physiological manifestations of the total activity of the organism.

For Berzelius the question of the wholeness of the animal organism was ostensibly paramount. What Berzelius had to say on this issue will be recalled.[j] The power to live belonged not to the constituent parts of organisms but resulted from their mutual interaction. Further he assumed that the vital force ensured the wholeness of the animal organism through the integrating activity of the nervous system. In relation to the latter, chemical activities appeared to be operating at lower levels. In this scheme physiological functions depending on chemical processes represented higher phenomena. What appeared to be foremost in the mind of Berzelius was the relation of the separate activities to the organic whole. It is here, then, that we must look for a clue to Berzelius's anti-reductionist reproach to chemists who failed to recognize that the anatomical element was part and parcel of the physiological phenomena whose chemical aspects they undertook to illuminate.

Turning now to Liebig, we find that his world outlook was not free from religious influences either.[k] Liebig's commitment to the fundamental validity of religion may certainly be discerned in what he wrote in *Animal Chemistry* about man's inherent incapacity to apprehend what life was. This situation was not to be altered by discoveries of 'laws of vitality', however numerous they might be.

On the one hand, 'life' was incomprehensible, similar to the Kantian unknowable 'thing in itself'. On the other hand, plant and animal activities were governed by laws accessible to scientific knowledge. Like Berzelius, Liebig was conscious of this contradiction but he chose to solve it along different lines.

We have already noted that Berzelius held firmly to the notion of vital force although it represented to him a mere word with no affixed idea. This enabled him to pursue chemical studies of products of organisms without having to refer to it constantly. In contrast Liebig turned to it and conferred on it a particular role, in his quest to establish the laws of vital motion. He regarded it as the specific cause of vital activity producing the characteristic vegetable and animal forms and structures. As to the nature of vital force, Liebig thought that it combined the properties of agencies causing chemical and electrical changes. As to its mode of operation, Liebig assumed that the vital force was behind the breaking down of food constituents into their elements and the building them up into the living tissues. As to laws governing these processes, Liebig proceeded on the supposition that they exclusively obeyed the laws of mechanics, physics and chemistry.[1] Closely allied to this thinking were Liebig's doubts – as shown below – about the relevance of physiological techniques.

The main effect of this thinking was to obliterate the differences between non-vital and vital phenomena, on the assumption that the latter were to be explained purely in mechanical, physical and chemical terms. What this reductionist approach (of which Liebig was by no means a solitary defender) left untackled were the origins and nature of the specific qualitative features of vital phenomena, including those arising from the relation of parts to the whole.

J. Liebig, *Ann.*, *50*, 295. 1844

'Berzelius und die Probabilitätstheorien' (Berzelius and the probability theories)

pp. 304–5

[Berzelius] calls our results probability theories because we conceive of the heart as a pressure as well as a suction pump in the sense in which perhaps the eye is compared with a *camera obscura*; because of a printing error in one, and only one, place in the book it is written that urine is separated from the venous blood, because we believe that the arterial blood passes through the kidney and the venous blood through the liver . . .

But if it is assumed that these views are crass errors, was then the aim of the author's work to establish and defend them? When he strove to ascertain the composition of the bile, urea, uric acid, that of the blood and their formation, when he strove to investigate their relationships to nutrition and to the secretions, is it not for his purpose completely immaterial whether the urine is separated from the venous or arterial blood, whether the heart is or is not a pressure and suction pump?

When the chemist maintains that the blood is not produced from starch and sugar, that bile is not to be found in the faeces but is given off as gas [*Luftform*]. When he develops the view that the drugs which are products of organic life have a share in the processes of the animal organism similar to the one we have recognized with certainty regarding all substances which nourish the plant kingdom. When he upholds that uric acid and urea are products of metabolism and do not arise directly from food. When he calls attention to the close relation between nutrition, heat loss and force consumption [*Kraftverbrauch*], are then these really probability theories, views without any real foundation, in the wake of the works which preceded them? Should then all researches carried out through thirty years remain absolutely without results capable of a useful application?

pp. 309–11

The chemist says: if I know the weight of the tobacco and the ash, then I know how much has gone away in the smoke. The physiologist says: prove this to me! If the chemist had weighed the smoke and quite neglected the tobacco and the ash, then he would have held the result as much more correct; so great is with many the absurdity of their direction of thought! . . .

It is established that, besides through the urine, skin and lungs, the carbon taken in through the mouth and wanting in the faeces has absolutely no other way out of the body; that the carbon passes out of the skin and lungs in the form of carbonic acid, that urea and water signify nothing else than carbonic acid and ammonia. In a quite simple equation the unknown quantity may be now deduced from two unknown ones and it may be maintained that a grown healthy man, who exercises daily for four hours and drags a heavy load, daily burns about 13.5 ounces of carbon in his body.

This conclusion is just as true as the statement of the mechanic who empirically ascertained on one hundred thousand soldiers that on the average a healthy, fully grown man without detriment to his health cannot carry more than a 30 pound burden for 8 hours. He does not do it like the physiologist who holds this conclusion to be erroneous, because this or that consumptive can carry only 10 pounds, or a stronger man 50 or 100 pounds.

Thus is has been ascertained that the average length of life for man amounts to thirty and some years, and yet very few men die just in the thirtieth year. All these conclusions are as near the truth as one possibly may get, they are on this account taken for the truth itself and serve as the foundation for the calculation of the proceeds derived from investments, and in life insurance arrangements, or the fixing of the weight of weapons and the pack that a soldier may carry.

The orthodox physiologist is not satisfied with this; observations made on nature on this scale do not satisfy him. Without concerning himself whether the man or animal has beforehand taken a meal, whether with full or empty stomach, he confines them in a chamber [*Kasten*] and estimates now the oxygen amounts which they breathe in and the carbon dioxide amounts which they breathe out. Instead of weighing the ash and tobacco he weighs the smoke, as if the sources of error with the latter were not a thousand times greater than with the first method. Assumed that observation was absolutely accurate, then with it he knows no more and no less than what a man in a such a chamber (under certain not more closely investigated and not normal conditions) in a given time breathes out and in. How much carbon however this individual uses in 24 hours, this does not disclose. If the experimentalist had given this man a bottle of good wine in the chamber, or if the latter beforehand had enjoyed a proper portion of cod liver oil, then without doubt quite other relations would have resulted.

pp. 326–8

The physiologist to whose mental eyes the animal body is as transparent as a glass-house, who with the greatest certainty knows the changes which the air in the lungs experiences, it is he who still needs endless and worthless special experiments in order to get a point of view. He shakes blood with air, and finds a trace of carbonic acid in the air, and although he perceives not the slightest oxygen absorption in his experiments, yet this carbonic acid evolution suffices for him to strengthen his conviction about the respiration process, although a handful of damp sawdust, a leaf would have behaved exactly so. What then if the blood would have been so resistant as not to give off any carbonic acid *outside the body*?

Everywhere and in all cases where we have succeeded in establishing the natural conditions of a phenomenon our conclusions have a far higher value than they can obtain through experiments. An experiment can never contradict a truth deduced in this way. We are indeed in such a plight that only with expenditure of endless trouble and time are we in the state to return to the conditions in which the phenomenon shows itself. With knowledge of the conditions the work is accomplished. The simplest, most correct and safest way remains always that of questioning nature; further experiments serve then to show us that we have not erred, they give foundations for useful applications and for further conclusions.

Let us then not make the work impossible to execute through difficulties which our phantasy introduces. Those which reality presents are great enough for our powers.

Selection 166

However noteworthy, the points raised in the controversy between Berzelius and Liebig appeared to have had little immediate effect on the development of the chemistry of life. Of greater significance and playing a major part in the conflict between mechanists and vitalists were the issues regarding the chemical or biological nature of fermentation. These were brought to a head especially by the debate between Justus Liebig and Louis Pasteur (1822–95) during the early 1870s.[a]

In considering Pasteur's biological interpretation of fermentation,[b] it is vital not to lose sight of its links with his work on spontaneous generation.[c] Pasteur's approach had a precedent in the work of his older contemporary, Schwann,[d] whose researches he was well acquainted with and valued highly. The intermittent criticism of Pasteur that he did not acknowledge his debt to Schwann is difficult to sustain.[e] Normally, it was a part of Pasteur's methodology to trace the development of the subject he was undertaking to investigate. While concerned with the outward recognition of his own contributions he accepted, at least in principle, that they constituted links in an historical chain of investigations.[f]

With regard to the paper from which the excerpts below are taken, commentators have mostly concentrated on Pasteur's success in preserving such easily decomposable fluids as blood and urine by exposing them to germ-free air and without heating them. Certainly at the time Pasteur saw this achievement as decisively significant for his view that fermentation and putrefaction were related slow oxidation processes carried out by microscopic living beings with a mode of life of their own. At the same time, Pasteur took it as further evidence against, indeed a final blow to the belief that they could arise from decaying organic material. However, Pasteur's experimental work along these lines, as his contemporary F. Engels (1820–95) accurately observed, had no bearing on the question of the spontaneous

origin of life from non-living matter.[g] It is necessary to clarify Pasteur's
position regarding the possibility of 'abiogenesis'[h] and his own earlier
attempts to produce life artificially, which stemmed from his ideas on the
relation between molecular asymmetry and life.[i]

These aspects are not the only ones that make the paper interesting.
There is another connected with Pasteur's concern to tackle the large
question of microorganisms' place in Nature, assigned to them by an all-
wise God. In this scheme God created microorganisms for the purpose of
destroying dead plant and animal materials. If this were not so, their
components could not return to the atmosphere and the soil and the
continuance of life would become impossible.

This, then, was the approach Pasteur adopted in order to combine his
religious beliefs with the results of his scientific thought and practice. His
approach centred on the conception of an invisibly animated atmosphere
which played an indispensable part in the circulation of matter between the
non-living and living worlds. This was Pasteur's contribution to the
demystification of 'the mysterious circle of organic life upon the surface of
the globe', a belief aired by Dumas (and Boussingault) nearly two decades
previously.[j]

L. Pasteur, *Comp.Rend.*, 56, 734. 1863

'Examen du rôle attribué au gaz oxygène atmosphérique dans la destruction
des matières animales et végétales après la mort'[a] (Investigation into the
role attributed to atmospheric oxygen gas in the destruction of animal and
vegetable substances after death)

pp. 734–6

The most ordinary observation has at all times demonstrated that animal and
vegetable substances, exposed after death to contact with atmospheric air, or
buried in the earth, disappear, in consequence of various transformations.

Fermentation, putrefaction, and slow combustion, are the three phenomena
which concur in the accomplishment of this great fact of the destruction of
organic substances – a condition necessary for the maintenance of life on the
earth . . .

Dead substances that ferment or putrefy do not yield solely to forces of a
purely physical or chemical nature. It will be necessary to banish from science
the whole of that collection of preconceived opinions which consist in
assuming that a certain class of organic substances – the nitrogenous plastic
substances – may acquire, by the hypothetical influence of direct oxidation, an
occult power, characterised by an internal agitation, communicable to organic
substances supposed to have little stability[b] . . .

In every case, life, manifesting itself in the lowest forms of organization,
appears to me to be one of the essential conditions of these phenomena, but
life of a nature unknown hitherto; that is to say, without consumption of air or
of free oxygen.

I today endeavour to demonstrate experimentally that the slow combustion
which takes place in dead organic substances, when they are exposed to the

air has, in most cases, an equally intimate connection with the presence of the lowest forms of life. This leads to the general conclusion that life controls [*préside*] the work of death in all its phases, and that the three terms of that perpetual return to the atmosphere, and to the mineral kingdom, of the elements which vegetables and animals have abstracted from them, are correlative acts of development and of the multiplication of organised beings ...

On 25 May 1860, in the open air in a garden I broke off the narrow and sealed end of an exhausted flask of 250 cubic centimetres capacity, containing 80 cubic centimetres of sugar solution with yeast, which had been heated to boiling. Immediately after admitting the air, I sealed the point of the flask with an [alcohol] lamp ... It very often happens, for example, that the liquid in the flask subsequently does not give rise either to the production of infusoria or moulds and it retains its original limpidity, although common air was admitted to the flask when it was opened. This was precisely the case with the above-mentioned flask. The liquid was still unaltered on 5 February 1863, on the day I analysed the air in the flask. This air contained:

Oxygen	18.1
Carbonic acid	1.4
Nitrogen by difference	80.5
	100.0

Therefore it is obvious that during the period of three years the albuminous substances of the beer yeast water mixed [*associées*] with sugared water and exposed to ordinary air, but under conditions in which animalcules and or moulds were not developed, had absorbed 2.7 per cent of oxygen, which they had partially converted into carbonic acid. Direct oxidation, slow combustion was therefore barely perceptible. Notwithstanding that during the three years the flask had been kept in a stove at 25° to 30° for eighteen months.

On 22 March 1860 I filled a flask of 250 cubic centimetres capacity, containing 60 to 80 cubic centimetres of boiled urine ... with air deprived of germs through a higher temperature. The liquid was perfectly clear in January 1863. Its colour tended slightly into a reddish-brown tint. A crystalline sandy deposit of uric acid had separated in small quantity on the sides of the flask. Moreover there were some clusters of needles which I recognized as calcium phosphate crystals. The urine was still acid, but this acidity was reduced rather than increased. Its smell recalled exactly the odour of fresh urine after boiling. The air of the flask contained:

Oxygen	11.4
Carbonic acid	11.5
Nitrogen by difference	77.1
	100.0

Thus after three months there still remained 11 to 12 per cent of oxygen. Moreover, all absorbed oxygen is recovered in carbonic acid, less the difference that may always result from the coefficients of solubility of two gases in the experimental liquid.

Be that as it may, it is obvious how slow and difficult is the direct oxidation of urinary substances by atmospheric air, when this air is placed under conditions unsuitable for promoting the development of lower organized beings [*êtres organisés inférieurs*] . . .

pp. 737–40

To sum up, in studying the slow combustion of dead organic substances under the influence of atmospheric oxygen, it is found that this combustion is not in doubt and that it varies in intensity and in mode of action. [That is] according to the nature of the organic substances comparably to metals that are not oxidized by air, such as gold and platina, to some that are moderately oxidizable, such as copper and lead, finally to others that are very prone to oxidation, such as potassium and sodium.

But is worthy of remark, and this is precisely the principal fact to which I today wish to draw the attention of the Academy: the slow combustion of organic substances after death, however real, is barely perceptible when the air is deprived of the germs of lower organisms. It becomes rapid, considerable, not to be compared with the former situation, should the organic substances be covered by moulds, mucors, bacteria, monads.[c] These small beings are agents of combustion the energy of which varying in accordance with their specific nature is sometimes extraordinary, as strikingly exemplified by the oxidation of alcohol, acetic acid, sugar by mycodermas of which I have informed the Academy a year ago.

The proximate principles[d] of living bodies might in some way remain indestructible if from the assembly of beings created by God the smallest, on the surface the most useless ones, would be removed. And life would become impossible because the restoration of all that which has ceased to live, to the atmosphere and to the mineral kingdom, would be all of a sudden suspended.

For all that, if I restricted myself to the above experiments, I might open myself to a serious objection . . . I have constantly worked not only on organic substances that were dead but, moreover, they were previously heated to the boiling point of water. Now, it cannot be doubted that organic substances are profoundly modified by a temperature of 100°. It was therefore necessary to study the possibility of slow combustion of natural organic substances that have not been previously heated, that have in short remained unaltered.

. . . I have succeeded in exposing to air, free from germs, fresh liquids that are highly susceptible to putrefaction . . . blood and urine . . . at present what I should like to call attention is above all to the slight activity of slow combustion, of direct oxidation of the constituents [*principes*] of blood . . . Direct oxidation of urinary substances is equally imperceptible. After forty days I found in one of the flasks:

Oxygen	19.2
Carbonic acid	0.8
Nitrogen	80.0
	100.0

The conclusions to which I have been led by the first series of experiments are therefore applicable in any case to organic substances whatever their structural condition . . . Doubtless it will be superfluous to observe that the experiments regarding blood and urine which I have just talked about to the

Academy strike the last blow at the doctrines of spontaneous generation as well as at the modern theory of fermentation.

Selection 167

Among the most popular, enduring and pervasive terms in the development of biochemical, indeed biological, thought was 'protoplasm'. It appears that the term was first used by Purkyně on 16 January 1839 in his talk to the Silesian Patriotic Society in Breslau.[a] The theme of the talk was clearly related to the writings by Schwann on the analogies in the structural elements of the animal and plant organism which made their appearance during the previous year. But how did he come to employ this term? Various suggestions have been advanced, pointing out the fact that Purkyně, originally destined for holy orders, could have become familiar with the word *protoplastus* used in old Catholic hymns for Adam. Also the word *plasma* occurs in old Latin church texts.[b]

Be that as it may, the concept of protoplasm developed undoubtedly in connection with previously held views that life was associated with a special kind of *substance* which G.L.L. de Buffon (1707–88) called *matière vivante*[c] and G.R. Treviranus (1776–1837) *lebende Materie, Lebensstoff*, or *Lebensmaterie*,[d] i.e. living matter. There was also the concept of the formative ground substances, capable of transformation in stages into mature plants and animals which goes back to the classical contribution of C.F. Wolff (1733–94) on epigenetic development (1759).[e]

With all these ideas Purkyně and his collaborators in Breslau were thoroughly familiar in the 1830s, when they were engaged in the systematic microscopic examination of animal tissues. Purkyně was among the first to draw attention to the analogies and dissimilarities between the microscopic structure of plants and animals. From his earlier embryological work – he discovered among others the *vesicula germinativa* or the nucleus in the avian egg – he was only too well aware that the development of organisms from the embryonic into the adult state took place in stages. At first (1834) he seemed to have employed the already known botanical term *cambium* both for the plant and animal rudimentary living substance. In 1835 his close collaborator G.G. Valentin (1810–83) spoke of *Urstoff* or *Urmasse* (primordial substance or mass) to which the cytoblastema of Schwann was closely related.[f]

Like Schwann, Purkyně did not approach the structure of tissues from a purely morphological point of view. Purkyně and Valentin had already demonstrated in their work on ciliary motion that morphological physiology could achieve remarkable results in the elucidation of fundamental biological processes.[g] The same broad outlook is revealed in Purkyně's work on the structure of gastric glands which cannot be separated from his chemical and physiological investigations of what then was called 'artificial digestion', following closely in the footsteps of Eberle, J. Müller and Schwann.[h]

It was on 19 September 1837 at the Annual Assembly of the Society of

German Naturalists and Physicians which, as it happened, took place in Prague, that Purkyně reported on his work on the structure of gastric glands and artificial digestion. He employed the term enchyme for the contents of 'granules', and it was the granular appearance under the microscope of the liver, spleen, thymus, thyroid, kidney, pancreas and other organs which made him state that as animals consisted of 'granules' (with a central nucleus), so plants were composed of 'granules or cells'. The comparative study of tissue enchyme led Purkyně not only to express the idea of a morphological analogy in the microscopic structure of plants and animals but also the hope that their biochemical and physiological features would be studied. He clearly stated that each plant cell has its *vita propria*, and that it forms its own specific contents by absorbing suitable substances from, and discharging them into, the general sap. He visualized a similar process of formation and dissolution of animal enchymes. At the end of his communication he suggested that their examination would contribute to the understanding of embryonic development, pathogenic processes and *pseudoplasmas*.

It is not clear what Purkyně meant by the term *pseudoplasma* but it appears that he believed in the existence of different forms of 'plasms' or living matter, in various stages of development and differentiation. Thus he conceived 'cambium' and 'protoplasm' to be analogical embryonic living materials of plants and animals. 'Enchyme' was the living matter of mature glands and other animal tissues, not exclusively granular in nature. For Purkyně granules were only one of the basic organized states of living matter, and fibres could be another. All this was linked to the view that the state of living matter oscillated between the fluid and the solid. Protoplasmatic granules, i.e. animal embryonic cells, were jelly-like, neither solid nor fluid, and this state seemed to persist in animals when they became mature. On the other hand, with plants the permeation of the fluid and the solid states was of a short duration, limited to early stages of their existence. In the course of the development of the plant, due to the separation of the fluid from the solid, true plant cells and then vessels were eventually formed.[i]

Whereas Purkyně conceived of protoplasm as the embryonic living material of animals, Hugo von Mohl (1805–72) introduced it in 1846 into botany to denote the living substances which gave rise to structures like the cell nucleus and cell membrane, in fact to the new cell. This remarkable physiological capability of the jelly-like substance to form new cells impressed Hugo von Mohl so much that he considered Schleiden's term for it, *Schleim* (mucilage), inadequate and suggested, therefore, that it should be replaced by *protoplasm*.[j] Whether von Mohl arrived at this term independently of Purkyně, or made himself familiar with it after having read about it in K.B. Reichert's (1811–83) review of advances in microscopical anatomy in *Müllers Archiv* (1841) is difficult to decide.[k]

Most of the work on the cell in the 1840s and 1850s was carried out by botanists. They distinguished between cell sap and cell contents, the importance of which began to be recognized in the life of the vegetable cell as compared with the membrane. One of the consequences of this was that the older concept of the vegetable cell as a mere bounded cavity or space began to be undermined. Comparisons were drawn and similarities discovered in the vegetable cellular material and sarcode, the jelly-like contractile transparent nitrogenous living substance which constituted, according to Felix Dujardin (1801–60), the bodies of lower animals (1835).[l]

In 1861 Max Schultze (1825–74) suggested that this term should be dropped and replaced by the word protoplasm which began to be used for cell substance.[m] Because animal cells did not seem to be bounded, yet in general they did not coalesce and retained their integrity, Schultze was led at the same time to formulate a new definition of the animal cell. He declared that it constituted nothing but a clump or blob of nucleated protoplasm.[n]

Schultze was not the only person who voiced his doubts about the adequacy of the cell picture, essentially derived from Schwann. This can be gathered from Ernst Brücke's (1819–92) critical remarks which were published in the same year as Schultze's article. Brücke raised the question whether cells constituted the 'elementary organisms' as they appeared to do, or could be subdivided into still lower living units. He was concerned with the findings that the cell was usually not a space enclosed by a membrane, that its contents were not fluid, and that the presence of the nucleus could not always be confirmed.

Brücke considered the cell contents to be the most important part of the cell, and it was in connection with this that he made an interesting observation. He asserted that the application of the concepts of 'solid' and 'liquid', as known in physics, to the state of the cell substance, had no validity. The organic compound molecules contained in living cells arranged themselves in such a manner that the resulting structure deserved the special name of 'organization'.[o]

The emergence of the central importance of the viscous cell substance changed within two decades the concept of the cell so that its outline became, in the true sense of the word, blurred. The cell became, as it were, submerged into protoplasm, a complex form of matter, endowed with the fundamental attribute of contractility. It was widely believed that protoplasmic contractility could usefully be studied in muscle. It should be remembered that Schultze's paper was concerned in the first place with the genesis of the muscle fibre and its structure. Following out this historically, great significance attaches to W. Kühne's (1837–1900) efforts to find out something more specific about protoplasm and muscle contractility in terms of chemistry (1859, 1864).[p]

Among those who greatly appreciated Kühne's work was Thomas Henry Huxley (1825–95), a forceful critic of the cell theory. He expressed his reservations in 1853 in an essay review of twelve contributions, encompassing three centuries, to the study of the embryonic development of plants and animals, of their common structural features and the nature of the forces behind these phenomena.

It is Huxley's critical assessment of the usefulness of the cell theory in throwing light on the difference between living and dead matter, rather than his castigations bearing on its histological aspects, that interests us here.

Huxley rejected the notion that vitality was to be associated with the cell or its morphological features. Instead he regarded it as 'the faculty . . . of exhibiting definite cycles of change in form and composition . . . a property inherent in certain kinds of matter.' According to Huxley life, that is, the total activity of living things, had to be viewed as a unity of ascendancy and decline. The succession of morphological and chemical phenomena accompanying it was thought to rest on 'the constant cyclical change of that which was, at first, morphologically and chemically indifferent and homogenous.'

As to what caused the cycle, Huxley stressed, nothing was known. He had no objection to calling the cause *Archaeus, Bildungs-trieb, Vis Essentialis,* Vital Force, Cell Force or by any other name, as long as it was understood to be amenable to scientific investigation and explanation. Huxley thought that it was responsible for the invisible atomic and molecular arrangements underlying the visible systems of cells and organs. What Huxley meant may be seen from the following summarizing statement:

> . . . and we have therefore maintained the broad doctrine established by Wolff, that the vital phenomena are not necessarily preceded by organization, nor are in any way the result or effect of formed parts, but that the faculty of manifesting them resides in the matter of which living bodies are composed as such – or, *to use the language of the day* [our italics – M.T.], that the 'vital forces' are molecular forces.[q]

What has emerged from the foregoing is that for Huxley in 1853 the common denominator of all living phenomena was not the cell but a substance which in its primordial stage was structureless. Moreover, that, while the latter's modifications into particular forms of life could be ultimately traced to the interactions between the molecules which made it up, Huxley allowed for the temporary nature of this finding. Whether what he meant was that molecules represented merely conventional symbols, enabling the study of living matter, or transient categories which scientific advance would replace by new ones reflecting its complex reality more adequately, it not clear.

Be that as it may, it was very largely due to Huxley himself that the universal substance of which all living things were thought to consist, that is protoplasm, became the centre of attention for those inside and outside science who were concerned with basic questions about the nature of life and its origin. It was in his lecture *On the physical basis of life,* delivered in 1868 in Edinburgh, that Huxley carried forward the story which he had begun fifteen years earlier. It should be noted that he now called the matter, common to all living things, protoplasm, a term which he did not employ in the earlier contribution. But he still shrank from acknowledging cells and preferred instead to speak of 'corpuscles' ('particles') of a nucleated or non-nucleated mass of protoplasm. As to the chemical nature of protoplasm, Huxley took it for granted that it was complex. And inasmuch as the most complex (organic) compounds known were proteins, Huxley inclined to the conception that protoplasm was proteinic.

The notion that life could not be considered independently of matter was by no means widely accepted at that time and certainly not by the Presbyterian audience to which Huxley was determined to bring it home by rendering the term protoplasm into 'physical basis or matter of life'. The significance of Huxley's lecture for the history of the study of the origin of life cannot be limited to this single aspect. In order to place it in historical perspective it should be remembered that the climate of opinion in the late 1860s favoured the view that life could have originated only from life. This was especially true after Pasteur's work on spontaneous generation.

How Huxley argued for the alternative view of the possibility of life arising from inorganic substances is shown in the opening excerpts below. It has proved to be a lasting challenge to partisans and foes alike, particularly regarding different yardsticks applied to the explanation of water on the one hand, and of living material, formed from its constituent elements on the other. Under certain conditions hydrogen and oxygen combine to form water, but a knowledge of the properties of gaseous hydrogen and oxygen does not enable us to predict the properties of water. Nevertheless, according to Huxley, it is scientifically valid to attribute the properties of water to molecular changes, i.e. to the interaction of its molecular components. He then proceeds to argue that it is also scientifically sound to accept transition from the non-living state to the living state of matter, by visualizing that, under certain circumstances, lifeless carbon, oxygen, hydrogen, and nitrogen, brought together, lead to manifestations of life.

It has not long ago been asserted that at the centre of the 1868 lecture lies the concept that 'vital forces are molecular forces', stated by Huxley in 1853.[r] Regardless of how we interpret the views Huxley held earlier, an attentive examination of the later text reveals that he expressed himself more subtly in 1868 than he has been credited with. The cited statement, 'the "vital forces" *are* [our italics – M.T.] molecular forces' (1853), is not identical with (1868) propositions

that the properties of protoplasm *result* [our italics – M.T.] from the nature and disposition of its molecules . . . that all vital action may . . . be said to be the *result* [our italics – M.T.] of the molecular forces of the protoplasm which displays it.

Huxley was an evolutionist, and, therefore, not surprisingly in his considerations the inference regarding *production* of life from non-life (novelty) began to make itself felt, in contrast merely to the *reduction* of life to non-life (sameness).[s] Indeed, Huxley's formulations bear the character of 'ultimate questions' which have not lost their force even in the present stage of the discussion of the origins of life.[t]

Last but not least, it is essential to refer to a further purpose of the lecture. It was described by Huxley to Ernst Haeckel (1834–1919), who in a way was his counterpart in Germany, in a letter of 20 January 1869 as follows: 'it contains a criticism of materialism which I should like you to consider'.[u]

As the last excerpt shows, Huxley was indeed intent on allaying fears about the implications of the materialist approach to the problem of the origin of life. According to Huxley, materialism is indispensable to the practice of science but insufficient to satisfy the 'higher' needs of men. This is where 'spiritualism', that is, idealism, comes into its own. Clearly Huxley's design was to demonstrate the onesidedness of both the materialist and idealist view of the world and the need to reconcile them. In keeping with what he wrote to Haeckel five days after speaking in Edinburgh (13 November 1868): 'We do not much mind heterodoxy, if it does not openly proclaim itself as such.'[v]

T.H. Huxley, *The Fortnightly Review*, 5, 129. 1869

'On the physical basis of life'[1]

pp. 135–6

Protoplasm, simple or nucleated, is the formal basis of all life. It is the clay of the potter: which, bake it and paint it as he will, remains clay, separated by artifice, and not by nature, from the commonest brick or sun-dried clod.

Thus it becomes clear that all living powers are cognate, and that all living forms are fundamentally of one character. The researches of the chemist have revealed a no less striking uniformity of material composition in living matter.

In perfect strictness, it is true that chemical investigation can tell us little or nothing directly, of the composition of living matter, inasmuch as such matter must needs die in the act of analysis – and upon this very obvious ground, objections, which I confess seem to me to be somewhat frivolous, have been raised to the drawing of any conclusions whatever respecting the composition of actually living matter, from that of the dead matter of life, which alone is accessible to us. But objectors of this class do not seem to reflect that it is also, in strictness, true that we know nothing about the composition of any body whatever, as it is. The statement that a crystal of calc-spar consists of carbonate of lime, is quite true, if we only mean that, by appropriate processes, it may be resolved into carbonic acid and quicklime. If you pass the same carbonic acid over the very quicklime thus obtained, you will obtain carbonate of lime again; but it will not be calc-spar, not anything like it. Can it, therefore, be said that chemical analysis teaches nothing about the chemical composition of calc-spar? Such a statement would be absurd; but it is hardly more so than the talk one occassionally hears about the uselessness of applying the results of chemical analysis to the living bodies which have yielded them.

One fact, at any rate, is out of reach of such refinements, and this is, that all the forms of protoplasm which have yet been examined contain the four elements, carbon, hydrogen, oxygen, and nitrogen, in very complex union, and that they behave similarly towards several reagents. To this complex combination, the nature of which has never been determined with exactness, the name of Protein has been applied. And if we use this term with such caution as may properly arise out of our comparative ignorance of the things for which it stands, it may be truly said, that all protoplasm is proteinaceous; or, as the white, or albumen, of an egg is one of the commonest examples of a nearly pure proteine matter, we may say that all living matter is more or less albuminoid.

Perhaps it would not yet be safe to say that all forms of protoplasm are affected by the direct action of electric shocks; and yet the number of cases in which the contraction of protoplasm is shewn to be effected by this agency increases every day.

Nor can it be affirmed with perfect confidence, that all forms of protoplasm are liable to undergo that peculiar coagulation at a tempernature of 40°–50° centigrade, which has been called 'heat-stiffening', though Kühne's beautiful researches have proved this occurrence to take place in so many and such diverse living beings, that it is hardly rash to expect that the law holds good for all.

Enough has, perhaps, been said to prove the existence of a general uniformity in the character of the protoplasm, or physical basis, of life, in whatever group of living beings it may be studied. But it will be understood that this general uniformity by no means excludes any amount of special modifications of the fundamental substance. The mineral, carbonate of lime, assumes an immense diversity of characters, though no one doubts that under all these Protean changes it is one and the same thing.

pp. 138–40

But it will be observed, that the existence of the matter of life depends on the pre-existence of certain compounds, namely, carbonic acid, water, and ammonia. Withdraw any one of these three from the world and all vital phenomena come to an end. They are related to the protoplasm of the plant, as the protoplasm of the plant is to that of the animal. Carbon, hydrogen, oxygen, and nitrogen are all lifeless bodies. Of these, carbon and oxygen unite in certain proportions and under certain conditions, to give rise to carbonic acid; hydrogen and oxygen produce water; nitrogen and hydrogen give rise to ammonia. These new compounds, like the elementary bodies of which they are composed, are lifeless. But when they are brought together, under certain conditions they give rise to the still more complex body, protoplasm, and this protoplasm exhibits the phenomena of life.

I see no break in this series of steps in molecular complication, and I am unable to understand why the language which is applicable to any one term of the series may not be used to any of the others. We think fit to call different kinds of matter carbon, oxygen, hydrogen, and nitrogen, and to speak of the various powers and activities of these substances as the properties of the matter of which they are composed.

When hydrogen and oxygen are mixed in a certain proportion, and an electric spark is passed through them, they disappear, and a quantity of water, equal in weight to the sum of their weights, appears in their place. There is not the slightest parity between the passive and active powers of the water and those of the oxygen and hydrogen which have given rise to it. At 32° Fahrenheit, and far below that temperature, oxygen and hydrogen are elastic gaseous bodies, whose particles tend to rush away from one another with great force. Water, at the same temperature, is a strong though brittle solid, whose particles tend to cohere into definite geometrical shapes, and sometimes build up frosty imitations of the most complex forms of vegetable foliage.

Nevertheless we call these, and many other strange phenomena, the properties of the water, and we do not hesitate to believe that, in some way or another, they result from the properties of the component elements of the water. We do not assume that a something called 'aquosity' entered into and took possession of the oxide of hydrogen as soon as it was formed, and then guided the aqueous particles to their places in the facets of the crystal, or amongst the leaflets of the hoar-frost. On the contrary, we live in the hope and in the faith that, by the advance of molecular physics, we shall by-and-by be able to see our way as clearly from the constituents of water to the properties of water, as we are now able to deduce the operations of a watch from the form of its parts and the manner in which they are put together.

Is the case in any way changed when carbonic acid, water, and ammonia disappear, and in their place, under the influence of pre-existing living protoplasm, an equivalent weight of the matter of life makes its appearance?

It is true that there is no sort of parity between the properties of the components and the properties of the resultant, but neither was there in the case of the water. It is also true that what I have spoken of as the influence of pre-existing living matter is something quite unintelligible; but does anybody quite comprehend the *modus operandi* of an electric spark, which traverses a mixture of oxygen and hydrogen? . . .

If scientific language is to possess a definite and constant signification whenever it is employed, it seems to me that we are logically bound to apply to the protoplasm, or physical basis of life, the same conceptions as those which are held to be legitimate elsewhere. If the phenomena exhibited by water are its properties, so are those presented by protoplasm, living or dead, its properties.

If the properties of water may be properly said to result from the nature and disposition of its component molecules, I can find no intelligible ground for refusing to say that the properties of protoplasm result from the nature and disposition of its molecules.

But I bid you beware that, in accepting these conclusions, you are placing your feet on the first rung of a ladder which, in most people's estimation, is the reverse of Jacob's, and leads to the antipodes of heaven. It may seem a small thing to admit that the dull vital actions of a fungus, or a foraminifer, are the properties of their protoplasm, and are the direct results of the nature of the matter of which they are composed. But if, as I have endeavoured to prove to you, their protoplasm is essentially identical with, and most readily converted into, that of any animal, I can discover no logical halting-place between the admission that such is the case, and the further concession that all vital action may, with equal propriety, be said to be the result of the molecular forces of the protoplasm which displays it. And if so, it must be true, in the same sense and to the same extent, that the thoughts to which I am now giving utterance, and your thoughts regarding them, are the expression of molecular changes in that matter of life which is the source of our other vital phenomena.

p. 145

Why trouble ourselves about matters of which, however important they may be, we do know nothing, and can know nothing? We live in a world which is full of misery and ignorance, and the plain duty of each and all of us is to try to make the little corner he can influence somewhat less miserable and somewhat less ignorant than it was before he entered it. To do this effectually it is necessary to be fully possessed of only two beliefs: the first that the order of nature is ascertainable by our faculties to an extent which is practically unlimited; the second, that our volition counts for something as a condition of the course of events.

Each of these beliefs can be verified experimentally, as often as we like to try. Each, therefore, stands upon the strongest foundation upon which any belief can rest; and forms one of our highest truths. If we find that the ascertainment of the order of nature is facilitated by using one terminology, or one set of symbols, rather than another, it is our clear duty to use the former; and no harm can accrue, so long as we bear in mind, that we are dealing merely with terms and symbols.

In itself it is of little moment whether we express the phenomena of matter in terms of spirit; or the phenomena of spirit, in terms of matter; matter may be regarded as a form of thought, thought may be regarded as a property of matter – each statement has a certain relative truth. But with a view to the progress of science, the materialistic terminology is in every way to be preferred. For it connects thought with the other phenomena of the universe, and suggests inquiry into the nature of those physical conditions, or concomitants of thought, which are more or less accessible to us, and a knowledge of which may, in future, help us to exercise the same kind of control over the world of thought, as we already possess in respect of the material world; whereas, the alternative, or spiritualistic, terminology is utterly barren, and leads to nothing but obscurity and confusion of ideas.

Thus there can be little doubt that the further science advances the more extensively and consistently will all the phenomena of nature be represented by materialistic formulæ and symbols.

But the man of science, who, forgetting the limits of philosophical inquiry, slides from these formulæ and symbols into what is commonly understood by materialism, seems to me to place himself on a level with the mathematician, who should mistake the x's and y's, with which he works his problems, for real entities – and with this further disadvantage, as compared with the mathematician, that the blunders of the latter are of no practical consequence, while the errors of systematic materialism may paralyse the energies and destroy the beauty of a life.

Selection 168

Was the cell or protoplasm the 'common denominator' of living things? We have seen in the previous selection something of the change of emphasis during 1840–70, from the cell towards protoplasm, when it came to a consideration of what constituted the underlying unity that encompassed all phenomena of life. It is in this context that we can usefully turn to Eduard Pflüger's (1829–1910) researches, begun around 1865, into physiological combustion. The 1872 and 1875 papers are of particular interest. In them he elaborated the principles of cellular respiration (including the fact that it is common to diverse forms of life) which have already been referred to.[a]

The effect of demonstrating that respiration is an intracellular process was to rehabilitate, so to speak, the cellular basis of living tissues from the functional point of view. Thus the cell emerged as the site not only for respiration but, gradually, for all chemical transformations taking place in plants and animals.

Pflüger was greatly praised later for his decisive contributions to the knowledge of the respiratory process. But he was also sharply criticized for his speculations on the nature of the chemistry of life. Essentially, he associated it with the decomposition and restitution of the unstable 'organized' or 'living' protein (*Organeiweiss*) which, he believed, virtually constituted the material substratum of cells (*Zellsubstanz*). In contrast to this, Pflüger regarded the unassimilated food proteins (*Nahrungseiweiss*) as 'dead' and visualized that only when they were incorporated into the living cells and took up oxygen did they begin 'to breathe, to live'. Pflüger maintained that the chemical difference between the two classes of proteins depended on the presence of cyanogen (CN) radicals in the 'living' protein.

In focusing attention on cyanogen, Pflüger was moved by an assumption which was widely shared by contemporary students of the chemistry of life in their efforts to deal with metabolism. Namely, that a knowledge of the structure of the vital products of organisms, including their artificial preparation, had a direct bearing on the understanding of the chemical processes occurring within them. Pflüger started from the known fact that animals excrete nitrogen principally in the form of urea and uric acid. Since both these compounds contained cyanogen radicals, according to Pflüger, it was reasonable to see in them the basic chemical group on which the 'living' state of the proteins in the cell turned. As evidence for this view, Pflüger pointed out that decomposition products from proteins regarded by him as non-living (e.g. egg white, seed proteins) did not contain cyanogen radicals.

It was observations of frogs surviving under anaerobic conditions that led him to speculate along these lines to surmise, as L. Hermann (1838–1914) did before him, that oxygen bound to protein molecule in tissue made it highly decomposable.[b] The production of carbonic acid by the respiring cell was explained by Pflüger as due to intramolecular heat bringing about the decomposition of the 'living' protein molecule into carbonic acid, water and, by way of intermediate cyanogen-containing products, amidic compounds.

Pflüger's conjectures also involved a supposition about the way in which living matter increased in size. He explained it by suggesting the matrix or ground substance (*Grundstoff*) of different types of cells was similar but not identical because of modifications essentially rooted in isomerism. The cellular materials varied because it was made of 'living' protein isomers that, in turn, were capable of 'growing' through polymerization, forming 'giant' molecules in the process.

In connection with the praise and criticism heaped on Pflüger, it has hardly been pointed out that they apply to interrelated work. As shown below, it included also a theory of origins of life, with cyanogen radicals playing a major role, still discussed by serious students in the field eighty to ninety years later.[c]

E. Pflüger, *Pflügers Arch.*, *10*, 251. 1875

'Beiträge zur Lehre von der Respiration. I. Ueber die physiologische Verbrennung in den lebendigen Organismen' (Contribution to the doctrine on respiration. I. On the physiological combustion in living organisms)

pp. 339–40

If one thinks about the beginning of organic life, one must not fix the eyes primarily on carbonic acid and ammonia. For both are the end of life not the beginning because, in the highest degree, they represent stable molecules, so far as one can speak of stability in connection with dynamic equilibrium. As the life process essentially presumes the possibility of *carbonic acid formation*, so life can take no departure from carbonic acid.

The beginning lies far more in *cyanogen* – How does cyanogen arise?

The fresh nitrogen of the air is capable of forming potassium cyanide if it is added to a strongly glowing mixture of potash and carbon, or to a mixture of potash or potassium carbonate and charcoal heated to a white glow. – The oxygen compounds of nitrogen – nitric acid, for example is indeed formed during storms – under similar conditions yield cyanogen compounds far more easily. – Further: ammonia led over glowing charcoal forms ammonium cyanate; similarly a mixture of carbonic oxide [carbon monoxide] and ammonia in contact with glowing platinum sponge. – Further: if ammonia is led over a glowing mixture of potassium carbonate and charcoal, or if sal ammoniac [ammonium chloride] is ignited with potassium carbonate and potash one obtains potassium cyanide. – Further: if carbonic oxide gas is heated with potassium hydrate [hydroxide] for a longer time potassium formate is formed, which with an ammonium salt can become ammonium formate. Ammonium formate on heating alone or with dehydrating substances yields with loss of water: hydrocyanic acid, cyanogen or cyanic acid.[1]

Accordingly, there is nothing clearer than the possibility of the formation of cyanogen compounds when the Earth was still *wholly* or *partially* in an igneous or heated state. I imagine that one must think that the cooling of the Earth's surface happened not uniformly and that single districts, which had cooled, also could again be heated, and so on.

In the same way in principle, the origin of the other essential constituents of the protein molecule has to be understood (what no chemist will deny) namely innumerable hydrocarbons and alkyl radicals [*Alkoholradikale*] respectively, without any intervention of living material through synthetic formations.

pp. 341–2

One sees how quite extraordinary and remarkably all the facts of chemistry indicate to us fire as the force which has produced the constituents of protein through synthesis. Life descends thus from the fire and is in its fundamental conditions generated [*angelegt*] at a time when the Earth was still a glowing ball of fire.

If one considers now the immeasurably long time in which the cooling of the Earth's surface took place – infinitely slowly, so cyanogen and the compounds which contain hydrocyans [*Cyanwasserstoffe*] and hydrocarbons, had sufficient long time and opportunity to follow their great inclination towards transformation and formation of polymers in the most extensive and different ways, and with participation of oxygen and later of water and the salts to go over into that self-decomposing [*selbstzersetzliche*] protein which is living matter.

I think thus that an intermediate stage leads from the lifeless to living nature.

Also today there is a glowing heavenly body, the Sun, which by way of the light sends into the far distance to the plants of the Earth the force which produces the constituents of the protein in them.

It appears to me that this is not incomprehensible if one assumes (for which much speaks) that the protein in the plant arises in no other way than that the already present living protein molecule becomes bigger at the cost of definite radicals or molecules offered to it, that is to say '*grows*'; for the protein formation in the plant is there, where it grows, where the *living protein is*.

One sees the 'growth' of organic matter convincingly in the almost endless carbon chains with their highly varied arrangements, as they are formed in the body of the plant. These chains originated from quite separate carbon atoms which were earlier contained in carbonic acid. Carbon has thus in the living molecules a great tendency to condition a growth through chain formation. But cyanogen possesses this tendency also in a high degree and, indeed, especially towards cyanogen again. But also, if need be, the ammonia does not lack it either. Thus the essential elements of the living protein have most decidedly the tendency to attract similar radicals and in this way to produce always larger molecules, that is to grow.

That now the living protein is in a specially favourable condition continuously to insert new elements of the same kind into its molecule proceeds from my theory, as the carbon and cyanogen radicals in their oscillations must always move into phases approaching the state which we in chemistry designate as *status nascens*.

Accordingly I would say that the first protein which appeared was immediately living material, endowed with the property in all its radicals to attract with great strength and preference especially similar constituents, in order to insert them chemically into the molecule and so to grow *in infinitum*. According to this conception the living protein needs to have no constant molecular weight, because it is a giant molecule engaged in continuous, never ending formation and itself again breaking down, which is probably related to the ordinary chemical molecules as the Sun is to the smallest meteor.

If one investigates fluid protein, one has to deal mostly with torn off bits of that giant molecule, which may well be as large as a whole creature. These bits need have no constant composition unless one produces equally large molecules through chemical intervention, that is [through] formation of decomposition products with a definite molecular weight.

In the plants thus the living protein proceeds to do what it always did since its first formation, that is to say continuing itself in all its parts through attraction [of entities] of the same kind to regenerate or to grow, on which account I believe that all protein present in the world today originates directly from that first. On this account I doubt [the existence of] *generatio spontanea* at the present time; also comparative biology points unmistakably to this, that all life has taken its origin from a single root only.

If in conclusion I have to summarize my hypothesis, then I say: '*The life process is the intramolecular heat of highly decomposable and, through dissociation, of decomposing protein molecules formed in the cell substance (essentially with the production of carbonic acid, water and amidic bodies) that continuously regenerate and also grow through polymerization.*'

Selection 169

The space of this section allows us to dwell on only a few key concepts which linked the chemical and biological fields at the time when the foundations of the modern chemistry of life were laid, and which formed part of its theoretical framework. Whatever the gaps, Claude Bernard's

(1813–78) contribution to this movement through his enquiries into methodical and theoretical principles of physiology cannot be ignored. The object here is not to introduce another topic of Bernard's scientific research but to concentrate on his general physiological viewpoint, which transcended the specialist features of his work.[a]

But which of the texts from his pen conveys it reliably? Although it is relevant to various Selections in this *Documentary History* as a whole, the question is particularly pertinent when it comes to presenting Bernard's general perception of the world of life. The intricacy of the problem has been emphasized by F.L. Holmes, the author of the scholarly work on Bernard's early research efforts. Holmes has found Bernard's laboratory notebooks, kept between 1842 and 1848, particularly helpful in this context:

Bernard is revealed in his early notebooks as an extraordinarily complex man, whose scientific approach is not easily summarized. It was richer and more capricious than the idealized version he made famous in his *Introduction to the Study of Experimental Medicine*.[b]

For good or ill, the excerpts below are from Bernard's penultimate book, *Leçons sur les phénomènes de la vie communs aux animaux et aux végétaux*, the proofs of which he was correcting at the time of his death, and which appeared posthumously (1878).[c] According to the publishers: 'In this volume M. Claude Bernard has gathered the whole of his precepts [*Doctrines*] and it is the most complete and methodical work that he leaves to the world of science.'[d]

On the face of it the volume reproduces, apart from the opening lecture, the course of nine lectures on general physiology which Bernard gave at the Museum of Natural History from the summer of 1870 and to which he brought an awareness of the historical dimension. Moreover, the volume reveals Bernard's aspiration to demonstrate that the experimental scientific method applied to the study of the world of life was 'self-sufficient', and hence need have no recourse to philosophy or, for that matter, to theology to sustain it.

Before introducing the excerpts themselves, a comment may be added on Bernard's conception of 'general physiology'. This term, corresponding to Bichat's 'general anatomy',[e] was employed by Bernard, as we have seen, in 1857 when he compared the formation of sugar in plants and animals and identified in it a chemical activity common to both. That work was to lead Bernard, during the 1870s, to the viewpoint that the real field of general physiology was the study of mechanisms, properties and conditions common to all beings. By contrast, he regarded the study of the phenomena peculiar to individual forms of living things as the subject of descriptive physiology, either special or comparative.[f]

It is worth recalling that by the beginning of the 1870s the accent in research into the phenomena of life was on specialist studies in natural history, botany, zoology and physiology. Hence Bernard's zeal in marking off physiological study from other biological studies in the opening lecture of the course:

It is necessary therefore to understand well the general movement now taking place before our eyes, which leads to the emancipation of physiology and its constitution in a definitive form.[9]

As to the excerpts, they focus attention on four of the major threads that run through the work, containing Bernard's last published work on how to approach and explain the phenomena of life. They can be grouped under the following headings: A. Definition and characteristics of life; B. *Milieu intérieur*; C. Cell, protoplasm and the chemistry of life; D. Physiological determinism. They will be considered in turn.

A. Bernard's thesis was as follows: 'It is possible to characterize life but not to define it.' It was prompted by Bernard's historical appreciation that scientific knowledge evolving in stages is in a position to obtain only approximate answers that at no time can be regarded as definitive. Thus attempts at defining life, Bernard argued, were doomed to failure because they presupposed complete knowledge of it which was not available. To pursue such an aim was simply putting the cart before the horse.

B. The formulation of the concept of the *milieu intérieur* or 'internal environment' has been widely recognized as Bernard's main contribution to the theoretical study and knowledge of living things. In the year of his death Bernard would probably have agreed with this assessment – whatever had been in his mind when he had fleetingly introduced it, in relation to the properties of blood – in 1857. In the jubilee year of 1963, Holmes summarized the evolution of Bernard's comprehension of the role of constancy of the internal environment as follows:

From 1857 until his death in 1878, the *milieu intérieur* was a persistently recurring theme in his writings. Believing that all courses in physiology should begin with the concept, he discussed it regularly in his yearly lectures . . . Through his long preoccupation with the concept Bernard gradually found in it a richness of meaning which he obviously had not foreseen when he rather casually first mentioned it. Only near the end of his career did he clearly perceive an implication now central to all discussions of the concept, that the internal environment can be maintained constant only by the continuous action of many complex regulatory mechanisms.[h]

Without seeking to place the *milieu intérieur* in its complete historical context, it is possible to identify at least two major components which went into its making. One derived from the cell theory: more specifically from Bernard's embracing Brücke's suggestion that cells were 'elementary organisms'.[i] Bernard regarded the cells that made up the body of an organism as infusoria-like small beings and the blood as their environment. He elaborated the view that the physical and chemical features of the blood respresented the internal environment of the organism on which its existence depended. The other component may be traced to Bernard's researches into the activities of the circulatory and nervous systems that pointed to mechanisms regulating and maintaining the internal environment.

The centrality of the *milieu intérieur* in Bernard's general thinking about the world of life is confirmed by a reading of *Phenomena of Life*. Bernard's audience was confronted with the term early in the course, that is, in the second lecture where he pointed to the crucial importance to life of the physicochemical conditions. Bernard differentiated between latent, variable (oscillating) and constant (free) states or forms of life, defined according to the degree of reciprocity between a living oganism and its surroundings. Bernard regarded an organism endowed with latent life as lacking interchange with the environment, and he offered the life of a grain as an example. He connected oscillating life with the life of plants and cold-blooded animals, and identified it as directly depending on the external environment. He associated free life with the life of higher animals and man who, shielded from the outside world through their internal environment, attained a high degree of autonomy. It was in discussing this last, third, form of life that Bernard produced his fullest exposition of the *milieu intérieur*.

C. It is hardly surprising that the issue whether life was fundamentally bound up with the cell or with protoplasm also engaged Bernard's attention. As mentioned, he adopted the idea that the cell was an organism. In his view the cell already constituted too complex a living system, taking on varied forms to serve as the starting point of life. Bernard reserved this role to amorphous, that is non-cellular, living matter or protoplasm, which he looked upon as a unique, complicated chemical substance of unknown composition that was identical in plants and animals. Regarding it as the 'repository of life', Bernard stated:

It ïs in the protoplasm, this active, working matter, that we should seek the explanation of life, the chemical phenomena of nutrition as well as higher vital reactions of sensibility and of movement.[j]

In this connection it should be appreciated that Bernard emphatically rejected the idea that the laws of general chemistry did not apply to the chemical and physicochemical reactions taking place in the plant and animal organism. However, it should also be noted that, drawing on his work on the glycogenic function of the liver, Bernard distinguished between non-vital 'destructive' (fermentative, i.e. enzymatic) and vital 'creative' (synthetic) chemical activities in living organisms.

Bernard sought to account for the peculiar capacity of organisms to synthesize by relating it to the as yet not comprehended activity of formed, that is, cellular protoplasm from which organs of living beings were made. At the root of it all was the power of chemical synthesis inherent to formless protoplasm, the consequence of which, in Bernard's opinion, was the conversion of the latter into morphologically differentiated cellular protoplasm. In retrospect it would seem that Bernard's relevant point regarding the correlation of chemical and morphological aspects in living things (already present in Schwann's work)[k] was virtually ignored.

D. To complete the introduction, a few remarks regarding Bernard's philosophical theory, which he called 'scientific determinism', are needed. It is possible to assume, on the basis of *Phenomena of Life*, that scientific determinism grew out of Bernard's dissatisfaction with what he took to be the one-sidedness of both the idealist and materialist conceptions of life that were current at the time. One the one hand, Bernard was unable to see how a solitary immaterial agency could engender phenomena of life. On the other hand, he thought it equally inconceivable that inert matter alone could give rise to the manifoldness of living forms and account for their purposive behaviour and continuity.

For Bernard, life without an immaterial principle was just as unthinkable as life without matter. If this was so, life encompassed two basic entities which appeared to contradict each other. Bernard's solution of the problem was subtle in that he differentiated between a non-antagonistic and an antagonistic type of contradiction denoted as 'conflict' and 'struggle'. Bernard argued that in the case of the conflict the factors, despite being at variance, were non-antagonistic and therefore capable of acting in conjunction with each other and engendering the characteristic properties of living systems. He contended that in the case of the struggle conditions were hostile to the emergence and existence of vital activity because the antagonistic and mutually exclusive factors did not make for a harmonious interplay. It is in this light that one has to understand Bernard's conception of life as a conflict. That is he viewed it essentially as a non-antagonistic contradiction between external conditions (i.e. environment) and internal conditions (i.e. genetic constitution of the organism). His treatment of the environment – organism complexity was to all intents and purposes dialectical.

Bernard's general philosophy of physiology and scientific determinism both derived from and shaped his opinion that the complex relations between the environment and living beings ultimately *determined* the manifold phenomena of the world of life. It was *the* task of physiology to study the relations but not to search for either their first or final cause. These were subjects for religion and theology but not for science.

C. Bernard, *Phenomena of Life Common to Animals and Vegetables* [Paris, 1878] (Dundee, 1974)

A.

pp. 11–12

Nothing is definable in natural sciences; all attempts at definition produce only a simple hypothesis. The knowledge of objects is acquired only successively from different and differing points of view. It is not at the beginning of these sciences that one can have of them integral and complete knowledge as any definition would suppose; it is only at the end and as an ideal term inaccessible to the study.

The method, which would consist of defining and deducing all from a definition, might be suitable to the abstract sciences, but is totally opposed to the spirit of the experimental ones.

That is why there is no need for defining life in physiology. When one talks about life, one is agreed on the subject without any difficulty, and that is enough to justify the use of the term without any ambiguity. It is quite sufficient to have a common understanding of the word 'life', to be able to use the term, but at the same time, one must know, that it is illusory, vain, even contrary to the spirit of science to seek for it an absolute definition. What one should try however, is to pinpoint its characteristics in their natural order of significance.

Nowadays it is very important to disembarrass general physiology thoroughly of the illusions that have agitated it for so long a time. It is an experimental science and does not have to offer definitions *'a priori'*.

pp. 14–15

We do not reject all hypotheses in science; they are, in any case only its scaffoldings. It is the facts, which constitute a science; but it does move and go forward with the help of hypotheses!

Let us now examine what are the general characteristics of living beings. We can summarise them under five headings: Organisation, Generation, Nutrition, Evolution, Senescence, Illness and Death.

A. Organisation results from a blending of complex substances acting on one another. This to us means an arrangement which gives rise to properties immanent in living matter, an arrangement inordinate and very complex, which nevertheless obeys the general laws of the chemistry of matter. Vital properties therefore are physico-chemical properties of organised matter.
B. The faculty of reproduction or generation, that is an action by which living beings spring from one another, is a chief and indisputable characteristic. Every being comes from his parents and after a certain time, he, in turn, becomes a parent, that is, gives origin to other beings.
C. Evolution is possibly the most remarkable feature of the living organism and thereby of life itself. A living being is born, grows, declines and dies. It is in a state of continuous change. It is subject to a death. It comes out of a seed, an egg or a grain, reaches through successive differentiation a certain degree of development, forms organs, some of them having a short duration and later disappearing, others which last as long as itself and then gets destroyed.

B.
p. 29

The manner of the relationship between living beings and the surrounding cosmic conditions allows us to consider three forms of life according to whether it is closely dependent on external conditions, whether it is less so, or whether the dependence is only relative. These three forms of life are: 1) *Life latent*; that is life non-manifest. 2) *Life oscillating*; that is life with manifestations variable and dependent on the external milieu. 3) *Life constant*; that is life having free manifestations and independent of the external milieu.

p. 48

The third form of life is constant or free life. It belongs to animals highest in the scale of organisation. Life in it is not suspended in any conditions: it flows constant and apparently indifferent to the variations of the cosmic environment and to the changes of material conditions surrounding the animal. Organs, tissues, and internal apparatus function noticeably evenly, and their activity is not subjected to such considerable variations as experienced by the animals with oscillating life. The reason for this is that the *inner environment* enveloping the organs, tissues and elements of tissues does not change. The atmospheric variations do not penetrate, so that it is true to say, that for a higher animal *physical conditions of the milieu* stay constant; the environment surrounding it is invariable, giving it a sort of atmosphere of its own in the forever changing cosmic milieu. It is an organism which has put itself into a glasshouse. Perpetual changes of the cosmic environment do not affect it; it is not tied to them, but free and independent.

I believe, that I was the first to insist on the idea that for an animal there are really two environments. The *external one*, in which the organism is placed, and the *internal* one, in which live the elements of the body tissues. The existence of the being is not lived in the external environment such as air is for the bird or the water sweet or salty for an aquatic animal, but in the *liquid internal milieu* formed by circulating organic fluids which envelope and bathe all anatomical elements of the tissues; it is the lymph or plasma, the liquid part of the blood which, in higher animals, permeates the tissues and pools all interstitial fluids nourishing local regions, being thus the source and the confluent of all elementary exchanges. This complex organism should be considered as a union of simple organisms which are anatomical elements living in the fluid internal environment.

The stability or fixity of the internal milieu is a condition of life free and independent. The mechanism responsible for it is the same as which ensures in the internal environment the maintenance of all conditions necessary to the life of the elements. This way we can understand why there could be no free and independent life for simpler organisms, whose constitutive elements are in direct contact with the outer cosmic environment and why, by contrast, this form of life is reserved exclusively for the beings that have reached the summit of complexity or organic differentiation.

The stability of the environment pre-supposes such a degree of perfection in the organism that external variations are instantly compensated and equilibrated. Consequently, far from being indifferent to the external world, the higher animal is actually in a very close and elaborate relationship with it, so that its equilibrium results from a continual and delicate compensatory adjustment which operates like the most sensitive of balances.

Conditions necessary to the life of elements which must be assembled and kept constant in the internal environment for the functioning of free life, are those we know already; water, oxygen, heat, chemical substances otherwise called the reserves.

These conditions are the same as are needed for the life of simple organisms, only in a perfected animal with independent life, there is a nervous system which regulates the conditions in order that they may work in harmony.

C.
pp. 95–6

We, for our part, do not accept vital force as an executive force. However, we do recognize that there are in living beings vital phenomena and chemical composites which belong to them. How then are we to understand their happening?

The chemistry of the laboratory and the chemistry of the living body are subject to the same laws. There are no two chemistries. Lavoisier has seen and said so. Only, laboratory chemistry is executed with the help of agents, of apparatus created by the chemist, whereas chemistry of the living being is performed with the help of agents and apparatuses, created by the organism. We have abundantly demonstrated the truth of this in relation to the agents of analysis or organic destruction. The chemist, for example, transforms starch into sugar by means of an acid which he himself has fabricated; he saponifies the fats using caustic potash, concentrated sulphuric acid, superheated water vapour, all agents he himself has created. An animal, on the other hand, as well as a germinating seed, transforms starch into sugar without any acid, with the help of one substance only, the diastase, which is an organic product. Fats in the animal get saponified in the intestine without any caustic potash, without superheated water vapour, but with the aid of pancreatic juice, a product of the secretion of a gland. Every laboratory has therefore its own special agents, but the chemical phenomena are in essence the same: the conversion of starch into sugar, the break-up of the fats into a fatty acid and glycerol, are in both cases produced by an identical chemical process.

It must be the same for the phenomena of organic creation. Laboratory chemistry, can effectuate syntheses just like a living body and already has produced many of them. Chemists have made essences, oils, fats, acids, which living organisms themselves manufacture. Here, however, we can affirm, that the agents of a synthesis differ. We may as yet not know, which these agents of the living syntheses are, but they certainly exist. We have mentioned various hypotheses formulated on the subject and we, ourselves, have arrived, by certain facts which we shall publish later, to attribute a certain role not only to the protoplasm but more so to the nucleus of the cells.

It can be said, that a chemist in his laboratory and the living organism with its own apparatus both work in the same way, but each with its own tools. The chemist may be able to fabricate the products of the living organism, but will never be able to make its tools, because these are the result of organic morphology, which is, as we shall soon see, outside chemistry proper; and under this aspect the chemist cannot make the simplest ferment any more than he can manufacture a whole living being.

In recapitulation, we see how obscure still are all those questions of the syntheses, of vital creations, in spite of all efforts devoted to the study of this subject.

For our part we do not think that the solution of these complex problems could be found by trying to get at them in their origin. On the contrary, we believe, that through observation of the facts nearest to us, we shall be able to advance step by step till we finally reach the determinism of these fundamental pheonomena.

Today we can only say, that the synthesis of complex substances such as of albuminoid bodies or of fatty matter, are completely unknown to us. The only one, on which we are a little better informed is the amylaceous or glycogenic synthesis in animals.

pp. 97–8

In animals and vegetables sugar then is formed from starch. The formation of this starch in both kingdoms is considered as an act of organic creation, a synthesis. By contrast, the formation of sugar is an organic destruction, hydration of the starch causing its conversion into dextrin and glucose; this substance, in turn, gives rise to lactic and carbonic acids by a series of operations resulting in the destruction of the sugar by means equivalent to the phenomena of oxidation.

Thus we find in animal and vegetable glycogeneses two phases characteristic of the great vital phenomena: 1. *Organic creation*: Synthesis of starch, synthesis of glycogen. 2. *Organic destruction*: Transformation of starch or glycogen into dextrin and sugar, then the destruction of the sugar by means analogous to combustions.

Unfortunately up to now we know sufficiently only about the phenomena of destruction of amylaceous substances. We know that in animals and in vegetables they take place under influence of *ferments*, diastase, the lactic ferment, these being the chemical agents specific to the organism. We also know, that in both kingdoms these phenomena engender heat while being accomplished.

As for the creation, the synthesis of the starch or glycogen, is still surrounded by great obscurities and that is true of vegetables as of animals.

p. 102

Cellular protoplasm is necessary only for the first phase, that is to say for the synthetic genesis of the basic product, but the destructive combustion can operate without the intervention of the protoplasm. Proofs of this are abundant. Glycogenic matter is one example. For its production there can be no substitute for the protoplasm animal or vegetal. By contrast, destruction is a chemical phenomenon which does not necessarily require the intervention of a living cellular agent and which can continue after death, or outside the organism. Of this the experiment of *washed liver* is a decisive proof. A jet of water is passed through the liver freshly cut out of the animal and therefore away from any vital influence; the sugar it contains is washed out; if we leave this liver for some time, we shall find in it a new lot of sugar. This operation can be successfully repeated a number of times, until all store of glycogenic matter is exhausted. Thus in this dead organ isolated from any physiological or vital influence, glycogenic matter continues to destroy itself as it does in life, but it does not renew itself.

How then does the cellular protoplasm act to form the basic element? This is a question to which alas there is as yet no answer. It could be supposed that glycogen does not appear through a true synthesis in the chemical sense, but through the break-up of the protoplasmic matter. It is for the future and perhaps not too distant future, to solve these problems that today we can only indicate, but of which we have already succeeded to analyse the principal conditions.

D.
p. 150

Various attempts that have been made in the history of science to find a definition of life led mostly to consider it either as a special principle, or as a resultant of general forces of nature, that is to say, to the concept vitalist or materialist. Both concepts appear to us badly formulated: the vitalist doctrine, because the so-called vital principle is not capable of acting by itself and thereby not able to explain anything by itself; on the contrary, it needs a host of general agents, physical and chemical. – The materialist doctrine is also faulty, as general agents of physical nature capable of bringing about separate vital phenomena do not explain their ordinance, their consensus and their linking.

From the point of view of organic interplay, we should perhaps say, that vital properties are at once the resultant and the principle. Superior vital faculties such as irritability, sensibility, intelligence, could, in fact, be considered as the result of the physico-chemical phenomena of nutrition; in that case it must also be admitted that these faculties become forms or principles of direction and manifestations of all the phenomena in the organism, no matter what their nature.

However, considering the question in an absolute way, we must say that life is neither a principle nor a resultant. It is not a principle, because being somewhat dormant or expectant, this principle is incapable of acting on its own. Neither is life a resultant of the forces of nature, because physico-chemical conditions which cause it to appear, do not impress it with a direction nor give it a specific form.

Neither of the two factors, the principle directing the phenomena any more than a whole set of material conditions for the manifestation of life, can explain it separately. They have to be united. In consequence, as we see it, life is a conflict. Its manifestations result from a close and harmonious relationship between the *conditions and the constitution* of the organism. These are the two prerequisites which are found present and we could say, in collaboration in each vital act. They are:

1. *Determined external physico-chemical conditions* governing the appearance of the phenomena, and
2. *Organic conditions or pre-established laws* regulating the succession, the concert, the harmony of all phenomena. These organic or morphological conditions derive from the atavism of anterior beings and form a heritage which they transmit to the actual living world.

p. 174

After what has been said in the preceding paragraphs, is it possible for us to attach ourselves to any system of philosophy? We might be tempted to count ourselves among materialists or physical chemists. But we belong to neither. For looking at the actual state of things we do affirm that there is a *special modality* in the physico-chemical phenomena of the organism. Are we then among the vitalists? Again no, as we do not accept an executive force outside physico-chemical forces. Are we then those empirical experimentalists who believe with Magendie,[a] that a fact has its own sufficiency, and that experimentation does not need a doctrine to direct itself? Not either; for we, on the contrary, are convinced that now especially, it is necessary to have a sound criterion for forming an opinion and a doctrine to unite all the facts acquired by science.

What then is this doctrine? It is *determinism*. To pretend that we can reach at the causes of the phenomena through the mind or matter is illusory. Neither the mind nor matter are the causes. Phenomena have no causes, specially not the phenomena of life, and in those that have an evolution, the concept of a cause disappears, because the idea of a constant succession does not carry with it the idea of dependence. The phenomena of evolution are a linked chain of a rigorous order; nevertheless, we know that an antecedent event does not command the following one. The obscure concept of a cause should be referred to the origin of things; it has no significance except as the first cause or the final cause. In a science it should make room for the concept of relationship or conditions. Determinism fixes the conditions of the phenomena. It allows us to foresee their happening and to provoke it when within our reach. It does not explain Nature to us, it makes us its masters.

Determinism is therefore the only scientific philosophy possible. It does in fact, forbid us the search for the 'why', as this 'why' is illusory. In return it exempts us from doing as did Faust, who from affirmation threw himself into negation. Like those religious orders where men mortify their flesh by privations, we too, in order to perfect our mind, must mortify it by excluding certain questions and admitting our inability to answer them. While we may think, or rather feel, that there is something beyond our scientific prudence, we should still wholeheartedly embrace determinism. So that, if at times, we allow our mind to be rocked gently by the wind of the unknown and the sublimities of ignorance, we will at least be able to recognise what science is, from what it is not.

Selection 170

Disciplinary contexts of the chemistry of life have been touched on in this section and elsewhere in the volume. The reason for dealing with the subject more specifically at this stage is connected with the appearance of the first issue of the journal *Zeitschrift für physiologische Chemie* in 1877.

The preamble by its editor, F. Hoppe-Seyler (1825–95),[a] constitutes an important historical document – the full translation is given below – because it contained the claim that the discipline called 'biochemistry' did attain the rank of an independent branch of science. This was all the more curious since the new journal carried a name associating it with physiological chemistry.

Implicit in this ambiguity was the opinion that physiological chemistry was not really to be viewed as a chemical arm of physiology. Undoubtedly, Hoppe-Seyler referred to physiological chemistry as 'biochemistry' because of a number of developments in the scientific study of the chemistry of life, during the period 1840–80, that transcended the scope of work on chemical aspects of physiological phenomena. On the one hand, the exploration of the chemical constituents and processes of living things was extended to topics frequently connected with problems of agricultural, medical and industrial practice. On the other hand, there was a growing recognition among investigators who concerned themselves with these problems that they formed the subject of specialized enquiries separate from physiological studies and having their own methods and concepts. This recognition reflected their awareness of the changing status of physiological chemistry: by 1880 they were pushing for its separation from physiology and, in effect, had begun to transform it into biochemistry.

The term 'biochemistry' itself was not new.[b] It was certainly employed by the Austrian V. Kletzinsky[c] as early as 1858 when he called his textbook on the subject: *Compendium der Biochemie*. Without going into detail, it should be said that Kletzinsky did not then consider biochemistry to be an independent discipline. Instead, he regarded it as one of three components of a unified science of life or biology; biophysics and biomorphology (i.e. anatomy and histology) being the other two.

As to Hoppe-Seyler's advocacy of the separation of physiological chemistry (biochemistry) from physiology, it was deprecated in no uncertain terms by Pflüger, the editor of the *Archiv für die gesammte Physiologie des Menschen und der Thiere*.[d] His starting point was that the task of physiology was the study of the life process (*Lebensprocess*). Physiology was *the* science of living matter (*lebendige Materie*) and, therefore, no less an autochtonous branch of scientific enquiry than the sciences of inanimate matter, physics and chemistry. Pflüger claimed that from the practical point of view it was difficult to differentiate between physical and chemical activities because they were so interwoven. Taking this claim as proved, Pflüger employed it ingeniously to argue for the rejection of the division of physiology into physiological physics and physiological chemistry. On practical grounds the division was intractable because of the difficulty of disentangling the intertwined state of physical and chemical forces, even on the supposition that they brought about all physiological processes including mental ones. But Pflüger also objected to the division on philosophical grounds, inasmuch as he held integrality to be the essence of life, depending as much on causal as on teleological relations.[e] Pflüger believed, in stark contrast to Hoppe-Seyler, that this kind of theorizing ('Speculation') had a role to play in understanding the phenomena of life and had, therefore, a legitimate place in his *Archiv*.

F. Hoppe-Seyler, *Z. physiol. Chem.*, 1, I, 1877–8

'Vorwort' (Preamble)

pp. I–III

The upsurge which organic chemistry has experienced in the last four decades enables it to pursue the biological problems (as already attempted earlier) not only in the analytical direction but also, by applying experimental procedures to the living organism, to subject the chemical life processes to full investigation. The synthetic results as well as the insight obtained through them into the structure of the chemical substances and its transformations through chemical processes, on which the most recent period of time must pride itself, have given the means and the ways to explore the causes of the phenomena of life in the structure and the relationship of the substances active in the organism with a certainty not anticipated earlier.

Through this biochemistry has grown from its first natural and necessary analytical beginnings to a science, which has not only placed itself alongside biophysics as of equal rank, but in activity and results competes with it; a glance in the annual reports on the achievements of these sciences suffices to place this beyond doubt. Although now scarcely anyone will deny the high significance of physiological chemistry, it has obviously still penetrated too little into the consciousness of those most closely affected by this scientific activity and its advances, and furthered in their own endeavours. At most German universities still now physiological chemistry, as a special science, is taught in effect insufficiently or not at all, and lectures on it are only seldom held.

The chief cause of this lamentable position appears to lie in the union of the physiological sciences still firmly adhered to. Of any exponent of science in a university one rightly requires that he, in his scientific sphere, possesses not only the knowledge to impart systematic instruction to the students but also knows the means and ways to pursue his own reliable researches. What physiologist now may well be able to boast as being so complete an expert in anatomy, physics and chemistry as to be able to advance into the field of physiology with success, following for the most part the completely different methods of these sciences. The sciences are certainly not different from art and handicrafts in that only that person may achieve in them something important who knows thoroughly the material to be worked on and understands how to use his tools. With the development which the natural sciences have reached in our time it is only very seldom that a quite specially gifted man will equally succeed in becoming sufficiently conversant with the anatomical, physical and chemical methods of research and the correspondingly different outlooks, to perform productive and reliable investigations in every direction. A separation is necessary here, namely a separation corresponding to those natural sciences the methods of which find application in the promotion of the knowledge of organisms and their life.

In close and easily recognizable connection with the unsatisfactory state described stands the lack of physiologico-chemical journals, a lack which has been sharply felt by physiological chemists for a long time. Out of the different journals of chemistry, physiology, practical medicine, hygiene and agriculture must be laboriously sought the works which the physiological chemist must know, in order to work further in his field, and if a work is executed ready for publication there comes to the author the question often difficult to decide, where he should send it in order that it first becomes known in general to professional colleagues and is not placed between microscopic, physical or indeed speculative papers.

In agreement with the persons who appear on the title page of this issue, whose names[a] give complete guaranteee for its strictly scientific direction, I have undertaken the publication of this journal for physiological chemistry and hope that it will fulfil its task to bring about a better assembly of the researches newly carried out in this field, and through this to show itself beneficial to science itself. I am strengthened in this hope by the approval which is expressed to me from many sides very decisively.
Strassburg, June 1877.

b. From *c*. 1880 to *c*. 1940: biochemistry as an independent discipline. Static and dynamic biochemistry. Unity of biochemistry. Origin and nature of life. Biochemistry and morphology

Selection 171

In the controversy over the connection of physiological chemistry with physiology, reported in the previous selection, it was Hoppe-Seyler's judgement, rather than Pflüger's, which eventually carried the day. An essential factor that paved the way for the establishment of biochemistry as an independent discipline was the expansion of enzyme chemistry after 1880.[a]

It would be misleading, however, to believe that the older views yielded easily to the new. Of historical interest in this respect is the attempt by the German physiologist, Max Verworn (1863–1921), to sponsor a theory of metabolism that placed enzymes effectively in the category of 'giant' molecules called 'biogens'. The term was introduced in 1895 by Verworn in his book, *Allgemeine Physiologie*, to express his doubts about the validity of Pflüger's calling the hypothetical labile compound of living substance (which supposedly conditioned life and from which life professedly derived) 'living' protein molecule.[b]

Verworn pointed out that no protein was known to be as unstable as the assumed living protein. Instead, he proposed that the concept of the living protein molecule was to be replaced by the biogen molecule, the chemical structure of which Verworn imagined to be built on the lines advanced by Paul Ehrlich (1854–1914). It will be recalled that in 1885 Ehrlich had suggested that the protoplasmic giant molecule envisaged by Pflüger consisted of a central nucleus and side-chains.[c] In effect Verworn attempted to unite the main features of the ideas of Pflüger and Ehrlich.

As shown below, the biogen molecule set out by Verworn in 1903 was a very complex carbon compound with a benzene ring as the nucleus for a number of side-chains. On the one hand, these contained nitrogen and perhaps iron atoms and functioned as receptors of oxygen. On the other hand, there were also present atomic groups (e.g. aldehydic) which served as the substrate for the oxidative dissociation of the biogen molecule.

After Ludimar Hermann's (1838–1914) inogen[d] and Pflüger's 'living' protein molecule, Verworn's biogen is the third of the class of giant molecules that we have encountered. They were not, as often assumed, thought of as units of living matter. Hermann was primarily concerned with the explanation of muscle contractility, Pflüger with tissue oxidation and the origin of life, while Verworn looked for a general metabolic mechanism.

Biogen was not so much a molecular unit of living matter as a carrier of life, undergoing metabolic changes within protoplasm. Protoplasm was not conceived of as a unitary substance either chemically or morphologically. The cell was the only form of living matter capable of continuous existence.[e]

Following the demonstration of cell-free fermentation in 1897,[f] the role of intracellular enzymes appeared in a new light. Enzymes rang the death-knell for biogens, but Verworn battled hard to save the biogen hypothesis of metabolic change. In the course of his analysis he made some acute observations, envisaging, for instance, that the enzyme was also a substrate.[g] He noted that in many respects enzymes were supposed to operate in a similar way to biogens. Hence those who believed that there were points of contact between both conceptions could consider the biogen hypothesis as an improved form of the enzyme hypothesis. Verworn himself, however, considered this approach an idle one. The enzyme hypothesis was no less hypothetical than the biogen hypothesis. But whereas the biogen hypothesis assumed that metabolic changes could all be explained from the changes of one compound, the enzyme hypothesis 'requires a great number of various enzymes in each cell, acting in a coordinated manner and each performing only its special function.'[h] In conclusion, Verworn maintained that where there was a conflict between the principles of singularity (i.e. biogen) and plurality (i.e. enzymes); the rules of scientific research demanded that the first should be given priority.

As shown below, the biogen hypothesis, formulated by Verworn at the turn of the century, tried to save the concept of the 'living' molecule by incorporating new developments emerging from contemporary studies of enzyme action.

M. Verworn, *Die Biogenhypothese* (Jena, 1903)

(The biogen hypothesis)

pp. 68–70

The central point of the biogen hypothesis is the assumption that in the living substance a complicated compound exists, the biogen, which itself undergoes an incessant metabolism, through rearrangement of the atoms at definite points of its large molecule continously dissociating and thereafter again reconstituting itself. This dissociation and restitution of the biogen molecule are made possible by complicated auxiliary mechanisms that are apparently only realized in the cellular formation of the living substance.

Concerning the chemical constitution of the biogen one can perhaps form the following general ideas. The biogen molecule is a very complex nitrogen-containing carbon compound and possesses, attached to the benzene ring as nucleus, diverse side-chains. Of which the one kind are nitrogen-containing or perhaps iron-containing chains of an aldehyde nature and yields the fuel for the oxidative dissociation of the biogen molecule.

The functional oxidation processes take place in the biogen molecule itself and not in its breakdown products. Through the intramolecular insertion of oxygen into the receptor group the already in itself very labile molecule attains the point of its greatest decomposability. With the functional dissociation oxygen goes over from the receptor group to the aldehyde group of the carbon chain, and passes out with the carbon atom of the same as carbonic acid. With this functional dissociation of the biogen molecule the essential energetic performances of the living substances are linked.

With the restitution comes about, on the one hand, a new taking up and binding of oxygen to the oxygen-transmitting side-chain that operates like an oxidase, and on the other hand, the released affinities in the carbon chain are again bound immediately by appropriate carbon-containing groups. This restitution of the biogen residue runs under ordinary conditions approximately as quickly as the functional breakdown.

Besides the function dissociation in which the whole nitrogen-containing part of the biogen molecule remains preserved, there occurs continuously in smaller degree and independently of the functional claim of the living substance a destructive breakdown. In it the biogen molecule as a result of its great lability suffers a profound decomposition with which nitrogen excretion is bound.

The new formation of biogen molecules and with this the growth of the living substance follows only with assistance from already present biogen molecules through polymerization of the single atom groups. The polymer biogen molecules arising in this way, when opportunity presents itself, break apart into the simple basic molecules. A lasting holding together of the polymer biogen molecules and growth to giant molecules is not to be assumed.

For the processes of restitution after the functional breakdown and the new formation of biogen through polymerization the arrangement of the cells and its differentiations create the necessary conditions. Through these it is taken care of that the necessary building stones are always in suitable form and sufficient amount in the appropriate place. The raw material for the making of the appropriate building stones is yielded in the first place by substances taken in from outside (oxygen and nourishment). But for times of need moreover reserve depots of oxygen and nourishment are present in the cells and indeed the reserve supply of nourishment always predominates quite considerably the reserve supply in oxygen.

The preparation and working up of the nourishment for building stones suitable for the restitution processes is essentially the work of the enzymes, the activity of which is spontaneously regulated by the state at the time and the conditions of the cells. As an integrating step in the chain of the preparatory processes in each cell the cell nucleus is inserted. In this respect in the different cell forms there play an indispensable role besides also special differentiations (for example the chlorophyll bodies in the plant cells).

Thus the centre of all happenings in the living substance is formed by the continuous building up and breaking down of the biogen, and all other processes are supporting auxiliary mechanisms in the service of the biogen.

Selection 172

Despite their initially common origins, embodied in the work of Schwann, cell morphology and cell chemistry for a long time developed side by side, more or less independently of each other. One of the effects of the work on enzymes and enzyme action, at the turn of the century, was to stimulate a new way of looking at chemically organized cell life which, in its turn, led to a new sub-microscopic picture of protoplasm. The new approach proceeded from the view that enzymes belonged to colloids and that they were catalysts conceivably capable of participating in reversible reactions.

The problem was clearly perceived and formulated in 1901 by Franz Hofmeister in a contribution which has the character of a classic and from which the excerpts below are taken.[a] He thought of cells as highly differentiated but, at the same time, integrated systems based on ordered chemical reactions effected by colloidal ferments. These he regarded as the most important instruments the cell possessed for the nature and sequence of chemical reactions within it.

As shown in the first excerpt, Hofmeister rejected the fashionable simple analogy between the burning of coal in the steam engine and the chemical changes taking place in the world of living things. He was well aware of the pitfall in equating the 'machinery' of the cell or the complete living organism to a steam engine. Instead of approaching it from the thermodynamical point of view, Hofmeister likened the cell to a physicochemical machine potentially self-regulated by the reversible nature of enzyme-catalyzed reactions, where it mattered which type of molecules were involved.

Hofmeister was very concerned to justify the belief that there was nothing improbable in the suggestion of a multitude of chemical reactions simultaneously taking place in a space that possessed the size range of the liver cell.[b] Indeed it is clear from the other two excerpts that Hofmeister suggested that the structural organization of the cell was the consequence of its chemical activities. From Hofmeister's description the possibility of a co-ordinated sub-microscopical physicochemical world underlying the visible morphological structures in the cell began to emerge.

F. Hofmeister, *Naturwissenschaftliche Rundschau*, *16*, 581. 1901

'Die chemische Organisation der Zelle' (The chemical organization of the cell)

p. 581

Here [in the organism] the nutrient material introduced as source of energy, before it is transformed into definite end products, undergoes a whole series of alterations which proceeding in parallel and in sequence, can be of very diverse chemical nature, and of very unequal energetic significance. Further while with the steam engine only the heat formed from the chemical energy is activated, so that it is quite immaterial from what fuel it comes, for the animal machine the substances that go to make nutrients are of the greatest significance. For these serve it not merely as heat source but simultaneously as building material which it needs for automatic repair of damaged parts, for replacement of any parts lost, and for production of new, similar machinery. Add to this that the nutrient material, even if it also has to serve merely as source of energy, must be transformed into definite intermediates depending on the functions that come into consideration. This in order to meet the special requirements of muscle contraction, nerve excitation, formation of secretions so that the result is a multiplicity of chemical processes taking place in the animal body which, in spite of keen work by individuals, can be no means be surveyed as yet.

To investigate these transformations is, however, not at all easier if it concerns itself not with a very complicated organism, such as the vertebrate animal, but with a protista or simple cell. For the advantage which appears to be given by the simplification of the anatomical structure is more than offset by the circumstance that a series of functions, which in the higher animal are divided amongst different organs and so are accessible for separate investigation, here appear compressed into the smallest space.

p. 613

But what still remain to be considered are the devices that ensure the undisturbed course of vital reactions. Is the cell as a whole a vessel, filled with a homogeneous solution, in which all chemical processes take place, or does it enclose a number of separated vessels, designed to ensure the undisturbed course of the single reactions side by side or in a logical sequence? . . . Most of the vital reactions taking place in the cell are linked with a colloidal substrate or at least with the mediation of a colloidal reagent, [that is] a ferment. They can therefore have quite well in the colloidal structure of the protoplasm a definite localization. Of the intracellular proferments and ferments especially is to be expected that, lacking diffusibility, they remain where they originated in the cell; there they root as it were, and only are activated when the adequate material is passed to them. Such a notion, however, presupposes the existence of numerous colloidal partition walls in the protoplasm, which does not appear strange to somebody at least who knows about the extraordinary tendency of many colloidal bodies to form membranes at the slightest provocation, thus especially where all the surfaces come into contact. Also the presence of definite organs, recognizable by the eye, the nucleus, the chromatophores and so on, the appearance of enclosures and secretions in vacuoles, the pigments in definite localities among others demonstrate the chemical heterogeneity [*Ungleichwerthigkeit*] and the complicated structure of the protoplasm.

However, even if one had not found so many indications for it, one would be obliged on *a priori* grounds to make such an assumption. First it would be otherwise very difficult to understand that in the protoplasm side by side quite different, partly chemically antagonistic processes, hydration and dehydration, oxidation and reduction processes can proceed without interference. But then, on assuming a single uniform reaction space in the cell, we would renounce a very important explanation possibility. In the protoplasm building up and breaking down of different substances goes on through a series of intermediate stages, whereby in no way is always the same kind of chemical involved but usually a series of reactions of different kinds. Thus we cannot picture the breakdown of glycocoll to urea without a setting free of the NH_2-group from a part of the glycocoll molecule, an oxidation of the rest, then a coming together of the fragments. These reactions however must proceed in a definite sequence otherwise as little urea can arise from them as, for example, aniline from benzene when one, reversing the usual course of the reaction, first treats benzene with reducing agents and then nitrates. A regulated [*gesetzmässige*] sequence of chemical reactions in the cell presupposes, however, separate work of the individual chemical agents and a definite direction of movement of the formed products, in short, a chemical organization which does not agree with the conception of a ubiquitous equivalence of the protoplasm throughout. On the other hand, [chemical organization] makes the promptitude and reliability with which [the protoplasm] functions even more understandable . . .

Now how can one picture this spatial separation of the chemical processes in the protoplasm I have already indicated. One needs only to think of the colloidal reagents as separated by impermeable partition walls. With the many-sided nature of the chemical processes therefore the demand arises for an extensive vacuolation, possibly too small to be observed by the naked eye, and so one may add also physiological considerations for the existence of a foam structure to the grounds, which were advanced by an outstanding authority on morphological aspects.[a] So it can be understood also that life, as we know it, is always tied to colloidal substrate, for only this makes possible with sufficient permeability for non-colloids a complicated building up in the smallest space.

p. 614

If on the one side the morphologist strives to elucidate the structure of the protoplasm up to its finest details, on the other hand the biochemist endeavours, with his apparently grosser and yet more penetrating aids, to ascertain the chemical performance of the same protoplasm, on the whole this still concerns only two different sides of the same coin. The one has in mind as final aim a horizontal and vertical projection, as detailed as possible, of the protoplasm structure. The other the description of the total in protoplasm operating processes by means of a connected chain of chemical and physical formulae. To make, so far as the facts permit, these two widely divergent sets of ideas compatible with each other will be the laborious but not unrewarding task of the future. However, already today one may say that the consideration of the cell as a machine working with chemical and physicochemical means nowhere leads to problems which unavoidably give rise to the assumption of other than known forces. And that so far as one can see, there is no reason for that resignation, which once expressed itself in an '*Ignorabimus*',[b] another time in vitalistic conclusions.

Selection 173

A significant feature of the study of the chemistry of life after 1880 was the rapidly increasing scale of the work which was done upon it. The expansion of research had brought into being, by 1910, specialized periodicals,[a] textbooks,[b] monographs[c] and professional organizations,[d] all of them reflecting the fact that the subject, now often called 'biochemistry'[e] was acquiring a separate disciplinary identity.

While biochemical research was, in effect, becoming an independent branch of science, there was reluctance on the part of those who held power (academic, political, financial) to recognize this development by establishing separate, full, chairs of biochemistry at universities, colleges, and other institutions of higher education. It was deemed sufficient for students, following a course of study in subjects such as human and veterinary medicine, pharmacy, agriculture or industrial food and drink production, to gain the necessary minimum biochemical knowledge through lectures devoted centrally to non-biochemical fields. Biochemical knowledge (or rather parts of it) entered into the curricula of medical chemistry, agricultural chemistry, organic chemistry, physiology, pharmacology, pathology and brewing science.

It is against this background that Kossel's[f] advocacy of the creation of separate chairs of biochemistry in Germany has to be viewed. For this distinguished biochemist, who held a Chair of Physiology at Heidelberg, this question was close to his heart. Kossel raised it in a speech dealing with the nature, scope and future of biochemistry, which he gave as *Prorektor* of the university to its assembly (21 November 1908) and from which the excerpts below are taken.

In arguing the case for the separateness of biochemistry as a branch of science, Kossel focused attention on delineating its subject matter. He regarded it as comprising two areas of study: one enquiring into the chemical composition of living matter and the other into the chemical processes occurring in it. Kossel stated that there were two approaches to the chemistry of life which corresponded to this division: one along statical ('descriptive') lines and another along dynamic ('experimental') ones. It should be noted that Kossel was certainly not the first to draw this line of demarcation in the ways of approaching the chemical aspects of the living world: it is one which seems to have been widely acknowledged by 1910.

Underlying and reflecting this standpoint were methodical and methodological considerations. On the one hand, they furthered the legitimization of biochemistry as a research and teaching discipline in its own right. On the other hand, they helped to shed light on the relationship of biochemistry to disciplines such as anatomy, chemistry and physiology.

Take static biochemistry and anatomy. They were both concerned with the study of the structure of living bodies, albeit at different levels. Indeed their close affiliation produced the characterization of statical biochemistry as 'chemical anatomy'. In his speech, interestingly and stimulatingly, Kossel pointed to the value of anatomy for the consideration of the chemical structure of constituents of living matter and, in turn, indicated the significance of the latter for the study of the developmental processes (p. 9):

Now an anatomical structure can be considered from two standpoints: development and function. Both these approaches have to be taken also into account where it concerns the understanding of *chemical* form. However, the enlargement of biochemistry in the developmental, ontogenetic and phylogenetic, direction has hardly been tackled.

As to dynamic biochemistry, Kossel believed that progress here would very much depend on studies of the synthetic action of enzymes. He also believed that studies of the chemical equilibrium established by enzymic synthesis and hydrolysis coming into balance would be particularly important. But beyond that, what is notable is Kossel's awareness of the potential contribution by dynamic biochemistry to the understanding, in chemical terms, of the common features of the world of life.

It is not without interest that Kossel envisaged an impetus to the study of the chemistry of life would come from the disclosures of novel, subatomic, phenomena which had been made since the turn of the century.

Last but not least, Kossel's unambiguous reductionist repudation of the vitalist position merits attention.

A. Kossel, *Die Probleme der Biochemie* (Heidelberg, 1908)

(The Problems of Biochemistry)

pp. 4–5

The scientific investigation of living beings may be carried out in two directions. First, the description of the *resting* body, its properties, its structure, its form; and secondly, the examination of the *active* organism, its life processes, for example, movement, nutrition. This double approach also applies to biochemistry. The subject of this science is twofold. On the one hand, the investigation and description of chemical products that form the body of animals and plants and, on the other hand, the detection of chemical processes that go on in these bodies. Thus the first part of biochemistry is, like anatomy, a descriptive science, we wish to call it descriptive biochemistry; the second pursues the same aim as physiology: the observation and explanation of living processes, it may be designated as experimental biochemistry.

Now the question may be raised about the special features of biochemistry with respect to anatomy on the one hand, and to physiology on the other hand? It is not the chemical investigation, transcending the capability of the microscope, a continuation of the anatomical one? Is experimental biochemistry not a branch of physiology? Indeed, if one and the same investigator could have the ability to master, at the same time, the chemical along with the physical, anatomical and especially physiological methods and approaches, so there would be no reason for separating these three fields (anatomy, biochemistry, and physiology) from each other. At the time when Tiedemann at this university performed his excellent anatomical investigations, when he wrote his great textbook of physiology and, in collaboration with Leopold Gmelin, successfully carried out investigations into the chemistry of digestive processes,[a] such a combination was still possible – not any more today. With the widening of our knowledge, also in these fields the intellectual and technical demands on the investigator have grown and in their totality surpass the capacity of the best brains today. This is how it has come about that in the *field of research* a complete separation between biochemistry and anatomy and physiology has taken place, expressed in the creation of laboratories, journals and textbooks. In Germany, however, the step towards a separation of the *chairs* has not been taken yet.

This development must not be designated, perhaps, as a deplorable specialization or a partition of an organic connection. On the one hand, the thread apparently is loosened but, on the other hand important new bonds are formed. The connection of experimental with descriptive biochemistry is being preserved, new relations are being established with botany, bacteriology, pathology and practical medicine as well as with other sciences.

At the beginning of all physiological investigations, there is the question regarding the common phenomena of life or organisms, animals and plants, their forms and the gross-mechanical [*grobmechanischen*] conditions of their living activity are fundamentally different. If here one wants to discover the inner connection, it is necessary to go back to elementary components and elementary functions, and hence for these researches the most likely successes are to be expected from a science dealing with smallest particles, the molecules and atoms. And indeed biochemical work has also helped to bridge the gap between animals and plants, between higher and lower organisms. Where excitation physiology [*Reizphysiologie*][b] was able to find only a few links, biochemistry has disclosed an extensive analogy in chemical structure and chemical functions. It has taught, on the basis on common chemical standpoints, to comprehend the breaking and building up of organic substances, respiration and assimilation, in the most diverse organisms. Through such work biochemistry has created for itself an independent sphere of ideas, which touches on the fundamental questions of biology, and has developed in a way which the universities will have to cope with by conceding it its own working space and by granting it a substantial influence on teaching. The mould into which the physiological sciences were cast around the middle of the past century does not satisfy any more today.

pp. 18–19

There is no doubt that the exploration of living matter with the methods of structural and theoretical chemistry will vouchsafe results of paramount significance. But besides tasks accessible to such treatment, there are other problems about which we do not know whether they could be elucidated on the basis of present-day chemical conceptions. To them belong processes of apparently chemical nature, which relate to phenomena of excitation and of responses to excitation [*Reizerfolge*], to formation of chemical antibodies, to morphogenesis, to chemical symmetry, heredity and so on. It is very probable that the treatment of these physiological processes will provide chemistry with new impulses and inspire new ideas. The development of organic chemistry as a whole, after all, proceeded from the examination of physiological subject matters.

The chemists of the 19th century succeeded in singling out, from the numerous substances produced by living beings, in the first place the simplest ones and in arranging them into a system. With them chemistry has studied the natural history of carbon, and through them it gained the idea of the possibilities for carbon, hydrogen, oxygen, nitrogen and other elements to combine, which could not have been deduced from the observation of non-living nature. In investigating these subjects, structural chemistry has developed its theories. The theories resulted in the evolution of synthetic chemistry and this synthetic work has exereted profound influence on the elucidation of physiologico-chemical phenomena. It is not a rather daring hope to expect that the process which had taken place in the history of the 19th century, will start over again this century. [That is] new ideas and theories of chemistry, derived from the consideration of living beings, originating from the field of biochemistry, and the repercussion of these theories on the doctrine of life [*Lehre von Leben*].

The physical and chemical works of the last decades have come to know phenomena which have completely changed our basic ideas about the nature of matter. Newly discovered radiations, the behaviour of radioactive substances, transformation of elements are processes the exploration of which places our ideas about the nature of atoms upon a new basis. Through these discoveries we are taught anew that the foundation on which we have to build our explanation of phenomena of life is only in the process of formation. But equally also the knowledge of living processes itself is still very imperfect.

We only know fragments of the world of phenomena, which have to be elucidated and we still stand at the beginnings of the development of the means which can serve [their] elucidation.

Despite this situation, already now the vitalist hypothesis assumes that to reduce all phenomena of life to physical and chemical processes will be impossible also in the future. Whether this assumption is meeting with more or less assent – science is going to proceed as if a limit to knowledge is not set.

Selection 174

The distinction between static and dynamic biochemistry (referred to in the foregoing selection) amounted to the delineation but not to the formulation of the subject matter of biochemistry. This was rightly taken to be the contribution of (Sir) Frederick Gowland Hopkins (1861–1947)[a] who in a number of public addresses examined the nature of biochemistry more deeply and farsightedly than anyone before him. It is appropriate, therefore, to devote this selection to what is probably his most celebrated discourse, the presidential address to the Physiology Section of the British Association for the Advancement of Science at Birmingham (11 September 1913).

Here Hopkins's purpose was to ventilate the position and aims of bio-chemistry as a separate borderland discipline. In doing so, he discussed the interrelations between biochemistry – strictly speaking animal chemistry – and biology and chemistry. Hopkins went beyond the view which sought to equate biochemistry with the application of the principles of chemistry to biological problems. Underscoring the immediate relevance of biological thought to biochemical practice, Hopkins came to regard biochemistry as a branch of biological sciences. However, he found that biologists, while paying some attention to physical chemistry, possessed little understanding or knowledge of biochemistry. At the same time, Hopkins observed that organic chemists widely distrusted the results of biochemical work on account of its being 'amateurish' and 'inexact'. He, in turn, reproached them for neglecting the colloidal state, catalysis and analysis of complex mixtures.

These were all topics of central interest to a biochemist. Hopkins traced their neglect to the organic chemist's obsessive belief

that the really significant happenings in the animal body are concerned in the main with substances of such high molecular weight and consequent vagueness of molecular structure as to make their reactions impossible of study by his available and accurate methods.

The repudiation of this approach, based on an insistence that metabolism dealt with simple molecules, constituted the core of Hopkins's argument. He was able to base it, of course, on his own investigations showing the part played by lactic acid in muscle contraction[b] and that of particular amino acids in animal nutrition.[c] As further support, Hopkins also invoked the work of other investigators, too numerous to be listed here.

What distinguished Hopkins from them was his belief that the starting point of investigations into chemical operations carried out by living organisms was not protoplasm, as such, but the chemical interaction of the molecules of its components, effected and governed by specific intracellular enzymes. In holding this view Hopkins was by no means unique at that time. Instance of this outlook was given earlier in this section by presenting Hofmeister's approach to the chemistry of the cell. Indeed, Hopkins's biochemical thinking flowed very much in the direction pointed out by Hofmeister. Nevertheless, Hopkins, before the Frist World War at any rate, did not imply as clearly as Hofmeister that the gap between the biochemical and morphological fields could be bridged. For Hopkins the living cell was not so much an enzymic physicochemical machine as a colloidal system of phases existing in a dynamic equilibrium. Reactions catalyzed by enzymes, he thought, took place in heterogeneous systems of phases. He also paid attention to the mode of action of enzymes, in the hope that this would throw light on the nature of organized living material (animal tissues) as chemical systems. Consequently the long-standing view (to the eradication of which Hopkins devoted much effort) of the chemistry of life in terms of the decomposition and restitution of this or that 'living' molecule was abandoned.

In assessing Hopkins's contribution to biochemical thought at this stage it may be said that he was concerned to supply it with the independent theoretical framework which it lacked, and hoped thereby to enable it to come fully into its own.

F.G. Hopkins, *Rep. Brit. Ass.*, **652. 1913**

'The dynamic side of biochemistry'

p. 652

My main thesis will be that in the study of the intermediate processes of metabolism we have to deal, not with complex substances which elude ordinary chemical methods, but with simple substances undergoing comprehensible reactions. By simple substances I mean such as are of easily ascertainable structure and of a molecular weight within a range to which the organic chemist is well accustomed. I intend also to emphasise the fact that it is not alone with the separation and identification of products from the animal that our present studies deal; but with their reactions in the body; with the dynamic side of biochemical phenomena.

pp. 657–8

There remains, I find, pretty widely spread, the feeling – due to earlier biological teaching – that, apart from substances which are obviously excreta, all the simpler products which can be found in cells or tissues are as a class mere dejecta, already too remote from the fundamental biochemical events to have much significance. So far from this being the case, recent progress points in the clearest way to the fact that the molecules with which a most important and significant part of the chemical dynamics of living tissues is concerned are of a comparatively simple character ... So long as there were any remains of the instinctive belief that the carbonic acid and urea which leave the body originate from oxidations occurring wholly in the vague complex of protoplasm, or at least that any intermediate products between the complex and the final excreta could only be looked for in the few substances that accumulate in considerable amount in the tissues (for instance, the creatin of muscle), the idea of seriously trying to trace within the body a series of processes which *begin* with such simple substances as tryosin [tyrosin] or leucin was as foreign to thought as was any conception that such processes could be of fundamental importance in metabolism. However vaguely held, such beliefs lasted long after there was justification for them; their belated survival was due, it seems to me, to a certain laziness exhibited by physiological thought when it trenched on matters chemical; they disappeared only when those accustomed to think in terms of molecular structure turned their attention to the subject. But it should be clearly understood that the progress made in these matters could only have come through the work and thought of those who combined with chemical knowledge trained instinct and feeling for biological possibilities. Our present knowledge of the fate of amino-acids, as of that of other substances in the body, has only been arrived at by the combination of many ingenious methods of study. It is easy in the animal, as in the laboratory, to determine the end-products of change; but, when the end result is reached in stages, it is by no means easy to determine what are the stages, since the intermediate products may elude us. And yet the whole significance of the processes concerned is to be sought in the succession of these stages. In animal experiments directed to the end under consideration, investigators have relied first of all upon the fact that the body, though the seat of a myriad reactions and capable perhaps of learning, to a limited extent and under stress of circumstances, new chemical accomplishments, is in general able to deal only with what is customary to it.

pp. 659–60

The earliest attempts at tracing the intermediate processes of metabolism looked for information to the products which accumulate in the tissues, but it seemed to be always tacitly assumed that only those few which are quantitatively prominent could be of importance to the main issues of metabolism. It is obvious, however, upon consideration, that the degree to which a substance accumulates is by itself no measure of its metabolic importance; no proof as to whether it is on some main line of change, or a stage in a quantitatively unimportant chemical bypath. For, if one substance be changing into another through a series of intermediate products, then, as soon as dynamical equilibrium has been established in the series, and to such equilibrium tissue processes always tend, the rate of production of any one intermediate product must be equal to the rate at which it changes into the next, and so throughout the series. Else individual intermediate products would accumulate or disappear, and the equilibrium be upset. Now the rate of chemical change in a substance is the product of its efficient concentration and the velocity constant of the particular reaction it is undergoing. Thus the relative concentration of each intermediate substance sharing in the dynamic equilibrium, or, in other words, the amount in which we shall find it at any moment in the tissue, will be inversely proportional to the velocity of the reaction which alters it. But the successive velocity constants in a series of reactions may vary greatly, and the relative accumulation of the different intermediate products must vary in the same degree. It is certain that in the tissues very few of such products accumulate in any save very small amount, but the amount of a product found is only really of significance if we are concerned with any function[1] which it may possibly possess. It is of no significance as a measure of the quantitative importance of the dynamical events which give rise to it.

pp. 662–3

Now interest in the chemical events . . . may still be damped by the feeling that, after all, when we go to the centre of things, to the bioplasm, where these processes are initiated and controlled, we shall find a milieu so complex that the happenings there, although they comprise the most significant links in the chain of events, must be wholly obscure, when viewed from the standpoint of structural organic chemistry. I would like you to consider how far this is necessarily the case.

The highly complex substances which form the most obvious part of the material of the living cell are relatively stable. Their special characters, and in particular the colloidal condition in which they exist, determine, of course, many of the most fundamental characteristics of the cell: its definite yet mobile structure, its mechanical qualities, including the contractility of the protoplasm, and those other colloidal characters which the modern physical chemist is studying so closely. For the dynamic chemical events which happen within the cell, these colloid complexes yield a special milieu, providing, as it were, special apparatus, and an organised laboratory. But in the cell itself, I believe, simple molecules undergo reactions of the kind we have been considering. These reactions, being catalysed by colloidal enzymes, do not occur in a strictly homogeneous medium, but they occur, I would argue, in the aqueous fluids of the cell under just such conditions of solution as obtained when they progress under the influence of enzymes *in vitro*.

There is, I know, a view which, if old, is in one modification or another still current in many quarters. This conceives of the unit of living matter as a definite, if very large and very labile molecule, and conceives of a mass of living matter as consisting of a congregation of such molecules in that definite sense in which a mass of, say, sugar is a congregation of molecules, all like to one another. In my opinion, such a view is as inhibitory to productive thought as it is lacking in basis. It matters little whether in this connection we speak of a 'molecule' or, in order to avoid the fairly obvious misuse of a word, we use the term 'biogen', or any similar expression with the same connotation. Especially, I believe, is such a view unfortunate when, as sometimes, it is made to carry the corollary that simple molecules, such as those provided by foodstuffs, only suffer change after they have become in a vague sense a part of such a giant molecule or biogen. Such assumptions became unnecessary as soon as we learnt that a stable substance may exhibit instability after it enters the living cell, not because it loses its chemical identity, and the chemical properties inherent in its own molecular structure, by being built into an unstable complex, but because in the cell it meets with agents (the intracellular enzymes) which catalyse certain reactions of which its molecule is normally capable.

Exactly what sort of material might, in the course of cosmic evolution, have first come to exhibit the elementary characters of living stuff, a question raised in the Presidential Address[a] which so stirred us last year, we do not, of course, know. But it is clear that the living cell as we now know it is not a mass of matter composed of a congregation of like molecules, but a highly differentiated system; the cell, in the modern phraseology of physical chemistry, is a system of co-existing phases of different constitutions.[2] Corresponding to the difference in their constitution, different chemical events may go on contemporaneously in the different phases, though every change in any phase affects the chemical and physico-chemical equilibrium of the whole system. Among these phases are to be reckoned not only the differentiated parts of the bioplasm strictly defined (if we can define it strictly) the macro- and micro-nuclei, nerve fibres, muscle fibres, &c., but the material which supports the cell structure, and what have been termed the 'metaplasmic' constituents of the cell. These last comprise not only the fat droplets, glycogen, starch grains, aleurone grains, and the like, but other deposits not to be demonstrated histologically. They must be held, too – a point which has not been sufficiently insisted upon – to comprise the diverse substances of smaller molecular weight and greater solubility, which are present in the more fluid phases of the system – namely, in the cell juices. It is important to remember that changes in any one of these constituent phases, including the metaplasmic phases, must affect the equilibrium of the whole cell system, and because of this necessary equilibrium relation it is difficult to say that any one of the constituent phases, such as we find *permanently* present in a living cell, even a metaplasmic phase, is less essential than any other to the 'life' of the cell, at least when we view it from the standpoint of metabolism. It is extremely difficult and probably impossible by any treatment of the animal to completely deprive the liver of its glycogen deposits, so long as the liver cells remain alive. Even an extreme variation in the quantity is in the present connection without significance because, as we know, the equilibrium of a polyphasic system is independent of the mass of any one of the phases; but I am inclined to the bold statement that the integrity of metabolic life of a liver cell is as much dependent on the co-existence of metaplasmic glycogen, however small in amount, as upon the co-existence of the nuclear material itself; so in other cells, if not upon glycogen, at least upon other metaplasmic constituents.

Now we should refuse to speak of the membrane of a cell, or of its glycogen store, as living material. We should not apply the term to the substances dissolved in the cell juice, and, indeed, would hardly apply it to the highly differentiated parts of the bioplasm if we thought of each detail separately. We are probably no more justified in applying it, when we consider it by itself, to what, as the result of microscopic studies, we recognise as 'undifferentiated' bioplasm. On ultimate analysis we can hardly speak at all of living matter in the cell; at any rate, we cannot, without gross misuse of terms, speak of the cell life as being associated with any one particular type of molecule. Its life is the expression of a particular dynamic equilibrium which obtains in a polyphasic system. Certain of the phases may be separated, mechanically or otherwise, as when we squeeze out the cell juices, and find that chemical processes still go on in them; but 'life,' as we instinctively define it, is a property of the cell as a whole, because it depends upon the organisation of processes, upon the equilibrium displayed by the totality of the co-existing phases.

pp. 664–6

It is clear that a special feature of the living cell is the organisation of chemical events within it. So long as we are content to conceive of all happenings as occurring within a biogen or living molecule all directive power can be attributed in some vague sense to its quite special properties.

But the last fifteen years have seen grow up a doctrine of a quite different sort which, while it has difficulties of its own, has the supreme merit of possessing an experimental basis and of encouraging by its very nature further experimental work. I mean the conception that each chemical reaction within the cell is directed and controlled by a specific catalyst. . . . We have arrived, indeed, at a stage when, with a huge array of examples before us, it is logical to conclude that all metabolic tissue reactions are catalysed by enzymes, and, knowing the general properties of these, we have every right to conclude that all reactions may be so catalysed in the synthetic as well as in the opposite sense. If we are astonished at the vast array of specific catalysts which must be present in the tissues, there are other facts which increase the complexity of things. Evidence continues to accumulate from the biological side to show that, as a matter of fact, the living cell can acquire *de novo* as the result of special stimulation new catalytic agents previously foreign to its organisation.

It is certain, from very numerous studies made upon the lower organisms, and especially upon bacteria, that the cell may acquire new chemical powers when made to depend upon an unaccustomed nutritive medium. . . .

If we are entitled to conceive of so large a part of the chemical dynamics of the cell as comprising simple metaplasmic reactions catalysed by independent specific enzymes, it is certain that our pure chemical studies of the happenings in tissue extracts, expressed cell juices, and the like, gain enormously in meaning and significance. We make a real step forward when we escape from the vagueness which attaches to the 'bioplasmic molecule' considered as the seat of all change. But I am not so foolish as to urge that the step is one towards obvious simplicity in our views concerning the cell. For what indeed are we to think of a chemical system in which so great an array of distinct catalysing agents is present or potentially present; a system, I would add, which when disturbed by the entry of a foreign substance regains its equilibrium through the agency of new-born catalysts adjusted to entirely new reactions? Here seems justification enough for the vitalistic view that events in the living cell are determined by final as well as by proximate causes, that its constitution has reference to the future as well as the past. But how can we conceive that any event called forth in any system by the entry of a simple molecule, an event related qualitatively of the structure of that molecule, can be of other than a chemical nature? The very complexity, therefore, which is apparent in the catalytic phenomena of the cell to my mind indicates that we must have here a case of what Henri Poincare[b] has called *la simplicité cachée*. Underlying the extreme complexity we may discover a simplicity which now escapes us. If so, I have of course no idea along what lines we are to reach the discovery of that simplicity, but I am sure the subject should attract the contemplative chemist, and especially him who is interested and versed in the dynamical side of his subject. If he can arrive at any hypothesis sufficiently general, to direct research he will have opened a new chapter of organic chemistry – almost will he have created a new chemistry.

pp. 667–8

But I would urge upon any young chemist who thinks of occupying himself with biological problems, the necessity for submitting for a year or two to a second discipline. If he merely migrate to a biological institute, prepared to determine the constitution of new products from the animal and study their reactions *in vitro*, he will be a very useful and acceptable person, but he will not become a bio-chemist. We want to learn how reactions run in the organism, and there is abundant evidence to show how little a mere knowledge of the constitution of substances, and a consideration of laboratory possibilities, can help on such knowledge. The animal body usually does the unexpected.[c]

But if the organic chemist will get into touch with the animal, it is sure that the possession of his special knowledge will serve him well. Difficulties and peculiarities in connection with technique may lead the professor of pure chemistry to call his work amateurish, and certainly his results, unlike those of the physical chemist, will not straightway lend themselves to mathematical treatment. He may himself, too, meet from time to time the spectre of Vitalism, and be led quite unjustifiably to wonder whether all his work may not be wide of the mark. But if he will first obtain for us a further supply of valuable qualitative facts concerning the reactions in the body, we may then say to him, as Tranio said to his master:

> 'The mathematics and the metaphysics
> Fall to them as you find your stomach serves you.'

All of us who are engaged in applying chemistry and physics to the study of living phenomena are apt to be posed with questions as to our goal, although we have but just set out on our journey. It seems to me that we should be content to believe that we shall ultimately be able at least to describe the living animal in the sense that the morphologist has described the dead; if such descriptions do not amount to final explanations, it is not our fault. If in 'life' there be some final residuum fated always to elude our methods, there is always the comforting truth of which Robert Louis Stevenson gave perhaps the finest expression, when he wrote:

> 'To travel hopefully is better than to arrive,
> And the true success is labour.'

Selection 175

On writing on Hopkins shortly after his death, two workers who were close to him hailed him as the *Fundator et Primas Abbas* of biochemistry in England.[a] If there is a biochemical heaven, Hopkins's deeds and behaviour entitle him to be canonized. However, it was not a call from the Lord to Hopkins to establish his see in Cambridge that brought him there in 1898. What happened was that he accepted an invitation from the Professor of Physiology (Sir) Michael Foster (1836–1907) to develop at Cambridge 'teaching and research in the chemical side of physiology'.[b]

It is idle to speculate on the course events might have taken if Foster had not suggested that Hopkins should move from London, where he had worked at Guy's Hospital during the day and in a privately run clinical laboratory in the evenings.[c] Or what might have happened if Hopkins had not accepted the offer and had stayed in London. What is certain is that Hopkins acquitted himself brilliantly in his task, in the face of continuous financial hardships from which he suffered in two ways. His personal income was highly unsatisfactory and this situation did not change when he became the first (unpaid) Professor of Biochemistry at Cambridge in 1914. But this development had a considerable importance for the discipline because it led to the formation of the Department of Biochemistry and its institutional separation from physiology (i.e. Physiological Laboratory). Neither the material nor financial conditions for carrying out the work connected with biochemical teaching and research were adequate until funding from outside the university became available. It allowed not only for the housing of the Department in a new building (formally opened on 9 May 1924) but also for the endowment of four teaching positions (Professorship, Readership, two Lectureships).[d]

Fifty years later one of the members of the Department who had been present at the opening of the building (the Sir William Dunn Institute of Biochemistry), Malcolm Dixon (1899–1985), looked back and discussed how the design of the building was affected by the kind of work being done at that time:[e]

You could not buy chemicals then – you had to make them; so the isolation of aminoacids [sic] was very important. So the building had two hot rooms (for running tryptic digests) but no cold room; many incubators but (as far as I can remember) not a single refrigerator! When we wanted ice we sent out to the fishmonger for a great block about two feet high, which had to be attacked with hammer and chisel. It was thought that we should be doing a lot of ether extractions; so (for distilling off without fire risk) each bench had a steam-heated water bath, and steam was laid on throughout the building from a large boiler in the basement. Our distilled water was got simply by condensing steam from the pipes. But in the evening, when the steam was turned off, the water in the baths and any dirt therein was sucked back into the pipes and in due course appeared in the distilled water!

In the old days we tended to work with whole animals (dogs and rats) and with whole perfused organs. Mitochondria were unknown, as were ribosomes and even pure enzymes. We had none of the modern facilities; most apparatus had to be home-made. There were no high-speed centrifuges (we had one which held nearly a litre and could sometimes be coaxed up to 2,000 rpm without breaking the tubes), no spectrophotometres, no O_2 electrodes, no electronics, no recording instruments, no chromatography, no glass electrodes for pH, no computers, no liquid air available, no ultramicroscopes, no Warburg manometers (although there were one or two adapted blood-gas manometers). We knew nothing of NAD, flavins, cytochromes, mitochondria or even ATP. Some of us worked on things like gamma-glucose, pnein, physin and thio-X whose very names are now forgotten.[f] Though we did know about lactic acid and purine metabolism, it was largely a different world.

Looking back over the fifteen years between the opening of the Dunn Institute and the outbreak of the Second World War, it is difficult not to agree with the view that during this period the Department of Biochemistry was at its peak. That belief has been expressed by one of Hopkins's out-of-the-ordinary pupils, Norman Wingate Pirie (b.1907), who explained the nature of his teacher's influence on the members of the Department as follows:[g]

He subtly directed our interests towards processes rather than substances. The word *subtly* should be stressed. He did not direct research but influenced it by his obvious interest in some aspects rather than others.

Among other achievements, he contributed to the widening of the scope of biochemistry by extending the study of the chemical processes in the animal body to those occurring in plants and micro-organisms. The appreciation of an underlying unity of some chemical reactions in diverse living forms began to be reflected in the growth of 'general biochemistry'. Whether Hopkins coined the term or not, he certainly was one of the first to conceptualize it and the work at Cambridge was one of the major factors in its gaining an independent footing in the scientific study of the world of life between the two World Wars.

The allegiance which biochemists accorded to Hopkins transcended the boundaries of Cambridge, which effectively became a Mecca for them. The esteem may be judged from the fact that when asked to discuss the subject of biological oxidations at the twelfth International Physiological Congress at Stockholm (1926), he first perhaps surprisingly provided his audience with a different treat. With an extensive knowledge of the position of biochemical teaching and research elsewhere in the world but speaking in the light of his own hard-won concrete Cambridge experience, Hopkins took up the cudgels, us it were *ex cathedra*, on behalf of what he aptly called the 'institutional needs of biochemistry'. At that time, as shown below, biochemistry's battle for a clearly defined institutionalized position in the spectrum had yet to be won.

F.G. Hopkins, *Skand.Arch.Physiol.*, *49*, 33. 1926

'On current views concerning the mechanisms of biological oxidation (with a Foreword on the institutional needs of biochemistry)'

pp. 33–7

I ought to say at once that I am among those who believe that independent Institutes of Biochemistry with specialised staffs for teaching and research should in every University stand by the side of the existing Institutes of Physiology. I know however that at some European academic centres the separation involved is still viewed with misgiving . . . Though it is fifty years since Hoppe-Seyler printed his appeal for the recognition of biochemistry as an independent discipline,[a] such recognition cannot be said to have yet arrived in his own country. In proportion to its outstanding academic equipment modern Germany provides but little institutional freedom for the discipline in question. It is under the circumstances difficult to see how she can continue to lead along the path which for a long time she trod almost alone. There is indeed as yet no considerable movement towards the separate endowment of biochemical studies anywhere in continental Europe; while in America it has been rapid. There are several outstanding European Chairs of course. In France we have the Chair occupied successively by such distinguished biochemists as Duclaux[b] and Georges Bertrand,[c] and a few others concerned in the main with medical teaching. In Italy there is a separate Institute at Rome and also at Naples. There are doubtless others elsewhere. Sweden in proportion to her academic resources has served the subject well. The existence of the old established Chair at Upsala so long associated with the name of Hammarsten,[d] a name by which biochemists conjure; the activities of the School at Lund, and the fact that the distinguished Professor of Organic Chemistry[e] at Stockholm is now a biochemist in all but name; these and other circumstances show that the country is alert to the importance of this branch of Science. We are now steadily increasing the endowment and equipment of biochemistry in Great Britain.

Academic centres in Europe however are, as I have said, behind those in America in their general encouragement of biochemical science. The initiation of the proportionately greater interest which is undoubtedly taken in the subject in the United States is, I believe, generally attributed to the influence of Chittenden[f] who a generation ago recrossed the Atlantic with an inspiration acquired in the laboratory of Kühne.[g]

Even America however lacks, I think, Institutes of a kind which I will now be bold enough to advocate.

It does not seem to me enough that the biochemist, hitherto housed in an Institute of animal Physiology, and therefore mainly occupied with animal studies, should migrate to a separate building and yet retain his preoccupations. The greatest need of biochemistry at the moment in my opinion is equipment which shall make possible the study under one roof (of course, from its own special standpoint alone) of all living material. No full understanding of the dynamics of life as a whole, no broad and adequate views of metabolism, can be obtained save by studying with equal concentration the green plant and micro-organisms as well as the animal. In leaving an Institute supposed to deal with the whole of animal physiology for one equipped for more embracing studies of the chemistry of living organisms as a whole, the biochemist would be far from narrowing his interests. If his intellectual frontiers would shrink on one side, they would extend in other and for him, more logical directions. The Institute I have in mind would steal something from the activities (usually, however, minor activities) of various existing biological Institutes, but would justify the theft by a highly profitable combination and coordination of the stolen materials. I am in fact only pleading for one of those reconstellations in intellectual pursuits which the progress of knowledge often makes desirable or necessary. I am sure that the one in question is now desirable.

Such Institutes of General Biochemistry would have to be equipped for teaching as well as for research, for we have to prepare a future generation to bear burdens greater than our own. Modern physical chemistry, and modern organic chemistry, reacting as they already are – even the latter – to the newer concepts of Physics, are to provide entirely new concepts for biochemistry. There must be experts to understand and apply them; a task which will be as impossible for those who must continue to concern themselves with the rapid growth on other sides of physiology and biology, as it will be for the equally preoccupied pure chemist. I do not undervalue the difficulties of equipping and staffing an Institute of general biochemistry such as I am picturing, but I must not stop to discuss them. I will only state that we have, in not too ambitious fashion, attempted the task in Cambridge. In addition to our ordinary medical teaching of chemical physiology in the older sense, my colleagues and I take a few advanced students each year over a much wider ground. Apart from hearing lectures on general biochemistry, they deal, in practical classes, with the chemical activities of bacteria, yeast and the green plant, and do not neglect even the invertebrate animal. Only then do they give attention to vertebrate metabolism, including their own. I think our experiment is a success, for our students undoubtedly acquire a scientific outlook which at present is somewhat unusual, and, so far, those who preside over other biological departments in Cambridge have viewed our proceedings in a friendly spirit. It is perhaps not possible, and may be not desirable, that such teaching should be provided at every academic centre, but I am sure it will justify itself wherever it is found.

Selection 176

A direct consequence of the presence of an immense variety of living forms has been the search for an underlying unity or for unifying principles encompassing the diverse chemical facets of their existence. For this, in this volume, we have the evidence in concepts such as vital force, fermentation, cycle, protoplasm, cellular respiration, enzyme and general biochemistry. Corresponding to the stages of conceptual development, the terms expressed a recognition of the elements common to the chemistry of life with all its many diverse aspects.

At this stage, it will be convenient to consider the efforts of the Dutch worker, Albert Jan Kluyver (1888–1956), in the mid-1920s to throw light on what constituted the common basis of chemical changes in living organisms. Kluyver taught at the Technological University of Delft, where he became Professor of General and Applied Microbiology in 1921, succeeding the eminent microbiologist M.W. Beijerinck (1851–1931). This was a surprising appointment because Kluyver's theoretical and practical knowledge of the field was slight, he having been employed mainly in the study of copra-fibre industry in Java between 1916 and 1921.

On taking up his new post, Kluyver felt that his first task was to acquaint himself with the situation in microbiological studies. To him, the 'outsider', the field appeared around 1925 to be in a state of confusion because of a lack of theoretical clarity. Not weighed down by considerations which, at times, befog the eye of the 'insider' he then boldly went on to proclaim the need to grasp the idea of the essential unity of the vast range of chemical activities occurring in the world of life – in Kluyver's words: 'From elephant to butyric acid bacterium'.[a]

Kluyver did not confine himself to diagnosis. Collaborating with H.J.L. Donker (b.1890), he attempted to show that there existed a general oxido-reductive mechanism of metabolic processes involving a catalytic transfer of hydrogen. Their work was based on utilizing and bringing together two contemporary developments. One derived from interpreting biological oxidation reactions in terms of hydrogen activation of substrates by dehydrogenases (Wieland,[b] Thunberg[c]). The other related to studies of hydrogen ion concentration effects in biological systems (Sørensen,[d] Michaelis,[e] Clark,[f] J. and D. Needham[g]).

As shown below, Kluyver and Donker advanced the view that the material substance of cell bodies (protoplasm) catalyzed the chemical changes taking place in living systems. Oxidation ensued by a stepwise catalytic removal of hydrogen from substrate molecules (dehydrogenation substrates) in sequence until the completion of the combustion to carbon dioxide and water. In this connection it is of interest that they found congenial Thunberg's notion that 'hydrogen has to be regarded as the elementary fuel of cells.'[h] It corresponded to their beliefs that there was a unifying way of looking at biochemical changes. With regard to hydrogen ion concentration, they saw in it the device delicately governing the equilibrium of oxido-reduction phenomena within protoplasm.

Before leaving the subject, a few remarks may be made concerning their view that protoplasm rather than enzymes played the part of the catalyst. The clue to this is to be found in the problem of ranges of enzyme specificity that aroused a good deal of debate at that time.[i] Kluyver and Donker belonged to the camp which found it difficult to visualize biochemical reactions as due to innumerable specific enzymes operating within a given cell. If protoplasm was the catalyst, they thought, it was possible to dispense with such an assumption. 'It is hardly necessary to emphasize,' they wrote, 'how thereby the idea of the biochemical working of the living cell will be simplified.'

A.J. Kluyver and H.J.L. Donker, *Chem.Zelle Gew.*, *13*, 134, 1926

'Die Einheit der Biochemie' (The unity of biochemistry)

pp. 187–8

The catalytic effect of the agent – which we provisionally have identified simply with the protoplasm of the cell – should be conditioned by the free affinity of the protoplasm for hydrogen or oxygen. For each biologically specific cell [*artspezifische*] this affinity would have a definite value and precisely this value (optimal for certain catalytic reactions) could determine the direction of dissimilation . . . This hypothesis, however, would have to be very plausibly extended to accepting that the said affinity of the cell for hydrogen is not an absolute constant but that it varies with the hydrogen ion concentration within the region of concentrations which are admissible for the cell. The endless diversity in biochemical behaviour of various biologically specific cells would thereby be traceable to a difference in a single property of the protoplasm – the affinity for hydrogen or oxygen respectively – to their dependence on hydrogen ion concentrations, typical and allowable for each species. The fact that also the tendency of substrates to be dehydrogenated and the tendency of acceptors to be hydrogenated are, as known, a function of hydrogen ion concentrations, contributes essentially to the enlargement of the external diversity of biochemical events in spite of their internal unity.

Additionally let us take into account that the oxidative or fermentative dissimilation of sugars always leads to intermediary or definitive formation of acids, the dissimilation of hydrolytic protein decomposition products to formation of ammonia. Then it is possible to arrive at the opinion (in order to substantiate the adaptation of the diverse cells to these most important substrate groupings) that the protoplasm of these cells has acquired a suitable affinity for hydrogen corresponding to appropriate hydrogen ion concentrations.

Finally we have meant to indicate that not only dissimilatory but also assimilatory processes essentially could be traced to the catalytic transference of hydrogen.

Selection 177

We noted earlier the attitudes of Pasteur,[a] Huxley[b] and Pflüger[c] to the problem of the chemical origins of life. Although the interest in it had never died out during the period 1880–1920, it was not until the third decade of this century that two independent attempts were made to bring it back into serious discussion. The first was due to Aleksandr Ivanovich Oparin (1894–1980) who treated the subject in a booklet in Russian that appeared in 1924. The second consisted in an 8-page essay written by John Burdon Sanderson Haldane (1892–1964) that was published in 1929 in the *Rationalist Annual*, a magazine devoted to the cause of free thought. However, the significance and the relatedness of their ideas on the origins of life were not grasped till well into the 1940s. Once this happened, the names of the authors began to be linked through the expression 'Oparin–Haldane hypothesis' which provided a new impulse to the study of an abiding problem. It is because of this close connection in the evolution of the subject that this selection includes excerpts from the two historically relevant works of both writers.

Now to introduce these presents some knotty problems. Why did Oparin and Haldane apparently make no appreciable impression in the short run, although they offered coherent albeit tentative biochemical solutions, and set them out clearly? Was it because the situation was not 'ripe' for that kind of hypothesizing in the 1920s? But then how can we explain the independent advent of two contributions, along corresponding lines, both dealing with the problem of the origins of life and appearing within five years or so of each other?

As far as Oparin is concerned it could be maintained that he failed to be noticed abroad because he was an unknown and relatively young biochemist, publishing in a language which the overwhelming majority of scientists outside the Soviet Union did not read. As to whether Oparin's ideas on the origins of life were appreciated in his homeland in the 1920s and 1930s, we have no direct information. We know that in 1935 Oparin was involved in the organization of the newly established Institute of Biochemistry of the Academy of Sciences under the directorship of A.N. Bach (whose former pupil he was).[d] In this connection it is not out of place to refer to contemporary reports on Soviet science by J.G. Crowther. As a writer and journalist specializing in scientific matters, Crowther visited the

Soviet Union seven times between 1927/8 and 1934/5. In his sympathetic accounts of the world of science and technology in the Soviet Union the name of Bach occurred but not that of Oparin.[e] Even if this might have been due to Crowther's relative neglect of the biomedical fields, it would be safe to conjecture that the problem of the origins of life, in general, and Oparin's outlook on it in particular, did not count as a major scientific issue in the Soviet Union in the period under review.

Be that as it may, in the case of Haldane the idea that he was not an academically established scientist in 1929 does not apply. He was Reader in Biochemistry in Cambridge (from 1923) and at the same time in charge of genetical investigations at the John Innes Horticultural Research Station (from 1927). He was already well known for his versatile researches in genetics and evolution, biochemistry and physiology. Moreover, by then he had made quite a name for himself in alerting the general public to the benefits and dangers of science in his popular writings and lectures on controversial subjects such as test-tube babies or gas warfare. But there was hardly any response even inside the ranks of science to his thoughts, published in 1929, on how life began on Earth. Thus J.D. Bernal (1901–71), disclosing that he had acquired his own lifelong interest in the subject from Haldane,[f] admitted:[g]

I knew he had written on it, but I had not then read his short essay in the Rationalist Annual (of 1929) . . . at the time, Haldane's ideas were dismissed as wild speculations. Only the clarity and beauty of his style and the attractiveness of his character ensured that they were read, albeit in small circles.

The problem has been enduringly attractive for intrinsic scientific reasons as well as for its religious and philosophical aspects. In this respect the situation in the 1920s was not different from previous times. What did change on the biochemical side was the attitude to the value of arriving at opinions on the origins of life through speculation. Sensing rather than perceiving the essential barrenness of the mechanist-vitalist debate, the bulk of biochemists – believers, agnostics and atheists alike – escaped into strict empiricism as a way out of this situation. Whatever the origins of life, it was agreed that the problem was virtually inaccessible to biochemical – or, for that matter, any other scientific – study. What was accessible was the empirical knowledge of the physical and chemical properties of living matter, provided one assumed the validity of physical and chemical laws in living systems. Whether this knowledge explained what life is was not a scientific but a non-scientific issue, belonging to the spheres of religion and philosophy. This was something separate from biochemists' concern with the chemical composition of living systems and from their studies of the chemical changes in them. The majority of biochemists shared the sentiments expressed in the preface to the first edition (1929) of what became a standard textbook of biochemistry:[h]

It is rather generally agreed among the scientists that the actions of a biological organism are expressions of the energy relationships due to chemical and physicochemical processes taking place within the cells and tissues which comprise the organism.

The biological organism can be looked upon as a complex system of chemical constituents composed mainly of proteins, carbohydrates, fats and lipids, mineral elements, and water, which are organized by the mysterious forces which we call 'life', and the actions and reactions of this protoplasmic mass are in turn determined by the energy interchanges of molecular transformations and surface and interfacial forces.

As to whether the moment was 'ripe' for Oparin and Haldane to attempt to carry the problem of the origins of life a stage further, the answer is 'yes' and 'no'. 'Yes' in the sense that two individuals independently and almost simultaneously were led to review the contemporary scientific evidence and concluded from it that the case for the historicity and naturalness of the formation of life from inorganic substances could be substantiated. 'No' in the sense that the full meaning and consequences of these two initiatives in raising the study of the chemical origins of life to a plausibly higher level was virtually not recognized.

The excerpts below indicate the common features but also the differences in their approaches. In trying to visualize how life originated from inorganic substances, Oparin went back to Pflüger's ideas on the possibility of the formation of cyanogen (cyan) compounds when the Earth was still in an igneous state and to D.I. Mendeleev's (1834–1907) hypothesis of the inorganic origin of petroleum centred on hydrolysis of metallic carbides. The third main element which went into Oparin's conceptual framework of the chemical beginnings of life was furnished by the study of the colloid state of matter.

In comparison with Oparin, Haldane took into account rather more recent scientific developments. These included photosynthetic studies of naturally occurring compounds by E.C.C. Baly (1871–1948)[i] and his collaborators at Liverpool, and the phenomenon of bacteriophagy first observed by F.W. Twort (1877–1950) and recognized as such by F. d'Herelle (1873–1949).[j] Of central interest to Haldane was the theme of the anaerobic nature of the environment within which life made its appearance.

Last but not least, because their later writings were influenced by Marxist philosophical thought, questions arise about its importance for the elaboration of their biochemical approaches to the origins of life in the 1920s. Regarding Oparin, the situation is that he made no conscious attempts to introduce it into his first study whereas he did this in his subsequent writings on the subject.[k] As to Haldane's philosophical basis, he certainly did not draw on Marxist philosophy then. It could be usefully asked, however, what part the early studies of Oparin and Haldane on the origins of life played in their embracing Marxism at a later stage.

A.I. Oparin, *The Origin of Life* (Moscow, 1924).[a]

pp. 222–5

Thus, at the period of existence of the Earth which we are considering, when it was a red star about to become extinguished, masses of carbides of iron and other metals which had formerly been concealed in its depths were being extruded on to its surface through cracks formed in its crust. Here, on the surface, they encountered the atmosphere of that time which differed in many respects from that of today. Water vapour was specially abundant in it. All the water in all the seas and oceans now on the Earth then existed in the form of superheated steam in the atmosphere. The carbides which flowed out on to the surface encountered this steam.

If we treat carbides of metals with superheated steam we obtain what are known as hydrocarbons, that is to say compounds consisting of carbon and hydrogen. These compounds must also have arisen when the carbides and steam met on the surface of the Earth. Of course some of these must immediately have been burnt, being oxidized by the oxygen of the air. However, under the conditions then prevailing this combustion must have been far from complete. Only a certain part (and a comparatively small one) of the hydrocarbons were fully oxidized, being converted to carbonic acid and water. A further part, owing to incomplete oxidation, gave rise to carbon monoxide and oxygen derivatives of hydrocarbons, while, finally, a certain proportion of the hydrocarbons completely escaped oxidation and was given off into the upper, cooler layers of the atmosphere without any alteration. The more the Earth cooled the lower became the temperature at which the interaction between the carbides and the water vapour took place, and less carbonic acid and more unoxidized hydrocarbons were formed.

Thus, the theoretical considerations put forward above lead us to the belief that at a particular time in the existence of the Earth compounds of carbon with hydrogen and oxygen were formed in its atmosphere. Let us see whether there are not facts in our natural surroundings supporting this idea.

Since the period under discussion is that during which the Earth passed through the stage of being a red star it would be quite appropriate to consider first what we know about these stars. Spectroscopic studies of the darkest red stars which are about to go out, carried out by the astronomer Vogel, led him to the conclusion that the atmospheres of these stars contain hydrocarbons. This fact was soon confirmed by several other workers.

Hydrocarbon lines have also been found in the spectra of comets, those heavenly bodies which from time to time pass through our solar system from interplanetary space. Furthermore, thanks to the studies of several scientists it has been found that cyan (a compound of carbon and nitrogen) and carbon monoxide are present in the gases which form the tails of comets.

By origin comets are related to a further class of heavenly bodies, the meteorites. We have already discussed these in an earlier chapter. Meteorites are specially interesting because, in falling, some of them reach the surface of our Earth in a more or less undamaged state in the form of red-hot stones from the sky. Thus they are accessible to direct chemical examination. They are, so to speak, lumps of matter, samples reaching us from the boundless region of interstellar space. Analysis of meteorites and study of their composition gives us the opportunity of getting a direct knowledge of some of the materials of which the stars are made. Most meteorites consist of native iron, partly combined with carbon and sometimes containing carbon in such quantities that it has been possible to isolate it from certain falling stars in the form of diamond dust. This composition of the meteorites is an extra confirmation of the correctness of the view that carbon exists on heavenly bodies in the form of mixtures or chemical compounds with metals.

In meteorites, however, other carbon compounds have also been found. By causing samples from meteorites to incandesce by means of an electric spark, scientists have managed to show that the hydrocarbon lines are certainly present in their spectra. It has even proved possible to isolate from some meteorites a considerable amount of hydrocarbons and to establish their nature by chemical studies.

Thus, we can demonstrate beyond doubt the presence of hydrocarbons on a number of heavenly bodies. This fact gives full support to the conclusions we had already drawn. There came a time in the life of the Earth at which the carbon which had been set free from its combination with metal and had combined with hydrogen formed a number of hydrocarbons. These were the first 'organic' compounds on the Earth.

Although only two elements, carbon and hydrogen, enter into the composition of these compounds these elements can join together in the most varied combinations and give rise to the most varied hydrocarbons. Organic chemists can now list a very large number of such compounds.

As the properties of hydrocarbons have been studied in great detail, a study of the conditions prevailing on the Earth when these compounds came into being makes it possible to put forward some suggestions as to which hydrocarbons were in fact formed. Without going into details we can only say that everything points to the view that it was the 'unsaturated' (free radical) hydrocarbons which were formed first, that is to say the most unstable members of the class we are discussing, having very large stores of chemical energy and great chemical potentialities, compounds which combine very easily both with one another and with other elements.

If these compounds could avoid oxidation at the time of their formation, then, during their stay in the hot, wet atmosphere of the Earth, they must certainly have combined with oxygen and given rise to the most varied substances composed of carbon, hydrogen and oxygen in various proportions (alcohols, aldehydes, ketones and organic acids).

Thus, all the considerations and facts which we have put forward above convince us that, even if not all the carbon, at least a great part of it, first appeared on the surface of the Earth, not in the form of the chemically inert carbon dioxide as had been thought, but in the form of unstable organic compounds capable of further transformation.

Let us leave these compounds for a while and take a look at what happened, during the period in the existence of the Earth which we are discussing, to the fourth element which enters into the composition of living things, namely nitrogen. At high temperatures nitrogen can form compounds with oxygen (the technical production of nitric acid depends on this). We are therefore justified in expecting the appearance of these compounds in the atmosphere of the Earth where the two elements involved were mixed. However, oxides of nitrogen are somewhat unstable. At temperatures of about 1,000°C these compounds break down and give off free nitrogen. Compounds of nitrogen with metals obtained industrially under conditions of white heat are far more stable. Such compounds must also have been formed in the atmosphere of the Earth by interaction between nitrogen and the incandescent vapour of the lighter metals. Later these compounds of metals with nitrogen, of a similar nature to carbides, were submitted to the action of superheated steam and formed ammonia,[1] which is a compound of nitrogen and hydrogen. Ammonia could also have been formed primarily at a far earlier stage in the existence of the Earth by the condensation of hydrogen and nitrogen in the upper layers of the atmosphere.

Furthermore, we cannot exclude the possibility of the formation of compounds of nitrogen and carbon. In this case the material obtained would be cyan, a substance with which we have already become acquainted when we were discussing the spectra of comets. Its presence in the gases surrounding these heavenly bodies confirms the possibility that it might also have been formed on the Earth. Thus, the atmosphere of the Earth at a certain period of its existence must have contained compounds of nitrogen in the form of ammonia and cyan as well as oxygen derivatives of hydrocarbons.

Although, from our point of view, cyan can hardly have played any important part in the further transformations of organic substances, we mention it because it is interesting to us in another way. It forms the starting-point of the extremely far-reaching and well-thought-out theory of the origin of life put forward by the well-known German physiologist Pflüger.

pp. 226–9

Finally the time came when the temperature of the surface layers of the Earth fell to 100°C. It became possible for water to exist in the form of liquid drops. Continuous downpours of rain fell upon the surface of the Earth from the moist atmosphere. They inundated it and formed a cover of water in the form of the original boiling ocean.

The first organic substances which had hitherto remained in the atmosphere were now dissolved in the water and fell to the ground with it. What were these substances?

We have already remarked on their main property at the end of the last section. They were substances having a large store of chemical energy and possessing great chemical potentialities. While still in the terrestrial atmosphere they had begun to combine with one another to give rise to very complicated compounds. In addition, they combined with oxygen and ammonia to give hydroxy and amino-derivatives of hydrocarbons (i.e. compounds of hydrocarbons with oxygen and nitrogen respectively).

When these substances fell from the atmosphere into the primaeval ocean they did not stop interacting with one another. Individual components of organic substances floating in the water met and combined with one another. Thus ever larger and more complicated particles were formed.

We can easily create a fairly accurate picture for ourselves of this process of aggregation (polymerization) of organic substances on the Earth by studying it in our chemical laboratories. In fact, the conditions in which organic substances existed in the stage of development of the Earth which we are dealing with can be achieved comparatively easily in our present-day laboratories. If we submit such substances as hydrocarbon radicals to the conditions described above and leave them to themselves we shall find the whole chain of reactions set out above taking place. The hydrocarbon radicals will be oxidized at the expense of the oxygen in the water and air to give the greatest variety of derivatives (alcohols, aldehydes, acids, etc.). This process takes place specially quickly at high temperatures and in the presence of iron and other metals.

Oxidized hydrocarbons readily combine with one another to form more complicated compounds. Many of these substances can also combine with ammonia and give rise to the development of the most varied nitrogen derivatives.

The process of aggregation of organic substances usually occurs rather slowly it is true. However, this is not very important. Whether it takes several months or several years, we still get, as a result of these processes, a mixture of various substances having a very complicated structure.

In these mixtures we may even find, among others, compounds of the nature of carbohydrates[2] and proteins. Both of these types of compounds play an important part in the structure of living material. We find them in all animals and plants without exception. In combination with other and yet more complicated substances they are, as it were, the foundation of life.

Of course, the substances which we produce artificially are not exactly the ones which can be isolated from living organisms. However, they are, if we may express it so, related to these compounds. The elementary composition, the structure of the particles and the chemical properties are almost the same in the one as in the other. The difference is only in detail.

The substances obtained by the method described above can serve as good nutrient material. They are specially nutritious for micro-organisms such as bacteria and moulds. This fact is specially important and we shall give a little more time to it . . . at present what are lacking above all are those substances containing much chemical energy which are the only things from which life could develop and which, themselves, could only be formed at extremely high temperatures. However, even if such substances were formed now in some place on the Earth, they could not proceed far in their development. At a certain stage of that development they would be eaten, one after the other. Destroyed by the ubiquitous bacteria which inhabit our soil, water and air.

Matters were different in that distant period of the existence of the Earth when organic substances first arose, when, as we believe, the Earth was barren and sterile. There were no bacteria nor any other micro-organisms on it, and the organic substances were perfectly free to indulge their tendency to undergo transformations for many, many thousands of years.

It is, of course, hard to say what these transformations were and what sort of substances resulted from them. The only thing that is certain is that these transformations were mainly directed towards an aggregation of material and the formation of more and more complicated and larger and larger particles.

However, we have seen in one of the preceding sections that substances with large and complicated particles have a great tendency to form colloidal solutions in water. Sooner or later such colloidal solutions of organic substances must have come into being in the watery covering of the Earth and once they had arisen they continued to exist, their molecules becoming more complicated and larger as time went on.

The state of colloidal solution is not, however, stable. For various, sometimes extremely slight causes, the dissolved substances come out of the colloidal solution in the form of precipitates, coagula or gels. It is impossible, incredible, to suppose that in the course of many hundreds or even thousands of years during which the terrestrial globe existed, the conditions did not arise 'by chance' somewhere which would lead to the formation of a gel in a colloidal solution. Such formation of aggregated pieces of organic material floating freely in the boundless watery spaces of the ocean which gave rise to them must certainly have occurred at some time in the existence of the Earth.

The moment when the gel was precipitated or the first coagulum formed, marked an extremely important stage in the process of the spontaneous generation of life. At this moment material which had formerly been structureless first acquired a structure and the transformation of organic compounds into an organic body took place. Not only this, but at the same time the body became an individual. Before this it had been inseparably fused with all the rest of the world, dissolved in it. Now, however, it separated itself out, though still very imperfectly, from that world and set itself apart from the environment surrounding it.

With certain reservations we can even consider that first piece of organic slime which came into being on the Earth as being the first organism. In fact it must have had many of those features which we now consider characteristic of life. It was composed of organic substances, it had a definite and complicated structure which was completely characteristic of it. It had a considerable store of chemical energy enabling it to undergo further transformations. Finally, even if it could not metabolize in the full sense of the word, it must certainly have had the ability to nourish itself, to absorb and assimilate substances from its environment, for this is present in every organic gel.

J.B.S. Haldane, *Rationalist Annual*, 3, 1929

'The origin of life'

pp 7–10

Within a few thousand years from its origin it probably cooled down so far as to develop a fairly permanent solid crust. For a long time, however, this crust must have been above the boiling-point of water, which condensed only gradually. The primitive atmosphere probably contained little or no oxygen, for our present supply of that gas is only about enough to burn all the coal and other organic remains found below and on the earth's surface. On the other hand, almost all the carbon of these organic substances, and much of the carbon now combined in chalk, limestone, and dolomite, were in the atmosphere as carbon dioxide. Probably a good deal of the nitrogen now in the air was combined with metals as nitride in the earth's crust, so that ammonia was constantly being formed by the action of water. The sun was perhaps slightly brighter than it is now, and as there was no oxygen in the atmosphere the chemically active ultra-violet rays from the sun were not, as they now are, mainly stopped by ozone (a modified form of oxygen) in the upper atmosphere, and oxygen itself lower down. They penetrated to the surface of the land and sea, or at least to the clouds.

Now, when ultra-violet light acts on a mixture of water, carbon dioxide, and ammonia, a vast variety of organic substances are made, including sugars and apparently some of the materials from which proteins are built up. This fact has been demonstrated in the laboratory by Baly of Liverpool and his colleagues. In this present world, such substances, if left about, decay – that is to say, they are destroyed by micro-organisms. But before the origin of life they must have accumulated till the primitive oceans reached the consistency of hot dilute soup. To-day an organism must trust to luck, skill, or strength to obtain its food. The first precursors of life found food available in considerable quantities, and had no competitors in the struggle for existence. As the primitive atmosphere contained little or no oxygen, they must have obtained the energy which they needed for growth by some other process than oxidation – in fact, by fermentation. For, as Pasteur put it, fermentation is life without oxygen.[b] If this was so, we should expect that high organisms like ourselves would start life as anaerobic beings, just as we start as single cells. This is the case. Embryo chicks for the first two or three days after fertilization use very little oxygen, but obtain the energy which they need for growth by fermenting sugar into lactic acid, like the bacteria which turns milk sour. So do various embryo mammals, and in all probability you and I lived mainly by fermentation during the first week of our pre-natal life. The cancer cell behaves in the same way. Warburg has shown that with its embryonic habit of unrestricted growth there goes an embryonic habit of fermentation.[c]

The first living or half-living things were probably large molecules synthesized under the influence of the sun's radiation, and only capable of reproduction in the particularly favourable medium in which they originated. Each presumably required a variety of highly specialized molecules before it could reproduce itself, and it depended on chance for a supply of them. This is the case to-day with most viruses, including the bacteriophage, which can grow only in presence of the complicated assortment of molecules found in a living cell.

The unicellular organisms, including bacteria, which were the simplest living things known a generation ago, are far more complicated. They are organisms – that is to say, systems whose parts co-operate. Each part is specialized to a particular chemical function, and prepares chemical molecules suitable for the growth of the other parts. In consequence, the cell as a whole can usually subsist on a few types of molecule, which are transformed within it into the more complex substances needed for the growth of the parts.

The cell consists of numerous half-living chemical molecules suspended in water and enclosed in an oily film. When the whole sea was a vast chemical laboratory the conditions for the formation of such films must have been relatively favourable; but for all that life may have remained in the virus stage for many millions of years before a suitable assemblage of elementary units was brought together in the first cell. There must have been many failures, but the first successful cell had plenty of food, and an immense advantage over its competitors.

It is probable that all organisms now alive are descended from one ancestor, for the following reason. Most of our structural molecules are asymmetrical, as shown by the fact that they rotate the plane of polarized light, and often form asymmetrical crystals. But of the two possible types of any such molecule, related to one another like a right and left boot, only one is found throughout living nature. The apparent exceptions to this rule are all small molecules which are not used in the building of the large structures which display the phenomena of life. There is nothing, so far as we can see, in the nature of things to prevent the existence of looking-glass organisms built from molecules which are, so to say, the mirror-images of those in our own bodies. Many of the requisite molecules have already been made in the laboratory. If life had originated independently on several occasions, such organisms would probably exist. As they do not, this event probably occurred only once, or, more probably, the descendants of the first living organism rapidly evolved far enough to overwhelm any later competitors when these arrived on the scene.

As the primitive organisms used up the foodstuffs available in the sea some of them began to perform in their own bodies the synthesis formerly performed haphazardly by the sunlight, thus ensuring a liberal supply of food. The first plants thus came into existence, living near the surface of the ocean, and making food with the aid of sunlight as do their descendants to-day. It is thought by many biologists that we animals are descended from them. Among the molecules in our own bodies are a number whose structure resembles that of chlorophyll, the green pigment with which the plants have harnessed the sunlight to their needs. We use them for other purposes than the plants – for example, for carrying oxygen – and we do not, of course, know whether they are, so to speak, descendants of chlorophyll or merely cousins. But since the oxygen liberated by the first plants must have killed off most of the other organisms, the former view is the more plausible.

The above conclusions are speculative. They will remain so until living creatures have been synthesized in the biochemical laboratory. We are a long way from that goal. It was only this year that Pictet for the first time made cane-sugar artifically.[d] It is doubtful whether any enzyme has been obtained quite pure. Nevertheless I hope to live to see one made artificially. I do not think I shall behold the synthesis of anything so nearly alive as a bacteriophage or a virus, and I do not suppose that a self-contained organism will be made for centuries. Until that is done the origin of life will remain a subject for speculation. But such speculation is not idle, because it is susceptible of experimental proof or disproof.

Some people will consider it a sufficient refutation of the above theories to say that they are materialistic, and that materialism can be refuted on philosophical grounds. They are no doubt compatible with materialism, but also with other philosophical tenets. The facts are, after all, fairly plain. Just as we know of sight only in connection with a particular kind of material system called the eye, so we know only of life in connection with certain arrangements of matter, of which the biochemist can give a good, but far from complete, account. The question at issue is: 'How did the first such system on this planet originate?' This is a historical problem to which I have given a very tentative answer on the not unreasonable hypothesis that a thousand million years ago matter obeyed the same laws that it does today.

This answer is compatible, for example, with the view that pre-existent mind or spirit can associate itself with certain kinds of matter. If so, we are left with the mystery as to why mind has so marked a preference for a particular type of colloidal organic substances. Personally I regard all attempts to describe the relation of mind to matter as rather clumsy metaphors. The biochemist knows no more, and no less, about this question than anyone else. His ignorance disqualifies him no more than the historian or the geologist from attempting to solve a historical problem.

Selection 178

We have seen that already before 1914 students of biochemistry such as Hofmeister[a] and Hopkins[b] recognized in the biochemical approach, in general, and in the study of enzymic systems, in particular, a means of gaining insight into the nature of the organization of living matter. However, this viewpoint had no appreciable effect in making the chemical organization of living matter a starting point of biochemical and biological research, and thus a meeting point between biochemistry and morphology until around 1930.

What Hofmeister and Hopkins were really facing was the problem of the relations of enzymic activity and cell structure, a question of some importance, for instance, in the history of cell respiration, yet relatively little discussed.

As noted before,[c] the discussion in the early 1920s centred on whether oxidation was due to hydrogen activation of substrates by dehydrogenases (Wieland-Thunberg) or oxygen activation by an iron-containing catalyst, the so-called *Atmungsferment* or respiratory ferment (Warburg). In addition, there was also the question of the part played by cellular structure in the oxidation process which was denied by Thunberg but stressed by Warburg. Compared with the stir created by the hydrogen–oxygen activation issue (resolved by 1924), the problem of the relations of enzymic activity and cell structure left the biochemists on the whole fairly cool. Only a few paid any attention to it, among them J.H. Quastel (1899–1987), who together with some colleagues examined bacterial systems and concluded that the activity of dehydrogenases was dependent on the integrity of the cell.[d] This idea did not find much favour with Thunberg, who believed Quastel went astray because he worked with bacteria. Thunberg wrote:[e]

In the case of such cells it is tempting to look upon the solid structure as essential for functions which in other cells are not necessarily bound to structure. If we work on larger cells and on tissues of softer consistency we are not so easily tempted to attach such a high importance to the cell structure for the dehydrogenation processes.

Another researcher who also found it necessary to relate the chemical activities inside the cell to its internal structure was (Sir) Rudolph Albert Peters (1889–1982). It was from such considerations that his 'co-ordinative' biochemical outlook arose.

Peters started his research career with the Cambridge physiologist J. Barcroft (1872–1947), before the First World War, on the nature of the union of oxygen with haemoglobin. His first doubts on the purely physical explanation of adsorption arose then and they were reinforced when he became familiar, in the later stages of the war, with the effects of war gases on living tissues, which he was asked to examine. He began to look for a chemical or physicochemical approach to permeability and adsorption. Like Quastel he found enough support in the work on monomolecular films of organic compounds at interfaces to press for a new understanding of integrated cell activity.[f]

Peters formulated this theoretical view for the first time in his Harben Lectures in 1929. He distinguished two types of surfaces, the external surface presented by the cell to its environment, and internal surfaces or interfaces within the cell, and suggested that 'a cell surface whether internal or external may be made of molecules so anchored that they constitute a chemical mosaic. This conception made co-ordinative biochemistry possible.'[g]

A year later, in a discussion on the structure of living matter arranged by the Faraday Society, Peters returned to his challenge of the current biochemical doctrine. Referring to 'normal architecture of the cell' and the 'dynamic equilibrium' as the two notions often employed to characterize life, Peters (see below) critically developed Hopkins's views on the colloidal conditions in which the dynamic equilibrium in the cell came to be established by bringing enzyme and surface chemistry together. However important enzymes were, Peters could not see his way to consider them as primarily responsible for the integrity of the cell. He made it clear that he was concerned with sub-microscopic cell organization, within which biochemical reactions could operate.[h]

It can hardly be said that Peters's communication was welcomed by all participants.[i] One of the leading physical chemists present, F.G. Donnan (1870–1956), felt that the expression 'chemical architecture of the cell' used by Peters was too vague. Viscosity experiments on living protoplasm did not offer support for structural conceptions. If the expression stood for molecular chain orientation then this view, although not experimentally confirmed, was not original. A much sharper attack came from one of the researchers working in the Cambridge laboratory, B. Woolf. He did not think that sub-visible structures had to be postulated for chemical reactions to proceed in the cell. Work with the yeast press juice and by Meyerhof with the cold water extract of muscle had shown that enzymes could operate in simple systems. This line of study should be continued by concentrating not only on the substrates of enzyme activity but also on the intermediate stages of metabolism. Peters's supposed invisible structure in protoplasm could hardly be tested in experimental practice and thus the suggestion was not very helpful for the advance of biochemical knowledge. Among those who came to the support of Peters, however, was J. Needham (b.1900), who believed that the speakers had tended to take a rather crude view of what Peters was trying to convey. Needham reminded the audience that certain results obtained by ultra-violet spectrophotometry of sea-urchin eggs upon cytolysis showed a difference in absorption protein curves, and asked: 'Does this not indicate in the living cell that proteins are associated with the other constituents in some radically different way from that which occurs in mixtures of compounds which are not organized?'

R.A. Peters, *Trans.Farad.Soc.*, *26*, 797. 1930

'Surface structure in the integration of cell activity'

pp. 797–8

The view which is presented here differs from most others in the stress which is laid upon architecture. Its keynote is the complete (or nearly complete) structural organisation of the cell. I believe this to be organised not only in respect to its grosser parts such as the nucleus, but also in regard to the actual chemical molecules of which it consists. Owing to the micro-heterogeneous nature of the system, surface effects take precedence over ordinary statistical, mass action relationships, and become in the ultimate limit responsible for the integration of the whole and therefore the direction of activities. It is believed that the directing effect of the internal surfaces is displayed predominantly by an organised network of protein molecules, forming a three dimensional mosaic extending throughout the cell. The enzymes would form part of this structure, their activity being largely controlled by the mosaic. This conception of living matter reconciles some difficulties; for instance independent chemical reactions could proceed simultaneously in various parts of the cell, but at the same time the mosaic could react as a whole to stimuli transmitted from the cell surface.

p. 798

During the pioneer developments of biochemistry, it has been usual to consider living matter in rather vague terms as a heterogeneous colloidal system in dynamic equilibrium, or to fall back upon conceptions of specific enzymes or of selective permeability. Investigation of these systems *in vitro* by chemical method must clearly form an essential step in our progress towards knowledge of the living stuff. But however valuable they may be in their own sphere, I feel that they no longer serve a useful purpose, when applied to the cell *in vivo* as general conceptions. I think even that they almost conceal some confusion of thought. Enzymes are essentially organic catalysts, capable by their very definition of accelerating chemical change. Specific enzymes may guide chemical change in the sense that an enzyme may determine that, out of a number of possible reactions, this one will take place rather than that; but enzymes are servants. They cannot determine what gets to them, nor can they be responsible for maintaining the integrity of structure which must exist; nor do we obtain much help from saying that the cell is in colloidal equilibrium. When liberated from the cell, by destroying the cell organisation, reactions run to completion; inside the cell they may not. They are under some kind of control.[1]

Selection 179

Until the late 1890s investigators studying microscopic, usually pathogenic, organisms did not distinguish between bacteria and viruses. The origins of the differentiation may be traced to the use of improved bacterial filtration techniques in the form of so-called 'candle filters', described in 1884 by one of Pasteur's chief collaborators, Charles Chamberland (1851–1908).[a] They were made use of by the Russian plant physiologist, Dmitri Iosifovich Ivanovsky (1864–1920), when he was asked in 1890 to investigate the mosaic disease that affected the tobacco plantations in the Crimea. About the results of this work, which appeared in print two years later, more is to come.

But before then a brief mention must be made of Adolf Mayer (1843–1942) who, apparently, was the first to investigate the tobacco mosaic disease and also gave it its name (1886). He believed that it was infectious and that the sap of diseased plants infected, on inoculation, healthy plants. Finding that the filtrate of such sap obtained by passing it through two layers of filter paper did not transmit the disease, Mayer concluded that it was caused by hitherto unknown bacteria.[b]

However, Ivanovsky thought that Mayer's experimental procedure pointed to a fungal rather than bacterial origin of the mosaic disease because the double paper prevented the passing of fungal spores but not bacteria. Ivanovsky found that the sap of diseased leaves, filtered through the

Chamberland filter, still transmitted the disease just as the unfiltered sap did, and he had no doubt that the disease was caused by bacteria.[c] But in a shorter paper in German, also in 1892, Ivanovsky interpreted the retention of the infectious properties of the filtrate rather more ambiguously:[d]

According to the opinions prevalent today, it seems to me that the latter is to be explained most simply by the assumption of a toxin secreted by the bacteria present, which is dissolved in the filtered sap. Besides this there is another equally acceptable explanation possible, namely, that the bacteria of the tobacco plant penetrated through the pores of the Chamberland filter-candles, even though before every experiment I checked the filter used in the usual manner and convinced myself of the absence of fine leaks and openings. (Added in footnote: It was impossible by means of a rubber bulb to press air through the filter-candles submerged in a cylinder of water.)

Six years later (1898) the explanation of observations along these lines was carried a stage further.

One came from Beijerinck of Delft who studied the tobacco mosaic disease and concluded that it was not due to a microbe but to a non-corpuscular living liquid virus. This was because he demonstrated the diffusibility of sap extracted from diseased leaves through solid agar. Moreover, layers of agar (about ½ mm thick) were used to inoculate healthy plants which produced signs of the disease. Further, Beijerinck pointed out that the virus only multiplied in the tissues of organs of growing tobacco plants and not outside them.[e]

The other suggestion was offered by two former associates of Robert Koch (1843–1910), who were asked by the German state authorities to deal with foot-and-mouth disease. They were F. Loeffler (1852–1915) and P. Frosch (1860–1928) who reported that the disease was transmitted to animals inoculated with filtered lymph just as easily as to animals (serving as controls) which were given corresponding amounts of unfiltered lymph. The German investigators were inclined to believe that the lymph contained unknown living organisms, rather than a toxin, which were capable of passing through the filter pores and must therefore be smaller than bacteria.[f]

This, then, was the beginning of the distinction between bacteria and 'filterable' or 'filter-passing' viruses as a separate group of infectious agents inducing diseases in plants and animals.[g] But from the outset questions were being asked about the living nature of viruses. The uncertainties pertained not only to the fluid or particulate nature of viruses (fuelled by the unsureness regarding their size) but also to the recognition that they seemed to multiply only in the presence of living hosts. In particular, from the discovery of bacteriophagy onwards (1915, 1917)[h] the opinion grew (shared by many workers in the field before the mid-1930s) that the viruses represented the lowest form of life and this, in the nature of things, placed them on the border between the non-living and living state.

Biochemical interest in the question was greatly increased from 1935 when Wendell Meredith Stanley (1904–71) of the Rockefeller Institute for Medical Research in New York announced that he had isolated a crystalline material, possessing the properties of the tobacco mosaic virus, from the

juice of Turkish tobacco plants that were infected with the virus. He identified the crystalline material as protein whose infectivity, chemical composition and optical rotation remained unchanged after ten crystallizations.
He regarded the virus 'as an autocatalytic protein, which for the present,
may be assumed to require the presence of living cells for multiplication'.[i]

There were those in the virus field who had their doubts about the purity
of Stanley's product, among them Frederick Charles Bawden (1908–72)
of the Rothamsted Experimental Station and N.W. Pirie still working at
the Cambridge Biochemical Laboratory. In 1937 they substantiated their
scepticism in a paper in which they described the isolation of liquid crystalline preparations from plants infected with tobacco mosaic. They obtained
them by combining precipitation methods employed in the study of proteins
with improved high-speed centrifugation. On the basis of the physical and
chemical properties of their products (except for the capacity for linear
aggregation) Bawden and Pirie identified them as nucleoproteins.

In connection with this paper the following retrospective observations by
Pirie may be quoted:[k]

The physical properties of the material we had isolated from virus-infected
plants excited considerable interest, partly because our biological outlook led
us, at the Royal Society soirée, to demonstrate anisotropy of flow by using a
goldfish and seahorse to stir dilute solution, placed between 'polaroid' sheets
rather than a more conventional stirring mechanism. The presence of nucleic
acid excited no interest at all; for two years it was strenuously opposed.
Nucleic acids were not fashionable at that date and no other authentic
nucleoprotein – that is to say, a preparation in which nucleic acid and protein
remained associated when in solution – was known. The 'tetranucleotide
hypothesis' was widely believed in. If nucleic acids had been tetranucleotides,
there would clearly be little scope for specificity in them. But anyone reading
the literature on nucleic acids at that date could see that the hypothesis was
baseless. Chemical evidence did no more than make it unnecessary to assume
the presence of more than four nucleotides, it had no bearing on the question
of whether four was the limit; it was obvious from their physical properties
that nucleic acids were large. We regarded the presence of nucleic acids as a
not unexpected fact rather than as a matter of philosophical import.

Why the majority of biochemists favoured the tetranucleotide hypothesis
and when they began to reject it is an historical problem deserving examination which cannot be undertaken here. But what attracts attention in an
ostensibly 'hard' biochemical paper is the scrutiny and repudiation of the
notion regarding the essential incompatibility between the crystalline and
living state.

Even though backed up by empirical evidence and the authority of
Hopkins (who communicated the paper), the appearance of a discussion
touching on the problem of the *nature* of life in an 'ordinary' scientific
communication published in the *Proceedings of the Royal Society* was
unusual. In a wider context the problem was most stimulatingly explored by
Pirie himself in a contribution from which the excerpts below are taken.
Pirie examined in turn the various qualities commonly associated with 'life'.
They included: adaptation, irritability, motion, metabolism, size and cellular

structure, chemical composition, growth and reproduction. He found 'that they are individually inadequate for even approximate definition' (p. 21).

Against their will so to speak, the biochemists were pushed by contemporary developments in virus research into concerning themselves with the seemingly 'metaphysical' question of drawing a sharp line between the non-living and living state. What they brought to light was that this was not possible. Indeed, it was the perception of this kind of dialectical relation that led them, or some of them at least, to examine the value of dialectical materialism in dealing with problems arising out of concrete scientific research.[1]

N.W. Pirie in *Perspectives in Biochemistry*, eds. J. Needham and D.E. Green (Cambridge, 1937), p. 11

'The meaninglessness of the terms life and living'

pp. 12–14

'Life' and 'living' are clearly words that the scientist has borrowed from the plain man. The loan has worked satisfactorily until comparatively recently, for the scientist seldom cared and certainly never knew just what he meant by these words, nor for that matter did the plain man. Now, however, systems are being discovered and studied which are neither obviously living nor obviously dead, and it is necessary to define these words or else give up using them and coin others. When one is asked whether a filter-passing virus is living or dead the only sensible answer is: 'I don't know; we know a number of things it will do and a number of things it won't and if some commission will define the word "living" I will try to see how the virus fits into the definition.' This answer does not as a rule satisfy the questioner, who generally has strong but unformulated opinions about what he means by the words living and dead. The virus worker may then be told, to take one example out of many equally foolish, that a certain virus cannot be living because its physical properties enable us to investigate it in some detail. It is for these reasons that it has seemed worth while to consider whether the words 'life' and 'living' can be defined at all. Rivers, discussing the same problem, has said: 'It is wise to leave the subject at this point, as further pursuit of it leads one into the sterile discussion of what life is, a problem still in the realm of metaphysics.'[a] A discussion is surely not sterile if it leads to an increase in the precision of language by deciding how life may be recognized, even if it should fail, metaphysically, to decide what it is. We have an aesthetic appreciation of the phenomena of life, and it cannot be stated dogmatically that there is here no underlying physical reality. This aesthetic appreciation is clearly what the original users of these words had in mind when they distinguished cats and rabbits from water and stones, but our experience is now wider, and more intermediate materials are known.

If it should be decided that there is no definable line between 'living' and 'non-living' matter the terms will still be useful for most purposes, in the same way that the words green and yellow are useful. For, although it is obvious that so far as our unaided vision is concerned there is a smooth transition between these two colours, so little of our colour experience falls in the region 'greenish-yellow' that it is convenient to name regions of the spectrum as we do and to imagine that they touch along legally definable lines. Similarly, in chemistry the terms 'organic' and 'inorganic' will serve as labels for the great majority of known compounds. It is possible that developments in certain aspects of chemistry, e.g. the chemistry of the co-ordination compounds, will lead to such a development of the border land that a large amount of chemical experience will have to be described without using these terms, but in chemistry as we now know it they are still useful. The terms have been preserved, although the modern definition of them is very different from that of Berzelius. As a third instance we may consider the words 'acid' and 'alkaline'. Here, as is often the case, increase in knowledge has enabled us to resolve the apparent antithesis and describe all the observed phenomena in terms of one quantity – hydrogen-ion concentration. In most regions of biology the old nomenclature is useless, for the words 'acid' and 'alkaline' do not have the proper connotations for describing fluids at, for example, pH 6 or 8. It is contended that there is a similar lack of meaning in attempts to describe many aspects of biology with the aid of the terms 'living' and 'non-living'. The words still have a very definite meaning when used by poets, knackers or soldiers, but little or none when used to describe the phenomena observed in tissue culture, virus research and kinetic studies on interrelated enzyme systems.

We may now discuss possible definitions of 'life' or, failing that, describe how to recognize systems which are best described as 'living'. The latter process, though logically less satisfactory, is fully as useful practically. The problem is, properly speaking, one for the philosophers of biology, but they, with an unusual unanimity, have neglected it or have produced such vague definitions that they amount merely to a pious hope that the reader, like the author, can recognize a living organism when he sees one. For the most part they have simply pointed out that, where measurements have been made, obviously living and obviously dead systems obey the same kind of laws; matter and energy are conserved, and there is no need to postulate entelechies or other agencies which are not a part of the stock in trade of some non-biological science.

pp. 20–1

As our knowledge of bacteriology increases it becomes possible to grow more and more bacteria in media that have been derived from 'living' materials but which are themselves generally looked on as 'non-living'. Few bacteria now remain to be cultivated in this way. Many protozoa and fungi and most viruses have, however, not been grown in the absence of 'living' cells; for this reason some writers have maintained that the viruses should not be called 'alive'. It is hard to see why the same terminological limitation is not imposed on the malaria parasite or the rust fungi. Our efforts to cultivate viruses *in vitro* are of such recent data that any positive statements about their necessary cultural requirements are absurdly premature. Even if it were proved that the presence of living cells is necessary for the multiplication of a virus there are many analogies with this state of affairs among the higher animals. A cow is not classified as 'not-living' because its existence in an environment in which grass is the principal foodstuff is dependent on the cellulase in its intestinal flora.

The case of a virus which has as yet only been shown to multiply inside a cell is more difficult and may in some ways resemble the multiplication of genes in the same environment. It has already been pointed out that the observed phenomena could be explained by looking on a virus as a crystal focus; this hypothesis, though not very plausible, is not necessarily false. Hardy long maintained the view that molecules inside a cell are in some way arranged or oriented, and the very frequent occurrence of birefringent structures in nature lends colour to this view, as do the experiments of Vlès & Gex on the difference between the ultraviolet absorption spectra of intact and cytolysed cells.[b] On this hypothesis the environments found by a virus inside and outside of a cell are fundamentally different, and it may be, to put it in a ludicrously simple form, that a small virus finds a protein molecule too large to be used unless it has already been oriented by the host cell. Many other hypotheses may be made to explain the difficulties found in the cultivation of viruses without assuming that their reproductive methods differ essentially from those of the pathogenic bacteria.

pp. 21–2

In the earlier part of this essay the transition from living to non-living was compared to the transition from green to yellow or from acid to alkaline. If this comparison were valid, it would be possible to lay down a precise but arbitrary dividing line. But as it has been shown that 'life' cannot be defined in terms of one variable as colours can, the comparison is not strictly valid and any arbitrary division would have to be made on a basis of the sum of a number of variables any one of which might be zero. The complexity of a definition like this is the main objection to it. Until a valid definition has been framed it seems prudent to avoid the use of the word 'life' in any discussion about border-line systems and to refrain from saying that certain observations on a system have proved that it is or is not 'alive'.

Selection 180

We referred already to J. Needham's concern with problems of cellular organization when he came to the support of Peters at the Faraday Society discussion on the structure of living matter.[a] As a pupil of Hopkins, Needham had learned from his teacher that the cell was a complex heterogenous system, endowed with a 'chemical geography', in which reactions between simple molecules were catalyzed by colloidal enzymes. Needham's deeper interest in cellular organization was aroused by listening to Peters's lectures on 'physicochemical cytology' which in turn owed so much to W.B. Hardy's stress on the need for obtaining more solid information on the colloidal state.[b] As a result of these influences Needham matured into a dedicated practitioner of physical biochemistry, especially in the fields of embryology and morphogenesis, which he made his own. For him the problem of biological organization was fundamental, and he believed it was within the possibilities of experimental science to take up the age-old questions and answer them meaningfully, if not exhaustively. It was for this reason that he opposed vitalism because he thought it inhibitory to further advance.

This, then, is the message of Needham's *Order and Life*, a book published in 1936 on the basis of his Terry Lectures at Yale University, delivered there the year before. Six years after Peters's first formulation, Needham discussed and developed the conception of co-ordinative biochemistry by which he meant 'the extension of morphology into biochemistry and the bridging of the gulf between the so-called sciences of matter and the so-called sciences of form'.[c] As one reads the account one sees more distinctly that the young Cambridge biochemist, in the light of knowledge existing in the mid-1930s, restated much more clearly ideas which Hofmeister had earlier proclaimed.

Needham reviewed the evidence for the thesis that morphological patterns were chemically conditioned and found it in diverse results and fields, such as changes in respiration upon cytolysis, relations between cellular structure and oxidation mechanism, the co-ordination of enzyme systems in muscular work and elsewhere. Above all, he showed a remarkable grasp of new knowledge of properties of large molecules derived from centrifugation, flow birefringence and X-ray studies of complex substances of biological origin. Possibilities were opening at last to give some substance to the idea previously expressed long before by investigators like Schwann, that morphogenesis was a process akin to crystallization. It became apparent that for the understanding of physiological processes such as muscle contraction knowledge of molecular organization was no less important than that of molecular reactions. This had important repercussions for the tackling of the fundamental problem of protoplasmic organization.

Many histologists and embryologists had tended to accept, since the last quarter of the previous century, one or the other 'histological' descriptions of protoplasmic structure such as granular, filar, fibrous, reticular, alveolar, foam-like and so on. It was believed that whatever the pattern, the structure constituted an elastic but fairly rigid framework. These views persisted well into the third and fourth decade of this century as shown, for instance, in the term 'cytosquelette' introduced by Wintrebert (1931).[d] Needham thought that it could prove useful to describe a markedly different situation where molecular events within the cell were responsible for morphological, dynamic patterns and brought in the concept of 'cyto-skeleton' or 'cell-skeleton' to cover it, as also the 'co-ordinative' model of Peters. The cell-skeleton was believed to be a fluid lattice with large molecules playing an important part in protoplasmic organization.[e] Did it mean that the once rejected 'giant' molecules were, in fact, being resurrected?

Although the impact of the giant molecule concept upon biological thought cannot be totally discounted, the history of the study of large molecules in the twentieth century can be traced rather to Nägeli's hypothesis, stated in the 1840s, that starch grains and cellular fibres are composed of sub-microscopic elongated crystalline particles, which he called 'micelles'. Some time ago J.S. Wilkie undertook a detailed study of Nägeli's work and concluded that Nägeli's views were progressively rejected in the period from 1882 to 1920.[f]

From the early 1920s, however, the study of large 'polymeric' molecules such as cellulose, starch, rubber, and proteins, began to receive a strong impulse from industry. There were not a few factories in existence for the manufacture of materials which were, in effect, only modifications of natural high polymers. Due to the endeavours of T. Svedberg, W.N. Haworth, H. Staudinger, W.H. Carothers, K.H. Meyer, H. Mark and others the study of large molecules advanced fairly rapidly.[g] An important role in the bringing together of biological and technical studies was the work on textile fibres by W.T. Astbury, J.B. Speakman and others. It is interesting to read in Astbury's book, *Fundamentals of Fibre Structure* (1933), what he had to say about Nägeli's work:[h]

The moral of this romance of research, and of many similar stories in the history of science, is again that of the profound unity underlying all natural phenomena. We need not suppose for a moment that Nägeli, as he pored over his plant cells and starch grains ever gave a thought to the welfare of the textile industry, but the fact remains that his problems were in essence also our problems, and therefore the gifts of his genius were as much for industry as for biology.

It was Astbury's work which led Needham to the aphorism that 'biology is largely the study of fibres'.[i] The notion that proteins were crystalline fibres provided a basis for picturing the cell-skeleton as a web-like structure, composed of chain-like molecules which was continuously built up and broken down. There could be no doubt that the concept of the cell-skeleton signified a substantial swing from the position of Hopkins, the great advocate of small molecules, who had little use for large molecules in cell-life. Needham recognized the change when he wrote:[j]

In general it may be said that in biochemistry the interest has in recent years been slowly shifting from chemical reactions as such to the chemical, indeed, 'organic', structure of the great molecules. How these great molecules come to be there, how they are formed and how replaced, brings us always back to the realm of metabolism. But for the unification of biochemistry and morphology, which every reflective biologist has at heart, this shift of interest is an excellent thing.

It speaks both for the teacher and the pupil that these sentences appeared in Needham's contribution to the *Hopkins-Festschrift* of which he was also one of the editors.

To comprehend the place of the concept of the cyto-skeleton in the history of ideas on the chemical organization of living matter, it is important to see that it was formulated by a biochemist concerned, above all, with the problem of organization:[k]

The enzymes involved in metabolism may be isolated and studied in relatively simple systems, analyses may be made of the substances entering and leaving the living body, even the blood and tissue fluids may be examined in relation to every conceivable bodily activity, change or disease – but all this avoids the main problem of biology, the origin, nature, and maintenance of specific organic structure. The building of a bridge between biochemical and morphological concepts is perhaps the most important task before biologists at the present time, and it may well be long before it is satisfactorily accomplished.

In order to do justice to the problems with which, at bottom, Needham was wrestling we have to view his scientific pursuits in a wider social context. In effect, since his cathartic experience of the General Strike in 1926, Needham has been concerned, in a nutshell, with disentangling and connecting what he had come to accept as the 'great forms of human experience; science, philosophy, religion, history and art', and with comprehending them as a whole.[1] This involved him in the study of Marxism and in coupling it with ideas which he had gleaned from writings on evolutionary naturalism, emergent evolutionism, holism, temporal realism and organic mechanism.[m] As a result, by the early 1940s, Needham was convinced that for the understanding of natural as well as social phenomena, including their relations, it was essential to regard nature and society as consisting of a series of dialectically connected levels of organization:[n]

We cannot consider nature otherwise than a series of levels of organisation, a series of dialectical syntheses. From ultimate physical particle to atom, from atom to molecule, from molecule to colloidal aggregate, from aggregate to living cell, from cell to organ, from organ to body, from animal body to social organisation, the series of organisational levels is complete. Nothing but energy (as we now call matter and motion) and the levels of organisation (or the stabilised dialectical syntheses) at different levels have been required for the building of our world. The consequences of this point of view are boundless. Social evolution is continuous with biological evolution, and the higher stages of social organisation, embodied in advanced ethics and in socialism, are not a pious hope based on optimistic ideas about human nature, but a necessary consequence of all foregoing evolution. We are in the midst of the dialectical process, which is not likely to stop at the bidding of those who sit, like Canute, with their feet in the water forbidding the flood of the tide.

From this point of view the cyto-skeleton embodied one of the transitional levels of protoplasmic organization the study of which could help to illumine the relation between morphology and biochemistry. All the same, plausible and attractive direct evidence for the existence of the cyto-skeleton was hard to come by. Needham admitted this in his Terry Lectures and also later in *Biochemistry and Morphogenesis* (1942), which included an extended version of the material and argument presented in *Order and Life*.

We now see that *Biochemistry and Morphogenesis* was practically Needham's last major contribution to biochemistry. By the time it was published his mind was made up to produce a work on the history of science and technology of China of which the first volume appeared in 1954. Here it is out of place to say anything, even in most general terms, about the immense significance of *Science and Civilisation in China*. What could be of benefit, however, to the reader of this *Documentary History* is to note that Needham's abiding interest in reductionism in biology played a part in his decision to occupy himself with Chinese thought, for his understanding of reductionism eventually made him aware of the traditionally anti-mechanical and organic philosophical features of China's culture. It is not an exaggeration to say that there were few biochemists, if any, during the 1930s who surpassed Needham in promoting and formulating the fact that the central problem of biology was the problem of organization (form).

In the enterprise with which he was concerned Needham was well aware that the point at issue was not only experimental but also philosophical. This emerges clearly from the 'General introduction' to *Biochemistry and Morphogenesis* from which the following selection is taken. Touching as it does on conceptual as well as disciplinary issues the selection provides perhaps the fitting ending to the last chapter of the present volume.

J. Needham, *Biochemistry and Morphogenesis* (Cambridge, 1942)

'General introduction'

p. xiv

The relation of biochemistry to morphology is not a banal matter. It has been implicit in countless discussions of theoretical biology and involves themes which have run right through the history of philosophy from the earliest speculations of Greek and Chinese thinkers through the sixteenth, seventeenth and eighteenth centuries, down to our own time. Here I wish only to state my position with regard to two of these, the 'irreducibility' of the biological, and the question of form and matter.

Morphologists, turning over the pages of the present book, will probably be inclined to remark that the author finds only 'a mass of substances' in the living and developing organism. Yet that is precisely what the organism is, containing in itself innumerable molecules of small size some of which are indeed the morphogenetic inductors, as well as the giant protein molecules which, in responding by new orientations, form the basis of the cellular and organic architecture. Nevertheless the biochemist approaching morphological problems does not forget that all these molecules are ordered and organised in a totally different way to anything which occurs outside living organisms. His difficulties arise rather from the vagueness and lack of quantitative precision common to so much morphological work and perhaps inseparable from it. But everything that can be done to remove this vagueness is being done, and after all, we have already a great many extremely well-defined histological and morphological factors which can be correlated with biochemical ones. Some may feel that the terminology of inductors, competences, etc., is too 'cut and dried', but the history of science abundantly shows that in every age, whether of the 'word-rectification' school in ancient China,[a] the Arabs, Albertus Magnus,[b] the early Royal Society, or the time of Haller and Boerhaave,[c] the invention and clarification of terms is one of the greatest limiting factors in scientific advance. A clear formulation, even though partial, is better than a vague one; and we must not, in avoiding rigidity, fall into obscurantism. Any who might be led to suppose that words such as 'inductor', 'evocator', etc., carry animistic undertones, would be under a complete misapprehension. They stand for definite chemical substances the nature of which is as yet not fully elucidated. They conceal no *archaeus*[a] or demon. In the use of the phrase 'oxidising agent' the chemist risks no confusion with 'passenger agent', and 'hydrogen-donators' head no subscription-lists. In this book, the word 'organiser' is always used to imply the presence of tissue, whether living or dead; 'evocator' refers only to a specific chemical substance; and 'inductor' is a looser word available for both these senses.[e]

pp. xiv–xvi

In the previous discussion the words 'organised' and 'ordered' are the keynotes. We cannot but consider the universe a series of levels of organisation and complexity, ranging from the sub-atomic level, through the atom, the molecule, the colloidal particle, the living nucleus and cell, to the organ and the organism, the psychological and sociological entity. It follows that the laws or regularities which we find at one level cannot be expected to appear at lower levels. The conditions for their appearance do not exist there. It is no use fishing in homogeneous solutions for the laws which hold good for the behaviour of liquid crystals, still less for those which hold for that of living organisms.[1]

Philosophers have often talked about the reducibility or irreducibility of biological facts to physico-chemical facts. These old controversies are unnecessary if we realise that we are dealing with a series of levels of organisation. We must seek to elucidate the regularities which occur at each of these levels without attempting either to force the higher or coarser processes into the framework of the lower or finer processes, or conversely to explain the lower by the higher. From this point of view, the regularities discovered by experimental morphology will always have their validity and will, in a sense, be unaffected by anything which either biochemistry on the one hand, or psychology on the other, may discover. The behaviour, for example, of an embryonic eye-cup isolated into saline solution, its capacities for self-differentiation, fusion with another eye-cup, lens-induction, regulation, etc., etc. will always remain the same however much our knowledge of biochemistry or biophysics may advance. This is the reason why prediction is possible at a level of organisation which, strictly speaking, we do not yet understand at all, for instance, genetics. But the important point is that although the regularities established at the level of experimental morphology are irrefragable, they will, in the absence of biochemical experimentation, remain for ever meaningless. Meaning can only be introduced into our knowledge of the external universe by the simultaneous prosecution of research at all the levels of complexity and organisation, for only in this way can we hope to understand how one is connected with the others.

This brings us to the ancient distinction between *forma* and *materia*, mainly due to the analytical mind of Aristotle – μορφή and εἶδοσ against ὕλη. This distinction has had an influence incalculably great on the historical development of biology.[2] Morphologists for many centuries past have devoted themselves to the study of living form without much consideration of the matter with which it is indissolubly connected. It is not surprising that the numerous devils of vitalism found a congenial abode in the mansions of empty *forma* thus suitably swept and garnished. Moreover, the morphological tradition[3] was to think of matter much too simply, ignoring the vast complexity of chemical structures and the unbroken line of sizes reaching from the hydrogen atom at one end (to say nothing of sub-atomic levels) to the virus molecule at the other. Only in the light of the conception of integrative levels can the saecular gulf between morphology and chemistry be bridged.

It is true that Aristotle held that there could be form without matter, though no matter without form. But according to him, the only entities which possessed form without matter, were the divine prime mover, the intelligent demiurges that moved the spheres, and perhaps the ψυχῆσ διανοητική (the rational soul) of man. Some of these are factors in which experimental science has never been very much interested. On the other hand, he maintained that there could be no matter without form, for however pure the matter was (even the chaotic primal menstrual matter which was the raw material of the embryo), it was always composed of the elements, that is to say, it was always hot, cold, dry or wet, and hence had a minimum of form. In its primitive way this mirrors the standpoint of modern science. Form is not the perquisite of the morphologist. It exists as the essential characteristic of the whole realm of organic chemistry, and cannot be excluded either from 'inorganic' chemistry or nuclear physics. But at that level it blends without distinction into order as such, and hence we should do well to give up all the old arguments about form and matter, replacing them by two factors more in accordance with what we now know of the universe, that is to say, Organisation and Energy. From this point of view there can be no sharp distinction between morphology and biochemistry, and we may have every hope that in the future we shall be able to see not only what laws the form of living organisms exhibits at its own level, but also how these laws are related to those which appear at lower and higher levels of organisation.[f]

Notes to Section VII

Selection 157

Introduction:

a. This contains a basic outline of the formulation of the vitalist–mechanist issue. As pointed out by Hilde Hein in a thoughtful survey of the subject: 'Neither mechanism or vitalism has remained static in its conceptualization of the nature of life.' See H. Hein, 'The endurance of the mechanism–vitalism controversy', *J.Hist.Biol.*, 5, 159–88. 1972 (p. 165). How the dispute affected the thought and work of individual scientists will be reflected in particular selections of this section.

b. H. Simmer, 'Zur Entwicklung der physiologischen Chemie', *CIBA Z.*, 8, 3013–44. 1958.

c. F.C. Medicus, *Von der Lebenskraft* (Man[n]heim, 1774). Medicus wrote on medical and botanical subjects and, among other positions, presided over the Palatinate Economic Society.

d. There is as yet no comprehensive treatment of Stahl which bears on his chemical *and* medical doctrines (phlogiston, *anima*) and their interrelationships. There is a perceptive study by Lester S. King which gives particular attention to Stahl's medical teachings, see *Dictionary of Scientific Biography* (New York, 1975), XII, 599–606. See also Teich, 'Circulation', Selection 1, Section I, note f.

e. For a biographical sketch of F. Hoffmann, the first Professor of Medicine at the University of Halle, see the article by Guenter B. Risse, *Dictionary of Scientific Biography* (New York, 1972), VI, 458–61. For a comparison of Stahl and Hoffmann consult L. S. King, 'Stahl and Hoffmann: a study in eighteenth century animism', *J.Hist.Med.*, 19, 118–30. 1964.

f. For information including further material on the man who held simultaneously the chairs of medicine, botany and chemistry at Leiden between 1718 and 1729 see G.A. Lindenboom, *Herman Boerhaave* (London, 1968).

g. For his many-sided scientific, literary and public activities, Haller occupies a prominent place in the history of the Swiss Enlightenment. His achievements are sketched by E. Hintzsche, *Dictionary of Scientific Biography* (New York, 1972), VI, 61–7.

Text:

1.   Grens Chemie, pt II §1395. p. 275 (second ed)
2.   I believe that we would give the least opportunity for misinterpretation if, instead of force we used the word *property of matter*. We would then in physiology first look at the general phenomena which organic matter has in common with inanimate nature, then we would consider those of its properties which are indeed specific to it, but belong to the whole animate kingdom of nature, then we would pass over to the special phenomena of the vegetable and animal matter. Regarding the animal matter we would consider its own kind of irritability and its modifications, nerve irritability, muscle irritability, gland irritability, etc., according to the differences in the composition of the matter in the special kinds of organ.
3.   The words *physical, chemical force*, etc. appear already to show that we have not always associated with them the correct concepts. Everything in the corporeal world acts physically: also the living organic matter and all forces can be traced in the end altogether to the differences in elements and to their single universal property, affinity.
4.   I have named the force of matter which characterizes the plant and animal kingdom the *vital force* and taken the word vital in its widest meaning. Perhaps others find the word *organic force* more suitable. I have not chosen it, however, because organization in ordinary usage denotes the formation of animate beings. But words are arbitrary signs of our concepts and it depends only how precisely we define the concept which we associate with a certain word.
5.   Schmidt, Empirische Psychologie Jena 1791, p. 413.
6.   I can therefore not assent to the definition of the vital force which Humboldt gives (in his Aphorismen aus der chemischen Physiologie der Pflanzen, Leipzig 1794. pp. 1–9). That inner force, he says, which loosens the bonds of chemical affinity and hinders the free combination of the elements in the body, we call vital force.
a.   Jacques de Vaucanson (1709–81), a French engineer interested in the construction of automatic mechanisms. Known for his construction of a mechanical loom.
b.   Farkas Wolfgang Kempelen (1734–1804), an Hungarian inventor of automatic devices, including a mechanical chess-player.
c.   Reference to the idea which conceived of acids and bases as compounds of the first order, and of salts and compounds of the second order. On these lines alum, for instance, represented a compound of the third order. Cf. J.R. Partington, *A History of Chemistry* (London, 1964), IV, p. 172.
d.   *Zellgewebe, tela cellulosa* – term introduced by Haller and held to comprise the loose connective tissue, but also the basis for a wide-ranging of tissues and organs. See Hintzsche, *Dictionary of Scientific Biography*, VI, p. 62.
e.   Cf. 'The association of breathing with life and its cessation with death had come down from the earliest historical times and beyond, and it is quite understandable that the functions of the body and all its organs and tissues should be thought of in terms of the movement and nature of a number of imperceptible breaths. This pneumatic proto-physiology and proto-biochemistry was common to all the civilizations of the Old World.' See 'Introduction' by J. Needham to *The Chemistry of Life. Eight Lectures on the History of Biochemistry* (Cambridge, 1970), p. ix, which he edited.
f.   In the original: *Hier ist nur das Ganze einer Maschine.*
g.   'Accident' – a category deriving from Aristotelian thought and referring to incidental but individuating properties of inanimate and animate matter such as size, temperature, colour, chemical combination.
h.   'Caloric' – a special sort of matter conceived to be the substance of heat. In 1787 Guyton de Morveau (1737–1816) spoke of 'matter of heat' as *calorique* in order to differentiate between the material cause and the sensation of heat to which it gave rise. See de Morveau, Lavoisier, Bertholet [*sic*], de Fourcroy, *Méthode de nomenclature chimique* (Paris, 1787), pp. 30–1.

Selection 158

Introduction:

a.   Brunnmark's translation was corrected by Thomas Young (1773–1829), renowned for his theory of colour vision. It has been pointed out that Young published a detailed extract from the translated text under his own name, *An Introduction to Medical Literature Including a System of Practical Nosology*, 1st–2nd edn (London, 1813, 1823). See N.G. Coley, 'The Animal Chemistry Club; assistant society to the Royal Society', *Not.Rec.R.S.*, 22, 173–185. 1967 (pp. 179, 184).

b. J.J. Berzelius, *Lehrbuch der Chemie*, 5th edn (Dresden and Leipzig, 1847), IV, pp. 5–7.

c. See Selection 9, Section I.

d. B.S. Jørgensen, 'More on Berzelius and the vital force', *J.Chem.Ed.*, 42, 394–6. 1965; idem, 'Berzelius und die Lebenskraft', *Centaurus*, 10, 258–81. 1964–5.

e. T.O. Lipman, 'More on Berzelius and the vital force', *J.Chem.Ed.*, 42, 396–7. 1965. There is virtually nothing on the subject in two relatively recent books in English on Berzelius: J.E. Jorpes, *Jac. Berzelius His Life and Work* (Uppsala, 1970); E.M. Melhado, *Jacob Berzelius. The Emergence of His Chemical System* (Uppsala, 1981).

Selection 159

Introduction:

a. F. Tiedemann and L. Gmelin, *Die Verdauung nach Versuchen* (Heidelberg and Leipzig, 1826), I, p. 3.

b. The blue colour produced by the action of iodine on starch was noted in Gay-Lussac's laboratory by [J.J] Colin and H.[F.] Gaultière de Claubry in 1814. Cf. 'Mémoire sur les combinaisons de l'iode avec les substances végétales et animales', *Ann.Chim.*, 90, 87–100. 1814.

c. G.S.C. Kirchhoff, 'Ueber die Zuckerbildung bei Malzen des Getreides und beym Bebrühen seines Mehls mit kochendem Wasser', *Schweiggers J.*, 14, 389–98. 1815. See also Selection 74, Section IV.

d. Tiedemann and Gmelin's work has been examined by N. Mani, 'Das Werk von Friedrich Tiedemann und Leopold Gmelin: 'Die Verdauung nach Versuchen', und seine Bedeutung für die Entwicklung der Ernährungslehre in der ersten Hälfte des 19. Jahrhunderts', *Gesnerus*, 13, 190–214. 1956 and F.L. Holmes, *Claude Bernard and Animal Chemistry* (Cambridge, Mass., 1974), pp. 149–59.

e. Cf. Holmes, *Bernard*, p. 150.

Text:

a. Named 'dextrin' by J.B. Biot and J.F. Persoz in 1853. See their 'Mémoire sur les modifications que la fécule et la gomme subissent sous l'influence des acides', *Ann.Chim.*, 52, 72–90. 1833.

b. Tiedemann and Gmelin speak of *speichelstoff-und osmazomartige Materie. Speichelstoff* was the name given by them to an ill-defined material constituent of saliva, see *Verdauung*, I, pp. 4–25. In 1840 Berzelius introduced the term 'ptyalin' for a similar though not identical body found in human saliva. In 1815 Thenard assigned the word *osmazôme* to an alcohol-soluble constituent of aqueous meat extracts. It was believed that the smell and taste of meat was derived from this substance. Later the term was applied to all alcohol-soluble nitrogeneous plant and animal extracts. Cf. Berzelius, *Lehrbuch*, IX, and Thenard, *Traité*, III; also Selection 10, Section I, notes d and c.

Selection 160

Introduction:

a. W. Prout, 'On the nature of the acid and saline matters usually existing in the stomachs of animals', *Phil.Trans.*, (I), 45–9. 1824.

b. F. Tiedemann and L. Gmelin, *Die Verdauung nach Versuchen* (Heidelberg und Leipzig, 1826) I, pp. 150–2.

c. A valuable guide to Prout is W.H. Brock's 'The life and work of William Prout', *Med.Hist.*, 9, 101–26. 1965. There is also his later contribution to the *Dictionary of Scientific Biography* (New York, 1975), XI, 172–4; also *From Protyle to Proton. William Prout and the Nature of Matter* (Bristol, 1985) (not seen).

d. The term was introduced in 1830 by J.J. Berzelius. See his 'Ueber die Zusammensetzung der Weinsäure und Traubensäure ... nebst allgemeinen Bemerkungen über solche

Körper, die gleiche Zusammensetzung, aber ungleiche Eigenschaften besitzen', *Ann.Phys.*, *19(95)*, 305–35. 1830.

Text:

1. I am indebted to my friend Mr Lunn[d] for this term.
2. When this subject first occupied my attention many years ago, I was at a loss to form any notion of the *modus operandi* of these minute admixtures of foreign bodies, except the mechanical one mentioned in the text, viz. that they operated by being interposed, as it were, among the essential elements of bodies, and thus by weakening or modifying their natural affinities. But the admirable Paper, published by Mr Herschel,[e] in the Philosophical Transactions for 1824, 'On certain motions produced in fluid conductors when transmitting the electric current', appeared to throw an entire new light on the subject. The facts brought forward in this Paper are of the most important kind, and seem to be to be evidently connected with a principle of a more general character, which when completely developed, will lead to the most unexpected results. 'That such minute proportions of extraneous matter,' says Mr H. 'should be found capable of communicating sensible mechanical motions and properties of a definite character to the body they are mixed with, is perhaps one of the most extraordinary facts that has yet appeared in chemistry. When we see energies so intense exerted by the ordinary forms of matter, we may reasonably ask what evidence we have for the imponderability of any of the powerful agents to which so large a part of the activity of material bodies seem to belong?'

   Any substance may be supposed capable of performing the part of a merorganizing body; but, in a certain point of view, *water* appears to constitute the *first* and *chief*, at least in organized substances.
a. Perceived as the chemistry of substances of plant or animal origin and, in this sense, perhaps first employed by J.J. Berzelius (1806). Cf. J.R. Partington, *A History of Chemistry* (London, 1964), IV, p. 233. See Introduction to Selection 165.
b. Egg white.
c. Uric acid.
d. According to Brock, Lunn could be the person who engraved for Prout.
e. John Herschel (1792–1871) – astronomer.

Selection 161

Introduction:

a. It is the reprint of the English translation from H.M. Leicester and H.S. Klickstein's *A Source Book in Chemistry 1400–1900* (New York, 1952; Cambridge, Mass., 1963), pp. 309–12. Contrary to some references in literature, 'Artificial formation of urea', in *The Quarterly Journal of Science, Literature and Art* (January–June), 491–2. 1828, is an abridged version ostensibly based on the full French translation from *Annales de chimie et de physique*, 37, 331–4. 1828, and not on the German original.
b. W.H. Warren, 'Contemporary reception of Wöhler's discovery of the synthesis of urea', *J.Chem.Ed.*, 5, 1539–57. 1928.
c. D. McKie, 'Wöhler's 'synthetic' urea and the rejection of vitalism: a chemical legend', *Nature*, 153, 608–10. 1944.
d. H. Kolbe, 'Beiträge zur Kenntniss der gepaarten Verbindungen', *Ann.*, 54, 145–88. 1845. Here, apparently, Hermann Kolbe (1818–84) employed the term 'synthesis' for the artificial preparation of organic compounds from its elements for the first time (p. 186).
e. Charles Gerhardt (1816–56). See Introduction to Selection 108, Section V.
f. Among those who referred approvingly to McKie's article were the historians of science, F.S. Bodenheimer, in *History of Biology* (London, 1958), p. 67, and G.S. Goodfield in her discussion of the mechanist–vitalist controversy as illustrated by problems of respiration and animal heat in *The Growth of Scientific Physiology* (London, 1960) p. 126. McKie's interpretation has been rejected by E. Campaigne in a short note reasserting that there is 'no chemical myth that Wöhler with his synthesis of urea in 1828 sounded the death-knell of vitalism'. See E. Campaigne, 'Wöhler and the overthrow of vitalism', *J.Chem.Ed.*, 32, 403. 1955. For a brief critique of McKie's as well as Campaigne's conclusions see L. Hartman, 'Wöhler and the vital force', *J. Chem.Ed.*, 34, 141–2. 1957.

McKie's opinion has been critically examined by M. Teich, 'On the historical foundations of modern biochemistry', *Clio Medica*, *1*, 41–57. 1965, on which these introductory comments draw. The paper – originally presented in 1961 – has been reprinted (with slight changes) in J. Needham (ed.), *The Chemistry of Life. Eight Lectures on the History of Biochemistry* (Cambridge, 1970), pp. 171–91. The view expressed here coincided to some extent with the independent evaluation of Wöhler's discovery offered in T.O. Lipman's 'Wöhler's preparation of urea and the fate of vitalism', *J.Chem.Ed.*, *41*, 452–8. 1964, and in J.H. Brooke's careful and painstaking study 'Wöhler's urea, and its vital force? – A verdict from the chemists', *Ambix*, *15*, 84–114. 1968. Lipman argued that vitalism 'arose because science was engaged in an attempt to explain organized bodies, not organic subtances. The distinction between organic bodies and organized bodies was subtle but real' (p. 457). Related to this are Brooke's observations regarding the uncertainty of what was meant by the terms 'body', 'organic', and 'organized' (pp. 103–4). In this connection it is not without interest that when Wöhler suggested to Liebig (13 October 1863) that he should write an article for the wider public on the synthesis of plant and animal substances from their elements, he particularly stressed the need to make clear 'the basic distinction between and organic and an organized body'. Wöhler's proposition came as a result of his dissatisfaction with the review and presentation of the developments in organic chemistry in M. Berthelot's (1827–1907) *Chimie organique fondée sur la synthèse* (1860), 'giving the impression that before him [i.e. Berthelot] there has been no scientific organic chemistry, that nothing has been known regarding the artificial formation of organic compounds from their elements'. Cf. A.W. Hofmann (ed.), *Aus Justus Liebigs und Friedrich Wöhler's Briefwechsel in den Jahren 1829–1873* (Braunschweig, 1888), II, pp. 145–6.

g. O. Wallach (ed.), *Briefwechsel zwischen J. Berzelius und F. Wöhler* (Leipzig, 1901), I, p. 206.

h. In a letter to Liebig (12 November 1863), Wöhler stressed his own lack of aptitude for philosophical thinking and mathematics. But Wöhler believed that he possessed the ability to observe – coupled with a kind of instinct to anticipate – real relations. Finding *Naturphilosophie* and Hegel's philosophy divorced from reality, he compared their effect on science to that of the mediaeval schoolmen. Cf. Hofmann, *Liebigs und Wöhlers Briefwechsel*, pp. 149–51.

i. Wallach, *Berzelius und Wöhler*, p. 208.

j. Wallach, *Berzelius und Wöhler*, p. 208.

k. J.J. Berzelius, *Lehrbuch der Chemie* (Dresden, 1831), IV (pt 1), p. 356.

l. J. Müller, *Handbuch der Physiologie des Menschen für Vorlesungen* (Coblenz, 1833), I (pt 1), p. 2.

m. Müller, *Handbuch*, p. 3.

n. Cf. Introduction to Selection 112, Section V.

Text:

1. Annals of Philosophy T. XI. p. 354.

2. The new atomic weights of Berzelius are used as a basis; accordingly N = 88.518, C = 76.437, H = 6.2398, O = 100.000, water ($\underline{\text{H}}$) = 112.479, cyanate of ammonia = $\underline{\text{NH}}^3$ + $\underline{\text{CNO}}$ and urea = $\underline{\text{NH}}^3$ + $\underline{\text{CNO}}$ + H.[e]

a. F. Wöhler, 'Ueber Cyan-verbindungen', *Ann.Phys.*, *3* (*79*), 177–82. 1825.

b. [J.L.] Proust, 'Faire pour la connaissance des urines et des calculs', *Ann.Chim.*, *14*, 257–88. 1820.

c. W. Prout, 'Observations on the nature of some of the proximate principles of the urine; with a few remarks on the means of preventing those diseases connected with a morbid state of that fluid', *Med.Chir.Trans.*, *8*, 526–49. 1817. An abridged version was published in *Ann.Phil.*, *11*, 352–6. 1818.

d. Reference to the discovery by M. Faraday (1791–1867) of a 'new carburet of hydrogen' (butylene) 'composed of carbon and hydrogen in the same proportion as in olefiant gas' (ethylene) but having 'double the density'. See Faraday, 'On new compounds of carbon and hydrogen, and on certain other products obtained during the decomposition of oil by heat', *Phil. Trans.*, pt 2, 440–66. 1825 (pp. 459–60).

e. For the significance of 'barred' symbols, see Selection 103, Section V, note f.

Selection 162

Introduction:

a. There is a brief reference to the paper by M. Florkin, *A History of Biochemistry* (Amsterdam, 1972), pts I, II, p. 119.
b. See Selection 74, Section IV.
c. See Selection 102, Section V.
d. In discussing oxidation and desoxidation of organic compounds Gmelin also considered their mechanism. Here, interestingly, he thought that the study of colour changes exhibited by dyes such as indigo or litmus *might* furnish a key. Among the first to take up the subject was the well-known Scottish chemist, Thomas Thomson (1773–1852), in 1820. He associated the formation of the blue colour in the indigo vat with absorption of oxygen making indigo an alkaline substance, and the loss of blue colour with deprivation of oxygen imparting to it acidity. 'Thus,' he wrote, 'indigo exhibits a striking refutation of the old notion that acidity is owing to the union of oxygen with an acidifiable basis.' See 'Indigo', *Ann.Phil.*, *15*, 465–8. 1820. Relating to this question, Gmelin's observations deserve to be introduced at this point. He thought there was uncertainty regarding the chemical side of the colour changes:

> For example, did desoxidants withdraw oxygen or add hydrogen; did oxidants remove carbon, or hydrogen, or both, or add oxygen; did the process repeat itself infinite times or was it accompanied each time by a decomposition of a certain quantity of the dye . . . (pp. 183–4).

Selection 163

Introduction:

a. Incorporated by J. Müller in his *Handbuch der Physiologie des Menschen für Vorlesungen* (Coblenz, 1837), II (pt 1), pp. 59–60.
b. See Selection 10, Section I.
c. Cf. Introduction to Selection 12, Section I.
d. Selection 12, Section I.
e. Th. Schwann, 'Ueber die Analogie in der Struktur und dem Wachsthume der Thiere und Pflanzen', *Neue Notizen aus dem Gebiete der Natur-und Heilkunde* (L.F. Froriep and R. Froriep), *5*, 33–6; 225–9. 1838; *6*, 21–3. 1838. Actually the articles were extracts from a letter to the well-known German physiologist, E.H. Weber (1795–1878).
f. The extracts below are from the English version, *Microscopical Researches into the Accordance in the Structure and Growth of Animals and Plants*, transl. H. Smith (London, 1847).
g. For a factual history of the cell theory, see J.R. Baker, 'The cell-theory: a restatement, history and critique', *Quart.J.Microscop.Sci.*, [3] *89*, 103–25. 1948; *90*, 87–108, 33. 1949; *93*, 157–90. 1952; *94*, 407–40. 1953; *96* 449–81. 1955. Cf. also A. Hughes, *A History of Cytology* (London and New York, 1959).
h. Jan Evangelista Purkyně (Purkinje) (1787–1869) – Czech physiologist. For his life and work consult H.J. John, *Jan Evangelista Purkyně Czech Scientist Patriot* (Philadelphia, 1959); V. Kruta, M. Teich, *Jan Evangelista Purkyně* (Prague, 1962).
i. See review of Schwann's book by J.E. Purkyně, *Opera omnia* (Prague, 1951), V, pp. 178–83, originally published in *Jahrb.wiss.Kritik*, 32–8, (2) 1840. See also Introduction to Selection 167.
j. Cf. Schwann's postscript defending his priority, *Microscopical Researches*, pp. 219–30.
k. Schwann, *Microscopical Researches*, pp. 190–1 (see below). the reference to 'elementary atoms', 'atoms of the second order', 'conglomerate molecules', has to be seen against the background of uncertainty regarding the concepts of equivalent, atom and molecule before the Congress at Karlsruhe in 1860. See also Introduction to Selection 108, Section V and Selection 157, note c.
l. M.J. Schleiden, 'Beiträge zur Phytogenesis', *Müllers Arch.*, 137–76. 1838.
m. Cf. M. Florkin, *Naissance et déviation de la théorie cellulaire dans l'oeuvre de Theodore Schwann* (Paris, 1960), 191f. The life and work of Schwann is examined by R. Watermann, *Theodor Schwann Leben und Werk* (Düsseldorf, 1960). This introduction draws on my article 'From "enchyme" to "cyto-skeleton" ', see Foreword, note i.

Text:

1. The word 'reciprocal action' must here be taken in its widest sense, as implying the preparation of material by one elementary part, which another requires for its own nutrition.

2. I could not avoid bringing forward fermentation as an example, because it is the best known illustration of the operation of the cells, and the simplest representation of the process which is repeated in each cell of the living body. Those who do not as yet admit the theory of fermentation set forth by Cagniard-Latour,[d] and myself, may take the development of any simple cells, especially of the spores, as an example; and we will in the text draw no conclusion from fermentation which cannot be proved from the development of other simple cells which grow independently, particularly the spores of the inferior plants. We have very conceivable proof that the fermentation-granules are fungi. Their form is that of fungi; in structure they, like them, consist of cells, many of which enclose other young cells. They grow, like fungi, by the shooting forth of new cells at their extremities; they propagate like them, partly by the separation of distinct cells, and partly by the generation of new cells within those already present, and the bursting of the parent-cells. Now, that these fungi are the cause of fermentation, follows, first, from the constancy of their occurrence during the process; secondly, from the cessation of fermentation under any influences by which they are known to be destroyed, especially boiling heat, arseniate of potass, etc.; and, thirdly, because the principle which excites the process of fermentation must be a substance which is again generated and increased by the process itself, a phenomenon which is met with only in living organisms. Neither do I see how any further proof can possibly be obtained otherwise than by chemical analysis, unless it can be proved that the carbonic acid and alcohol are formed only at the surface of the fungi. I have made a number of attempts to prove this, but they have not as yet completely answered the purpose. A long test-tube was filled with a weak solution of sugar, coloured of a delicate blue with litmus, and a very small quantity of yeast was added to it, so that fermentation might not begin until several hours afterwards, and the fungi, having thus previously settled at the bottom, the fluid might become clear. When the carbonic acid (which remained in solution) commenced to be formed, the reddening of the blue fluid actually began at the bottom of the tube. If at the beginning a rod were put into the tube, so that the fungi might settle upon it also, the reddening began both at the bottom, and upon the rod. This proves, at least, than an undissolved substance which is heavier than water gives rise to fermentation; and the experiment was next repeated on a small scale under the microscope, to see whether the reddening really proceeded from the fungi, but the colour was too pale to be distinguished, and when the fluid was coloured more deeply no fermentation ensued; meanwhile, it is probable that a reagent upon carbonic acid may be found which will serve for microscopic observation, and not interrupt fermentation. The foregoing inquiry into the process by which organized bodies are formed, may perhaps, however, serve in some measure to recommend this theory of fermentation to the attention of chemists.[e]

a. The development of the term is treated in F.C. Bing, 'The history of the word "metabolism"', *J.Hist.Med.*, *26*, 158–80. 1971.
b. 100°C.
c. Between 12.5° and 40°C.
d. See Selection 11, Section I.
e. For the dating of these experiments (August 1838), see M. Florkin, *A History of Biochemistry* (Amsterdam, 1972), pts I, II, pp. 331–2.

Selection 164

Introduction:

a. See Selection 53, Section III.
b. Cf. Selection 14, Section I.
c. F.L. Holmes, 'The *milieu intérieur* and the cell theory', *Bull.Hist.Med.*, *37*, 315–35. 1963 (p. 322).
d. Selection 162.

e.   Cf. Selection 1, Section I; Selection 31, Section II; Selection 137, Section VI.
f.   Jean Baptiste Boussingault (1802–87) pioneered investigations on the source of nitrogen in plants, and laid the groundwork for 'nitrogen balance' studies in animal nutrition. See Introduction to Selection 118, Section V, note b. A biography in book form is by F.W.J. McCosh, *Boussingault Chemist and Agriculturist* (Dordrecht and Boston, 1984).
g.   According to Dumas, the lecture was given on 20 August 1841 and immediately reproduced in many journals under the title *Essai de statique chimique des êtres organisés*, and as a separate booklet (constituting its second edition) in 1842. This draws on the article with the same name published by Dumas in *Annales de chimie et de physique*, [3] *4*, 115–26. 1842 in which he rejected Liebig's accusations of having appropriated ideas from the latter's *Organic Chemistry in its Applications to Agriculture and Physiology* (1840). Cf. also F.L. Holmes's 'Introduction' to the facsimile of the Cambridge (Mass.) edition of Liebig's *Animal Chemistry in its Application to Physiology and Pathology* (New York and London, 1964), pp. xxviii–xxxi.

Text:

a.   Nitrogen.
b.   Starchy matters.
c.   Dextrins. Cf. Selection 159, note a.

Selection 165

Introduction:

a.   For a brief but still valuable account of these issues see C. Schorlemmer, *The Rise and Development of Organic Chemistry*, rev. edn (London, 1894), chap. II. For a more recent treatment consult A.I. Ihde, *The Development of Modern Chemistry* (New York, Evanston, and London, 1964), chap. 7. See also S.C. Kapoor, 'Dumas and organic classification', *Ambix*, *16*, 1–65. 1969.
b.   J. Carrière (ed.), *Berzelius und Liebig. Ihre Briefe von 1831–1845* (Munich and Leipzig, 1893), pp. 210–11.
c.   See Introduction to Selection 108, Section V. For a perceptive discussion of problems connected with the integration of the two branches of chemistry see J.H. Brooke's 'Organic synthesis and the unification of chemistry – a reappraisal', *Brit.J.Hist.Sci.*, *5*, 363–92. 1970–1.
d.   J. Liebig, *Organic Chemistry in its Applications to Agriculture and Physiology* (London, 1840), pp. vi–vii.
e.   Readers are referred to the facsimile of the Cambridge (Mass.) English edition (1842) reprinted in 1964. The 'Introduction' by F.L. Holmes, which has repeatedly been mentioned in this volume, constitutes an indispensable guide to the work.
f.   Vance Hall suggests that Liebig's work, *Chemische Untersuchungen über das Fleisch* (Heidelberg, 1847), published in English as *Researches on the Chemistry of Food* (London, 1847), was intended as the third part of the report. Cf. V.M.D. Hall, 'The role of force or power in Liebig's physiological chemistry', *Med.Hist.*, *24*, 20–59. 1980 (p. 24, note 16). However plausible, this suggestion does not seem to be supported by what Liebig wrote in the preface of the work:

      The present little work contains the analytical details of my investigations on these subjects which in accordance with the plan [of a new edition] of the Animal Chemistry, could not be introduced into that work.

g.   Selection 9, Section I.
h.   Cf. 'But for the correct comprehension of the chemical phenomena occurring in the living body knowledge of its structure, that is of its anatomy, is indispensable'. J.J. Berzelius, *Lehrbuch der Chemie*, 4th edn (Dresden and Leipzig, 1840), 9, p. 3. See also J. Berzelius, 'Thierchemie', *Jahres-Ber.*, *22*, 535–6. 1842.
i.   J. Berzelius, 'Thierchemie', *Jahres-Ber.*, *23*, 575–83. 1843.
j.   Selection 158.
k.   The subject has been examined by O. Sonntag, 'Religion and science in the thought of Liebig', *Ambix*, *24*, 159–69. 1977.

l.   This is based on Chapters I and II of *Animal Chemistry*. For other treatments bearing on Liebig's ideas of vital force and vitality, with further bibliographic references, see T.O. Lipman, 'Vitalism and reductionism in Liebig's physiological thought', *Isis*, *58*, 167–85. 1967, and Hall, 'Liebig's physiological chemistry', *Med.Hist.*, *24*, 20–59. 1980.

Selection 166

Introduction:

a.   J. Liebig, 'Ueber die Gährung und die Quelle der Muskelkraft', *Ann.*, *153*, 1–47, 137–228. 1870; L. Pasteur, 'Note sur un memoir de M. Liebig, relatif aux fermentations', *Comp.Rend.*, *73*, 1419–24. 1871.

b.   Selection 16, Section I.

c.   L. Pasteur, 'Mémoir sur les corpuscles organisés qui existent dans l'atmosphère, examen de la doctrine des generations spontanées', *Ann.Sci.Nat.(Zool.)*, [4] *16*, 5–98. 1861.

d.   Selection 12, Section I.

e.   Cf., for example, 'Mais le travail de Schwann renferme des observations précieuses qui jettent beaucoup de jour sur l'origine des fermentations spontanées et qui permittent d'interpreter autrement que ne l'a fait Gay-Lussac les expériences d'Appert'. See L. Pasteur, 'Mémoires sur la fermentation alcoolique', *Ann.Chim.*, [3] *58*, 323–426. 1860. (p. 370). See also quotation from Pasteur's letter to Schwann on 15 June 1878 in W. Bulloch, *The History of Bacteriology*, repr. (London, 1960), p. 96.

f.   Cf. 'Toutes les fois qu'on le peut faire, il est utile de montrer la liaison des faits nouveaux avec les faits antérieurs de même ordre. Rien de plus satisfaisant pour l'esprit que de pouvoir suivre une découverte dès son origine jusqu'à ses derniers développements'. Pasteur, 'Fermentation alcoolique', p. 371.

g.   Cf. 'Pasteur's attempts in this direction are useless; for those who believe in this possibility he will never be able to prove their impossibility by these experiments alone, but they are important because they furnish much enlightenment on these organisms, their life, their germs, etc.' F. Engels, *Dialectics of Nature* (London, 1940), p. 189.

h.   The terms 'biogenesis' and 'abiogenesis' appear to originate with T.H. Huxley who employed them in his 'Address', when he was President to the British Association for hte Advancement of Science at Liverpool in 1870. Cf. 'And thus the hypothesis that living matter always arises by the agency of preexisting living matter . . . I shall call it the hypothesis of *Biogenesis*; and I shall term the contrary doctrine – that living matter may be produced by not living matter – the hypothesis of *Abiogenesis*', *Rep.Brit.Ass.*, lxxv–lxxvi. 1870.

i.   See J. Farley and G.L. Geison, 'Science, politics and spontaneous generation in nineteenth-century France: the Pasteur-Pouchet debate', *Bull.Hist.Med.*, *48*, 161–98. 1974 (pp. 172–9). The authors have attempted to deal with the political and religious contexts of Pasteur's endeavours in this area. For a critique of this effort, see N. Roll-Hansen, 'Experimental method and spontaneous generation: the controversy between Pasteur and Pouchet, 1859–64', *J.Hist.Med.*, *34*, 273–92. 1979.

j.   Selection 164. Also in this connection, one cannot help thinking of the scheme worked out by Stahl around 1720 which involved the circulation of phlogiston connecting the mineral, vegetable and animal realms. Stahl considered the function of the atmosphere (naturally devoid of life) as so significant in this process that he came to think of it as a separate, fourth, natural kingdom. The subject is dealt with by Teich in 'Circulation', see Introduction to Selection 1, Section I, note f.

Text:

a.   The translation draws in part on the incomplete and not too precise English version of the article that appeared under the title 'The function of atmospheric oxygen in the destruction of animal and vegetal substances after death', *Chem.News*, *7*, 280–282. 1863.

b.   This clearly is a reference to Liebig's chemical explanation of fermentation, see Selection 14, Section I. The term 'nitrogenous plastic substances' may also be traced to Liebig's division of the food of man into two classes, 'the nitrogenized and the non-nitrogenized'. He called the former the 'plastic elements of nutrition' and the latter 'elements of respiration'. Cf. J. Liebig, *Animal Chemistry or Organic Chemistry in its Application to Physiology and Pathology* (Cambridge, Mass., 1842), p. 92.

c. Pasteur's taxonomy was apparently imprecise. This was noted by his contemporaries and explained by his lack of real competency in the microscopy of living things. Cf. Bulloch, *Bacteriology*, p. 187.

d. On 'proximate analysis', see Selection 3, Section I, note d.

Selection 167

Introduction:

a. J.E. Purkyně, 'Ueber die Analogien in den Struktur -Elementen des thierischen und pflanzlichen Organismus', *Opera omnia* (Prague, 1937), II, pp. 90–1. Breslau is now in Poland and known under its Polish name, Wrocław.

b. F.K. Studnička, 'Noch einiges über das Wort Protoplasma', *Protoplasma*, 27, 619–25. 1936–7.

c. [G.L.L. de] Buffon, *Histoire naturelle, générale et particulière*, 2nd edn (Paris, 1750), II, pp. 30–40.

d. G.R. Treviranus, *Biologie oder Philosophie der lebenden Natur für Naturforscher und Aerzte* (Göttingen, 1803), II, pp. 311, 403–4.

e. C.F. Wolff, *Theoria generationis* (Hallae ad Salam, 1759).

f. Studnička, *Protoplasma*, 621.

g. M. Teich, 'Purkyně and Valentin on ciliary motion: an early investigation in morphological physiology', *Brit.J.Hist.Sci.*, 5, 168–77. 1970.

h. See Selection 10, Section I.

i. J.E. Purkyně, 'Über den Bau der Magen – Drüsen und über die Natur des Verdaungsprocesses', *Opera omnia*, (Prague, 1939), III, pp. 43–5.

j. H.v. Mohl, 'Ueber die Saftbewegung im Innern der Zellen', *Bot.Ztg.*, 4, 73–8; 89–94. 1846 (pp. 75–6).

k. K.B. Reichert, 'Bericht über dei Fortschritte der mikroskopischen Anatomie in den Jahren 1839 und 1840', *Müllers Arch.*, clxii–ccxxvii. 1841 (p. clxiii); Studnička, *Protoplasma*; 621–2.

l. F. Dujardin, 'Recherches sur les organismes inférieurs', *Ann.Sci.Nat.(Zool.)*, [2], 4, 343–77. 1835.

m. The term protoplasm was reintroduced into zoology by Robert Remak (1815–65) when he was engaged in his fundamental studies of egg cleavage. In 1852 he equated the yolk to the protoplasm of the egg cell. Three years later, after discussing the varying meanings of 'yolk' in the literature, and its role in holoblastic and meroblastic cleavage, he made it clear that it was necessary to differentiate between protoplasm or *zooplasm* as living matter and yolk as food material. See R. Remak, 'Ueber extracellulare Entstehung thierischer Zellen und über die Vermehrung derselben durch Theilung', *Müllers Arch.*, 47–57. 1852 (p. 50); idem, *Untersuchungen über die Entwicklung der Wirbelthiere* (Berlin, 1855), pp. 81–2.

n. M. Schultze, 'Ueber Muskelkörperchen und das, was man eine Zelle zu nennen habe', *Müllers Arch*, 1–27, 1861.

o. E. Brücke, 'Die Elementarorganismen', *Sitzsber (Wien)*, 4 (2. Abth), 381–406. 1861.

p. See Introduction to Selection 52, Section III.

q. T.H. Huxley, 'The cell-theory', *Brit.For.Med.Chir.Rev.*, 12, 285–314. 1853. The reader is reminded that *Archaeus* – a personalized spiritual entity conceived by Paracelsus (1493–1543) – was thought to govern non-living as well as living phenomena. *Bildungstrieb* or *nisus formativus* was an immaterial morphogenetic agency postulated by the German comparative anatomist and physiologist J.F. Blumenbach (1752–1840). In contrast to these apparently non-immanent powers, C.F. Wolff seemed to have identified in *vis essentialis* a built-in force accounting for morphogenetic movements in living forms.

r. M. Florkin, *A History of Biochemistry* (Amsterdam, 1975), pt III, p. 5.

s. Cf. T.H. Huxley, 'Address', *Rep.Brit.Ass.*, lxxxiii–lxxxiv. 1870.

and if it were given to me to look beyond the abyss of geologically recorded time to the still more remote period when the earth was passing through physical and chemical conditions, which it can no more see again than a man can recall his infancy, I should expect to be a witness of the evolution of living protoplasm from not living matter . . .

t. As to Charles Darwin (1809–82), even on the basis of a few scattered remarks in the published correspondence it is clear that he was aware that his own work on the origin of species had a direct bearing on the subject of the origin of life, but that it was then (1863) as hard to deal with it as to answer the question conerning the origin of matter itself. Cf. F. Darwin (ed.), *More Letters of Charles Darwin* (London, 1903), I, p. 419; idem, *The Life and Letters of Charles Darwin* (London, 1887), III, p. 18. In the early 1870s Darwin contemplated the possibility of life having arisen or even arising from non-living matter. While, so to speak, hoping against hope for an experimental test, he understood that it could not be carried out easily. See *Life and Letters*, III, pp. 168–9. Apart from the problem of reproducing the historical conditions, that is, 'some warm little pond, with all sorts of ammonia and phosphoric salts, light, heat, electricity, &c.', which Darwin visualized as possibly having given rise to the first living matter on the Earth, he also assumed that 'such matter would be instantly devoured or absorbed'. See *Life and Letters*, III, p. 18. Beyond this scientific concern, there was Darwin's perennial worry of how to come to terms theologically and philosophically with chance and necessity. Thus on 12 July 1870 he wrote to his botanist friend, J.D. Hooker (1817–1911): 'My theology is a simple muddle . . . Spontaneous generation seems almost as great a puzzle as preordination.' See *More Letters*, I, p. 321.

u. L. Huxley (ed.), *Life and Letters of Thomas Henry Huxley*, (London, 1903), I, p. 431.

v. Huxley, *Life and Letters*, I, p. 440.

Text:

1. The substance of this paper was contained in a discourse which was delivered in Edinburgh on the evening of Sunday, the 8th of November, 1868 – being the first of a series of Sunday evening addresses upon non-theological topics, instituted by the Rev. J. Cranbrook. Some phrases, which could possess only a transitory and local interest, have been omitted; instead of the newspaper report of the Archbishop of York's address, his Grace's subsequently-published pamphlet 'On the Limits of Philosophical Inquiry', is quoted; and I have, here and there, endeavoured to express my meaning more fully and clearly than I seem to have done in speaking – if I may judge by sundry criticisms upon what I am supposed to have said, which have appeared. But in substance, and, so far as my recollection serves, in form, what is here written corresponds with what was there said.

Selection 168

Introduction:

a. Selection 53, Section III. See also Pflüger's addendum, 'Ueber die physiologische Verbrennung in den lebendigen Organismen', *Pflügers Arch.*, *10*, 641–4. 1875.

b. Selection 52, Section III.

c. A.I. Oparin, *The Origin of Life on Earth*, 3rd edn (London, 1957), pp. 82–4; J.D. Bernal, *The Origin of Life*, (London, 1967), p. 225.

Text:

1. See August Kekulé, Organische Chemie, 1867. Vol. I. p. 309.

Selection 169

Introduction:

a. Cf. Selection 77, Section IV and Selection 139, Section VI.

b. F.L. Holmes, *Claude Bernard and Animal Chemistry* (Cambridge, Mass., 1974), p. xv.

c. The quotations and excerpts are from the English version *Phenomena of Life Common to Animals and Vegetables*, transl. by R.P. and M.A. Cook, (Dundee, 1974).

d. Bernard, *Phenomena of Life*, p. vii.

e.   M.F.X. Bichat (1771–1802), author of the influential *Anatomie générale* (1801). Working without a microscope, he identified twenty-one tissues as the anatomical elements, fundamental to the normal activity in the human body. For a review of the history of general physiology see J. Schiller, *Physiology and Classification* (Paris, 1980), chap. XIII.

f.   Bernard, *Phenomena of Life*, p. 152. Cf. also Selection 77, Section IV.

g.   Bernard, *Phenomena of Life*, p. 3.

h.   F.L. Holmes, 'Claude Bernard and the milieu intérieur', *Arch.Int.Hist.Sci.*, *16*, 369–76. 1963 (p. 369). Surprisingly not much has been written about the subject. In addition to this short but valuable review, including information about the historical background, see the longer paper by the same author, 'The *milieu intérieur* and the cell theory', *Bull.Hist.Med.*, *37*, 315–35. 1963 and J. Schiller, *Claude Bernard et les Problèmes Scientifiques de Son Temps* (Paris, 1967), chap. XI.

i.   See Introduction to Selection 167, note c.

j.   Bernard, *Phenomena of Life*, p. 85.

k.   Selection 163.

Text:

a.   François Magendie (1783–1855) – Bernard's teacher who pioneered experimental physiology in France.

Selection 170

Introduction:

a.   Hoppe-Seyler was Professor of Physiological Chemistry (and Hygiene) at Strassburg (Strasbourg) from 1872. He had come from Tübingen where he had formally held the Chair of Applied Chemistry since 1861, although the university apparently possessed Germany's oldest independent institution (founded in 1845) for laboratory work in physiological chemistry.

b.   It has been stated that A.I. Khodnev (Kharkov University) published a 'textbook on biochemistry' in 1847. See the editorial 'Role of the USSR Academy of Sciences in the development of native biochemistry', *Biokhimiya*, *39*, 243–57. 1974 (p. 243, in Russian).

c.   Vinzenz Kletzinsky (1826–82). Originally trained as a doctor, he taught chemistry at the newly opened *Oberreal-und Gewerbeschule* at Wieden (Vienna) from 1855. He was a chemist with very broad interests in applied and theoretical chemistry. Thus he was the first in Vienna to lecture on Kekulé's structural concepts in organic chemistry. See the obituary by A.E. Haswell in *Ber.chem.Ges.*, *15*, 3310–15. 1882.

d.   E. Pflüger, 'Die Physiologie und ihre Zukunft', *Pflügers Arch.*, *15*, 361–65. 1877. See also his 'Wesen und Aufgaben der Physiologie', ibid., *18*, 427–42. 1878.

e.   Cf. E. Pflüger, 'Die teleologische Mechanik der lebendigen Natur', *Pflügers Arch.*, *15*, 57–103. 1877.

Text:

a.   E. Baumann (1846–1891) – Berlin; K. Gäthgens (1839–1915) – Rostock; E.(von) Gorup-Besanez (1817–78) – Erlangen; G. Hüfner (1840–1908) – Tübingen; K. Huppert (1832–1904) – Prague; R. Maly (1839–91) – Graz; E. Salkowski (1844–1923) – Berlin.

Selection 171

Introduction:

a.   Cf. Section Ib.

b.   Selection 171.

c.   Selection 56, Section III.

d.   Selection 52, Section III.

e.   M. Verworn, *General Physiology*, transl. 2nd German edn 1897 (London, 1899), pp. 80, 484.

f.   Selection 20, Section I.

g.   M. Verworn, *Die Biogenhypothese* (Jena, 1903), p. 15.
h.   Verworn, *Biogenhypothese*, p. 113.

Selection 172

Introduction:

a.   It was intended, but due to illness not given, as a lecture to the plenary session of the Annual Assembly of the Society of German Naturalists and Physicians at Leipzig in 1901. Apart from appearing in *Naturwissenschaftliche Rundschau*, the text was published separately as a booklet by Vieweg (Braunschweig, 1901). We have come across Franz Hofmeister (1850–1922) in connection with his experimental and theoretical work on proteins (Selections 112 and 115, Section V). Although a major figure in the history of biochemistry, his name is not included in the very helpful *Dictionary of Scientific Biography* (New York, 1970   ). In 1896 he left Prague, where he held the Chair of Pharmacology at the medical faculty of the German branch of the University, for Strassburg (Strasbourg) where he was to succeed Hoppe-Seyler as Professor of Physiological Chemistry until 1918. For an obituary see J. Pohl and K. Spiro, 'Franz Hofmeister sein Leben und Wirken', *Ergeb.Physiol.*, *22*, 1–31. 1923. There is a list of Hofmeister's and his pupils' publications on pp. 39–50.
b.   He envisaged that about 2×10^{13} water, protein, lipoid, crystalloid and salt molecules were involved in a space of $8,000 \mu^3$. This was the assumed volume of a liver cell on the basis that it was a cube the side of which measured $20 \mu.^c$ Cf. F. Hofmeister, 'Vom chemisch-morphologischen Grenzgebiet', *Z.Morph.Anthr.*, *18*, 717–24. 1914.
c.   1 micron $(\mu) = 10^{-4}$ cm.

Text:

a.   This refers to the idea of Otto Bütschli (1848–1920) regarding the 'honeycombed' or 'alveolate' structure of protoplasm. Cf. his *Untersuchungen über mikroskopische Schäume und das Protoplasma*, (Leipzig, 1892).
b.   A reference to a celebrated passage in the lecture by the Berlin physiologist, Emil du Bois-Reymond (1818–96), delivered at the 45th Assembly of the Society of German Naturalists and Physicians at Leipzig on 14 August 1872. It proclaimed that the essence of matter, force and human thought will remain an insoluble riddle forever. Cf. E. du Bois-Reymond, ed. Estelle du Bois-Reymond, *Reden*, (Leipzig, 1912), I, p. 473. Clearly, Hofmeister disagreed with this view.

Selection 173

Introduction:

a.   *Beiträge zur chemischen Physiologie und Pathologie. Zeitschrift für die gesamte Biochemie* (1902); *Biochemisches Centralblatt* (1902–3); *Journal of Biological Chemistry* (1905); *Biochemische Zeitschrift* (1906); *Biochemical Journal* (1906).
b.   By then textbooks appeared that took account of the division of the subject matter of biochemistry into static and dynamic, of which more will be said in a moment. Cf., for example, S. Fränkel, *Descriptive Biochemie* and *Dynamische Biochemie* (Wiesbaden, 1907, 1911).
c.   Here the authoritative series, *Monographs on Biochemistry*, edited by R.H.A. Plimmer and F.G. Hopkins and published by Longmans, Green and Co. has to be mentioned. Starting with W.M. Bayliss, *The Nature of Enzyme Action* (1908), each monograph contained the editorial *General Preface* which because of its historical interest is reproduced in full:

> The subject of Physiological Chemistry, or Biochemistry, is enlarging its borders to such an extent at the present time that no single text-book upon the subject, without being cumbrous, can adequately deal with it as a whole, so as to give both a general and a detailed account of its present position. It is, moreover, difficult, in the case of the larger text-books, to keep abreast of so rapidly growing a science by means of new editions, and such volumes are therefore issued when much of their contents has become obsolete.

For this reason, an attempt is being made to place this branch of science in a more accessible position by issuing a series of monographs upon the various chapters of the subject, each independent of and yet dependent upon the others, so that from time to time, as new material and the demand therefor necessitate, a new edition of each monograph can be issued without reissuing the whole series. In this way, both the expenses of publication and the expense to the purchaser will be diminished, and by a moderate outlay it will be possible to obtain a full account of any particular subject as nearly current as possible.

The editors of these monographs have kept two objects in view: firstly, that each author should be himself working at the subject with which he deals; and, secondly, that a *Bibliography*, as complete as possible, should be included, in order to avoid cross references, which are apt to be wrongly cited, and in order that each monograph may yield full and independent information of the work which has been done upon the subject.

d. American Society of Biological Chemists (1906); Biochemical Society (1911). In Britain there was an earlier, albeit unsuccessful, attempt to unify chemistry and physiology institutionally between 1808 and 1825. See N.G. Coley, 'The Animal Chemistry Club: assistant society to the Royal Society', *Not.Rec.R.S.*, *22*, 173–85. 1967.
e. Also spelled 'bio-chemistry'. In addition the term 'biological chemistry' came into use.
f. Albrecht Kossel (1853–1927). Received the Nobel Prize in Physiology or Medicine in 1910 for contributions to knowledge of the chemistry of the cell nucleus.

Text:

a. See Selection 159.
b. Presumably this meant physiology of the muscles and nerves.

Selection 174

Introduction:

a. M. Stephenson, 'Sir F.G. Hopkins' teaching and scientific influence' in J. Needham and E. Baldwin (eds), *Hopkins and Biochemistry* (Cambridge, 1949), p. 36. This commemorative volume contains Hopkins's 'Autobiography' and also a selection of his addresses.
b. Selection 81, Section IV.
c. Selection 120, Section V.

Text:

1. A product of metabolism can only be said to have a 'function' in a cell or in the body when, being the end-product of one reaction, it initiates or modifies reactions in another milieu.
2. See in this connection the very able exposition of the views developed by Zwaardemaker and others, by Botazzi in Winterstein's *Handbuch*, vol. i.
a. E.A. Schäfer, 'Address', *Rep.Brit.Ass.*, 3–36. 1912.
b. H. Poincaré (1854–1912) – distinguished French mathematician who also published widely known works on scientific methodology and the philosophy of science. Cf. *La science et l'hypothèse* (1906); *Science et méthode* (1908); *La valeur de la science* (1913).
c. The insistence on the inadequacy of a purely chemical approach to biology threads its way through all Hopkins's thinking and emerges as one of its hallmarks. See Hopkins's remarks on Krebs's discovery of the ornithine cycle, Introduction to Selection 129, Section V, note a.

Selection 175

Introduction:

a. J. and D. Needham, 'Sir Frederick, Gowland Hopkins. O.M. P.R.S.', *Brit.Med.Bull.*, *5*, 299–301. 1948 (p. 300). Reprinted, with additions, in J. Needham and E. Baldwin (eds), *Hopkins and Biochemistry*, (Cambridge, 1949), pp. 113–19 (p. 115).
b. 'Autobiography of Sir Frederick Gowland Hopkins', in *Hopkins and Biochemistry*, p. 20. The institutional rise of physiology in Cambridge is the subject of G.L. Geison's *Michael Foster and the Cambridge School of Physiology* (Princeton, 1978).

c. 'Autobiography', p. 19.
d. The events involving the complex relations between the world of science and contemporary life that led to the establishment of the Dunn Institute of Biochemistry were examined by R.E. Kohler, 'Walter Fletcher, F.G. Hopkins and the Dunn Institute of Biochemistry: a case study in the patronage of science', *Isis*, *69*, 331–55. 1979.
e. Professor Malcolm Dixon in the internally published volume, *Sir William Dunn Institute of Biochemistry 1924–74*, commemorating the Fiftieth Anniversary of the opening of the Institute (pp. 10–11). Information on laboratory equipment in the 1930s will be found in a short article by N.W. Pirie, 'The research environment', *Biologist*, *27*, 6–8. 1980.
f. For 'pnein' see Selection 59, Section III. Volumes of *Biochemical Journal* between 1911 and 1926 appear not to contain any reference to the other terms.
g. N.W. Pirie, 'Sir Frederick Gowland Hopkins (1861–1947)', in G. Semenza (ed.), *A History of Biochemistry. Selected Topics in the History of Biochemistry Personal Recollections. I.* (Amsterdam, 1983), p. 124. This is volume 35 of *Comprehensive Biochemistry*, edited by A. Neuberger and L.L. M. Van Deenen (formerly edited by M. Florkin and E.H. Stotz).

Text:

a. See Selection 170.
b. Émile Duclaux (1840–1904) – Professor at the Sorbonne who taught at the Institut Pasteur.
c. Gabriel Bertrand (1867–1962) – Duclaux' successor at the Sorbonne. He coined the term 'coferment'. See Introduction to Selection 22, Section I, note a.
d. Olof Hammarsten (1841–1932).
e. Hans (von) Euler-Chelpin (1873–1964) – Nobel Prize in Chemistry (1929).
f. Russel Henry Chittenden (1856–1943) – Professor at Yale University. He was known for his advocacy of low protein diet. See Selection 119, Section V, note c.
g. For Kühne's contribution – he was Professor of Physiology at Heidelberg – to enzyme and muscle chemistry, see Selection 17, Section I and Introduction to Selection 52, Section III.

Selection 176

Introduction:

a. On Kluyver, see A.F. Kamp, J.W.M. La Rivière and W. Verhoeven (eds), *Albert Jan Kluyver. His Life and Work* (Amsterdam, 1959). On the contribution of Dutch workers, including Kluyver, to the development of microbiology, see C.B. van Niel, ' "The Delft School" and the rise of general microbiology', *Bact.Revs.*, *13*, 161–74. 1949.
b. Selection 60, Section III.
c. Selection 63, Section III.
d. Selection 23, Section I.
e. Selection 24, Section I.
f. W.M. Clark, *The Determination of Hydrogen Ions* (Baltimore, 1923).
g. J. and D.M. Needham, 'The hydrogen-ion concentration and the oxidation-reduction potential of the cell-interior and cleavage: a micro-injection study', *Proc.Roy.Soc.*, (*B*), *98*, 259–86. 1925; 'The hydrogen-ion concentration and oxidation-reduction potential of the cell-interior before and after fertilisation: a micro-injection study on marine eggs', ibid., *99*, 173–99. 1925–6. In this later paper it is interesting how the authors attempted to visualize the morphology of the protoplasmic matrix in the light of its pH and rH. It will be remembered that while $pH = \log_{10}\dfrac{1}{[H^+]}$, $rH = \log_{10}\dfrac{1}{[H]}$:

The constancy of the pH values under all conditions investigated is exceedingly striking. Neither moderately severe pathological changes, nor ontogenetic changes in the individual, nor phylogenetic changes as between one group of animals and another lead to any difference in the result. For the rH the constancy for the individual is similarly seen, but small variations between the different families exist. One is left with a picture of protoplasm as a sort of wax, out of which a diversity of morphological patterns are modelled: the wax itself remaining unchanged until an extremity of pathogenic influence may cause it to melt.

h.   T. Thunberg, 'Besteht ein genetischer Zusammenhang zwischen dem eingeatmeten Sauerstoff und dem Sauerstoff der ausgeatmeten Kohlensäure?', *Naturwiss.*, *10*, 417–20. 1922. See also Thunberg's earlier paper 'Zur Kenntnis des intermediären Stoffwechsels und der dabei wirksamen Enzyme', *Skand.Arch.Physiol.*, *40*, 1–91. 1920.
i.   A.J. Kluyver and H.J.L. Donker, 'The catalytic transference of hydrogen as the basis of the chemistry of dissimilation processes', *Proc. (Amsterdam)*, *28*, 605–18. 1925; S. Kostytschew, 'Über die Nichtexistenz einiger Fermente', *Z.physiol.Chem.*, *154*, 262–75. 1926; A.J. Kluyver, 'Über die Nichtexistenz einiger Fermente', *Z.physiol.Chem.*, *158*, 111–12. 1926. See also Selection 28, Section I.

Selection 177

Introduction:

a.   Selection 166.
b.   Selection 167.
c.   Selection 168.
d.   For brief biographical information on Oparin, see *Great Soviet Encyclopedia*, 3rd edn (New York–London, 1978), *18*, 463–4.
e.   Cf. J.G. Crowther, *Science in Soviet Russia* (London, 1930), pp. 68–70; *Soviet Science* (London, 1936), p. 21.
f.   Haldane's life and work is discussed by N.W. Pirie in *Biog.Mem.F.R.S.*, *12*, 219–49. 1966 (with the assistance of Sewall Wright, Motoo Kimura, L.S. Penrose, M.S. Bartlett and F.N. David). Consult also the commemorative volume edited by K.R. Dronamraju, *Haldane and Modern Biology* (Baltimore, 1968). There is an informative though less strictly scientific biography by R. Clark, *J.B.S. The Life and Work of J.B.S. Haldane* (London, 1968).
g.   J.D. Bernal, *The Origin of Life* (London, 1967), pp. ix, 251.
h.   R.A. Gortner, jun. and W.A. Gortner (eds), *Outlines of Biochemistry*, 3rd edn (New York–London, 1949), p. xi.
i.   E.C.C. Baly, *A Course of Three Lectures in Photochemistry. The Rice Institute Pamphlet*, XII (1), 1925, pp. 65–104; *Photosynthesis* (London, 1940), chap. VI.
j.   F.W. Twort, 'An investigation on the nature of ultra-microscopic viruses', *Lancet*, (II), 1241–3. 1915; F. d'Herelle, 'Sur un microbe invisible antagoniste de bacilles dysentériques', *Compt.Rend.*, *165*, 373–5. 1917. See also his *The Bacteriophage and its Behaviour*, transl. by G.H. Smith (London, 1926).
k.   A.I. Oparin, *The Origin of Life* (New York, 1938); *The Origin of Life on the Earth*, 3rd revised and enlarged edn (Edinburgh and London, 1957); *Genesis and Evolutionary Development of Life* (New York and London, 1968).

Text (i):

1.   As everyone knows, spirits of hartshorn is a solution of ammonia in water.
2.   Carbohydrates are organic substances composed of carbon, hydrogen and oxygen, the hydrogen and oxygen being present in the same proportions as in water. The various sugars such as glucose, sucrose and fructose as well as starch and cellulose are typical examples of carbohydrates.

a.   Translation by Ann Synge of A.I. Oparin (1924), *Proiskhozhdenie zhizny*. Moscow Izd. Moskovskĭ Rabochĭi. Published in J.D. Bernal, *The Origin of Life* (London, 1967), pp. 199–244.

Text (ii):

b.   Selection 16, Section I.
c.   O. Warburg, K. Posener and E. Negelein, 'Über den Stoffwechsel der Carcinomzelle', *Biochem.Z.*, *152*, 309–44. 1924.
d.   See A. Pictet and H. Vogel, 'Synthèse du saccharose', *Helv.Chim.Acta*, *11*, 436–42. 1928. This work attracted widespread interest but could not be repeated. Cf. G. Barger, 'The Pictet Memorial Lecture Amé Pictet (1857–1937)', *J.Chem.Soc.*, 11113–25. 1937.

Selection 178

Introduction:

a. Selection 172.
b. Selection 174.
c. Section IIIb.
d. Selection 28, Section I. See also J.H. Quastel, 'The mechanism of bacterial action', *Trans.Farad.Soc.*, *26*, 853–61. 1930.
e. T. Thunberg, 'The hydrogen-activating enzymes of the cells', *Quart.Rev.Biol.*, *5*, 318–47. 1930 (p. 333).
f. Cf. W.B. Hardy, *Collected Scientific Papers* (Cambridge, 1936); W.D. Harkins, *The Physical Chemistry of Surface Films* (New York, 1952); I. Langmuir, *Surface Phenomena* published in *The Collected Works* (Oxford, 1961), 9; N.K. Adam, *The Physics and Chemistry of Surfaces* (Oxford, 1930). (The second and third editions appeared in 1938 and 1942 respectively.)
g. The Harben Lectures (1929) were reprinted in Sir R.A. Peters, *Biochemical Lesions and Lethal Synthesis* (Oxford, 1963), pp. 216–305. The quoted passage is on p. 221.
h. It is not without historical interest that Peters developed his critical remarks in the presence of Hopkins who acted as Chairman of the meeting. It would seem as if Hopkins responded to them in two public addresses which he gave shortly afterwards. They were the *33rd Robert Boyle Lecture* 'The problems of specificity in biochemical catalysis' (1931) and the *Second Purser Memorial Lecture* 'Some aspects of biochemistry. The organising capacities of specific catalysts' (1932). Both were reprinted in J. Needham and E. Baldwin (eds), *Hopkins and Biochemistry 1861–1947* (Cambridge, 1949), pp. 211–24; 225–41. See also Introduction to Selection 174.
i. See 'General discussion', *Trans.Farad.Soc.*, *26*, 815–17, 819. 1930.

Text:

1. See also Hardy, VI., *Coll.Symp.Mono.*, for emphasis of this point.

Selection 179

Introduction:

a. Ch. Chamberland, 'Sur un filtre donnant de l'eau physiologiquement pure', *Comp.Rend.*, *99*, 247–8. 1884.
b. A. Mayer, 'Ueber die Mosaikkrankheit des Tabaks', *Landwirt. Versuchs.-St*, *32*, 451–67. 1886. Translated by J. Johnson in *Phytopathological Classics 7* (published by the *American Phytopathological Society*, 1942; repr. 1968), pp. 11–24.
c. Dm. Ivanovsky, *O dvukh boleznyakh tabaka* (St Petersburg, 1892), p. 17.
d. *Phytopathological Classics 7*, pp. 27–30. The quotation is to be found on p. 30.
e. M.W. Beijerinck, 'Ueber ein Contagium vivum fluidum als Ursache der Fleckenkrankheit der Tabaksblätter', *Verhandelingen (Amsterdam)*, No. 5, 1898.
f. [F.] Loeffler and [P.] Frosch, 'Berichte der Kommission zur Erforschung der Maul-und Klauenseuche bei dem Institut für Infektionskrankheiten in Berlin', *Centralbl.f. Bakt. und Parasit.*, *23*, 371–91. 1898.
g. It would appear that the usage of 'viruses' without the adjectives 'filterable' or 'filter-passing' arose during the 1930s.
h. Introduction to Selection 177, note j.
i. W.M. Stanley, 'Isolation of a crystalline protein possessing the properties of tobacco-mosaic virus', *Science, 81*, 644–5. 1935.
j. F.C. Bawden and N.W. Pirie, 'The isolation and some properties of liquid crystalline substances from solanaceous plants infected with three strains of tobacco mosaic virus', *Proc.Roy.Soc. (B)*, *128*, 274–320. 1937.
k. N.W. Pirie, 'F.C. Bawden', *Biog.Mem.F.R.S.*, *19*, 19–63. 1973 (p. 41).
l. See Introduction to Selection 177.

Text:

a. T.M. Rivers, *Filterable Viruses* (London, 1928), p. 19.
b. F. Vlès and M. Gex, 'Recherches sur le spectre ultra-violet de l'oeuf d'Oursin (*Paracentrotus lividus Lk.*) et de ses constituants', *Arch.Phys.Biol.*, *6*, 255–86. 1928.

Selection 180

Introduction:

a. Selection 178.
b. W.B. Hardy, *Collected Scientific Papers* (Cambridge, 1936).
c. J. Needham, *Order and Life*, repr. edn (Cambridge, 1968), p. 138. The reprint contains an informative 'Author's foreword' where the circumstances in which the book was written and the developments that have occurred since are discussed briefly.
d. P. Wintrebert, 'La rotation immédiate de l'oeuf pondu et la rotation d'activation chez *Discoglossus pictus* Otth.', *Comp.Rend.Soc.Biol.*, *106*, 439–42. 1931 (p. 441).
e. Needham, *Order and Life*, pp. 149f.
f. J.S. Wilkie, 'Nägeli's work on the fine structure of living matter' – IIIa, *Ann.Sci.*, *16*, 209. 1960. Cf. Introduction to Selection 127, Section V, note b.
g. H. Mark, 'Polymers review and preview', *Impact of science on society*, *18*, 27–33. 1968. T. Svedberg (1884–1971) – Nobel Prize in Chemistry (1926) for work on disperse systems; W.N. Haworth (1883–1950) – Nobel Prize in Chemistry (1937) for investigations on carbohydrates and vitamin C; H. Staudinger (1881–1965) – Nobel Prize in Chemistry (1953) for discoveries in the field of macromolecular chemistry; W.H. Carothers (1896–1937) – in charge of research in organic chemistry at the du Pont Company; K.H. Meyer (1883–1952) – Professor at the University of Geneva; H.F. Mark (1895–●) – Professor and Director of Institute of Polymer Research at the Polytechnic Institute of New York (Brooklyn).
h. W.T. Astbury, *Fundamentals of Fibre Structure* (London, 1933), pp. 100–1. W.T. Astbury (1898–1961) – Professor of Biomolecular Structure at Leeds University; J.B. Speakman (1898–1969) – Professor of Textile Industries at Leeds University.
i. Needham, *Order and Life*, pp. 146, 152.
j. J. Needham, 'Chemical aspects of morphogenetic fields' in J. Needham and D.E. Green (eds), *Perspectives in Biochemistry* (Cambridge, 1937), p. 78.
k. J. Needham, 'Integrative levels; a revaluation of the idea of progress (Herbert Spencer Lecture at Oxford University, 1937)' in *Time: The refreshing river (Essays and Addresses, 1932–1942)* (London, 1943), p. 245.
l. Cf. J. Needham, 'Metamorphoses of scepticism. Introductory essay (1941)', *Time: The refreshing river*, p. 8.
m. In this respect the works which Needham refers to repeatedly as having influenced him were: R.W. Sellars, *Evolutionary Naturalism* (Chicago, 1922); C. Lloyd-Morgan, *Emergent Evolution* (London, 1923) and *Life, Mind and Spirit* (London, 1926); J.C. Smuts, *Holism and Evolution* (London, 1926); S. Alexander, *Space, Time and Deity* (London, 1920, 1927); A.N. Whitehead, *Science and the Modern World* (Cambridge, Mass., 1925).
n. Needham, 'Metamorphoses', in *Time: The refreshing river*, p. 15.

Text:

1. For a fuller discussion of integrative levels, see the present author's Herbert Spencer lecture at Oxford, *Integrative levels; a revaluation of the idea of Progress*. The general picture there drawn has relations with evolutionary naturalism, emergent evolutionism, and dialectical materialism.
2. For a brilliant summary of Aristotle's views on these subjects, with convenient references to the texts themselves see W.D. Ross' *Aristotle*.
3. Originating from the conception of becoming as the privation of one form and the donation of another.
a. A reference to the old Confucian *Chêng ming* doctrine (*c.* 500 BC), which insisted on the attachment of specific unvarying meanings to words, as for example the calling of a spade

a spade, no matter what powerful influences might wish to call it a shovel. (I owe this comment to Joseph Needham.)

b.   Albertus Magnus (*c.* 1193–1280) – prolific author of commentaries on Aristotle and other scientific topics.

c.   Cf. Introduction to Selection 157, notes f and g.

d.   Cf. Introduction to Selection 167, note q.

e.   Cf. 'Glossary', *Biochemistry and Morphogenesis*, pp. 681–9.

f.   For Needham's reappraisement of the book in the light of the changes that have come since it was published, see J. Needham, 'Organizer phenomena after four decades: a retrospect and prospect' in K.R. Dronamraju (ed.), *Haldane and Modern Biology* (Baltimore, 1968), pp. 277–98. (This was also printed at the beginning of the 1966 edition of *Biochemistry and Morphogenesis*.)

Afterword

As stated in the Foreword, the object of this volume has been to collect historical material regarding the rise and development of the science of the chemistry of life from about 1770 to about 1940. Though not all-inclusive, it should be representative enough to guide readers to an understanding of the growth and nature of the discipline known as 'modern biochemistry' during the period under review. Whether they have made themselves familiar with all the material in the volume or have only dipped into certain Selections, they may find these few concluding reflections useful.

The rise of modern biochemistry: c. 1770–c. 1880

Discussing the theme of the historical foundations of modern biochemistry more than a quarter century ago, I proposed that there were three main stages of development: 1. A gradual separation of the science of the chemistry of life from the still unified body of chemistry into organic chemistry (c. 1800–c. 1840); 2. The linking of organic chemistry with physiology to form physiological chemistry (c. 1840–c. 1880); 3. The transformation of physiological chemistry into biochemistry, a process which began around 1880.[a]

Considering the evidence set forth in this book, there are grounds enough for maintaining that during the period 1800–80 the science of the chemistry of life was put on a firm foundation and that the suggested stages are not merely historical abstractions. At the same time, looking at the matter today, it is clearer now than it was then that the rise of modern biochemistry, treated in the (a) subdivisions of the seven thematic Sections which deal with particular topics, can be better understood by going back in history to the last three decades of the eighteenth century. It was from the 1770s that the theoretical framework of the doctrine of phlogiston, within which the chemistry of plants, animals and minerals had been coherently presented, began to be affected by the movement which eventually brought about its replacement by the new theoretical system centred on oxygen. This was an event in the history of chemistry comparable to that of the establishment of the theory of gravitation in the history of astronomy. What is less appreciated are its links with interdependent and interacting efforts to illumine fermentation and respiration in terms of chemical relations, including developments arising out of the discovery of the process that later became known as photosynthesis. As we have seen, on the observational and experimental side a major role was played by men such as Priestley and Lavoisier, even though, theoretically, they stood on opposite, phlogistic and antiphlogistic, sides.

Lavoisier's theory of matter and the chemistry of life

Concerning the chemistry of life it may be worthwhile here to dwell briefly on a very rarely noted passage which occurs in Lavoisier's *Elements of*

Chemistry (1789). In view of Lavoisier's premature death and the all-embracing, ordered manner in which this publication expounded the facets of chemistry in terms of the new oxidation theory, it represents the high point in Lavoisier's activities as a chemist: not least because it contains the formulation of the principle of conservation of matter, the key to the understanding of all chemical transformation in the inanimate as well as the animate world. It was largely due to the comprehensiveness of this treatise that chemistry entered the nineteenth century as a unified science, equally applicable to the study of chemical properties and to the relationships between substances of earthly, plant and animal origin.

Reference was made to Lavoisier's list of five simple substances 'belonging to all the kingdoms of nature, which may be considered as the elements of bodies': light, caloric, oxygen, azote (nitrogen) and hydrogen. It was also stated that this might be taken as an example of Lavoisier's not having broken with the past completely.[b] Lavoisier attempted, to account for the material diversity in nature by tracing it to the interaction of a fixed number of simple substances or elements. What is of particular interest to us here is that his list is headed by light and caloric, ostensibly designed to occupy the roles played by aether and phlogiston (fire) in the preceding conceptions of the compositions of the inorganic and organic bodies.

Lavoisier was aware of the suppositional features of his theory of matter, associated with the concept of caloric and especially of light. At any rate, he imagined that the gaseous state of substances resulted from their union with a sufficient amount of caloric which was presumed to surround and permeate them constantly. But however conjectural this approach to the gaseous state was, Lavoisier admitted that even less was known when it came to considering the nature of the action of light on substances which included compounds formed by their combination. Nevertheless, Lavoisier put light at the head of his list of elements. Why? In order to appreciate this approach, one can not do better than go back to Lavoisier and read what he had to say in this context:

Experiments upon vegetation give reason to believe that light combines with certain parts of vegetables, and that the green of their leaves, and the various colours of their flowers, is chiefly owing to this combination. This much is certain, that plants which grow in darkness are perfectly white, languid; and unhealthy, and that to make them recover vigour, and to acquire their natural colours, the direct influence of light is absolutely necessary. Somewhat similar takes place even upon animals: Mankind degenerate to a certain degree when employed in sedentary manufactures, or from living in crowded houses, or in the narrow lanes of large cities; whereas they improve in their nature and constitution in most of the country labours which are carried on in the open air. Organization, sensation, spontaneous motion, and all the operations of life, only exist at the surface of the earth, and in places exposed to the influence of light. Without it nature itself would be lifeless and inanimate. By means of light, the benevolence of the Deity hath filled the surface of the earth with organization, sensation, and intelligence. The fable of Promotheus [*sic*] might perhaps be considered as giving a hint of this philosophical truth, which had even presented itself to the knowledge of the ancients. I have intentionally avoided any disquisitions relative to organized bodies in this work, for which reason the phenomena of respiration, sanguification, and animal heat, are not considered; but I hope, at some future time, to be able to elucidate these curious subjects.[c]

Now it is easier to answer the question why Lavoisier's list of elements starts off with light. It was bound up with his deistic interpretation of the creative act described in the bible as God saying: 'Let there be light.' It further emerges from the passage that Lavoisier contemplated *all* forms of life (plant, animal and human beings) as absolutely depending on light. Apparently, there is no direct historical connection between Lavoisier's observation on the centrality of light for the world of life and its substantiation by later work on photosynthesis, arguably *the* life-sustaining reaction.

Be that as it may, historically, Lavoisier's unified conception of the chemical composition of mineral, plant and animal substances, including his interpretation of respiration as slow combustion, was of enormous influence. Taken for granted after his death, Lavoisier's approach affected the character of the chemistry of life pervasively and decisively. Yet for all this, Lavoisier did embrace an elemental theory of matter which in some ways still gravitated towards the past. Believing that there were five common ingredients of the stuff from which non-living and living things are ultimately made of, Lavoisier placed them in a special category. Thus the chemist, who has been universally credited with having revolutionized his field, proceeded far less revolutionarily than assumed, centainly in the maner in which he conceived and classified the 'elementary' status of his 'simple substances'. An instructive reminder that a critical history of science has to allow, without becoming relativist, for the real but ever-moving and permeating boundaries between 'old' and 'new' scientific knowledge.[d]

The coming of modern biochemistry into its own: c. 1880–1940

Now let us look overall at the topics set out in the (b) subdivisions of each Section, which cover approximately the period 1880–1940 and see whether it is possible to draw the multiple threads together, that is, with a view to saying something about their bearing on the transformation of physiological chemistry into biochemistry.

Obviously, the subject matter of those subdivisions still carried the stamp of the historical connection with physiology and the results obtained partly in response to the needs and problems of medicine, agriculture and industry would be directly or indirectly valuable for them. Nevertheless, during those sixty years, the study of the chemistry of life underwent a conspicuous metamorphosis which involved its separation from physiology and made it, as 'biochemistry', a recognizable and a recognized independent branch of science. It became recognizable through the study of enzymes which, historically, owing as it did its origins to preoccupations with phenomena of fermentation and digestion, eventually stood out among biochemistry's wide range of interests as the topic peculiar to it. It became recognized because both practitioners of biochemistry and outsiders finally came to accept its independent status and place in the spectrum of sciences.

Let us pursue the discussion a little further on the basis of subdividing the period 1880–1940, effectively into two phases. It enables us to gain a more detailed insight into the developments that brought enzymes into the foreground of the biochemical research scene during those years.

The earlier phase, 1880–1920, saw the beginning of enzyme kinetics and of the knowledge of enzyme specificity. Apart from establishing cell-free fermentation, other important steps during this phase were the discoveries of the reversibility of enzyme action and the influence of hydrogen ion concentration on enzyme activity. Here it should be noted that despite all the indications that these advances were derived from research undertaken in the study of enzymes *per se*, their origins as well as their impact cannot be seen solely in enzymic perspective (sugar structure, physicochemical properties of proteins, fermentation practice). But by the eve of the Great War enzyme chemistry had arrived – it was to draw the chemistry of life away from the current static thinking towards a dynamic one and, indeed, enzyme chemistry helped to give dynamic thinking reality.

The dynamic standpoint was advocated before 1914 by leaders of biochemical thought such as Hofmeister and Hopkins, but work on these lines did not get off the ground until the later phase 1920–40. The delay was, in part, due to the 'outer' circumstances of war. But the actual history indicates that it was due even more to the 'inner' circumstances bound up with theoretical concepts and technical procedures.

After 1920 this situation began to change with the further expansion of the study of enzymes, in particular in relation to biological oxidation. In the preceding sections we have seen how experimentally significant proved to be the spread of the methylene blue reduction procedure in vacuum tubes (T. Thunberg), the employment of microspectroscopy (D. Keilin) and the development of manometry, tissue slice technique and spectrophotometry (O. Warburg). Other important steps during 1920–40 were efforts at purifications (R. Willstätter, M. Dixon) and crystallizations of enzymes (J.B. Sumner, J.H. Northrop). A striking achievement in this area was the crystallization of enzymes of fermentation, carried out during the late 1930s (O. Warburg and associates). The studies of intermediate metabolism began to profit from the use of radioactive isotopes from the mid-1930s (R. Schoenheimer and co-workers).

These techniques played a vital part in throwing light on the chemical processes associated with respiration, fermentation and photosynthesis, thus contributing concretely to a fuller understanding of the dynamics of the chemistry of life. While the findings were revealing the flux underlying the chemical systems in the world of life, they were also pointing to the ordered catalytic involvement of enzymes in (almost) all of them.

Taken in all, it was through the rise and expansion of enzyme chemistry that the science of the chemistry of life found itself ready to come into its own. Between 1880 and 1940 biochemistry emerged conceptually and institutionally as an independent discipline, with the overall task of comprehending the dynamic aspects of the chemistry of life. Nevertheless, it is only proper to point out that the subject of enzymes did not carry as much weight in biochemical education as it did in research. In this respect, even by 1946, the status of enzymes was described criticaly by D.E. Green (*b.*1910), an American leader in the field, as follows:

Most modern textbooks of biochemistry often treat the subject of enzymes much in the manner that feathers and scales are dealt with in textbooks of anatomy. If this were merely another instance of the lag of textbooks behind developments in research there would be no cause for concern. But more appears to be involved than the traditional lag. The implications of enzyme chemistry have yet to be more generally understood; and until that time arrives the textbooks will continue to regard enzymes as chemical oddities, if not to ignore them altogether.[e]

What was it that Green took amiss? Though not explicitly stated, he was suggesting that biochemists were to be made fully aware of what might be called 'enzymic reductionism', the belief that life ultimately could be reduced to enzyme chemistry.

The domain of biochemistry and its history

Here we touch on the question of the chemistry of life as an interdiscipline. We have seen how its practitioners, from Lavoisier to Needham, were openly or tacitly wrestling with it and produced answers which, fundamentally, turned on how to come to terms with its two aspects or components, one chemical, the other biological. All these searchers agreed that the laws of physics and chemistry also held good in the world of life and therefore provided the avenue for its scientific exploration.

Brought up on and influenced by religious notions, as they were, they faced the more fundamental issue of whether the laws of physics and chemistry could account adequately for the manifold phenomena of life. How did they deal with that? One conclusion that arises from looking through the material of this volume is that they opted, by and large, for a position that allowed the coexistence of vitalist and mechanist conceptions. As a result, considerable scope remained for different types of materialistic and idealistic explanations of the origins and nature of life.

As a further consequence of this situation, individual scientists' opinions about the theoretical and experimental handling of the chemical side of life varied considerably. From about 1840 onwards they led, at times, to vigorous controversies over the subordination of the biological to the chemical side, or vice versa. At the centre of these disputes was the effort to decide between the chemical and biological explanations of alcoholic fermentation. Viewed against this background we can see from E. Buchner's report in 1897 that he had observed cell-free fermentation was a critical contribution to the resolution of this issue. This achievement has been hailed, especially by biochemists, as the date of the birth of modern biochemistry. A kindred claim regarding modern organic chemistry has been made for Wöhler's laboratory synthesis of urea in 1828.

In this volume these dates have been taken to be important landmarks on the road to comprehending that although the laws of physics and chemistry were fully applicable to the world of life, the latter was govered by laws of biology. That understanding was reflected in the gradual but discernible shift which removed the chemistry of life (organic chemistry, physiological chemistry, biochemistry) from the orbit of chemistry and brought it within the sphere of biology. Crucial for this change of direction was the establishment of the cell theory.

This claim is supported by the abundance of material in the *Documentary History* which shows that the historical relations between cell morphology and cell chemistry played an essential role in advancing contacts between biology and chemistry and, therefore, contributed crucially to the development of the chemistry of life from about 1840. The cross-fertilization of the two fields, differing at times in approach and the use of techniques, eventually produced a theoretical awareness of the need to link structural and dynamic outlooks in the study of the chemical organisation of living matter, although little experimental work had been done in this context by 1940. Hence, presumably, the startling effect of the discovery by Engelhardt and Lyubimova in 1939 of the enzymic activity of myosin. It must be regarded as a landmark, and not only in the history of muscle biochemistry. Undoubtedly, a major historical factor behind the sporadic attempts to examine, for example, the relations of enzymic activity and cell structure was the biochemists' preoccupation with work on pure enzymes. But as H.A. Krebs pointed out the

purification of enzymes may modify their properties and that it is therefore essential to investigate the behaviour of enzymes not only in the purified state but also in their natural environment. The study of crude enzymes may bring to light important characteristics which may escape attention in the examination of the pure enzymes. This follows from the consideration that living cells are systems whereby the whole is more than the sum of its components. The integration of the parts to a unit involves an arrangement where the component parts influence each other's behaviour. In the terms of chemistry this interlocking implies that cell constituents modify each other's chemical reactivity, an interplay which is an essential part of the regulatory mechanisms. It distinguishes a complex unit from a complex mixture.[f]

This is in line with the warnings of N.W. Pirie[g] and A. Tiselius[h] who have emphasized that biochemists have to be concerned just as much with the 'impure', the 'non-uniform', the 'dissimilar' and the 'heterogeneous' as they are with the 'pure', the 'uniform', the 'identical' and the 'homogeneous'.

When R.A. Peters proposed in 1929 and 1930 to take into account the inner architecture of the cell on the basis of a 'co-ordinative biochemical' approach, his views were apparently not acceptable to the pure colloid chemists, though the more thoughtful biochemists began to appreciate the necessity of linking biological form and chemistry together, as reflected in the ideas of the cell micro-morphology put forward by J. Needham.

The story is complex because in the actual development of biochemistry, as indicated, for example, by the history of fermentation chemistry, vital staining, or the knowledge of the nature of fibres, the 'external' and the 'internal' factors do not operate separately. The study of the history of biochemistry has to bear in mind the social and philosophical, including religious, aspects just as it has to take the relations of structural and dynamic approaches into consideration against the background of important biological generalizations such as the cell and evolutionary theories.[i] In this way, perhaps, an adequate comprehension and presentation of the intellectual and technical difficulties in the development of modern biochemistry and its place in history can be attained.[j]

Notes to Afterword

a.   M. Teich, 'Historical foundations', Introduction to Selection 161, Section VII, note f.

b.   Introduction to Selection 46, Section III, note d.

c.   A.-L. Lavoisier, *Elements of Chemistry* (Edinburgh, 1790), pp. 183–4. Due to his untimely and tragic death, Lavoisier was unable to achieve his objective. Engulfed by the march of revolutionary events, he was tried for his membership of the detested *Ferme générale* – the private body with the responsibility for collecting Government revenue – and guillotined in 1794. For the hitherto most comprehensive treatment of Lavoisier's activities in this area, see the already referred to work of F.L. Holmes: *Lavoisier and the Chemistry of Life An Exploration of Scientific Creativity* (Madison, 1985). See also his 'Lavoisier and Krebs. The individual scientist in the near and deeper past', *Isis*, 75, 131–42. 1984.

d.   R.C. Jennings makes the point that Lavoisier's 'acceptance of the Stahlian theory of phlogiston and the theory that fire has weight, have not been adequately noted by historians. Historical studies have concentrated on his rejection of the former and his virtual rejection of the latter; but to place Lavoisier's developing views in a proper historical context it is important to note that he did initially hold these views'. See his article 'Lavoisier's views on phlogiston and the matter of fire before about 1770', *Ambix*, 27, 206–9. 1981 (p. 208).

e.   D.E. Green, 'Biochemistry from the standpoint of enzymes' in D.E. Green (ed.), *Currents in Biochemical Research* (New York, 1946), p. 149.

f.   H.A. Krebs, 'Enzymic activity and cellular structure' in M. Kasha and B. Pullman (eds), *Horizons in Biochemistry* (New York–London, 1962), p. 291.

g.   N.W. Pirie, 'Criteria of purity used in the study of large molecules of biological origin', *Biol.Revs.*, 15, 377–404. 1940; 'Patterns of assumption about large molecules', *Arch.Biochem.Biophys.Suppl.*, 1, 21–9. 1962.

h.   A. Tiselius, 'Reflections from both sides of the counter', *Ann.Rev.Biochem.*, 37, 1–24, 1968. (pp. 14–15). Arne Tiselius (1902–71) – Swedish recipient of the Nobel Prize in Chemistry (1948) in particular for work on heterogeneity of serum proteins.

i.   For a recent discussion of the significance of the 'evolutionary axiom' in biochemistry see J.E. Baldwin and H. Krebs, 'The evolution of metabolic cycles', *Nature*, 221, 381–2. 1981.

j.   This draws on Teich, 'From "enchyme" to "cytoskeleton"', Foreword, note i.

Select bibliography

Section I

Ahrens, F.B., 'Das Gärungsproblem', *Sammlung Chemischer und chemisch-technischer Vorträge*, 7, no. 8. 1902.

Bayliss, W.M., *The Nature of Enzyme Action*, 1st–5th edn (London: Longmans, Green and Co., 1908–25).

Dixon, M. and E.C. Webb, *Enzymes*, 1st–2nd edn (London: Longmans, 1958, 1964).

Duclaux, E., *Traité de Microbiologie* (Paris: Masson et Cie, 1898–1901), vols I–IV.

Euler, H., *General Chemistry of the Enzymes*, translated by T.H. Pope (New York: John Wiley and Sons; London: Chapman and Hall, 1912).

Euler, H.v., *Chemie der Enzyme*, pt I, *Allgemeine Chemie der Enzyme*, (Munich-Wiesbaden: J.F. Bergmann, 1920).

Green, J.R., *The Soluble Ferments and Fermentation*, 1st–2nd edn (Cambridge, England: At the University Press, 1899, 1901).

Haldane, J.B.S., *Enzymes*, (London: Longmans, Green and Co., 1930).

Harden, A., *Alcoholic Fermentation*, 1st–4th edn (London: Longmans, Green and Co., 1911–32).

Northrop, J.H., *Crystalline Enzymes. The Chemistry of Pepsin, Trypsin, and Bacteriophage* (New York: Columbia University Press, 1939).

Oppenheimer, C., *Die Fermente und ihre Wirkungen Nebst einem Sonderkapitel von Dr R. Kuhn*, 5th edn (Leipzig: Georg Thieme, 1925), vol. I.

Schadewaldt, H., 'Zur Geschichte des Fermentbegriffes', reprint from *Festschrift der Kali-Chemie Aktiengesellschaft Hannover* (n.d.).

Segal, H.L., 'The development of enzyme kinetics' in P.D. Boyer, H. Lardy, K. Myrbäck (eds), *The Enzymes*, 2nd edn (New York: Academic Press, 1959), vol. 1.

Sumner, J.B. and G.F. Somers, *Chemistry and Methods of Enzymes*, 3rd printing (New York: Academic Press, 1946).

Tauber, H., 'The chemical nature of enzymes', *Chem.Revs.*, 15, 99–121. 1934.

Walden, P., 'Aus der Entwicklungsgeschichte der Enzymologie von ihren Anfängen bis zum Anbruch des zwanzigsten Jahrhunderts', *Erg.Enzymf.*, 10, 1–64. 1949.

Section II

Fogg, G.E., *Photosynthesis* (London: The English Universities Press, 1968).

Hill, R. and C.P. Whittingham, *Photosynthesis*, 2nd edn (London: Methuen, 1957).

Hill, R., 'Historical outline photosynthesis and the chloroplast', reprint from M. Gibbs (ed.), *Structure and Function of Chloroplasts* (Printed in Germany: Springer-Verlag, 1971). Not in trade.

Hill, R., 'Days of visual spectroscopy', *Ann.Rev.Plant Physiol.*, 26, 1–11. 1975.

Spoehr, H.A., *Photosynthesis* (New York: Chemical Catalog Co., 1926).

Stiles, W., *Photosynthesis. The Assimilation of Carbon by Green Plants*, (London: Longmans, Green and Co., 1925).

Section III

Batelli, F. and L. Stern, 'Die Oxydationsfermente', *Erg.Physiol.*, 12, 96–268. 1912.

Dakin, H.B., *Oxidations and Reductions in the Animal Body*, 1st–2nd edn (London: Longmans, Green and Co., 1912, 1922).

Dixon, M., 'Respiratory carriers' in J. Needham and D.E. Green (eds), *Perspectives in Biochemistry* (Cambridge: At the University Press, 1937), pp. 114–26.

Dixon, M., 'Biological oxidations and reductions', *Ann.Rev.Biochem.*, *8*, 1–36. 1939.

Green, D.E., *Mechanism of Biological Oxidations*, (Cambridge: At the University Press, 1940).

Krebs, H.A., 'Intermediary hydrogen-transport in biological oxidations', *Perspectives in Biochemistry*, pp. 150–4.

Scheer, B.T., 'The development of the concept of tissue respirations', *Ann.Sci.*, *4*, 295–305. 1939.

Section IV

Armstrong, E.F., *The Simple Carbohydrates and the Glucosides*, 1–4th edn (London: Longmans, Green and Co., 1910, 1912, 1919, 1924). The following is the 5th edn published in two separate volumes.

Armstrong, E.F. and K.F. Armstrong, *The Glycosides* (London: Longmans, Green and Co., 1931).

Armstrong, E.F. and K.F. Armstrong, *The Carbohydrates*, (London: Longmans, Green and Co., 1934).

Bell, D.J., 'Recent accomplishments in carbohydrate chemistry', *Perspectives in Biochemistry*, pp. 187–200.

Cori, C.F., 'Phosphorylation of glycogen and glucose', *Symp.Soc.Exp.Biol.*, *5*, 131–40. 1941.

Cori, C.F., 'The role of lactic acid in the development of biochemistry' in A. Kornberg, B.L. Horecker, L. Corundella and J. Oró (eds), *Reflections on Biochemistry* (Oxford: Pergamon Press, 1976), pp. 17–26.

Fischer, E., *Gesammelte Werke Untersuchungen über Kohlenhydrate und Fermente* 1884–1908; 1908–19 (Berlin: J. Springer, 1909, 1929), 2 vols.

Lohmann, K., 'The chemistry and metabolism of the compounds of phosphorus', *Ann.Rev.Biochem.*, *7*, 125–42. 1938.

Meyerhof, O., *Chemical Dynamics of Life Phenomena* (Philadelphia: J.B. Lippincott, 1924).

Needham, D.M., *The Biochemistry of Muscle* (London: Methuen, 1932).

Needham, D.M., *Machina Carnis* (Cambridge: At the University Press, 1971).

Section V

Adair, G.S., 'The chemistry of the amino acids and proteins', *Ann.Rev.Biochem.*, *6*, 163–87. 1937.

Cathcart, E.P, *The Physiology of Protein Metabolism*, 1st–2nd edn (London: Longmans, Green and Co., 1912, 1921).

Chibnall, A.C., 'Second Procter Memorial Lecture. The contribution of the analytical chemist to the problem of protein structure', *Journal of the International Society of Leather Trades Chemists*, *30*, 1–20. 1946.

Cohn, E.J., 'Die physikalische Chemie der Eiweisskörper', *Erg.Physiol.*, *33*, 781–882. 1931.

Edsall, J.T., 'Chemistry of the proteins and amino acids', *Ann.Rev.Biochem.*, *11*, 151–82. 1942.

Fantini, B., 'Chemical and biological classification of proteins', *Hist.Phil.Life Sci.*, *5*, 3–32. 1983.

Fischer, E., *Gesammelte Werke Untersuchungen über Aminosäuren, Polypeptide und Proteine* 1899–1906; 1907–19 (Berlin: J. Springer, 1906, 1923), vol. I, II.

Hunter, A., *Creatine and Creatinine* (London: Longmans, Green and Co., 1928).

Osborne, T.B., *The Vegetable Proteins*, 1st–2nd edn (London: Longmans, Green and Co., 1909, 1924).

Plimmer, R.H.A., *The Chemical Constitution of the Proteins*, 1st–3rd edn (London: Longmans, Green and Co., 1908, 1912, 1913), vol. I, II.

Robertson, T.B., *The Physical Chemistry of Proteins* (London: Longmans, Green & Co., 1918).

Schryver, S. B., *The General Characters of the Proteins* (Longmans, Green and Co., 1909).

Tracey, M.V., *Proteins and Life* (London: The Pilot Press, 1948).

Vickery, H.B., 'The contribution of the analytical chemist to protein chemistry', *Ann.N.Y.Acad.Sci.*, 47, 63–94. 1946.

Vickery, H.B. and T.B. Osborne, 'A Review of hypotheses of the structure of proteins', *Physiol.Revs.*, 8, 393–446. 1928.

Vickery, H.B. and C.A.L. Schmidt, 'The history of the discovery of the amino acids', *Chem.Revs.*, 9, 169–318. 1931.

Werner, E.A., *The Chemistry of Urea. The Theory of its Constitution, and Mode of its Formation in Living Organisms* (London: Longmans, Green and Co., 1923).

Section VI

Bloor, W.R., *Biochemistry of the Fatty Acids, their Compounds, the Lipids* (New York: Reinhold Publishing Co., 1943).

Breusch, F.L., 'The biochemistry of fatty acid metabolism', *Adv.Enzym.*, 8, 343–423. 1948.

Dam, H., 'Historical introduction' in R.P. Cook (ed.), *Cholesterol, Chemistry, Biochemistry, and Pathology* (New York: Academic Press, 1958).

Kleinzeller, A., 'Synthesis of lipides', *Adv.Enzym.*, 8, 299–341. 1948.

Leathes, J.B., *The Fats* (London: Longmans, Green and Co., 1910).

Leathes, J.B., and H.S. Raper, *The Fats*, 2nd edn (London: Longmans, Green and Co., 1925).

MacLean, H., *Lecithin and Allied Substances. The Lipines* (London: Longmans, Green and Co., 1918).

MacLean, H. and I. Smedley MacLean, *Lecithin and Allied Substances*, 2nd edn (London: Longmans, Green and Co., 1927).

Schoenheimer, R., *The Dynamic State of Body Constituents* (Cambridge, Mass.: Harvard University Press, 1942).

Section VII including Foreword and Afterword

Abir-Am, P., 'From biochemistry to molecular biology: DNA and the acculturated journey of the critic of science E. Chargaff', *Hist.Phil.Life Sci.*, 2, 3–60. 1980.

Abir-Am, P., 'The biotheoretical gathering,' *Hist Sci.*, 25, 1–70. 1987.

Allen, G.E., *Life Science in the Twentieth Century* (New York: John Wiley and Sons, 1975).

Arrhenius, S., *Quantitative Laws of Biological Chemistry* (London: G. Bell and Sons, 1915).

Bacon, J.S.D., 'Life outside the cell (or, what the biologist saw)' in W. Bartley, H.L. Kornberg, J.R. Quayle (eds), *Essays in Cell Metabolism* (London: Wiley-Interscience, 1970), pp. 45–66.

Baldwin, E., *Dynamic Aspects of Biochemistry* (Cambridge: At the University Press, 1947).

Bayertz, K., 'Naturwissenschaften und Philosophie Drei Gründe für ihre Differenzierung im frühen 19.Jahrhundert' in M. Hahn and H.J. Sandkühler (eds), *Die Teilung der Vernuft* (Cologne: Pahl-Rugenstein, 1982), pp. 106–20 [discussion, pp. 120–7].

Bernal, J.D., *The Physical Basis of Life* (London: Routledge and Kegan Paul, 1951).

Bernal, J.D., *Science in Industry in the Nineteenth Century* (London: Routledge and Kegan Paul, 1951).

Bernal, J.D., *Science in History*, 1st–3rd edn (London: Watts, 1954, 1957, 1965).

Bertalanffy, L.v., *Theoretische Biologie* (Berlin: Gebrüder Bornträger, 1932), vol. I.

Berntalanffy, L.v., *Modern Theories of Development*, translated and adapted by J.H. Woodger (London: Oxford University Press, 1933).

Bertalanffy, L.v., *Problems of Life. An Evaluation of Modern Biological Thought* (London: Watts, 1952).

Brooks, C.McC. and P.F. Cranefield (eds), *The Historical Development of Physiological Thought* (New York: The Hafner Publishing Company, 1959).

Bukharin, N.I. et al., Science at the Crossroads, 2nd edn with a new Foreword by J. Needham and a new Introduction by P.G. Werskey (London: Frank Cass and Co., 1971).

Büttner, J. (ed.), *History of Clinical Chemistry* (Berlin-New York: de Gruyter, 1983).

Bynum, W.F., E.J. Browne, R. Porter, (eds), *Dictionary of the History of Science* (London-Basingstoke: Macmillan, 1981).

Canguilhem, G., *Études d'Histoire et de Philosophie des Sciences* (Paris: J. Vrin, 1968).

Canguilhem, G., *La Connaissance de la Vie*, 2nd edn (Paris: J.Vrin, 1969).

Chittenden, R.H., *The Development of Physiological Chemistry in the United States* (New York: Chemical Catalog Co., 1930).

Coleman, W., *Biology in the Nineteenth Century: Problems of form, function and transformation* (New York: John Wiley and Sons, 1971).

Culotta, C.A., 'Tissue oxidation and theoretical physiology: Bernard, Ludwig, and Pflüger', *Bull.Hist.Med.*, *44*, 103–40. 1970.

Debru, C., *L'Esprit des Protéines: Histoire et philosophie biochimiques* (Paris: Hermann, 1983).

Driesch, H., *The Science and Philosophy of Organism* (London: Adam and Charles Black, 1908), vols I, II.

Edsall, J.T., 'A historical sketch of the Department of the Physical Chemistry, Harvard Medical School: 1920–1950', *American Scientist*, *38*, 580–93. 1950.

Edsall, J.T., 'The evolution of biochemistry', *Science*, *180*, 606–8. 1973.

Edsall, J.T., 'The Journal of Biological Chemistry after seventy-five years', *J.Biol.Chem.*, *255*, 8939–51. 1980.

Edsall, J.T., 'Progress in our understanding of biology' in G.A. Almond, M. Chodorow and R.H. Pearce (eds), *Progress and its Discontents* (Berkeley: University of California Press, 1982), pp. 135–60.

Farley, J., *The Spontaneous Generation Controversy from Descartes to Oparin* (Baltimore and London: Johns Hopkins University Press, 1977).

Florkin, M., *A History of Biochemistry*, vols 30–3A,B in M. Florkin and E.H. Stotz (eds), *Comprehensive Biochemistry* (Amsterdam: Elsevier, 1972–5).

Franklin, K.J., 'A short history of the International Congresses of Physiologists', *Ann.Sci.*, *3*, 241–335. 1938.

Fruton, J.S., 'The place of biochemistry in the university', *Yale J.Biol.and Med.*, *23*, 305–10. 1951.

Fruton, J.S., 'The Rockefeller Institute for Medical Research. An essay review', *J.Hist.Med.*, *21*, 71–7. 1966.

Fruton, J.S., Molecules and life Historical Essays on the Interplay of Chemistry and Biology (New York: Wiley-Interscience, 1972).

Fruton, J.S., 'The emergence of biochemistry', *Science*, *192*, 327–34. 1976.

Fruton, J.S., 'The interplay of chemistry and biology at the turn of the century' in

C.G. Bernhard, E. Crawford and P. Sörbom (eds), *Science Technology and Society in the Time of Alfred Nobel* (Oxford: Pergamon Press, 1982), pp. 74–96.

Fruton, J.S., 'Contrasts in scientific style. Emil Fischer and Franz Hofmeister: their research groups and their theory of protein structure', *Proc.Am.Phil.Soc.*, *129*, 313–70. 1985.

Geison, G.L., 'The protoplasmic theory of life and the vitalist–mechanist debate', *Isis*, *60*, 273–92. 1969.

Geison, G.L., *Michael Foster and the Cambridge School of Physiology: the Scientific Enterprise in Late Victorian Society* (Princeton: Princeton University Press, 1978).

Gray, J., 'The mechanical view of life', *Rep.Brit.Ass.*, 81–92. 1933.

Gregory, F., *Scientific Materialism in Nineteenth Century Germany* (Dordrecht-Boston: D. Reidel, 1977).

Goodfield, G.J., *The Growth of Scientific Physiology* (London: Hutchinson, 1960).

Haldane, J.B.S., *The Marxist Philosophy and the Sciences* (London: George Allen and Unwin, 1938).

Haldane, J.B.S., 'Protoplasm', *The Modern Quarterly*, *2*, 128–35. 1939.

Haldane, J.B.S., 'The scientific work of J.S. Haldane', *Penguin Science Survey* (2), 11–33. 1961.

Haldane, J.B.S., 'Life and mind as physical realities', *Penguin Science Survey* (B), 224–38. 1963.

Haldane, J.B.S., 'An autobiography in brief', *Persp.Biol.and Med.*, *9*, 476–81. 1965–66.

Haldane, J.S., *The Sciences and Philosophy*, (London: Hodder and Stoughton, n.d.).

Haldane, J.S., *The Philosophical Basis of Biology* (London: Hodder and Stoughton, 1931).

Haldane, J.S., *The Philosophy of a Biologist*, 2nd edn (Oxford: At the Clarendon Press, 1936).

Hall, T.S., 'Life as opposed transformation', *J.Hist.Med.*, *20*, 262–75. 1965.

Hall, T.S., *Ideas of Life and Matter. Studies in the History of General Physiology 600 B.C.–1900 A.D. From the Enlightenment to the End of the Nineteenth Century* (Chicago and London: University of Chicago Press, 1969), vol. II.

Hall, V.M.D., 'Biochemistry' in C.A. Russell (ed.), *Recent Developments in the History of Chemistry* (London: The Royal Society of Chemistry, 1985), chap. 9.

Haraway, D.J., *Crystals, Fabrics, and Fields: Metaphors of organicism in twentieth century developmental biology* (New Haven-London: Yale Unviersity Press, 1976).

Harrow, B., *Textbook of Biochemistry*, 1st–3rd edn (Philadelphia: W.B. Saunders Company, 1938, 1940, 1943).

Henderson, L.J., *The Fitness of the Environment. An Inquiry into the Biological Significance of the Properties of Matter*, Introduction by G. Wald, (Boston: Beacon Press, 1958) [1st edn 1913].

Hickel, E., 'Der Apothekerberuf als Keimzelle naturwissenschaftlicher Berufe in Deutschland', *Medizinhistorisches Journal*, *13*, 259–76. 1978.

Hickel, E., 'Die industrielle Arzneimittelforschung am Ende des 19.Jahrhunderts und die Durchsetzung einer reduktionistischen Biologie', *Argument-Sonderband AS 54*, 132–54. 1981.

Hiebert, E.N., 'The problem of organic analysis' in *L'Aventure de la Science (Melanges Alexandre Koyré)*, introduced by I.B. Cohen and R. Taton (Paris, 1964), pp. 303–25.

Hiebert, E.N., 'The uses and abuses of thermodynamics in religion', *Daedalus*, *95*, 1046–80. 1966.

Hodgkin, A.L. et al., *The Pursuit of Nature. Informal Essays on the History of Physiology* (Cambridge: Cambridge University Press, 1977).

Hogben, L., *Nature of Living Matter* (London: Kegan Paul, Trench, Trubner and Co., 1930).

Holmes, F.L., 'Biochemistry and the Historian', *Hist.Sci.*, *13*, 114–21. 1975.

Hopkins, F.G., 'Biochemistry: present position and outlook', *The Lancet*, 1247–52. 1924.

Jacob, F., *The Logic of Living Systems. A History of Heredity*, translated by B.E. Spillman (London: Allen Lane, 1974).

Janko, J., 'Chemistry in the synthetic treatments of physiology in the 2nd half of the 19th and the beginning of the 20th century', *Acta historiae rerum naturalium necnon technicarum Special Issue 7*, 185–208. 1974.

Janko, J., 'From physiological chemistry to biochemistry', *Acta historiae rerum naturalium necnon technicarum Special Issue 9*, 223–79. 1977.

Jevons, J.R., *The Biological Approach to Life*, 1st–2nd edn (London: George Allen and Unwin, 1964, 1968).

Kahane, E., *La Vie N'Existe Pas!*, Foreword by A.I. Oparin (Paris: Éditions rationalistes, 1962).

Kahane, E., 'Biochemistry–its development and its tasks', *Impact*, *14*, 101–18. 1964.

Kamminga, H., *Studies in the History of Ideas on the Origin of Life*, Ph.D. thesis, University of London, 1980.

Karlson, P., '100 Jahre Biochemie im Spiegel von Hoppe-Seyler's Zeitschrift für Physiologische Chemie', *Z.physiol.Chem.*, *358*, 717–52. 1977.

Knight, B.C.J.G., 'The growth of microbiology. The Fifth Marjory Stephenson Lecture', *J.Gen.Microbiol.*, *27*, 357–72. 1962.

Kohler, jun., R.E., 'The enzyme theory and the origin of biochemistry', *Isis*, *64*, 181–96. 1973.

Kohler, R.E., 'The history of biochemistry: a survey', *J.Hist.Biol.*, *8*, 275–318. 1975.

Kohler, R.E., *From Medical Chemistry to Biochemistry. The Making of a Biomedical Discipline* (Cambridge: Cambridge University Press, 1982).

Larson, J., 'Vital forces: regulative principles or constitutive agents? A strategy in German physiology, 1786–1802', *Isis*, *70*, 235–49. 1979.

Leicester, H.M., *Development of Biochemical Concepts from Ancient to Modern Times* (Cambridge, Mass.: Harvard University Press, 1974).

Lenoir, T., 'Teleology without regrets. The transformation of physiology in Germany',. *Stud.Hist.Phil.Sci.*, *12*, 293–354. 1981.

Lenoir, T., *The Strategy of Life* (Dordrecht-Boston: D. Reidel, 1982).

Levy, H., 'The fallacy of mechanism', *The Modern Quarterly*, *1*, 5–14. 1938.

Löw, R., *Pflanzenchemie zwischen Lavoisier und Liebig* (Straubing-Munich: Donau-Vlg, 1977).

McCay, C.M., *Notes on the History of Nutrition Research*, edited by F. Verzár (Berne: Hans Huber Publishers, 1973).

Mendelsohn, E., 'Cell theory and the development of general physiology', *Arch.Int.Hist.Sci.*, *6*, 419–29. 1963.

Mendelsohn, E., *Heat and Life: The development of the theory of animal heat* (Cambridge, Mass.: Harvard Unviersity Press, 1964).

Mendelsohn, E., 'The biological sciences in the nineteenth century: some problems and sources', *Hist. Sci.*, *3*, 39–59. 1964.

Mendelsohn, E., 'Physical models and physiological concepts: explanation in nineteenth-century biology', *Brit.J.Hist.Sci.*, *2*, 201–19. 1965.

Monod, J., *Chance and Necessity. An Essay on the National Philosophy of Modern Biology*, translated by A. Wainhouse (Glasgow: Collins Fontana Books, 1974).

Morgan, N., 'The development of biochemistry in England through botany and the brewing industry (1870–1890)', *2*, 141–66. 1980.

Morgan, N., 'William Dobinson Halliburton, F.R.S. (1860–1931) pioneer of British biochemistry?', *Not.Rec.R.S.*, *38*, 129–45. 1983.

Morton, R.A., *The Biochemical Society. Its History and Activities* (London: The Biochemical Society, 1969).

Multhauf, R.P., 'J.B. van Helmont's reformation of the Galenic doctrine of digestion', *Bull.Hist.Med.*, *29*, 154–63. 1955.

Needham, J., 'The philosophical basis of biochemistry', *Monist*, *35*, 27–48. 1925.

Needham, J., 'Mechanistic biology and the religious consciousness', in J. Needham (ed.), *Science, Religion and Reality* (London: The Sheldon Press, 1935), pp. 219–57.

Needham, J., 'Biochemical aspects of form and growth' in L.L. Whyte (ed.), *Aspects of Form and Growth*, 2nd edn (London: Lund Humphries, Publishers, 1968), pp. 77–86.

Needham, J., (ed) *The Chemistry of Life Lectures on the History of Biochemistry, with an Introduction by J. Needham* (Cambridge: At the University Press, 1970).

Needham, J., 'The making of an Honorary Taoist' in M. Teich and R. Young (eds), *Changing Perspectives in the History of Science* (London: Heinemann Educational Books, 1973), pp. 1–20 (under pseudonym Henry Holorenshaw).

Needham, J., *Moulds of Understanding. A Pattern of Natural Philosophy*, edited and introduced by G. Werskey (London: George Allen and Unwin, 1976).

Olby, R., *The Path to the Double Helix*, Foreword by Francis Crick (London and Basingstoke: Macmillan, 1974).

Olby, R.C., 'The significance of the macromolecules in the historiography of molecular biology', *Hist.Phil.Life Sci.*, *1*, 185–179. 1979.

Pagel, W., *Joan Baptista Van Helmont. Reformer of Science and Medicine* (Cambridge: Cambridge University Press, 1982).

Parascandola, J., 'Organismic and holistic concepts in the thought of L.J. Henderson', *J.Hist.Biol.*, *4*, 63–118. 1971.

Parascandola, J., 'L.J. Henderson and the theory of buffer action', *Medizinhistorisches Journal*, *6*, 297–309. 1971.

Parascandola, J., 'Reflection on the history of pharmacology', *Pharmacy in History*, *22*, 131–40. 1980.

Pickstone, J.V., 'Globules and coagula: concepts of tissue formation in the early nineteenth century', *J.Hist.Med.*, *28*, 336–56. 1973.

Pirie, N.W., 'The position of stereoisomerism in argument about the origins of life', *Trans. Bose Res. Inst.*, 111–20.

Pirie, N.W., 'Retrospect on the biochemistry of plant viruses' in T.W. Goodwin (ed.), *British Biochemistry Past and Present* (London–New York: Academic Press, 1970), pp. 43–56.

Pirie, N.W., 'Infection from worms to nucleotides', *Interdiscipl.Sci.Revs.*, *5*, 312–19. 1980.

Plimmer, R.H.A., *The History of the Biochemical Society 1911–1949* (Cambridge: Printed for the Society at the University Press, 1949).

Prenant, M., *Biology and Marxism*, translated by C.D. Greaves with a Foreword by J. Needham (London: Lawrence and Wishart, 1938).

Prigogine, I., 'Es gibt keine wirkliche Evolution, wenn alles gegebenist', *Dialektik Beiträge zu Philosophie und Wissenschaften*, *5*, 121–133. 1982.

Regelmann, J.-P., 'Thesen zum beruflichen Hintergrund der biochemischen Forschung in Deutschland im 19.Jahrhundert' in M. Hahn and H.J. Sandkühler (eds), *Die Teilung der Vernunft* (Cologne: Pahl-Rugenstein, 1982), pp. 141–6. Discussion, pp. 146–7.

Roe, S.A., *Matter, Life, and Generation. Eighteenth-Century Embryology and the Haller-Wolff Debate* (Cambridge: Cambridge University Press, 1981).

Roll-Hansen, N., 'Louis Pasteur – a case against reductionist historiography', *Brit.J.Phil.Sci.*, *23*, 347–61. 1972.

Roll-Hansen, N., *Reductionism in Biological Research. Three Historical Case Studies*

(Oslo: Haakon Arnesen A.s., 1979).

Roll-Hansen, N., 'E.S. Russel and J.H. Woodger: the failure of two twentieth-century opponents of mechanistic philosophy', *J.Hist.Biol.*, *17*, 399–428. 1984.

Rose, S. with C. Sanderson, *The Chemistry of Life*, 2nd edn (Harmondsworth: Penguin Books, 1979).

Rossiter, M.W., *The Emergence of Agricultural Science. Justus Liebig and the Americans* (New Haven and London: Yale University Press, 1975).

Rothschuh, K.E., *Physiologie der Wandel ihrer Konzepte, Probleme und Methoden vom 16.bis 19.Jahrhundert* (Freiburg: Vlg. Karl Albert, 1968).

Rothschuh, K.E., *History of Physiology*, translated by G.B. Risse (Huntington, NY: Robert Krieger, 1973).

Rousseau, G.S. and R. Porter, (eds) *The Ferment of Knowledge. Studies in the Historiography of Eighteenth-Century Science* (Cambridge: Cambridge University Press, 1980).

Russel, E.J., 'Rothamsted and its experimental station', *Agric.Hist.*, *16*, 161–83. 1942.

Russel, E.J., *A History of Agricultural Science in Great Britain, 1620–1954* (London: George Allen and Unwin, 1966).

Russell, E.S., *Form and Function. A Contribution to the History of Animal Morphology* (London: John Murray, 1916).

Russell, E.S., *The Study of Living Things* (London: Methuen, 1924).

Russell, E.S., *The Interpretation of Development and Heredity. A Study in Biological Method* (Oxford: At the Clarendon Press, 1930).

Schäfer, E.A., 'Address', *Rep.Brit.Ass.*, 3–36. 1912.

Schiller, J., *La Notion d'Organisation dans l'Histoire de la Biologie* (Paris: Maloine, 1978).

Schrödinger, E., *What is Life? The Physical Aspect of Life* (Cambridge: At the University Press, 1944).

Schröer, H., *Carl Ludwig Begründer der messenden Experimentalphysiologie, 1816– 1895* (Stuttgart: Wissenschaftliche Verlagsgesellschaft, 1967).

Sharpey-Shafer, E., *History of the Physiological Society during its First Years 1876– 1926* (London: Cambridge University Press, 1927).

Semenza, G (ed.) *Selected Topics in the History of Biochemistry. Personal Recollections. I, II.*, vols 35, 36 in A. Neuberger and L.L.M. Van Deenen (eds), *Comprehensive Biochemistry* (Amsterdam: Elsevier, 1983, 1986).

Simmer, H., 'Aus den Anfängen der physiologischen Chemie in Deutschland', *Sudhoffs Arch.*, *39*, 216–36. 1955.

Staudinger, H., *From Organic Chemistry to Macromolecules*, translated by J. Fock and M. Fried with a Foreword by H.F. Mark (New York: Wiley-Interscience, 1970).

Stephenson, M., *Bacterial Metabolism* (London: Longmans, Green and Co., 1930).

Stoll, A., *Ein Gang durch biochemische Forschungsarbeiten* (Berlin: J. Springer, 1933).

Ströker, E., *Theoriewandel in der Wissenschaftsgeschichte Chemie im 18.Jahrhundert* (Frankfurt on Main: Klostermann, 1982).

Štrbáňová, S., 'Biochemical journals and their profile in 1840–1930', *Acta historiae rerum naturalium necnon technicarum Special Issue*, *16*, 149–95. 1981.

Teich, M., 'The history of modern biochemistry: the first phase', *Actes du XIe Congrès International d'Histoire des Sciences* (Warsaw-Cracow, 1965), vol. V. pp. 223–6.

Teich, M., 'The history of modern biochemistry: the second phase: *c.* 1920–1940/45', *Actes du XIIe Congrès International d'Histoire des Sciences* (Paris, 1968), vol. VIII, pp. 199–203.

Teich, M., 'A single path to the double helix?', *Hist.Sci.*, *13*, 264–83. 1975.

Teich, M., 'Book review' of D.J. Haraway, *Crystals, Fabrics, and Fields: metaphors of organicism in twentieth-century developmental biology* in *Brit.J.Hist.Sci.*, *11*, 92–4. 1978.

Teich, M., 'Book review' of R.E. Kohler, *From Medical Chemistry to Biochemistry. The Making of a Biomedical Discipline* in *Brit.J.Hist.Sci.*, *17*, 239–41. 1984.

Teich, M., 'Die Suche der Biochemie nach Identität', in E. Hickel (ed.), *Biochemische Forschung im 19.Jahrhundert – mit einer Bibliographie der Quellen* (Braunschweig: Braunschweiger Veröffentlichungen zur Geschichte der Pharmazie und der Naturwissenschaften, vol. 32, 1989).

Temkin, O., *The Double Face of Janus and Other Essays in the History of Medicine* (Baltimore–London: Johns Hopkins University Press, 1977).

Trim, A.R., 'British biochemistry', *Brit.Med.Bull.*, *5*, 357–61. 1948.

Uschmann, G., 'Haeckel's biological materialism', *Hist.Phil.Life Sci.*, *1*, 101–18. 1979.

Waddington, C.H., 'Some European contributions to the prehistory of molecular biology', *Nature*, *221*, 318–21. 1969.

Werskey, G., *The Visible College* (London: Allen Lane, 1978).

Woodger, J.H., *Biological Principles A Critical Study* (London: Kegan Paul, Trench, Trubner and Co., 1929).

Woodger, J.H., 'The "concept of organism" and the relation between embryology and genetics', *Quart.Rev.Biol.*, *5*, 1–22. 1930.

Yoxen, E.J., 'Where does Schroedinger's "What is life?" belong in the history of molecular biology?', *Hist.Sci.*, *17*, 17–52. 1979.